Introduction to Finite Element Analysis

Using Creo Simulate 4.0®

Randy H. Shih
Oregon Institute of Technology

SDC Publications

SDC Publications
P.O. Box 1334
Mission, KS 66222
913-262-2664
www.SDCpublications.com
Publisher: Stephen Schroff

ISBN-13: 978-1-63057-108-5
ISBN-10: 1-63057-108-3

Printed and bound in the United States of America.

Preface

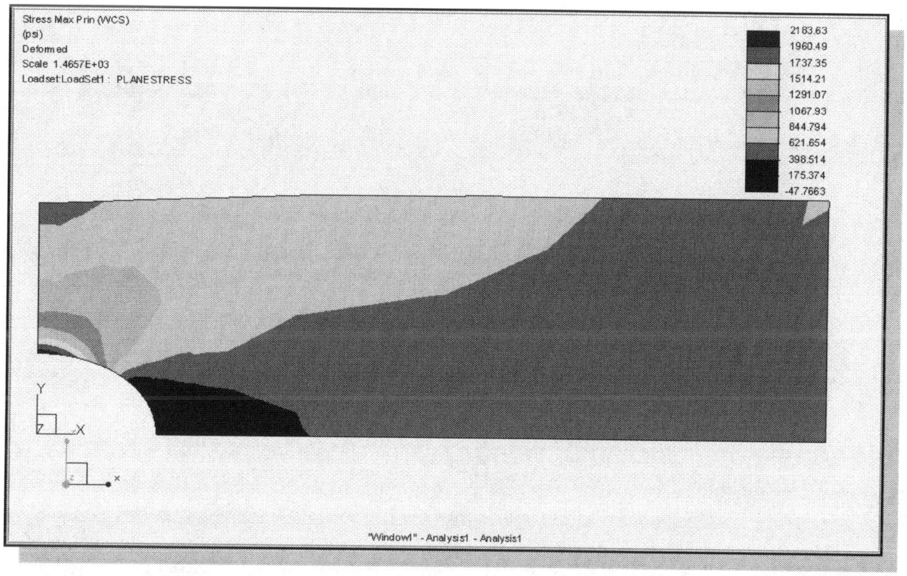

The primary goal of ***Introduction to Finite Element Analysis using Creo Simulate*** is to introduce the aspects of finite element analysis that are important to engineers and designers. Theoretical aspects of finite element analysis are also introduced as they are needed to help better understand the operation. The primary emphasis of the text is placed on the practical concepts and procedures for using *Creo Simulate* in performing *Linear Static Stress Analysis*; but the basic modal analysis procedure is also covered. This text is intended to be used as a training guide for students and professionals. This text covers *Creo Simulate* and the lessons proceed in a pedagogical fashion to guide you from constructing basic truss elements to generating three-dimensional solid elements from solid models. This text takes a hands-on, exercise-intensive approach to all the important *Finite Element Analysis* techniques and concepts. This textbook contains a series of ten tutorial style lessons designed to introduce beginning FEA users to *Creo Simulate*. This text is also helpful to *Creo Simulate* users upgrading from a previous release of the software. The finite element analysis techniques and concepts discussed in this text are also applicable to other FEA packages. The basic premise of this book is that the more designs you create using *Creo Simulate*, the better you learn the software. With this in mind, each lesson introduces a new set of commands and concepts, building on previous lessons. This book does not attempt to cover all of *Creo Simulate's* features, only to provide an introduction to the software. It is intended to help you establish a good basis for exploring and growing in the exciting field of **Computer Aided Engineering**.

Acknowledgments

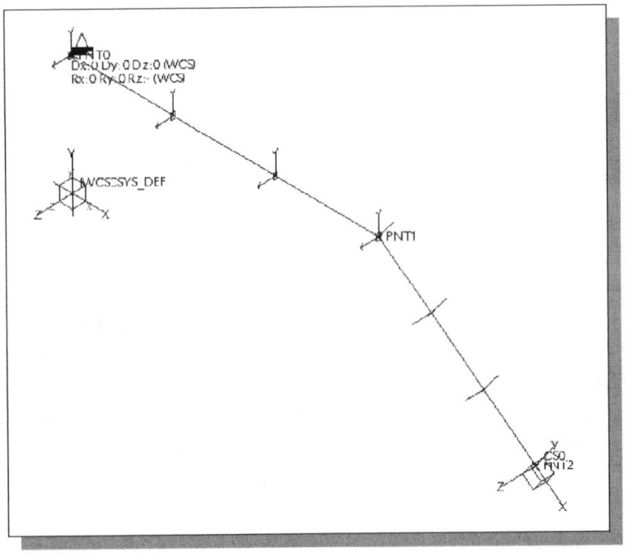

This book would not have been possible without a great deal of support. First, special thanks to two great teachers, Prof. George R. Schade of University of Nebraska-Lincoln and Mr. Denwu Lee, who taught me the fundamentals, the intrigue, and the sheer fun of Computer Aided Engineering.

The effort and support of the editorial and production staff of SDC Publications is gratefully acknowledged. I would especially like to thank Stephen Schroff and Mary Schmidt for their support and helpful suggestions during this project.

I am grateful that the Mechanical and Manufacturing Engineering and Technology Department of Oregon Institute of Technology has provided me with an excellent environment in which to pursue my interests in teaching and research.

Finally, truly unbounded thanks are due to my wife Hsiu-Ling and my daughter Casandra for their understanding and encouragement throughout this project.

Randy H. Shih
Klamath Falls, Oregon
Summer 2017

Table of Contents

Chapter 2
Truss Elements in Two-Dimensional Spaces

Chapter 3
2D Trusses in MS Excel and the Truss Solver

Chapter 4
Creo Simulate Two-Dimensional Truss Analysis

Chapter 5
Three-Dimensional Truss Analysis

Chapter 6
Basic Beam Analysis

Chapter 7
Beam Analysis Tools

Chapter 10
Three-Dimensional Solid Elements

Chapter 11
Axisymmetric and Thin Shell Elements

Chapter 12
Dynamic Modal Analysis

Introduction

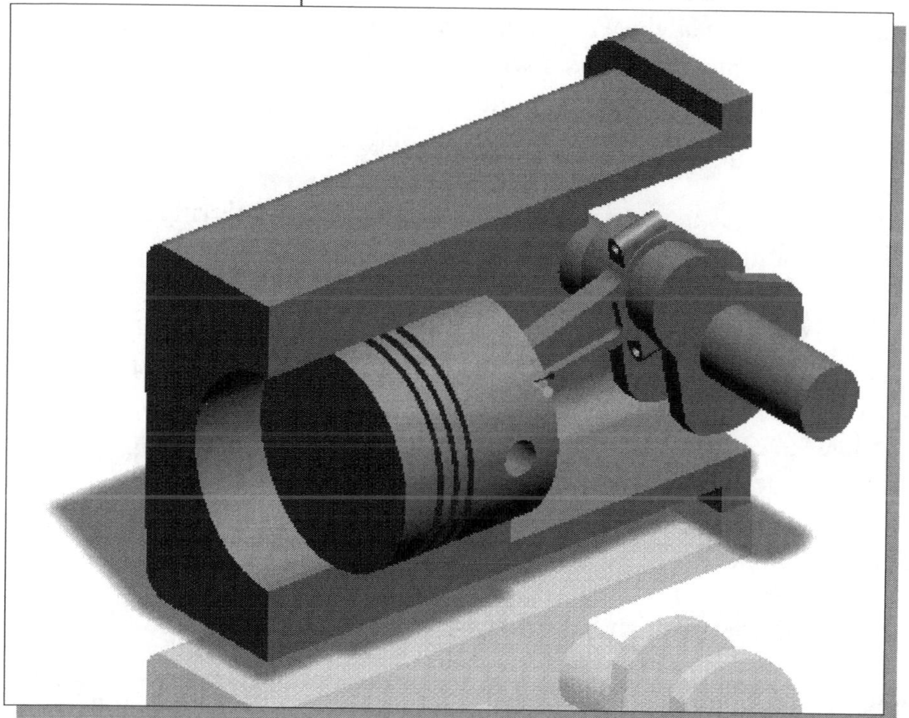

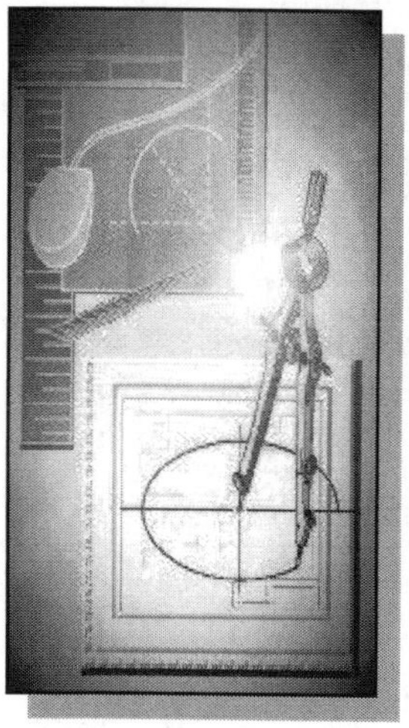

Learning Objectives

- Development of Finite Element Analysis.
- FEA Modeling Considerations.
- Finite Element Analysis Procedure.
- Getting Started with Creo Parametric.
- Startup Options and Units Setup.
- Creo Parametric Screen Layout.
- Mouse Buttons.
- Creo Parametric On-Line Help.

Introduction

Design includes all activities involved from the original concept to the finished product. Design is the process by which products are created and modified. For many years, designers sought ways to describe and analyze three-dimensional designs without building physical models. With the advancements in computer technology, the creation of three-dimensional models on computers offers a wide range of benefits. Computer models are easier to interpret and easily altered. Simulations of real-life loads can be applied to computer models and the results graphically displayed.

Finite Element Analysis (FEA) is a numerical method for solving engineering problems by simulating real-life-operating situations on computers. Typical problems solved by finite element analysis include structural analysis, heat transfer, fluid flow, soil mechanics, acoustics, and electromagnetism. *Creo Parametric* is an integrated package of mechanical computer aided engineering software tools developed by **Parametric Technology Corporation (PTC)**. *Creo* is a suite of programs, including the *Finite Element Analysis* module ***(Creo Simulate)***, which is used to facilitate a concurrent engineering approach to the design, analysis, and manufacturing of mechanical engineering products. This text focuses on basic structural analysis using the integrated *Creo Parametric* and *Creo Simulate.*

Development of Finite Element Analysis

Finite element analysis procedures evolved gradually from the work of many people in the fields of engineering, physics, and applied mathematics. The finite element analysis procedure was first applied to problems of stress analysis. The essential ideas began to appear in publications during the 1940s. In 1941, Hrenikoff proposed that the elastic behavior of a physically continuous plate would be similar to a framework of one-dimensional rods and beams, connected together at discrete points. The problem could then be handled by familiar methods for trusses and frameworks. In 1943, Courant's paper detailed an approach to solving the torsion problem in elasticity. Courant described the use of piecewise linear polynomials over a triangularized region. Courant's work was not noticed and soon forgotten, since the procedure was impractical to solve by hand.

In the early 1950s, with the developments in digital computers, Argyris and Kelsey converted the well-established "framework-analysis" procedure into matrix format. In 1956, Turner, Clough, Matin, and Topp derived stiffness matrices for truss elements, beam elements and two-dimensional triangular and rectangular elements in plane stress. Clough introduced the first use of the phrase "finite element" in 1960. In 1961, Melosh developed a flat, rectangular-plate bending-element, followed by development of the curved-shell bending-element by Grafton and Strome in 1963. Martin developed the first three-dimensional element in 1961 followed by Gallagher, Padlog and Bijlaard in 1962 and Melosh in 1964.

From the mid-1960s to the end of the 1970s, finite element analysis procedures spread beyond structural analysis into many other fields of application. Large general purpose FEA software began to appear. By the late 1980s, FEA software became available on microcomputers, complete with automatic mesh-generation, interactive graphics, and pre-processing and post-processing capabilities.

In this text, we will follow a logical order, parallel to the historical development of the finite element analysis procedures, in learning the fundamental concepts and commands for performing finite element analysis using *Creo Parametric* and ***Creo Simulate***. We will begin with the one-dimensional truss element, beam element, and move toward the more advanced features of *Creo Simulate*. This text also covers the general procedures of performing two-dimensional and three-dimensional solid FE analyses. The concepts and techniques presented in this text are also applicable to other FEA packages. Throughout the text, many of the classic strength of materials and machine design problems are used as examples and exercises, which hopefully will help build up the user's confidence on performing FEA analyses.

FEA Modeling Considerations

The analysis of an engineering problem requires the idealization of the problem into a mathematical model. It is clear that we can only analyze the selected mathematical model, and that all the assumptions in this model will be reflected in the predicted results. We cannot expect any more information in the prediction than the information contained in the model. Therefore, it is crucial to select an appropriate mathematical model that will most closely represent the actual situation. It is also important to realize that we cannot predict the response exactly because it is impossible to formulate a mathematical model that will represent all the information contained in an actual system.

As a general rule, finite element modeling should start with a simple model. Once a mathematical model has been solved accurately and the results have been interpreted, it is feasible to consider a more refined model in order to increase the accuracy of the prediction of the actual system. For example, in a structural analysis, the formulation of the actual loads into appropriate models can drastically change the results of the analysis. The results from the simple model, combined with an understanding of the behavior of the system, will assist us in deciding whether and at which part of the model we want to use further refinements. Clearly, the more complicated model will include more complex response effects, but it will also be more costly and sometimes more difficult to interpret the solutions.

Modeling requires that the physical action of the problem be understood well enough to choose suitable kinds of analyses. We want to avoid the waste of time and computer resources caused by over-refinement and badly shaped elements. Once the results have been calculated, we must check them to see if they are reasonable. Checking is very important because it is easy to make mistakes when we rely upon the FEA software to solve complicated systems.

Types of Finite Elements

The finite element analysis method is a numerical solution technique that finds an approximate solution by dividing a region into small sub-regions. The solution within each sub-region that satisfies the governing equations can be reached much more simply than that required for the entire region. The sub-regions are called ***elements*** and the elements are assembled through interconnecting a finite number of points on each element called ***nodes***. Numerous types of finite elements can be found in commercial FEA software and new types of elements are being developed as research is done worldwide. Depending on the dimensions, finite elements can be divided into three categories:

1. **One-dimensional** line elements: Truss, beam, and boundary elements.

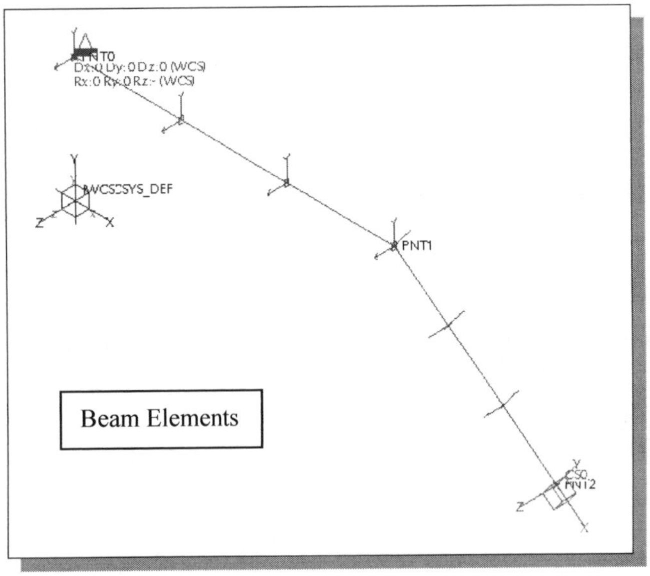

2. **Two-dimensional** plane elements: plane stress, plane strain, axisymmetric, membrane, plate, and shell elements.

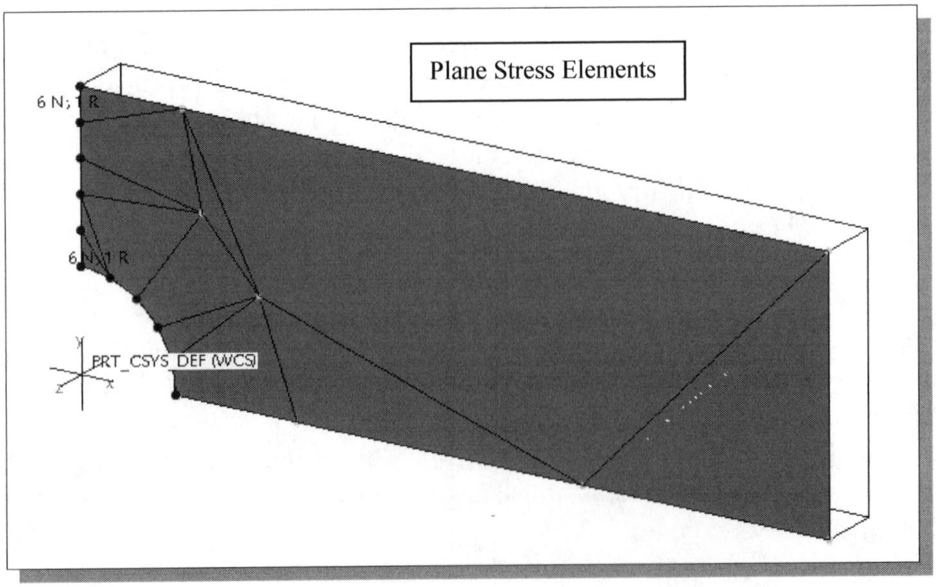

3. **Three-dimensional** volume elements: tetrahedral, hexahedral, and brick elements.

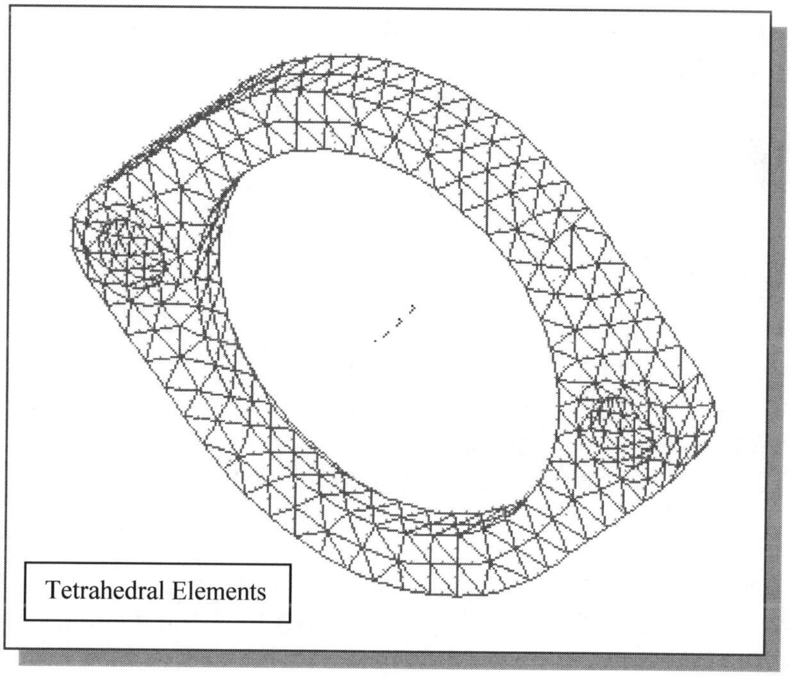

Typically, finite element solutions using one-dimensional line elements are as accurate as solutions obtained using conventional truss and beam theories. It is usually easier to get FEA results than doing hand calculations using conventional theories. However, very few closed form solutions exist for two-dimensional elements and almost none exist for three-dimensional solid elements.

In theory, all designs could be modeled with three-dimensional volume elements. However, this is not practical since many designs can be simplified with reasonable assumptions to obtain adequate FEA results without any loss of accuracy. Using simplified models greatly reduces the time and effort in reaching FEA solutions.

Finite Element Analysis Procedure

Prior to carrying out the finite element analysis, it is important to do an approximate preliminary analysis to gain some insights into the problem and as a means of checking the finite element analysis results.

For a typical linear static analysis problem, the finite element analysis requires the following steps:

1. Preliminary Analysis.

2. Preparation of the finite element model:
   a. Model the problem into finite elements.
   b. Prescribe the geometric and material information of the system.
   c. Prescribe how the system is supported.
   d. Prescribe how the loads are applied to the system.

3. Perform calculations:
   a. Generate a stiffness matrix of each element.
   b. Assemble the individual stiffness matrices to obtain the overall, or global, stiffness matrix.
   c. Solve the global equations and compute displacements, strains, and stresses.

4. Post-processing of the results:
   a. Viewing the stress contours and the displaced shape.
   b. Checking any discrepancy between the preliminary analysis results and the FEA results.

Matrix Definitions

The use of vectors and matrices is of fundamental importance in engineering analysis because it is with the use of these quantities that complex procedures can be expressed in a compact and elegant manner. One need not understand vectors or matrices in order to use FEA software. However, by studying the matrix structural analysis, we develop an understanding of the procedures common to the implementation of structural analysis as well as the general finite element analysis. The objective of this section is to present the fundamentals of matrices, with emphasis on those aspects that are important in finite element analysis. In the next chapter we will introduce the derivation of matrix structural analysis, the stiffness matrix method. *Matrix Algebra* is a powerful tool for use in programming the FEA methods for electronic digital computers. Matrix notation represents a simple and easy-to-use notation for writing and solving sets of simultaneous algebraic equations.

A **matrix** is a rectangular array of elements arranged in rows and columns. Applications in this text deal only with matrices whose elements are real numbers. For example,

$$[\mathbf{A}] = \begin{bmatrix} A_{11} & A_{12} & A_{13} \\ A_{21} & A_{22} & A_{23} \end{bmatrix}$$

[A] is a rectangular array of two rows and three columns, thus called a 2 × 3 matrix. The element A_{ij} is the element in the ith row and j^{th} column.

- **Column Matrix (Row Matrix)** A column (row) matrix is a matrix having one column (row). A single-column array is commonly called a column matrix or *vector*. For example:

$$\{F\} = \begin{Bmatrix} F_1 \\ F_2 \\ F_3 \end{Bmatrix}$$

- **Square Matrix** A square matrix is a matrix having equal numbers of rows and columns.

- **Diagonal Matrix** A diagonal matrix is a square matrix with nonzero elements only along the diagonal of the matrix.

- **Addition** The addition of matrices involves the summation of elements having the same "address" in each matrix. The matrices to be summed must have identical dimensions. The addition of matrices of different dimensions is not defined. Matrix addition is associative and commutative.

For example:

$$[\mathbf{A}] + [\mathbf{B}] = \begin{bmatrix} \textcircled{1} & 2 \\ 3 & \boxed{4} \end{bmatrix} + \begin{bmatrix} \textcircled{2} & 4 \\ 6 & \boxed{8} \end{bmatrix} = \begin{bmatrix} \textcircled{3} & 6 \\ 9 & \boxed{12} \end{bmatrix}$$

- **Multiplication by a Constant** If a matrix is to be multiplied by a constant, every element in the matrix is multiplied by that constant. Also, if a constant is factored out of a matrix, it is factored out of each element. For example:

$$\mathbf{3 \times [A] = 3 \times} \begin{bmatrix} 1 & 2 \\ 3 & 4 \end{bmatrix} = \begin{bmatrix} 3 & 6 \\ 9 & 12 \end{bmatrix}$$

- **Multiplication of Two Matrices** Assume that [C] = [A][B], where [A], [B], and [C] are matrices. Element Cij in matrix [C] is defined as follows:

$$C_{ij} = A_{i1} \times B_{1j} + A_{i2} \times B_{2j} + \cdots + A_{ik} \times B_{kj}$$

For example:

$$[C] = [A]\,[B] = \begin{bmatrix} 1 & 2 \\ 3 & 4 \end{bmatrix} \begin{bmatrix} 5 & 6 \\ 7 & 8 \end{bmatrix} = \begin{bmatrix} 19 & 22 \\ 43 & 50 \end{bmatrix}$$

$$C_{11} = 1 \times 5 + 2 \times 7 = 19, \quad C_{12} = 1 \times 6 + 2 \times 8 = 22$$
$$C_{21} = 3 \times 5 + 4 \times 7 = 43, \quad C_{22} = 3 \times 6 + 4 \times 8 = 50$$

- **Identity Matrix** An identity matrix is a diagonal matrix with each diagonal element equal to unity.

For example:

$$[I] = \begin{bmatrix} 1 & 0 & 0 & 0 \\ 0 & 1 & 0 & 0 \\ 0 & 0 & 1 & 0 \\ 0 & 0 & 0 & 1 \end{bmatrix}$$

- **Transpose of a Matrix** The transpose of a matrix is a matrix obtained by interchanging rows and columns. Every matrix has a transpose. The transpose of a column matrix (vector) is a row matrix; the transpose of a row matrix is a column matrix.

For example:

$$[A] = \begin{bmatrix} 1 & 2 & 3 \\ 4 & 5 & 6 \end{bmatrix} \qquad [A]^T = \begin{bmatrix} 1 & 4 \\ 2 & 5 \\ 3 & 6 \end{bmatrix}$$

- **Inverse of a Square Matrix** A square matrix *may* have an inverse. The product of a matrix and its inverse matrix yields the identity matrix.

$$[A]\,[A]^{-1} = [A]^{-1}\,[A] = [I]$$

The reader is referred to the following techniques for matrix inversion:

1. Gauss-Jordan elimination method
2. Gauss-Seidel iteration method

These techniques are popular and are discussed in most texts on numerical techniques.

Getting Started with Creo Parametric

Creo Parametric is composed of several application software modules (these modules are called *applications*), all sharing a common database. In this text, the main concentration is placed on the solid modeling modules used for part design. The general procedures required in creating solid models, engineering drawings, and assemblies are illustrated.

Starting Creo Parametric

How to start *Creo Parametric* depends on the type of workstation and the particular software configuration you are using. With most *Windows* and *UNIX* systems, you may select **Creo Parametric** on the *Start* menu or select the **Creo Parametric** icon on the desktop. Consult your instructor or technical support personnel if you have difficulty starting the software.

The program takes a while to load, so be patient. The tutorials in this text are based on the assumption that you are using *Creo Parametric's* default settings. If your system has been customized for other uses, some of the settings may appear differently and not work with the step-by-step instructions in the tutorials. Contact your instructor and/or technical support personnel to restore the default software configuration.

Creo Parametric Screen Layout

The default *Creo Parametric drawing screen* contains the *Quick Access* toolbar, the *Ribbon* toolbar, the *Navigator* area, the *Web Browser*, the *message area*, the *Status bar,* and the *Folder Tree.* A line of quick text appears next to the icon as you move the *mouse cursor* over different icons. You may resize the *Creo Parametric* drawing window by clicking and dragging at the edges of the window, or relocate the window by clicking and dragging at the window title area.

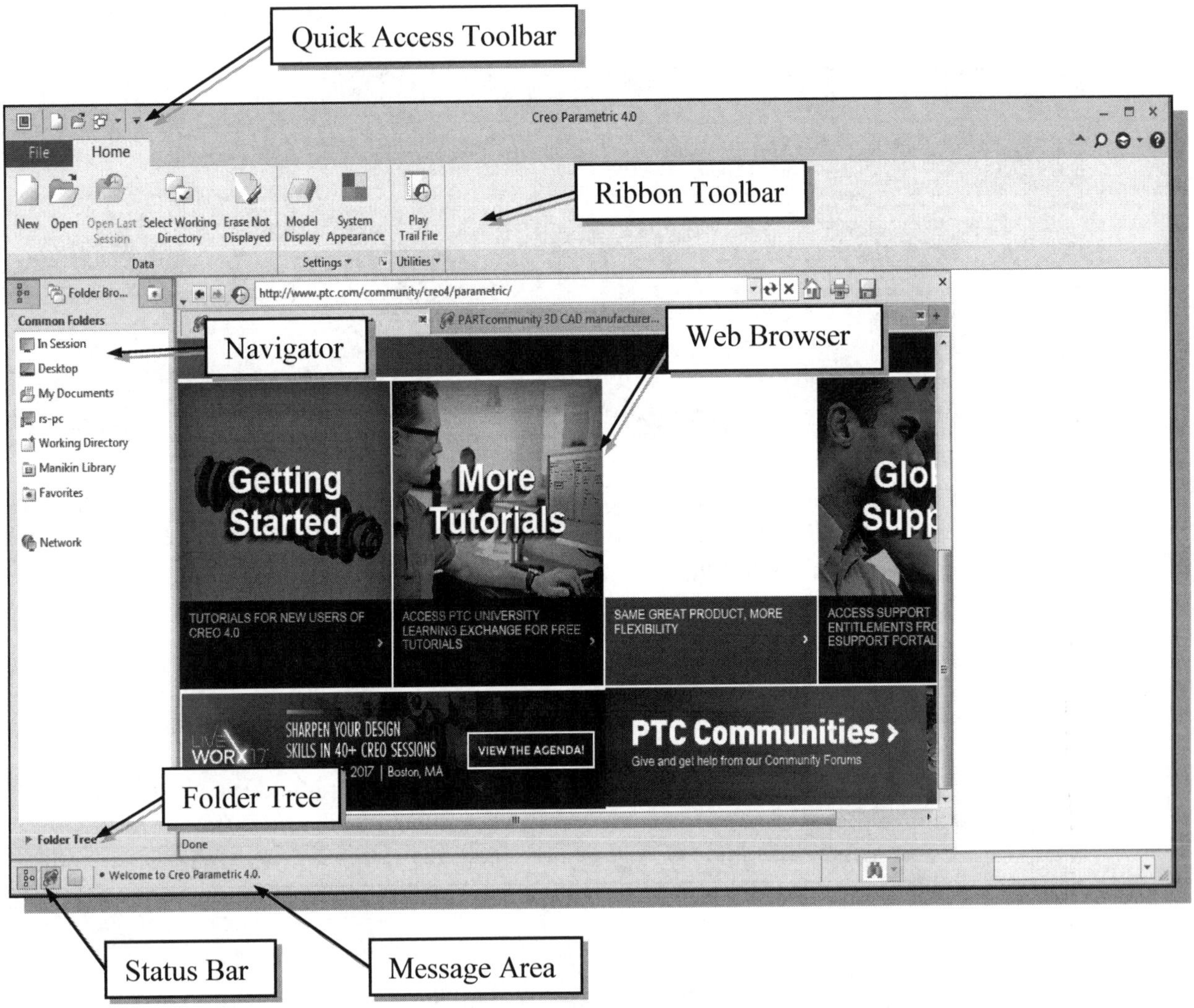

❖ *Creo Parametric* uses the **context sensitive menus** approach, which means additional selection menus and option windows will only be available when they are applicable to the current task. The appearance of the *Creo Parametric* main window and toolbars, as shown in the figure above, remains the same on the screen throughout the different phases of modeling. Note that the menu items and icons of the non-applicable options are grayed out, which means they are temporarily disabled.

- **Ribbon toolbar**
  The *Ribbon* toolbar at the top of the main window contains operations that you can use for all modes of the system. The *Ribbon* toolbar contains command buttons organized in a set of tabs. On each tab, the related buttons are grouped. You can minimize the ribbon to make more space available on your screen. You can customize the ribbon by adding, removing, or moving buttons.

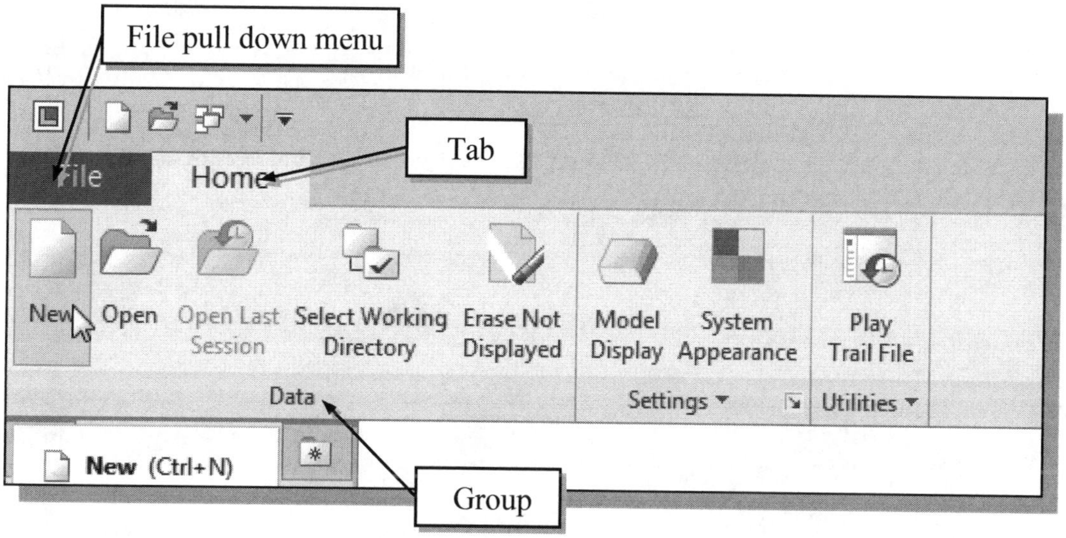

- **Quick Access toolbar**
  The *Standard* toolbar located at the top of the main window, below the pull-down menus, allows us quick access to frequently used commands. For example, the *view related* commands, such as Zoom, Pan, and Shaded Image, are tools to help manipulate the viewing of graphical objects. We can customize the toolbar by adding and removing sets of options or individual commands.

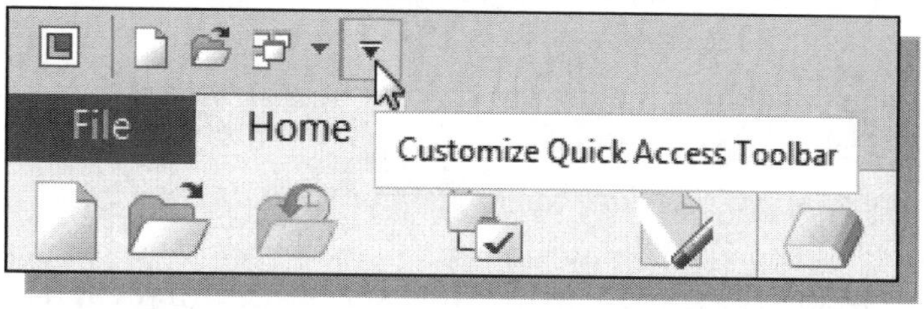

- **Message area**
  The *message* area provides both an interface to the commands and also a line of quick help on the activated command.

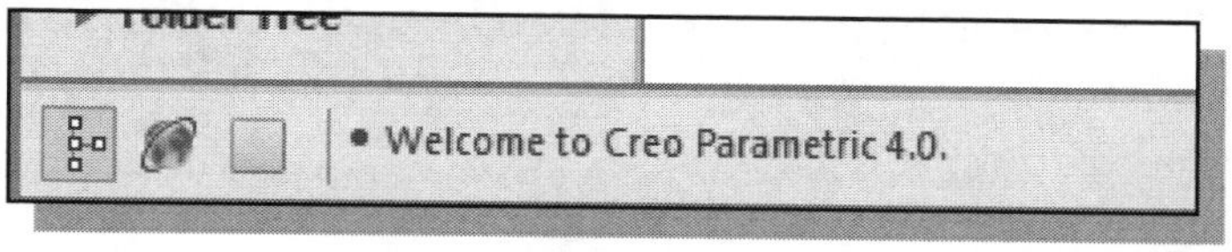

- **Graphics Display area**
  The *graphics display* area is the area where parts, assemblies, and drawings are displayed. The model's display is controlled by the *environment settings*. The display control toolbar is also located near the top of the graphics area.

- **Navigator**
  The tabs across the top of the *Navigator* provide direct access to four groups of options. The initial content of the *Navigator* is the *Folder* information where *Creo Parametric* is launched. The *Navigator* includes the *Model Tree*, *Layer Tree*, *Detail Tree*, *Folder browser*, and *Favorites*.

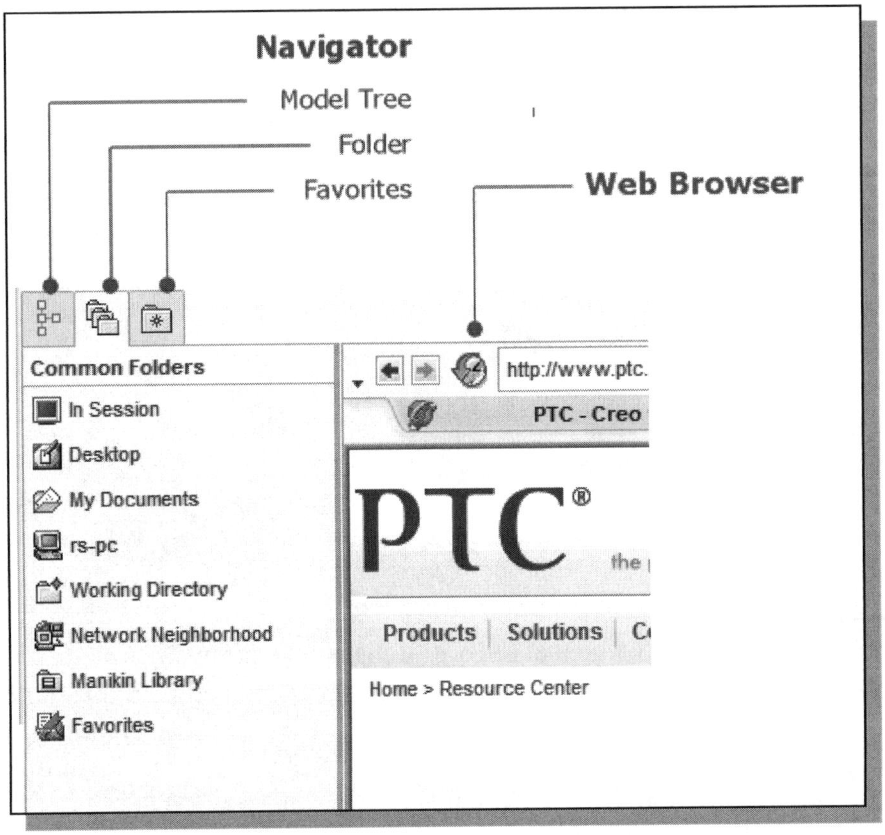

- **Web Browser**
  The *Creo Parametric Web Browser* allows the user to quickly access **model information** and **on-line documentation**. It also provides general web browsing capabilities. The initial content of the *Creo Parametric* web browser is the *Creo Parametric Web Tools*. The web browser will close automatically when you open a model; it can also be opened or closed by toggling the *Browser Sash* with the left-mouse-button.

- **Navigator Display Controls**
  The *Navigator* and the *Web Browser* can be opened or closed by toggling the *Display Controls* with the left-mouse-button.

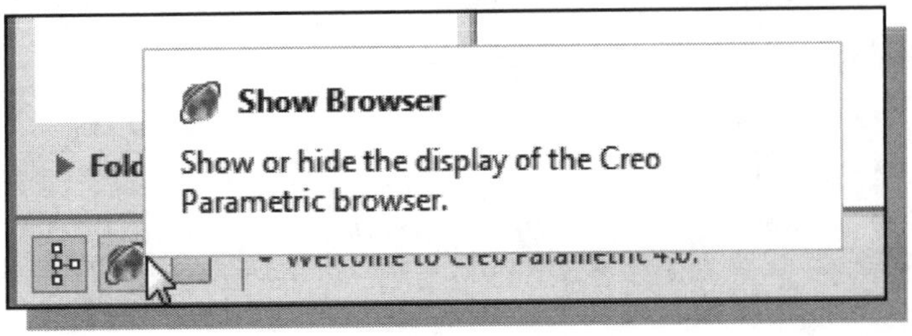

Basic Functions of Mouse Buttons

Creo Parametric utilizes the mouse buttons extensively. In learning *Creo Parametric's* interactive environment, it is important to understand the basic functions of the mouse buttons. A three-button mouse is highly recommended with *Creo Parametric* since the package uses all three buttons for various functions. If a two-button mouse is used instead, some of the operations, such as the *Dynamic Viewing* functions, will not be available.

- **Left mouse button**
  The left mouse button is used for most operations, such as selecting menus and icons, or picking graphic entities. One click of the button is used to select icons, menus and form entries, and to pick graphic items.

- **Middle mouse button**
  The middle button is often used to accept the default setting of a prompt or to end a process. It is also used as the shortcut to perform the *Dynamic Viewing* functions: Rotate, Zoom, and Pan.

- **Right mouse button**
  The right mouse button is used to query a selection and is also used in the *Creo Parametric Sketcher* to create tangent arcs.

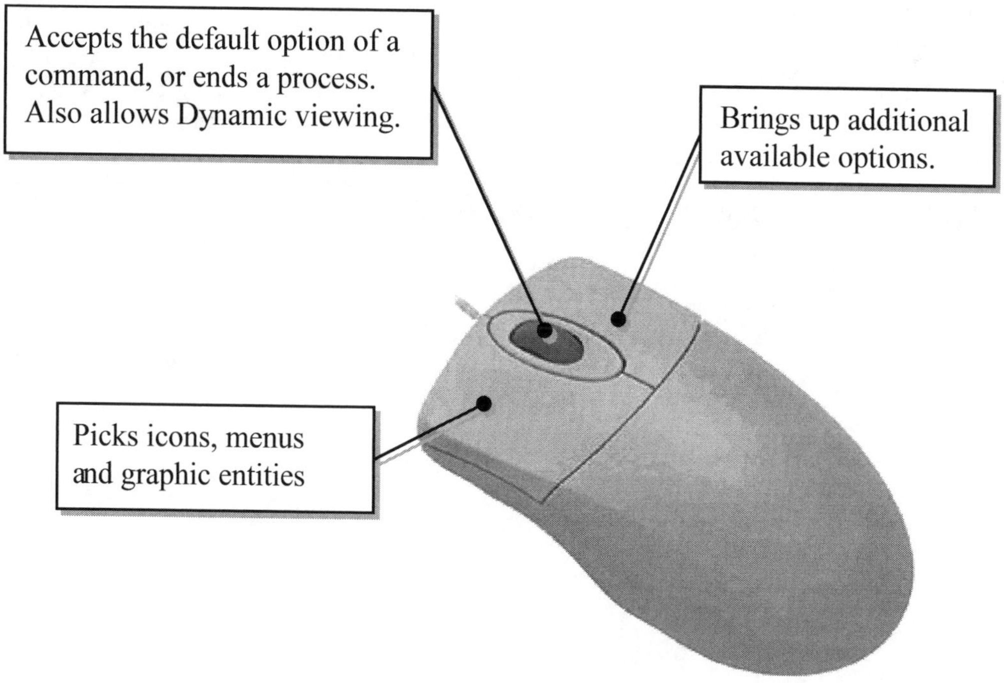
Accepts the default option of a command, or ends a process. Also allows Dynamic viewing.
Brings up additional available options.
Picks icons, menus and graphic entities

Model Tree Window and Feature Toolbars

The *Creo Parametric* ***Model Tree*** window and ***Ribbon*** toolbars are two of the most important *Creo Parametric* interfaces; they will appear on the screen when we begin a modeling session.

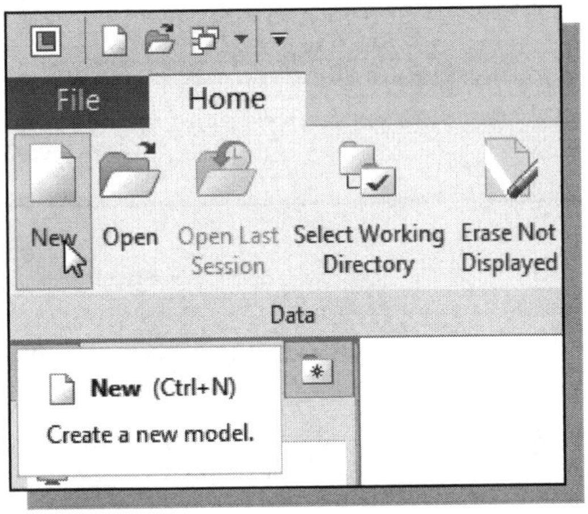

1. Click on the **New** icon, the first icon in the *Ribbon* toolbar as shown.

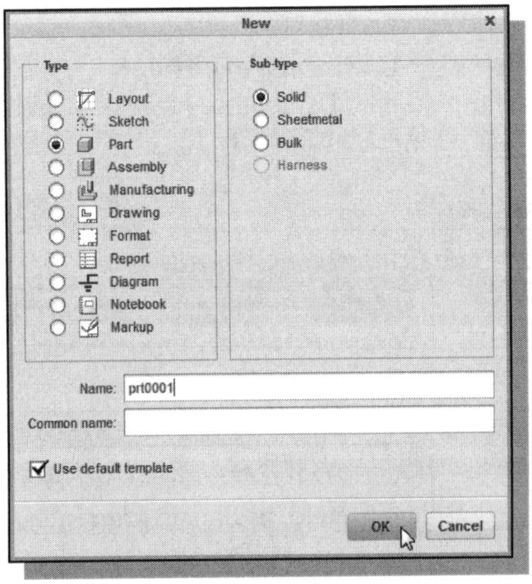

2. In the *New* dialog box, confirm the model's Type is set to **Part** (and the Sub-type is set to Solid).

3. Click on the **OK** button to accept the default settings and enter the *Creo Parametric Part Modeling* mode.

❖ The ***Navigator*** and the ***Ribbon*** toolbar now display commands related to the current part modeling task.

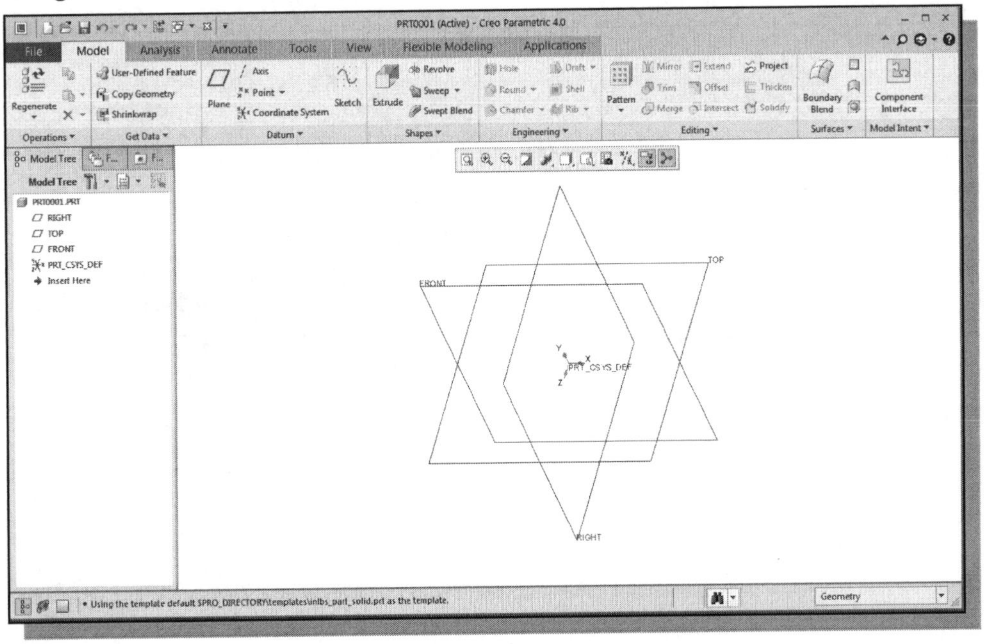

Online Help

Several types of online *Help* are available at any time during a *Creo Parametric* session. *Creo Parametric* provides many help functions, such as:

- ***Creo Parametric Help system***: The multiple Creo help system can be accessed through the File tab in the Ribbon of the *Creo Parametric* window. Note that many of the Help options use the *Creo Parametric Web Tools*, which allow us to access the *Creo Parametric* resource center available at the PTC website through the internet.

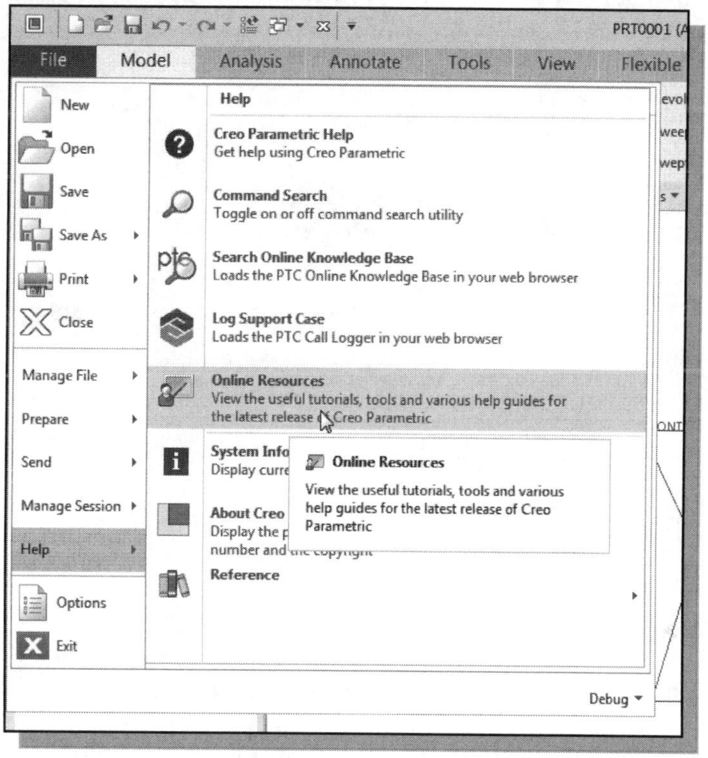

- ***Creo Parametric Help Center***: Click on the **Help** or the **Command Search** option in the *Ribbon* toolbar to access the *Creo Parametric* ***Help Center***.

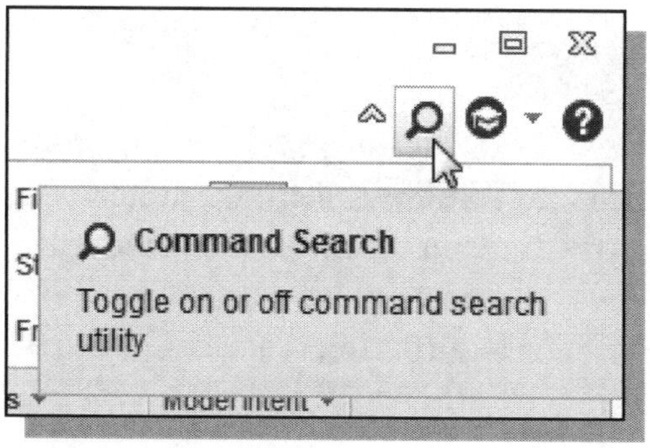

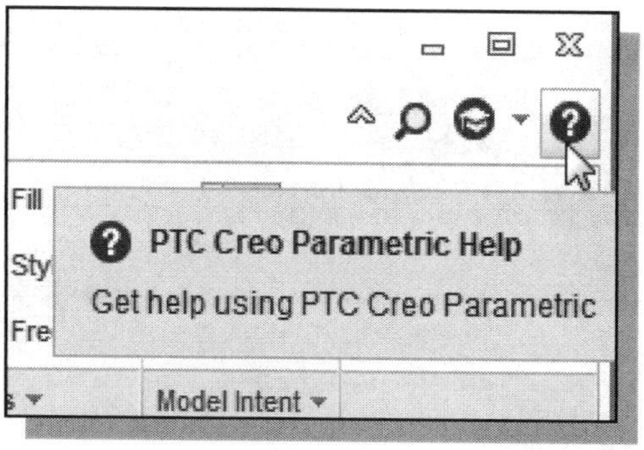

- ***Context Sensitive Help***: The context sensitive help can be used to quickly access the Help menu on specific tasks or icons displayed on the screen.

Leaving Creo Parametric

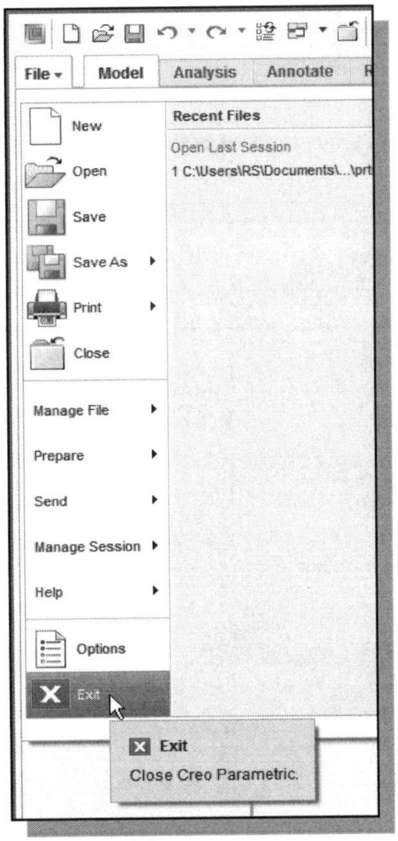

1. To leave *Creo Parametric*, use the left-mouse-button and click on **File** tab in the *Creo Parametric Ribbon* toolbar, then choose **Exit** from the pull-down menu.

2. In the confirmation dialog box, click on the **Yes** button to exit *Creo Parametric*.

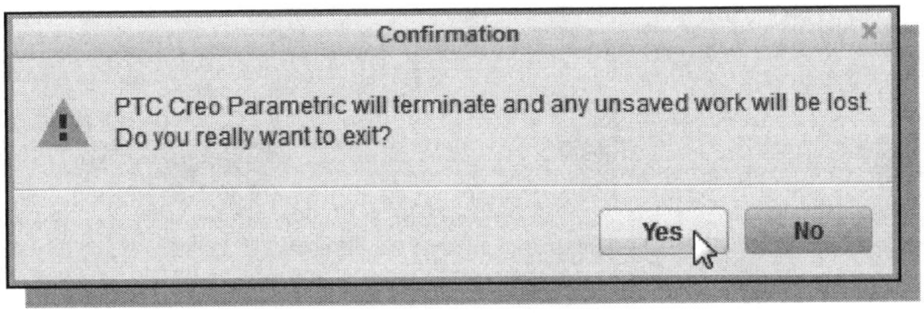

Creating a CAD files folder

It is a good practice to create a separate folder to store your CAD files. You should not save your CAD files in the same folder where the *Creo Parametric* application is located. It is much easier to organize and back up your project files if they are in a separate folder. Making folders within this folder for different types of projects will help you organize your CAD files even further. When creating CAD files in *Creo Parametric*, it is strongly recommended that you *save* your CAD files on the hard drive.

To create a new folder with most *Windows* systems:

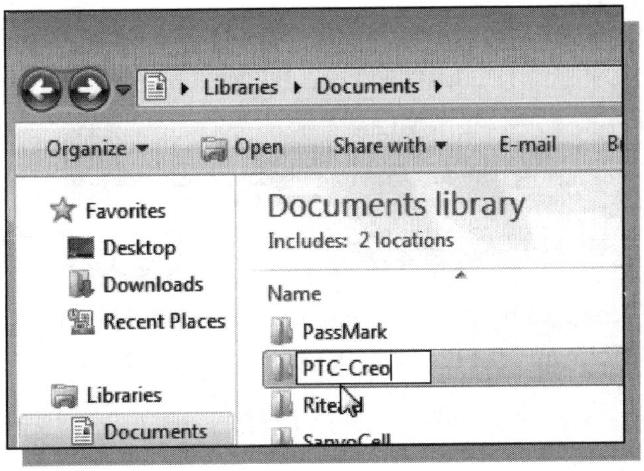

1. In *My Computer*, or start the **Windows Explorer** under the *Start* menu, open the **Documents** folder.

2. **Right-click** once and select **New-Folder**.

3. Enter the new name of the folder, and press **ENTER**.

Chapter 1
The Direct Stiffness Method

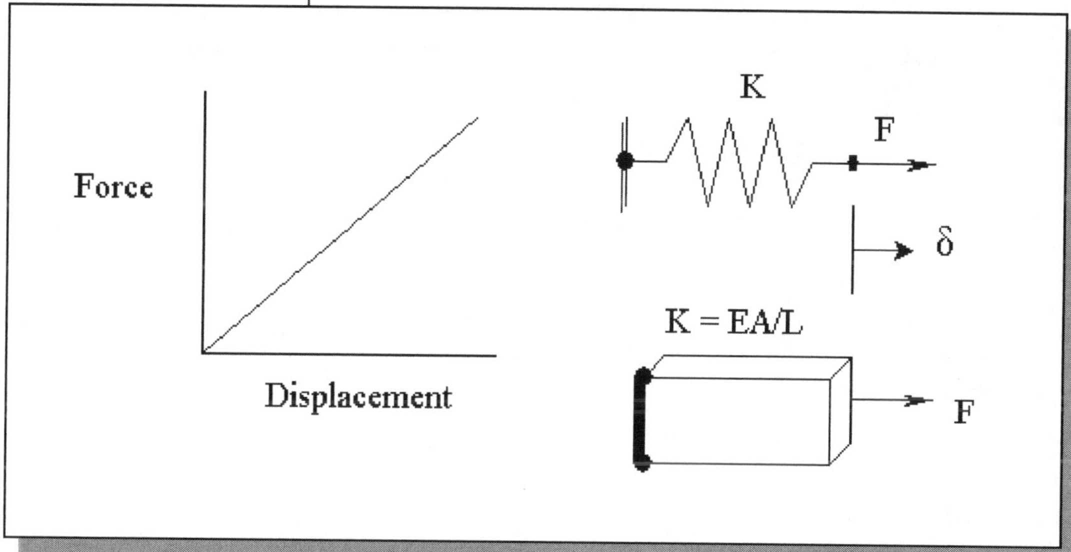

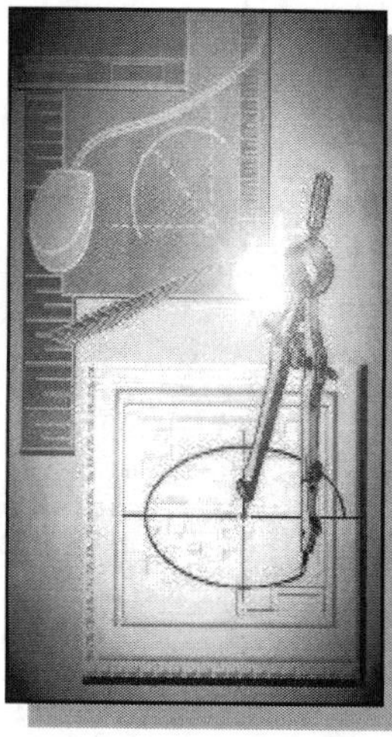

Learning Objectives

- ♦ **Understand system equations for truss elements.**
- ♦ **Understand the setup of a Stiffness Matrix.**
- ♦ **Apply the Direct Stiffness Method.**
- ♦ **Create an Extruded solid model using Creo Parametric.**
- ♦ **Use the Display Viewing commands.**
- ♦ **Use the Creo Parametric 2D Sketcher.**
- ♦ **Create Cutout features.**
- ♦ **Use the Basic Editing commands.**

Introduction

The direct stiffness method, developed in the 1940s, is generally considered the origin of finite element analysis. The **direct stiffness method** is used mostly for *Linear Static analysis*. Linear Static analysis is appropriate if deflections are small and vary only slowly. Linear Static analysis omits time as a variable. It also excludes plastic action and deflections that change the way loads are applied. The direct stiffness method for Linear Static analysis follows the *laws of Statics* and the *laws of Strength of Materials*.

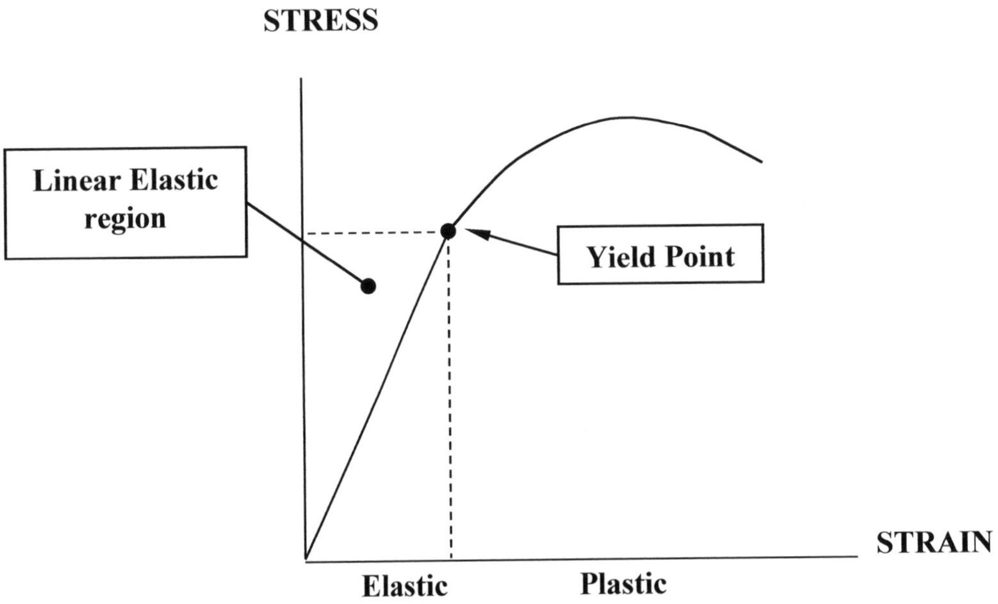

Stress-Strain diagram of typical ductile material

This chapter introduces the fundamentals of finite element analysis by illustrating an analysis of a one-dimensional truss system using the direct stiffness method. The main objective of this chapter is to present the classical procedure common to the implementation of structural analysis. The direct stiffness method utilizes *matrices* and *matrix algebra* to organize and solve the governing system equations. Matrices, which are ordered arrays of numbers that are subjected to specific rules, can be used to assist the solution process in a compact and elegant manner. Of course, only a limited discussion of the direct stiffness method is given here, but we hope that the focused practical treatment will provide a strong basis for understanding the procedure to perform finite element analysis with *Creo Simulate*.

The later sections of this chapter demonstrate the procedure to create a solid model using *Creo Parametric*. The step-by-step tutorial introduces the *Creo Parametric* user interface and serves as a preview to some of the basic modeling techniques demonstrated in the later chapters.

One-dimensional Truss Element

The simplest type of engineering structure is the truss structure. A truss member is a slender (the length is much larger than the cross-section dimensions) **two-force** member. Members are joined by pins and only have the capability to support tensile or compressive loads axially along the length. Consider a uniform slender prismatic bar (shown below) of length L, cross-sectional area A, and elastic modulus E. The ends of the bar are identified as nodes. The nodes are the points of attachment to other elements. The nodes are also the points for which displacements are calculated. The truss element is a two-force member element; forces are applied to the nodes only, and the displacements of all nodes are confined to the axes of elements.

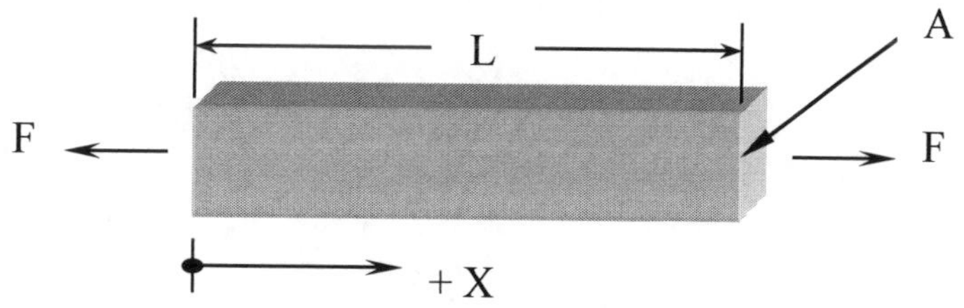

In this initial discussion of the truss element, we will consider the motion of the element to be restricted to the horizontal axis (one-dimensional). Forces are applied along the X axis and displacements of all nodes will be along the X axis.

For the analysis, we will establish the following sign conventions:

1. Forces and displacements are defined as positive when they are acting in the positive X direction as shown in the above figure.

2. The position of a node in the undeformed condition is the finite element position for that node.

If equal and opposite forces of magnitude F are applied to the end nodes, from the elementary strength of materials, the member will undergo a change in length according to the equation:

$$\delta = \frac{FL}{EA}$$

This equation can also be written as $\delta = F/K$, which is similar to *Hooke's Law* used in a linear spring. In a linear spring, the symbol K is called the **spring constant** or **stiffness** of the spring. For a truss element, we can see that an equivalent spring element can be used to simplify the representation of the model, where the spring constant is calculated as **K=EA/L**.

Force-Displacement Curve of a Linear Spring

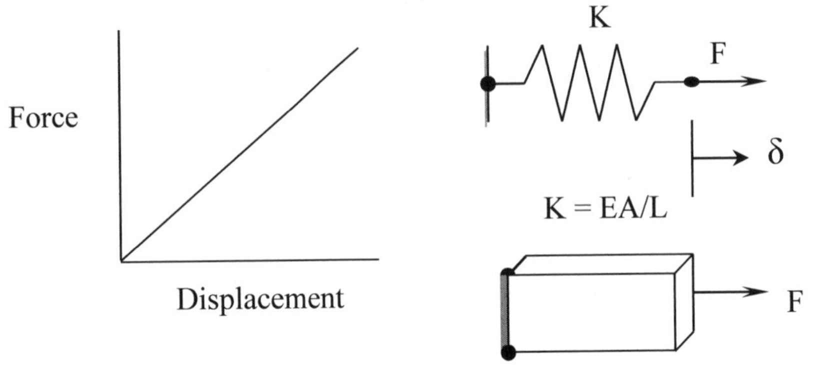

We will use the general equations of a single one-dimensional truss element to illustrate the formulation of the stiffness matrix method:

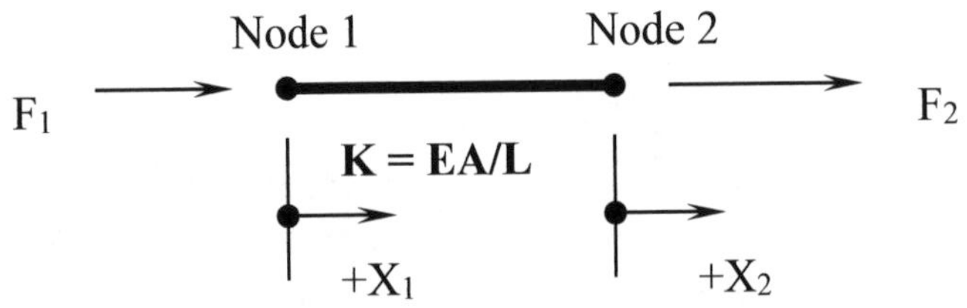

By using the *Relative Motion Analysis* method, we can derive the general expressions of the applied forces (F_1 and F_2) in terms of the displacements of the nodes (X_1 and X_2) and the stiffness constant (K).

1. Let $X_1 = 0$,

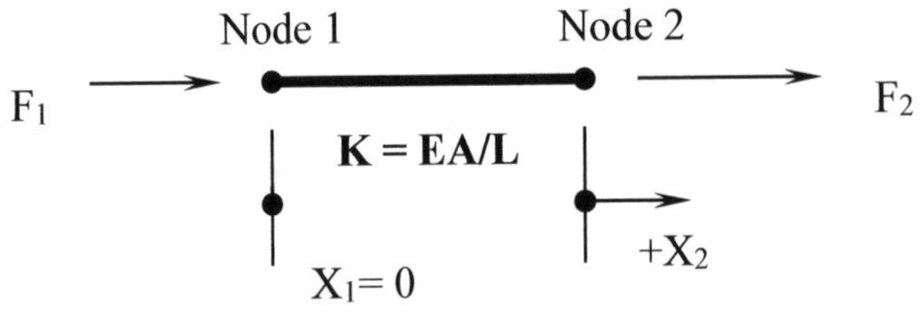

Based on Hooke's law and equilibrium equation:

$$\begin{cases} F_2 = K\,X_2 \\ F_1 = -\,F_2 = -\,K\,X_2 \end{cases}$$

2. Let $X_2 = 0$,

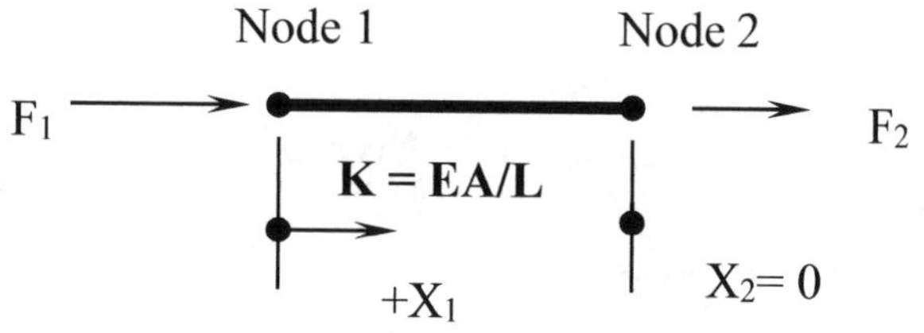

Based on *Hooke's Law* and equilibrium:

$$\begin{cases} F_1 = K\,X_1 \\ F_2 = -\,F_1 = -\,K\,X_1 \end{cases}$$

Using the *Method of Superposition,* the two sets of equations can be combined:

$$\begin{aligned} F_1 &= \;\;K\,X_1 - K\,X_2 \\ F_2 &= -\,K\,X_1 + K\,X_2 \end{aligned}$$

The two equations can be put into matrix form as follows:

$$\begin{Bmatrix} F_1 \\ F_2 \end{Bmatrix} = \begin{bmatrix} +K & -K \\ -K & +K \end{bmatrix} \begin{Bmatrix} X_1 \\ X_2 \end{Bmatrix}$$

This is the general force-displacement relation for a two-force member element, and the equations can be applied to all members in an assemblage of elements. The following example illustrates a system with three elements.

Example 1.1:

Consider an assemblage of three of these two-force member elements. (Motion is restricted to one-dimension, along the X axis.)

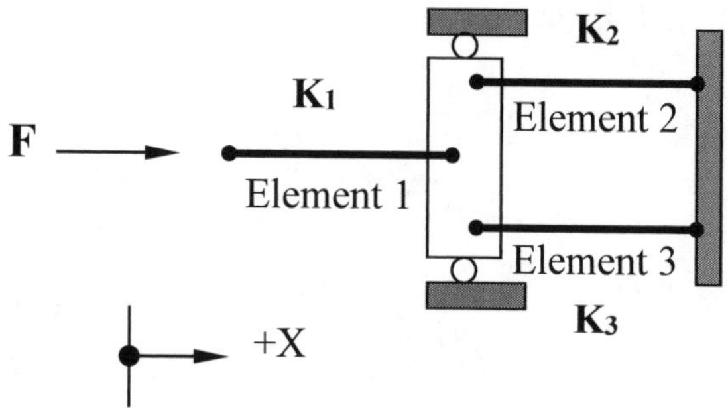

The assemblage consists of three elements and four nodes. The *Free Body Diagram* of the system with node numbers and element numbers labeled:

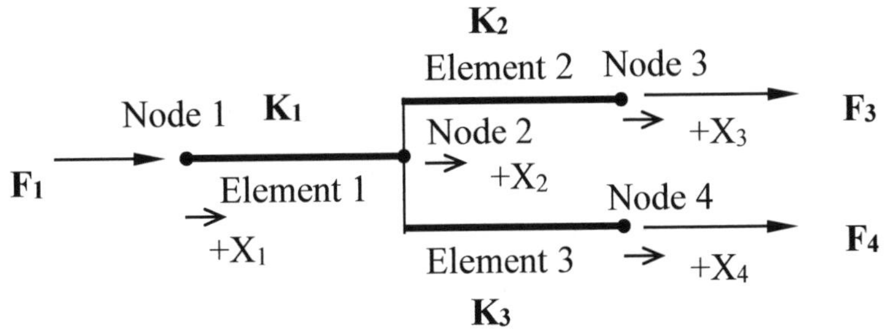

Consider now the application of the general force-displacement relation equations to the assemblage of the elements.

Element 1:

$$\begin{Bmatrix} F_1 \\ F_{21} \end{Bmatrix} = \begin{bmatrix} +K_1 & -K_1 \\ -K_1 & +K_1 \end{bmatrix} \begin{Bmatrix} X_1 \\ X_2 \end{Bmatrix}$$

Element 2:

$$\begin{Bmatrix} F_{22} \\ F_3 \end{Bmatrix} = \begin{bmatrix} +K_2 & -K_2 \\ -K_2 & +K_2 \end{bmatrix} \begin{Bmatrix} X_2 \\ X_3 \end{Bmatrix}$$

Element 3:

$$\begin{Bmatrix} F_{23} \\ F_4 \end{Bmatrix} = \begin{bmatrix} +K_3 & -K_3 \\ -K_3 & +K_3 \end{bmatrix} \begin{Bmatrix} X_2 \\ X_4 \end{Bmatrix}$$

Expanding the general force-displacement relation equations into an *Overall Global Matrix* (containing all nodal displacements):

Element 1:

$$\begin{Bmatrix} F_1 \\ F_{21} \\ 0 \\ 0 \end{Bmatrix} = \begin{bmatrix} +K_1 & -K_1 & 0 & 0 \\ -K_1 & +K_1 & 0 & 0 \\ 0 & 0 & 0 & 0 \\ 0 & 0 & 0 & 0 \end{bmatrix} \begin{Bmatrix} X_1 \\ X_2 \\ X_3 \\ X_4 \end{Bmatrix}$$

Element 2:

$$\begin{Bmatrix} 0 \\ F_{22} \\ F_3 \\ 0 \end{Bmatrix} = \begin{bmatrix} 0 & 0 & 0 & 0 \\ 0 & +K_2 & -K_2 & 0 \\ 0 & -K_2 & +K_2 & 0 \\ 0 & 0 & 0 & 0 \end{bmatrix} \begin{Bmatrix} X_1 \\ X_2 \\ X_3 \\ X_4 \end{Bmatrix}$$

Element 3:

$$\begin{Bmatrix} 0 \\ F_{23} \\ 0 \\ F_4 \end{Bmatrix} = \begin{bmatrix} 0 & 0 & 0 & 0 \\ 0 & +K_3 & 0 & -K_3 \\ 0 & 0 & 0 & 0 \\ 0 & -K_3 & 0 & +K_3 \end{bmatrix} \begin{Bmatrix} X_1 \\ X_2 \\ X_3 \\ X_4 \end{Bmatrix}$$

Summing the three sets of general equations: (Note $F_2=F_{21}+F_{22}+F_{32}$)

$$\begin{Bmatrix} F_1 \\ F_2 \\ F_3 \\ F_4 \end{Bmatrix} = \begin{bmatrix} K_1 & -K_1 & 0 & 0 \\ -K_1 & (K_1+K_2+K_3) & -K_2 & -K_3 \\ 0 & -K_2 & K_2 & 0 \\ 0 & -K_3 & 0 & +K_3 \end{bmatrix} \begin{Bmatrix} X_1 \\ X_2 \\ X_3 \\ X_4 \end{Bmatrix}$$

Overall Global Stiffness Matrix

Once the *Overall Global Stiffness Matrix* is developed for the structure, the next step is to substitute boundary conditions and solve for the unknown displacements. At every node in the structure, either the externally applied load or the nodal displacement is needed as a boundary condition. We will demonstrate this procedure with the following example.

Example 1.2:

Given:

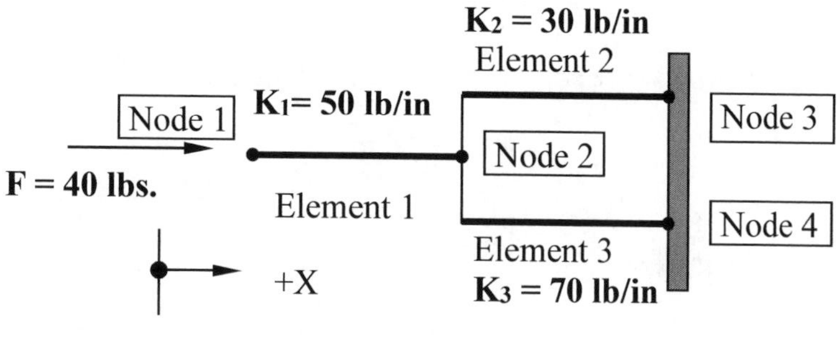

Find: Nodal displacements and reaction forces.

Solution:

From example 1.1, the overall global force-displacement equation set:

$$\begin{Bmatrix} F_1 \\ F_2 \\ F_3 \\ F_4 \end{Bmatrix} = \begin{bmatrix} 50 & -50 & 0 & 0 \\ -50 & (50+30+70) & -30 & -70 \\ 0 & -30 & 30 & 0 \\ 0 & -70 & 0 & 70 \end{bmatrix} \begin{Bmatrix} X_1 \\ X_2 \\ X_3 \\ X_4 \end{Bmatrix}$$

Next, apply the known boundary conditions to the system: the right-ends of element 2 and element 3 are attached to the vertical wall; therefore, these two nodal displacements (X_3 and X_4) are zero.

$$\begin{Bmatrix} F_1 \\ F_2 \\ F_3 \\ F_4 \end{Bmatrix} = \begin{bmatrix} 50 & -50 & 0 & 0 \\ -50 & (50+30+70) & -30 & -70 \\ 0 & -30 & 30 & 0 \\ 0 & -70 & 0 & 70 \end{bmatrix} \begin{Bmatrix} X_1 \\ X_2 \\ 0 \\ 0 \end{Bmatrix}$$

The two displacements we need to solve the system are X_1 and X_1. Remove any unnecessary columns in the matrix:

$$\begin{Bmatrix} F_1 \\ F_2 \\ F_3 \\ F_4 \end{Bmatrix} = \begin{bmatrix} 50 & -50 \\ -50 & 150 \\ 0 & -30 \\ 0 & -70 \end{bmatrix} \begin{Bmatrix} X_1 \\ X_2 \end{Bmatrix}$$

Next, include the applied loads into the equations. The external load at *Node 1* is 40 lbs. and there is no external load at *Node 2.*

$$\begin{Bmatrix} 40 \\ 0 \\ F_3 \\ F_4 \end{Bmatrix} = \begin{bmatrix} 50 & -50 \\ -50 & 150 \\ 0 & -30 \\ 0 & -70 \end{bmatrix} \begin{Bmatrix} X_1 \\ X_2 \end{Bmatrix}$$

The Matrix represents the following four simultaneous system equations:

$$\begin{aligned} 40 &= 50\,X_1 - 50\,X_2 \\ 0 &= -\,50\,X_1 + 150\,X_2 \\ F_3 &= 0\,X_1 - 30\,X_2 \\ F_4 &= 0\,X_1 - 70\,X_2 \end{aligned}$$

From the first two equations, we can solve for X_1 and X_2:

$$X_1 = 1.2 \text{ in.}$$
$$X_2 = 0.4 \text{ in.}$$

Substituting these known values into the last two equations, we can now solve for F_3 and F_4:

$$F_3 = 0\,X_1 - 30\,X_2 = -30 \times 0.4 = 12 \text{ lbs.}$$
$$F_4 = 0\,X_1 - 70\,X_2 = -70 \times 0.4 = 28 \text{ lbs.}$$

From the above analysis, we can now reconstruct the *Free Body Diagram* (*FBD*) of the system:

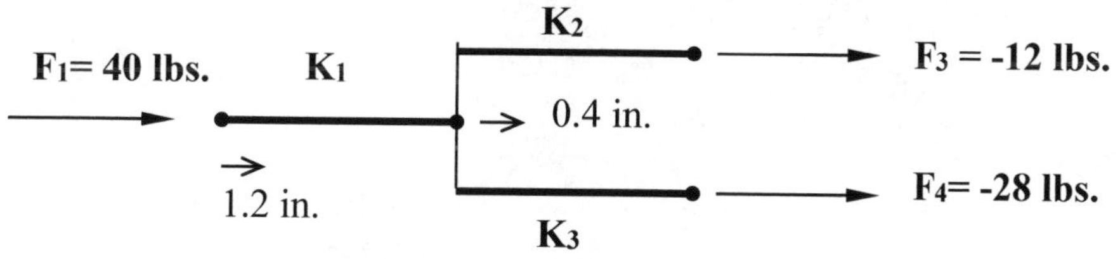

- The above sections illustrated the fundamental operation of the direct stiffness method, the classical finite element analysis procedure. As can be seen, the formulation of the global force-displacement relation equations is based on the general force-displacement equations of a single one-dimensional truss element. The two-force-member element (truss element) is the simplest type of element used in FEA. The procedure to formulate and solve the global force-displacement equations is straightforward, but somewhat tedious. In real-life applications, the use of a truss element in one-dimensional space is rare and very limited. In the next chapter, we will expand the procedure to solve two-dimensional truss frameworks.

The following sections illustrate the procedure to create a solid model using *Creo Parametric*. The step-by-step tutorial introduces the basic *Creo Parametric* user interface and the tutorial serves as a preview to some of the basic modeling techniques demonstrated in the later chapters.

Basic Solid Modeling using Creo Parametric

One of the methods to create solid models in *Creo Parametric* is to create a two-dimensional shape and then *extrude* the two-dimensional shape to define a volume in the third dimension. This is an effective way to construct three-dimensional solid models since many designs are in fact the same shape in one direction. This method also conforms to the design process that helps the designer with conceptual design along with the capability to capture the design intent. *Creo Parametric* provides many powerful modeling tools and there are many different approaches available to accomplish modeling tasks.

The Adjuster Design

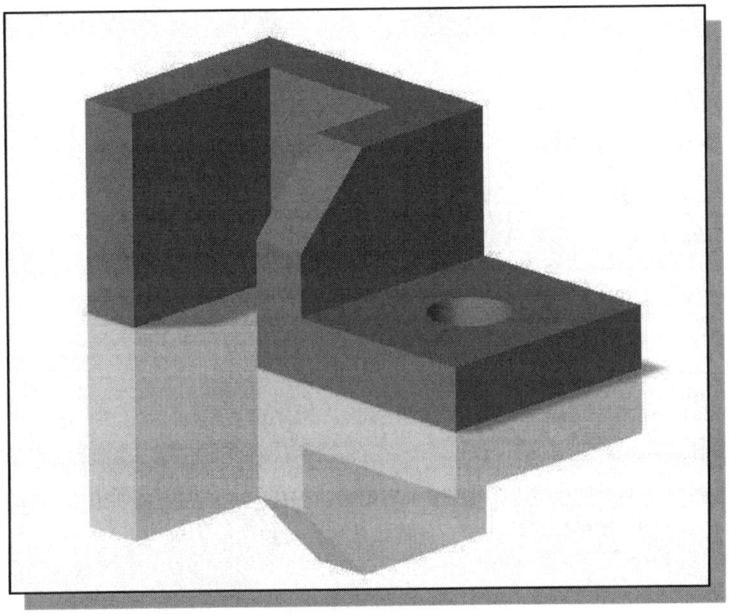

Starting Creo Parametric

How to start *Creo Parametric* depends on the type of workstation and the particular software configuration you are using. With most *Windows* and *UNIX* systems, you may select **Creo Parametric** on the *Start* menu or select the **Creo Parametric** icon on the desktop. Consult your instructor or technical support personnel if you have difficulty starting the software.

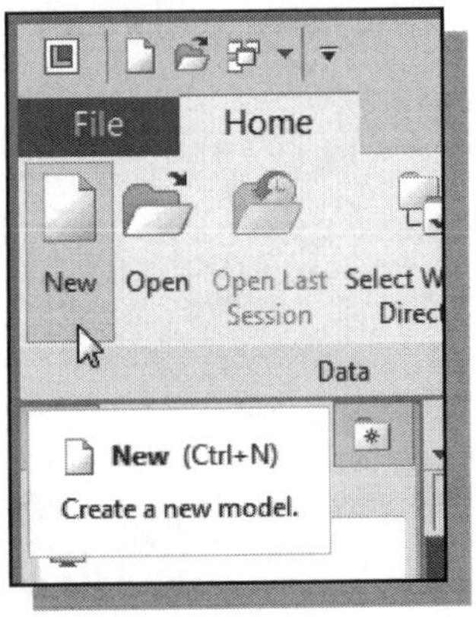

1. Select the **Creo Parametric** option on the *Start* menu or select the **Creo Parametric** icon on the desktop to start *Creo Parametric*. The *Creo Parametric* main window will appear on the screen.

2. Click on the **New** icon, located in the *Ribbon toolbar* as shown.

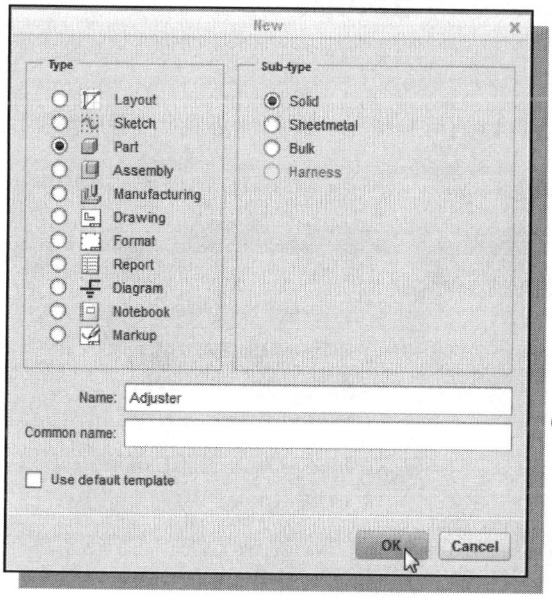

3. In the *New* dialog box, confirm the model's **Type** is set to **Part** (**Solid** Sub-type).

4. Enter ***Adjuster*** as the part **Name** as shown in the figure.

5. Turn *off* the **Use default template** option.

6. Click on the **OK** button to accept the settings.

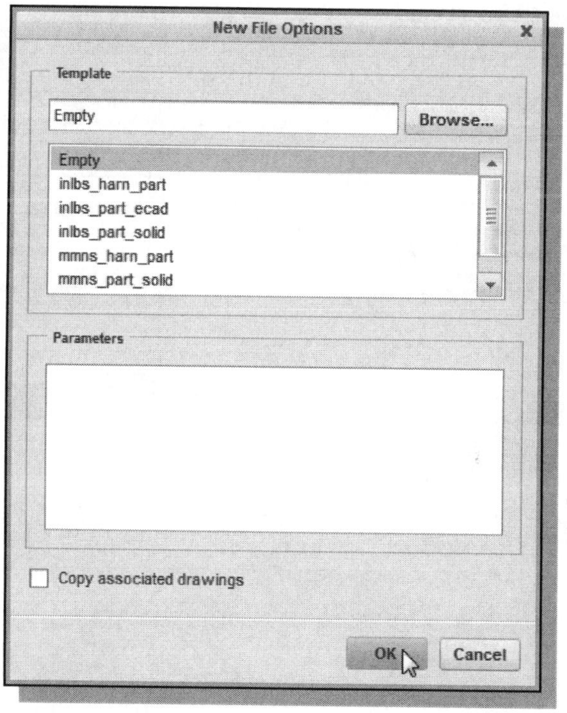

7. In the *New File Options* dialog box, select **EMPTY** in the option list to not use any template file.

8. Click on the **OK** button to accept the settings and enter the *Creo Parametric Part Modeling* mode.

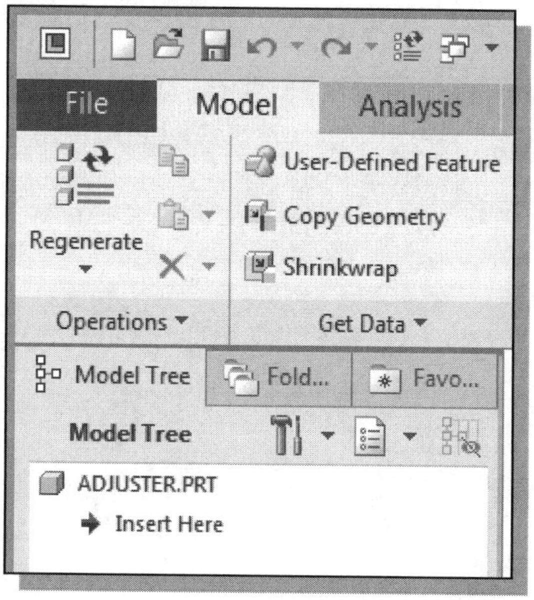

❖ Note that the part name, *Adjuster.prt*, appears in the *Navigator Model Tree* window and the title bar area of the main window.

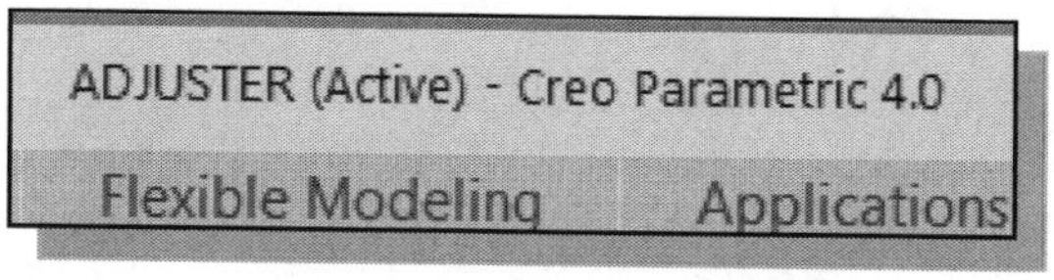

Step 1: Units and Basic Datum Geometry Setups

♦ **Units Setup**

When starting a new model, the first thing we should do is to choose the set of units we want to use.

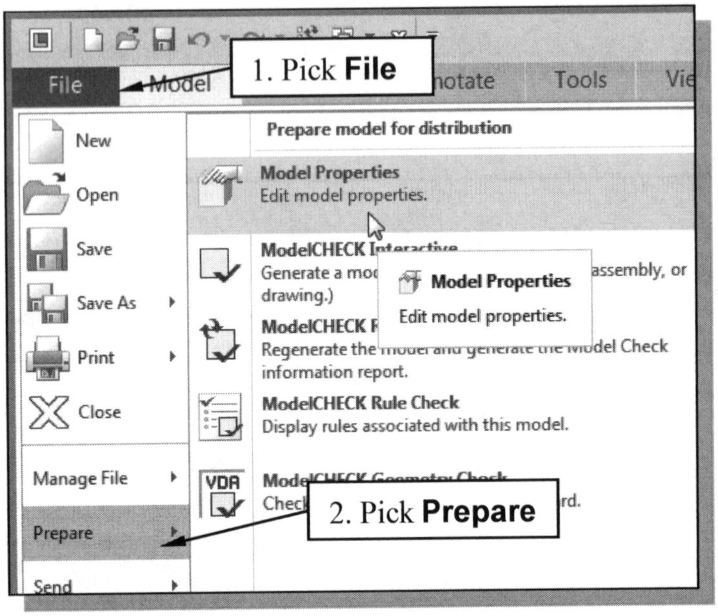

1. Use the left-mouse-button and select **File** in the pull-down menu area.

2. Use the left-mouse-button and select **Prepare** in the **pull-down list** as shown.

3. Select **Model Properties** in the ***expanded list*** as shown.

➢ Note that the *Creo Parametric* menu system is context-sensitive, which means that the menu items and icons of the non-applicable options are grayed out (temporarily disabled).

4. Select the **Change** option that is to the right of the **Units** option in the ***Model Properties*** window.

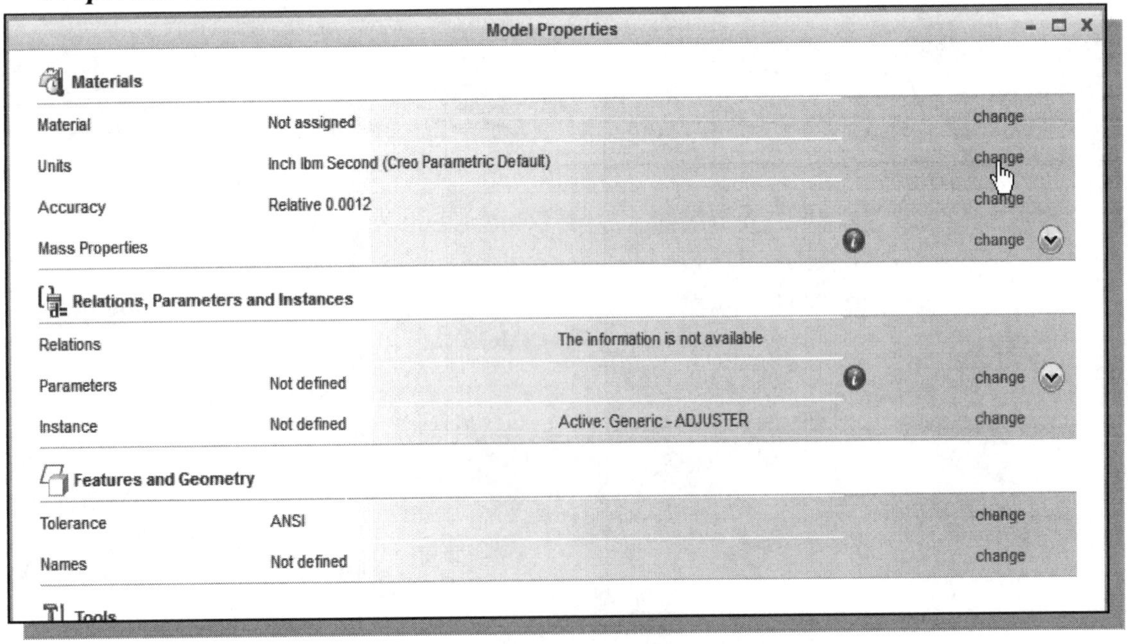

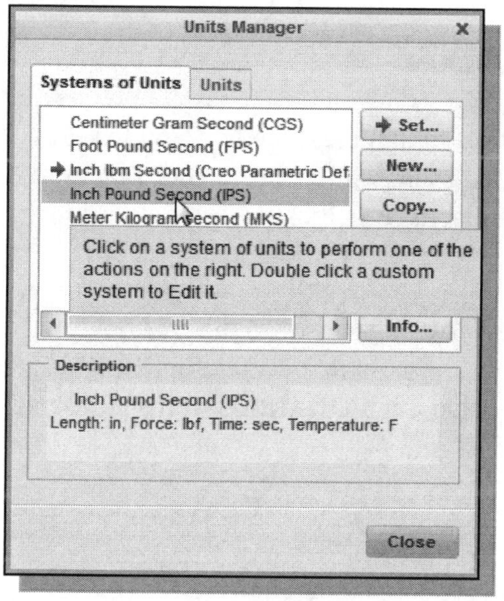

5. In the **Units Manager – Systems of Units** form, the *Creo Parametric* default setting **Inch lbm Second** is displayed. The set of units is stored with the model file when you save. Pick **Inch Pound Second (IPS)** by clicking in the list window as shown.

6. Click on the **Set** button to accept the selection. Notice the arrow in the Units list now points toward the **Inch Pound Second (IPS)** units set.

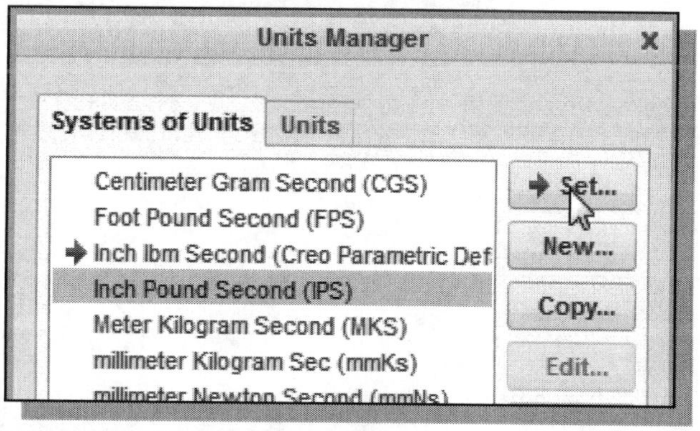

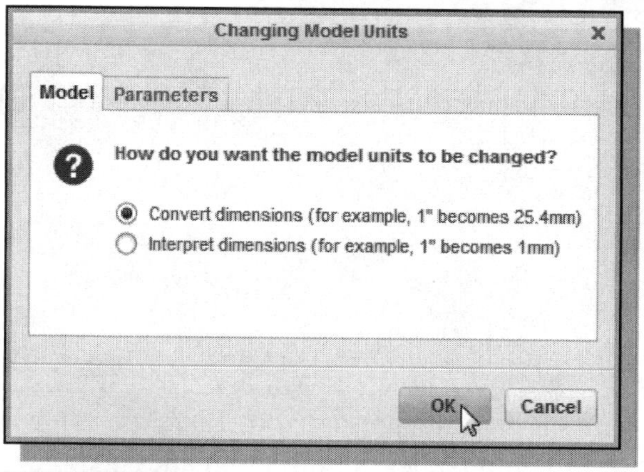

7. In the *Changing Model Units* dialog box, click on the **OK** button to accept the default option to change the units.

➢ Note that *Creo Parametric* allows us to change model units even after the model has been constructed; we can change the units by (1) *Convert dimensions* or (2) *Interpret dimensions*.

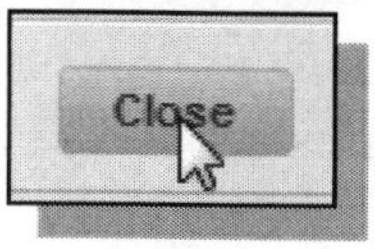

8. Click on the **Close** button to exit the *Units Manager* dialog box.

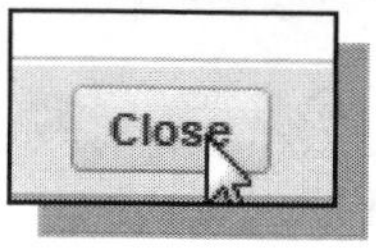

9. Pick **Close** to exit the ***Model Properties*** window.

♦ Adding the First Part Features – Datum Planes

❖ *Creo Parametric* provides many powerful tools for model creation. In doing feature-based parametric modeling, it is a good practice to establish three reference planes to locate the part in space. The reference planes can be used as location references in feature constructions.

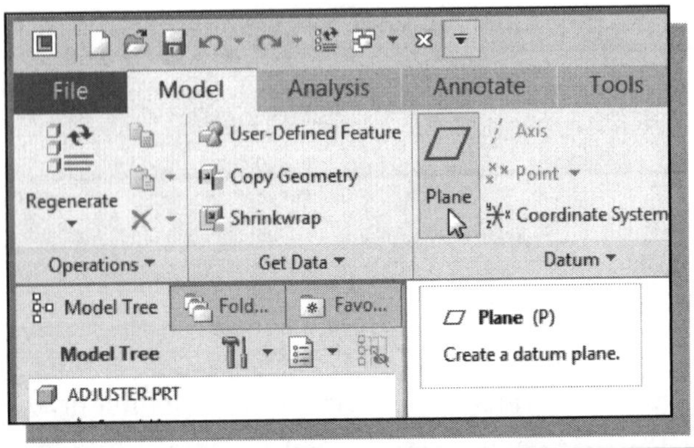

➢ Move the cursor to the ***Datum*** toolbar on the *Ribbon* toolbar and click on the **Datum Plane** tool icon as shown.

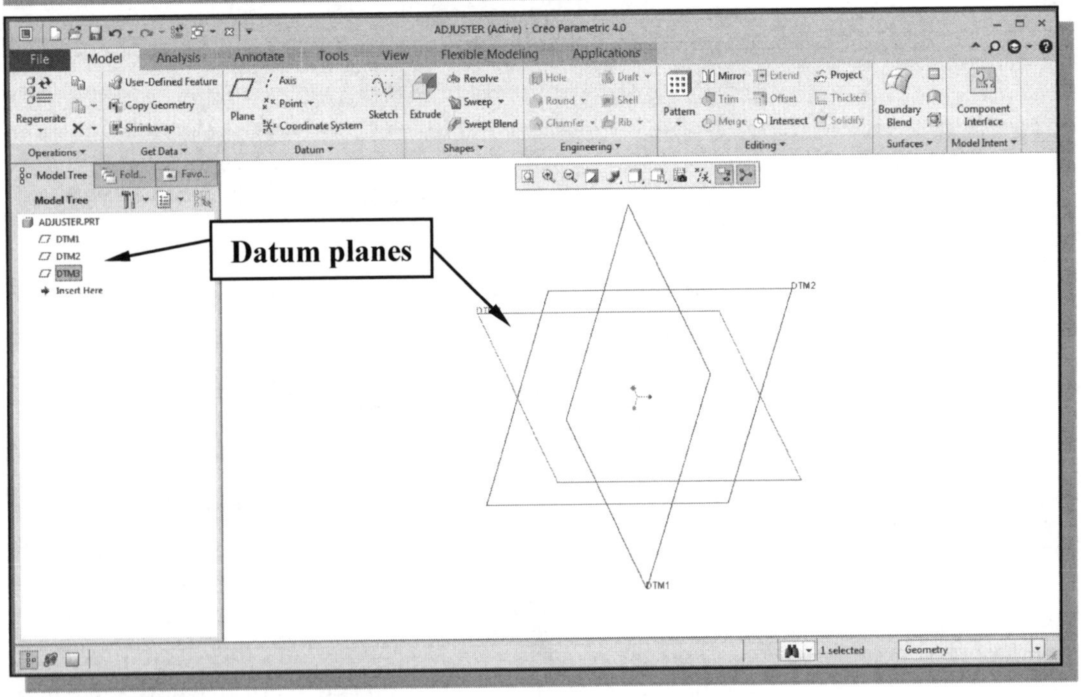

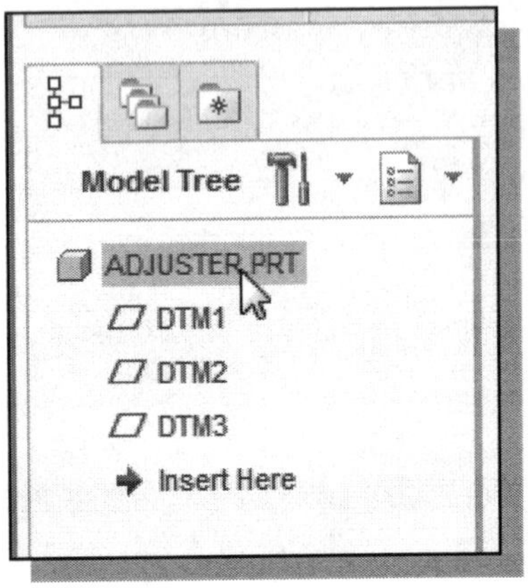

❖ In the *Navigator Model Tree* window and the display area, three datum planes represented by three rectangles are displayed. Datum planes are infinite planes and they are perpendicular to each other. We can consider these planes as XY, YZ, and ZX planes of a Cartesian coordinate system.

➢ Click the model name, ***Adjuster.prt***, in the *Navigator* window to **deselect** the last created feature.

♦ Switching On/Off the Plane Tag Display

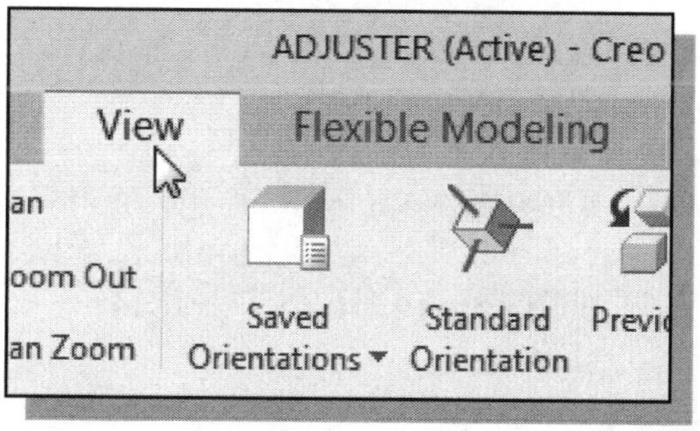

1. Click on the **View** tab in the Ribbon toolbar to show the view related commands in Creo.

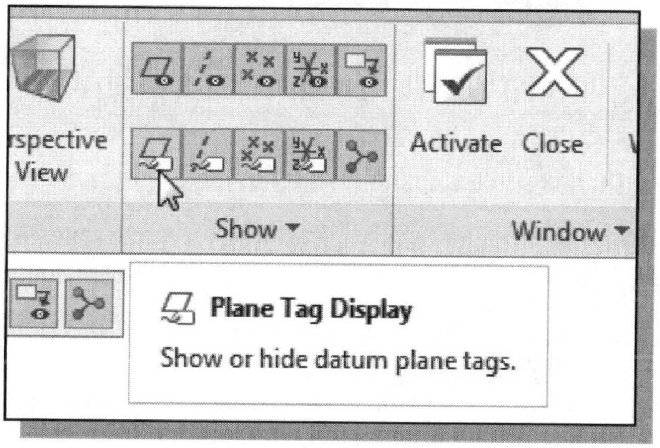

2. Click on the **Plane Tag Display** icon to toggle on/off the display of the plane tag.

♦ Note the other options available to display the **Axis Tag** and/or **Point Tag**.

♦ On your own, experiment with turning on/off the Plane Tag; set the datum planes to display with the associated names as shown before proceeding to the next section.

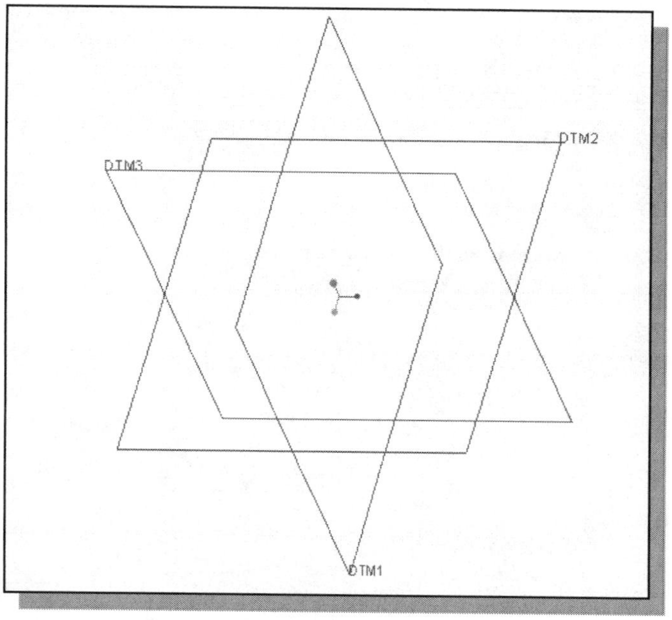

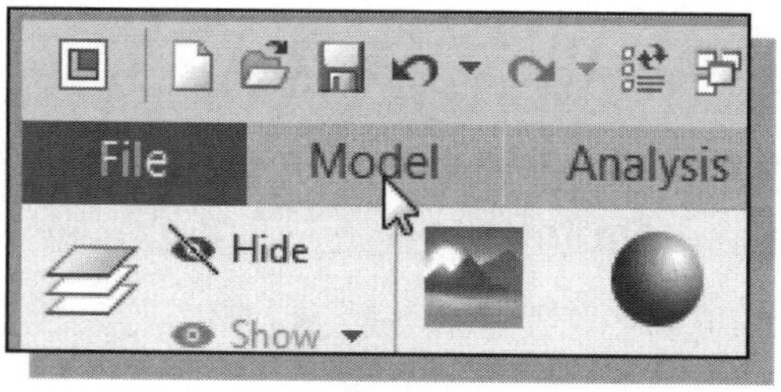

3. Click on the **Model** tab in the *Ribbon toolbar* to return to the Model toolbar as shown.

Step 2: Determine/Set up the Base Solid Feature

For the *Adjuster* design, we will create an extruded solid as the base feature.

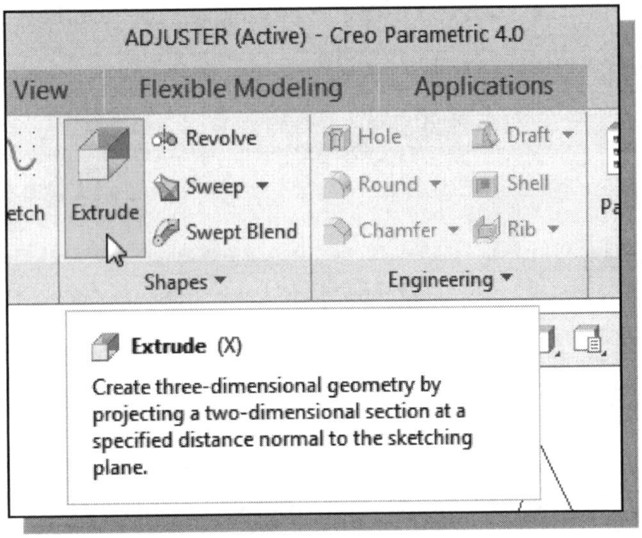

1. In the ***Shapes*** toolbar (the fourth group in the *Ribbon* toolbar), click on the **Extrude** tool icon as shown.

- The ***Feature Option Dashboard***, which contains applicable construction options, is displayed in the *Ribbon* toolbar of the *Creo Parametric* main window.

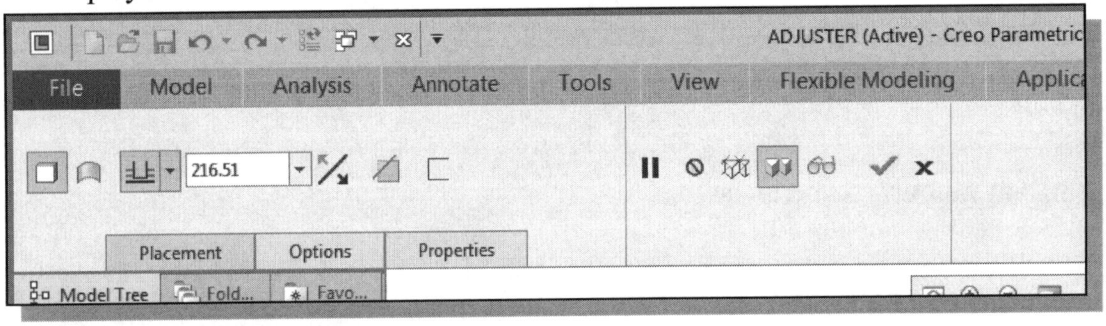

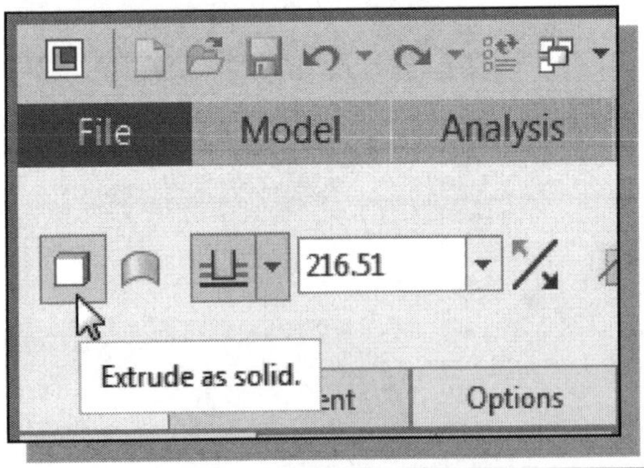

2. On your own, move the cursor over the icons and read the descriptions of the different options available. Note that the default extrude option is set to **Extrude as solid**.

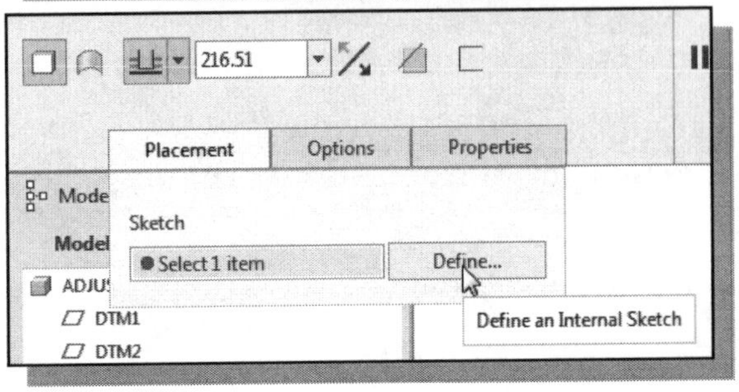

3. Click the **Placement** option and choose **Define** to begin creating a new *internal sketch*.

Sketching plane – It is an XY CRT, but an XYZ World

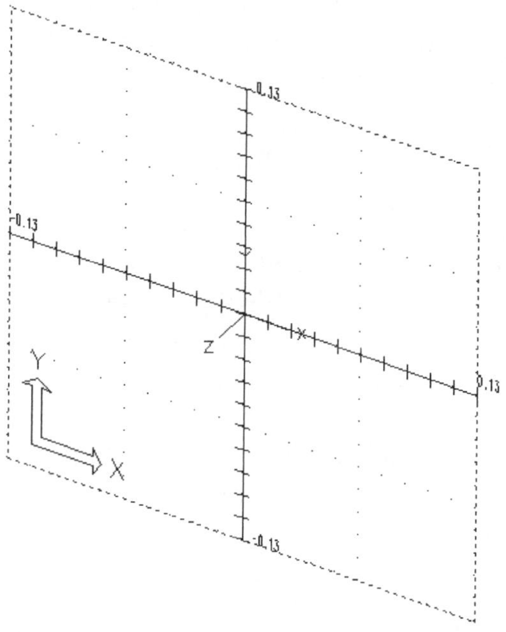

Design modeling software is becoming more powerful and user friendly, yet the system still does only what the user tells it to do. When using a geometric modeler, we therefore need to have a good understanding of what its inherent limitations are. We should also have a good understanding of what we want to do and what to expect, as the results are based on what is available.

In most 3D geometric modelers, 3D objects are located and defined in what is usually called **world space** or **global space**. Although a number of different coordinate systems can be used to create and manipulate objects in a 3D modeling system, the objects are typically defined and stored using the world space. The world space is usually a **3D Cartesian coordinate system** that the user cannot change or manipulate.

In most engineering designs, models can be very complex, and it would be tedious and confusing if only the world coordinate system were available. Practical 3D modeling systems allow the user to define **Local Coordinate Systems (LCS)** or **User Coordinate Systems (UCS)** relative to the world coordinate system. Once a local coordinate system is defined, we can then create geometry in terms of this more convenient system.

Although objects are created and stored in 3D space coordinates, most of the geometric entities can be referenced using 2D Cartesian coordinate systems. Typical input devices such as a mouse or digitizer are two-dimensional by nature; the movement of the input device is interpreted by the system in a planar sense. The same limitation is true of common output devices, such as CRT displays and plotters. The modeling software performs a series of three-dimensional to two-dimensional transformations to correctly project 3D objects onto the 2D display plane.

The *Creo Parametric* ***sketching plane*** is a special construction approach that enables the planar nature of the 2D input devices to be directly mapped into the 3D coordinate system. The *sketching plane* is a local coordinate system that can be aligned to an existing face of a part or a reference plane.

Think of the sketching plane as the surface on which we can sketch the 2D sections of the parts. It is similar to a piece of paper, a white board, or a chalkboard that can be attached to any planar surface. The first sketch we create is usually drawn on one of the established datum planes. Subsequent sketches/features can then be created on sketching planes that are aligned to existing **planar faces of the solid part** or **datum planes.**

Defining the Sketching Plane

The *sketching plane* is a reference location where two-dimensional sketches are created. The *sketching plane* can be any planar part surface or datum plane. Note that *Creo Parametric* uses a two-step approach in setting up the selection and alignment of the sketching plane.

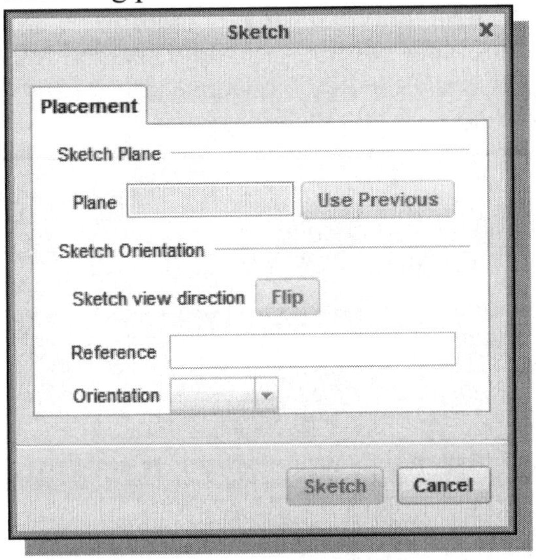

❖ In the *Section Placement* window, the selection of the sketch plane and the orientation of the sketching plane are organized into two groups as shown in the figure. The **Sketch Plane** can be set to any surfaces, including datum planes. The **Sketch Orientation** is set based on the selection of the Sketch plane.

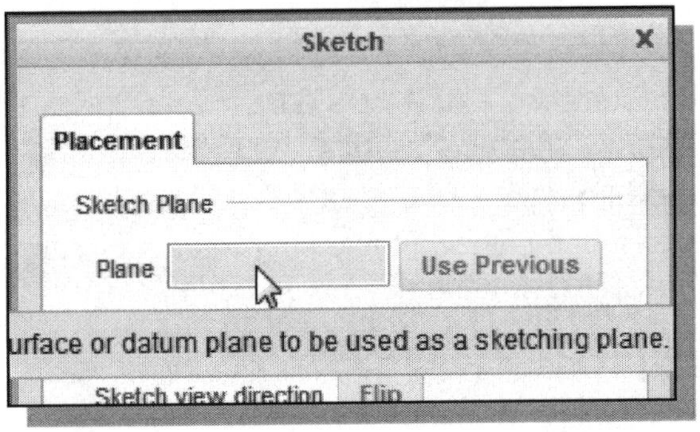

1. Notice the **Plane** option box in the *Sketch* window is activated, and the message "*Select a plane or surface to define sketch plane.*" is displayed in the quick help tip and also in the message area.

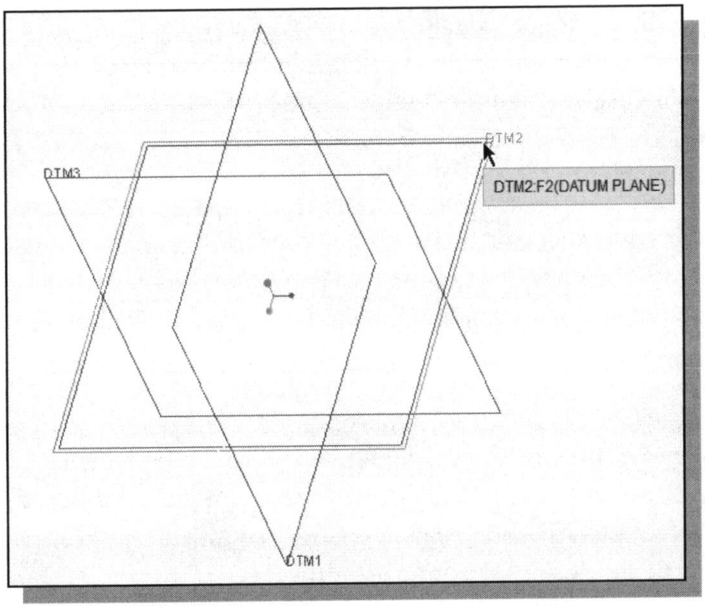

2. In the *graphic area*, select **DTM2** by clicking on any edge of the plane as shown.

❖ Notice an arrow appears on the left edge of DTM2. The arrow direction indicates the viewing aligned direction of the sketch plane. The viewing direction can be reversed by clicking on the **Flip** button in the *Sketch Orientation* section of the pop-up window.

Defining the Orientation of the Sketching Plane

Although we have selected the sketching plane, *Creo Parametric* still needs additional information to define the orientation of the sketch plane. *Creo Parametric* expects us to choose a reference plane (any plane that is perpendicular to the selected sketch plane) and the orientation of the reference plane is relative to the computer screen.

❑ **To define the orientation of the sketching plane, select the facing direction of the reference plane with respect to the computer screen.**

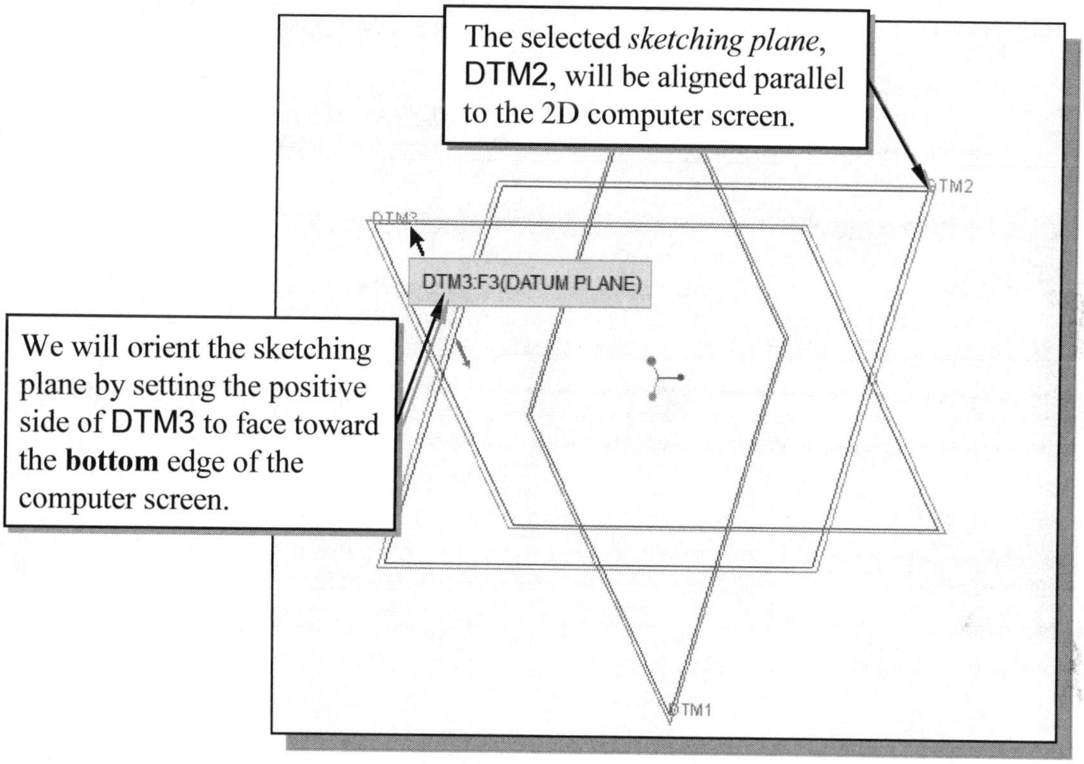

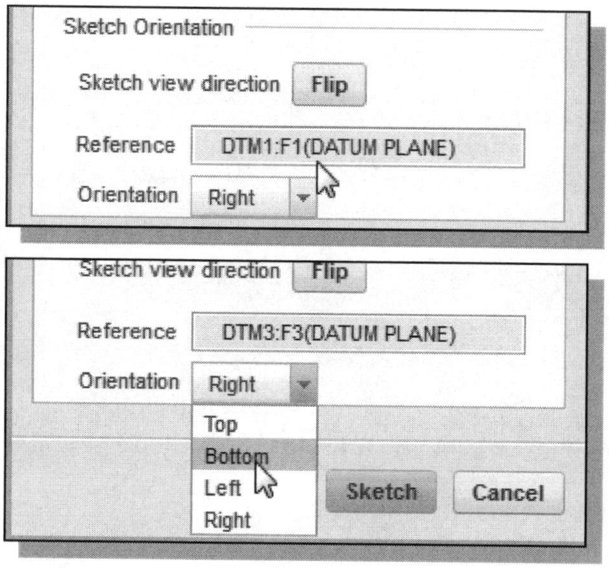

1. Notice the **Reference** option box in the Sketch Orientation window is now activated. The message "*Select a reference, such as surface, plane or edge to define view orientation.*" is displayed in the message area.

2. In the *graphic area*, select **DTM3** by clicking on one of the datum plane edges as shown in the above figure.

3. In the **Orientation** list, pick **Bottom** to set the orientation of the reference plane.

4. Pick **Sketch** to exit the *Section Placement* window and proceed to enter the *Creo Parametric Sketcher* mode.

5. To orient the sketching plane parallel to the screen, the **Sketch View** icon in the *Display View* toolbar is available as shown. Note the orientation of the sketching plane is adjusted based on the setup on the previous page.

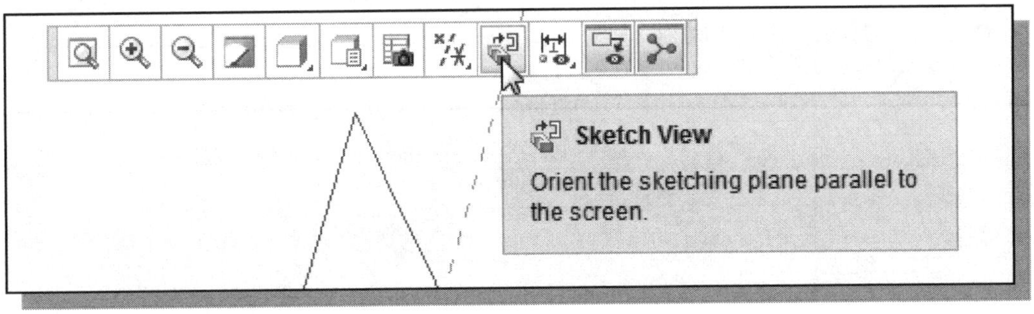

- *Creo Parametric* will now rotate the three *datum planes*: DTM2 aligned to the screen and the positive side of DTM3 facing toward the bottom edge of the computer screen.

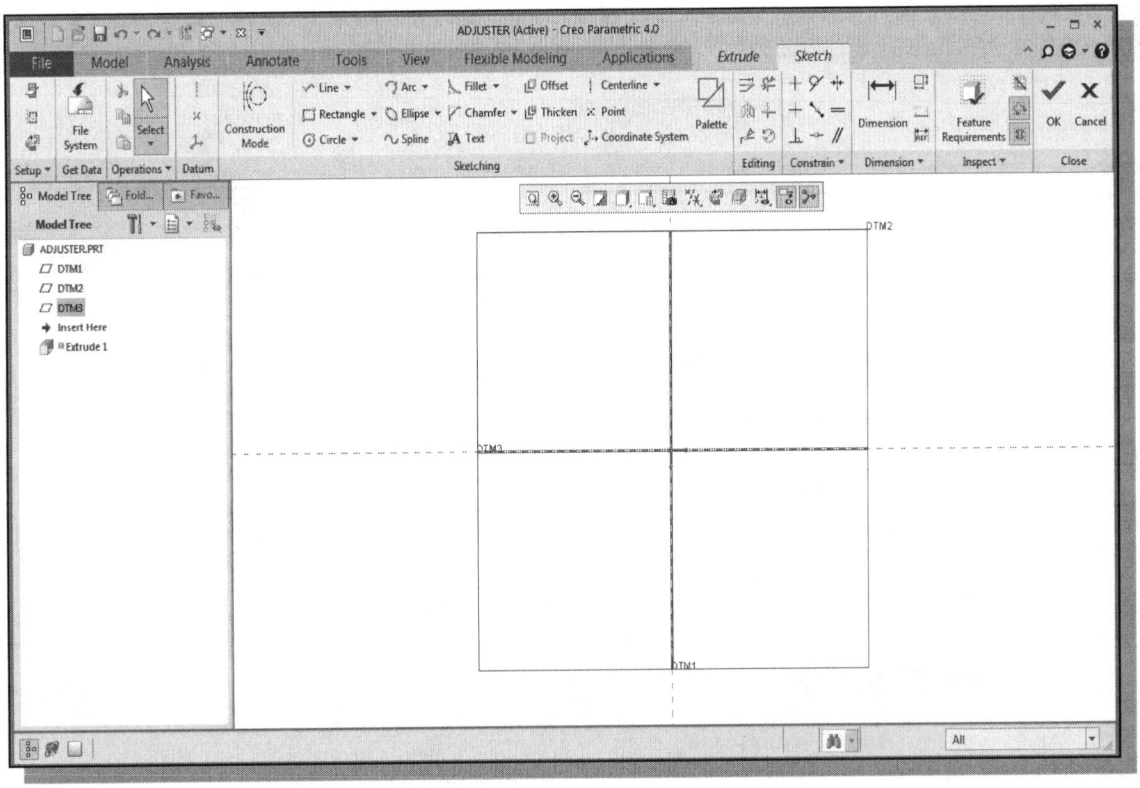

❖ The orientation of the *sketching plane* can be very confusing to new users. Read through this section carefully again to make sure you understand the steps involved.

Step 3: Creating 2D Rough Sketches

♦ Shape Before Size – Creating Rough Sketches

Quite often during the early design stage, the shape of a design may not have any precise dimensions. Most conventional CAD systems require the user to input the precise lengths and location dimensions of all geometric entities defining the design, and some of the values may not be available during the early design stage. With *parametric modeling*, we can use the computer to elaborate and formulate the design idea further during the initial design stage. With *Creo Parametric*, we can use the computer as an electronic sketchpad to help us concentrate on the formulation of forms and shapes for the design. This approach is the main advantage of *parametric modeling* over conventional solid-modeling techniques.

As the name implies, ***rough sketches*** are not precise at all. When sketching, we simply sketch the geometry so it closely resembles the desired shape. Precise scale or dimensions are not needed. *Creo Parametric* provides us with many tools to assist in finalizing sketches, known as ***sections***. For example, geometric entities such as horizontal and vertical lines are set automatically. However, if the rough sketches are poor, much more work will be required to generate the desired parametric sketches. Here are some general guidelines for creating sketches in *Creo Parametric*:

- **Create a sketch that is proportional to the desired shape.** Concentrate on the shapes and forms of the design.

- **Keep the sketches simple.** Leave out small geometry features such as fillets, rounds, and chamfers. They can easily be placed using the Fillet and Chamfer commands after the parametric sketches have been established.

- **Exaggerate the geometric features of the desired shape.** For example, if the desired angle is 85 degrees, create an angle that is 50 or 60 degrees. Otherwise, *Creo Parametric* might assume the intended angle to be a 90-degree angle.

- **Draw the geometry so that it does not overlap.** The sketched geometry should eventually form a closed region. *Self-intersecting* geometric shapes are not allowed.

- **The sketched geometric entities should form a closed region.** To create a solid feature, such as an extruded solid, a closed region section is required so that the extruded solid forms a 3D volume.

➢ **Note:** The concepts and principles involved in *parametric modeling* are very different, and sometimes they are totally opposite, to those of the conventional computer aided drafting systems. In order to understand and fully utilize *Creo Parametric's* functionality, it will be helpful to take a *Zen* approach to learning the topics presented in this text: **Temporarily forget your knowledge and experiences using conventional computer aided drafting systems.**

♦ The Creo Parametric SKETCHER and INTENT MANAGER

In previous generation CAD programs, construction of models relies on exact dimensional values, and adjustments to dimensional values are quite difficult once the model is built. With *Creo Parametric,* we can now treat the sketch as if it is being done on a napkin, and it is the general shape of the design that we are more interested in defining. The *Creo Parametric* part model contains more than just the final geometry. It also contains the *design intent* that governs what will happen when geometry changes. The design philosophy of "**shape before size**" is implemented through the use of the *Creo Parametric* ***Sketcher***. This allows the designer to construct solid models in a higher level and leave all the geometric details to *Creo Parametric.*

In *Creo Parametric,* previously known as Pro/ENGINEER, one of the more important functionalities is the ***Intent Manager*** in the *2D Sketcher.*

The ***Intent Manager*** enables us to do:

- Dynamic dimensioning and constraints
- Add or delete constraints explicitly
- Undo any *Sketcher* operation

The first thing that *Creo Parametric Sketcher* expects us to do, which is displayed in the *References* window, is to specify *sketching references*. In the previous sections, we created the three datum planes to help orient the model in 3D space. Now we need to orient the 2D sketch with respect to the three datum planes. At least two references are required to orient the sketch in the horizontal direction and in the vertical direction. By default, the two planes (in our example, DTM1 and DTM3) that are perpendicular to the sketching plane (DTM2) are automatically selected.

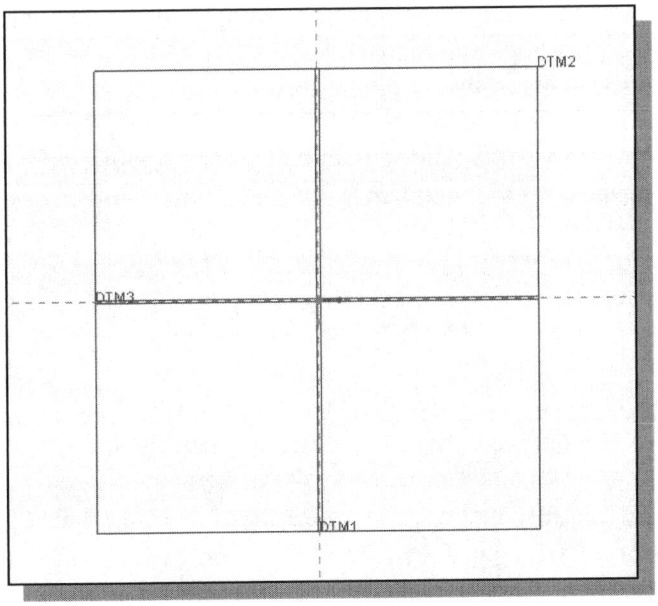

➢ Note that DTM1 and DTM3 are pre-selected as the sketching references. In the graphics area, the two references are displayed with two **dashed lines**.

- In *Creo Parametric*, a 2D sketch needs to be Fully Placed with respect to at least two references. In this case, DTM1 is used to control the horizontal placement of geometry, where DTM3 is used to control the vertical placements.

❖ Next, we will create a rough sketch by using some of the visual aids available, and then update the design through the associated control parameters.

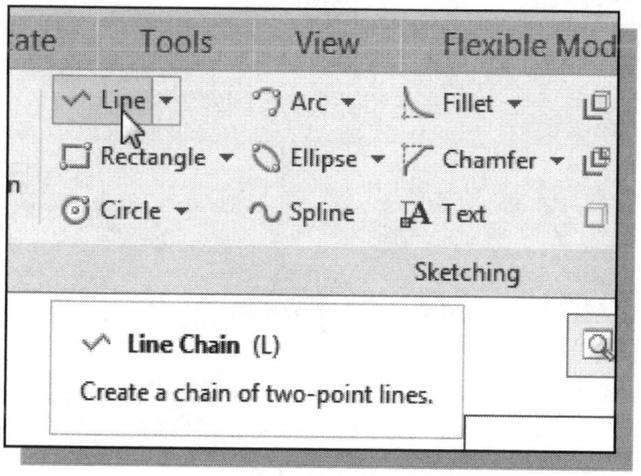

- Move the graphics cursor to the **Line** icon in the *Sketching* toolbar. A *help-tip* box appears next to the cursor to provide a brief description of the command.

❖ The *Sketching* toolbar provides tools for creating the basic 2D geometry that can be used to create features and parts.

Graphics Cursors

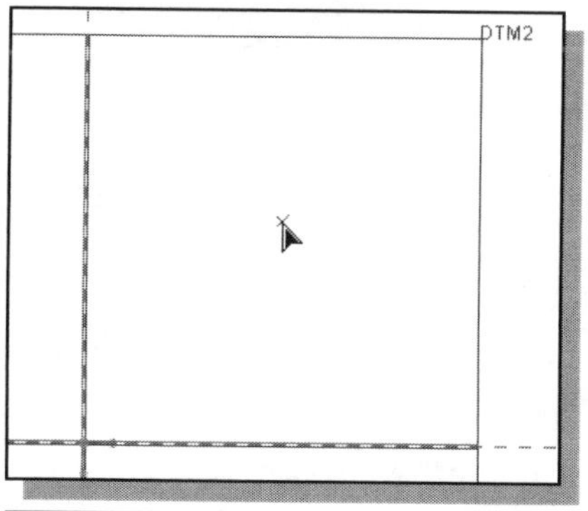

❖ Notice the cursor changes from an arrow to an arrow with a small crosshair when graphical input is expected.

1. As you move the graphics cursor, you will see different symbols appear at different locations.

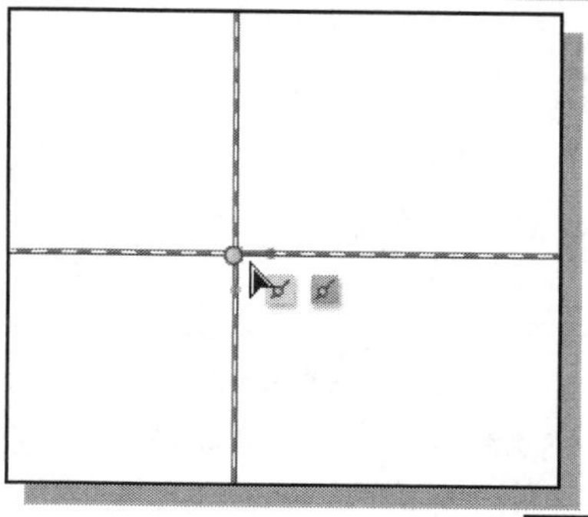

2. Move the cursor near the intersection of the two references, and notice that the small crosshair attached to the cursor will automatically snap to the intersection point. **Left-click** once to place the starting point as shown. Notice the small geometric constraint symbol next to the cursor indicating the first line endpoint is Coincident with the intersection point.

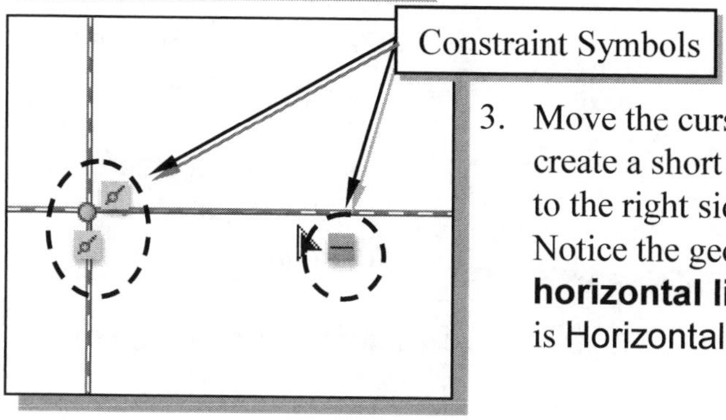

3. Move the cursor along the vertical reference and create a short horizontal line by clicking at a location to the right side of the starting point as shown. Notice the geometric constraint symbol, a **short horizontal line**, indicating the created line segment is Horizontal.

Geometric Constraint Symbols

Creo Parametric displays different visual clues, or symbols, to show you alignments, perpendicularities, tangencies, etc. These constraints are used to capture the *design intent* by creating constraints where they are recognized. *Creo Parametric* displays the governing geometric rules as models are built.

Vertical	indicates a line segment is vertical
Horizontal	indicates a line segment is horizontal
Equal Length	indicates two line segments are of equal length
or Equal Radii	indicates two curves are of equal radii
Tangent	indicates two entities are tangent to each other
Parallel	indicates a segment is parallel to other entities
Perpendicular	indicates a segment is perpendicular to other entities
Symmetry	indicates two points are symmetrical
Point on Entity	indicates the point is on another entity

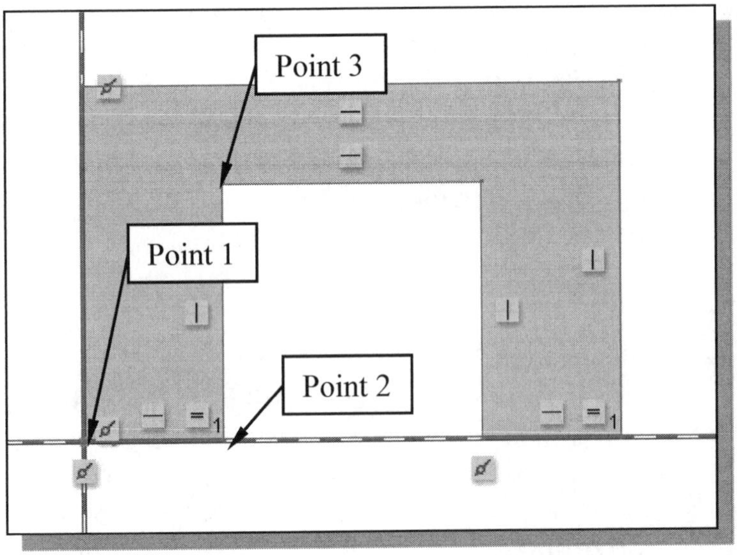

1. Complete the sketch as shown, a closed region ending at the starting point (***Point 1***). Watch the displayed constraint symbols while sketching, especially the **Equal Length** constraint, **=**, applied to the two short horizontal edges.

➢ Note that all segments are either *vertical* or *horizontal*.

2. Inside the *graphics area*, click **twice** with the **middle-mouse-button** to end the current line sketch.

❖ *Creo Parametric's* ***Intent Manager*** automatically places dimensions and constraints on the sketched geometry. This is known as the ***Dynamic Dimensioning and Constraints*** feature. Constraints and dimensions are added "on the fly." Do not be concerned with the size of the sketched geometry or the displayed dimensional values; we will modify the sketched geometry in the following sections.

Dynamic Viewing Functions

❖ *Creo Parametric* provides a special user interface, ***Dynamic Viewing,*** which enables convenient viewing of the entities in the display area at any time. The ***Dynamic Viewing*** functions are controlled with the combinations of the middle mouse button, the **[Ctrl]** key and the **[Shift]** key on the keyboard.

Zooming – Turn the **mouse wheel** or **[Ctrl]** key and **[middle-mouse-button]**

Use the **mouse wheel** to perform the zooming option; turning the wheel forward will reduce the scale of display. Hold down the **[Ctrl]** key and press down the middle-mouse-button in the display area. Drag the mouse vertically on the screen to adjust the scale of the display. Moving upward will reduce the scale of the display, making the entities display smaller on the screen. Moving downward will magnify the scale of the display.

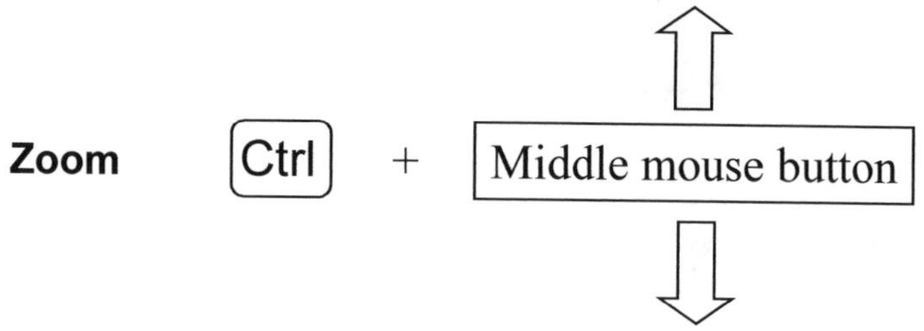

Panning – **[Shift]** key and **[middle-mouse-button]**

Hold down the **[Shift]** key and press down the middle-mouse-button in the display area. Drag the mouse to pan the display. This allows you to reposition the display while maintaining the same scale factor of the display. This function acts as if you are using a video camera. You control the display by moving the mouse.

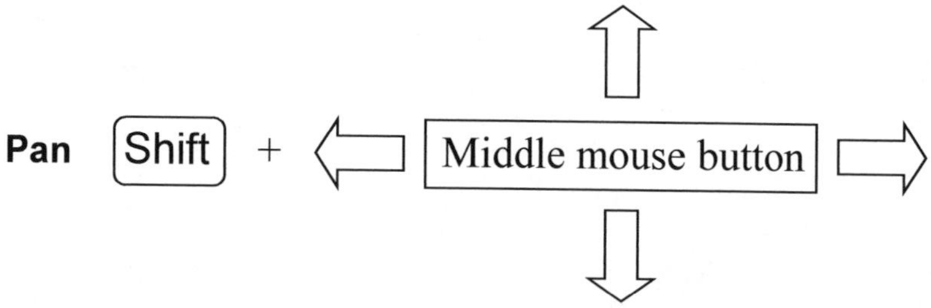

➢ On your own, use the *Dynamic Viewing* functions to reposition and magnify the scale of the 2D sketch to the center of the screen so that it is easier to work with.

Step 4: Apply/Modify Constraints and Dimensions

➢ As the sketch is made, *Creo Parametric* automatically applies geometric constraints (such as **Horizontal**, **Vertical** and **Equal Length**) and dimensions to the sketched geometry. We can continue to modify the geometry, apply additional constraints and/or dimensions, or define/modify the size and location of the existing geometry. It is more than likely that some of the automatically applied dimensions may not match with the design intent we have in mind. For example, we might want to have dimensions identifying the overall-height, overall-width, and the width of the inside-cut of the design, as shown in the figures below.

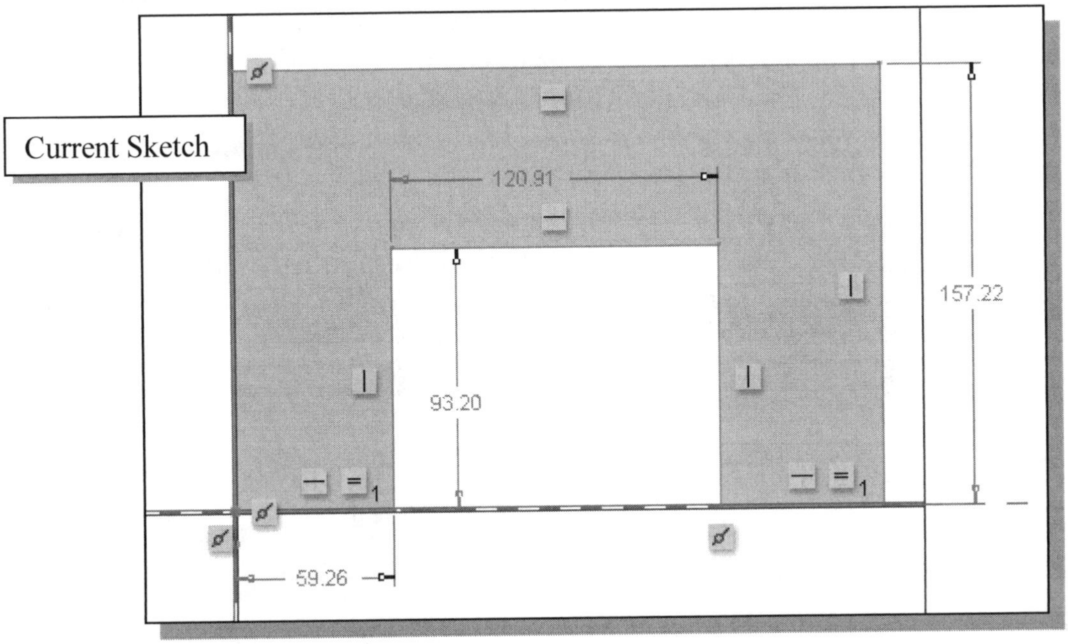

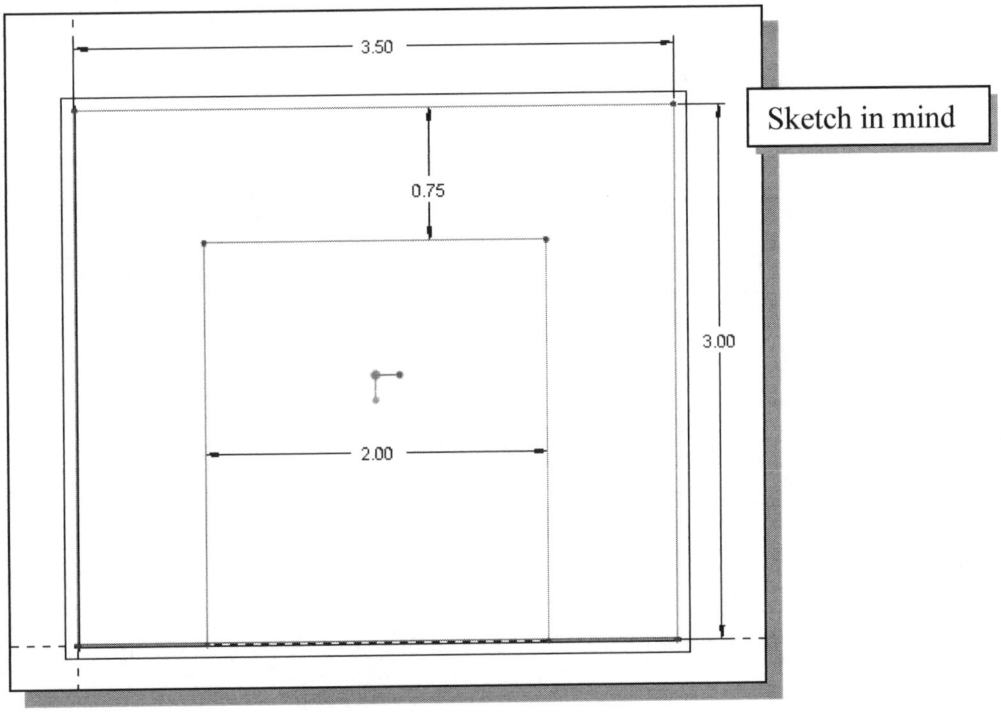

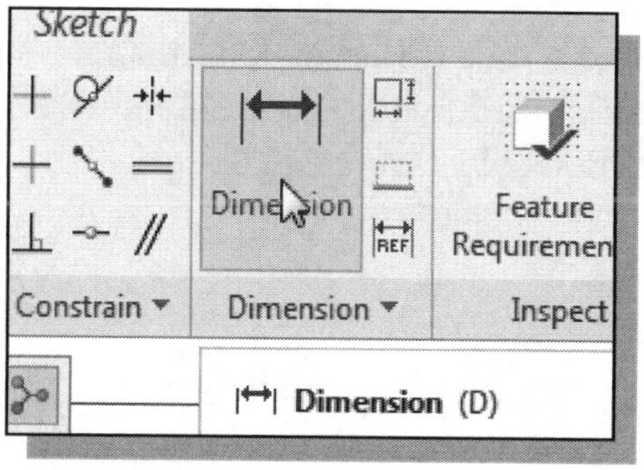
Sketch
Dimension
Feature Requiremen
Constrain
Dimension
Inspect
Dimension (D)

1. Click on the **Normal Dimension** icon in the *Sketching* toolbar as shown. This command allows us to create defining dimensions.

2. Select the **inside horizontal line** by left-clicking once on the line as shown.

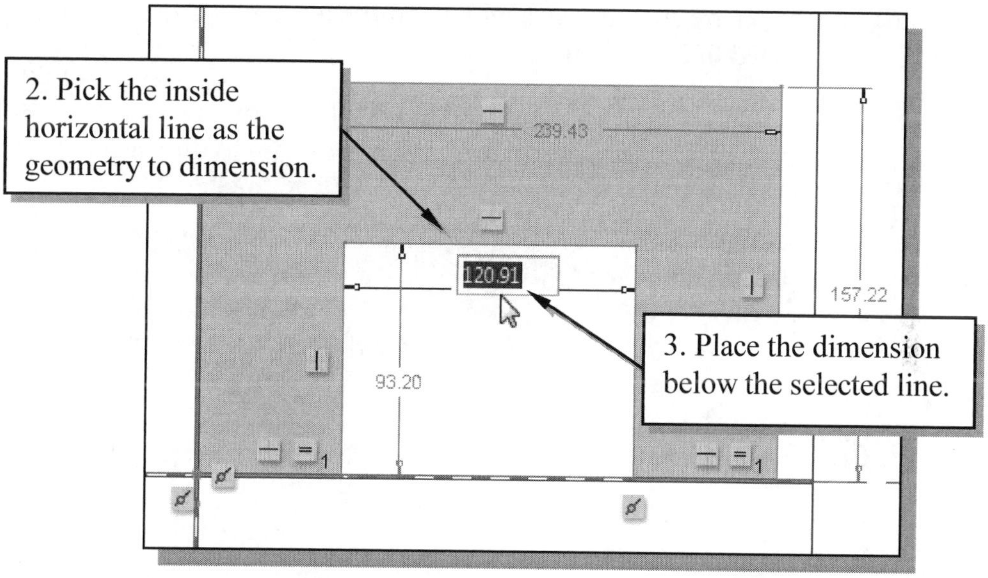
2. Pick the inside horizontal line as the geometry to dimension.
239.43
120.91
157.22
93.20
3. Place the dimension below the selected line.

3. Move the graphics cursor below the selected line and click once with the **middle-mouse-button** to place the dimension. (Note that the value displayed on your screen might be different than what is shown in the above figure.)

4. Click again with the **middle-mouse-button** to exit the *Edit Dimension* mode.

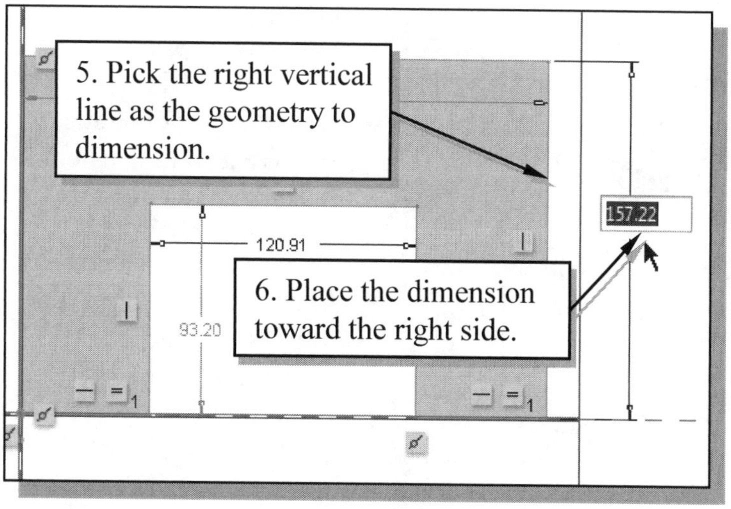
5. Pick the right vertical line as the geometry to dimension.
157.22
120.91
93.20
6. Place the dimension toward the right side.

5. Select the *right vertical line.*

6. Place the dimension, by clicking once with the **middle-mouse-button** at a location toward the right of the sketch.

❖ The Dimension command will create a length dimension if a single line is selected.

- Notice the overall-height dimension applied automatically by the *Intent Manager* is removed as the new dimension is defined.

❖ Note that the dimensions we just created are displayed with a different color than those that are applied automatically. The dimensions created by the *Intent Manager* are called **weak dimensions**, which can be replaced/deleted as we create specific defining dimensions to satisfy our design intent.

7. Select the **top horizontal line** as shown below.

8. Select the **inside horizontal line** as shown below.

9. Place the dimension, by clicking once with the **middle-mouse-button,** at a location in between the selected lines as shown below.

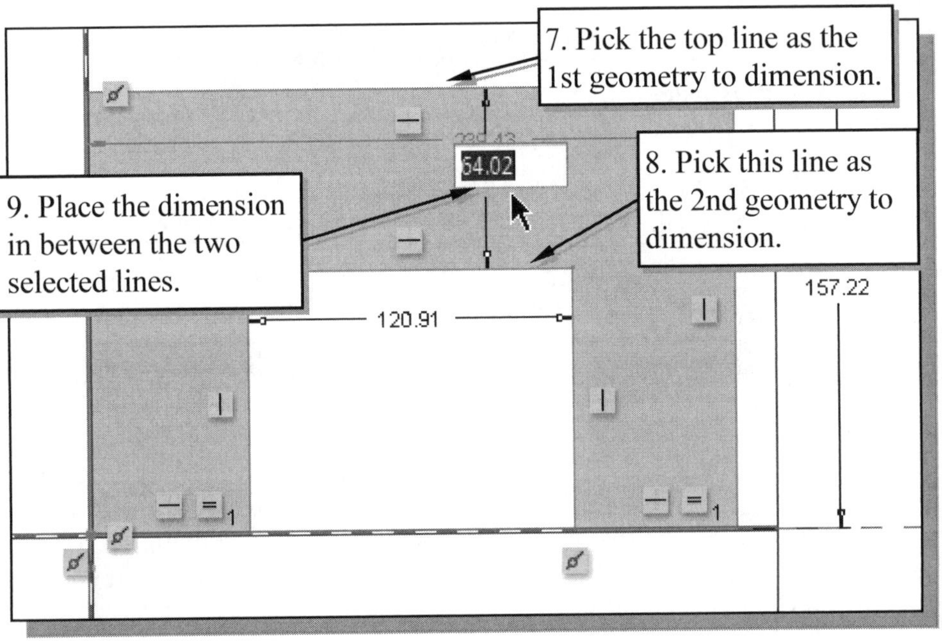
7. Pick the top line as the 1st geometry to dimension.
64.02
8. Pick this line as the 2nd geometry to dimension.
9. Place the dimension in between the two selected lines.
157.22
120.91

10. Click again with the **middle-mouse-button** to exit the *Edit Dimension* mode.

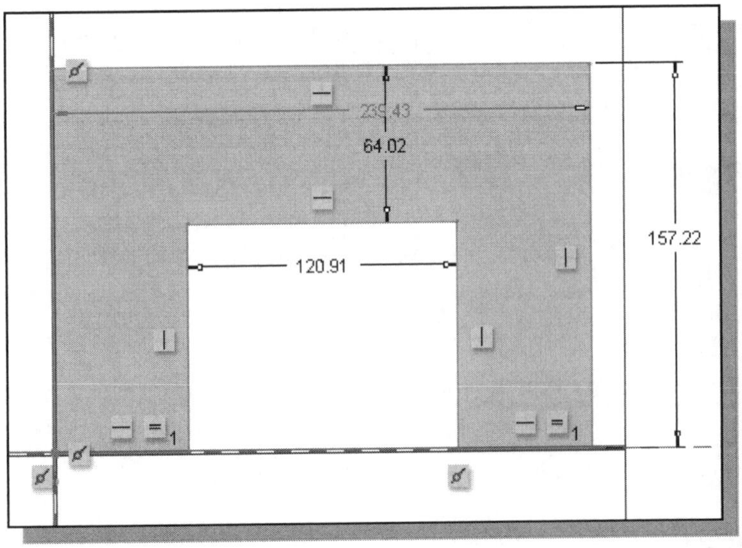
64.02
157.22
120.91

❖ When two parallel lines are selected, the Dimension command will create a dimension measuring the distance in between.

➢ Examine the established dimensions and constraints in the sketch. Is the sketch fully defined? Or should we add additional dimensions?

Modifying the Dimensions in a Sketch

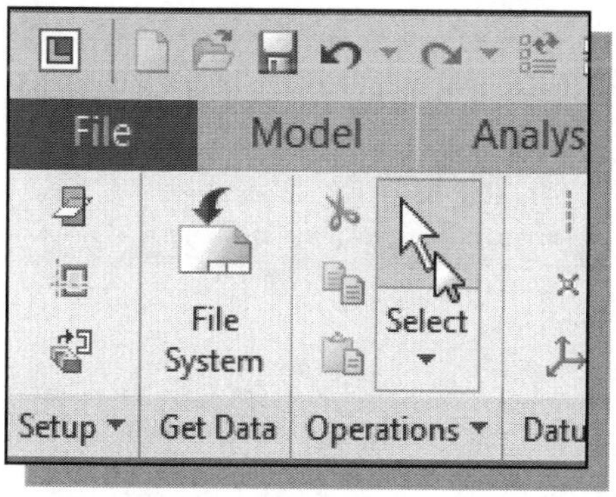

1. Click on the **Select** icon in the *Operations* toolbar as shown. The Select command allows us to perform several modification operations on both the sketched geometry and dimensions.

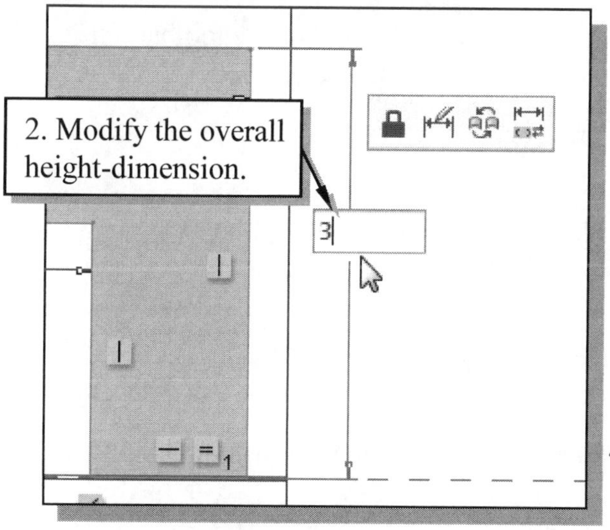

2. Select the overall height dimension of the sketch by **double-clicking** with the left-mouse-button on the dimension text.

3. In the *dimension value* box, the current length of the line is displayed. Enter **3** as the new value for the dimension.

4. Press the **ENTER** key once to accept the entered value.

➢ *Creo Parametric* will update the sketch using the entered dimension value. Since the other dimensions are much larger, the sketched shape becomes greatly distorted.

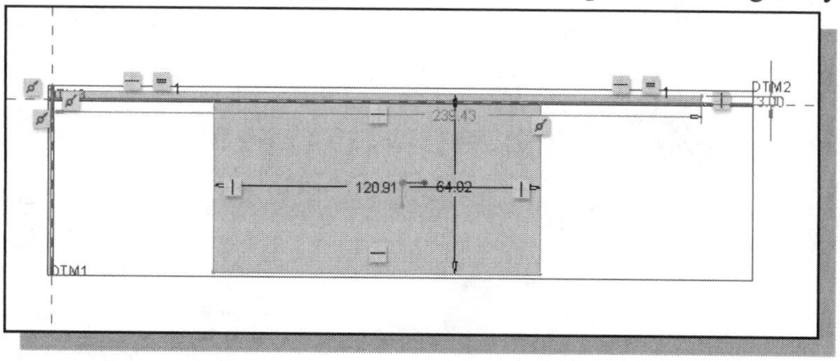

5. Click on the **Undo** icon in the *Quick Access* toolbar to undo the Modify Dimension performed.

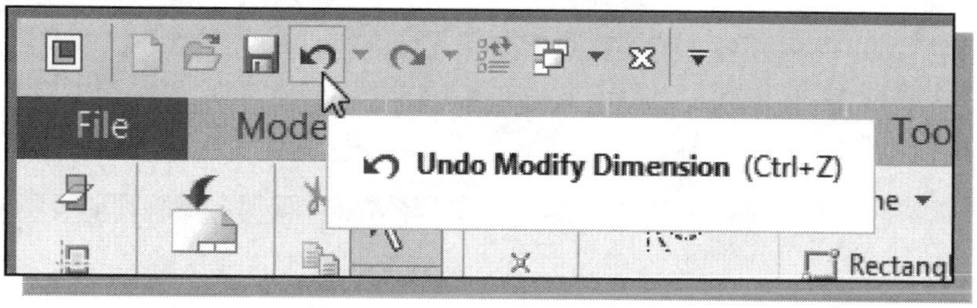

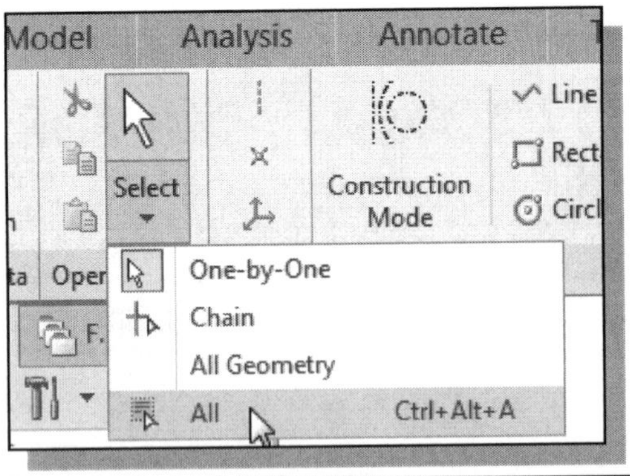

6. In the pull-down menu area, click on the down-arrow on **Select** to display the option list and select the **All** option as shown.

   (Note that using the hotkey combination **Crtl+Alt+A** can also activate this option.)

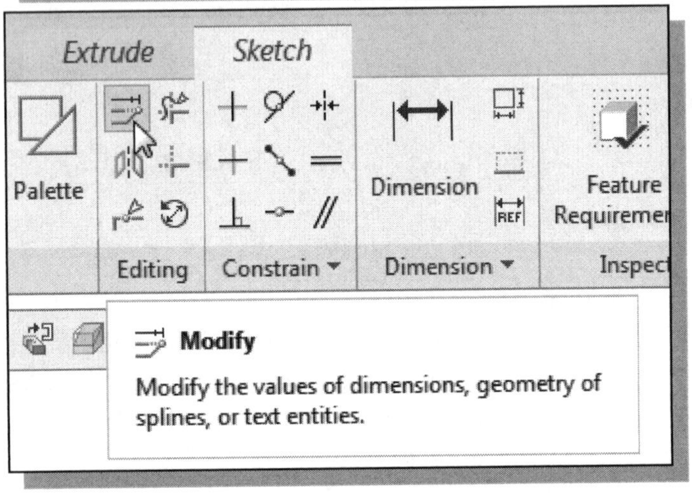

7. In the *Editing* toolbar, click on the **Modify** icon as shown.

- With the pre-selection option, all dimensions are selected and listed in the *Modify Dimensions* dialog box.

8. Turn ***off*** the **Regenerate** option by unchecking the option in the *Modify Dimensions* dialog box as shown.

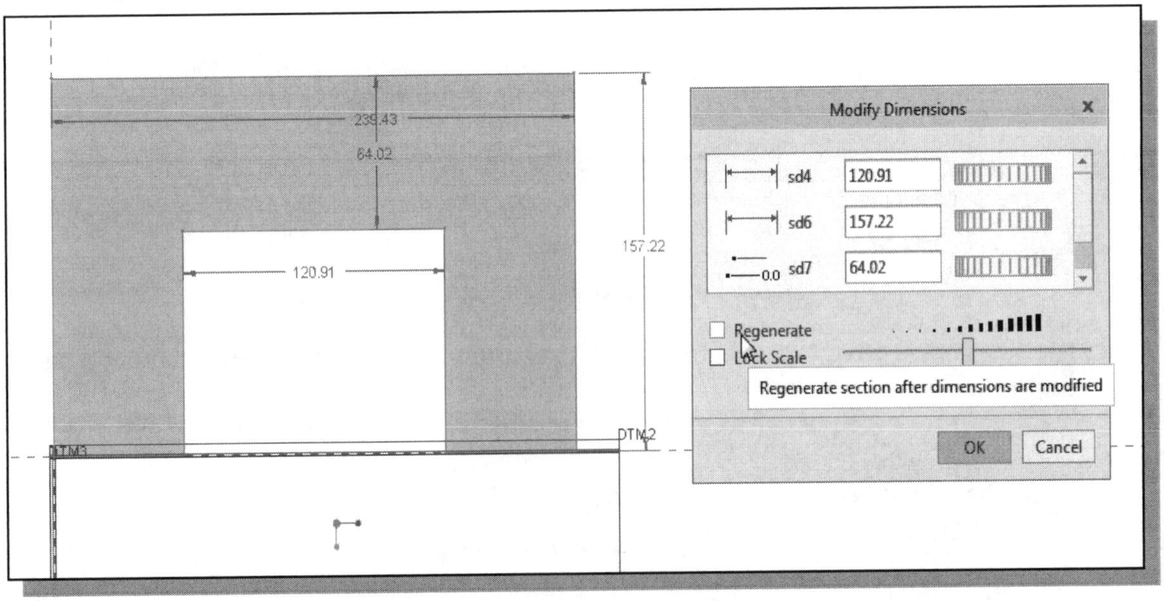

9. On your own, adjust the dimensions as shown below. Note that the dimension selected in the *Modify Dimensions* dialog box is identified with an enclosed box in the display area.

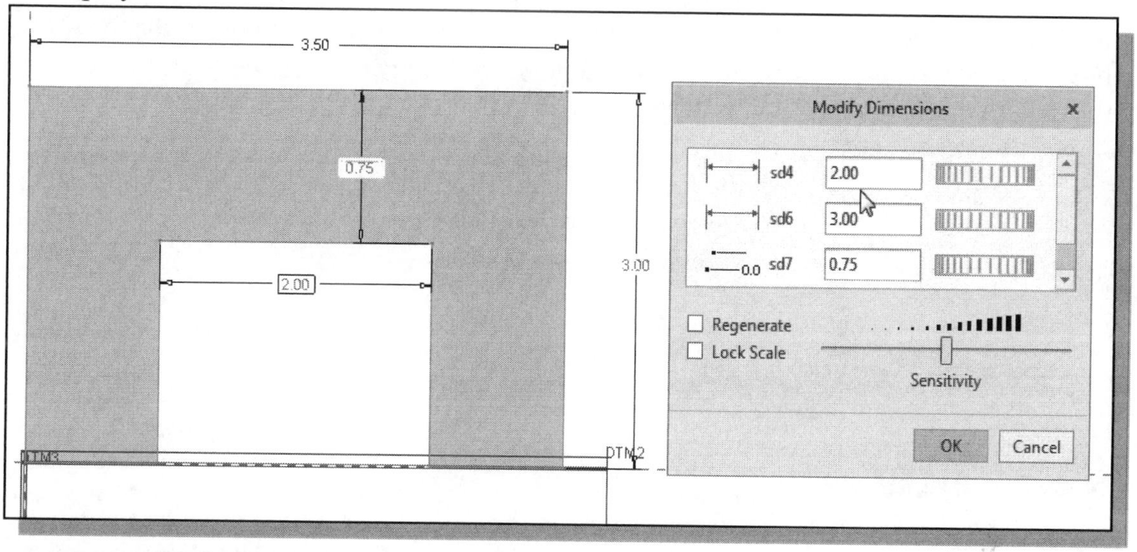

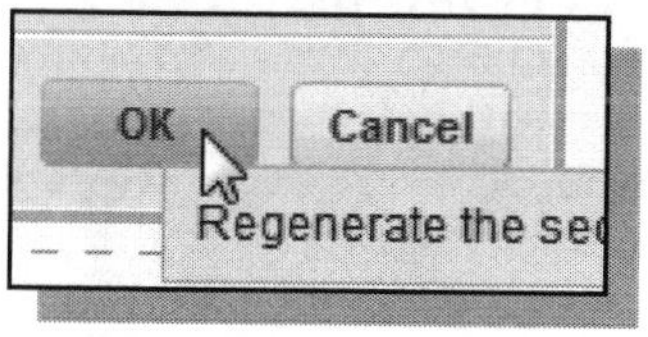

10. Inside the *Modify Dimensions* dialog box, click on the **OK** button to regenerate the sketched geometry and exit the **Modify Dimensions** command.

Repositioning Dimensions

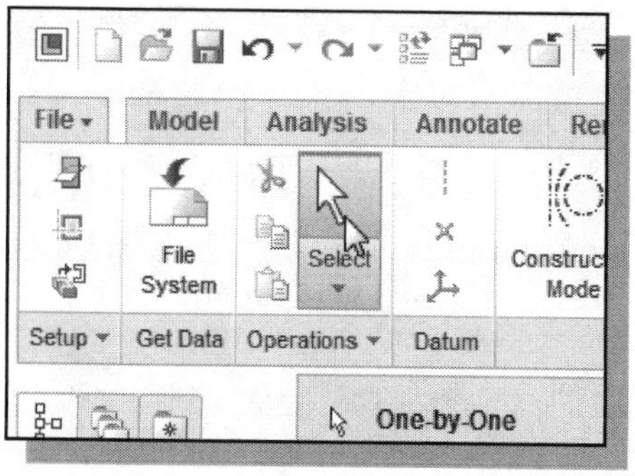

1. Confirm the **Select** icon, in the *Operations* toolbar, is activated as shown.

2. Press and hold down the **left-mouse-button** on any dimension text, then drag the dimension to a new location in the display area. (Note the cursor is changed to a moving arrow icon during this operation.)

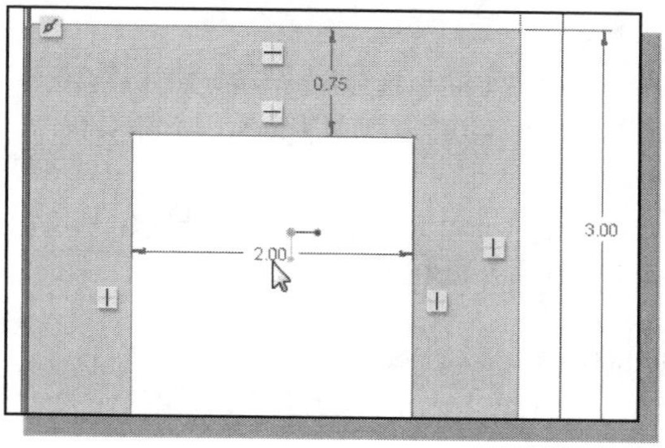

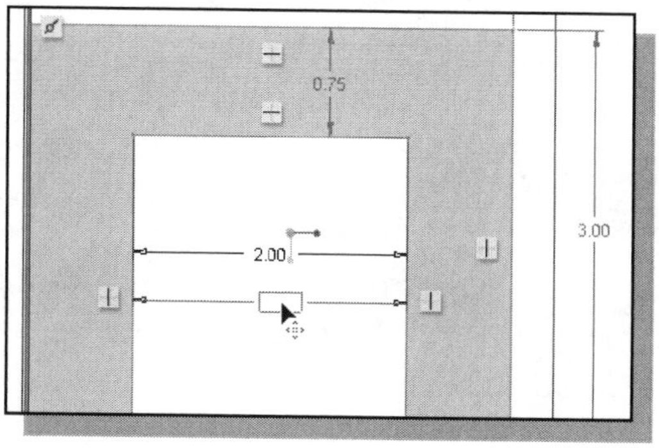

Step 5: Completing the Base Solid Feature

Now that the 2D sketch is completed, we will proceed to the next step: creating a 3D part from the 2D section. Extruding a 2D section is one of the common methods that can be used to create 3D parts. We can extrude planar faces along a path. In *Creo Parametric*, the default extrusion direction is perpendicular to the selected sketching plane, DTM2.

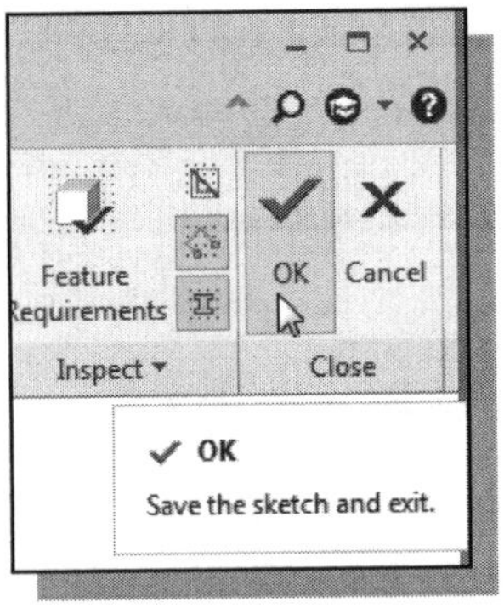

1. In the *Ribbon* toolbar, click **OK** to exit the *Creo Parametric 2D Sketcher.* The 2D sketch is the first element of the *Extrude* feature definition.

2. In the *Feature Option Dashboard*, confirm the *depth value* option is set as shown. This option sets the extrusion of the section by **Extrude from sketch plane by a specified depth value**.

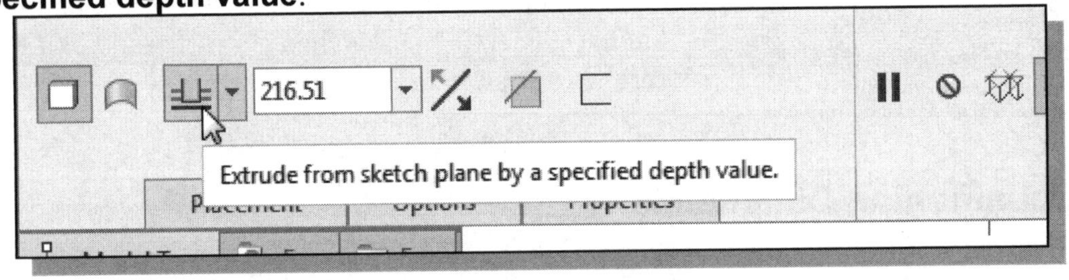

3. In the *depth value* box, enter **2.5** as the extrusion depth.

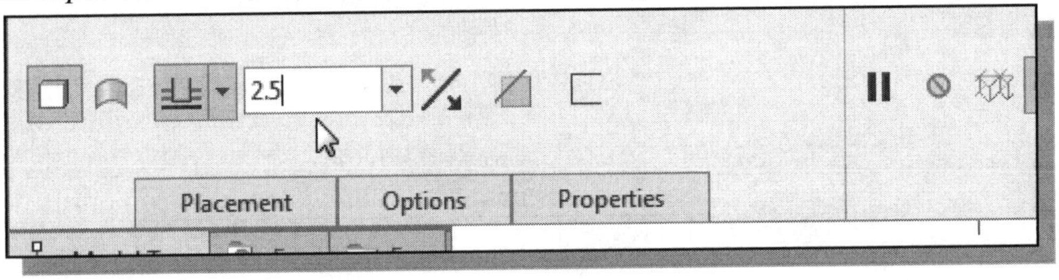

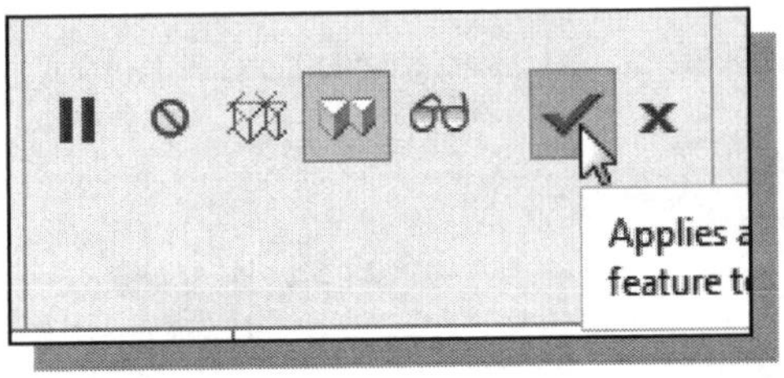

4. In the *Dashboard* area, click **Accept** to proceed with the creation of the solid feature.

5. Use the hotkey combination **CTRL+D** to reset the display to the default 3D orientation.

➢ Note that all dimensions disappeared from the screen. All parametric definitions are stored in the ***Creo Parametric* database**, and any of the parametric definitions can be displayed and edited at any time.

The Third Dynamic Viewing Function

3D Dynamic Rotation – [middle mouse button]

Press down the middle-mouse-button in the display area. Drag the mouse on the screen to rotate the model about the screen.

3D Rotation

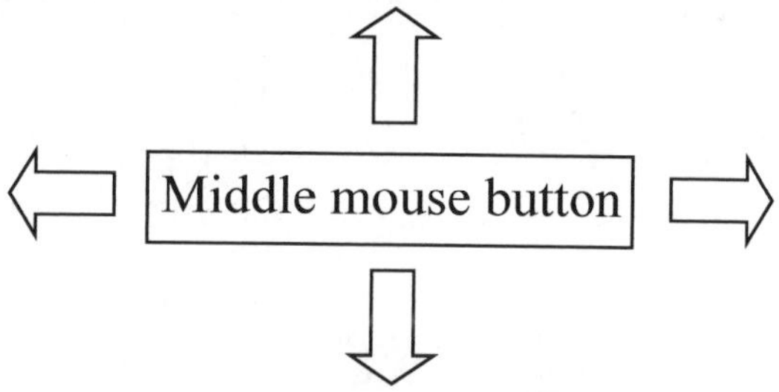

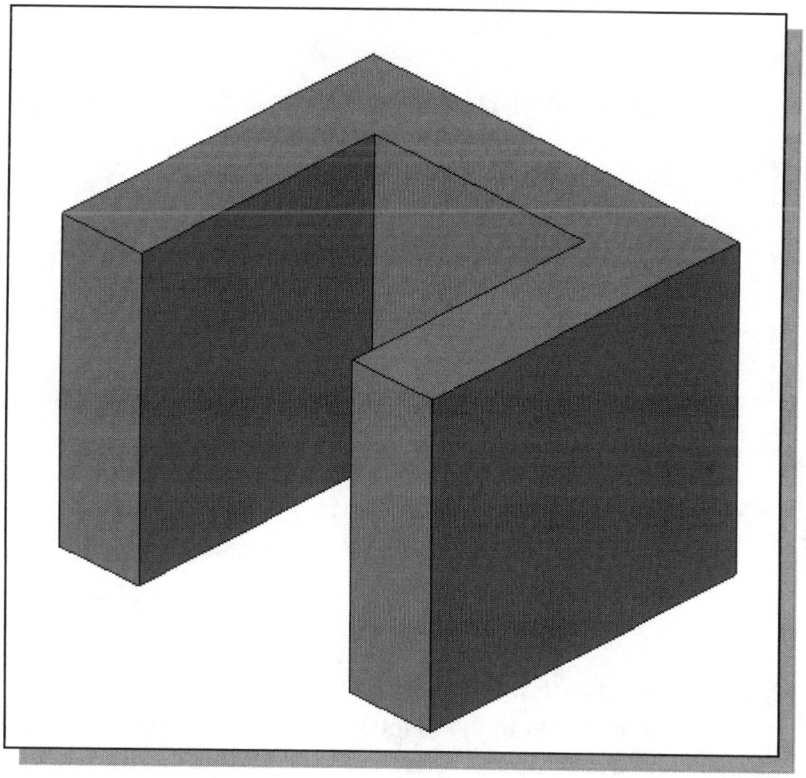

- On your own, practice the use of the ***Dynamic Viewing functions***; note that these are convenient viewing functions at any time.

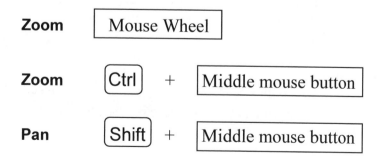

Zoom Mouse Wheel

Zoom Ctrl + Middle mouse button

Pan Shift + Middle mouse button

Display Modes: Wireframe, Shaded, Hidden Edge, No Hidden

The display in the graphics window has six display modes: Shading with Edges, Shading with Reflections, Shading, No Hidden lines, Hidden Line, and Wireframe image. To change the display mode in the active window, click on the display mode button in the *Display* toolbar to display the list as shown.

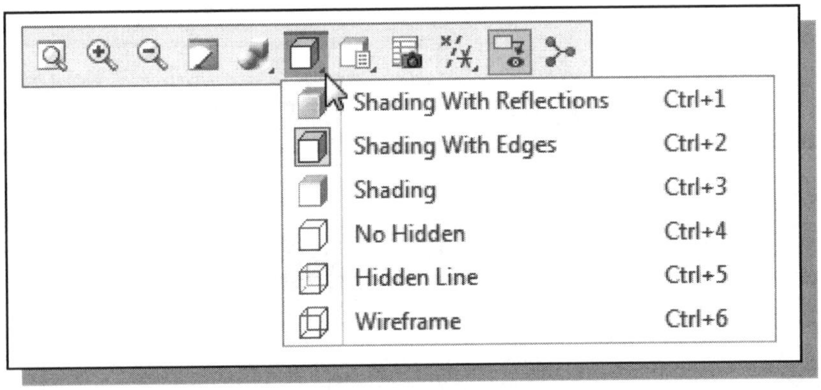

❖ **Shading With Reflections:**
The second icon in the display mode button group generates a more enhanced shaded image of the 3D object.

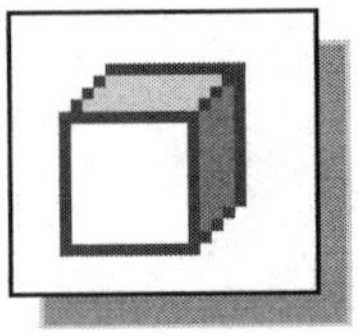

❖ **Shading With Edges:**
The first icon in the display mode button group generates a shaded image of the 3D object with edges highlighted.

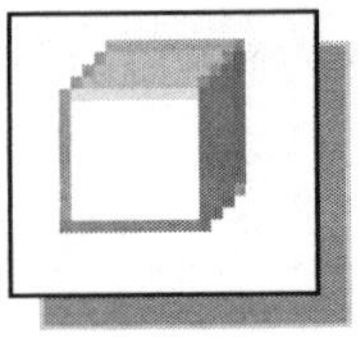

❖ **Shading:**
The third icon in the display mode button group generates a shaded image of the 3D object.

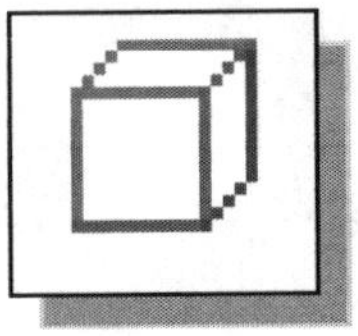

❖ **No Hidden:**
The fourth icon in the display mode button group can be used to generate a wireframe image of the 3D object with all the back lines removed.

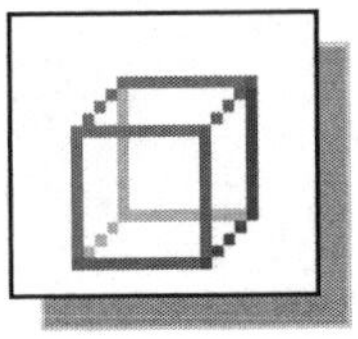

❖ **Hidden Line:**
The fifth icon in the display mode button group can be used to generate a wireframe image of the 3D object with all the back lines shown as hidden lines.

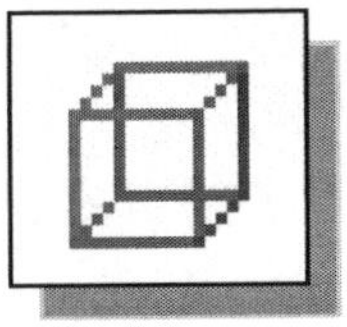

❖ **Wireframe:**
The sixth icon in the display mode button group allows the display of 3D objects using the basic wireframe representation scheme.

Step 6: Adding Additional Features

Next, we will create another extrusion feature that will be added to the existing solid object.

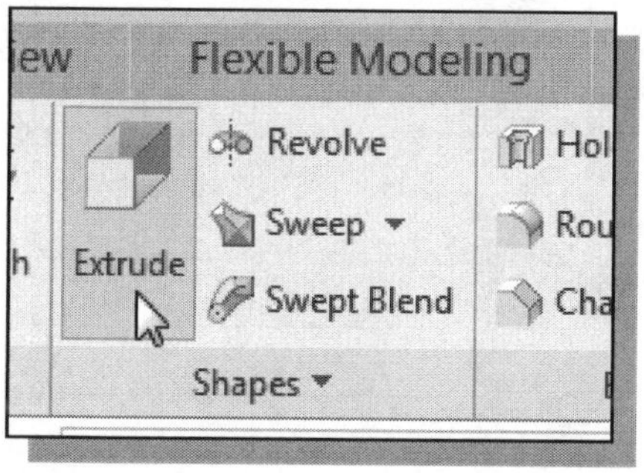

1. In the ***Shapes*** toolbar (the fourth toolbar in the *Ribbon* toolbar), select the **Extrude** tool option as shown.

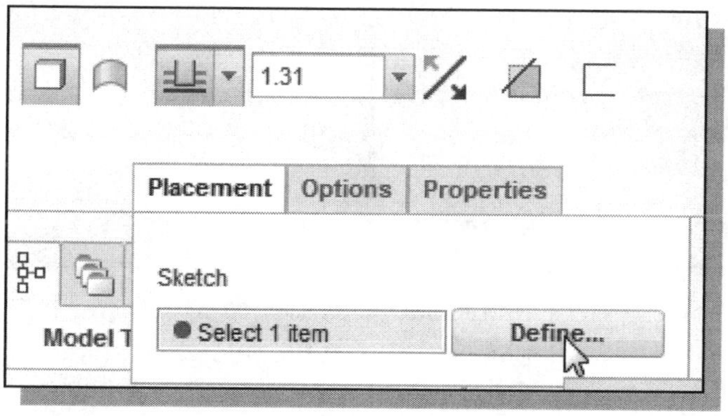

2. Click the **Placement** option and choose **Define** to begin creating a new *internal sketch.*

3. Pick the right vertical face of the solid model as the sketching plane as shown in the figure below.

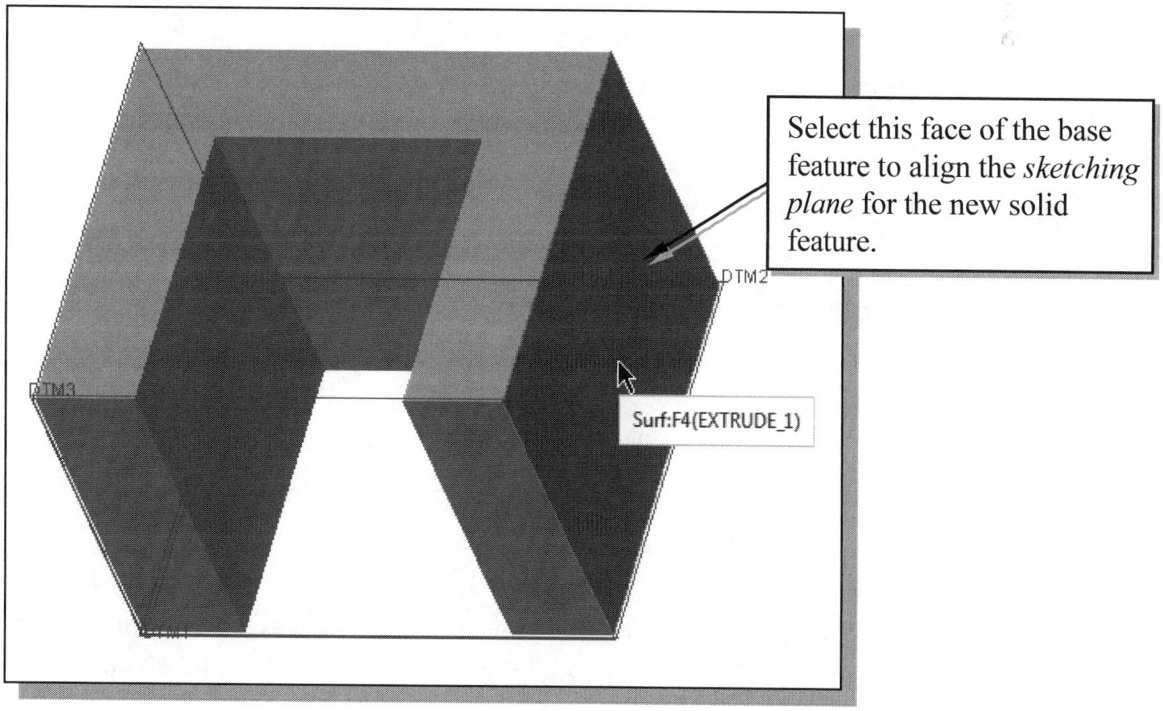

4. In the display area, pick the **top face** of the base feature as the reference plane as shown.

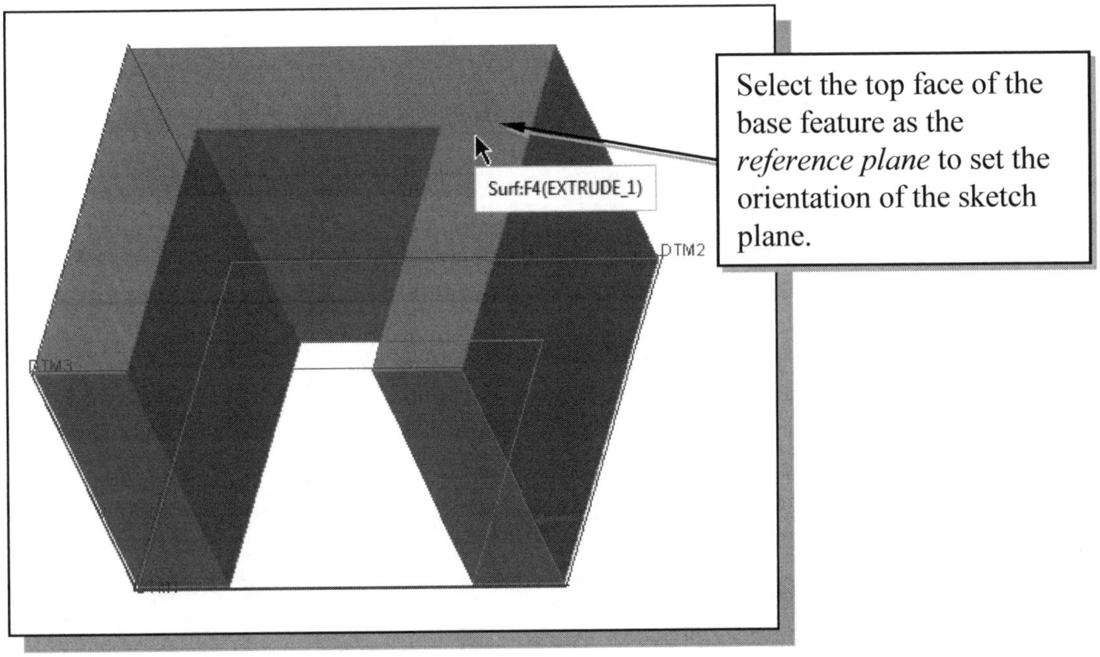

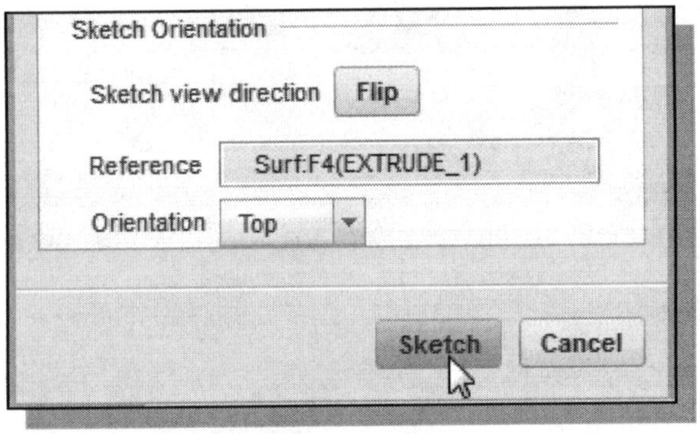

5. In the **Sketch Orientation** menu, pick **Top** to set the reference plane **Orientation**.

6. Pick **Sketch** to exit the *Section Placement* window and proceed to enter the *Creo Parametric Sketcher* mode.

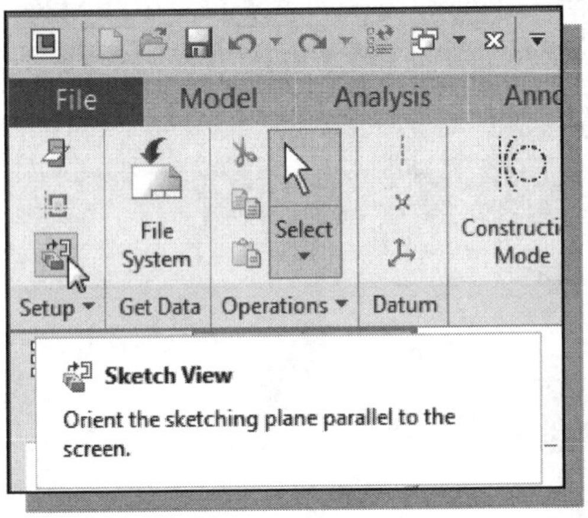

7. To orient the sketching plane parallel to the screen, click on the **Sketch View** icon in the *Setup* toolbar as shown.

8. Note that the top surface of the solid model and one of the datum planes are pre-selected as the sketching references to aid the positioning of the sketched geometry. In the graphics area, the two references are highlighted and displayed with two dashed lines.

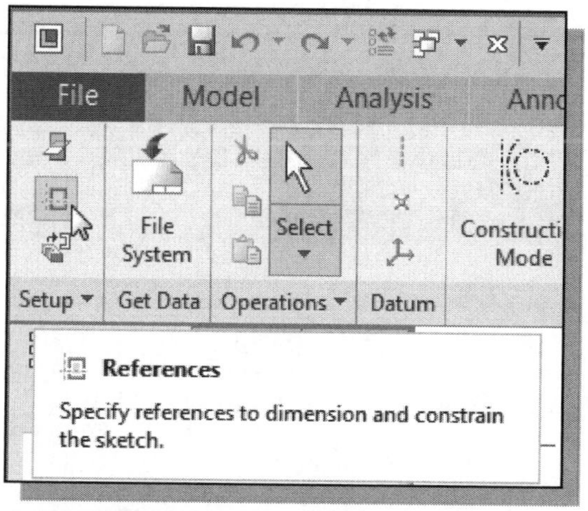

9. In the ***Setup*** toolbar, click on **References** to display the option list and select the **References** option.

➢ This will bring up the *References* dialog box.

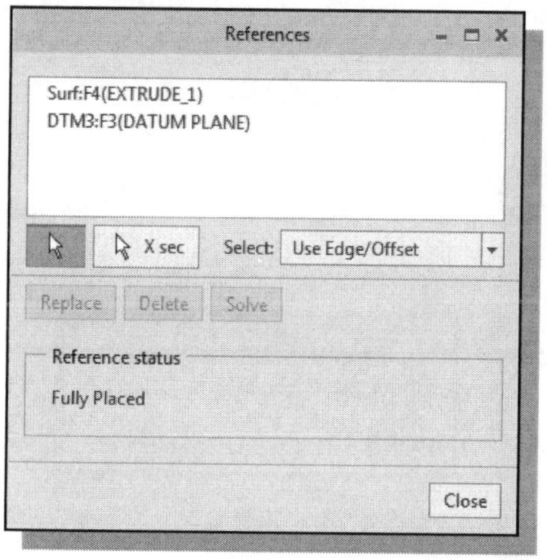

➢ Note that, in the *References* dialog box, the top surface of the solid model and **DTM3** are pre-selected as the sketching references as shown.

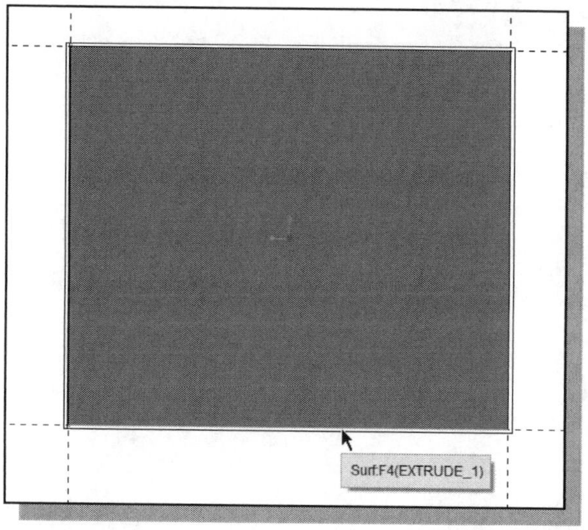

10. Select the **right edge** and the bottom edge of the base feature so that the four sides of the selected sketching plane, or corresponding datum planes, are used as references as shown.

11. Click **Solve** to apply the changes.

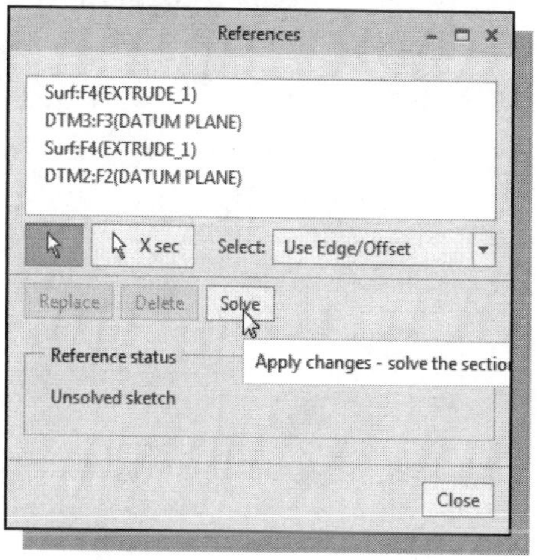

12. In the *References* dialog box, click on the **Close** button to accept the selections.

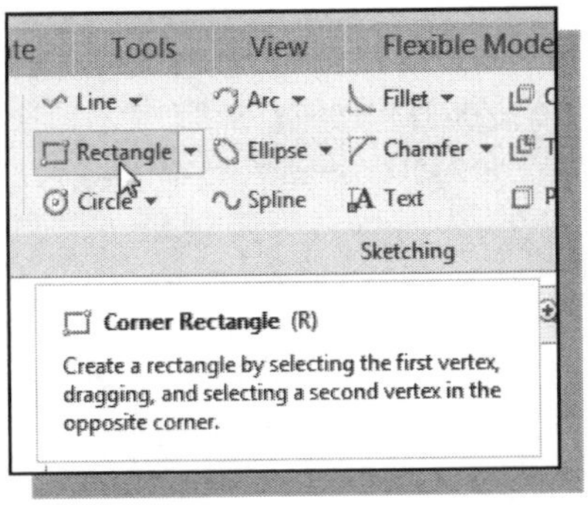

13. In the *Sketching* toolbar, click on the **Corner Rectangle** icon as shown to activate the Rectangle command.

14. Create a rectangle by clicking on the **lower left corner** of the solid model as shown below.

15. Move the cursor upward and place the opposite corner of the rectangle along the **right edge** of the base solid as shown below.

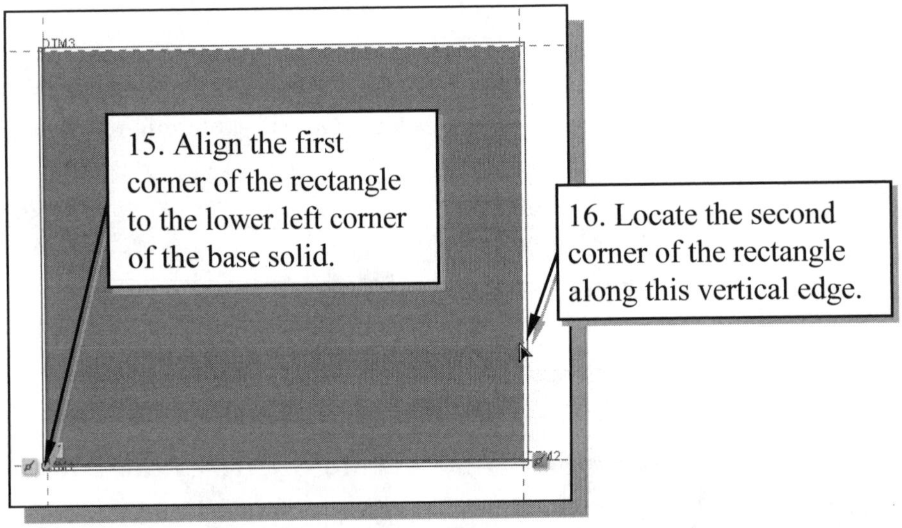

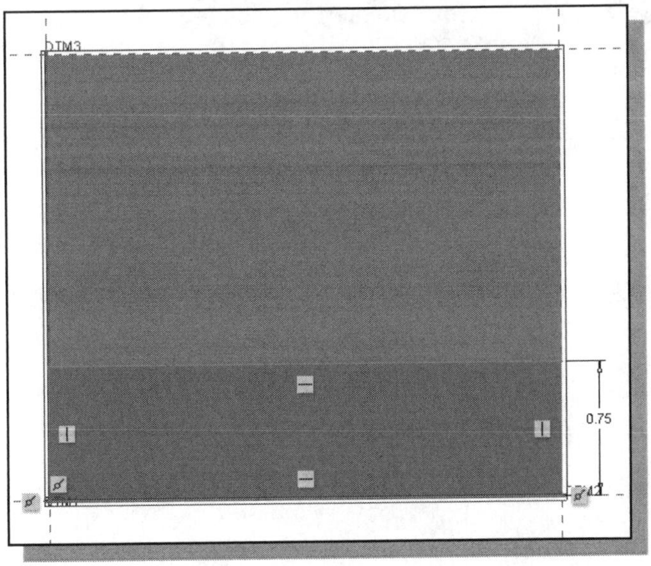

16. On your own, modify the height dimension to **0.75** as shown.

- Note that only one dimension, the height dimension, is applied to the 2D sketch; the width of the rectangle is defined by the references.

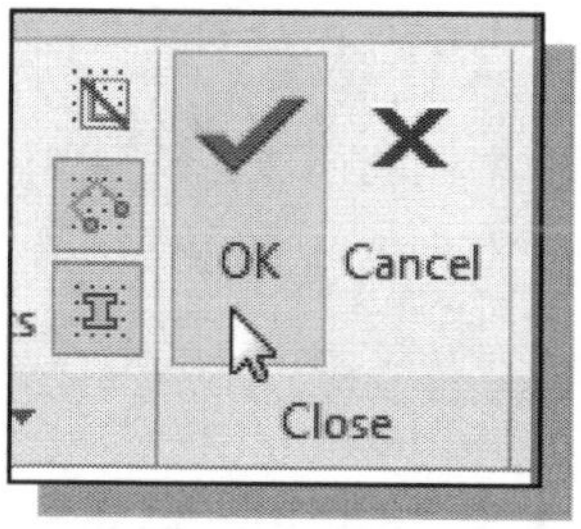

17. In the *Ribbon* toolbar, click on the **OK** icon to end the *Creo Parametric 2D Sketcher* and proceed to the next element of the feature definition.

18. In the *Feature Option Dashboard*, confirm the *depth value* option is set and enter **2.5** as the extrusion depth as shown.

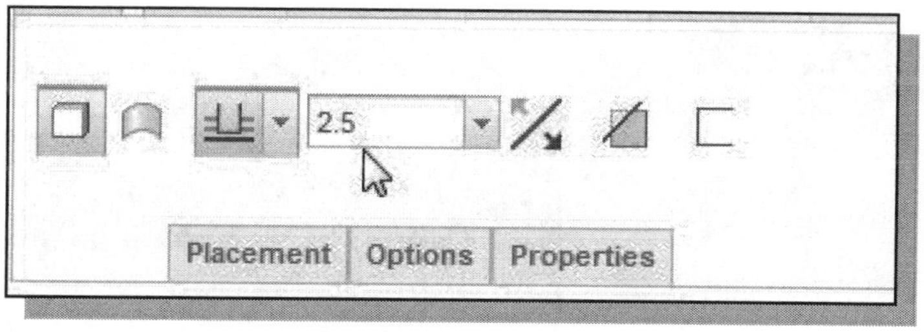

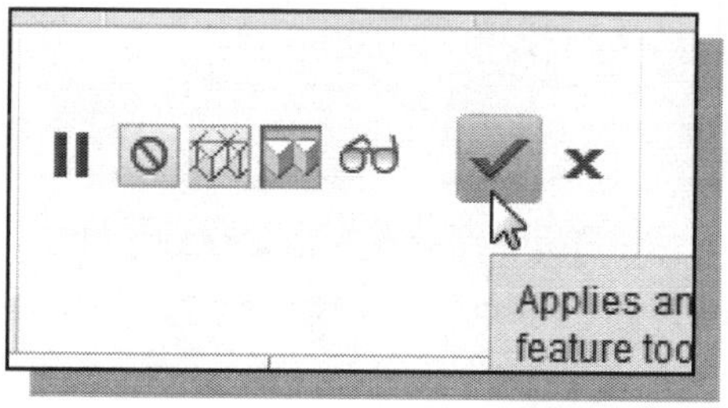

19. In the *dashboard* area, click **Accept** to proceed with the creation of the solid feature.

Creating a Cut Feature

We will create a circular cut as the next solid feature of the design. Note that the procedure in creating a *cut feature* is almost the same as creating a *protrusion feature.*

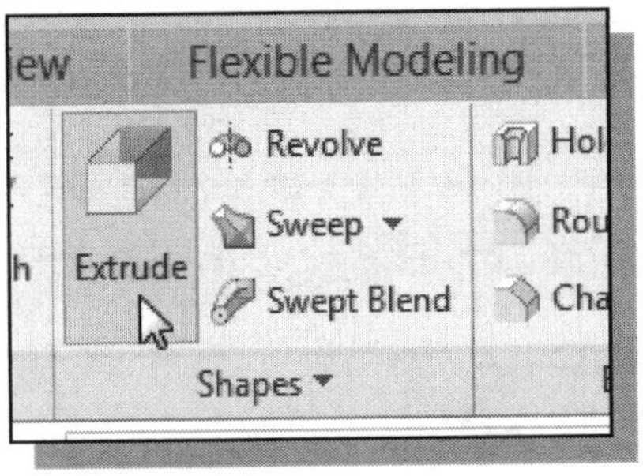

1. In the *Shapes* toolbar (the fourth toolbar in the *Ribbon* toolbar area), select the **Extrude** tool option as shown.

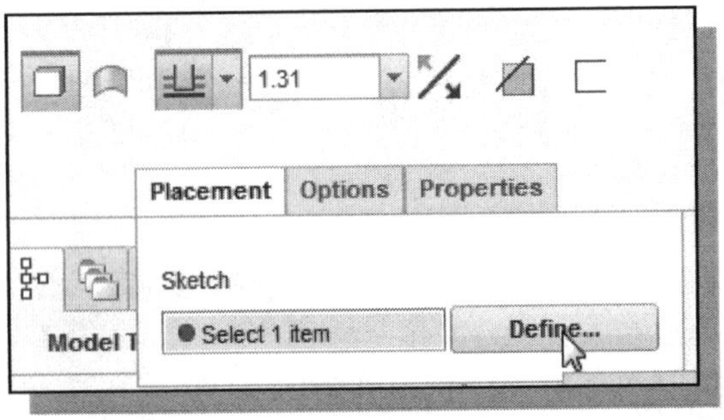

2. Click the **Placement** option and choose **Define** to begin creating a new *internal sketch.*

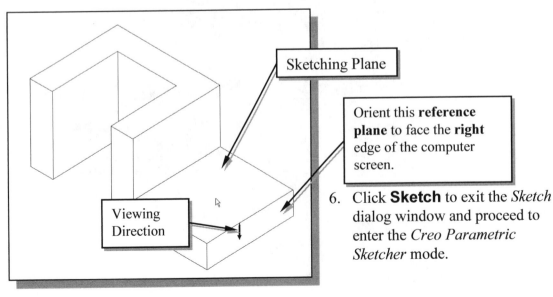

3. We will use the top surface of the last feature as the sketching plane. Click once, with the **left-mouse-button**, inside the top surface of the rectangular solid feature as shown in the figure below.

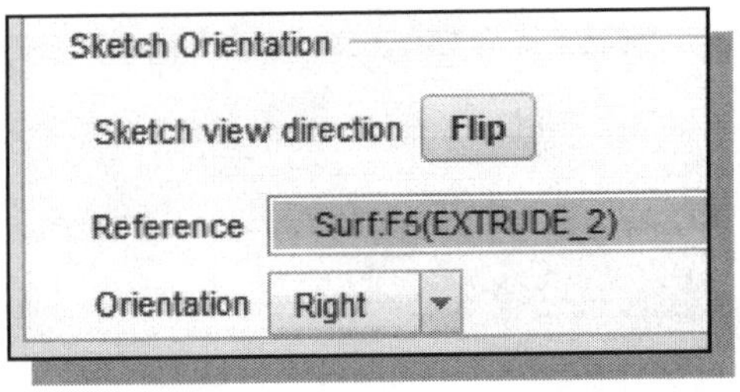

4. In the **Sketch Orientation** menu, confirm the reference plane **Orientation** is set to **Right**.

5. Pick the **right vertical face** of the second solid feature as the reference plane, which will be oriented toward the right edge of the computer screen.

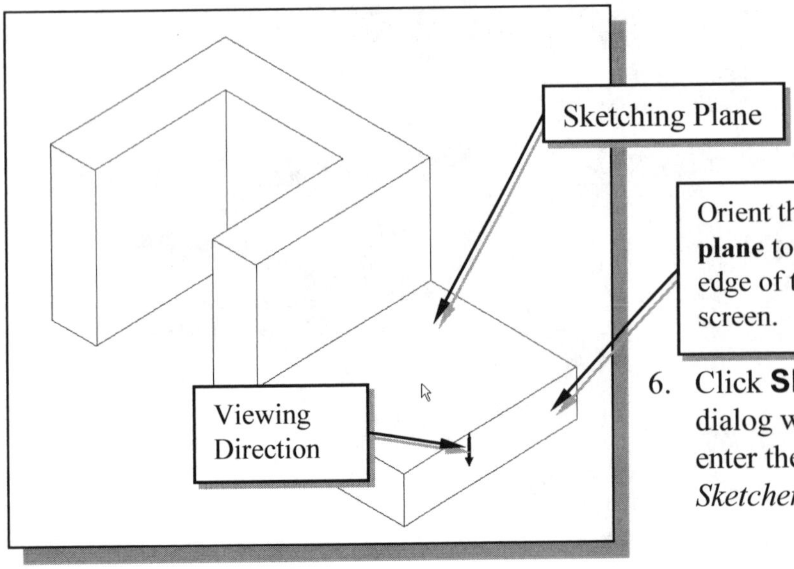

6. Click **Sketch** to exit the *Sketch* dialog window and proceed to enter the *Creo Parametric Sketcher* mode.

Creating the 2D Section of the Cut Feature

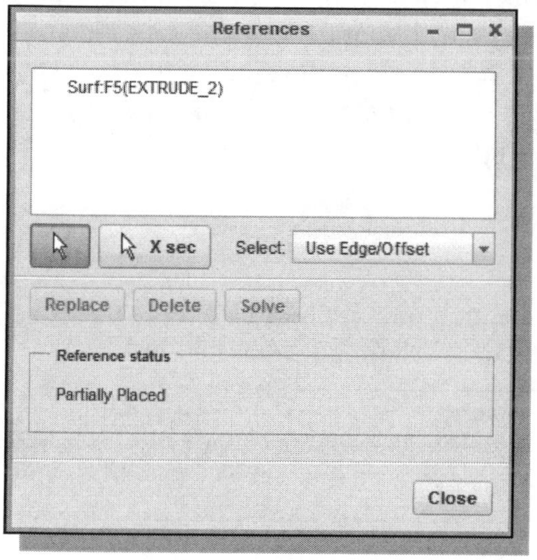

1. Note that the **right vertical plane** is pre-selected as a reference for the new sketch.

   - Note that at least one horizontal reference and one vertical reference are required to position a 2D sketch. We will need at least one more vertical reference for this sketch.

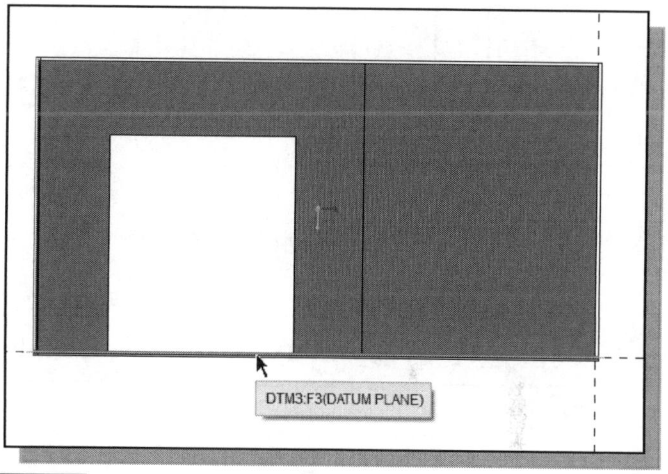

2. Select **DTM3** as the vertical sketching references as shown. In the graphics area, the two references are highlighted and displayed with two dashed lines.

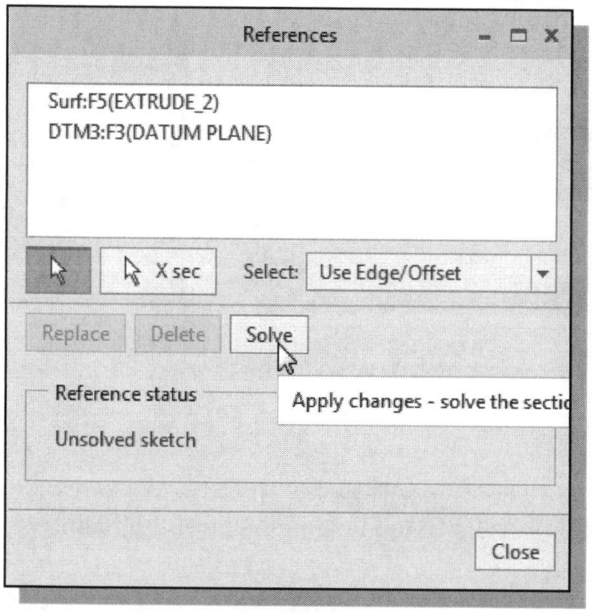

3. Click **Solve** to apply the changes.

4. Click on the **Close** button to accept the selected references and proceed to entering the *Creo Parametric Sketcher* module.

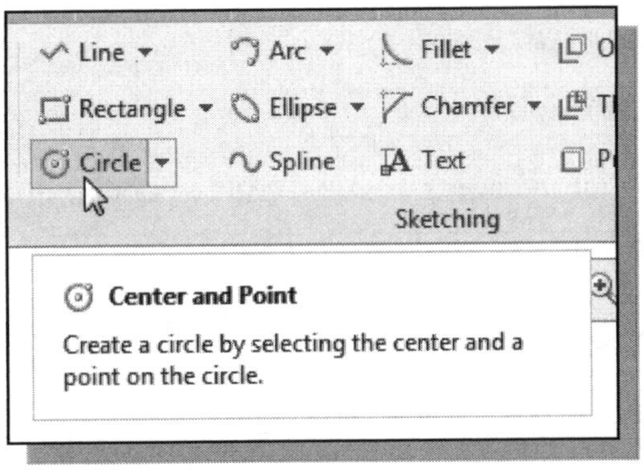

5. In the *Sketching* toolbar, select **Circle** as shown. The default option is to create a circle by specifying the center point and a point through which the circle will pass. The message "*Select the center of a circle*" is displayed in the message area.

6. On your own, create a circle of arbitrary size on the sketching plane as shown.

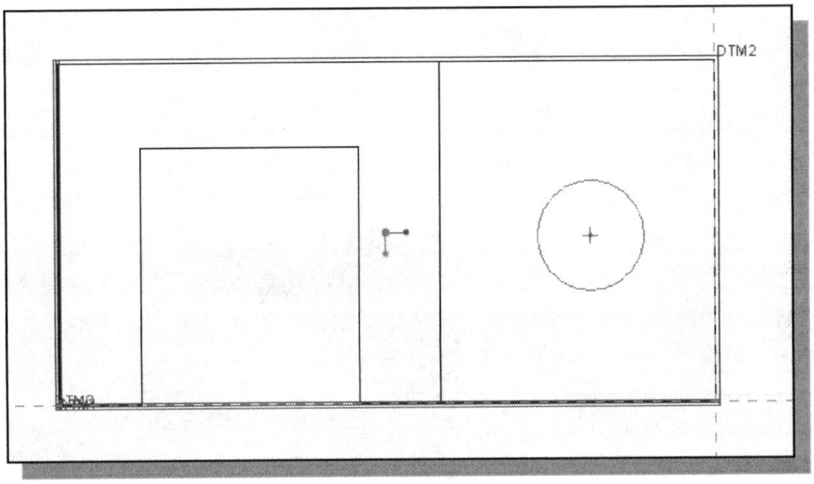

7. On your own, edit/modify the dimensions as shown.

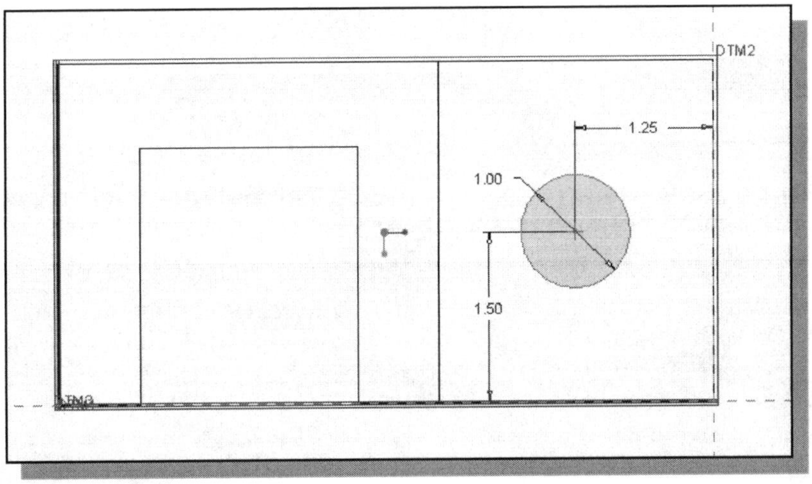

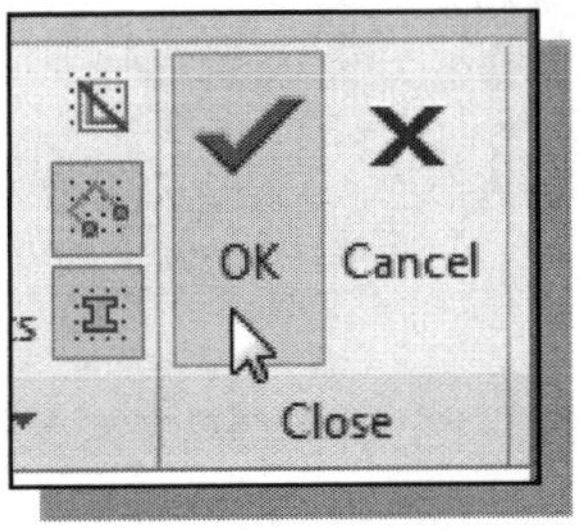

8. In the *Ribbon* toolbar, click **OK** to exit the *Creo Parametric 2D Sketcher* and proceed to the next element of the feature definition.

9. Switch on the **Remove Material** option as shown in the figure below.

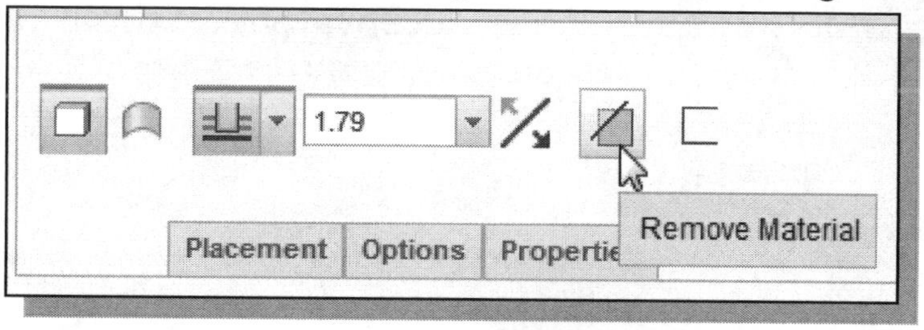

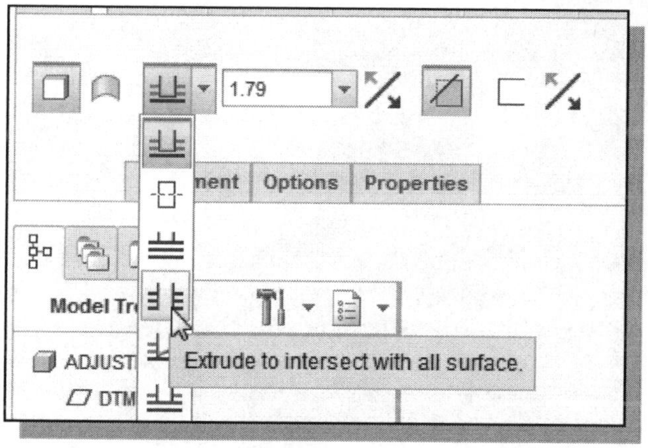

10. In the *Feature Option Dashboard*, select the **Extrude to intersect with all surface** option as shown.

- Note that this **Thru All** option does not require us to enter a value to define the depth of the extrusion; *Creo Parametric* will calculate the required value to assure the extrusion is through the entire solid model.

11. On your own, use the *Dynamic Rotate* function to view the feature.

12. Click on the **Flip direction** icon as shown in the figure below to set the cut direction.

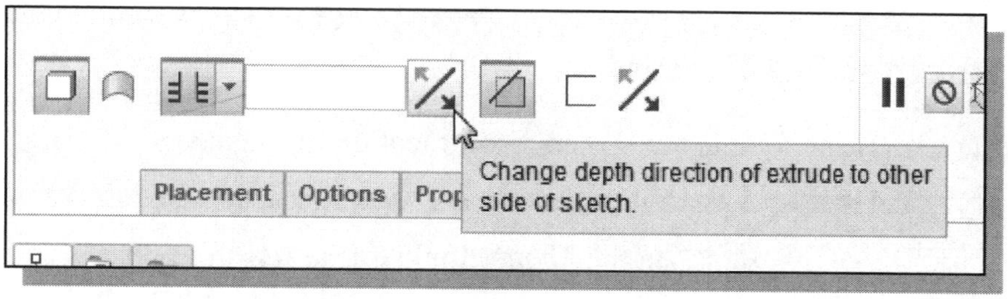

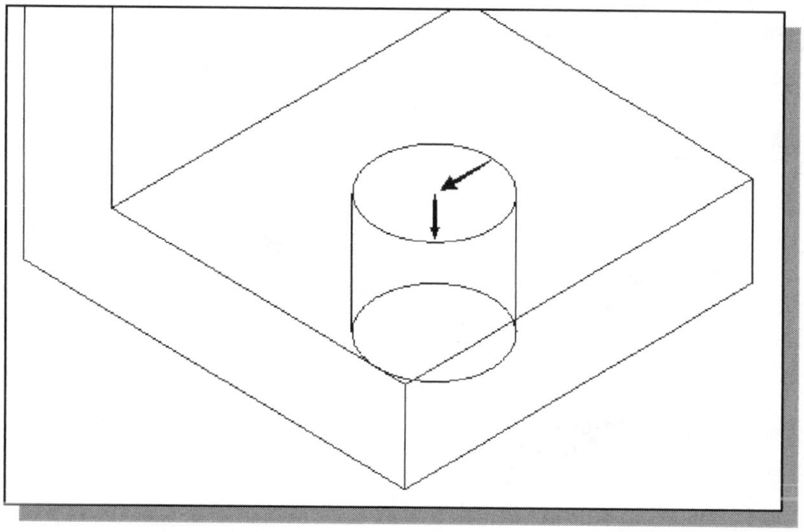

13. Click on **Accept** to proceed with the extrusion option.

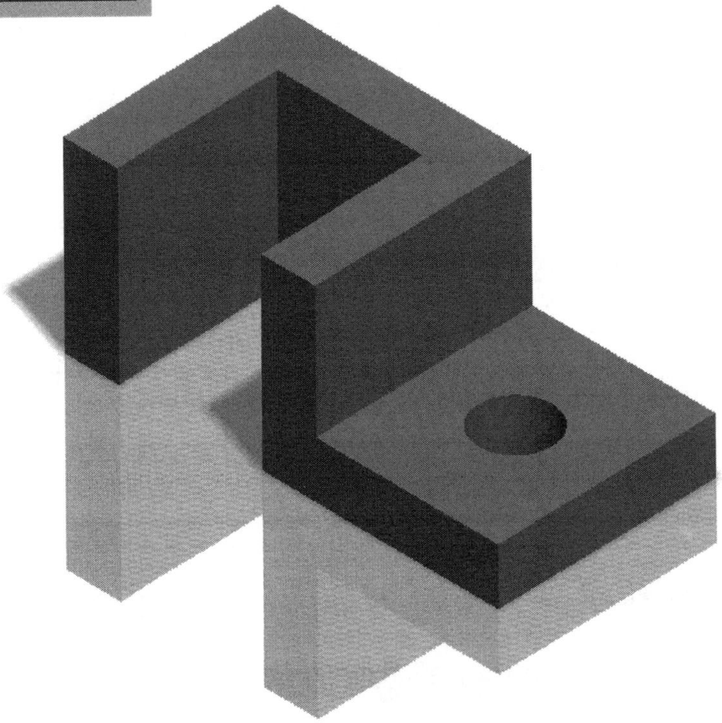

Creating another Cut Feature

We will next create a triangular cut as the next solid feature of the design.

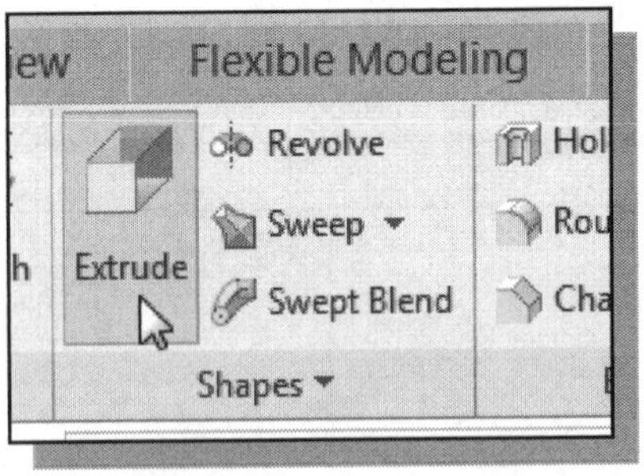

1. In the *Shapes* toolbar (the fourth toolbar in the *Ribbon* toolbar area), select the **Extrude** tool option as shown.

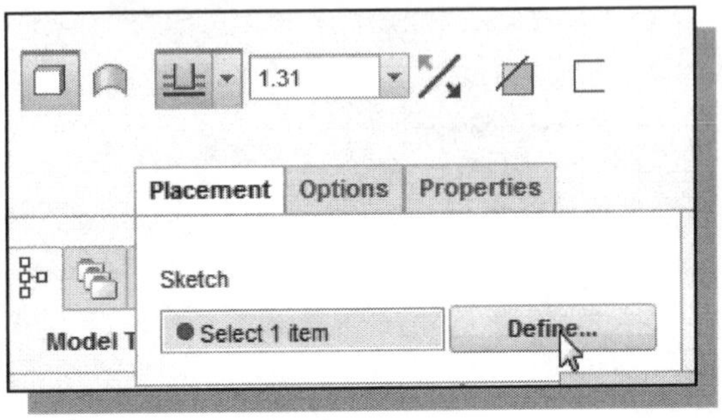

2. Click the **Placement** option and choose **Define** to begin creating a new *internal sketch.*

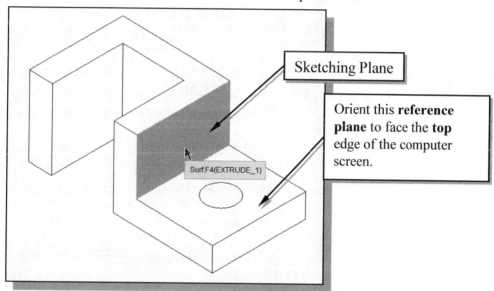

3. We will use the right vertical surface of the first solid feature as the sketching plane. Click once, with the **left-mouse-button**, inside the top surface of the rectangular solid feature as shown in the figure below.

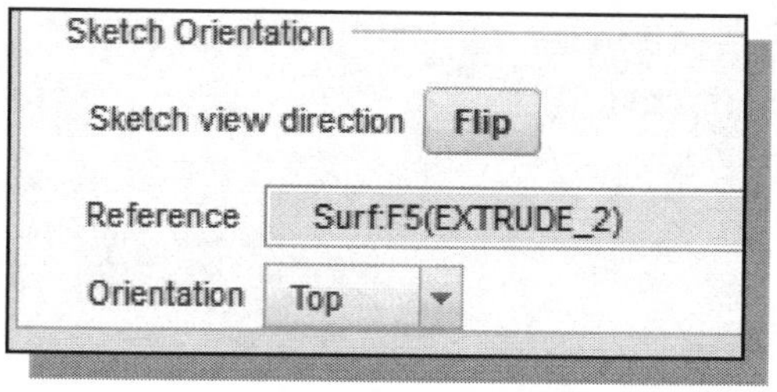

4. In the **Sketch Orientation** menu, set the reference plane **Orientation** to **top**.

5. Pick the **top horizontal face** of the second solid feature as the reference plane, which will be oriented toward the top edge of the computer screen.

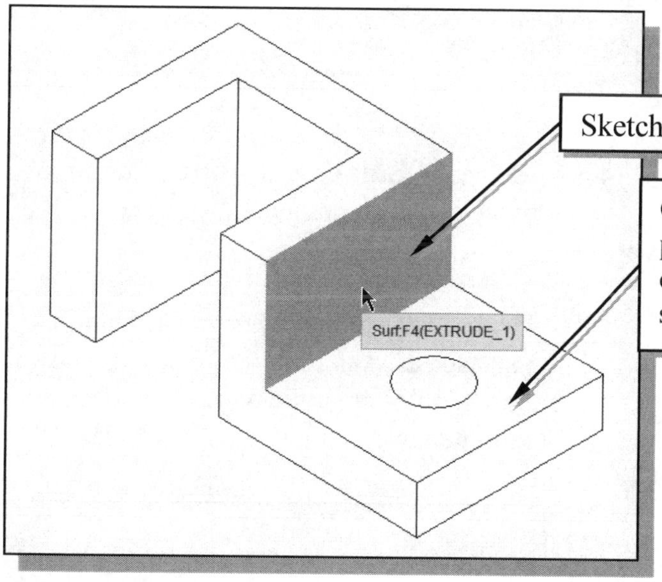

6. Click **Sketch** to exit the *Sketch* dialog window and proceed to enter the *Creo Parametric Sketcher* mode.

Delete/Select the Sketching References

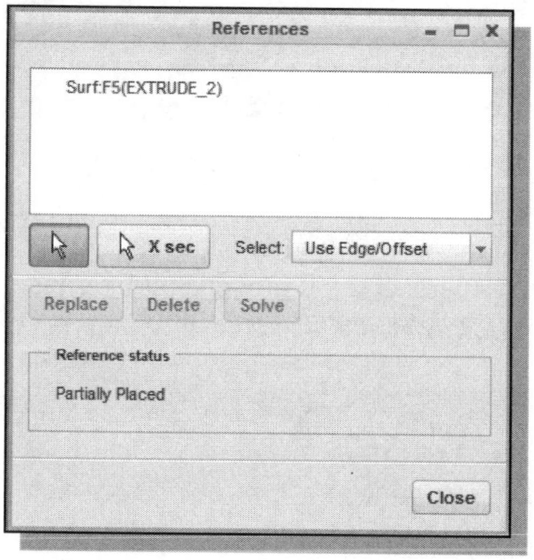

1. Note that currently the **top horizontal plane** of the second solid feature is pre-selected as a reference for the new sketch.

   - Note that at least one horizontal reference and one vertical reference are required to position a 2D sketch. We will need at least one more vertical reference for this sketch.

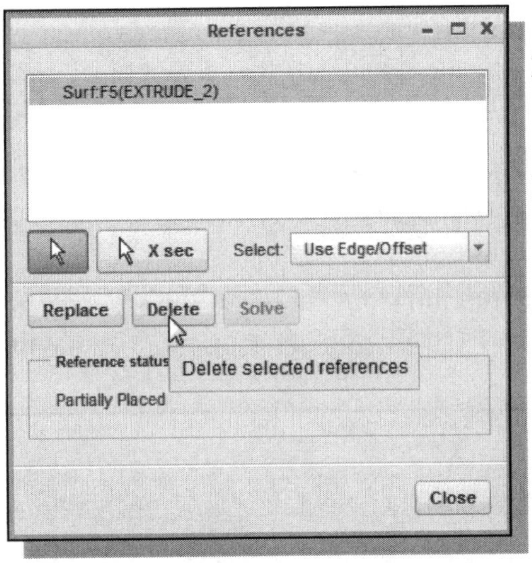

2. Select the **referenced surface** in the item list.

3. Click **Delete** to remove the item as a reference plane.

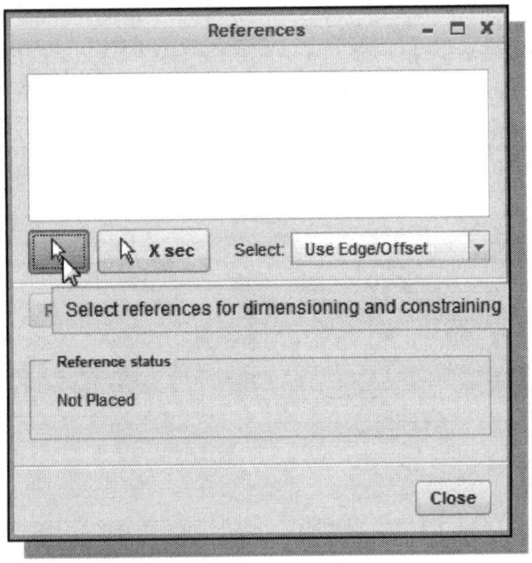

4. Click on the **Select Reference** button to activate select command.

5. Select the left vertical edge, **DTM3**, and the top surface as sketching references as shown. In the graphics area, the two references are highlighted and displayed with two dashed lines.

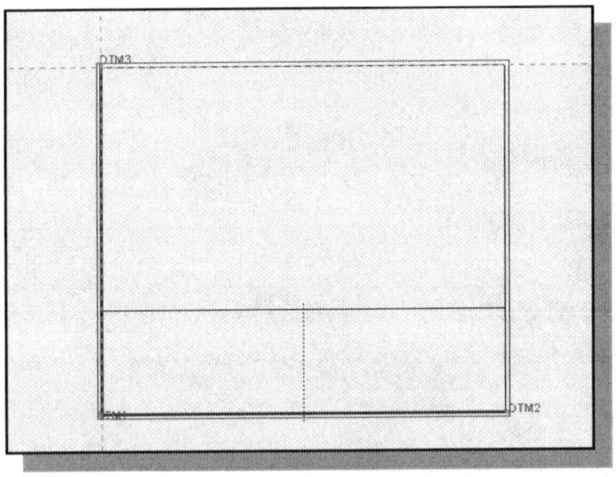

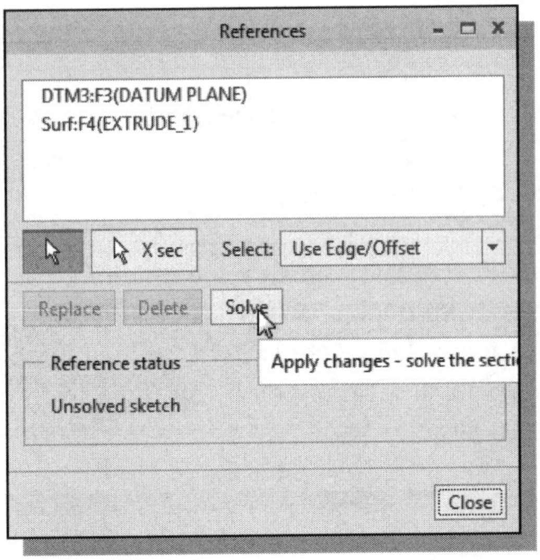

6. Click **Solve** to apply the changes.

7. Click on the **Close** button to accept the selected references and proceed to entering the *Creo Parametric Sketcher* module.

Create the 2D Section

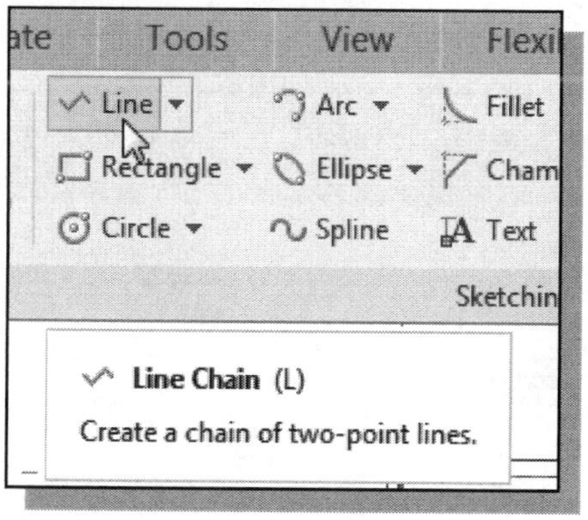

1. In the *Sketching* toolbar, select **Line** as shown. The default option is to create a chain of lines by specifying the endpoint of the line segments.

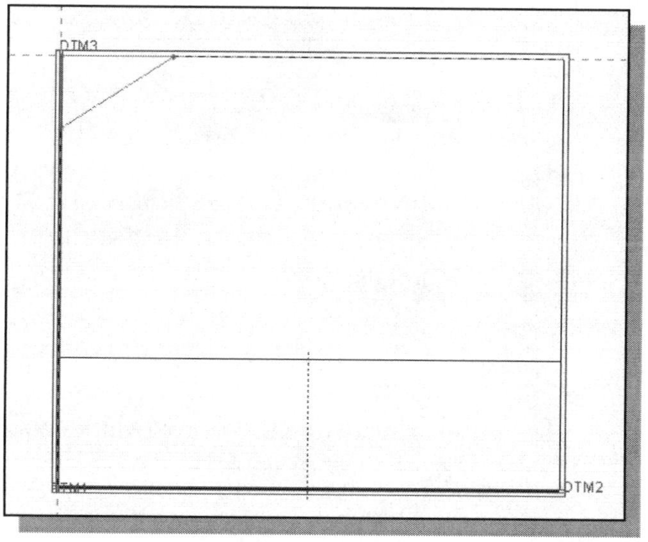

2. On your own, starting at the top left corner and create a triangle of arbitrary size on the sketching plane as shown.

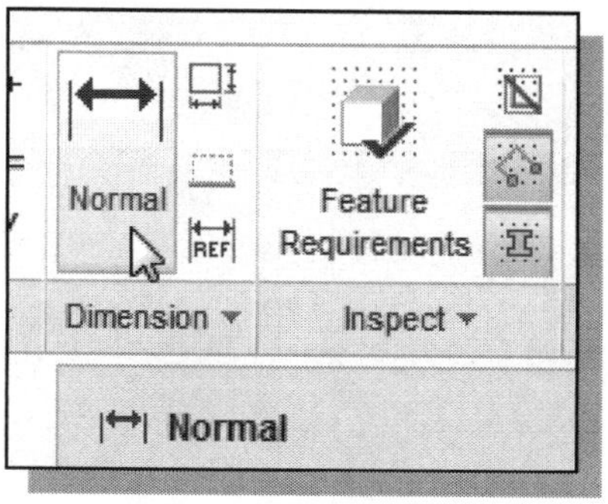

3. On your own, edit/modify the length dimension of the top edge of the triangle as shown.

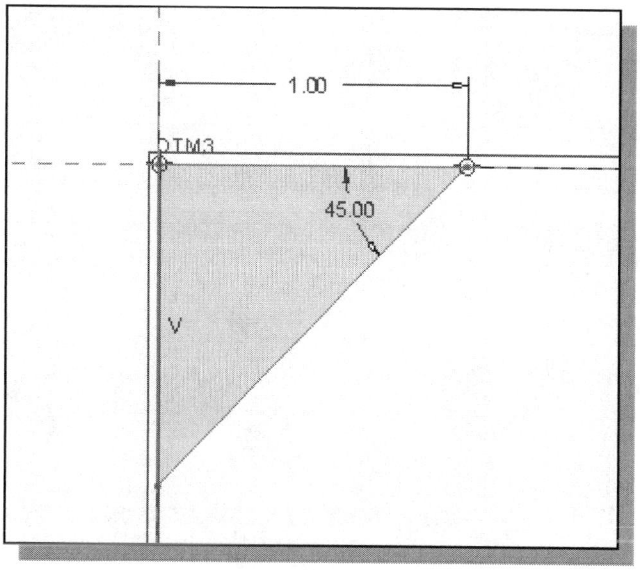

4. Create the angle dimension by selecting the two adjacent lines and place the angular dimension *inside* the *desired quadrant*.

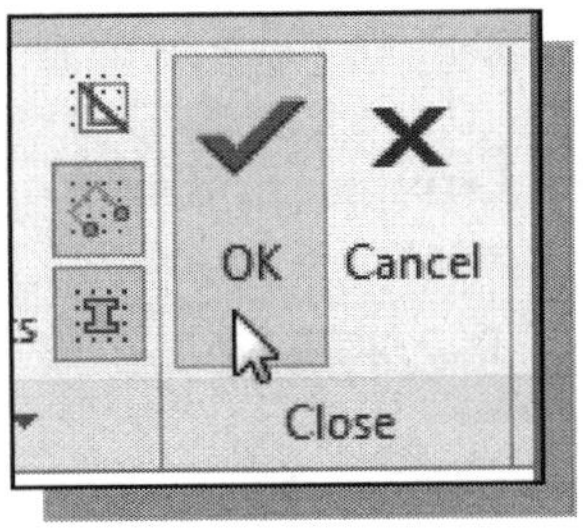

5. In the *Ribbon* toolbar, click **OK** to exit the *Creo Parametric 2D Sketcher* and proceed to the next element of the feature definition.

6. Switch on the **Remove Material** option as shown in the figure below.

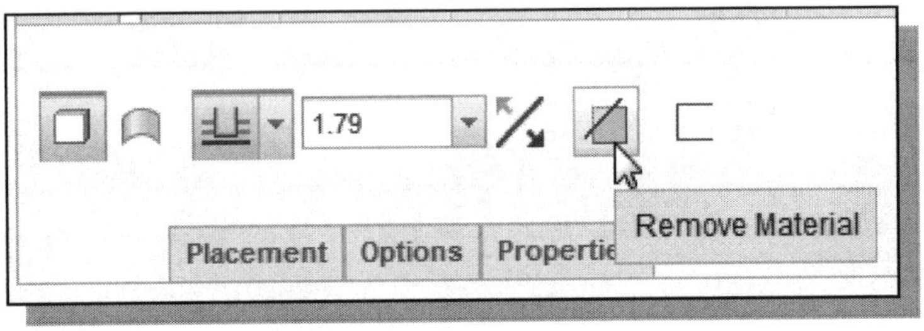

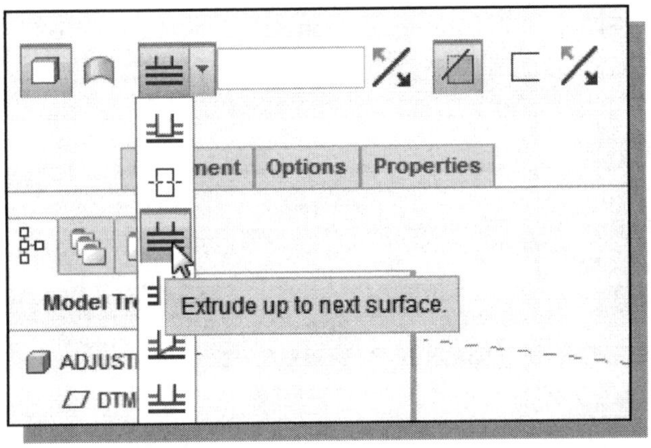

7. In the *Feature Option Dashboard*, select the **Extrude to Next surface** option as shown.

- Note that with the **Extrude to Next surface** option; *Creo Parametric* will calculate the required value to assure the extrusion is to the next surface.

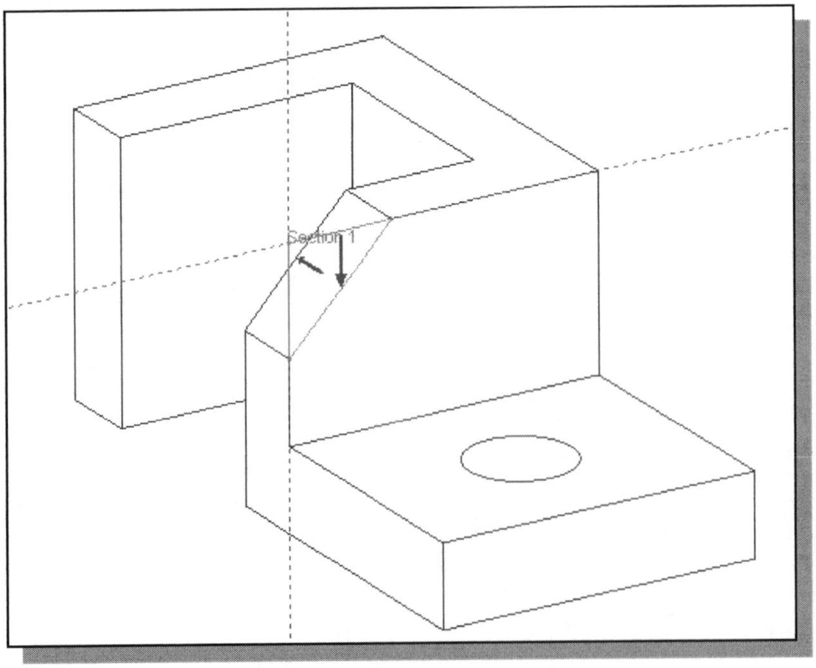

8. Click on **Accept** to proceed with the extrusion option.

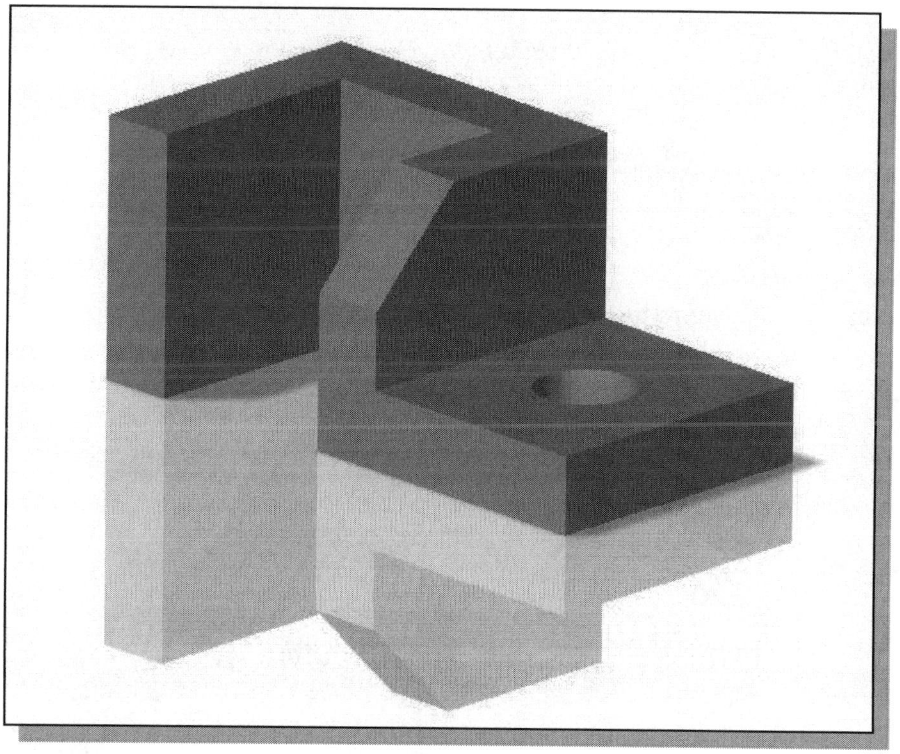

Save the Part

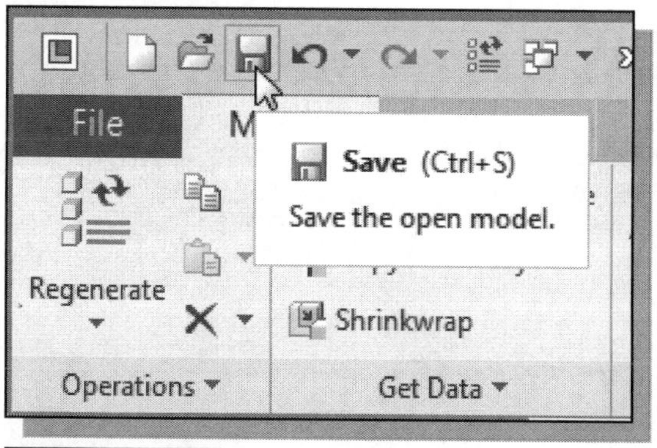

1. Select **Save** in the *Standard* toolbar, or you can also use the "**Ctrl-S**" combination (press down the **[Ctrl]** key and hit the **[S]** key once) to save the part.

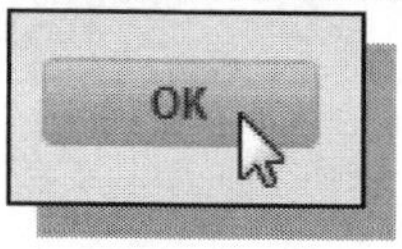

2. In the *message area*, the part name is displayed. Click on the **OK** button to save the file.

❖ It is a good habit to save your model periodically. In general, you should save your work onto the disk at an interval of every 15 to 20 minutes.

Review Questions

1. The truss element used in finite element analysis is considered as a two-force member element. List and describe the assumptions of a two-force member.

2. What is the size of the stiffness matrix for a single element? What is the size of the overall global stiffness matrix in example 1.2?

3. What is the first thing we should set up when building a new CAD model in *Creo Parametric*?

4. How does the *Creo Parametric Intent Manager* assist us in sketching?

5. How do we remove the dimensions created by the *Intent Manager*?

6. How do we modify more than one dimension at a time?

7. Identify and describe the following commands:

(a)

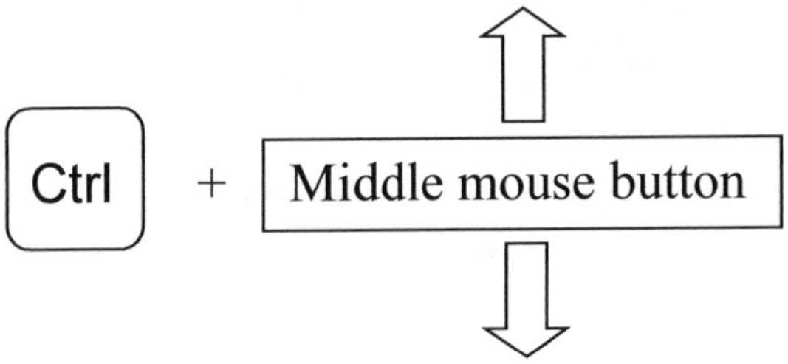

(b)

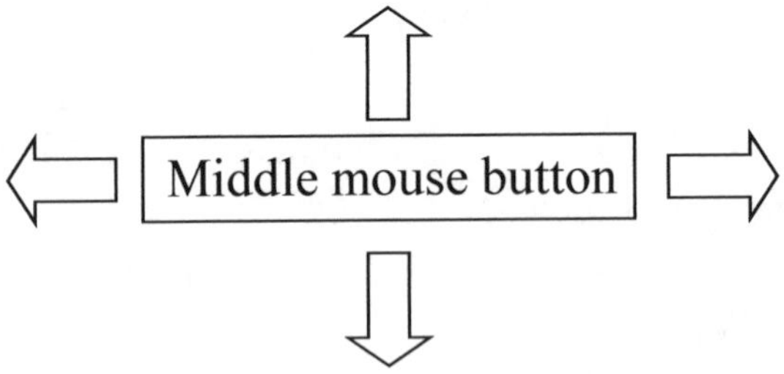

Exercises

1. For the one-dimensional 3 truss-element system shown, determine the nodal displacements and reaction forces using the direct stiffness method.

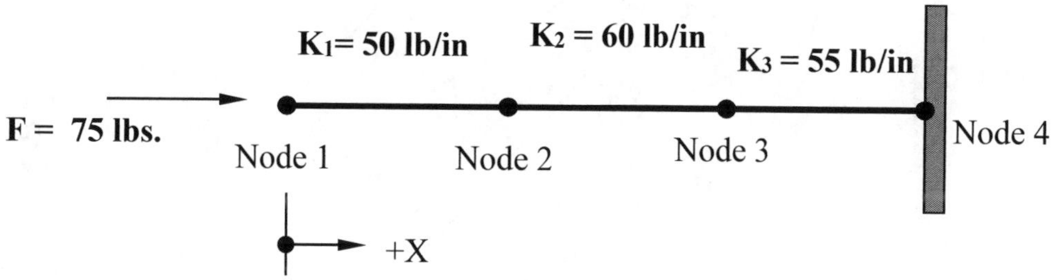

2. For the one-dimensional 4-member truss-element system shown, determine the nodal displacements and reaction forces using the direct stiffness method.

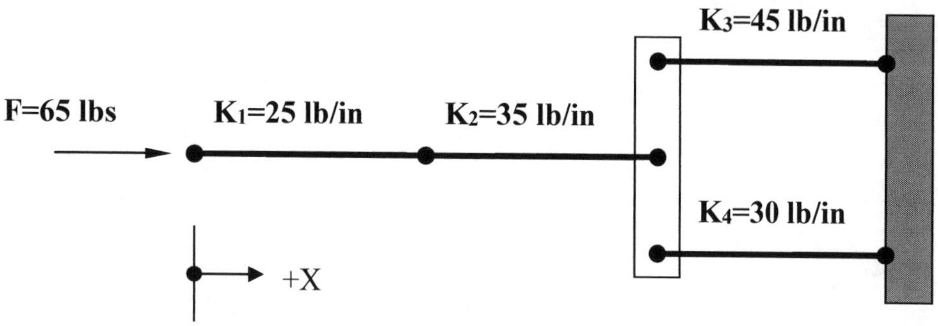

3.

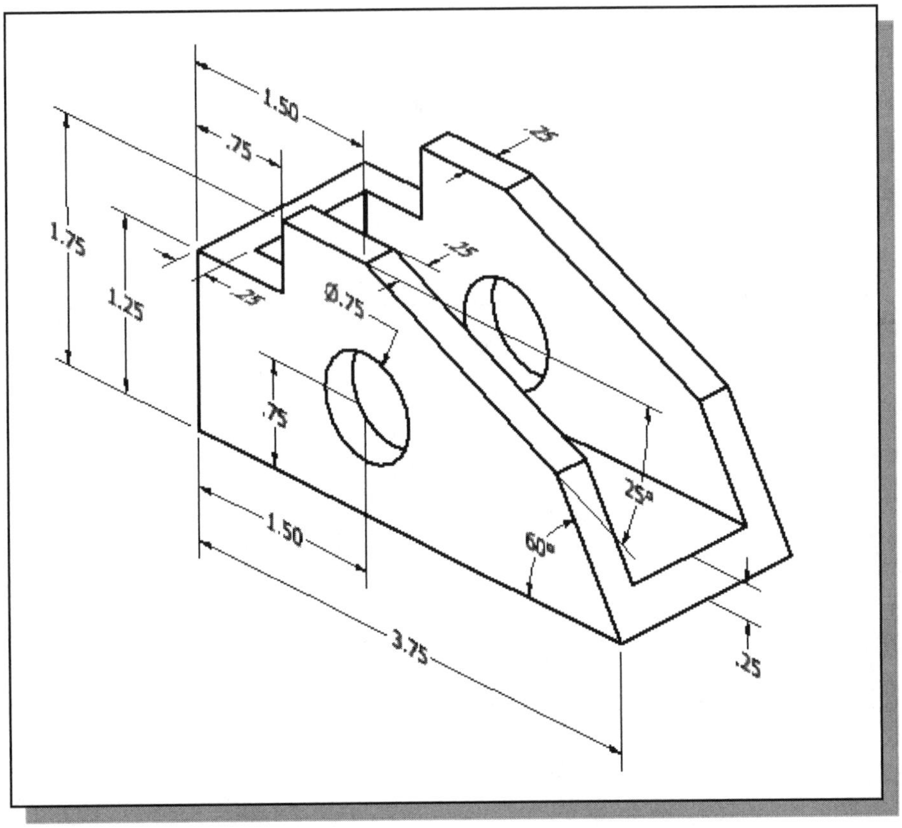

Chapter 2
Truss Elements in Two-Dimensional Spaces

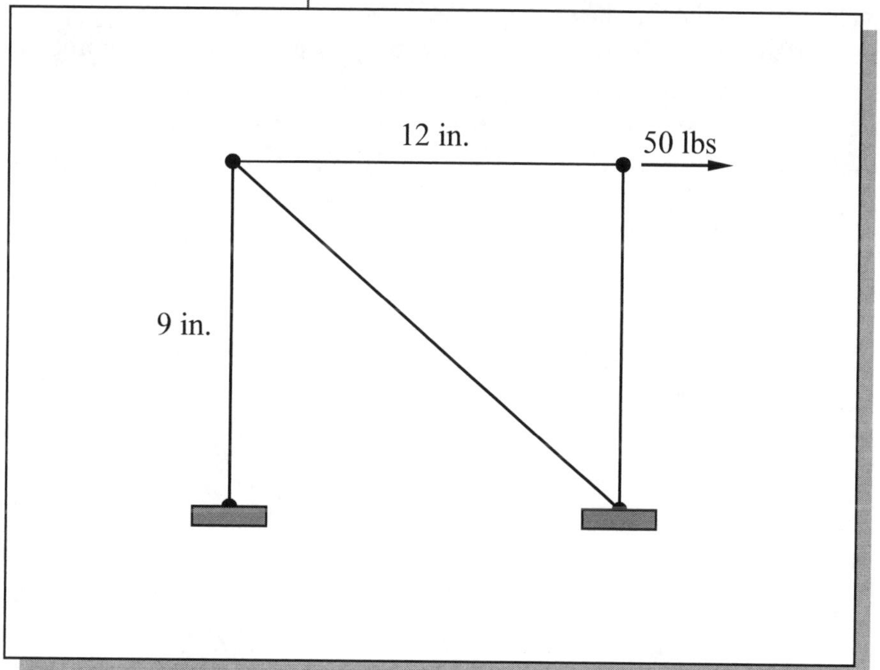

Learning Objectives

- Perform 2D Coordinates Transformation.
- Expand the Direct Stiffness Method to 2D Trusses.
- Derive the general 2D element Stiffness Matrix.
- Assemble the Global Stiffness Matrix for 2D Trusses.
- Solve 2D trusses using the Direct Stiffness Method.

Introduction

This chapter presents the formulation of the direct stiffness method of truss elements in a two-dimensional space and the general procedure for solving two-dimensional truss structures using the direct stiffness method. The primary focus of this text is on the aspects of finite element analysis that are more important to the user than the programmer. However, for a user to utilize the software correctly and effectively, some understanding of the element formulation and computational aspects are also important. In this chapter, a two-dimensional truss structure consisting of two truss elements (as shown below) is used to illustrate the solution process of the direct stiffness method.

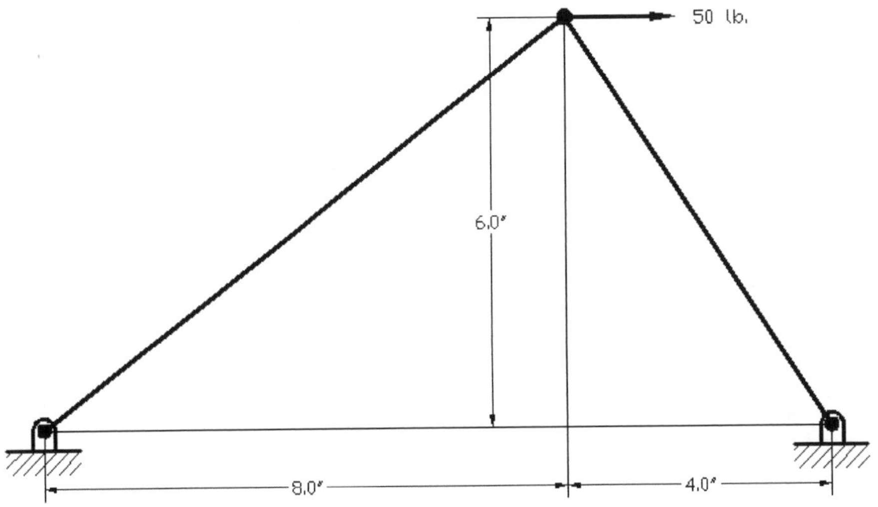

Truss Elements in Two-Dimensional Spaces

As introduced in the previous chapter, the system equations (stiffness matrix) of a truss element can be represented using the system equations of a linear spring in one-dimensional space.

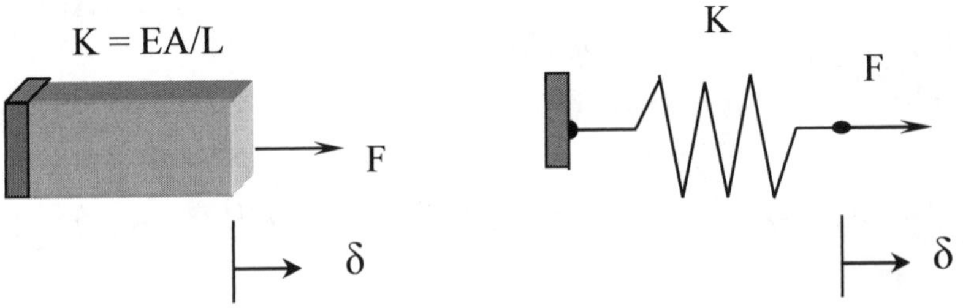

Free Body Diagram:

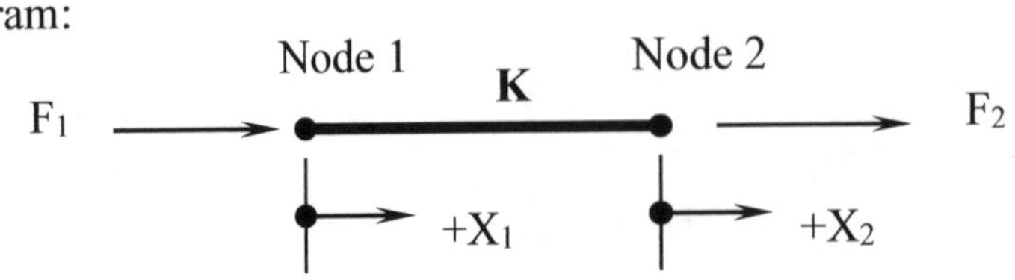

The general force-displacement equations in matrix form:

$$\begin{Bmatrix} F_1 \\ F_2 \end{Bmatrix} = \begin{bmatrix} +K & -K \\ -K & +K \end{bmatrix} \begin{Bmatrix} X_1 \\ X_2 \end{Bmatrix}$$

For a truss element, **K = EA/L**

$$\begin{Bmatrix} F_1 \\ F_2 \end{Bmatrix} = \frac{\mathbf{EA}}{\mathbf{L}} \begin{bmatrix} +1 & -1 \\ -1 & +1 \end{bmatrix} \begin{Bmatrix} X_1 \\ X_2 \end{Bmatrix}$$

For truss members positioned in two-dimensional space, two coordinate systems are established:

1. The global coordinate system (**X** and **Y** axes) chosen to represent the entire structure.
2. The local coordinate system (***X*** and ***Y*** axes) selected to align the ***X*** axis along the length of the element.

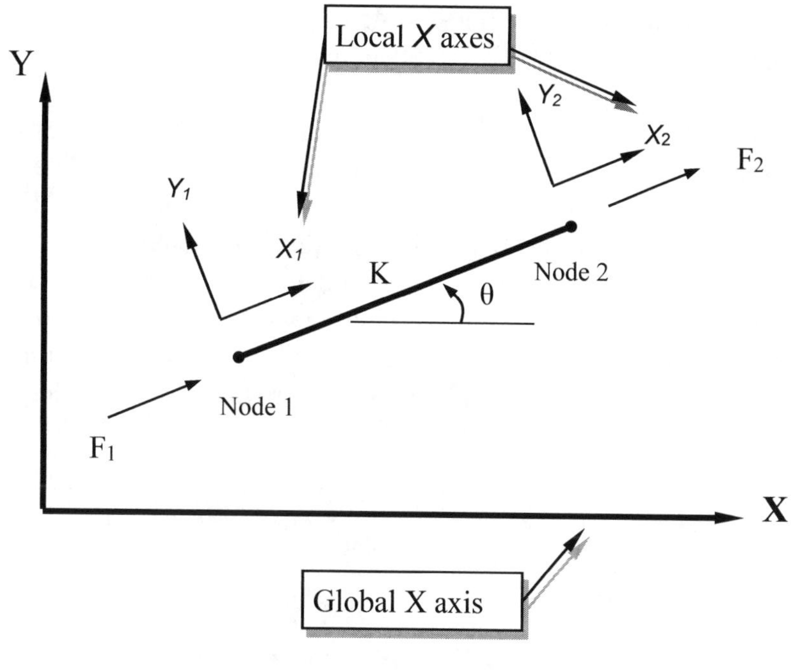

The force-displacement equations expressed in terms of components in the local *XY* coordinate system:

$$\begin{Bmatrix} F_{1X} \\ F_{2X} \end{Bmatrix} = \frac{\mathbf{EA}}{\mathbf{L}} \begin{bmatrix} +1 & -1 \\ -1 & +1 \end{bmatrix} \begin{Bmatrix} X_1 \\ X_2 \end{Bmatrix}$$

The above stiffness matrix (system equations in matrix form) can be expanded to incorporate the two force components at each node and the two displacement components at each node.

Force Components (Local Coordinate System)

$$\begin{Bmatrix} F_{1X} \\ F_{1Y} \\ F_{2X} \\ F_{2Y} \end{Bmatrix} = \frac{\mathbf{EA}}{\mathbf{L}} \begin{pmatrix} +1 & 0 & -1 & 0 \\ 0 & 0 & 0 & 0 \\ -1 & 0 & +1 & 0 \\ 0 & 0 & 0 & 0 \end{pmatrix} \begin{Bmatrix} X_1 \\ Y_1 \\ X_2 \\ Y_2 \end{Bmatrix}$$

Nodal Displacements (Local Coordinate System)

In regard to the expanded local stiffness matrix (system equations in matrix form):

1. It is always a square matrix.
2. It is always symmetrical for linear systems.
3. The diagonal elements are always positive or zero.

The above stiffness matrix, expressed in terms of the established 2D local coordinate system, represents a single truss element in a two-dimensional space. In a general structure, many elements are involved, and they would be oriented with different angles. The above stiffness matrix is a general form of a <u>SINGLE</u> element in a 2D local coordinate system. Imagine the number of coordinate systems involved for a 20-member structure. For the example that will be illustrated in the following sections, two local coordinate systems (one for each element) are needed for the truss structure shown below. The two local coordinate systems (X_1Y_1 and X_2Y_2) are aligned to the elements.

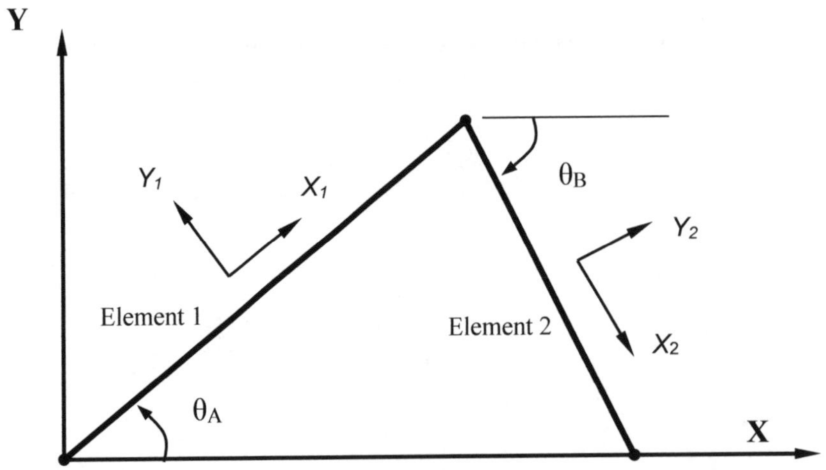

In order to solve the system equations of two-dimensional truss structures, it is necessary to assemble all elements' stiffness matrices into a **global stiffness matrix**, with all the equations of the individual elements referring to a common global coordinate system. This requires the use of *coordinate transformation equations* applied to system equations for all elements in the structure. For a one-dimensional truss structure (illustrated in chapter 2), the local coordinate system coincides with the global coordinate system; therefore, no coordinate transformation is needed to assemble the global stiffness matrix (the stiffness matrix in the global coordinate system). In the next section, the coordinate transformation equations are derived for truss elements in two-dimensional spaces.

Coordinate Transformation

A vector, in a two-dimensional space, can be expressed in terms of any coordinate system set of unit vectors.

For example,

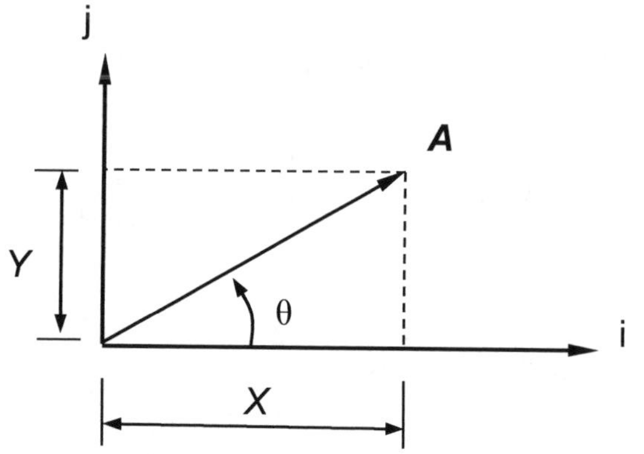

Vector ***A*** can be expressed as:

$$\mathbf{A} = X\,i + Y\,j$$

Where i and j are unit vectors along the X- and Y-axes.

Magnitudes of X and Y can also be expressed as:

$$X = A\cos(\theta)$$
$$Y = A\sin(\theta)$$

Where X, Y and A are scalar quantities.

Therefore,

$$\mathbf{A} = X\,i + Y\,j = A\cos(\theta)\,i + A\sin(\theta)\,j \quad \text{----------} \ (1)$$

Next, establish a new unit vector (u) in the same direction as vector $\boldsymbol{A}$.

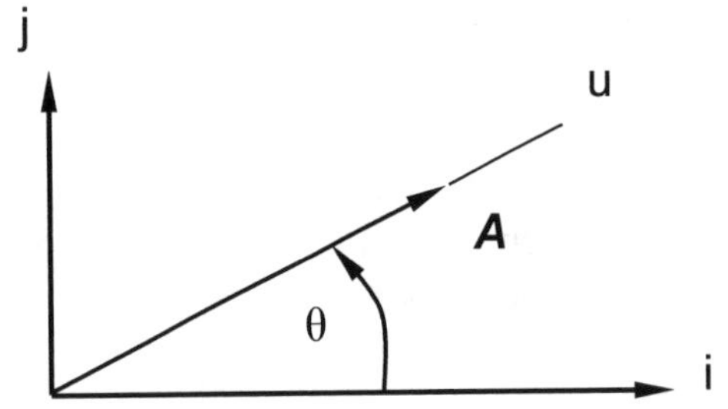

Vector A can now be expressed as: $\boldsymbol{A} = A\,u$ ------------ (2)

Both equations (the above (1) and (2)) represent vector $\boldsymbol{A}$:

$$\boldsymbol{A} = A\,u = A\cos(\theta)\,i + A\sin(\theta)\,j$$

The unit vector u can now be expressed in terms of the original set of unit vectors i and j:

$$\boxed{u = \cos(\theta)\,i + \sin(\theta)\,j}$$

Now consider another vector $\boldsymbol{B}$:

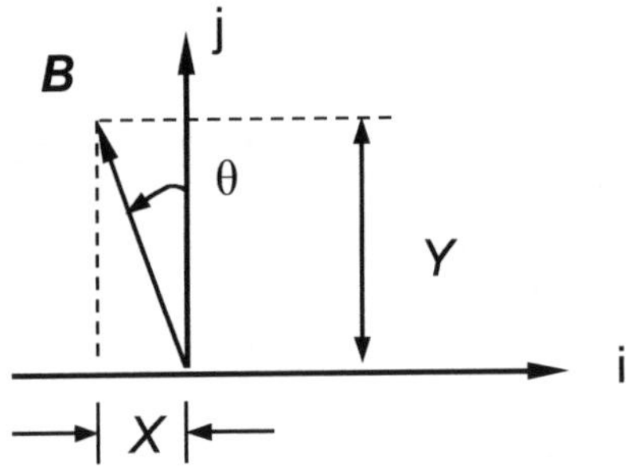

Vector $\boldsymbol{B}$ can be expressed as:

$$\boldsymbol{B} = -X\,i + Y\,j$$

Where i and j are unit vectors along the X- and Y-axes.

Magnitudes of X and Y can also be expressed as components of the magnitude of the vector:

$$X = B\sin(\theta)$$
$$Y = B\cos(\theta)$$

Where X, Y and B are scalar quantities.

Therefore,

$$\boldsymbol{B} = -X\,i + Y\,j = -B\sin(\theta)\,i + B\cos(\theta)\,j \text{ --------- (3)}$$

Next, establish a new unit vector (v) along vector $\boldsymbol{B}$.

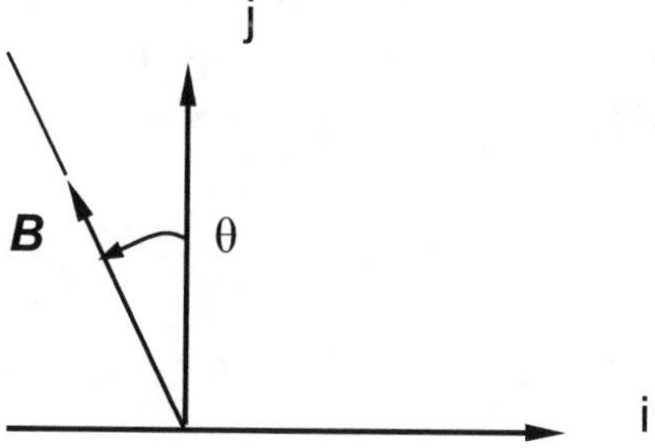

Vector $\boldsymbol{B}$ can now be expressed as: $\boldsymbol{B} = B\, v$ ------------- (4)

Equations (3) and (4) represent vector $\boldsymbol{B}$:

$$\boldsymbol{B} = B\, v = -\,B \sin(\theta)\, i + B \cos(\theta)\, j$$

The unit vector v can now be expressed in terms of the original set of unit vectors i and j:

$$v = -\sin(\theta)\, i + \cos(\theta)\, j$$

We have established the coordinate transformation equations that can be used to transform vectors from ij coordinates to the rotated uv coordinates.

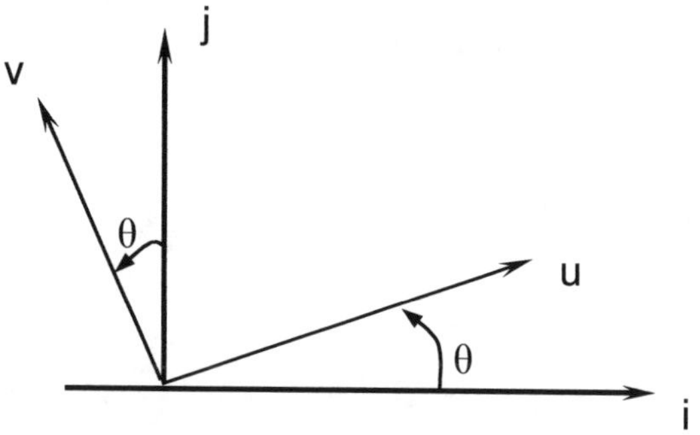

Coordinate Transformation Equations:

$$u = \cos(\theta)\, i + \sin(\theta)\, j$$
$$v = -\sin(\theta)\, i + \cos(\theta)\, j$$

Direction cosines

In matrix form,

$$\begin{Bmatrix} u \\ v \end{Bmatrix} = \begin{bmatrix} \cos(\theta) & \sin(\theta) \\ -\sin(\theta) & \cos(\theta) \end{bmatrix} \begin{Bmatrix} i \\ j \end{Bmatrix}$$

The above *direction cosines* allow us to transform vectors from the GLOBAL coordinates to the LOCAL coordinates. It is also necessary to be able to transform vectors from the LOCAL coordinates to the GLOBAL coordinates. Although it is possible to derive the LOCAL to GLOBAL transformation equations in a similar manner as demonstrated for the above equations, the *MATRIX operations* provide a slightly more elegant approach.

The above equations can be represented symbolically as:

$$\{a\} = [\,l\,]\,\{b\}$$

where {a} and {b} are direction vectors, [l] is the direction cosines.

Perform the matrix operations to derive the reverse transformation equations in terms of the above direction cosines:

$$\{b\} = [\,?\,]\,\{a\}.$$

First, multiply by $[\,l\,]^{-1}$ to remove the *direction cosines* from the right hand side of the original equation.

$$\{a\} = [\,l\,]\,\{b\}$$

$$[\,l\,]^{-1}\{a\} = [\,l\,]^{-1}\,[\,l\,]\,\{b\}$$

From matrix algebra, $[\,l\,]^{-1}\,[\,l\,] = [\,I\,]$ *and* $[\,I\,]\{b\} = \{b\}$.

The equation can now be simplified as

$$[\,l\,]^{-1}\{a\} = \{b\}$$

For *linear statics analyses*, the *direction cosine* is an *orthogonal matrix* and the *inverse of the matrix* is equal to the transpose of the matrix.

$$[\,l\,]^{-1} = [\,l\,]^{T}$$

Therefore, the transformation equation can be expressed as:

$$[\,l\,]^{T}\{a\} = \{b\}$$

The transformation equations that enable us to transform any vector from a *LOCAL coordinate system* to the *GLOBAL coordinate system* become:

LOCAL coordinates to the GLOBAL coordinates:

$$\begin{Bmatrix} i \\ j \end{Bmatrix} = \begin{bmatrix} \cos(\theta) & -\sin(\theta) \\ \sin(\theta) & \cos(\theta) \end{bmatrix} \begin{Bmatrix} u \\ v \end{Bmatrix}$$

The reverse transformation can also be established by applying the transformation equations that transform any vector from the *GLOBAL coordinate system* to the *LOCAL coordinate system*:

GLOBAL coordinates to the LOCAL coordinates:

$$\begin{Bmatrix} u \\ v \end{Bmatrix} = \begin{bmatrix} \cos(\theta) & \sin(\theta) \\ -\sin(\theta) & \cos(\theta) \end{bmatrix} \begin{Bmatrix} i \\ j \end{Bmatrix}$$

As it is the case with many mathematical equations, derivation of the equations usually appears to be much more complex than the actual application and utilization of the equations. The following example illustrates the application of the two-dimensional *coordinate transformation equations* on a point in between two coordinate systems.

Example 2.1

Given:

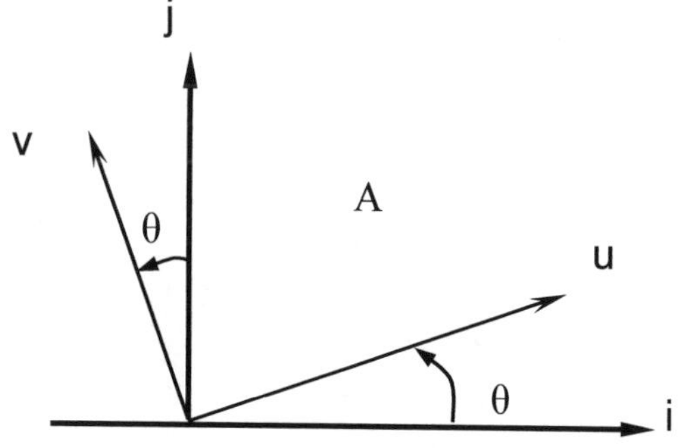

The coordinates of point A: (20 i, 40 j).

Find: The coordinates of point A if the local coordinate system is rotated 15 degrees relative to the global coordinate system.

Solution:

Using the coordinate transformation equations (GLOBAL coordinates to the LOCAL coordinates):

$$\begin{Bmatrix} u \\ v \end{Bmatrix} = \begin{bmatrix} \cos(\theta) & \sin(\theta) \\ -\sin(\theta) & \cos(\theta) \end{bmatrix} \begin{Bmatrix} i \\ j \end{Bmatrix}$$

$$= \begin{bmatrix} \cos(15^\circ) & \sin(15^\circ) \\ -\sin(15^\circ) & \cos(15^\circ) \end{bmatrix} \begin{Bmatrix} 20 \\ 40 \end{Bmatrix}$$

$$= \begin{Bmatrix} 29.7 \\ 32.5 \end{Bmatrix}$$

➢ On your own, perform a coordinate transformation to determine the global coordinates of point A using the *LOCAL* coordinates of (29.7,32.5) with the 15 degrees angle in between the two coordinate systems.

Global Stiffness Matrix

For a single truss element, using the coordinate transformation equations, we can proceed to transform the local stiffness matrix to the global stiffness matrix.

For a single truss element arbitrarily positioned in a two-dimensional space:

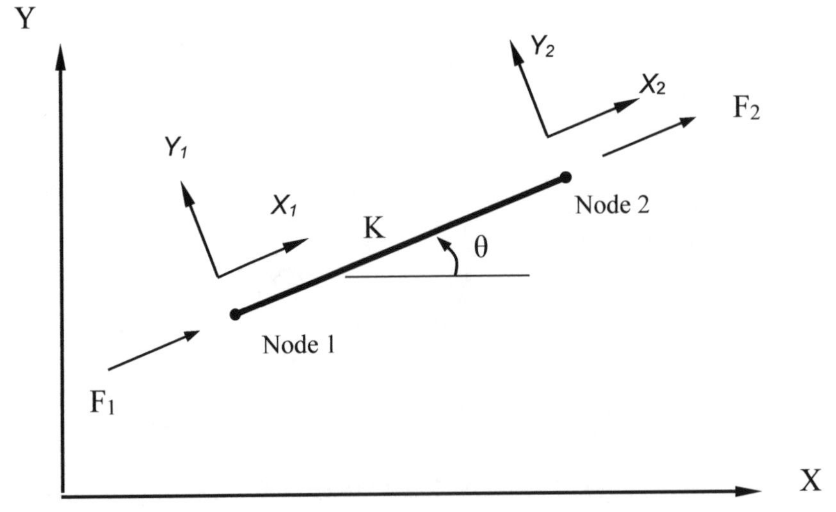

The force-displacement equations (in the local coordinate system) can be expressed as:

$$\begin{Bmatrix} F_{1X} \\ F_{1Y} \\ F_{2X} \\ F_{2Y} \end{Bmatrix} = \frac{\mathbf{EA}}{\mathbf{L}} \begin{pmatrix} +1 & 0 & -1 & 0 \\ 0 & 0 & 0 & 0 \\ -1 & 0 & +1 & 0 \\ 0 & 0 & 0 & 0 \end{pmatrix} \begin{Bmatrix} X_1 \\ Y_1 \\ X_2 \\ Y_2 \end{Bmatrix}$$

Local Stiffness Matrix

Next, apply the coordinate transformation equations to establish the general GLOBAL STIFFNESS MATRIX of a single truss element in a two-dimensional space.

First, the displacement transformation equations (GLOBAL to LOCAL):

$$\begin{Bmatrix} X_1 \\ Y_1 \\ X_2 \\ Y_2 \end{Bmatrix} = \begin{pmatrix} \cos(\theta) & \sin(\theta) & 0 & 0 \\ -\sin(\theta) & \cos(\theta) & 0 & 0 \\ 0 & 0 & \cos(\theta) & \sin(\theta) \\ 0 & 0 & -\sin(\theta) & \cos(\theta) \end{pmatrix} \begin{Bmatrix} \mathrm{X}_1 \\ \mathrm{Y}_1 \\ \mathrm{X}_2 \\ \mathrm{Y}_2 \end{Bmatrix}$$

Local | Global

The force transformation equations (GLOBAL to LOCAL):

$$\begin{Bmatrix} F_{1x} \\ F_{1y} \\ F_{2x} \\ F_{2y} \end{Bmatrix} = \begin{pmatrix} \cos(\theta) & \sin(\theta) & 0 & 0 \\ -\sin(\theta) & \cos(\theta) & 0 & 0 \\ 0 & 0 & \cos(\theta) & \sin(\theta) \\ 0 & 0 & -\sin(\theta) & \cos(\theta) \end{pmatrix} \begin{Bmatrix} \mathrm{F}_{1x} \\ \mathrm{F}_{1y} \\ \mathrm{F}_{2x} \\ \mathrm{F}_{2y} \end{Bmatrix}$$

Local | Global

The above three sets of equations can be represented as:

$\{F\} = [K]\{X\}$ ------- *Local force-displacement equation*

$\{X\} = [l]\{\mathrm{X}\}$ ------- *Displacement transformation equation*

$\{F\} = [l]\{\mathrm{F}\}$ ------- *Force transformation equation*

We will next perform *matrix operations* to obtain the GLOBAL stiffness matrix: Starting with the local force-displacement equation

$$\{F\} = [K]\{X\}$$

Local Local

Next, substituting the transformation equations for $\{F\}$ and $\{X\}$,

$$[l]\{\mathrm{F}\} = [K][l]\{\mathrm{X}\}$$

Multiply both sides of the equation with $[l]^{-1}$,

$$[l]^{-1}[l]\{\mathrm{F}\} = [l]^{-1}[K][l]\{\mathrm{X}\}$$

The equation can be simplified as:

$$\{\mathrm{F}\} = [l]^{-1}[K][l]\{\mathrm{X}\}$$

or

Global Global

$$\{\mathrm{F}\} = [l]^{T}[K][l]\{\mathrm{X}\}$$

The GLOBAL force-displacement equation is then expressed as:

$$\{\mathrm{F}\} = [l]^{T}[K][l]\{\mathrm{X}\}$$

or

$$\{\mathrm{F}\} = [\mathrm{K}]\{\mathrm{X}\}$$

The global stiffness matrix [**K**] can now be expressed in terms of the local stiffness matrix.

$$[\mathrm{K}] = [l]^{T}[K][l]$$

For a single truss element in a two-dimensional space, the global stiffness matrix is

$$[\mathrm{K}] = \frac{\mathbf{EA}}{\mathbf{L}} \begin{bmatrix} \cos^2(\theta) & \cos(\theta)\sin(\theta) & -\cos^2(\theta) & -\cos(\theta)\sin(\theta) \\ \cos(\theta)\sin(\theta) & \sin^2(\theta) & -\cos(\theta)\sin(\theta) & -\sin^2(\theta) \\ -\cos^2(\theta) & -\cos(\theta)\sin(\theta) & \cos^2(\theta) & \cos(\theta)\sin(\theta) \\ -\cos(\theta)\sin(\theta) & -\sin^2(\theta) & \sin(\theta)\cos(\theta) & \sin^2(\theta) \end{bmatrix}$$

- The above matrix can be applied to any truss element positioned in a two-dimensional space. We can now assemble the global stiffness matrix and analyze any two-dimensional multiple-elements truss structures. The following example illustrates, using the general global stiffness matrix derived above, the formulation and solution process of a 2D truss structure.

Example 2.2

Given: A two-dimensional truss structure as shown. (All joints are **Pin Joints.**)

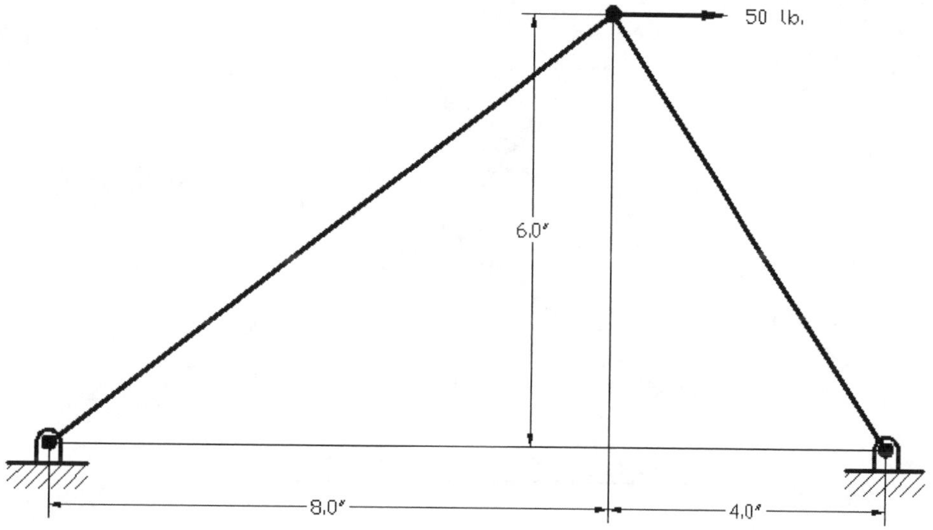

Material: Steel rod, diameter ¼ in.

Find: Displacements of each node and stresses in each member.

Solution:

The system contains two elements and three nodes. The nodes and elements are labeled as shown below.

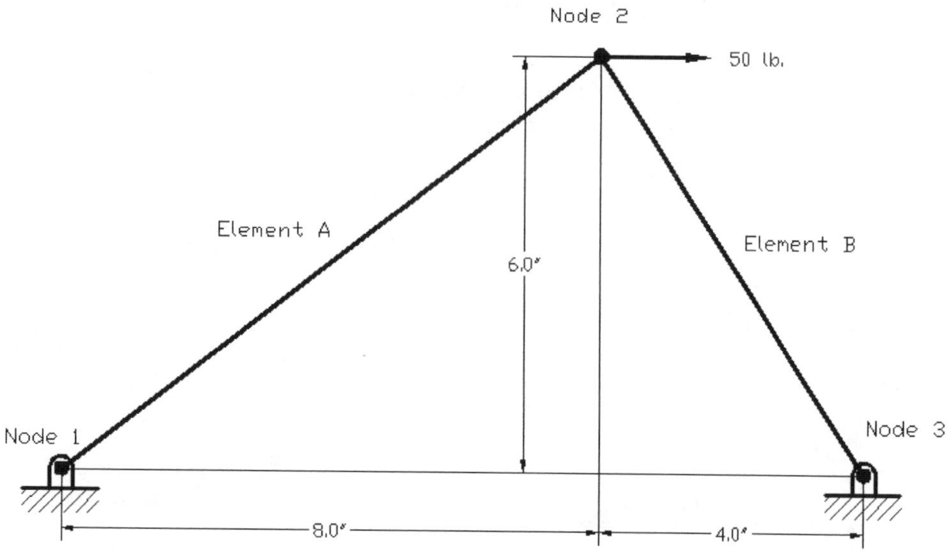

First, establish the GLOBAL stiffness matrix (system equations in matrix form) for each element.

Element A (Node 1 to Node 2)

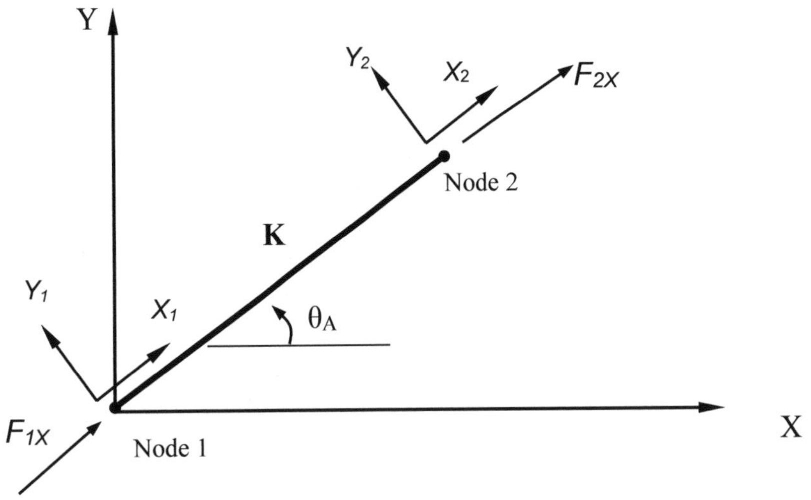

$\theta_A = \tan^{-1}(6/8) = 36.87°$,
E (Young's modulus) = 30×10^6 psi
A (Cross sectional area) = $\pi r^2 = 0.049$ in^2,
L (Length of element) = $(6^2 + 8^2)^{1/2} = 10$ in.

Therefore, $\dfrac{\mathbf{EA}}{\mathbf{L}} = 147262$ lb/in.

The LOCAL force-displacement equations:

Local coordinate system

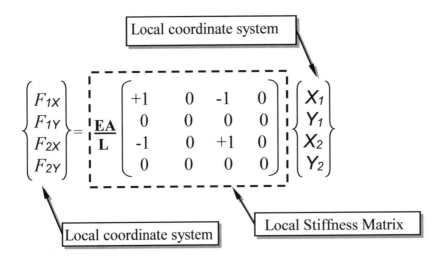

$$\begin{Bmatrix} F_{1X} \\ F_{1Y} \\ F_{2X} \\ F_{2Y} \end{Bmatrix} = \frac{\mathbf{EA}}{\mathbf{L}} \begin{bmatrix} +1 & 0 & -1 & 0 \\ 0 & 0 & 0 & 0 \\ -1 & 0 & +1 & 0 \\ 0 & 0 & 0 & 0 \end{bmatrix} \begin{Bmatrix} X_1 \\ Y_1 \\ X_2 \\ Y_2 \end{Bmatrix}$$

Local coordinate system

Local Stiffness Matrix

Using the equations we have derived, the GLOBAL system equations for *element A* can be expressed as:

$$\{ F \} = [K] \{ X \}$$

$$[K] = \frac{EA}{L} \begin{bmatrix} \cos^2(\theta) & \cos(\theta)\sin(\theta) & -\cos^2(\theta) & -\cos(\theta)\sin(\theta) \\ \cos(\theta)\sin(\theta) & \sin^2(\theta) & -\cos(\theta)\sin(\theta) & -\sin^2(\theta) \\ -\cos^2(\theta) & -\cos(\theta)\sin(\theta) & \cos^2(\theta) & \cos(\theta)\sin(\theta) \\ -\cos(\theta)\sin(\theta) & -\sin^2(\theta) & \sin(\theta)\cos(\theta) & \sin^2(\theta) \end{bmatrix}$$

Therefore,

$$\begin{Bmatrix} F_{1X} \\ F_{1Y} \\ F_{2XA} \\ F_{2YA} \end{Bmatrix} = \mathbf{147262} \begin{bmatrix} .64 & .48 & -.64 & -.48 \\ .48 & .36 & -.48 & -.36 \\ -.64 & -.48 & .64 & .48 \\ -.48 & -.36 & .48 & .36 \end{bmatrix} \begin{Bmatrix} X_1 \\ Y_1 \\ X_2 \\ Y_2 \end{Bmatrix}$$

Global (force vector) · Global Stiffness Matrix · Global (displacement vector)

Element B (Node 2 to Node 3)

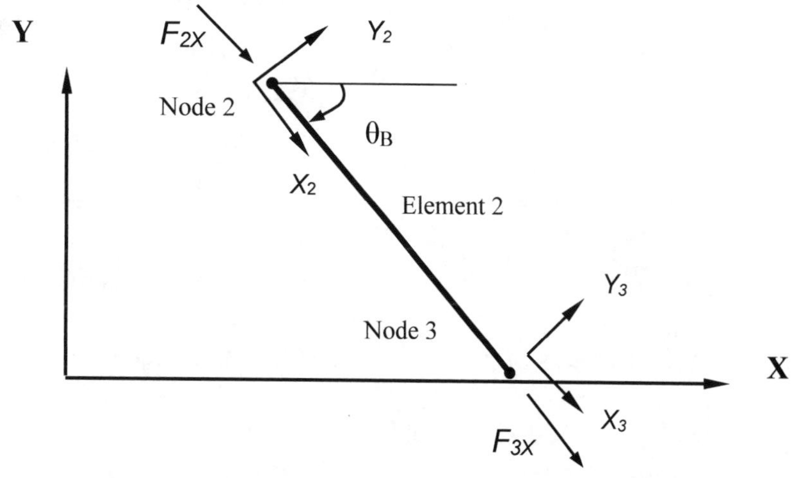

$\theta_B = -\tan^{-1}(6/4) = -56.31°$ (negative angle as shown)

E (Young's modulus) = 30×10^6 psi

A (Cross sectional area) = $\pi r^2 = 0.049$ in^2

L (Length of element) = $(4^2 + 6^2)^{1/2} = 7.21$ in.

$$\frac{EA}{L} = 204216 \text{ lb/in.}$$

$$\begin{Bmatrix} F_{2X} \\ F_{2Y} \\ F_{3X} \\ F_{3Y} \end{Bmatrix} = \frac{EA}{L} \begin{pmatrix} +1 & 0 & -1 & 0 \\ 0 & 0 & 0 & 0 \\ -1 & 0 & +1 & 0 \\ 0 & 0 & 0 & 0 \end{pmatrix} \begin{Bmatrix} X_2 \\ Y_2 \\ X_3 \\ Y_3 \end{Bmatrix}$$

Local system — Local system — Local Stiffness Matrix

Using the equations we derived in the previous sections, the GLOBAL system equations for *element B* are:

$$\{ F \} = [K] \{ X \}$$

$$[K] = \frac{EA}{L} \begin{pmatrix} \cos^2(\theta) & \cos(\theta)\sin(\theta) & -\cos^2(\theta) & -\cos(\theta)\sin(\theta) \\ \cos(\theta)\sin(\theta) & \sin^2(\theta) & -\cos(\theta)\sin(\theta) & -\sin^2(\theta) \\ -\cos^2(\theta) & -\cos(\theta)\sin(\theta) & \cos^2(\theta) & \cos(\theta)\sin(\theta) \\ -\cos(\theta)\sin(\theta) & -\sin^2(\theta) & \sin(\theta)\cos(\theta) & \sin^2(\theta) \end{pmatrix}$$

Therefore,

$$\begin{Bmatrix} F_{2XB} \\ F_{2YB} \\ F_{3X} \\ F_{3Y} \end{Bmatrix} = \mathbf{204216} \begin{pmatrix} 0.307 & -0.462 & -0.307 & 0.462 \\ -0.462 & 0.692 & 0.462 & -0.692 \\ -0.307 & 0.462 & 0.307 & -0.462 \\ 0.462 & -0.692 & -0.462 & 0.692 \end{pmatrix} \begin{Bmatrix} X_2 \\ Y_2 \\ X_3 \\ Y_3 \end{Bmatrix}$$

Global — Global — Global Stiffness Matrix

Now we are ready to assemble the overall global stiffness matrix of the structure.

Summing the two sets of global force-displacement equations:

$$\begin{Bmatrix} F_{1X} \\ F_{1Y} \\ F_{2X} \\ F_{2Y} \\ F_{3X} \\ F_{3Y} \end{Bmatrix} = \begin{pmatrix} 94248 & 70686 & -94248 & -70686 & 0 & 0 \\ 70686 & 53014 & -70686 & -53014 & 0 & 0 \\ -94248 & -70686 & 157083 & -23568 & -62836 & 94253 \\ -70686 & -53014 & -23568 & 194395 & 94253 & -141380 \\ 0 & 0 & -62836 & 94253 & 62836 & -94253 \\ 0 & 0 & 94253 & -141380 & -94253 & 141380 \end{pmatrix} \begin{Bmatrix} X_1 \\ Y_1 \\ X_2 \\ Y_2 \\ X_3 \\ Y_3 \end{Bmatrix}$$

Next, apply the following known boundary conditions into the system equations:

(a) Node 1 and Node 3 are fixed-points; therefore, any displacement components of these two node-points are zero (X_1, Y_1 and X_3, Y_3).

(b) The only external load is at Node 2: F_{2x} = 50 lbs.
Therefore,

$$\begin{Bmatrix} F_{1X} \\ F_{1Y} \\ 50 \\ 0 \\ F_{3X} \\ F_{3Y} \end{Bmatrix} = \begin{pmatrix} 94248 & 70686 & -94248 & -70686 & 0 & 0 \\ 70686 & 53014 & -70686 & -53014 & 0 & 0 \\ -94248 & -70686 & 157083 & -23568 & -628360 & 94253 \\ -70686 & -53014 & -23568 & 194395 & 94253 & -141380 \\ 0 & 0 & -62836 & 94253 & 62836 & -94253 \\ 0 & 0 & 94253 & -141380 & -94253 & 141380 \end{pmatrix} \begin{Bmatrix} 0 \\ 0 \\ X_2 \\ Y_2 \\ 0 \\ 0 \end{Bmatrix}$$

The two displacements we need to solve are X_2 and Y_2. Let's simplify the above matrix by removing the unaffected/unnecessary columns in the matrix.

$$\begin{Bmatrix} F_{1X} \\ F_{1Y} \\ 50 \\ 0 \\ F_{3X} \\ F_{3Y} \end{Bmatrix} = \begin{pmatrix} -94248 & -70686 \\ -70686 & -53014 \\ 157083 & -23568 \\ -23568 & 194395 \\ -62836 & 94253 \\ 94253 & -141380 \end{pmatrix} \begin{Bmatrix} X_2 \\ Y_2 \end{Bmatrix}$$

Solve for nodal displacements X_2 and Y_2:

$$\begin{Bmatrix} 50 \\ 0 \end{Bmatrix} = \begin{bmatrix} 157083 & -23568 \\ -23568 & 194395 \end{bmatrix} \begin{Bmatrix} X_2 \\ Y_2 \end{Bmatrix}$$

$\mathbf{X_2 = 3.24\ e^{-4}}$ **in.**
$\mathbf{Y_2 = 3.93\ e^{-5}}$ **in.**

Substitute the known X_2 and Y_2 values into the matrix and solve for the reaction forces:

$$\begin{Bmatrix} F_{1X} \\ F_{1Y} \\ F_{3X} \\ F_{3Y} \end{Bmatrix} = \begin{pmatrix} -94248 & -70686 \\ -70686 & -53014 \\ -62836 & 94253 \\ 94253 & -141380 \end{pmatrix} \begin{Bmatrix} \mathbf{3.24\ e^{-4}} \\ \mathbf{3.93\ e^{-5}} \end{Bmatrix}$$

Therefore,

$$\mathbf{F_{1X} = -33.33\ lbs.,\quad F_{1Y} = -25\ lbs.}$$
$$\mathbf{F_{3X} = -16.67\ lbs.,\quad F_{3Y} = 25\ lbs.}$$

To determine the normal stress in each truss member, one option is to use the displacement transformation equations to transform the results from the global coordinate system back to the local coordinate system.

Element A

$\{X\} = [l]\{X\}$ ------- *Displacement transformation equation*

$$\begin{Bmatrix} X_1 \\ Y_1 \\ X_2 \\ Y_2 \end{Bmatrix} = \begin{bmatrix} .8 & .6 & 0 & 0 \\ -.6 & .8 & 0 & 0 \\ 0 & 0 & .8 & .6 \\ 0 & 0 & -.6 & .8 \end{bmatrix} \begin{Bmatrix} 0 \\ 0 \\ X_2 \\ Y_2 \end{Bmatrix}$$

Local — Global

$$\boldsymbol{X_2 = 2.83\ e^{-4}}$$
$$\boldsymbol{Y_2 = -1.63\ e^{-4}}$$

The LOCAL force-displacement equations:

Local system

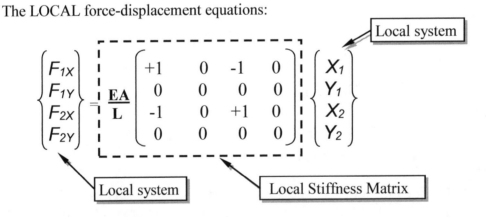

$$\begin{Bmatrix} F_{1X} \\ F_{1Y} \\ F_{2X} \\ F_{2Y} \end{Bmatrix} = \frac{EA}{L} \begin{bmatrix} +1 & 0 & -1 & 0 \\ 0 & 0 & 0 & 0 \\ -1 & 0 & +1 & 0 \\ 0 & 0 & 0 & 0 \end{bmatrix} \begin{Bmatrix} X_1 \\ Y_1 \\ X_2 \\ Y_2 \end{Bmatrix}$$

Local system — Local Stiffness Matrix

$$\boldsymbol{F_{1X} = -41.67\ lbs.,\quad F_{1Y} = 0\ lb.}$$
$$\boldsymbol{F_{2X} = 41.67\ lbs.,\quad F_{2Y} = 0\ lb.}$$

Therefore, the normal stress developed in *Element A* can be calculated as (41.67/0.049)=**850 psi.**

- On your own, calculate the normal stress developed in *Element B*.

Review Questions

1. Determine the coordinates of point A if the local coordinate system is rotated 15 degrees relative to the global coordinate system. The global coordinates of point A: (30,50).

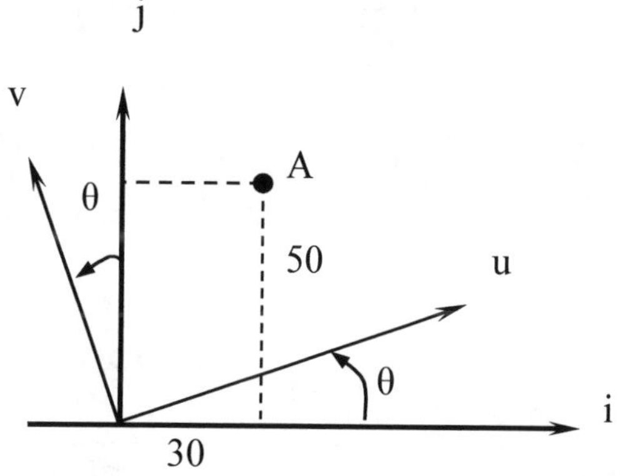

2. Determine the global coordinates of point B if the local coordinate system is rotated 30 degrees relative to the global coordinate system. The local coordinates of point B: (30,15).

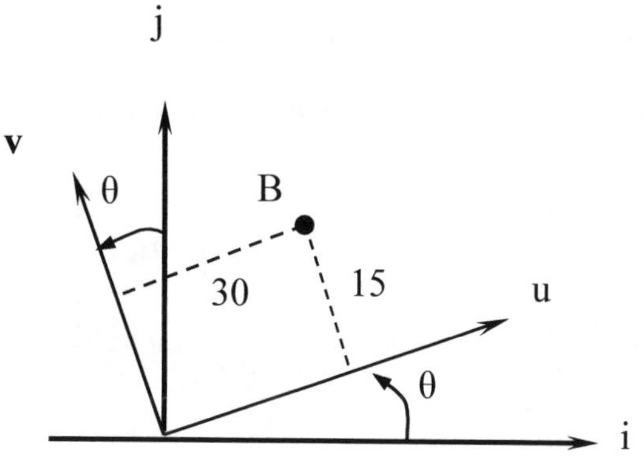

Exercises

1. Given: two-dimensional truss structure as shown. (All joints are pin joints.)

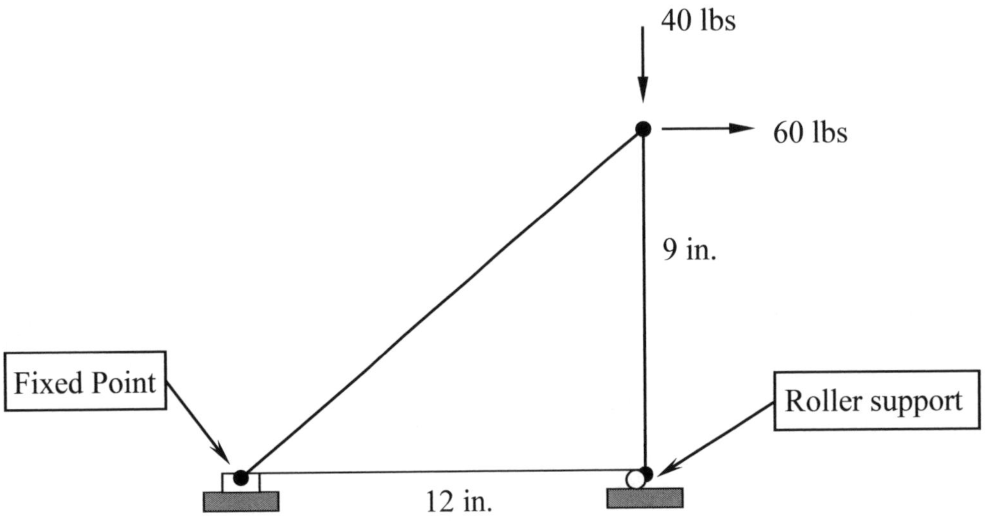

Material: Steel, diameter ¼ in.

Find: (a) Displacements of the nodes.
(b) Normal stresses developed in the members.

2. Given: Two-dimensional truss structure as shown. (All joints are pin joints.)

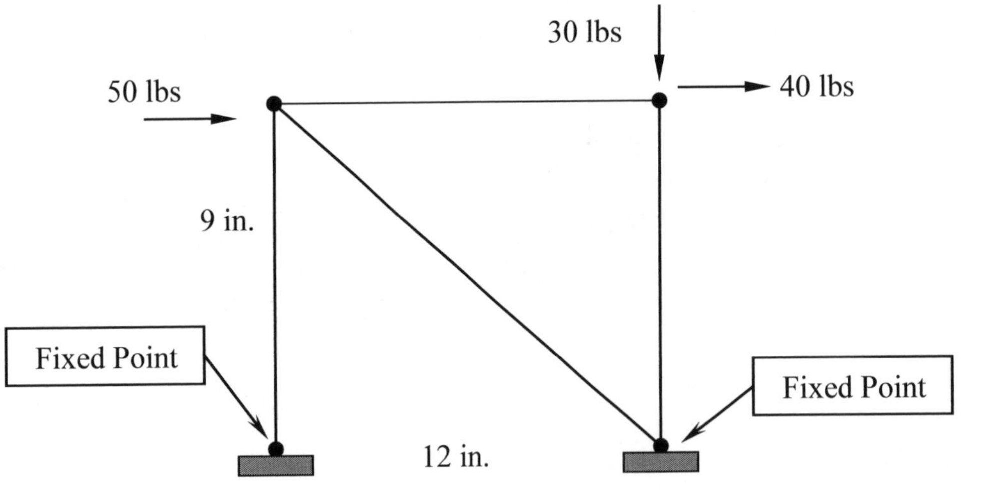

Material: Steel, diameter ¼ in.

Find: (a) Displacements of the nodes.
(b) Normal stresses developed in the members.

Chapter 3
2D Trusses in MS Excel and Truss Solver

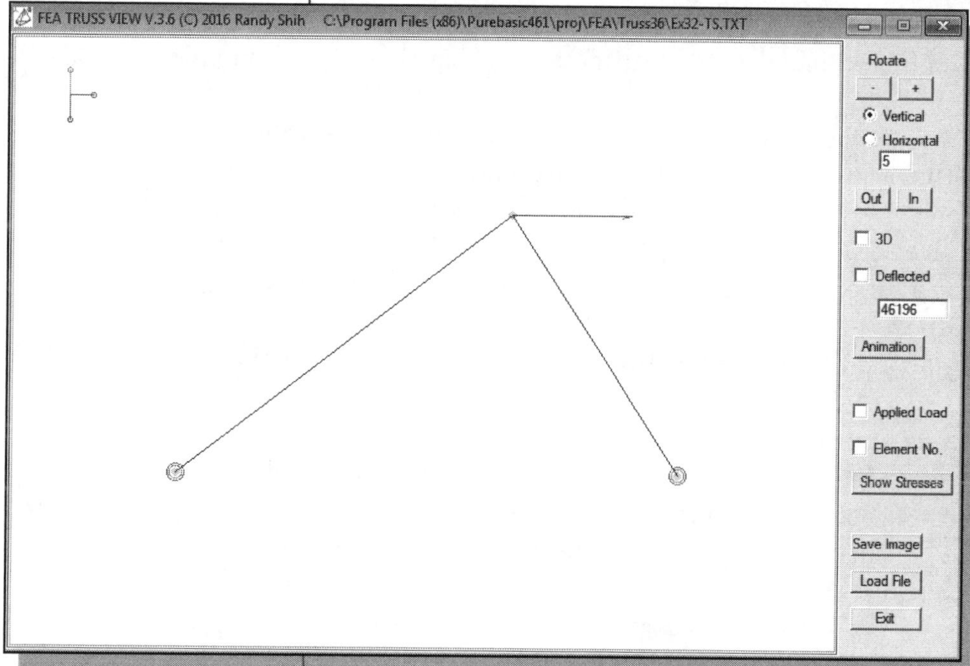

Learning Objectives

- ♦ **Perform the Direct Stiffness Matrix Method using MS Excel.**
- ♦ **Using MS Excel built-in commands to solve simultaneous equations.**
- ♦ **Solve 2D trusses using the Truss Solver program.**
- ♦ **Understand the general FEA procedure concepts.**

Direct Stiffness Matrix Method using Excel

Microsoft Excel is a very popular general purpose spreadsheet program, which was originally designed to do accounting related calculations and simple database operations. However, ***Microsoft Excel*** does have sufficient engineering functions to be very useful for engineers and scientists in performing engineering calculations as well.

.

In the case of implementing the direct stiffness matrix method, ***Excel*** provides several advantages:

- The set up of the direct stiffness matrix method using a spreadsheet, such as ***MS Excel***, helps us to gain a better understanding of the FEA procedure.
- The spreadsheet takes away the tedious number crunching task, which is usually prone to errors.
- Checking for errors and making corrections are relatively easy to do in a spreadsheet.
- The color and line features in ***MS Excel*** can be used to highlight and therefore keep tracks of the series of calculations performed.
- ***MS Excel*** also provides commonly used matrices operations, which are otherwise difficult to attain in the traditional calculator, pencil and paper approach.
- Duplicating and editing formulas are fairly easy to do in a spreadsheet. ***MS Excel*** can also recalculate a new set of values in an established spreadsheet automatically.
- The use of ***MS Excel*** provides us with additional insight into the principles involved in the finite element method, without getting bogged down with the programming aspects of the problem solution.

Example 3.1

Use ***MS Excel*** to perform the calculations illustrated in Example 2.2.
Material: Steel rod, diameter ¼ in.

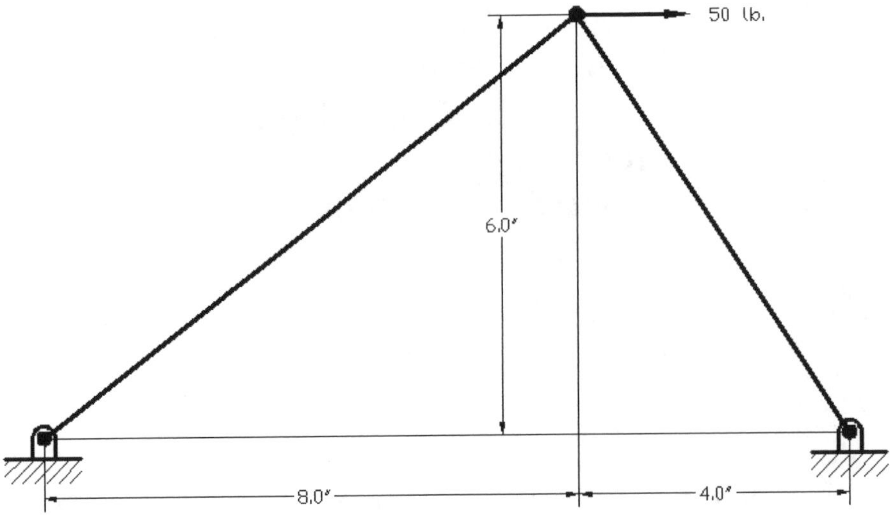

The system contains two elements and three nodes. The nodes and elements are labeled as shown below.

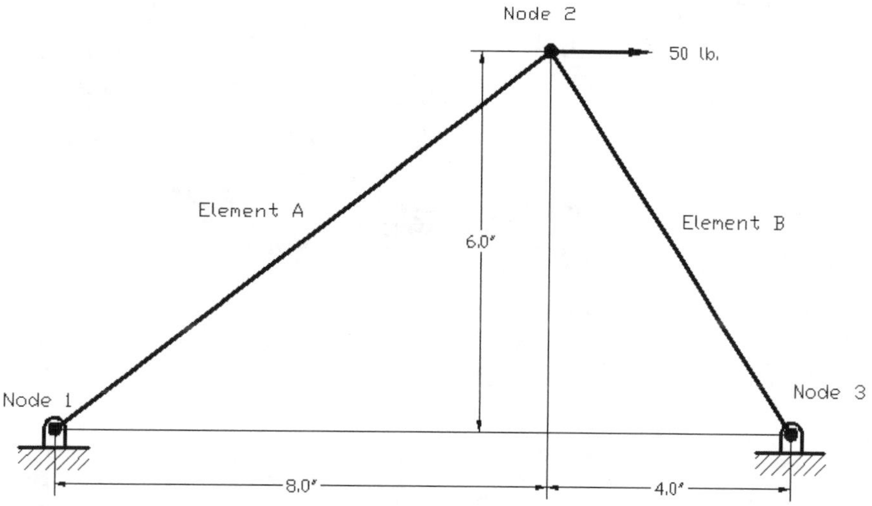

Establish the Global K Matrix for Each Member

1. Start *Microsoft Excel* through the desktop icon or the *Explorer* toolbar.

2. First enter the general problem information and element labels near the top of the spreadsheet.

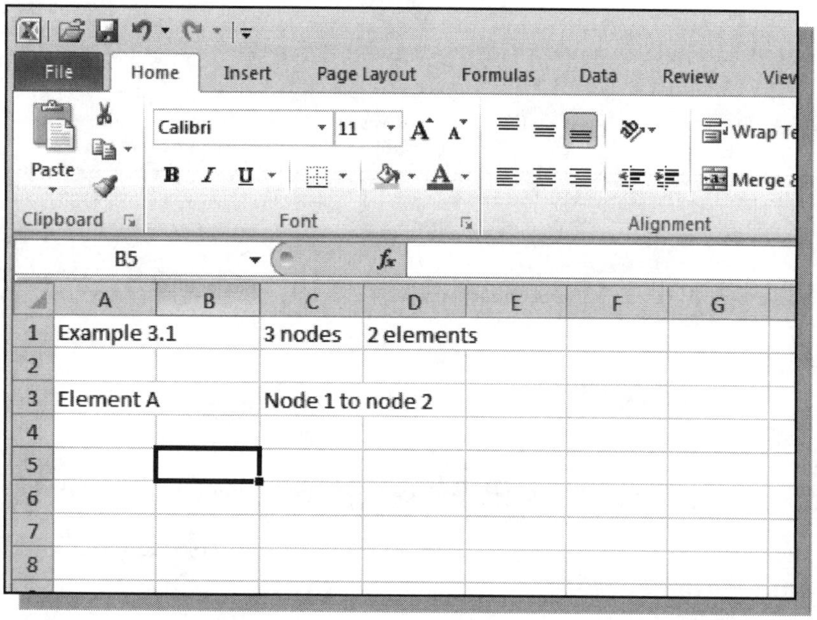

❖ Note that we will be following the same convention as outlined in Example 2.2.

3. To calculate the angle of *Element A*, enter the formula: **=ATAN2(8,6)**; the resulting angle is in **radians**.

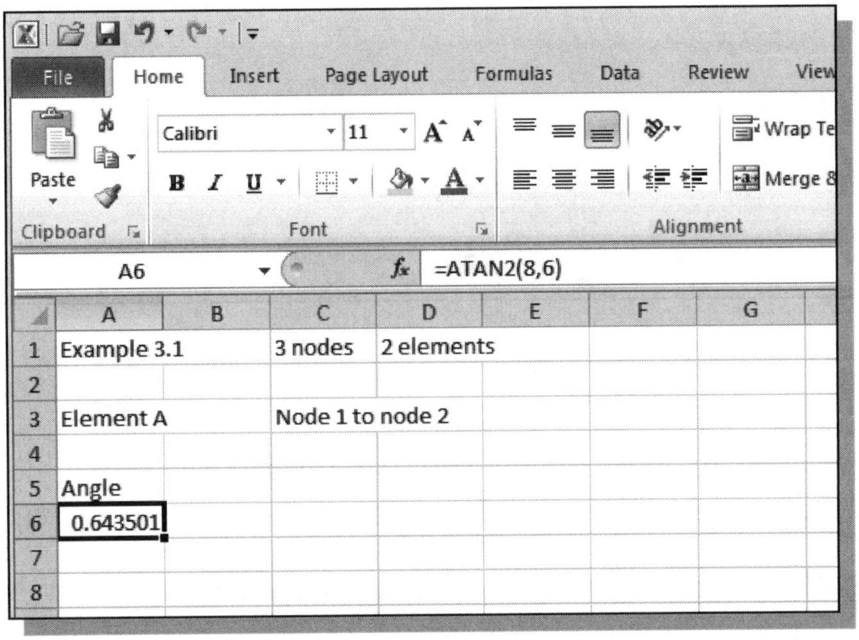

❖ Note the equal sign is used to identify the formula; the **ATAN2** function will calculate the arctangent function in all four quadrants.

4. To convert the radians to degrees, use the **Degrees** function as shown.

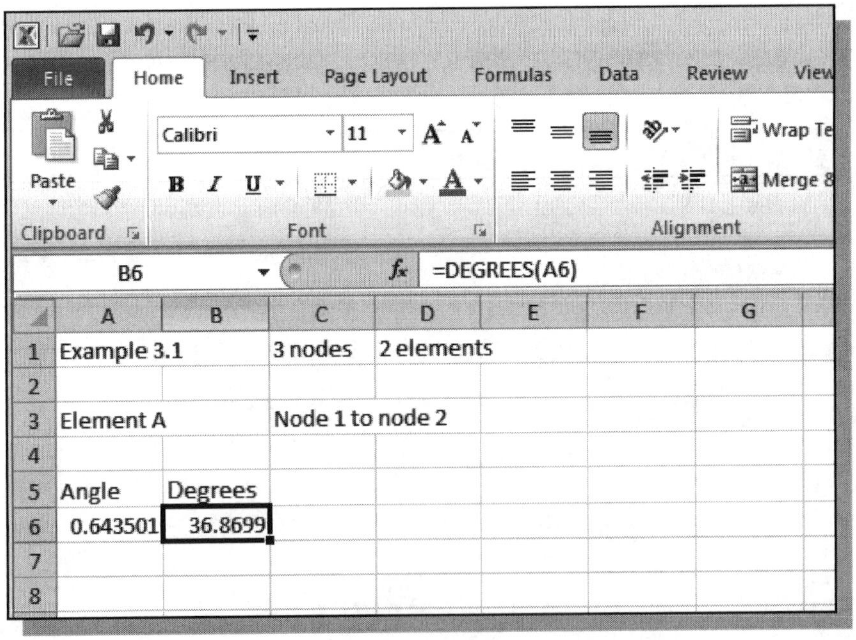

❖ Note that the conversion from radians to degrees can also be calculated using the following equation:

$$\textbf{Degrees} = \textbf{Radians} \times 180/\pi$$

5. Next enter the modulus of elasticity (E), and cross-section area based on the diameter of the truss member.

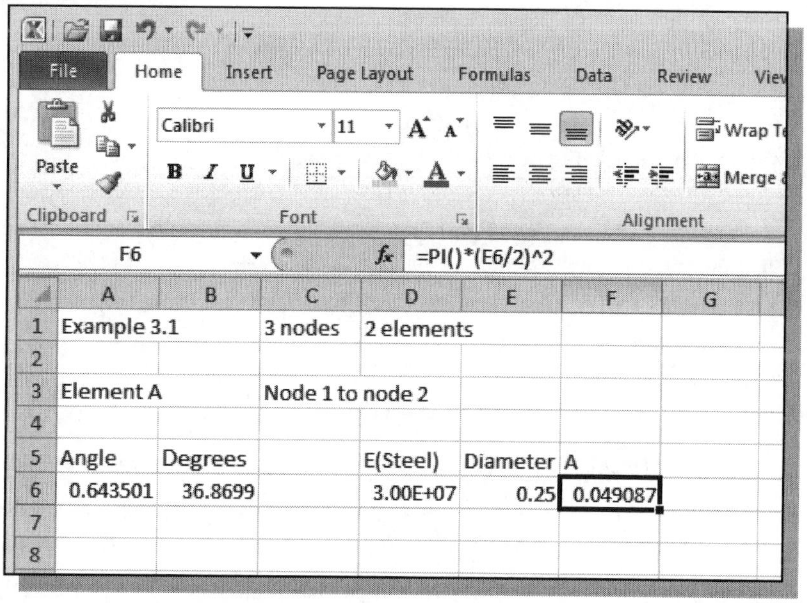

❖ Note the PI() function gives the π value in *Excel.*

6. To calculate the length of the member, enter the formula as shown.

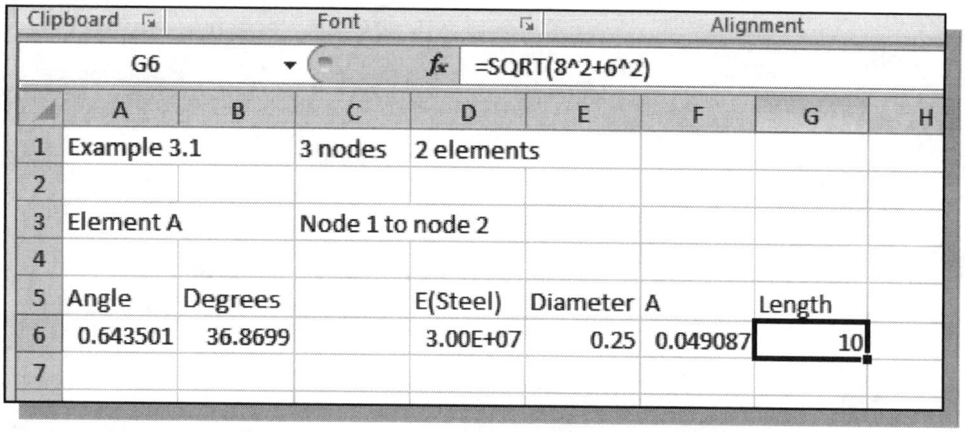

❖ The **SQRT** function can be used to calculate the square root value.

7. Now, we can calculate the k value. Note the equation is **k=EA/L.**

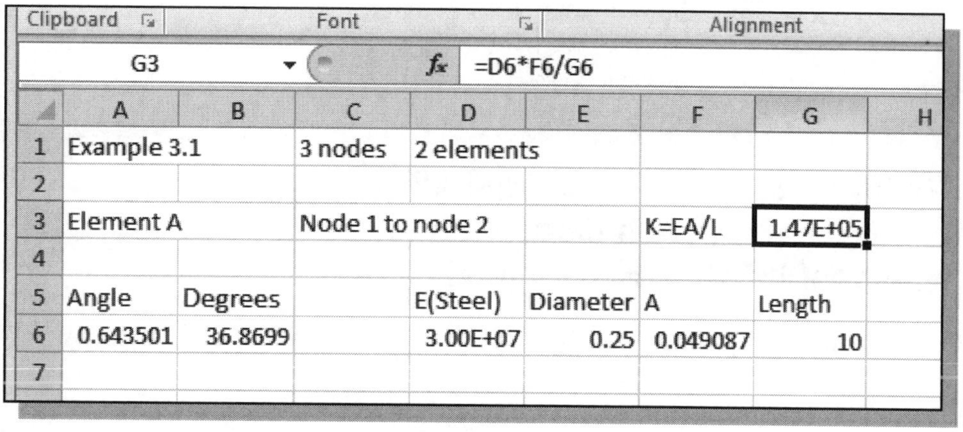

8. Next calculate the cosine and sine of the radian angle as shown.

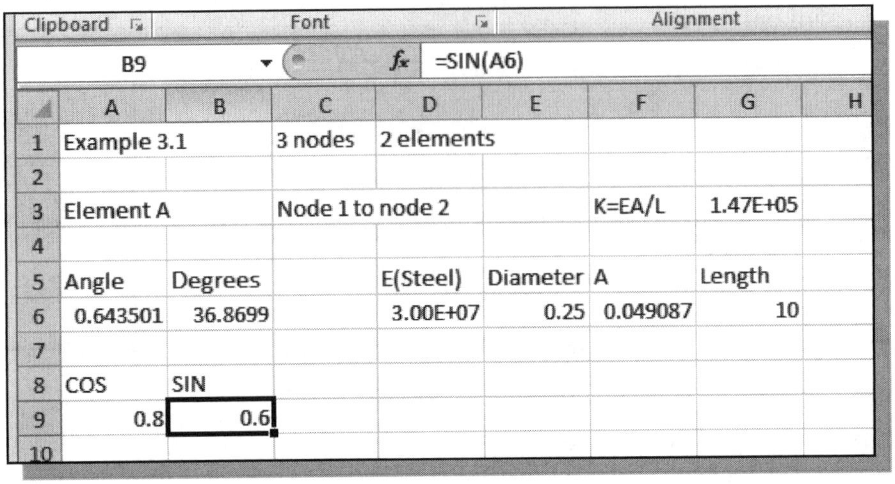

	A	B	C	D	E	F	G	H
1	Example 3.1		3 nodes	2 elements				
2								
3	Element A		Node 1 to node 2			K=EA/L	1.47E+05	
4								
5	Angle	Degrees		E(Steel)	Diameter	A	Length	
6	0.643501	36.8699		3.00E+07	0.25	0.049087	10	
7								
8	COS	SIN						
9	0.8	0.6						
10								

9. Next calculate the 4x4 global K matrix using the cosine and sine values.

$$[K] = \frac{EA}{L}\begin{pmatrix} \cos^2(\theta) & \cos(\theta)\sin(\theta) & -\cos^2(\theta) & -\cos(\theta)\sin(\theta) \\ \cos(\theta)\sin(\theta) & \sin^2(\theta) & -\cos(\theta)\sin(\theta) & -\sin^2(\theta) \\ -\cos^2(\theta) & -\cos(\theta)\sin(\theta) & \cos^2(\theta) & \cos(\theta)\sin(\theta) \\ -\cos(\theta)\sin(\theta) & -\sin^2(\theta) & \sin(\theta)\cos(\theta) & \sin^2(\theta) \end{pmatrix}$$

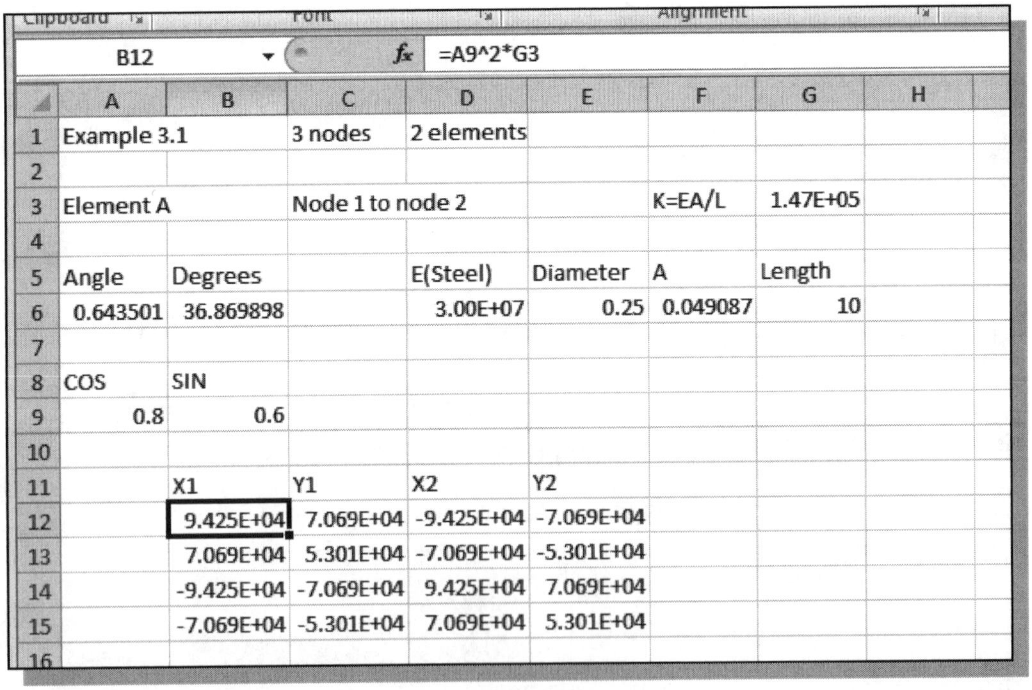

	A	B	C	D	E	F	G	H
1	Example 3.1		3 nodes	2 elements				
2								
3	Element A		Node 1 to node 2			K=EA/L	1.47E+05	
4								
5	Angle	Degrees		E(Steel)	Diameter	A	Length	
6	0.643501	36.869898		3.00E+07	0.25	0.049087	10	
7								
8	COS	SIN						
9	0.8	0.6						
10								
11		X1	Y1	X2	Y2			
12		9.425E+04	7.069E+04	-9.425E+04	-7.069E+04			
13		7.069E+04	5.301E+04	-7.069E+04	-5.301E+04			
14		-9.425E+04	-7.069E+04	9.425E+04	7.069E+04			
15		-7.069E+04	-5.301E+04	7.069E+04	5.301E+04			
16								

$K_{11}=k(\cos(\theta))^2$, $K_{12}=k\cos(\theta)\sin(\theta)$,
$K_{13}=-k(\cos(\theta))^2$, $K_{14}=-k\cos(\theta)\sin(\theta)$,
$K_{21}=k\cos(\theta)\sin(\theta)$, $K_{22}=k(\sin(\theta))^2$........

❖ On your own, compare the calculated values to the numbers shown on page 2-15.

10. Once we have confirmed the calculations are correct for *Element A*, we can simply ***copy and paste*** the established formulas for any other elements. Select the established formulas as shown.

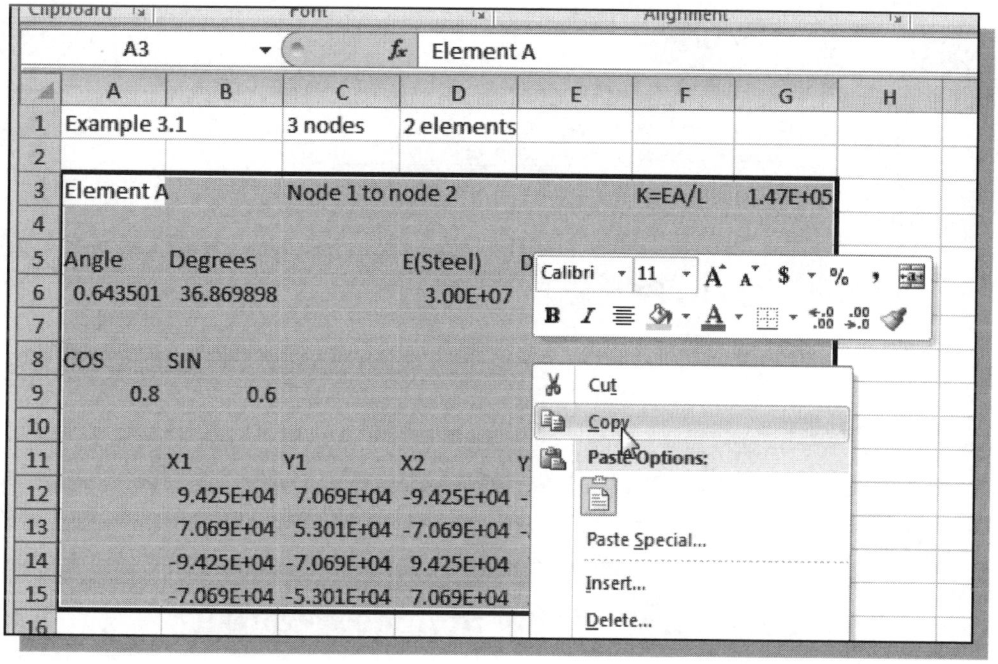

11. Click the **Copy** icon or right-mouse-click to bring up the option menu and choose Copy as shown in the above figure.

12. Move the cursor a few cells below the selected cells as shown.

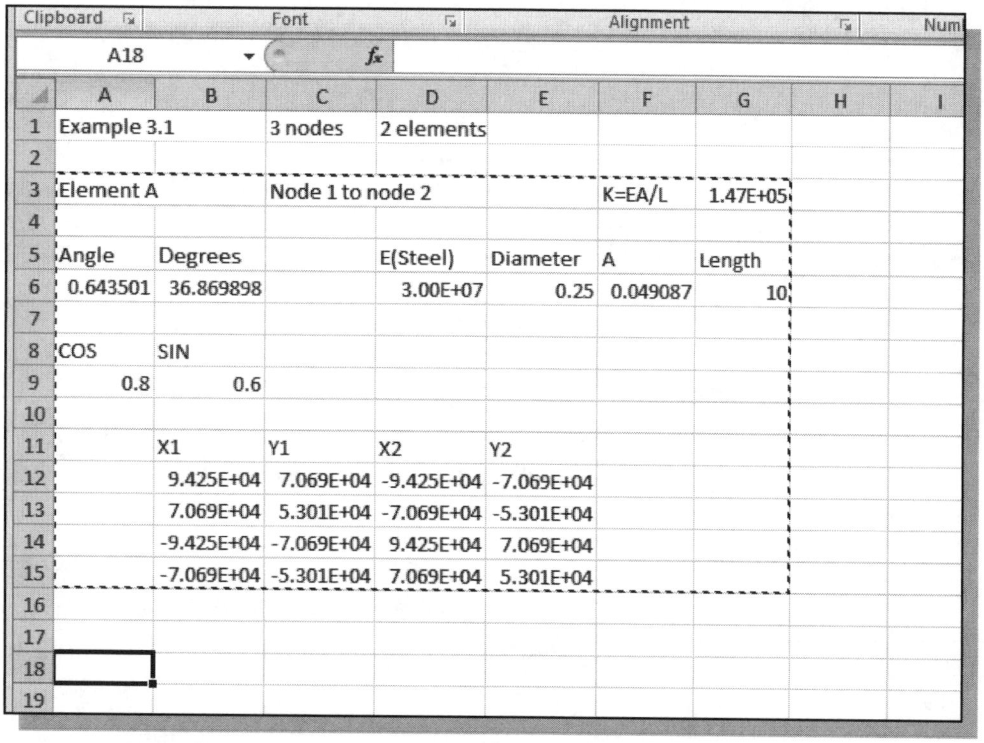

13. Press the **[ENTER]** key once to create a copy of the selected cells.

14. On your own, update the second set for *Element B* as shown in the figure below. Hint: Update the values by changing the values in the following two cells: **Radian Angle** and **Length**.

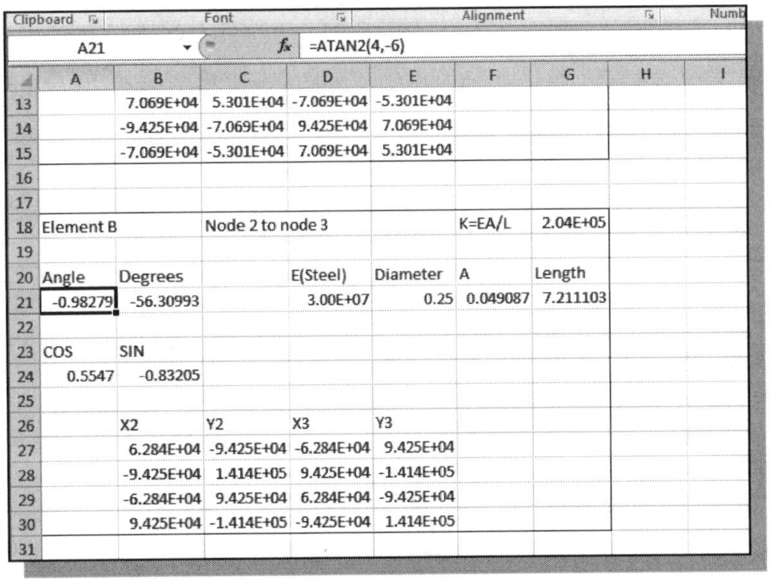

	A	B	C	D	E	F	G
13		7.069E+04	5.301E+04	-7.069E+04	-5.301E+04		
14		-9.425E+04	-7.069E+04	9.425E+04	7.069E+04		
15		-7.069E+04	-5.301E+04	7.069E+04	5.301E+04		
16							
17							
18	Element B		Node 2 to node 3			K=EA/L	2.04E+05
19							
20	Angle	Degrees		E(Steel)	Diameter	A	Length
21	-0.98279	-56.30993		3.00E+07	0.25	0.049087	7.211103
22							
23	COS	SIN					
24	0.5547	-0.83205					
25							
26		X2	Y2	X3	Y3		
27		6.284E+04	-9.425E+04	-6.284E+04	9.425E+04		
28		-9.425E+04	1.414E+05	9.425E+04	-1.414E+05		
29		-6.284E+04	9.425E+04	6.284E+04	-9.425E+04		
30		9.425E+04	-1.414E+05	-9.425E+04	1.414E+05		
31							

Assembly of the Overall Global Stiffness Matrix

1. On your own, set up the labels for the overall global stiffness matrix as shown. The two large rectangles indicate the locations for where the two element matrices will fit. You can also use different colors to highlight the cells.

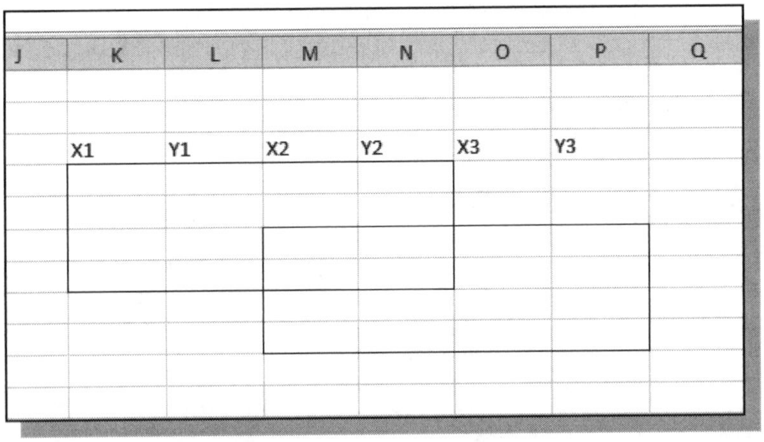

❖ Note that four cells from the two matrices will overlap each other. The four cells are marked with the small rectangle in the above figure.

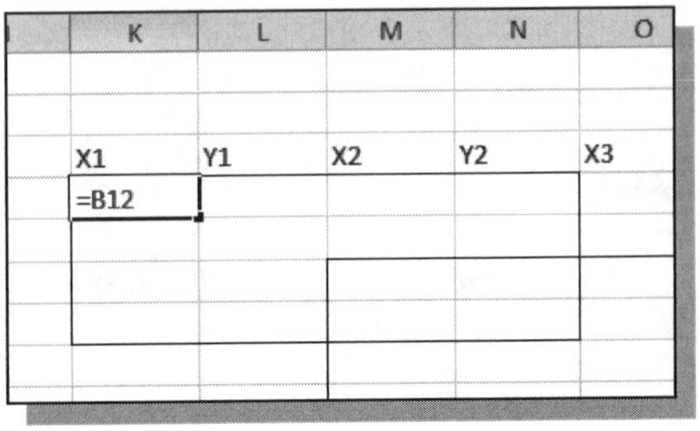

2. Reference the first cell to the first cell of the *Element A* matrix by entering first the equal sign and then the cell location:

=B12 [ENTER]

3. Repeat the last step to fill in the first rectangle, which are simply referencing the cells in the *Element A* matrix. (For the cell locations, refer to the image shown on page 3-6.)

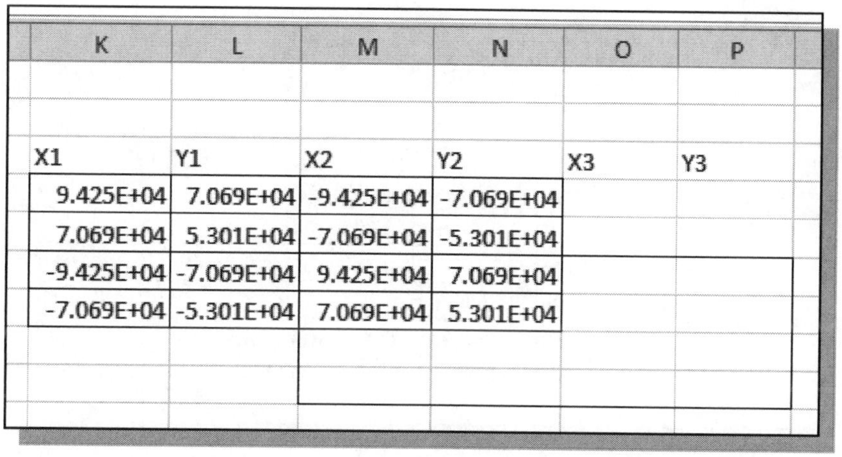

K	L	M	N	O	P
X1	Y1	X2	Y2	X3	Y3
9.425E+04	7.069E+04	-9.425E+04	-7.069E+04		
7.069E+04	5.301E+04	-7.069E+04	-5.301E+04		
-9.425E+04	-7.069E+04	9.425E+04	7.069E+04		
-7.069E+04	-5.301E+04	7.069E+04	5.301E+04		

4. Edit the third cell in the X2 column; this is where the first overlap occurs. Edit the cell, so that it includes the first cell of the *Element B* matrix: **=D14+B27**

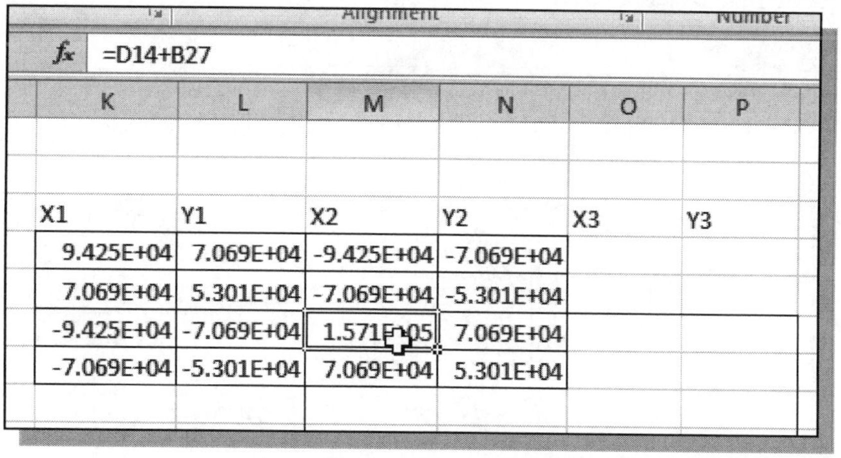

=D14+B27

K	L	M	N	O	P
X1	Y1	X2	Y2	X3	Y3
9.425E+04	7.069E+04	-9.425E+04	-7.069E+04		
7.069E+04	5.301E+04	-7.069E+04	-5.301E+04		
-9.425E+04	-7.069E+04	1.571E+05	7.069E+04		
-7.069E+04	-5.301E+04	7.069E+04	5.301E+04		

5. Repeat the above step and complete the four overlapped cells as shown.

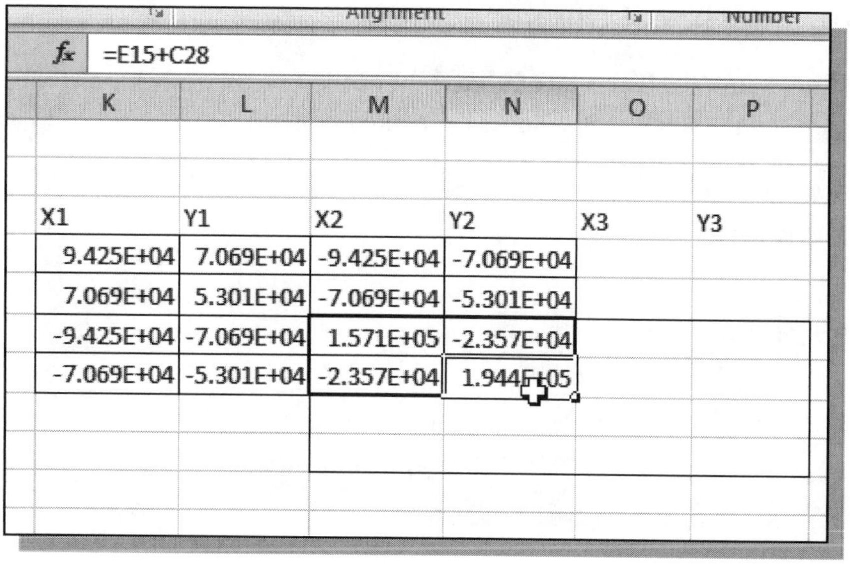

=E15+C28

K	L	M	N	O	P
X1	Y1	X2	Y2	X3	Y3
9.425E+04	7.069E+04	-9.425E+04	-7.069E+04		
7.069E+04	5.301E+04	-7.069E+04	-5.301E+04		
-9.425E+04	-7.069E+04	1.571E+05	-2.357E+04		
-7.069E+04	-5.301E+04	-2.357E+04	1.944E+05		

6. Fill in the rest of the second rectangle by referencing to the *Element B* matrix; the results should appear as shown.

f_x =E27

K	L	M	N	O	P
X1	Y1	X2	Y2	X3	Y3
9.425E+04	7.069E+04	-9.425E+04	-7.069E+04		
7.069E+04	5.301E+04	-7.069E+04	-5.301E+04		
-9.425E+04	-7.069E+04	1.571E+05	-2.357E+04	-6.284E+04	9.425E+04
-7.069E+04	-5.301E+04	-2.357E+04	1.944E+05	9.425E+04	-1.414E+05
		-6.284E+04	9.425E+04	6.284E+04	-9.425E+04
		9.425E+04	-1.414E+05	-9.425E+04	1.414E+05

7. Enter eight **zeros** in the remaining blank cells. Note that these zeros are necessary for the calculations of the reaction forces, which we will do once the global displacements are calculated.

f_x 0

K	L	M	N	O	P
X1	Y1	X2	Y2	X3	Y3
9.425E+04	7.069E+04	-9.425E+04	-7.069E+04	0	0
7.069E+04	5.301E+04	-7.069E+04	-5.301E+04	0	0
-9.425E+04	-7.069E+04	1.571E+05	-2.357E+04	-6.284E+04	9.425E+04
-7.069E+04	-5.301E+04	-2.357E+04	1.944E+05	9.425E+04	-1.414E+05
0	0	-6.284E+04	9.425E+04	6.284E+04	-9.425E+04
0	0	9.425E+04	-1.414E+05	-9.425E+04	1.414E+05

Solve for the Global Displacements

1. Label the global displacements to the right of the overall global matrix; enter the four zeros as shown. (Node 1 and Node 3 are fixed points.)

	X3	Y3			
04	0	0		X1	0
04	0	0		Y1	0
04	-6.284E+04	9.425E+04		X2=?	
05	9.425E+04	-1.414E+05		Y2=?	
04	6.284E+04	-9.425E+04		X3	0
05	-9.425E+04	1.414E+05		Y3	0

2. Label the global forces to the left of the overall global matrix; enter **50** and **0** to represent the loads that are applied at Node 2.

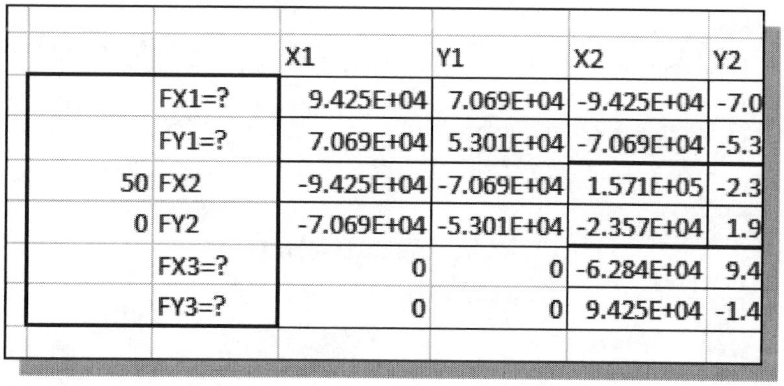

❖ We will solve the two unknown displacements, X2 and Y2, first. This can be done using the established matrix equations.
For $\{\mathbf{F}\} = [\,\mathbf{K}\,]\,\{\mathbf{X}\}$, $\{X\}$ can be solved by finding the inverse [K] matrix.
$\{\mathbf{X}\} = [\,\mathbf{K}\,]^{-1}\,\{\mathbf{F}\}$, where the $\{X\}$ vector represents the two unknowns.

3. Place a label for calculating the **Inverse K** as shown and select a 2x2 array as shown in the figure.

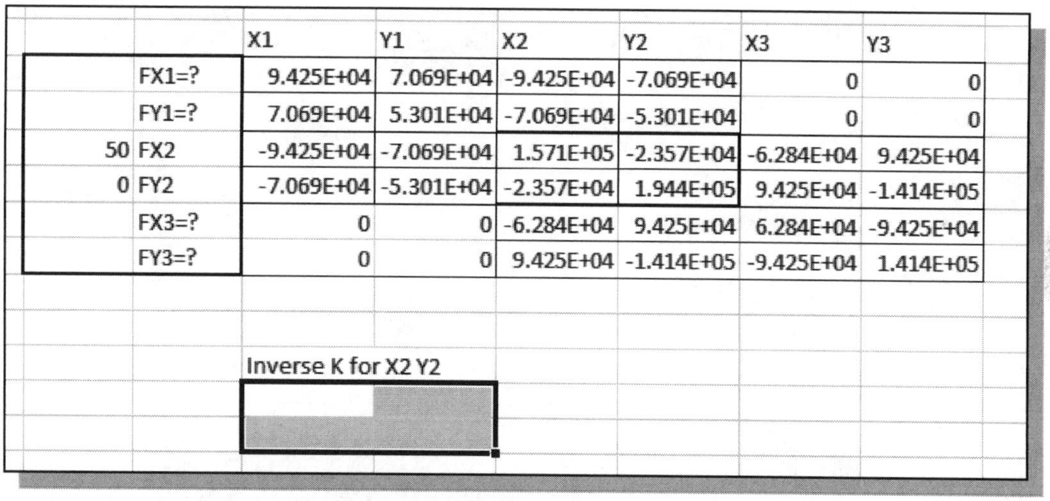

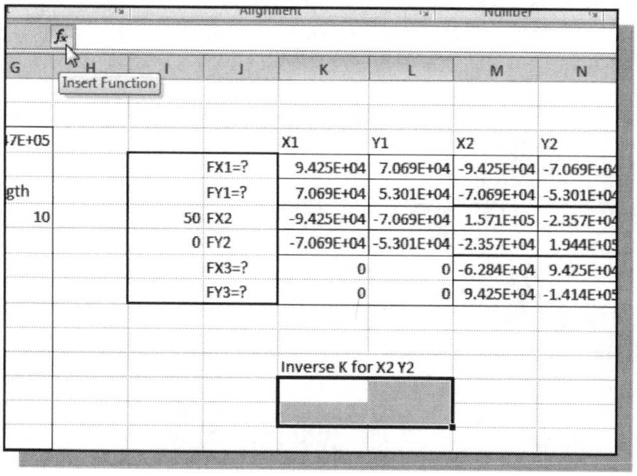

4. Click on the **Insert Function** icon that is located in front of the input edit box as shown in the figure.

❖ *Excel* provides a variety of functions, which include many of the more commonly used financial, math and trigonometric functions.

5. In the ***Insert Function*** dialog box, select **Math & Trig** as shown.

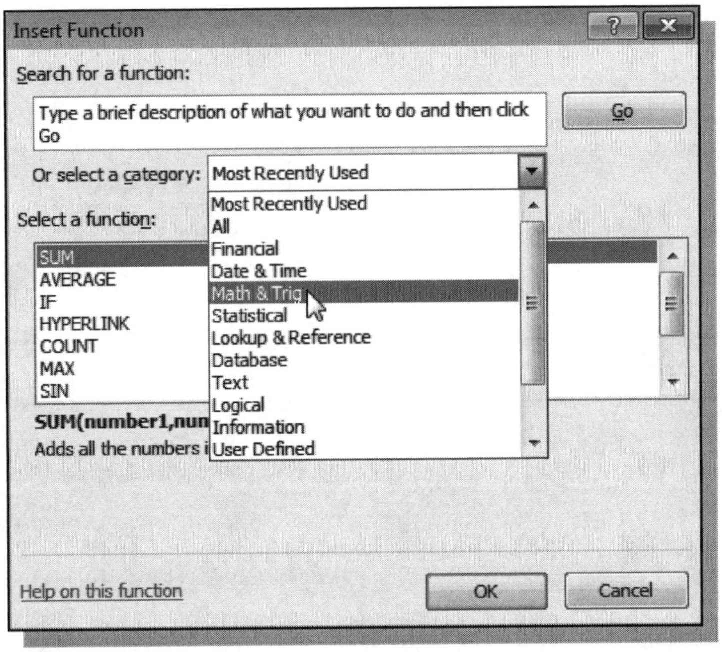

6. Select the **MINVERSE** command as shown.

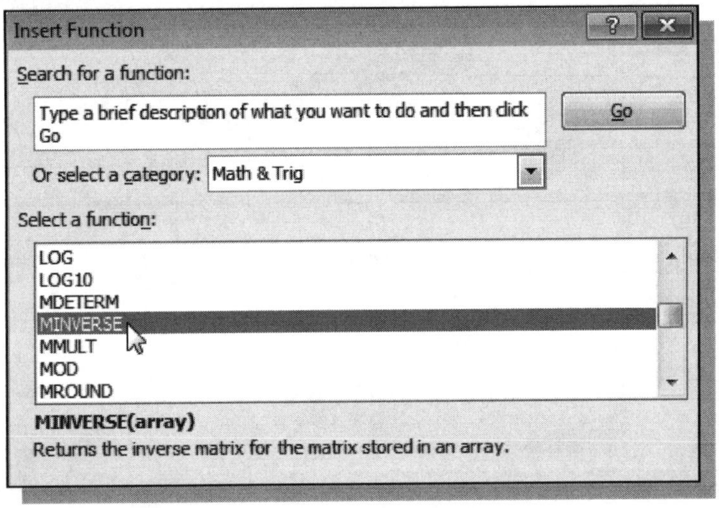

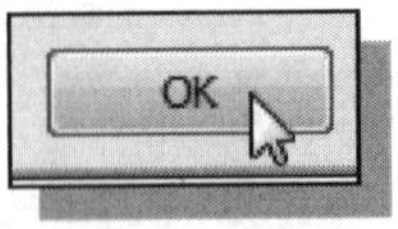

7. Click **OK** to proceed with the **MINVERSE** command.

8. The ***Function Arguments*** dialog box appears on the screen with a brief description of the function.

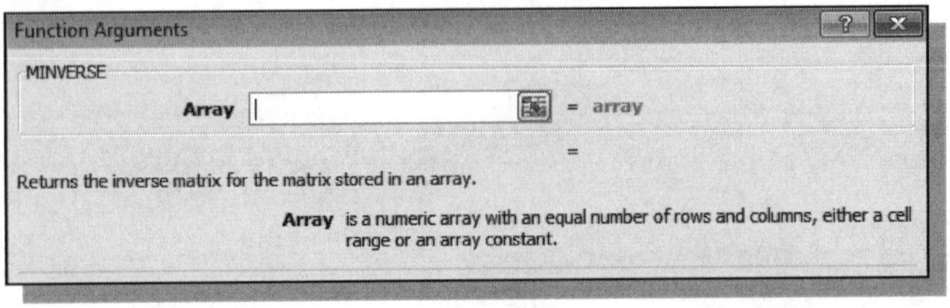

9. Select the 2x2 array as shown in the figure below.

X1	Y1	X2	Y2	X3	Y3
9.425E+04	7.069E+04	-9.425E+04	-7.069E+04	0	0
7.069E+04	5.301E+04	-7.069E+04	-5.301E+04	0	0
-9.425E+04	-7.069E+04	1.571E+05	-2.357E+04	-6.284E+04	9.425E+04
-7.069E+04	-5.301E+04	-2.357E+04	1.[illegible]4E+05	9.425E+04	-1.414E+05
0	0	-6.284E+04	9.425E+04	6.284E+04	-9.425E+04
0	0	9.425E+04	-1.414E+05	-9.425E+04	1.414E+05

10. It is important to note that when using the matrix commands in *Excel*, we must press **CTRL+SHIFT+ENTER** to perform the calculations. The Inverse K is calculated as shown in the figure below.

		X1	Y1	X2	Y2	X3	Y3
	FX1=?	9.425E+04	7.069E+04	-9.425E+04	-7.069E+04	0	0
	FY1=?	7.069E+04	5.301E+04	-7.069E+04	-5.301E+04	0	0
50	FX2	-9.425E+04	-7.069E+04	1.571E+05	-2.357E+04	-6.284E+04	9.425E+04
0	FY2	-7.069E+04	-5.301E+04	-2.357E+04	1.944E+05	9.425E+04	-1.414E+05
	FX3=?	0	0	-6.284E+04	9.425E+04	6.284E+04	-9.425E+04
	FY3=?	0	0	9.425E+04	-1.414E+05	-9.425E+04	1.414E+05

Inverse K for X2 Y2

6.484E-06	7.861E-07
7.8609E-07	5.239E-06

11. Label the two global displacements, and pre-select the 2x1 array as shown.

3=?	0	0	9.425E+04	-1.414E+05	-9.425E+04	1.414E+05

Inverse K for X2 Y2

6.484E-06	7.861E-07		X2	
7.8609E-07	5.239E-06		Y2	

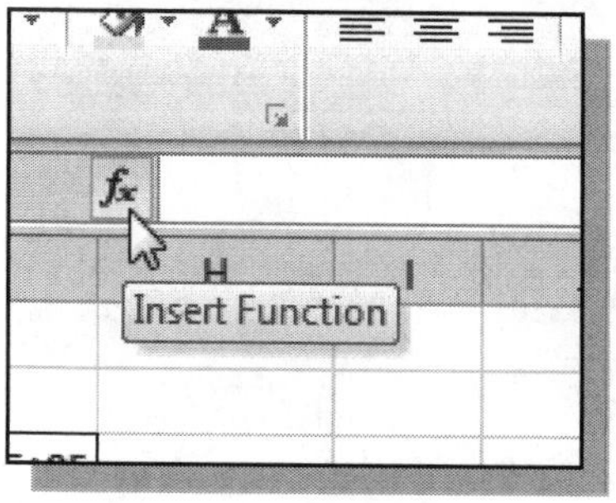

12. Click on the **Insert Function** icon that is located in front of the input edit box as shown in the figure.

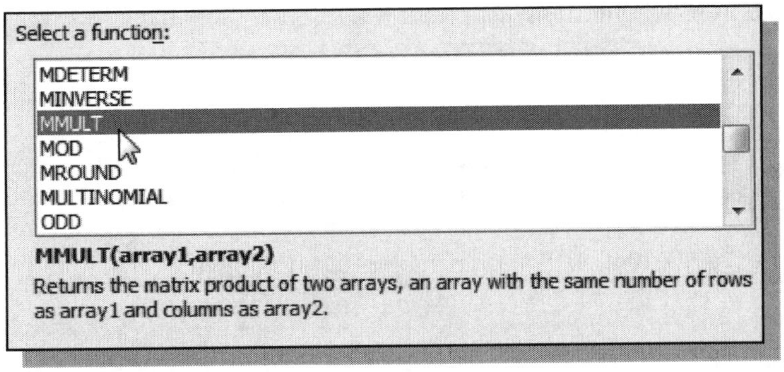

13. Select the **MMULT** command as shown.

14. Click **OK** to proceed with the **MMULT** command.

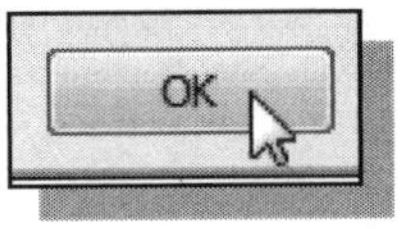

15. The ***Function Arguments*** dialog box appears on the screen with a brief description of the function. Note that two arrays are required for this function.

16. Select the calculated **Inverse [K] for X2 and Y2**, which is a 2x2 array, as shown.

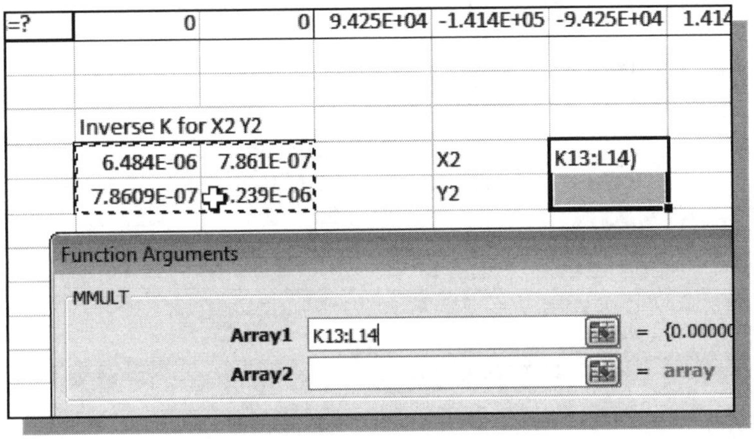

17. Hit the **Tab** key once to move the cursor to the next input box and select the two known forces at node 2 (*FX2* and *FY2*).

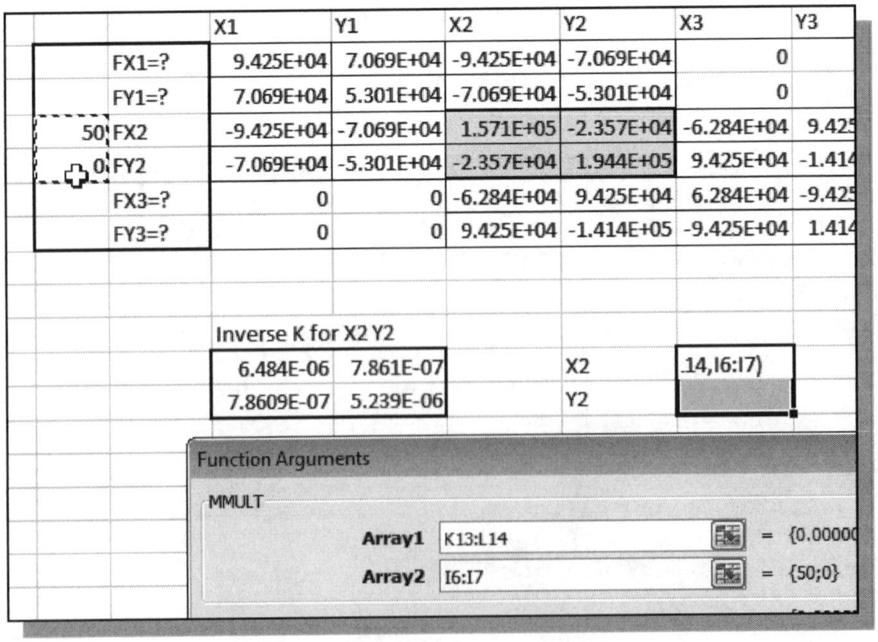

18. It is important to note that when using the matrix commands in *Excel*, we must press **CTRL+SHIFT+ENTER** to perform the calculations. The global displacements X2 and Y2 are calculated as shown in the figure below.

		X1	Y1	X2	Y2	X3	Y3
	FX1=?	9.425E+04	7.069E+04	-9.425E+04	-7.069E+04	0	0
	FY1=?	7.069E+04	5.301E+04	-7.069E+04	-5.301E+04	0	0
50	FX2	-9.425E+04	-7.069E+04	1.571E+05	-2.357E+04	-6.284E+04	9.425E+04
0	FY2	-7.069E+04	-5.301E+04	-2.357E+04	1.944E+05	9.425E+04	-1.414E+05
	FX3=?	0	0	-6.284E+04	9.425E+04	6.284E+04	-9.425E+04
	FY3=?	0	0	9.425E+04	-1.414E+05	-9.425E+04	1.414E+05
		Inverse K for X2 Y2					
		6.484E-06	7.861E-07		X2	0.000324199	
		7.8609E-07	5.239E-06		Y2	3.93046E-05	

- Note the calculation is based on the equation $\{\mathbf{X}\} = [\ \mathbf{K}\ ]^{-1}\ \{\mathbf{F}\}$, and in matrix multiplication, the order of the matrices are non-interchangeable. So, the inverse K matrix is selected first, followed by the Force matrix.

19. Set the X2 value, in the overall global displacement matrix, to reference the X2 value just calculated as shown.

	Y2	X3	Y3			
E+04	-7.069E+04	0	0		X1	0
E+04	-5.301E+04	0	0		Y1	0
E+05	-2.357E+04	-6.284E+04	9.425E+04		X2=?	=O13
E+04	1.944E+05	9.425E+04	-1.414E+05		Y2=?	
E+04	9.425E+04	6.284E+04	-9.425E+04		X3	0
E+04	-1.414E+05	-9.425E+04	1.414E+05		Y3	0
	X2	0.000324199				
	Y2	3.93046E-05				

E+04	-5.301E+04	0	0		Y1	0
E+05	-2.357E+04	-6.284E+04	9.425E+04		X2=?	0.000324
E+04	1.944E+05	9.425E+04	-1.414E+05		Y2=?	=O14
E+04	9.425E+04	6.284E+04	-9.425E+04		X3	0
E+04	-1.414E+05	-9.425E+04	1.414E+05		Y3	0
	X2	0.000324199				
	Y2	3.93046E-05				

20. Repeat the above step and set the Y2 value as shown as well.

Calculate the Reaction Forces

1. Pre-select the **six cells** in front of the global forces labels as shown.

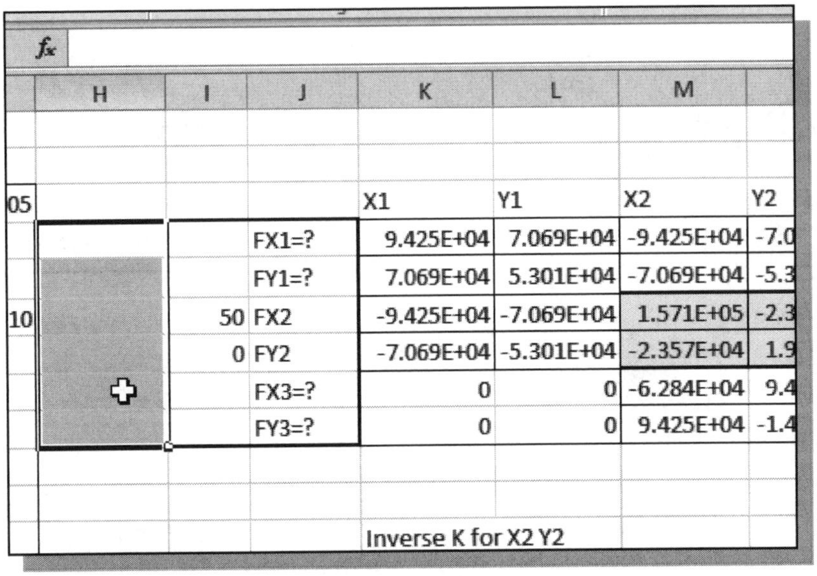

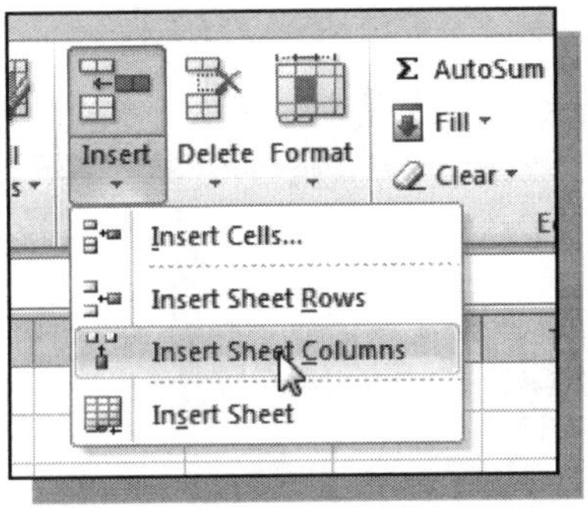

❖ Note the **Insert Sheet Columns** and the **Insert Sheet Rows** commands are also available to add additional columns and rows in between existing cells.

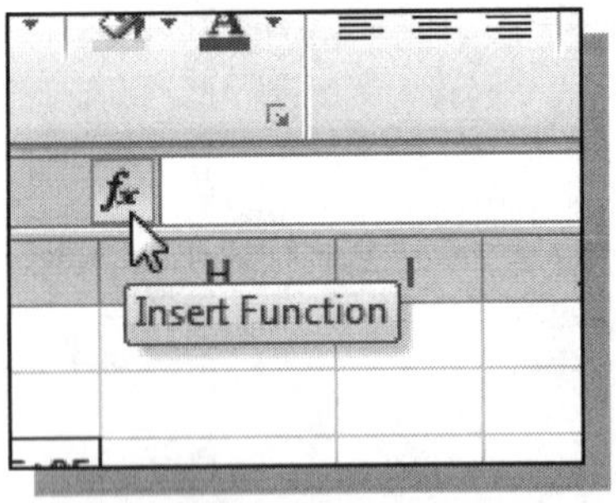

2. Click on the **Insert Function** icon that is located in front of the input edit box as shown in the figure.

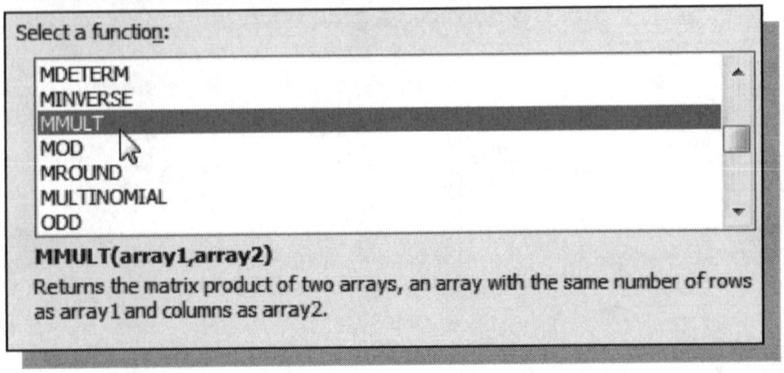

3. Select the **MMULT** command as shown.

4. Click **OK** to proceed with the **MMULT** command.

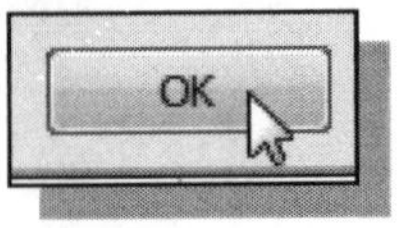

5. The ***Function Arguments*** dialog box appears on the screen with a brief description of the function. Note that two arrays are required for this function.

6. Select the **overall global [k] matrix**, which is the 6x6 array, as shown as the first array.

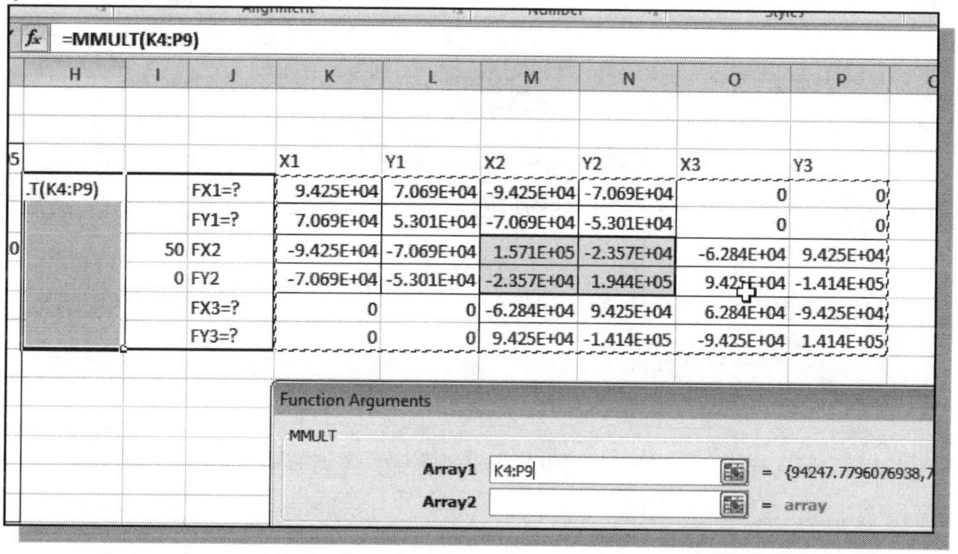

7. Hit the **Tab** key once to move the cursor to the second array input box.

8. Select the **overall global displacement matrix** as the second array. This is the 6x1 array as shown in the figure.

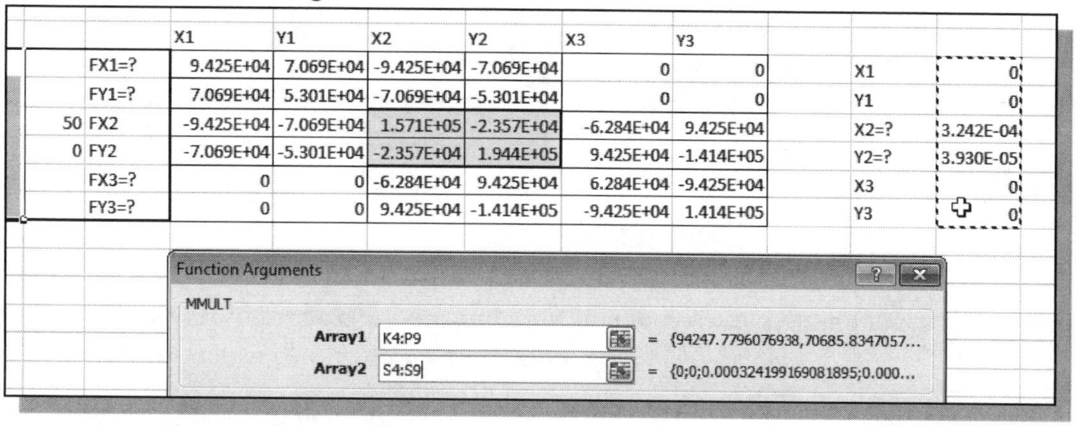

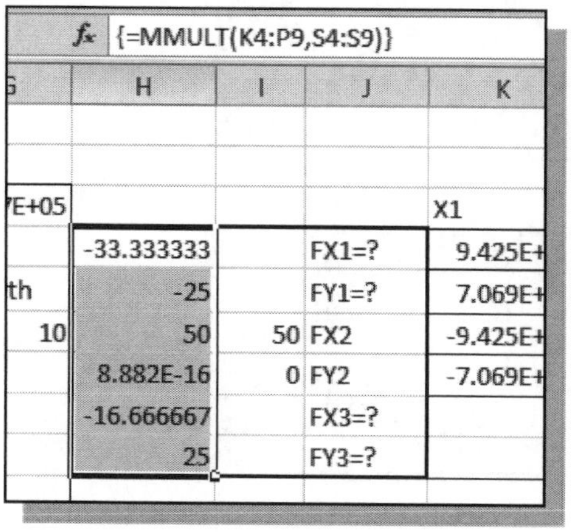

9. It is important to note that when using the matrix commands in *Excel*, we must press **CTRL+SHIFT+ENTER** to perform the calculations. The global forces are calculated as shown in the figure.

❖ Note this calculation also provides a quick check against the initial forces at node 2. The small FY2 represents the rounding error that exists in computer software.

Determine the Stresses in Elements

To determine the normal stress in each truss member, one option is to use the displacement transformation equations to transform the results from the global coordinate system back to the local coordinate system.

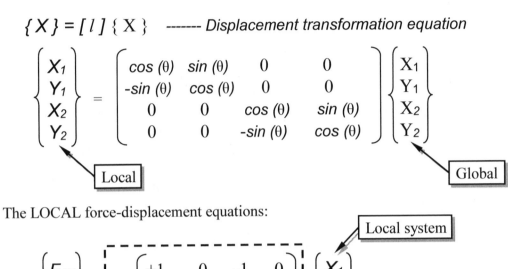

$\{X\} = [l]\{X\}$ ------- *Displacement transformation equation*

$$\begin{Bmatrix} X_1 \\ Y_1 \\ X_2 \\ Y_2 \end{Bmatrix} = \begin{bmatrix} \cos(\theta) & \sin(\theta) & 0 & 0 \\ -\sin(\theta) & \cos(\theta) & 0 & 0 \\ 0 & 0 & \cos(\theta) & \sin(\theta) \\ 0 & 0 & -\sin(\theta) & \cos(\theta) \end{bmatrix} \begin{Bmatrix} X_1 \\ Y_1 \\ X_2 \\ Y_2 \end{Bmatrix}$$

Local — Global

The LOCAL force-displacement equations:

$$\begin{Bmatrix} F_{1X} \\ F_{1Y} \\ F_{2X} \\ F_{2Y} \end{Bmatrix} = \frac{EA}{L} \begin{bmatrix} +1 & 0 & -1 & 0 \\ 0 & 0 & 0 & 0 \\ -1 & 0 & +1 & 0 \\ 0 & 0 & 0 & 0 \end{bmatrix} \begin{Bmatrix} X_1 \\ Y_1 \\ X_2 \\ Y_2 \end{Bmatrix}$$

Local system — Local Stiffness Matrix — Local system

Element A

1. We will begin to set up the solution in finding the normal stress of *Element A*. Below the inverse K matrix, label *Element A* and set up the direction cosines as shown. (Use the Cosine and Sine functions of the *Element A* angle.)

	7.8609E-07	5.239E-06		Y2	3.93
Element A					
	x=lX				
	0.8	0.6	0	0	
	-0.6	0.8	0	0	
	0	0	0.8	0.6	
	0	0	-0.6	0.8	

2. Pre-select the **four cells** in front of the local displacement labels as shown.

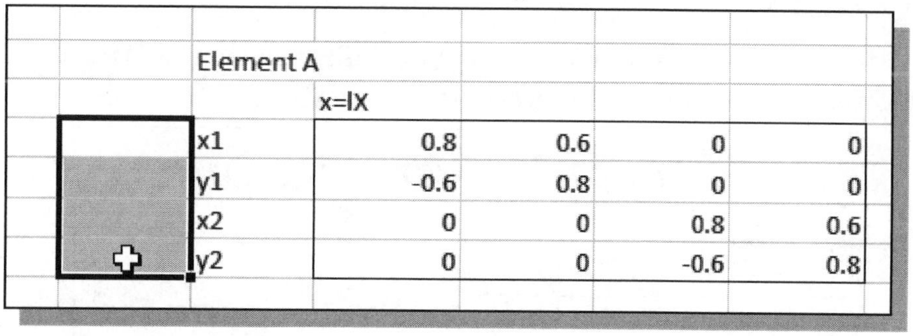

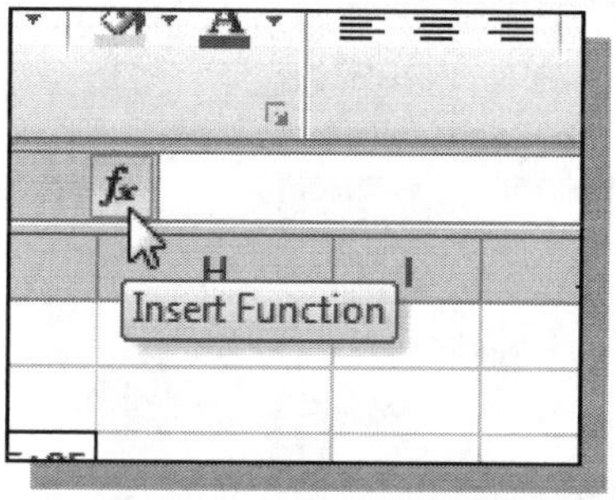

3. Click on the **Insert Function** icon that is located in front of the input edit box as shown in the figure.

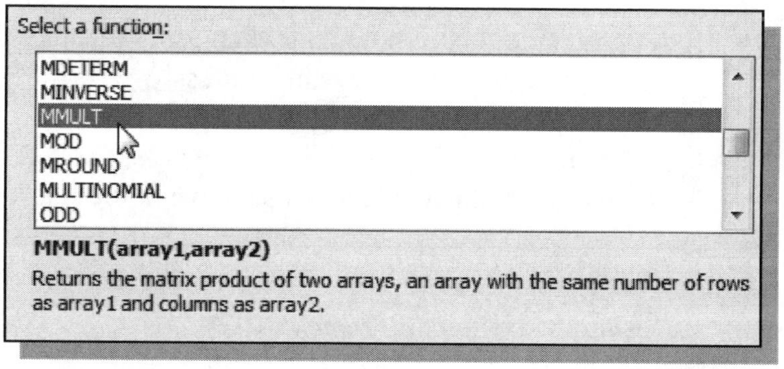

4. Select the **MMULT** command as shown.

5. Click **OK** to proceed with the **MMULT** command.

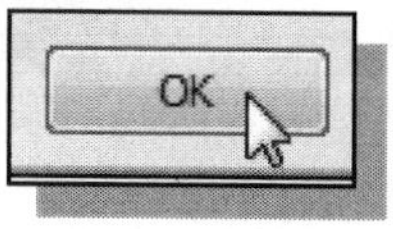

6. Select the **directional cosine matrix**, which is the 4x4 array, which we just set up as the first array.

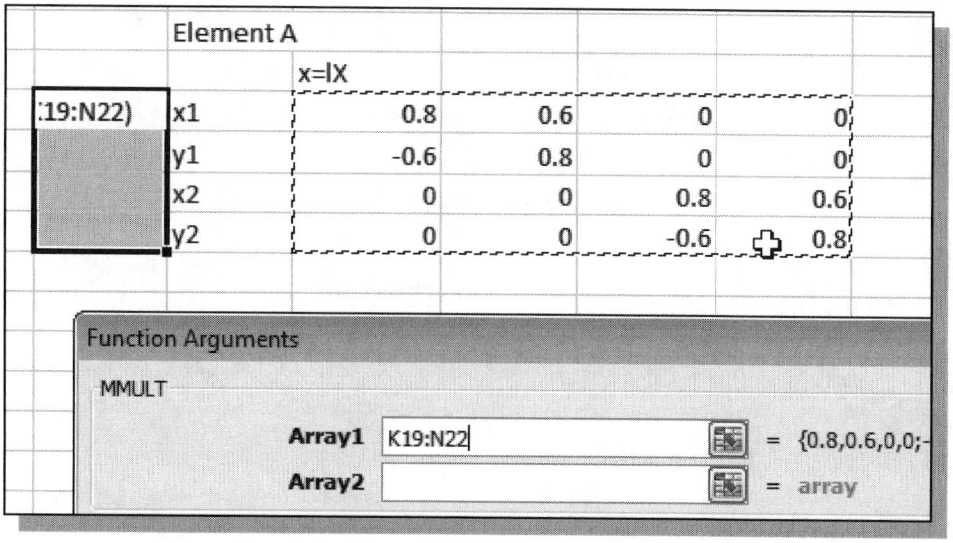

7. Hit the **Tab** key once to move the cursor to the second array input box.

8. Select the **first four cells** of the **overall global displacement matrix** as the second array. This is the 4x1 array as shown in the figure.

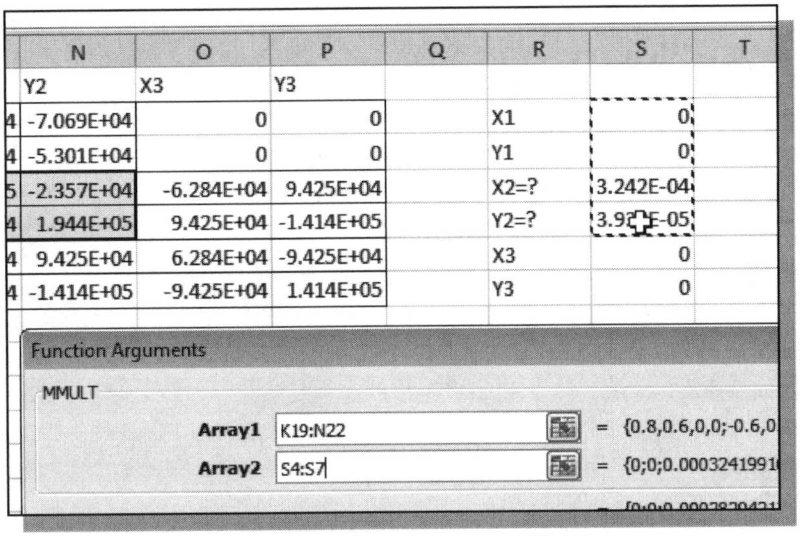
N
O
P
Q
R
S
T
Y2
X3
Y3
-7.069E+04 0 0 X1 0
-5.301E+04 0 0 Y1 0
-2.357E+04 -6.284E+04 9.425E+04 X2=? 3.242E-04
1.944E+05 9.425E+04 -1.414E+05 Y2=?
9.425E+04 6.284E+04 -9.425E+04 X3 0
-1.414E+05 -9.425E+04 1.414E+05 Y3 0
Function Arguments
MMULT
Array1 K19:N22 = {0.8,0.6,0,0;-0.6,0
Array2 S4:S7 = {0;0;0.0003241991

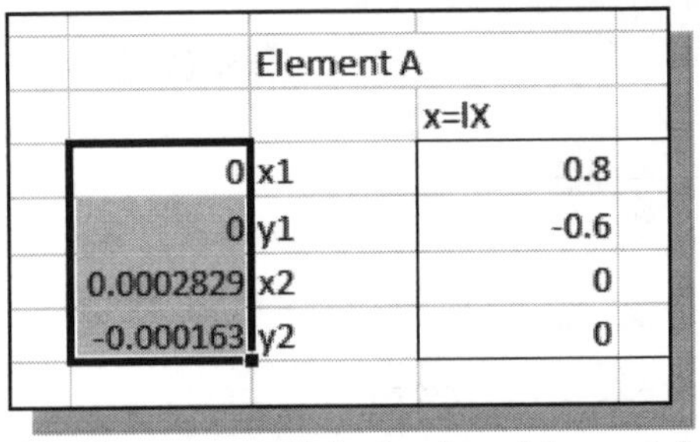
Element A
x=lX
0 x1 0.8
0 y1 -0.6
0.0002829 x2 0
-0.000163 y2 0

9. It is important to note that when using the matrix commands in *Excel*, we must press **CTRL+SHIFT+ENTER** to perform the calculations. The local displacements of node 1 and node 2 are calculated as shown in the figure.

10. Next establish the **local k matrix**, using the k value of *Element A*.

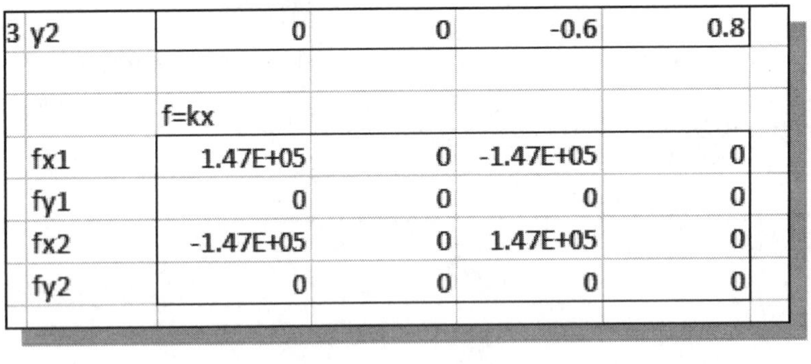
y2 0 0 -0.6 0.8
f=kx
fx1 1.47E+05 0 -1.47E+05 0
fy1 0 0 0 0
fx2 -1.47E+05 0 1.47E+05 0
fy2 0 0 0 0

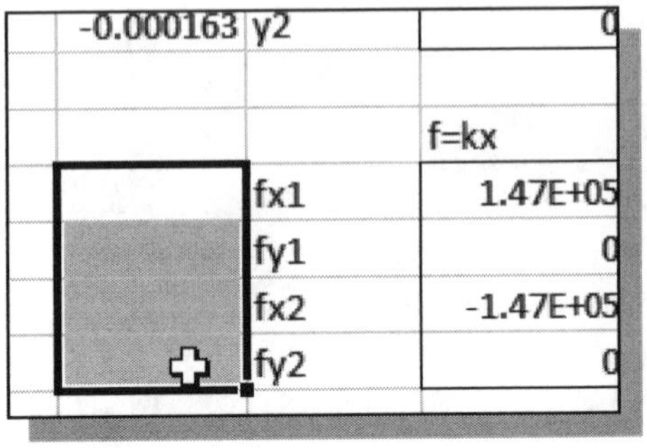
-0.000163 y2
f=kx
fx1 1.47E+05
fy1
fx2 -1.47E+05
fy2

11. Pre-select the **four cells** in front of the local force labels as shown.

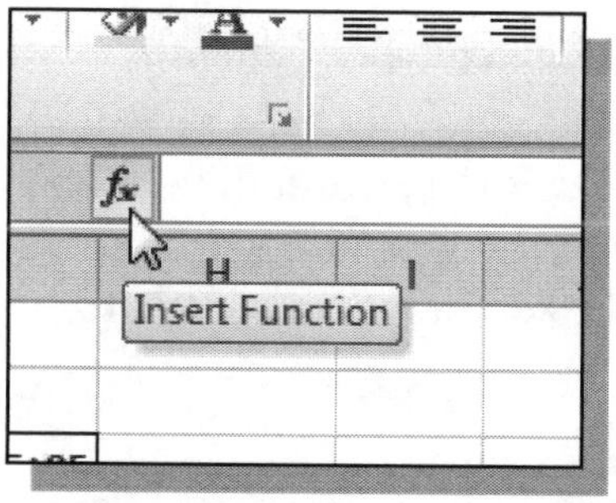

12. Click on the **Insert Function** icon that is located in front of the input edit box as shown in the figure.

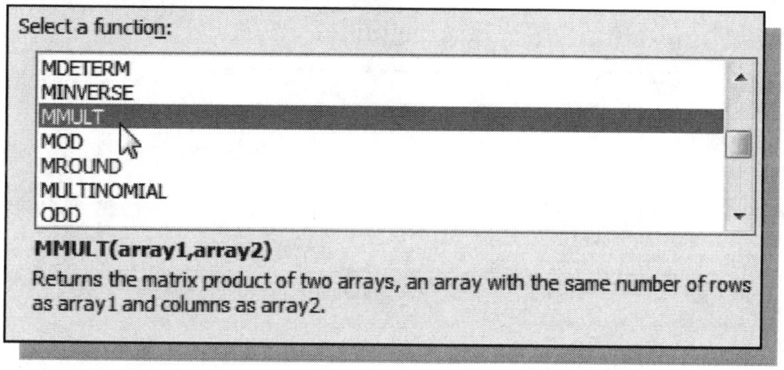

13. Select the **MMULT** command as shown.

14. Click **OK** to proceed with the **MMULT** command.

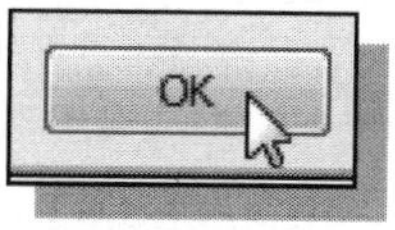

15. Select the **local k matrix**, which is the 4x4 array, which we just set up as the first array.

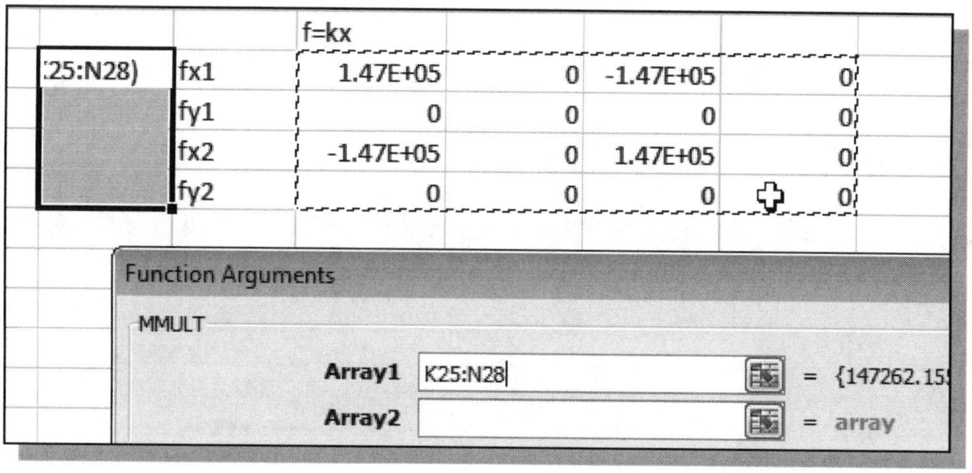

16. Select the **local displacement matrix** as the second array. This is the 4x1 array which we calculated in step 9.

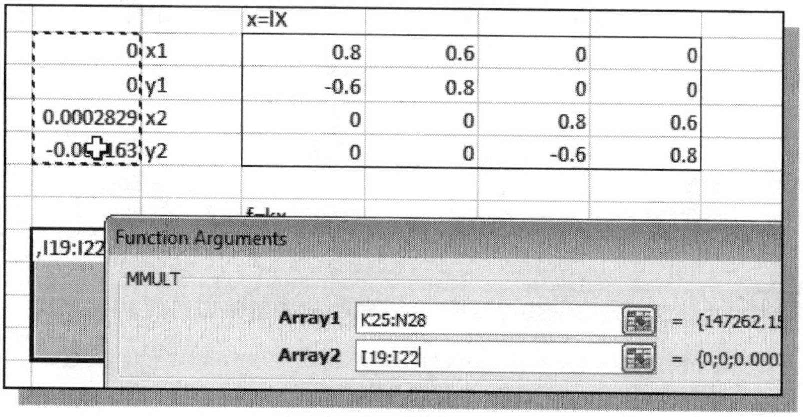

-0.000163	y2	0
		f=kx
-41.66667	fx1	1.47E+05
0	fy1	0
41.666667	fx2	-1.47E+05
0	fy2	0

17. It is important to note that when using the matrix commands in *Excel*, we must press **CTRL+SHIFT+ENTER** to perform the calculations. The local forces of node 1 and node 2 are calculated as shown in the figure.

18. To calculate the normal stress of *Element A*, use the equation σ = F/A.

			f=kx			
	-41.66667	fx1	1.47E+05	0	-1.47E+05	0
Stress	0	fy1	0	0	0	0
=I27/F6	41.666667	fx2	-1.47E+05	0	1.47E+05	0
	0	fy2	0	0	0	0

19. On your own, repeat the above steps and find the normal stress in *Element B*.

		Element A				
			x=lX			
	0	x1	0.8	0.6	0	0
	0	y1	-0.6	0.8	0	0
	0.0002829	x2	0	0	0.8	0.6
	-0.000163	y2	0	0	-0.6	0.8
			f=kx			
	-41.66667	fx1	1.47E+05	0	-1.47E+05	0
Stress	0	fy1	0	0	0	0
848.82636	41.666667	fx2	-1.47E+05	0	1.47E+05	0
	0	fy2	0	0	0	0
		Element B				
			x=lX			
	0.0001471	x2	0.5547002	-0.83205	0	0
	0.0002916	y2	0.83205029	0.5547002	0	0
	0	x3	0	0	0.5547002	-0.83205
	0	y3	0	0	0.8320503	0.5547002
			f=kx			
	30.046261	fx1	2.04E+05	0	-2.04E+05	0
Stress	0	fy1	0	0	0	0
-612.0974	-30.04626	fx2	-2.04E+05	0	2.04E+05	0
	0	fy2	0	0	0	0

❖ With the built-in functions and convenient editing tools of *Excel*, the direct stiffness matrix method becomes quite feasible to solve 2D truss problems.

The Truss Solver and the Truss View programs

The *Truss Solver* program is a custom-built *Windows* based computer program that can be used to solve 1D, 2D and 3D trusses. The *Truss Solver* program is based on the direct stiffness matrix method with a built-in editor and uses the *Cholesky* decomposition method in solving the simultaneous equations. The program is very compact, 62 Kbytes in size, and it will run in any *Windows* based system.

The *Truss View* program is a simple viewer program that can be used to view the results of the *Truss Solver* program. It simply provides a graphical display of the information that is stored in the output file of the *Truss Solver* program; no FEA analysis is done in the *Truss View* program.

❖ First download the *Truss36.zip* file, which contains the *Truss Solver* and the *Truss View* programs, from the *SDC Publications* website.

1. Launch your internet browser, such as the *MS Internet Explorer* or *Mozilla Firefox* web browsers.

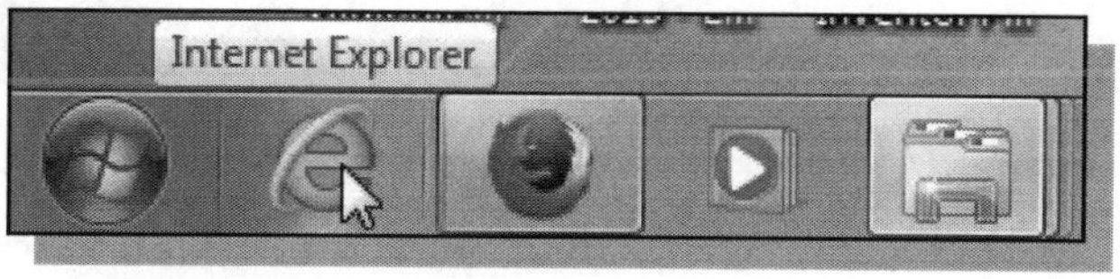

2. In the *URL address* box, enter www.SDCPublications.com/downloads/978-1-63057-108-5

3. Click the **Download Now** button to download the *Truss Solver* software (*Truss37.zip*) to your computer.

4. On your own, extract the content of the ZIP file to a folder or on the desktop. Two files are included in the ZIP file: (1) the *Truss Solver* program, *Truss37.exe,* and (2) the *Truss View* program, *TrusVW36.exe.*

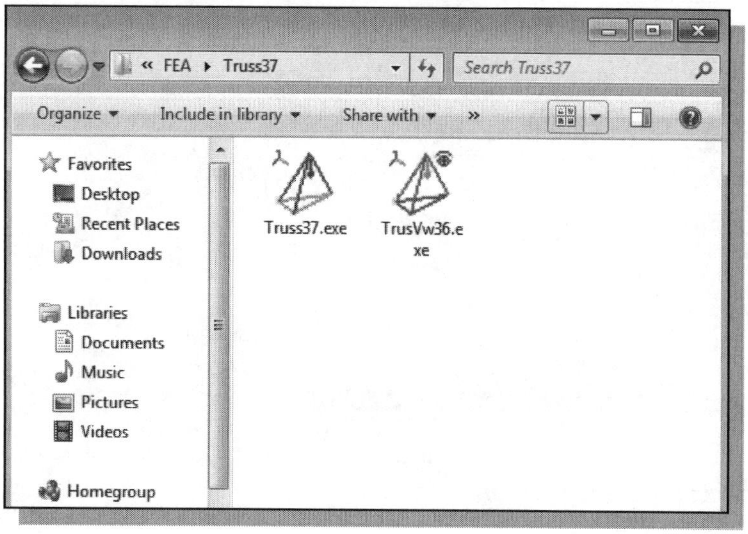

5. Start the ***Truss Solver*** by double-clicking on the **Truss37.exe** icon.

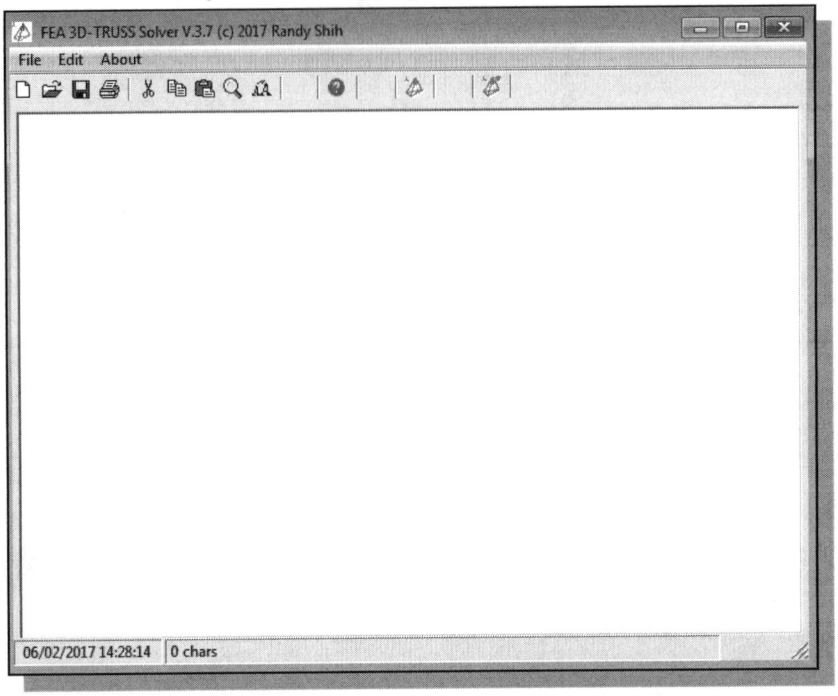

6. Hit the [**F1**] key once or click on the **Help** button to display the *Help* page.

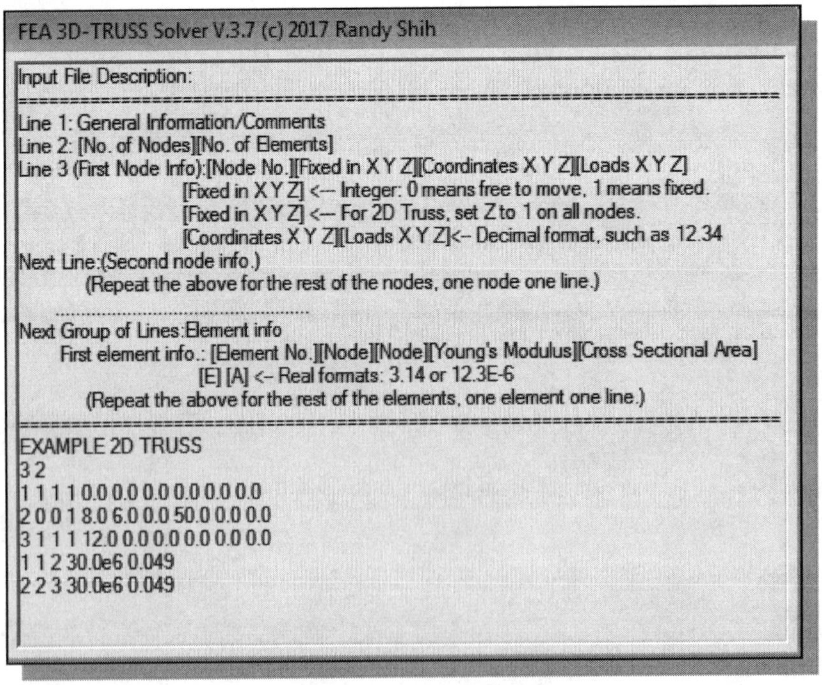

❖ The *Help* page shows the format and information needed to create the input file for the *Truss Solver* program.

7. The first line is a comment line, which can be used to describe any general information about the truss system being solved. For this example, enter: **Example 3.2, 2D truss**.

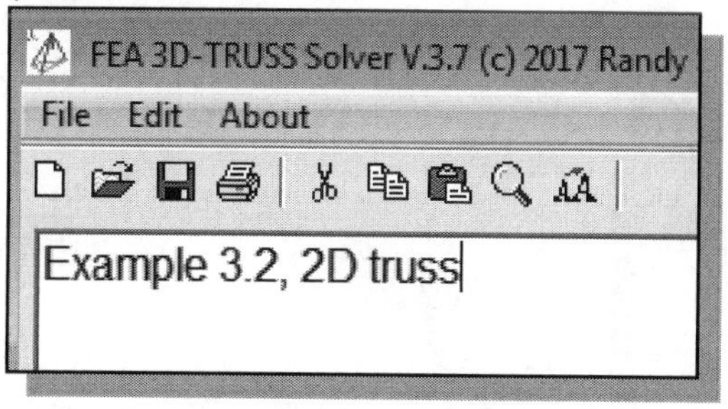

8. The second line contains two numbers: the number of nodes and the number of elements in the system. For our example, we enter **3 2** (the two numbers are separated by a space) to indicate the system has 3 nodes and 2 elements.

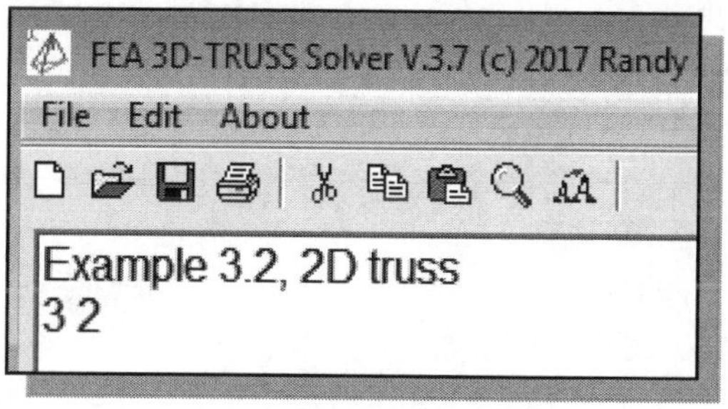

9. Since we have indicated there are three nodes, the next three lines will be describing the three nodes. Enter **1**, **2** and **3** as the first number of the three lines.

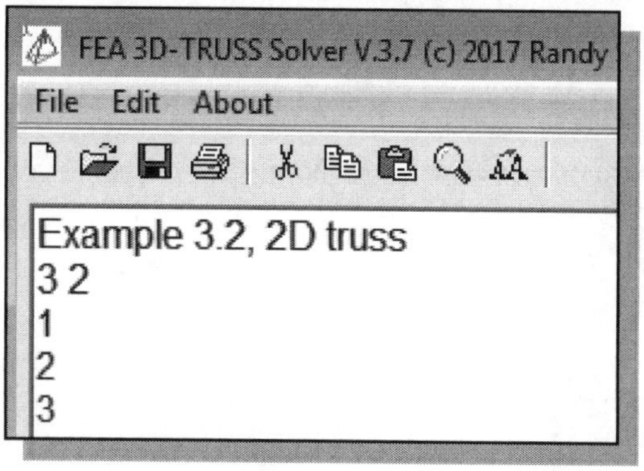

10. The next three numbers, behind the node number, are used to identify the restraints of the node in X, Y and Z directions. Use 1 to indicate it cannot move in that direction; use 0 if it is free to move in that direction. Enter **1 1 1** to indicate node 1 cannot move in all three directions.

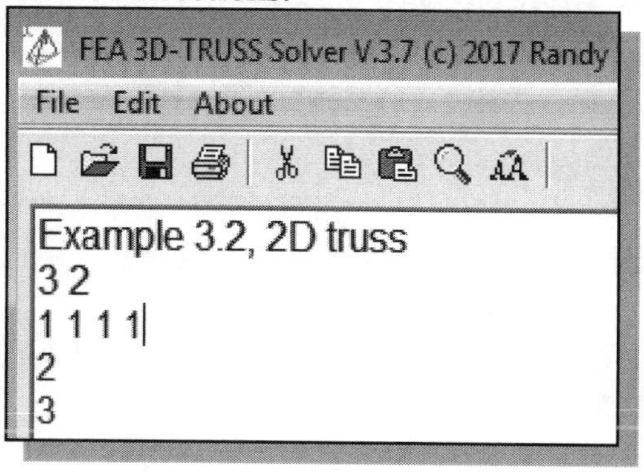

11. The next three numbers are the coordinates of the node; the coordinates need to be expressed in Real format. Enter **0.0 0.0 0.0** to identify the node is aligned to the origin of the coordinate system.

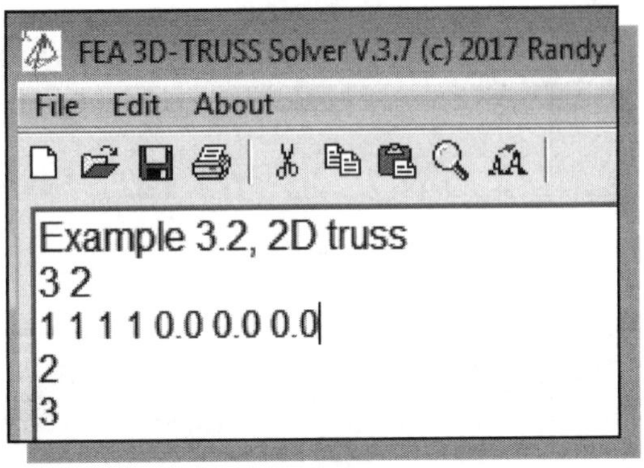

12. The last three numbers describe the external loads applied, components in the X, Y and Z directions, at the node. The numbers also need to be in Real format. Enter **0.0 0.0 0.0** to indicate no loads at node 1.

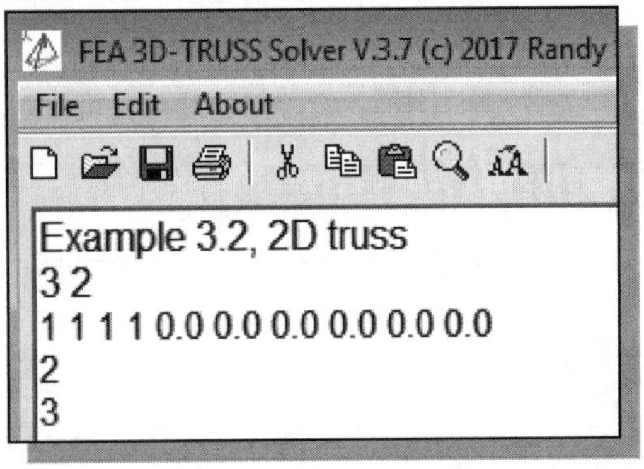

13. Repeat the above steps and enter the information for node 2 and node 3 as shown.

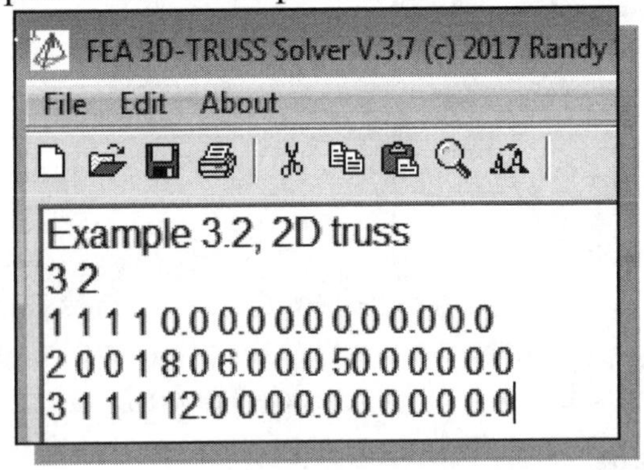

14. For the next two lines, we will describe the element information. Enter **1** and **2** as the element numbers as shown.

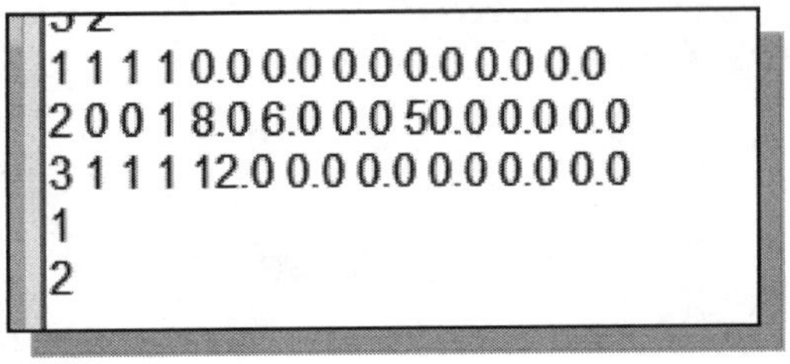

15. The two numbers after the element number are used to describe the two nodes connected to the element. Enter **1 2** to indicate the first element is connected to node 1 and node 2.

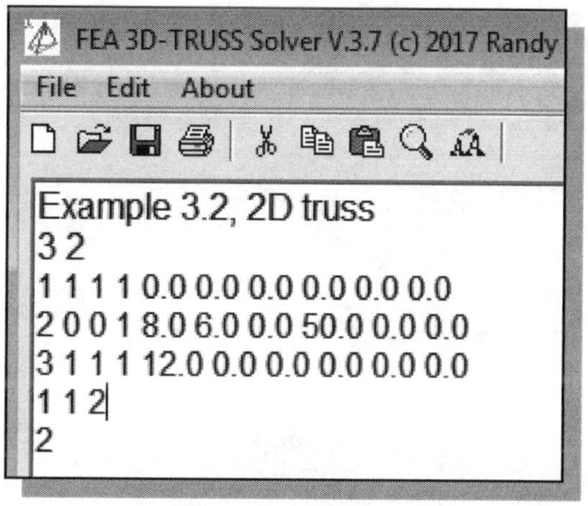

16. The next two numbers describe the **Modulus of elasticity** for the material used and the **cross-sectional area** of the first element.

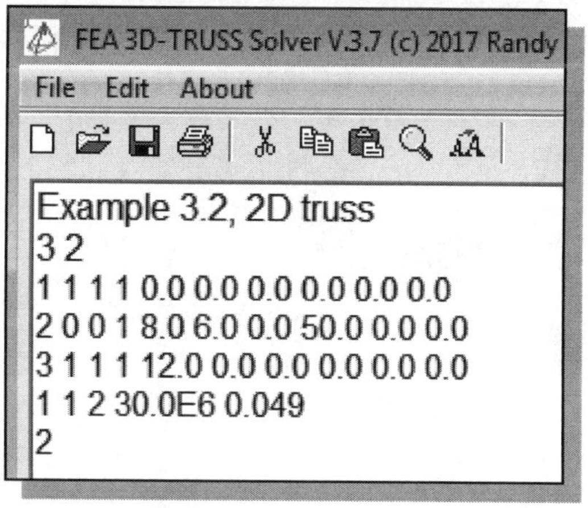

17. Repeat the above step and enter the material and cross-sectional information of the second element.

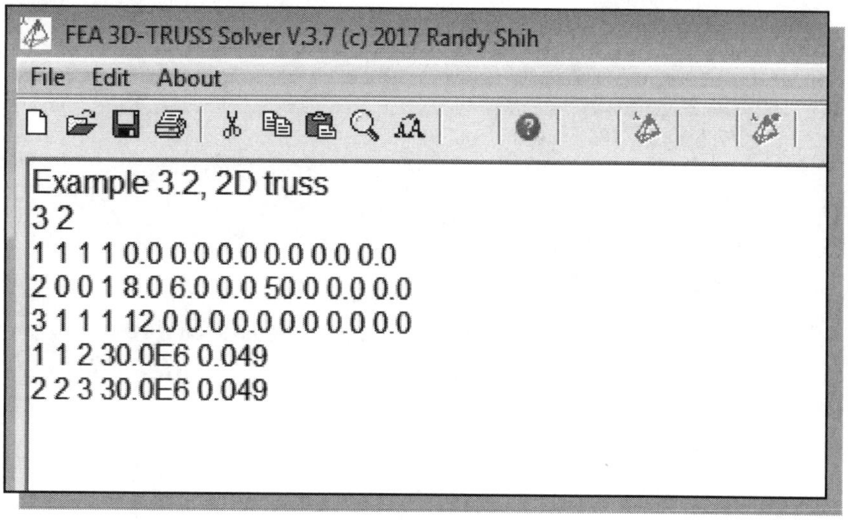

18. Notice the **Truss Solver** button is currently grayed-out, which indicates the command is unavailable. The *Truss Solver* will read the data from a disk file; therefore, it is necessary to save the data to disk.

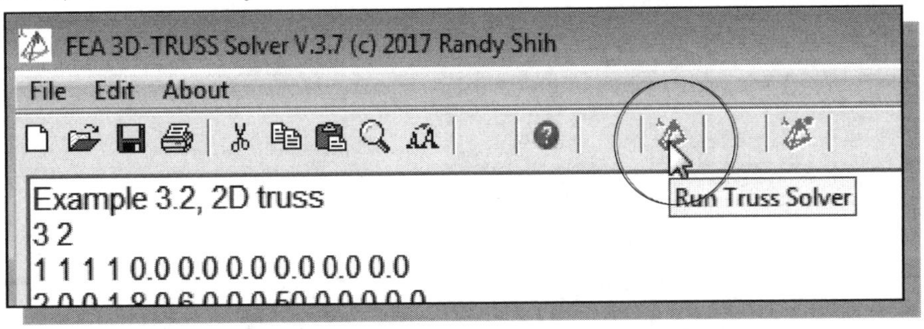

19. Click on the **Save** button as shown.

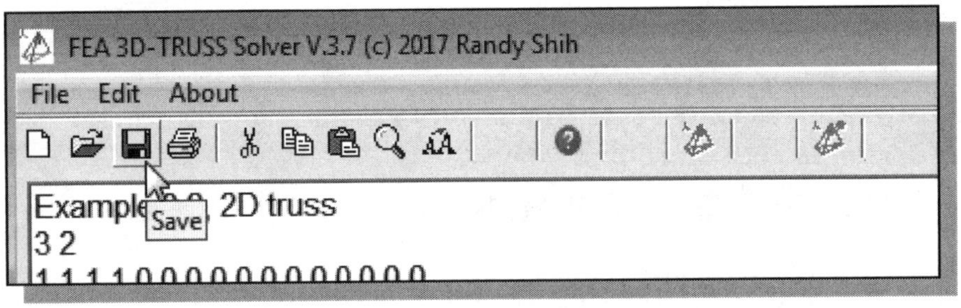

20. Enter **Ex32.txt** as the filename and click **Save** to store the file to disk.

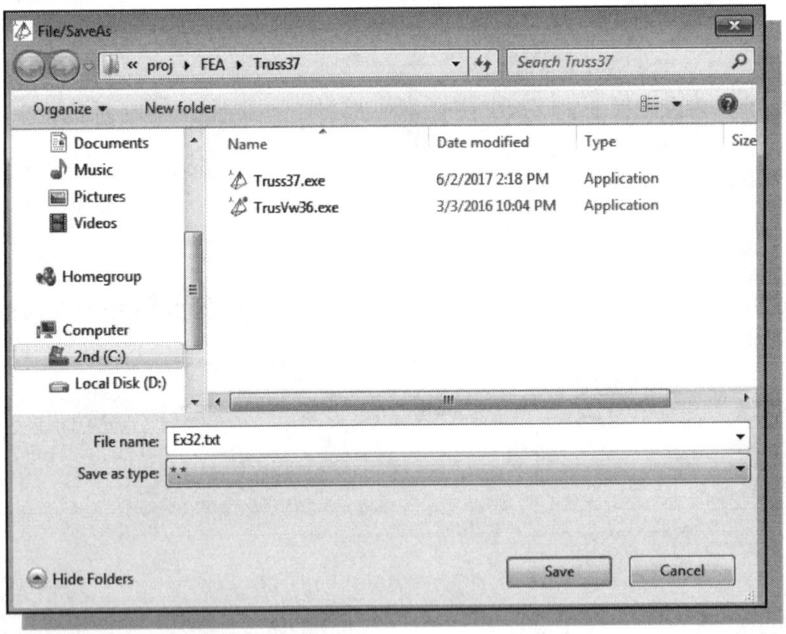

21. Notice the **Run Truss Solver** button is now available. Click on it to **run** the *Truss Solver*.

22. The *Truss Solver* will then process the data and the solution is displayed inside the built-in editor.

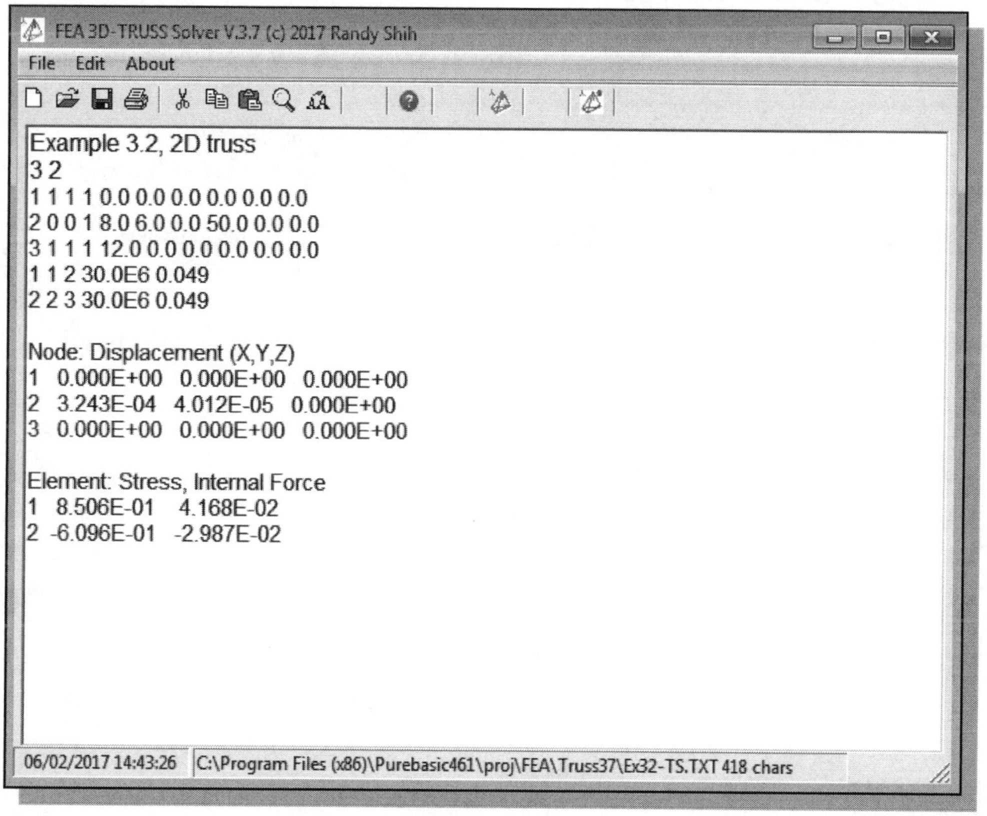

- The solution contains two sections: (1) displacements of the nodes in X, Y and Z directions, and (2) the stresses and internal forces of the elements.

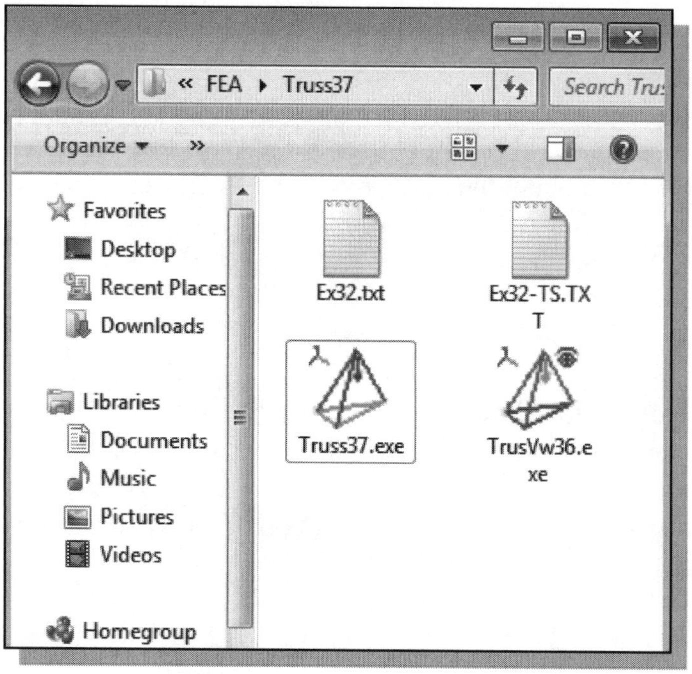

- ➢ Note the results of the *Truss Solver* are also stored in the *Ex32-TS.txt* file. Same filename as the input file, with a different filename extension.

- ❖ Also note that both *Ex32.txt* and *Ex32-TS.txt* are plain text files, which can be opened with any text editor, such as *Windows Notepad*.

The Truss View program

1. Launch the ***Truss View*** program by clicking on the **Launch Truss View** icon as shown.

   ❖ Note the FEA results stored in *Ex32-TS.txt* are transferred to the Truss View program.

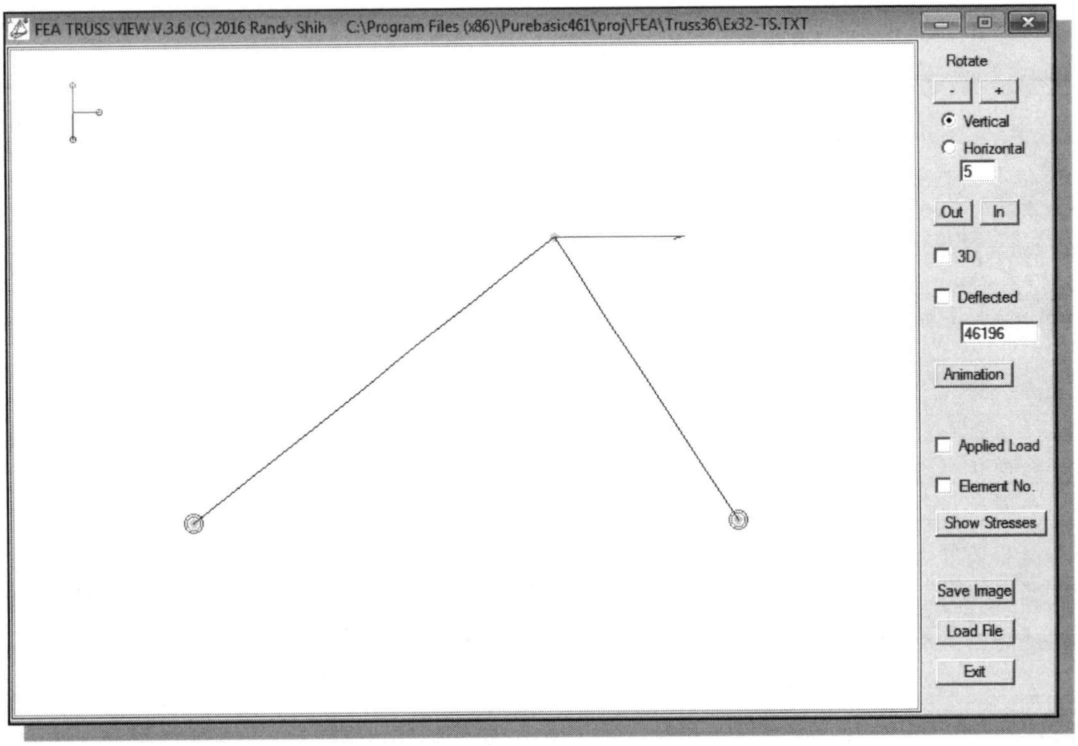

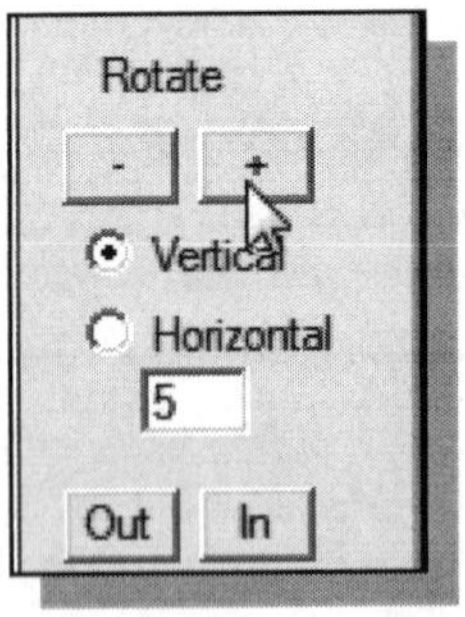

2. The **Rotate** command is generally used for **3D trusses**, which we can choose to rotate about the horizontal or vertical axes.

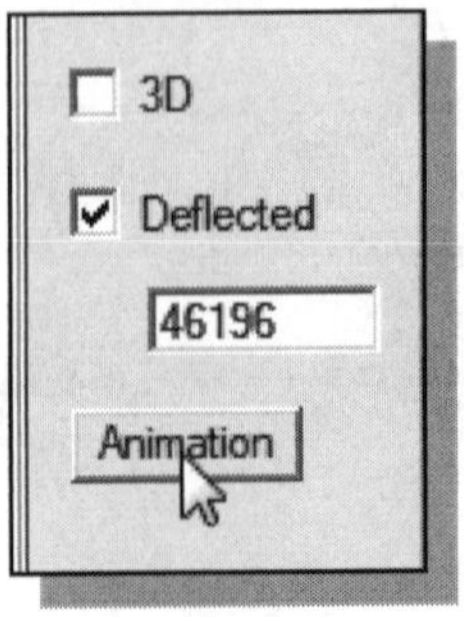

3. The **Deflected** option shows the deflected node location using the computer calculated scale factor that can be edited.

4. The **Animation** button allows the simulation of the system as the load is applied.

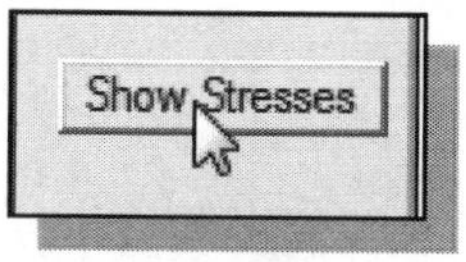

5. Click on the **Show Stresses** button to display the list of the stresses and forces for the elements.

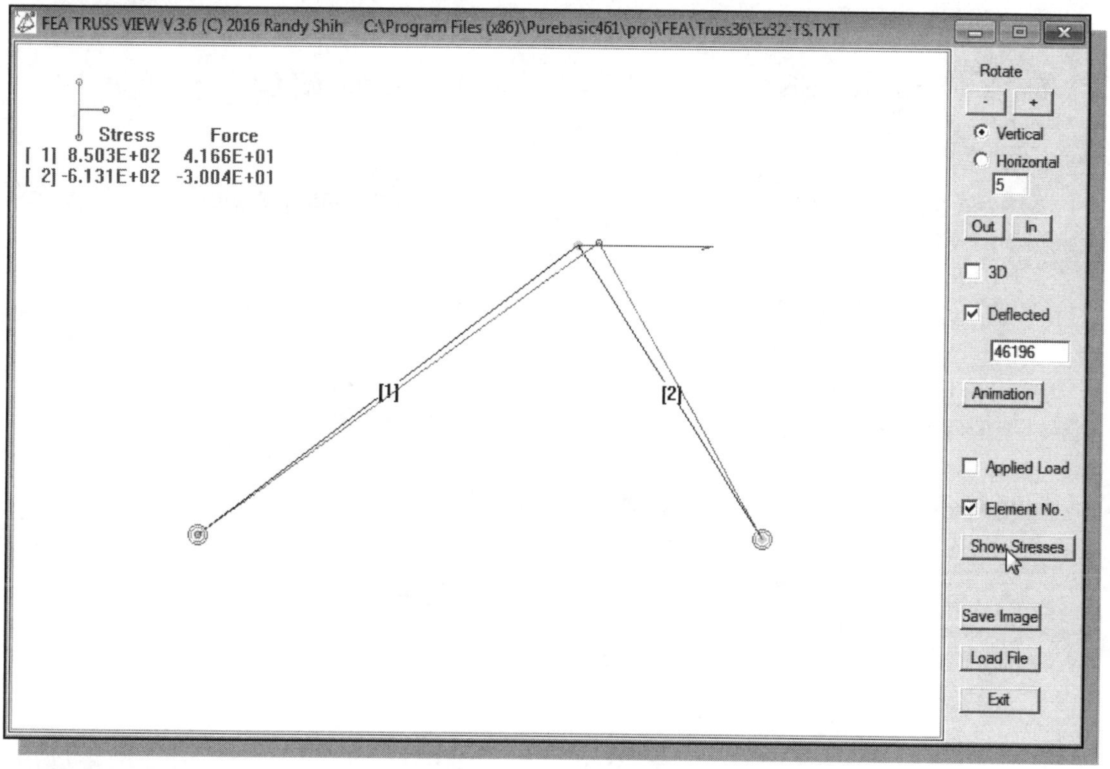

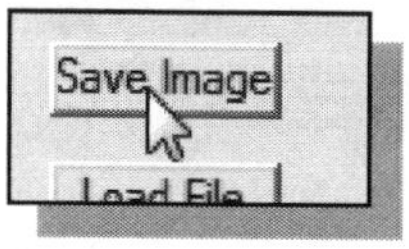

6. Click on the **Save Image** button to save a copy of the screen image as a jpg image file.

➢ The use of the *Truss View* program provides the user with a visual representation of the FEA results, which makes interpreting the FEA results much easier than reading the lengthy text based output file. The use of the *Truss Solver* and the *Truss View* programs also illustrates the basic three steps of the FEA procedure: (1) **Pre-processing**, where the user creates an input file that contains all of the system information; (2) **Solver**, the FEA procedure solving of the system; and (3) **Post-processing**, displaying and viewing the FEA results.

Review Questions

1. List and describe two of the commands in *MS Excel* to perform matrix operations that were used in the tutorial.

2. What is the difference between the **ATAN2** function and the **ATAN** function in *MS Excel*.

3. In *MS Excel*, the **trigonometric functions** require the associated angles to be in degrees or radians?

4. How do we convert from an angle measurement from radians to degrees?

5. What are the advantages of performing the direct stiffness matrix method to solve a 2D Truss problem using *MS Excel* versus using a calculator?

6. Using ***MS Excel***, determine the inverse matrix of matrix A:

$$\begin{pmatrix} 8 & 2 & 3 & 12 & 11 \\ 2 & 4 & 7 & 0.25 & 5 \\ 3 & 7 & 3 & 5 & 6 \\ 12 & 0.25 & 5 & 2 & 4 \\ 11 & 5 & 6 & 4 & 8 \end{pmatrix}$$

7. Using ***MS Excel***, determine the displacements of the following simultaneous equations:

$$\begin{Bmatrix} 25 \\ 13.25 \\ 18 \\ 19.25 \end{Bmatrix} = \begin{bmatrix} 8 & 2 & 3 & 12 \\ 2 & 4 & 7 & 0.25 \\ 3 & 7 & 3 & 5 \\ 12 & 0.25 & 5 & 2 \end{bmatrix} \begin{Bmatrix} X_1 \\ Y_1 \\ X_2 \\ Y_2 \end{Bmatrix}$$

Exercises

Solve the following problems using *MS Excel* and the *Truss Solver* program.

1. Given: two-dimensional truss structure as shown. (All joints are pin joints.)

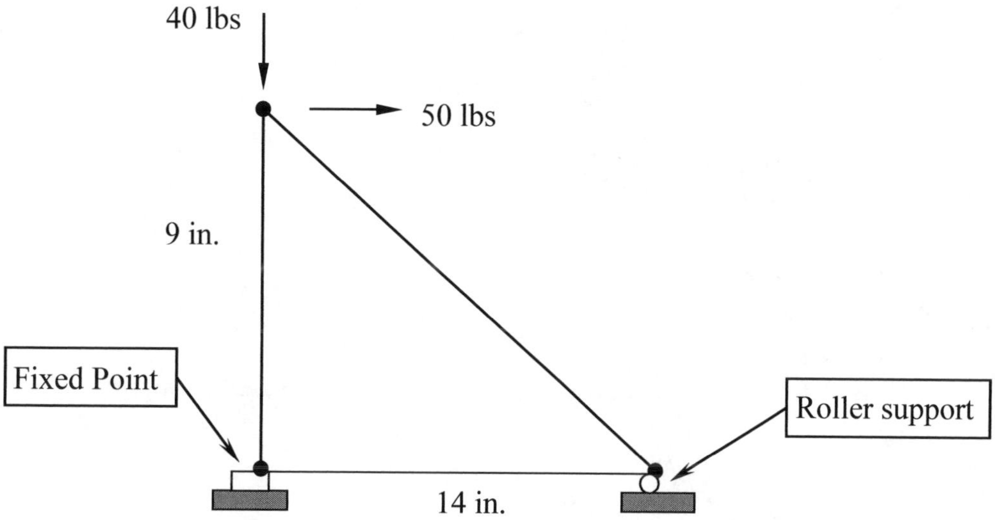

Material: Steel, diameter ¼ in.

Find: (a) Displacements of the nodes.
(b) Normal stresses developed in the members.

2. Given: Two-dimensional truss structure as shown. (All joints are pin joints.)

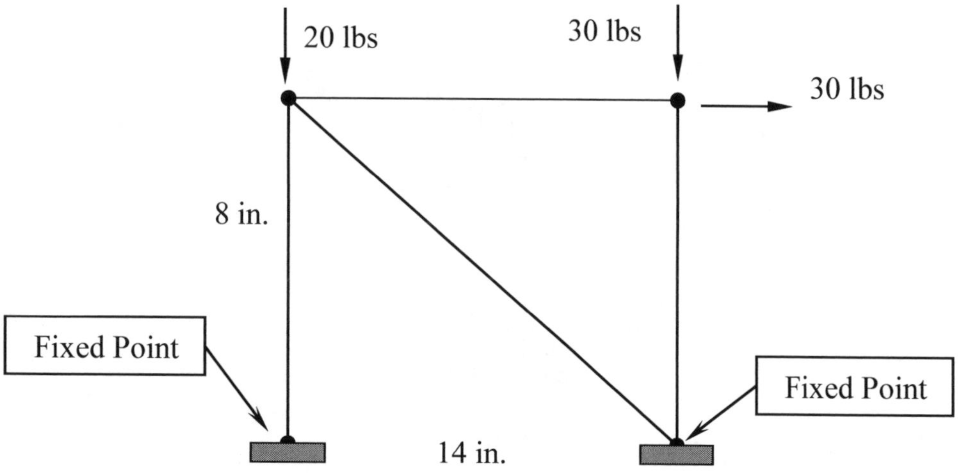

Material: Steel, diameter ¼ in.

Find: (a) Displacements of the nodes.
(b) Normal stresses developed in the members.

3. Given: two-dimensional truss structure as shown. (All joints are pin joints.)

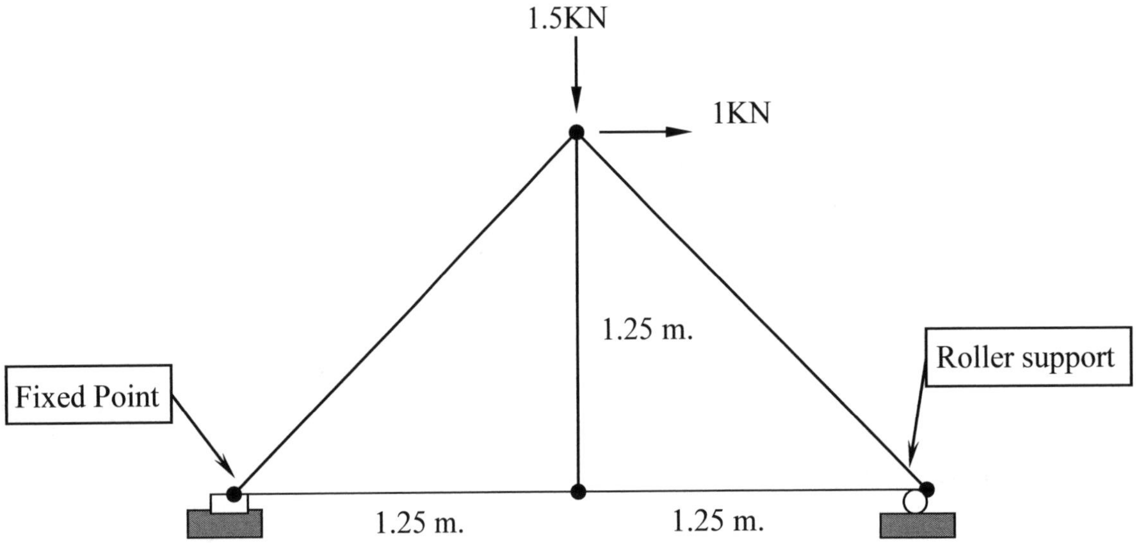

Material: Steel, diameter 25mm.

Find: (a) Displacements of the nodes.
(b) Normal stresses developed in the members.

Chapter 4
Creo Simulate Two-Dimensional Truss Analysis

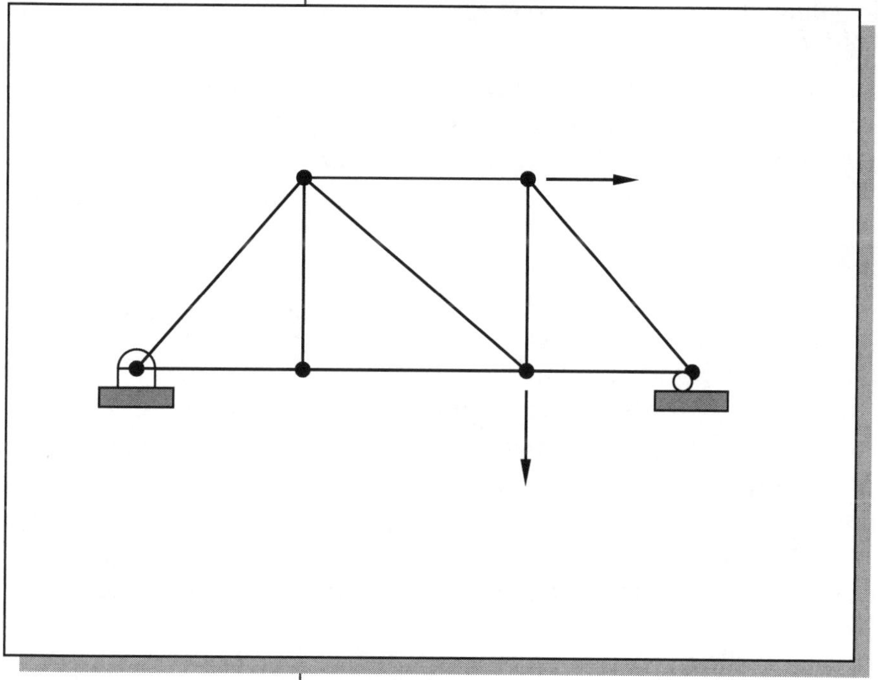

Learning Objectives

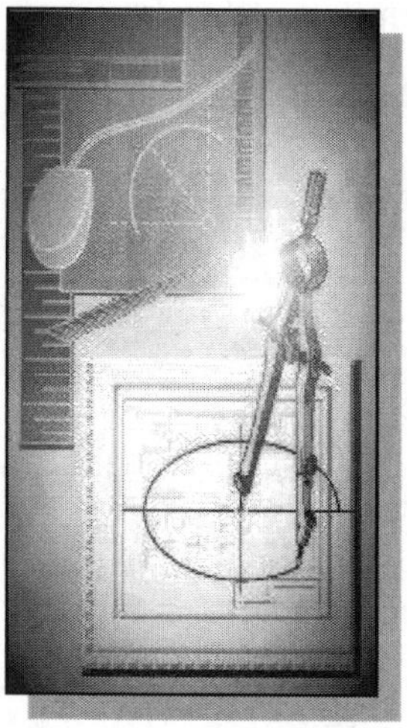

- Create Creo Simulate FE Truss Model.
- Create Nodes and Truss Elements in Creo Simulate.
- Apply Loads and Boundary conditions at Nodes.
- Run the Creo Simulate FEA Solver.
- View and Examine the Creo Simulate FEA Results.
- Understand the General Computer FEA Procedure.

Finite Element Analysis Procedure

While all real-life structures are three-dimensional in nature, solutions of many stress analyses are done on two-dimensional spaces and sometimes one-dimensional spaces. Quite often, the approximations represent the three-dimensional members very well and there is no need to do a three-dimensional analysis.

This chapter demonstrates the entire process of creating, solving, and viewing the results of a finite element analysis on a two-dimensional truss structure using the FEA application software, *Creo Simulate,* which is available in *Creo Parametric*. The following illustration follows the typical procedure of performing finite element analysis:

a. Preliminary analysis of the system:
   Perform an approximate calculation to gain some insights about the system.

b. Preparation of the finite element model:
   1. Prescribe the geometric and material information of the system.
   2. Prescribe how the system is supported.
   3. Determine how the loads are applied to the system.

c. Perform the calculations:
   Solve the system equations and compute displacements, strains and stresses.

d. Post-processing of the results:
   View the results of the FEA procedure and confirm the results of the analysis by comparing to the preliminary analysis.

➢ Before going through the tutorial, perform a preliminary analysis of the two-dimensional truss structure as shown below. First create a free body diagram of the entire structure to find the reactions at the supports. Then use either the classical *joint method* or the *section method* to find the forces in each of the members. Which member would you expect to have the highest stress? On your own, calculate the stress for the member you identified as the highest-stress member and compare to the computer solution at the end of this chapter.

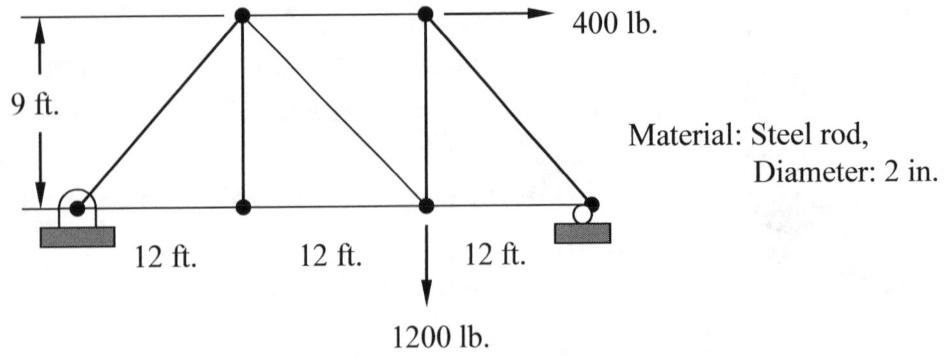

Preliminary Analysis

Determine the normal stress in each member of the truss structure shown.

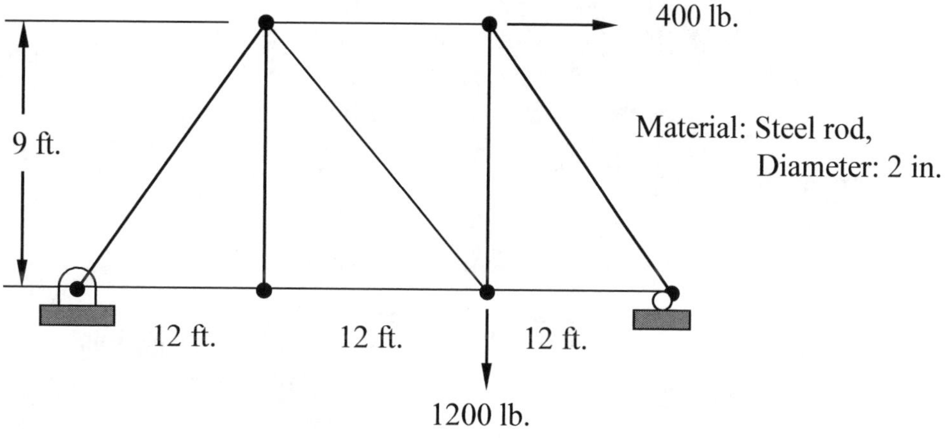

Prior to carrying out the finite element analysis, it is important to do an approximate preliminary analysis to gain some insights into the problem and as a means to check the finite element analysis results.

Free Body Diagram of the structure:

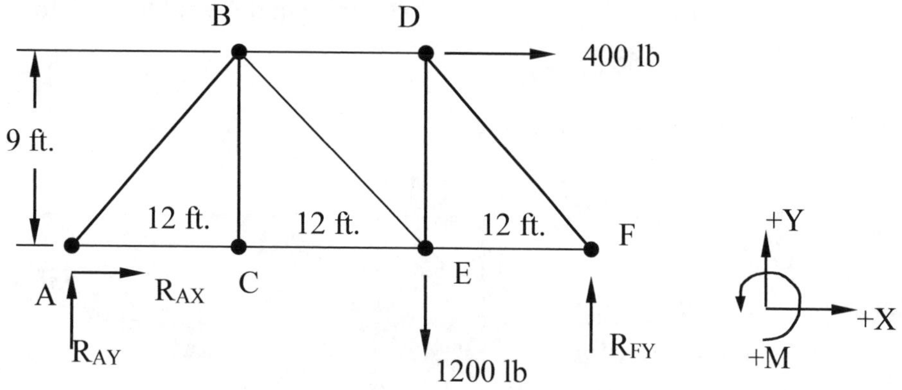

By inspection, member BC can be identified as a ZERO-FORCE member. Therefore, the stress in BC will be zero.

$\Sigma M_A = 36 \times R_{FY} - 24 \times 1200 - 9 \times 400 = 0$

Solving for R_{FY}:

$R_{FY} = 900$ lb.

Next, using the *Joint Method* (Conventional Statics analysis technique), solve for internal forces in members DF and EF.

FBD of point F:

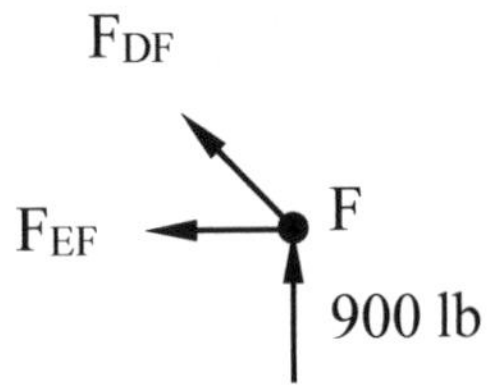

$$\Sigma F_X = -F_{EF} - F_{DF} \times \frac{4}{5} = 0$$

$$\Sigma F_Y = 900 + F_{DF} \times \frac{3}{5} = 0$$

Solving the two simultaneous equations with two unknowns:

F_{DF} = **- 1500 lb**.; therefore, normal stress $\sigma_{DF} = -1500/\pi =$ **-477.5 psi**

F_{EF} = **1200 lb**.; therefore, normal stress $\sigma_{EF} = 1200/\pi =$ **382 psi**

➢ It is not necessary to solve the entire problem by hand. We will compare the three results we have calculated so far to the computer solution in the following sections.

Starting Creo Parametric

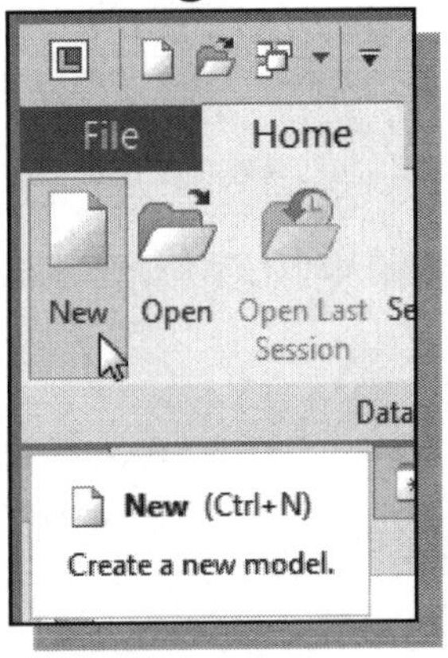

1. Select the **Creo Parametric** option on the *Start* menu or select the **Creo Parametric** icon on the desktop to start *Creo Parametric*. The *Creo Parametric* main window will appear on the screen.

2. Click on the **New** icon, located in the *Ribbon toolbar* as shown.

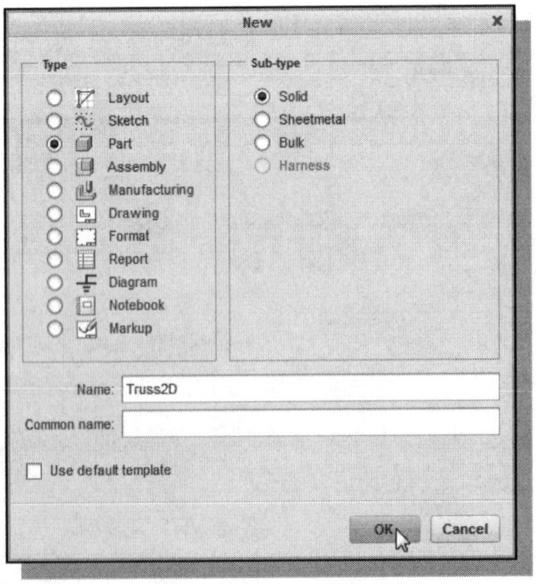

3. In the *New* dialog box, confirm the model's Type is set to **Part** (**Solid** Sub-type).

4. Enter **Truss2D** as the part Name as shown in the figure.

5. Turn ***off*** the **Use default template** option.

6. Click on the **OK** button to accept the settings.

7. In the *New File Options* dialog box, select **Empty** in the option list to not use any template file.

8. Click on the **OK** button to accept the settings and enter the *Creo Parametric Part Modeling* mode.

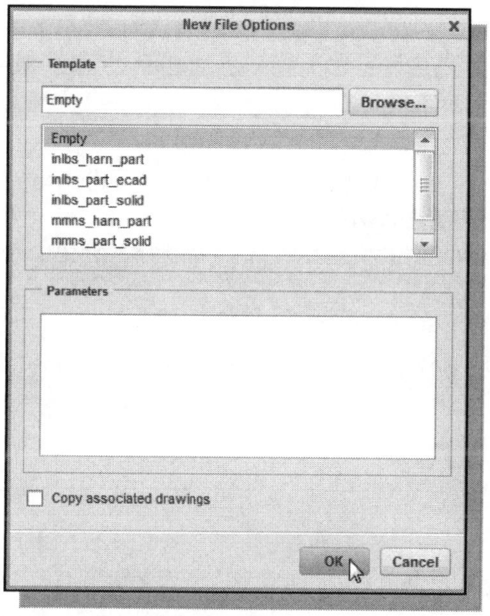

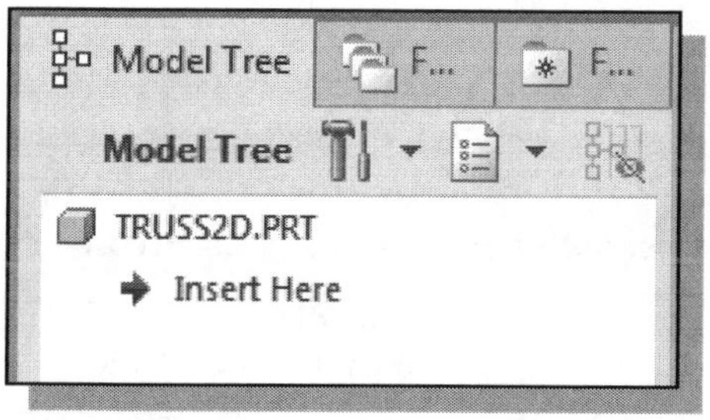

❖ Note that the part name, *Truss2D,* appears in the title area of the main window and in the *Navigator Model Tree* window.

Units and Basic Datum Geometry Setups

♦ **Units Setup**

When starting a new model, the first thing we should do is to choose the set of units we want to use.

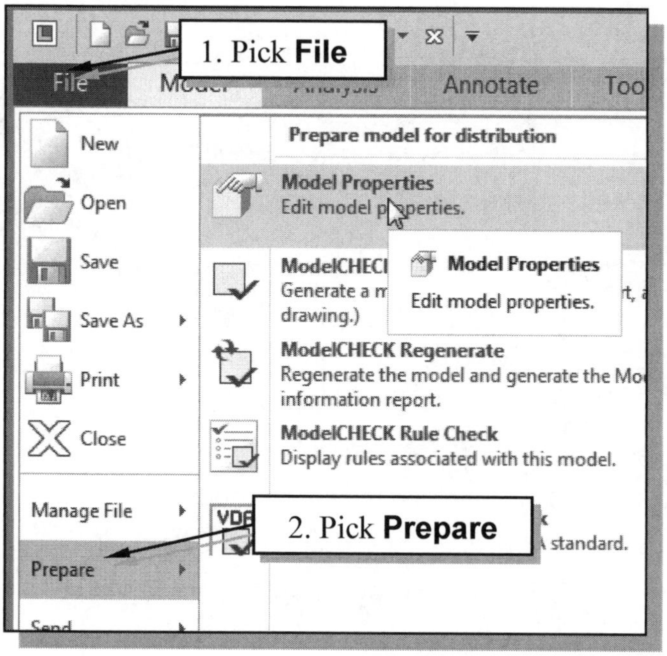

1. Use the left-mouse-button and select the **File** pull-down menu.

2. Use the left-mouse-button and select **Prepare** in the **pull-down list** as shown.

3. Select **Model Properties** in the *expanded list* as shown.

➢ Note that the *Creo Parametric* menu system is context-sensitive, which means that the menu items and icons of the non-applicable options are grayed out (temporarily disabled).

4. Select the **Change** option that is to the right of the **Units** option in the ***Model Properties*** window.

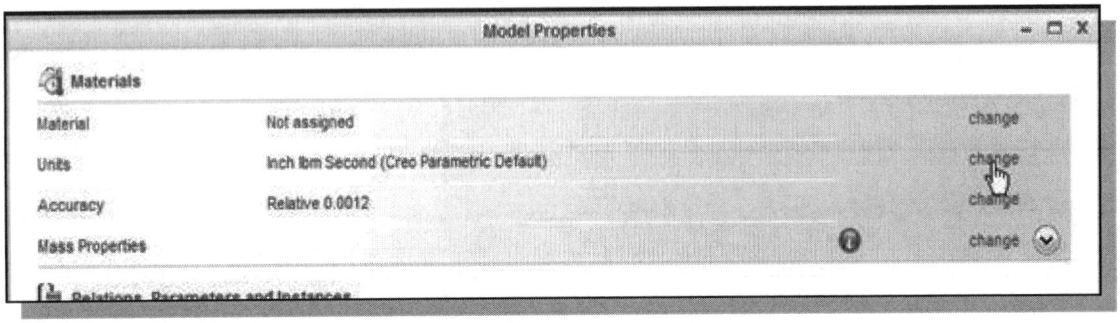

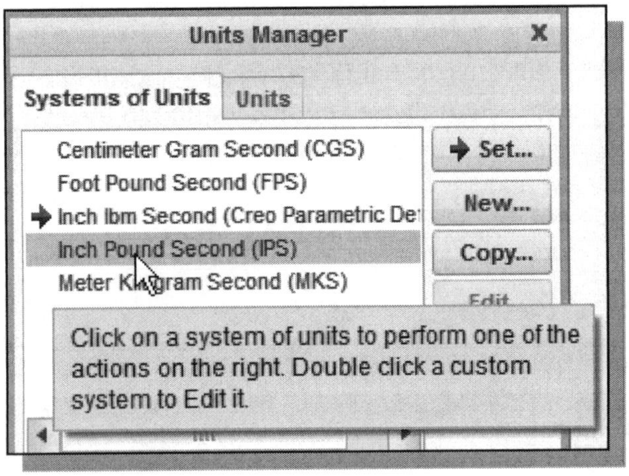

5. In the **Units Manager – Systems of Units** form, the *Creo Parametric* default setting **Inch lbm Second** is displayed. The set of units is stored with the model file when you save. Pick **Inch Pound Second (IPS)** by clicking in the list window as shown.

6. Click on the **Set** button to accept the selection.

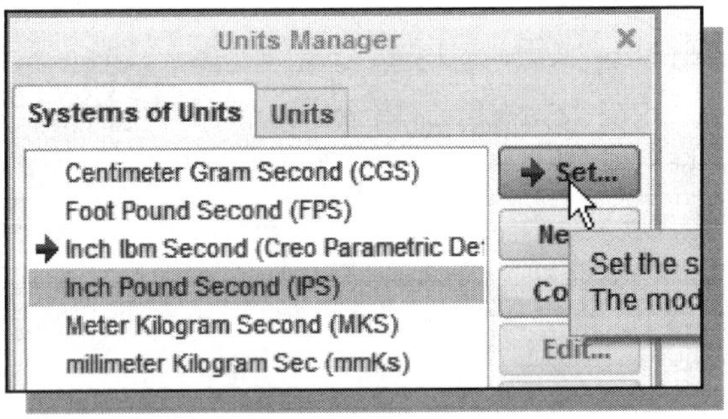

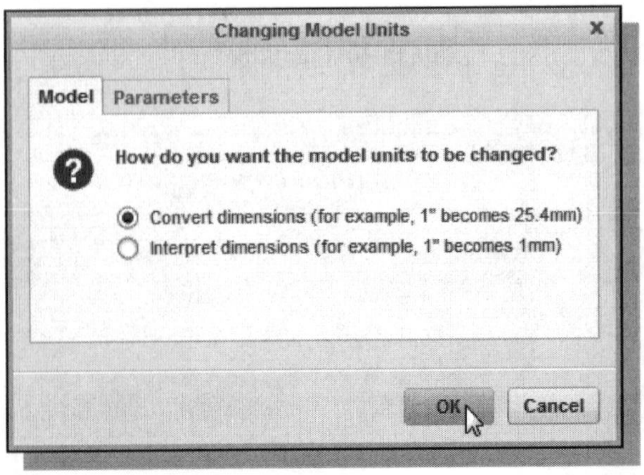

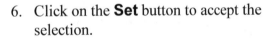

7. In the *Changing Model Units* dialog box, click on the **OK** button to accept the default option to change of the units.

8. Click on the **Close** button to exit the *Units Manager* dialog box.

9. Pick **Close** to exit the *Model Properties* window.

♦ Adding the First Part Features – Datum Planes

❖ *Creo Parametric* provides many powerful tools for model creation. In doing feature-based parametric modeling, it is a good practice to establish three reference planes to locate the part in space. The reference planes can be used as location references in feature constructions.

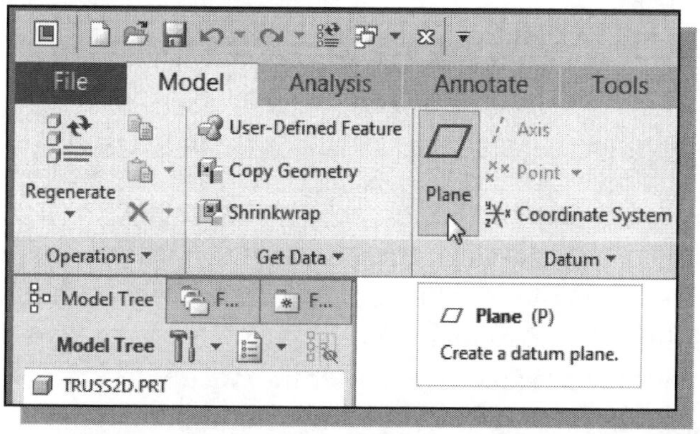

1. Move the cursor to the *Datum* toolbar in the *Ribbon* toolbar area and click on the **Datum Plane** tool icon as shown.

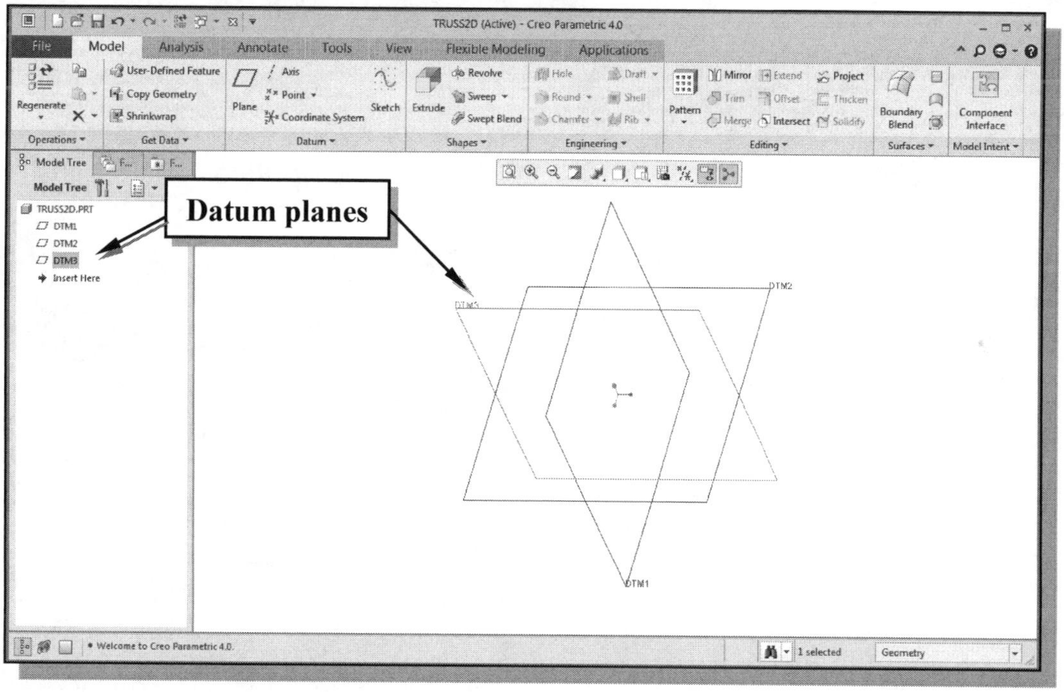

❖ In the *Navigator Model Tree* window and the display area, three datum planes represented by three rectangles are displayed. Datum planes are infinite planes and they are perpendicular to each other. We can consider these planes as XY, YZ, and ZX planes of a Cartesian coordinate system.

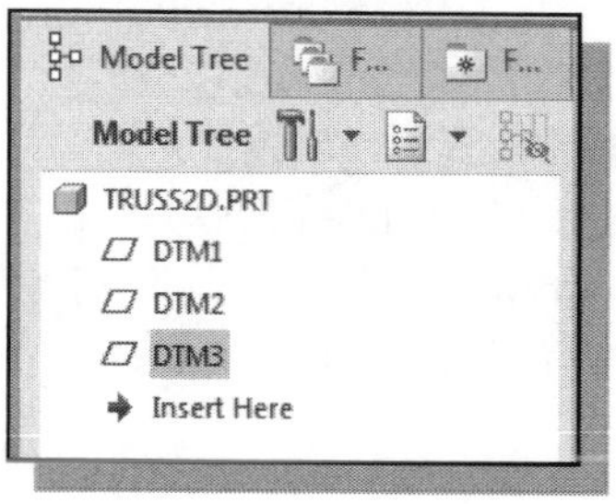

❖ Note that setting up the three mutually perpendicular datum planes provides a starting point to allow us to switch to the integrated mode of *Creo Simulate*. We will create the FEA Truss model directly inside *Creo Simulate*.

The Integrated Mode of Creo Simulate

Creo Simulate is a multi-discipline Computer Aided Engineering (CAE) tool that enables the user to simulate the physical behavior of a model, and therefore enable the user to improve the design. *Creo Simulate* can be used to predict how a design will behave in the real world by calculating stresses, deflections, frequencies, heat transfer paths, etc.

The *Creo Simulate* product line features three modules: ***Structure***, ***Thermal***, and ***Motion***. *Structure* focuses on the structural integrity of the design, *Thermal* evaluates heat-transfer characteristics, and *Motion* provides the kinematics details of moving parts in the design.

Creo Simulate is available in two modes: *Integrated* mode and *Independent* mode. In *Integrated* mode, all of the *Creo Simulate* functions are integrated as part of *Creo Parametric*. This version of *Creo Simulate* offers the convenience and power of *Creo Parametric's* parametric modeling capabilities. *Creo Simulate* is also available as an independent package (the *Independent* mode), where a separate user interface (UI) is used. Models can be imported or created using *Creo Simulate's* somewhat limited built-in FEA-geometry-creation commands.

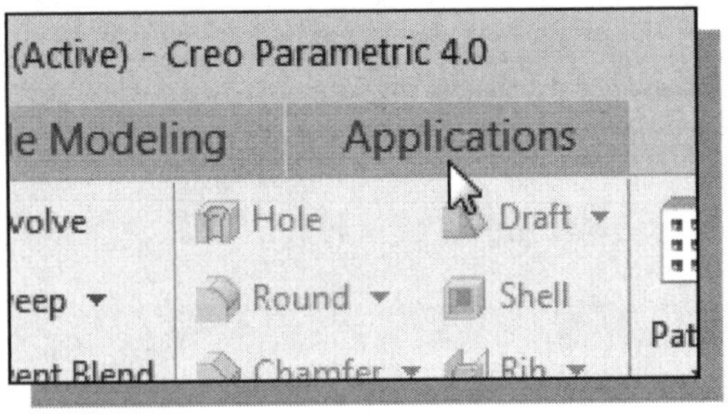

1. Start the *Integrated* mode of *Creo Simulate* by selecting the **Applications** tab as shown.

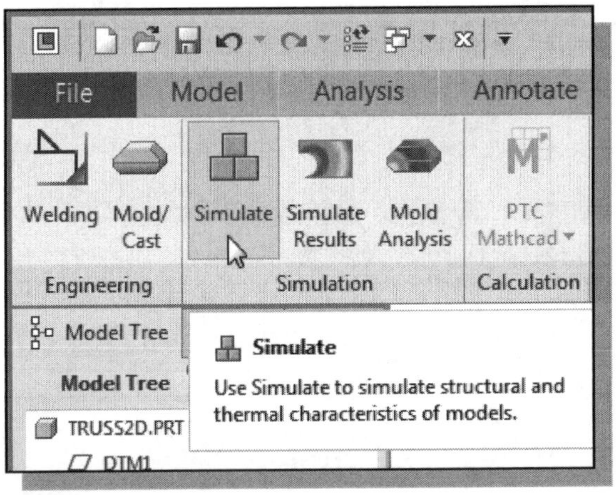

2. Click **Simulate** to activate the integrated *Creo Simulate* mode as shown.

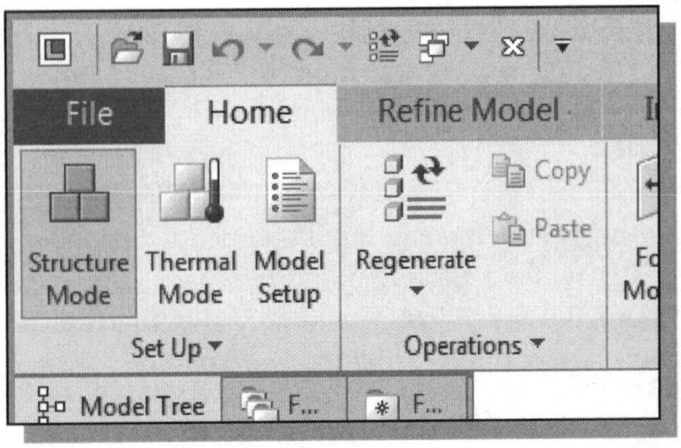

3. In the *Simulate Ribbon*, ***Structure*** mode is the default analysis type.

❖ Note that in integrated mode, all of the *Creo Simulate* functions are integrated as part of *Creo Parametric*. All of the *Creo Simulate* commands are accessible through the *Ribbon* toolbar area.

Create Datum Points as FEA Nodes

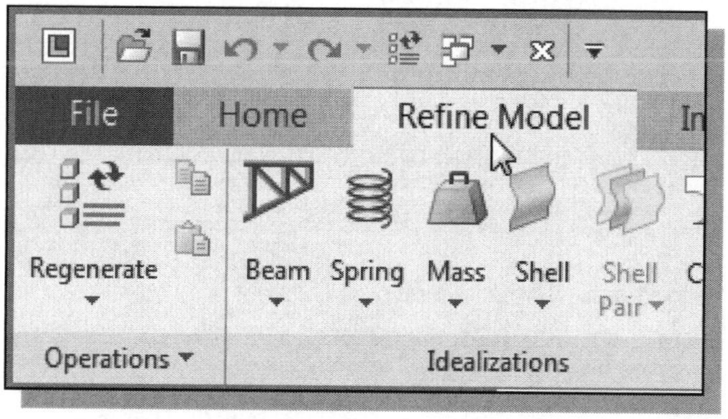

1. Click on the **Refine Model** tab to view the available commands.

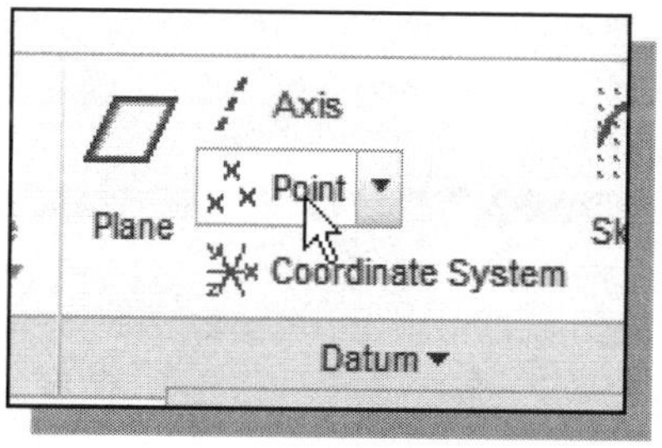

2. Choose **Datum Point** tool in the icon panel displayed on the *Ribbon* of the *Creo Parametric* main screen.

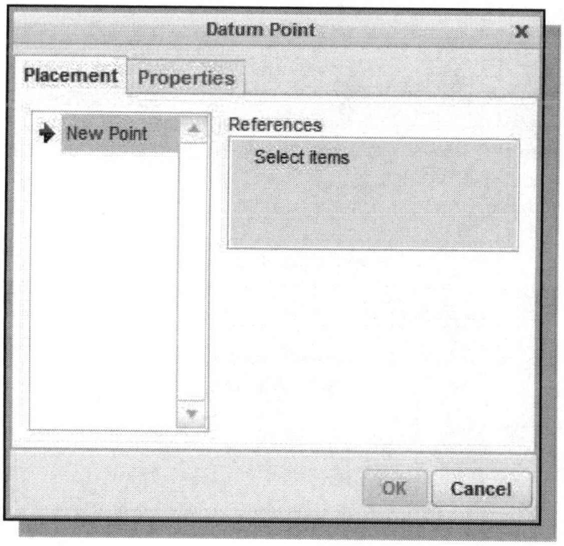

❖ In the ***Datum Point*** window, there is currently no datum point in the Placement list. We can create points by using existing objects as References.

3. Inside the *graphics area*, select **DTM3** by clicking on one of the edges as shown.

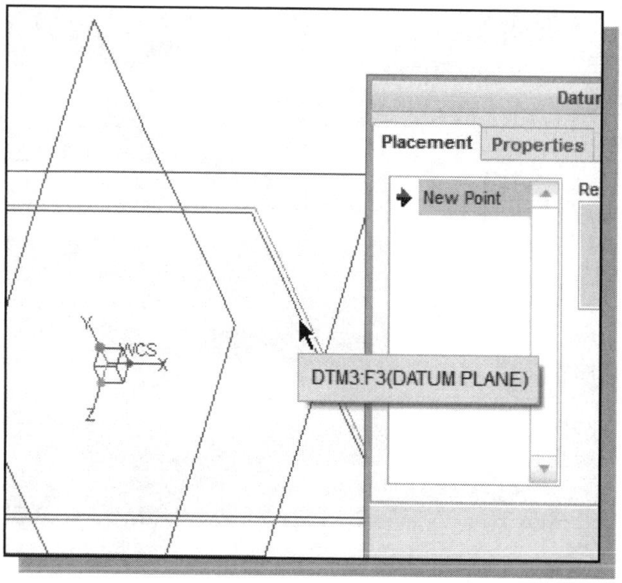

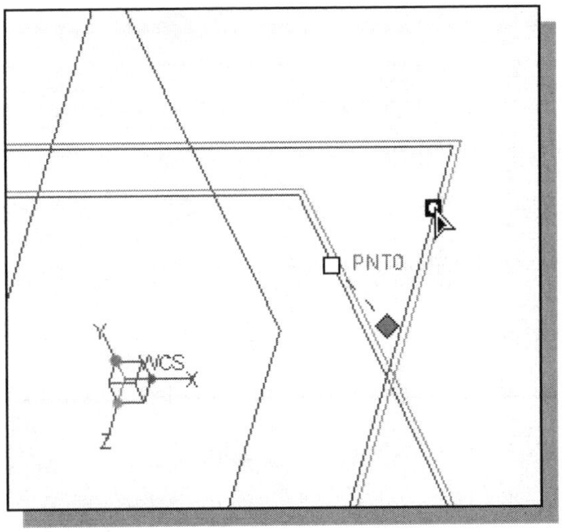

4. Grab one of the handle points and place it on datum plane **DTM2**, as shown. DTM2 is now being used as an Offset reference to position the point.

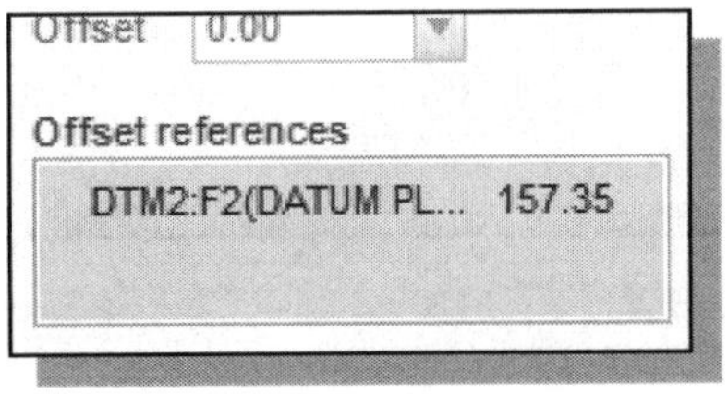

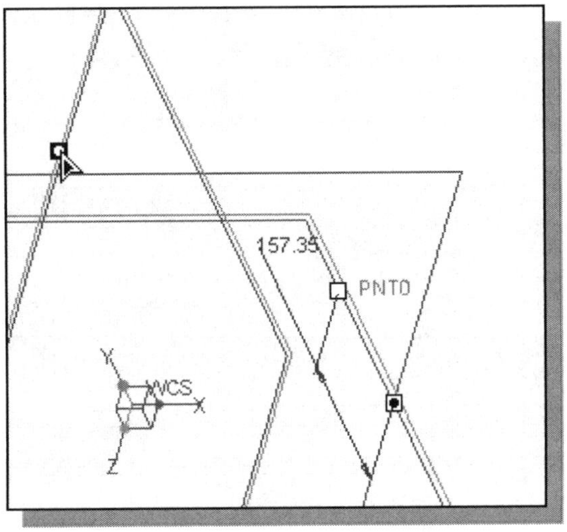

5. Repeat the above step and place the other handle point on **DTM1**.

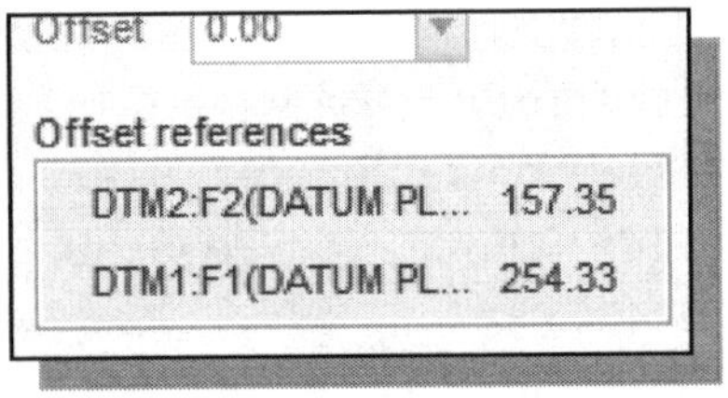

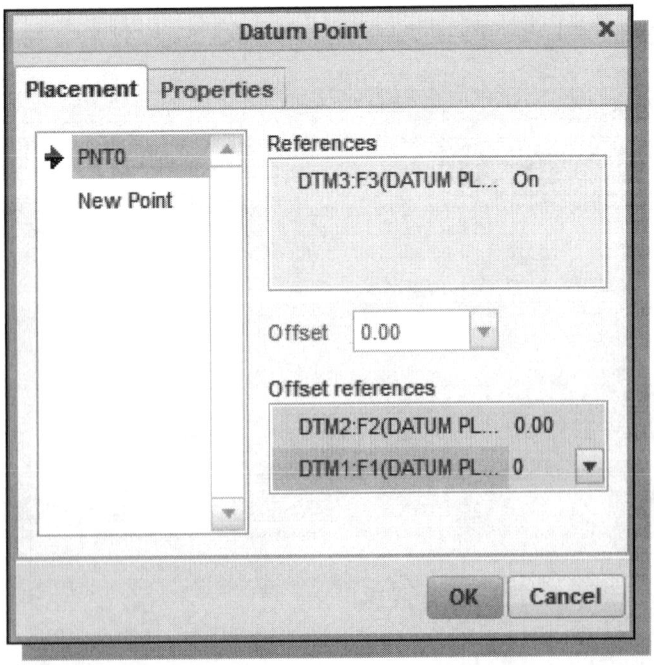

6. Click on one of the Offset values inside the **References** option box, in the *Datum Point* window, and change the value to **0.0** as shown.

7. Click on the second value and also adjust it to **0.0** as shown.

❖ We have created a datum point located exactly at the origin of the coordinate system.

8. In the *Datum Point* window, create a new datum point by clicking on the **New Point** option as shown.

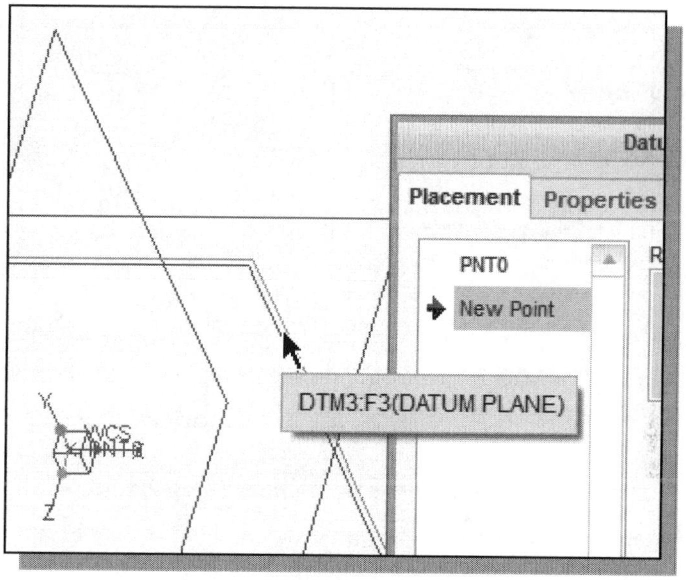

9. Inside the graphics area, select **DTM3** by clicking on one of the edges as shown.

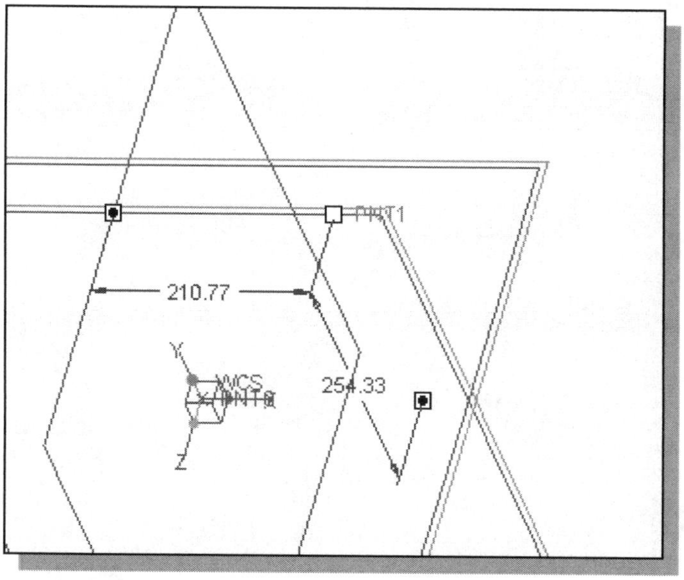

10. Grab the handle points and place them on the two datum planes, **DTM1** and **DTM2**, as shown.

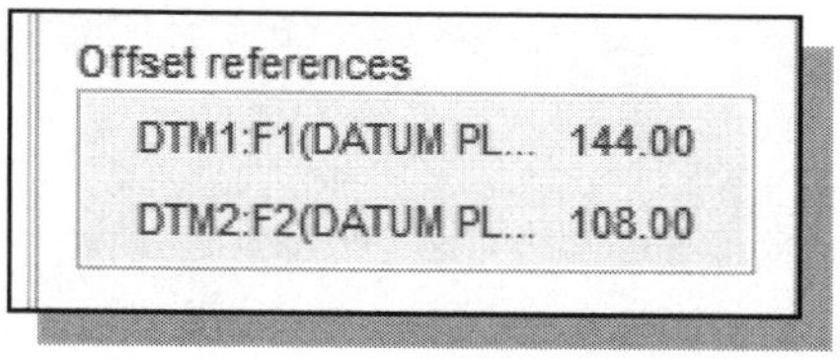

11. Adjust the Offset values of DTM1 to **144** and DTM2 to **108** as shown.

12. In the *Datum Point* window, create a new datum point by clicking on the New Point option.

13. On your own, create the additional four datum points as shown. Note the bottom four points are aligned to the horizontal reference.

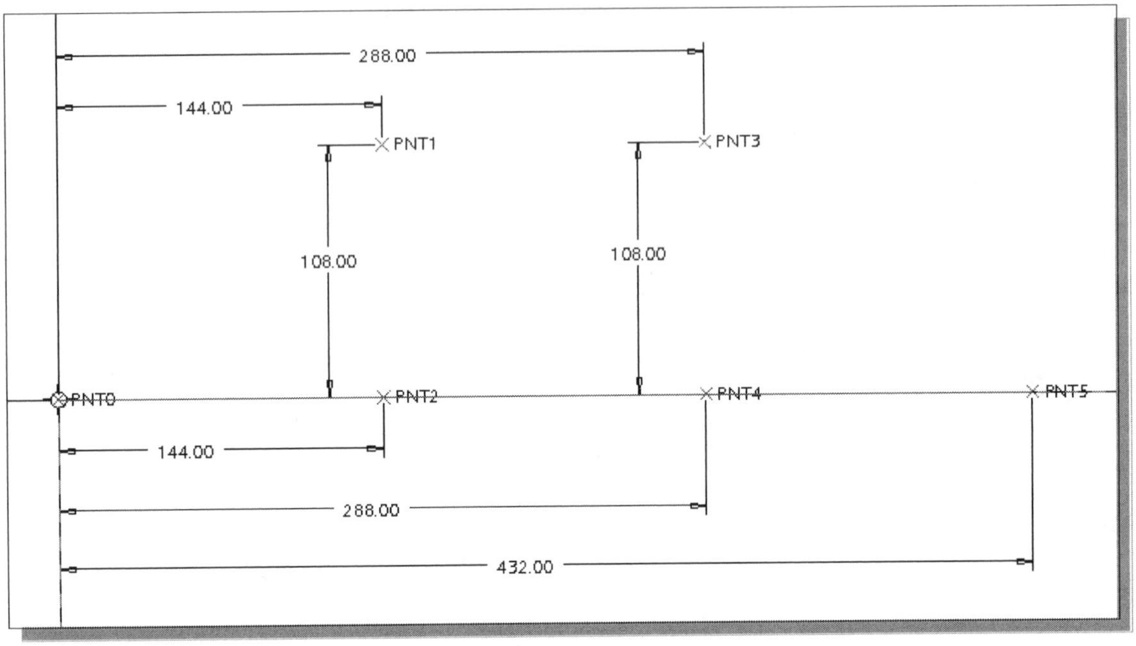

14. Click on the **OK** button to accept the creation of the datum points.

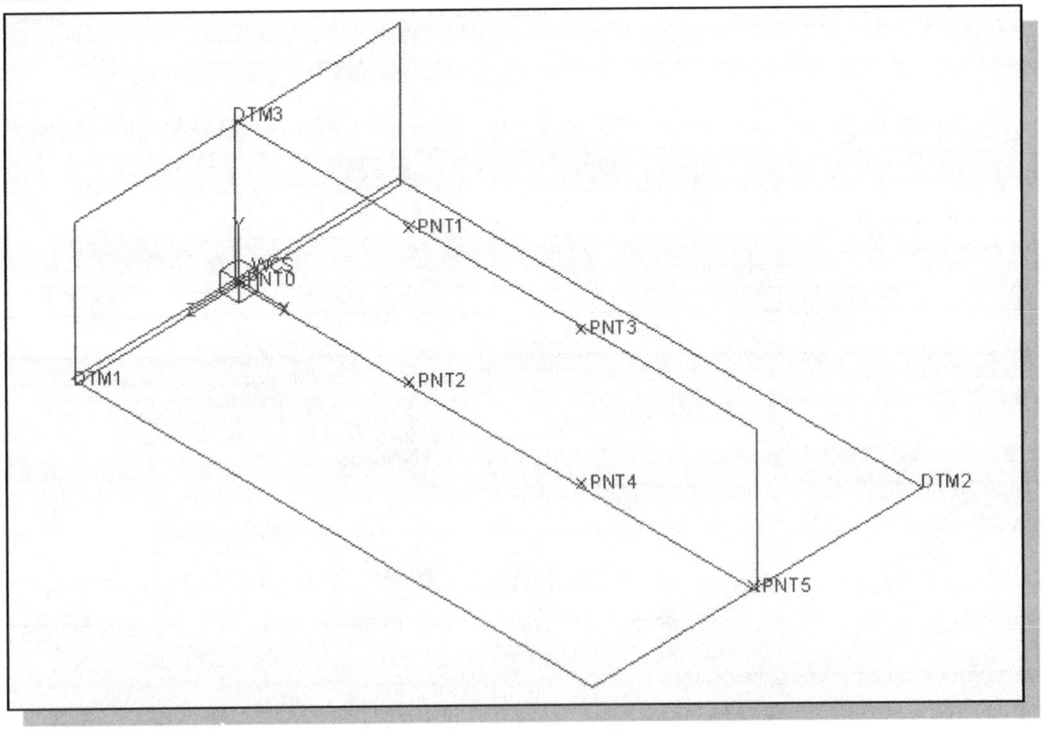

❖ Note the datum points created are labeled as **PNT0, PNT1, PNT2, PNT3, PNT4** and **PNT5** which correspond to the order they were created.

Set up an Element Cross Section

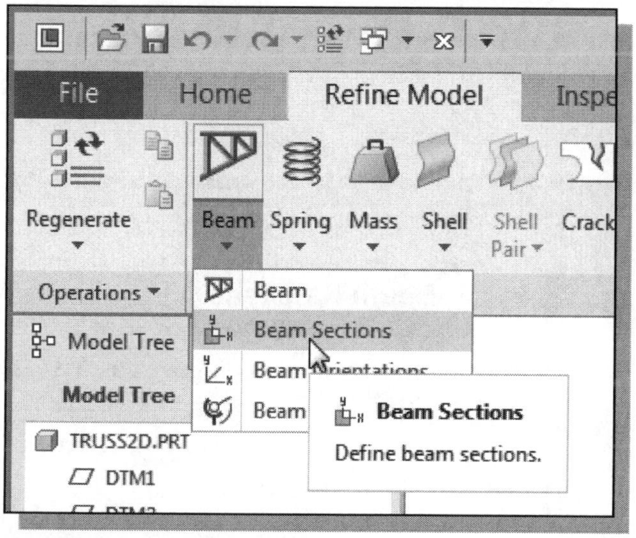

1. Choose **Beam Sections** in the **Refine Model** tab as shown.

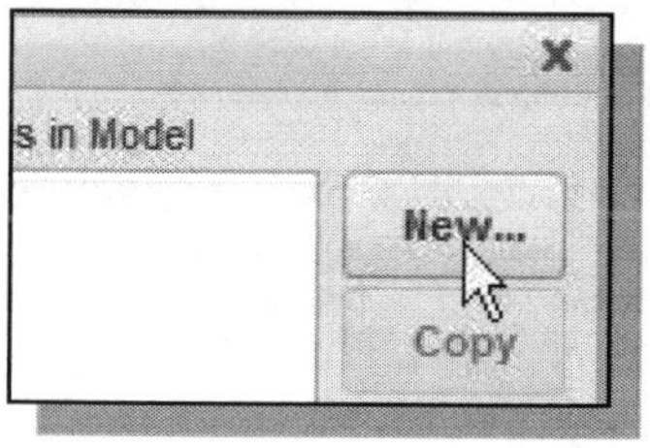

2. Click **New** to create a new beam cross-section.

3. Enter **CircularRod** as the Name of the new beam section.

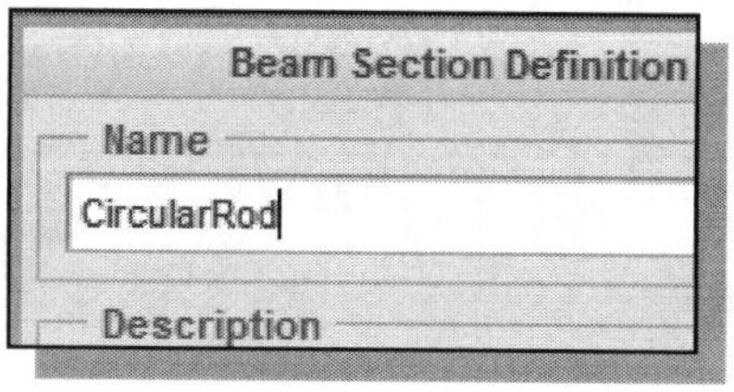

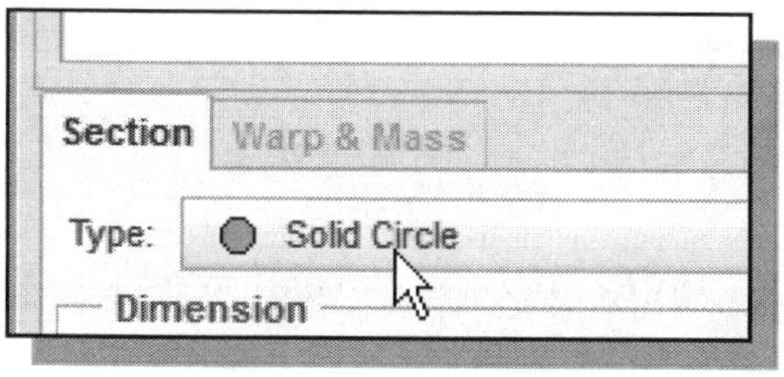

4. Choose **Solid Circle** as the cross-section type as shown.

5. Enter **1.0** as the radius of the circular beam cross section.

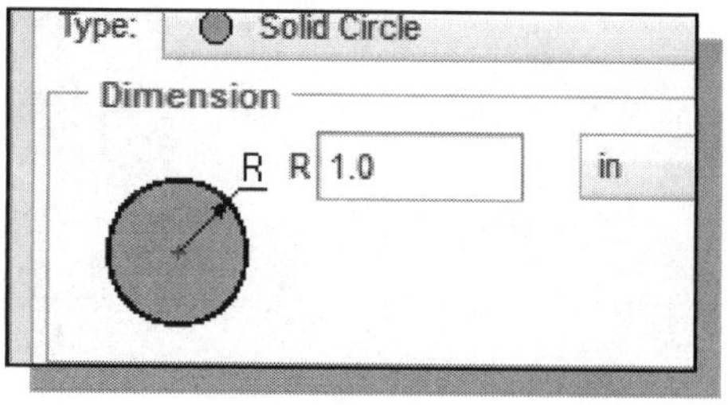

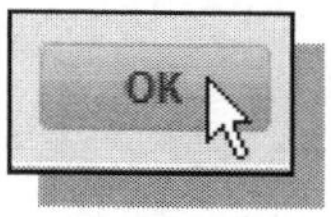

6. Click **OK** to accept the creation of the new beam section.

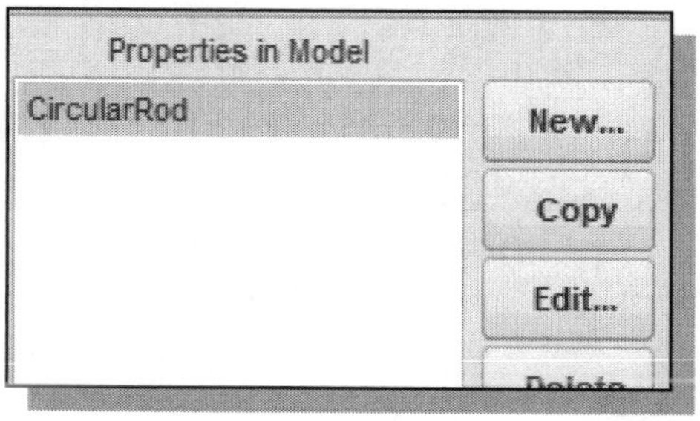

7. Click **Close** to accept the creation of the new beam section.

Set up Beam Element Releases

In *Creo Simulate*, the *truss element* is considered a special type of *beam element*. This is done by releasing the rotational degrees of freedom of the beam element.

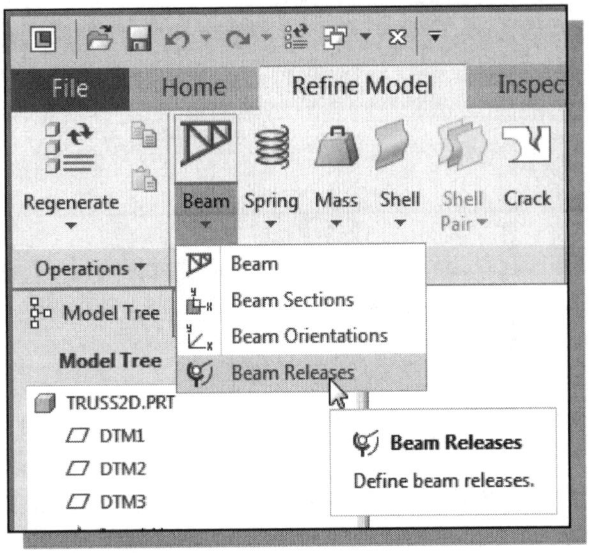

1. Choose **Beam Releases** from the *Idealizations* toolbar as shown.

❖ The **Beam Releases** command can be used to specify the degrees of freedom for a beam end or ends. By default, *Creo Simulate* assumes that all degrees of freedom are locked at the ends of the beam.

2. Click **New** to create a new beam cross-section.

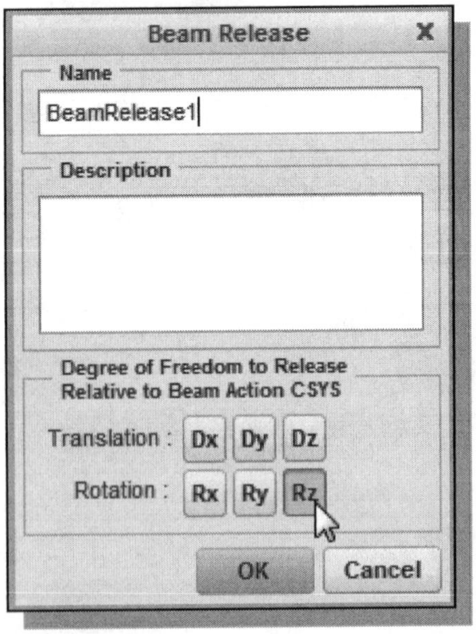

3. Click on the **Rz** icon to allow free rotation about the Z axis.

❖ A 2D truss member is free to rotate about the perpendicular axis of the truss structure.

4. Click **OK** to accept the definition of the new beam release.

5. Click **Close** to accept the creation of the new beam release.

Select and Examine the Element Material Property

Before creating the FEA elements, we will first set up the *Material Property* for the elements. The *Material Property* contains the general material information, such as *Modulus of Elasticity*, *Poisson's Ratio*, etc., that is necessary for the FEA analysis.

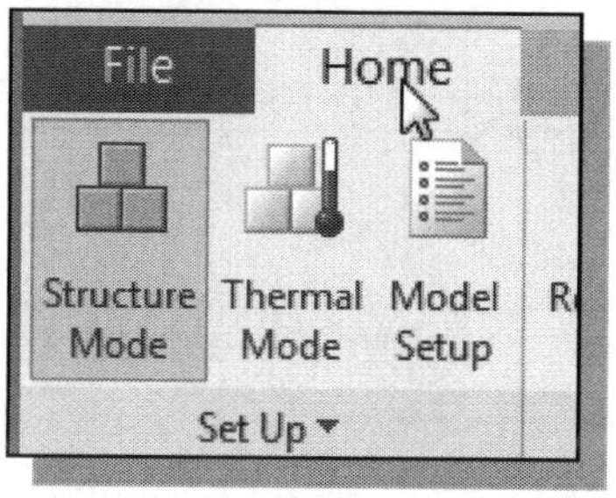

1. Click on the **Home** tab to return to the main *Creo Simulate* commands toolbar.

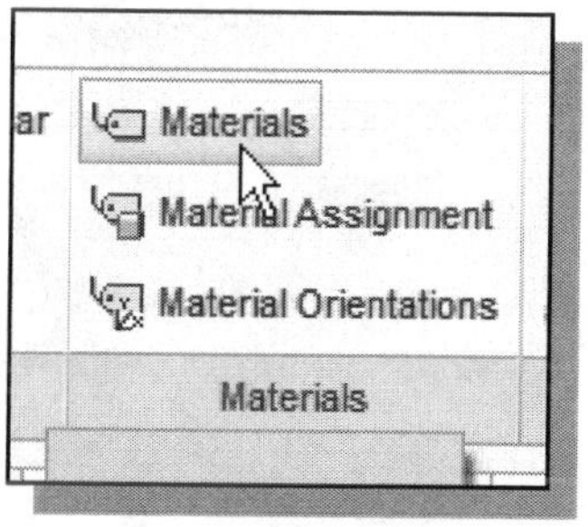

2. Choose **Materials** in the *Materials* group from the *Ribbon* toolbar as shown.

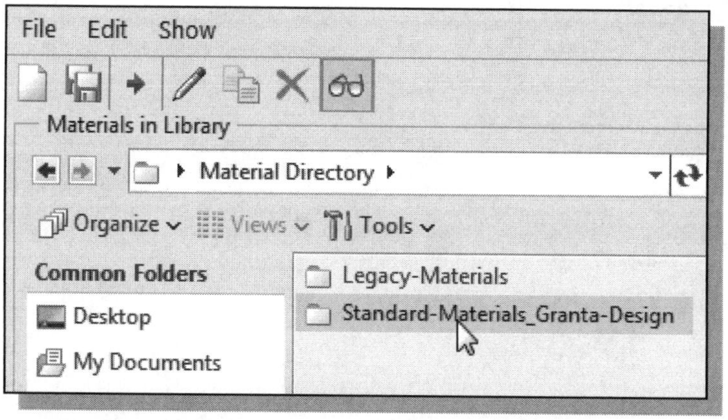

3. Select the *Standard-Materials_Granta-Design* library by **double-clicking** in the list area. The material library comes with many of the commonly used materials. Note that new material properties can also be added and all material properties can be edited.

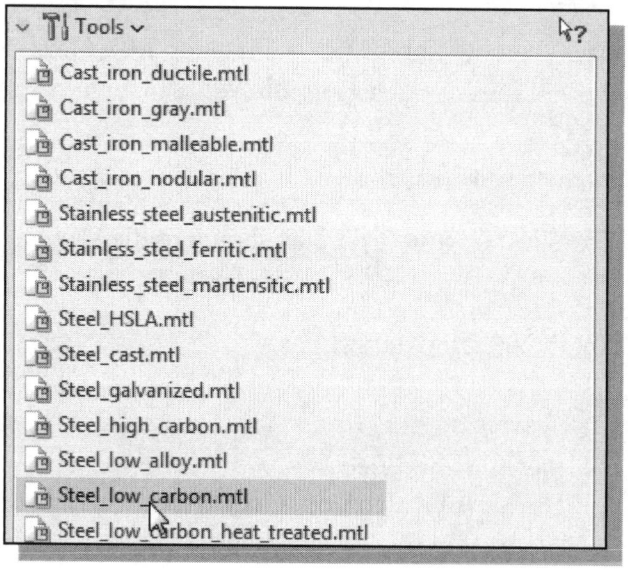

4. Choose **Ferrous-metals → Steel_low_Carbon** from the material library as shown.

5. In the *Materials* dialog box, click the **Edit** icon as shown.

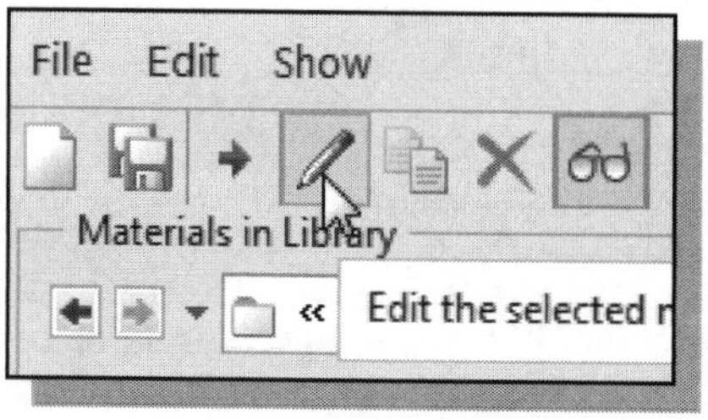

6. In the *Material Definition* dialog box, the material properties of **Steel_low_Carbon** are displayed and can be edited.

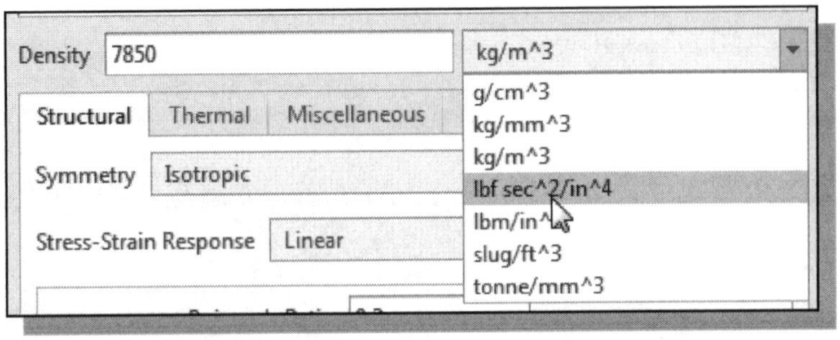

7. Click on the down arrow and choose **lbf sec^2/in^4** as the new Density **Units** as shown.

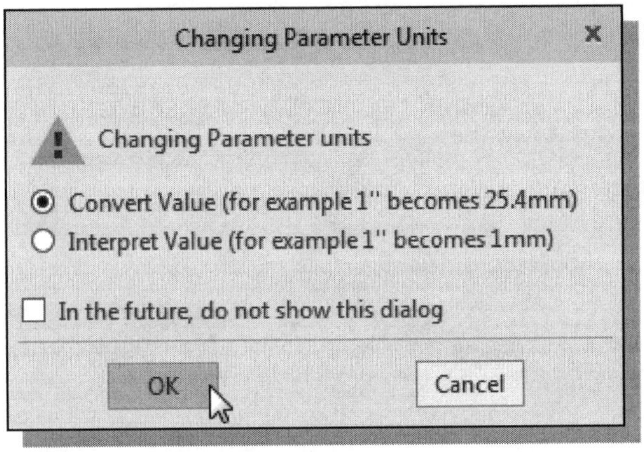

8. In the Changing Parameter Units dialog box, select Convert Value as shown.

9. Click **OK** to confirm the change settings as shown.

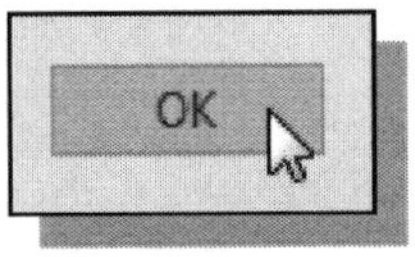

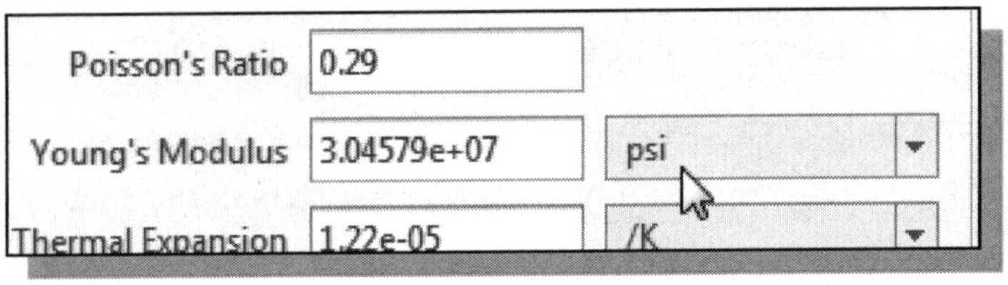

10. Set the Units of Young's Modulus to **psi** as shown.

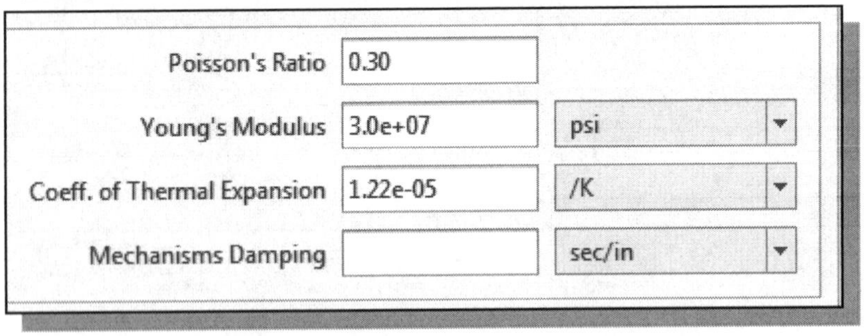

11. Click the value box of Young's Modulus and enter **3.0E+7** as shown.

12. Repeat the above steps and adjust the Poisson's Ratio to **0.3** as shown in the above figure.

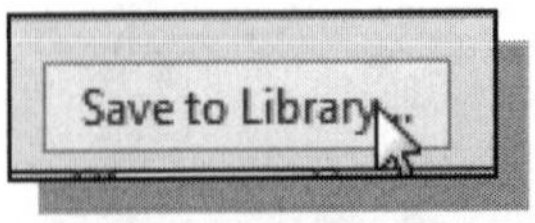

13. Click **OK** to accept the material definitions.

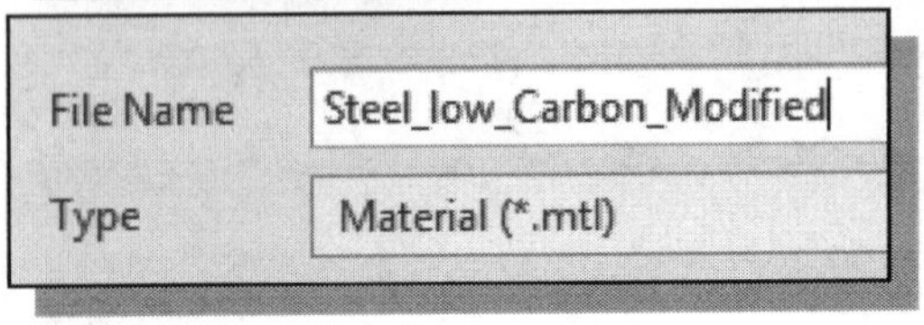

14. Save the modified *Material Definition* as **Steel_low_Carbon_Modified.**

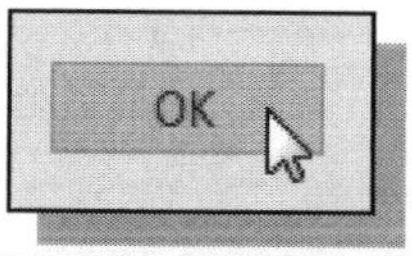

15. Click **OK** to accept the settings and save the modified material definition.

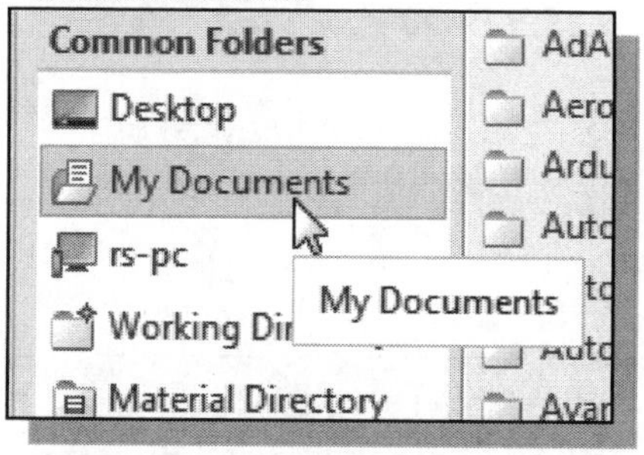

16. On your own, locate the folder where we just saved the modified material definition.

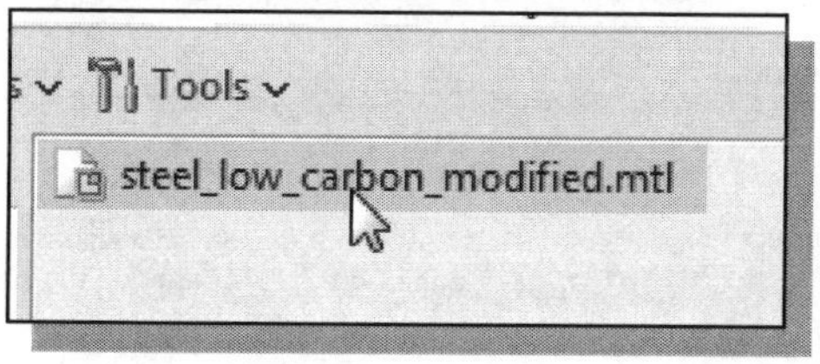

17. **Double-click** on the modified material definition.

18. In the *Message dialog box*, click **Yes** to convert all units in the selected material to match the units set for the model.

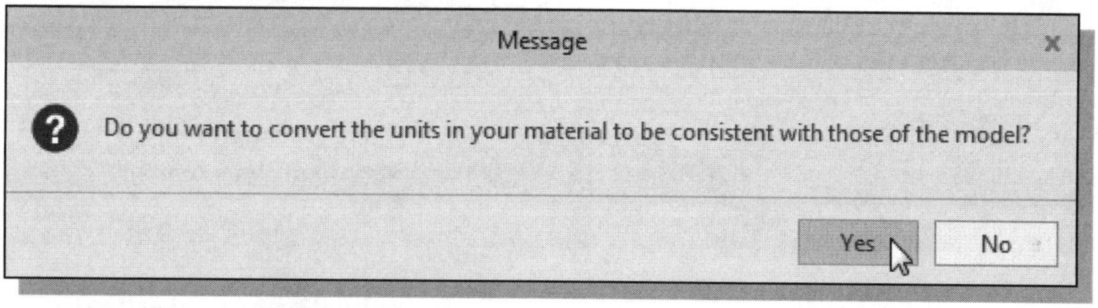

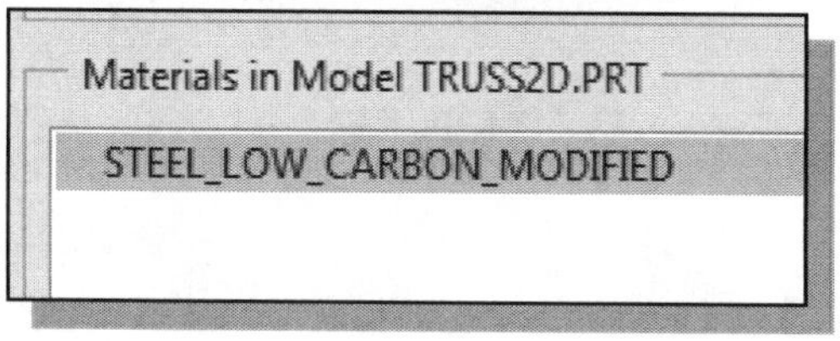

19. Notice the selected *Material definition* appears in the **Materials Box** available for use with the model.

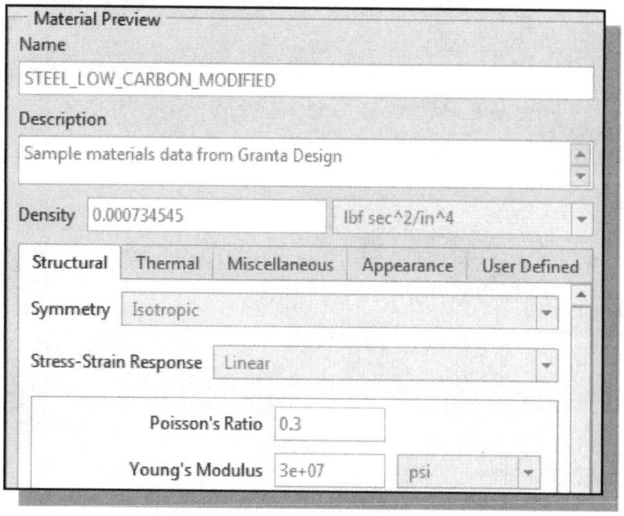

20. Examine the material information in the *Material Preview* area and click **OK** to accept the settings.

Create Elements

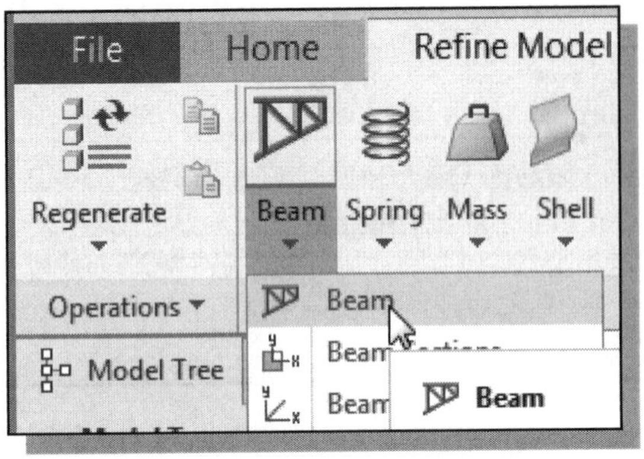

1. Click on the **Refine Model** tab to view the available commands.

2. Choose **Beam** in the icon panel. The *Beam Definition* window appears.

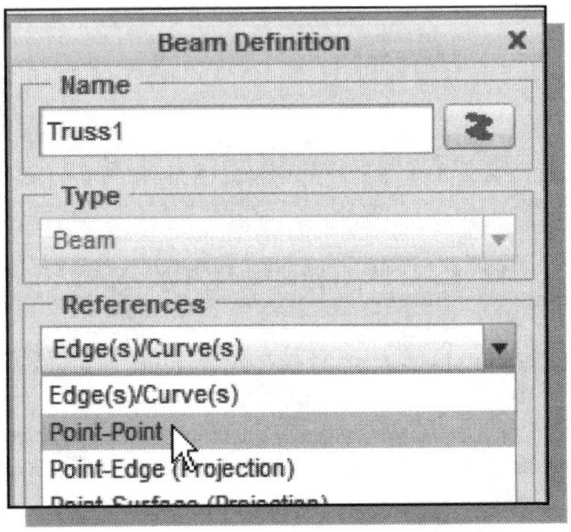

3. Enter **Truss1** as the Name of the first element to be created.

4. In the References option list, select **Point-Point** as shown.

5. Set the two Point options to **Single**.

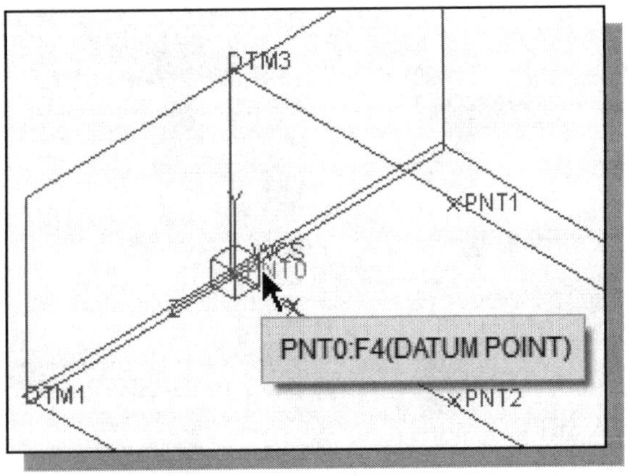

6. Click on the first datum point, **PNT0**, as the first point of the element.

7. Choose **PTN1** as the second point of the element.

❖ Note the selections are displayed in the *Beam Definition* window.

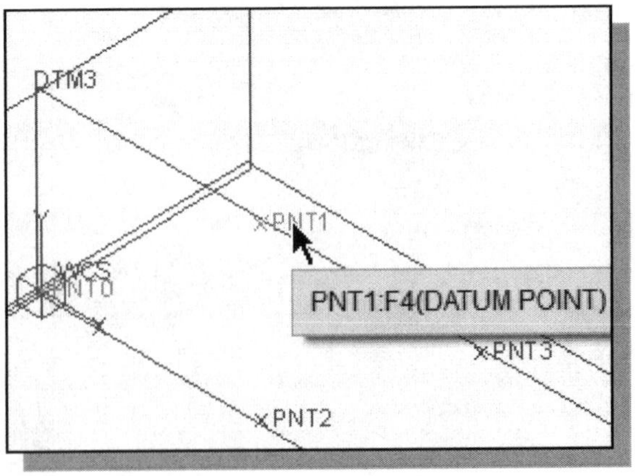

8. Select **Steel_low_Carbon_Modified** as the material of the element.

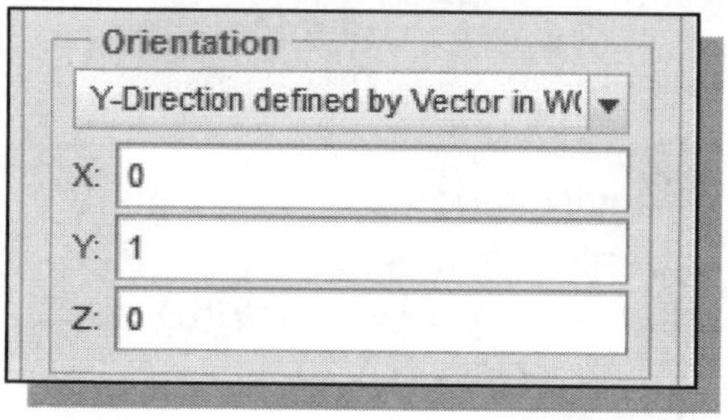

- In the **Properties** section, *Simulate* expects us to identify the Y-Direction of the beam. This option allows us to set the orientation of the beam in 3D space. The Y direction is referring to using the *Beam Action Coordinate System*.

Beam Action Coordinate System (BACS)

In the use of *beam elements*, it is necessary to set their orientations relative to the *World Coordinate System* (**WCS**). For the most general case, a beam orientation can require setting up three beam coordinate systems: the ***Beam Action Coordinate System (BACS)***, the ***Beam Shape Coordinate System (BSCS)***, and the ***Beam Centroidal Principal Coordinate System (BCPCS)***. However, for standard shapes that are built-in in *Creo Simulate*, the setting of the BACS relative to the WCS is generally sufficient and the procedure is very straightforward.

The origin of the BACS is identified by the order of selection of the datum points. The X axis of the BACS is always set to be parallel to the axis of the beam. Note that this concept is similar to the derivation of the 2D truss elements using the direct stiffness matrix method, as described in Chapter 2. The orientation of the BACS can be controlled by setting the Y direction relative to the WCS.

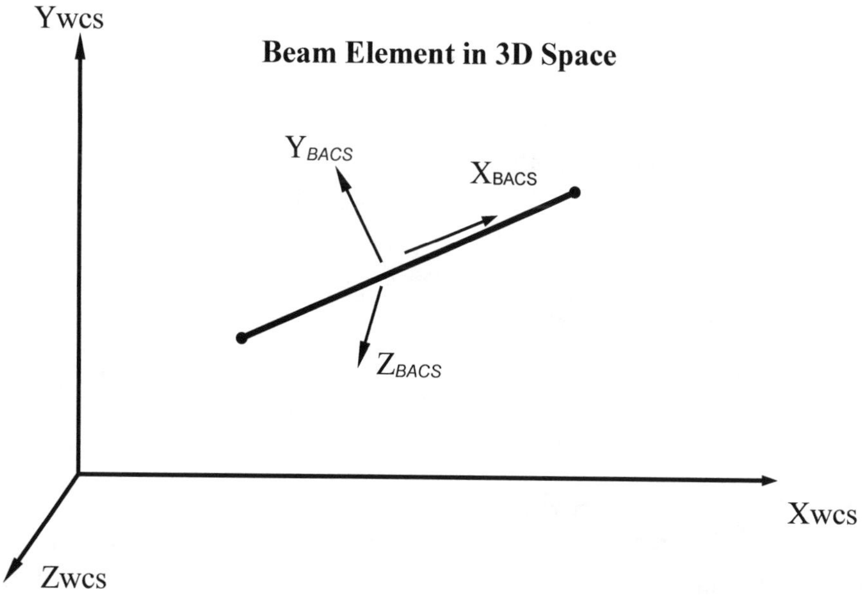

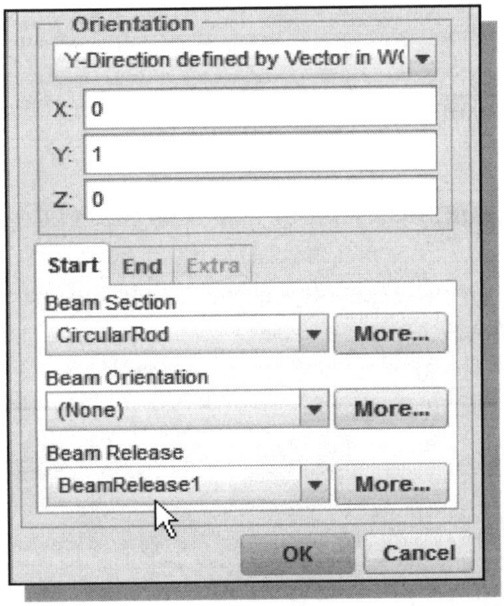

1. Confirm the Y-Direction is set to **WCS** with **0,1,0** as the X, Y and Z directions.

2. Set the *Beam Release* to **BeamRelease1** as shown.

3. Confirm the other two **Start** options are set as shown: Beam Section to **CircularRod** and Beam Orientation set to **(None)** as shown.

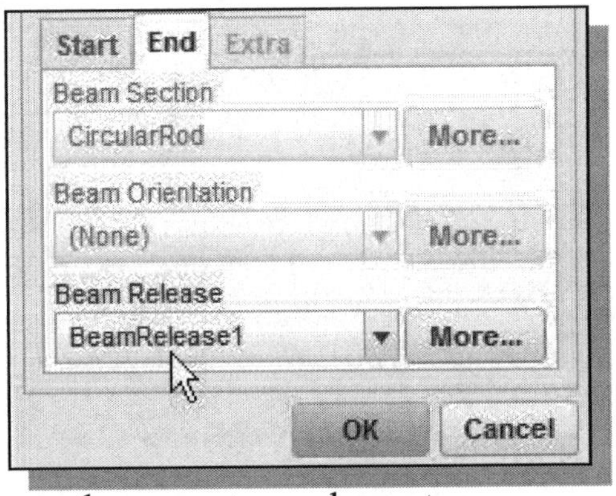

4. Click on the **End** tab and set the Beam Release to **BeamRelease1** as shown.

5. Click on the **OK** button to create the first *Beam Element* of the structure.

6. On your own, repeat the above procedure, and create the seven truss elements as shown. Note the BACS is automatically adjusted.

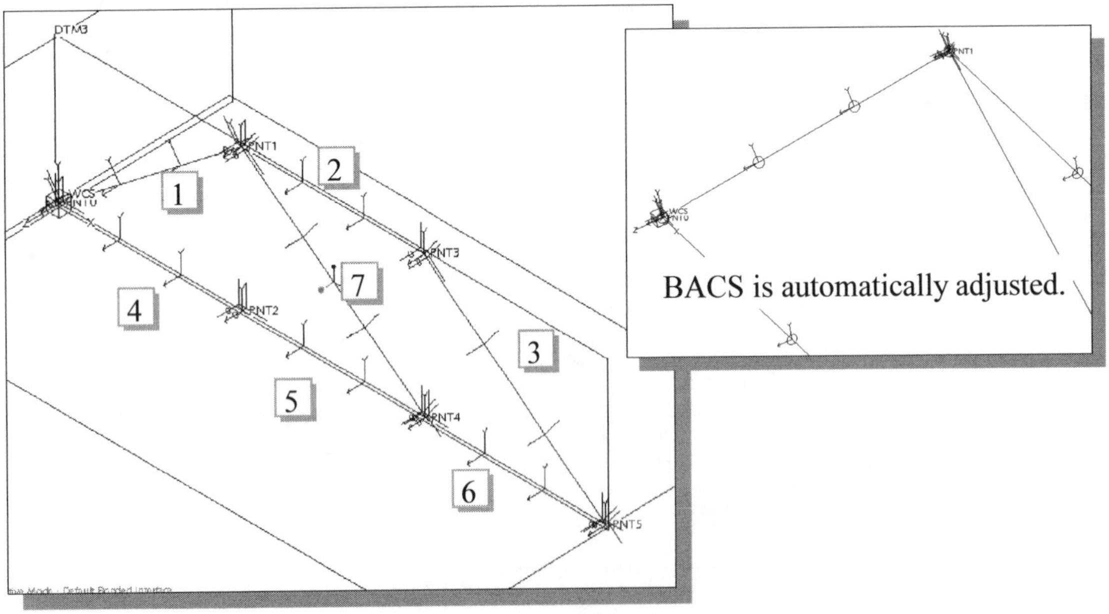

❖ Be sure to set the **Material** and **Beam Release** options on all the elements correctly. Missing any one of these settings will prevent the *Solver* from finding the correct FEA results.

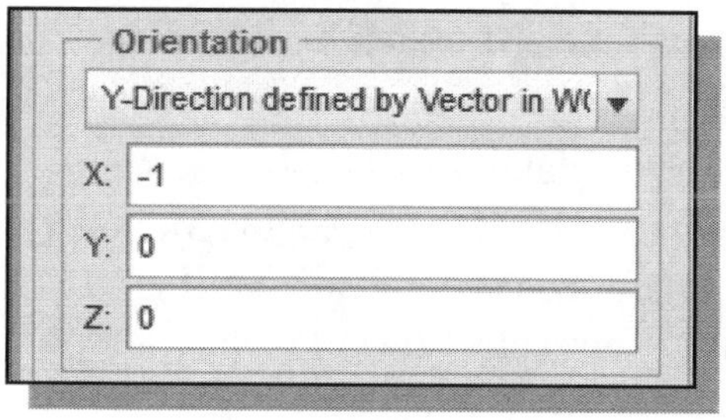

7. For the two remaining vertical elements, repeat the above procedure, but set the **Y Direction** to use **WCS** with **-1,0,0** as the **X**, **Y** and **Z** components as shown.

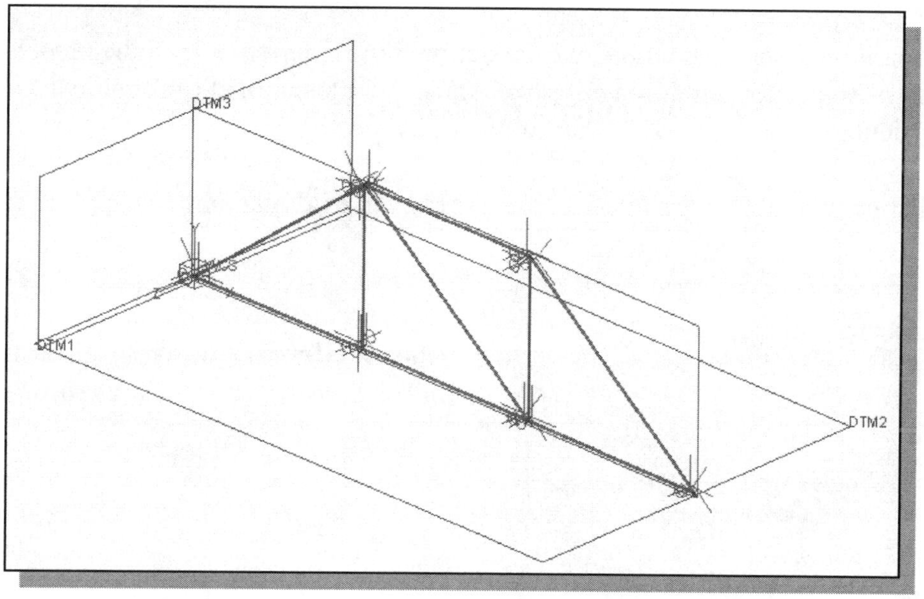

❖ Note that the BACS X direction of a truss member is always aligned to the member direction, and the Y direction is always set at 90° from the X direction.

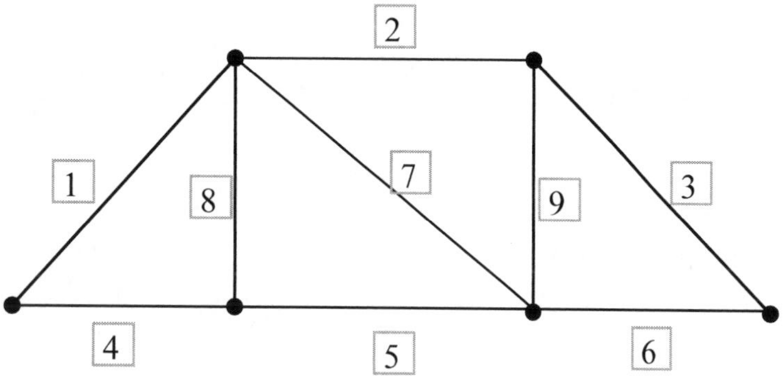

➢ For clarity purposes, we will refer to the different elements by the labels shown in the figure above.

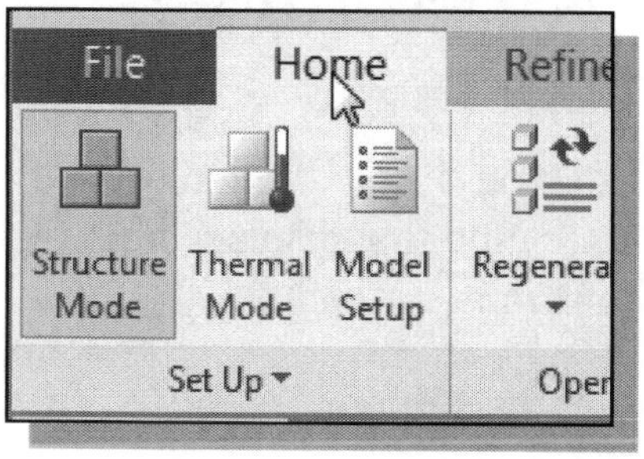

8. Click on the **Home** tab to return to the main *Creo Simulate* commands toolbar.

Apply Boundary Conditions – Constraints and Loads

In *Simulate* FEA analysis, the real-world conditions are simulated through the use of Constraints and Loads. In defining constraints for a *Simulate* model, the goal is to model how the real-life system is supported. Loads are also generalized and idealized. It is important to note that the computer models are idealized; our selections of different types of supports and loads will affect the FEA results.

In constraining a *Simulate* model, we are defining the extent to which the model can move in reference to a coordinate system. Thus, the translational or rotational movements of a particular support should be carefully considered.

Note that *truss members* are two-force members; the loads can only be applied at the nodes.

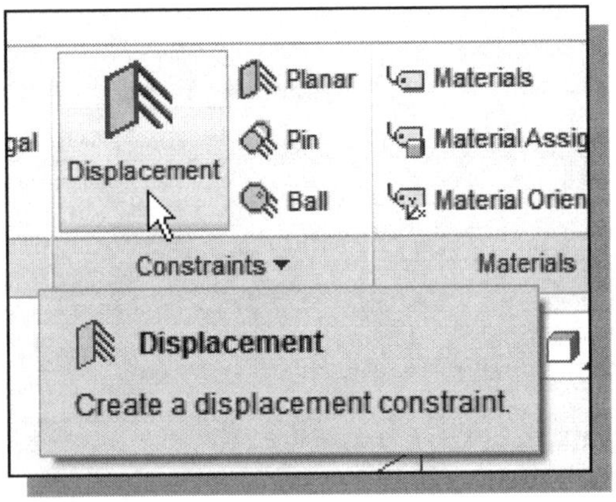

1. Choose **Displacement** by clicking the icon in the *Constraints* toolbar as shown.

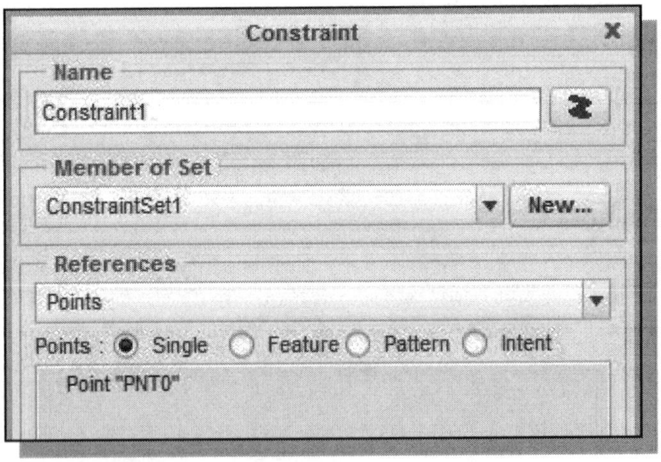

2. Set the References to **Points** and select **PNT0** as shown.

❖ PNT0 is a fixed support point; it cannot move in the X, Y or Z directions. For 2D truss systems, all joints are *pin-joints*; therefore, the Z rotational movement is free at PNT0.

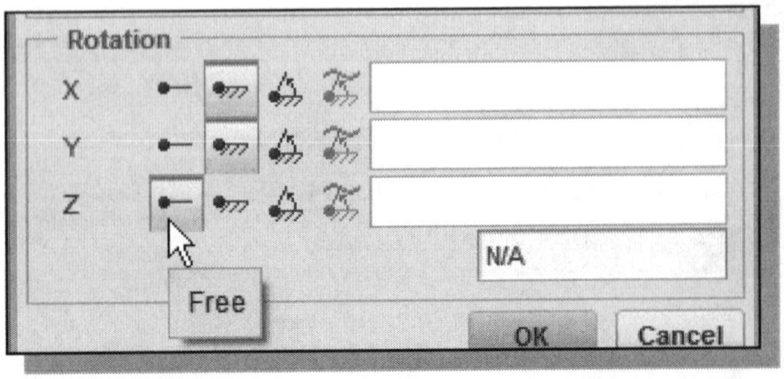

3. Set Z Rotation to **Free** and X, Y Rotations to **Fixed** as shown.

4. Click on the **OK** button to accept the first displacement constraint settings.

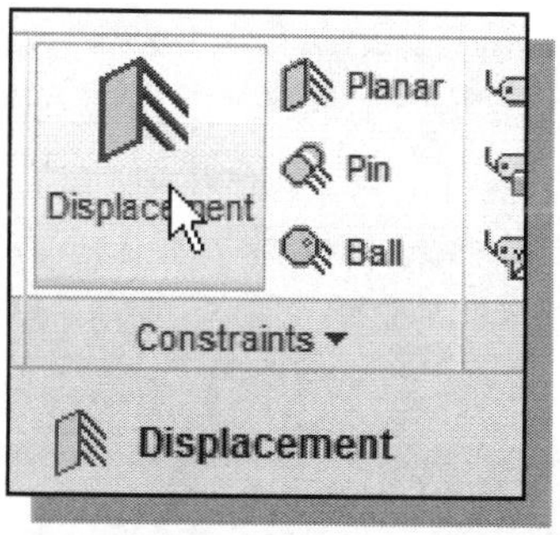

5. Choose **Displacement** by clicking the icon in the *Constraints* toolbar as shown.

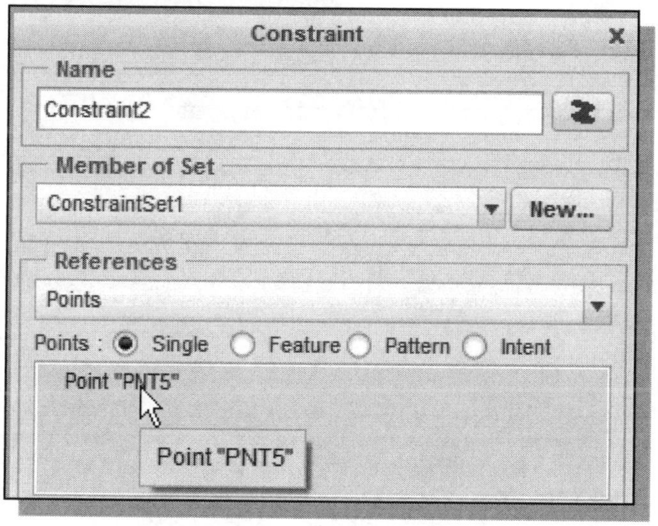

6. Confirm the *Constraint* set is still set to **ConstraintSet1**.

❖ Note that a constraint set can contain multiple constraints.

7. Set the References to **Points** and select **PNT5** as shown.

❖ PNT5 has a roller support; it can move in the X direction, but not in the Y or Z directions. For 2D truss systems, all joints are pin-joints; therefore, the Z rotational movement is also free at PNT5.

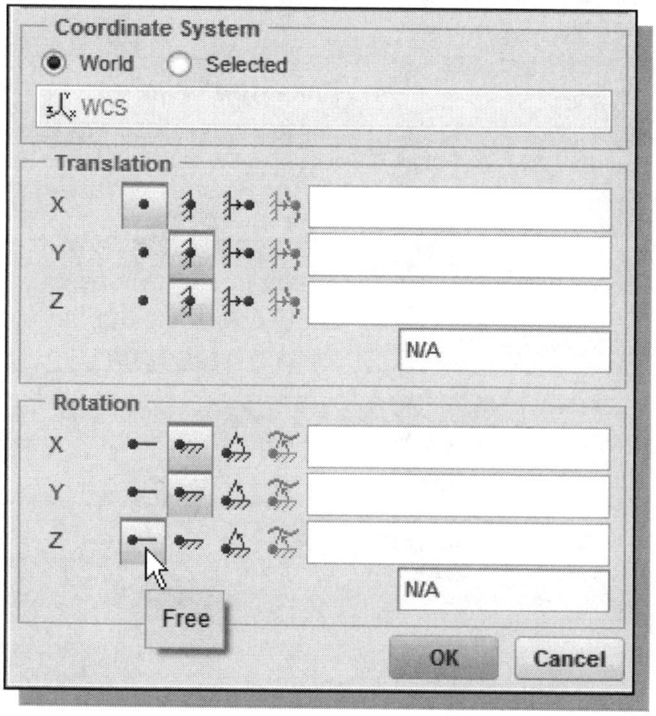

8. Set X Translation to **Free** as shown.

9. Set Z Rotation to **Free** as shown.

10. Click on the **OK** button to accept the Displacement constraint settings.

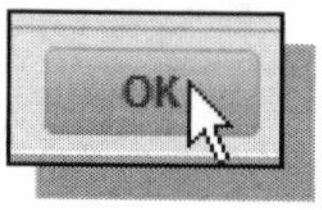

Apply External Loads

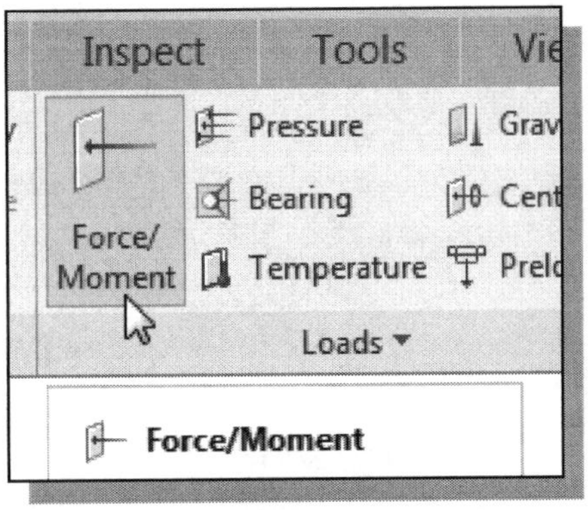

1. Choose **Force/Moment Load** by clicking the icon in the Loads toolbar as shown.

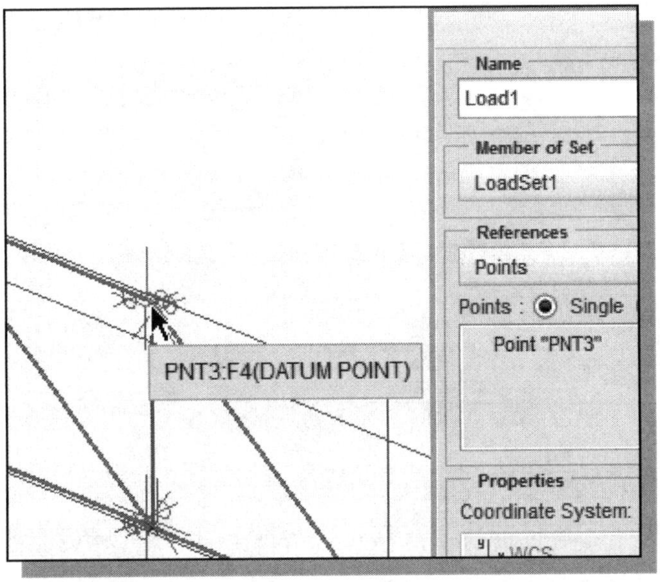

2. Set the References to **Points**.

3. Pick the **upper right node**, **PNT3**, of the truss system as shown.

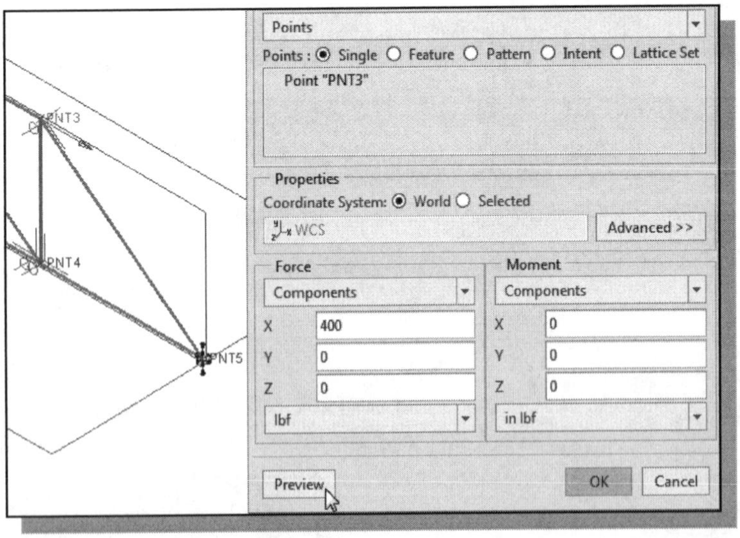

4. Enter **400** in the X component Force value box as shown.

5. Click on the **Preview** button to confirm the applied load is pointing in the correct direction.

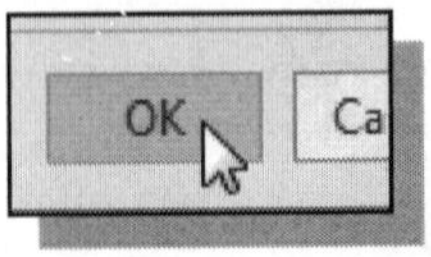

6. Click on the **OK** button to accept the first load settings.

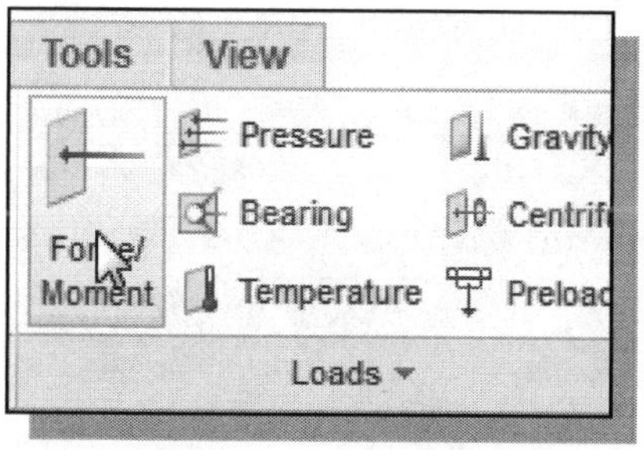

7. Choose **Force/Moment Load** by clicking the icon in the toolbar as shown.

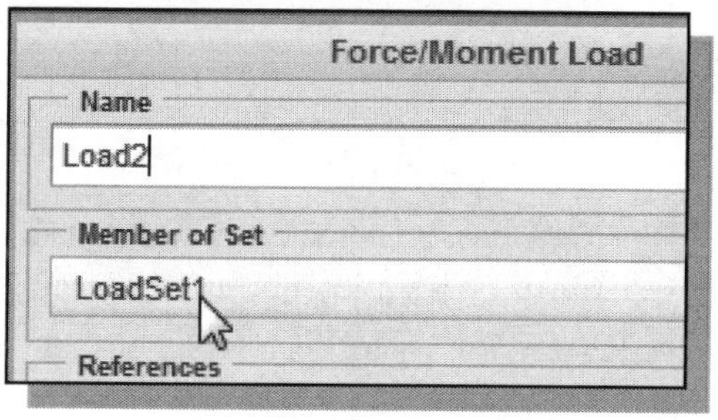

8. Confirm the *Load* set is still set to **LoadSet1**.

❖ Note that a load set can contain multiple loads.

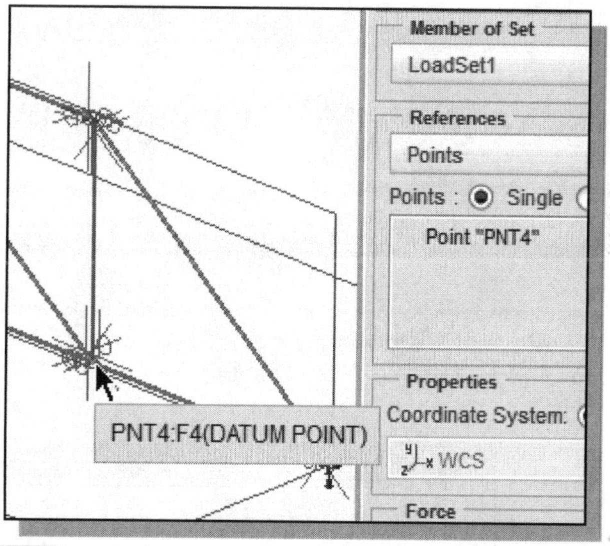

9. Set the References to **Points**.

10. Pick the 3rd bottom node, **PNT4**, of the truss system as shown.

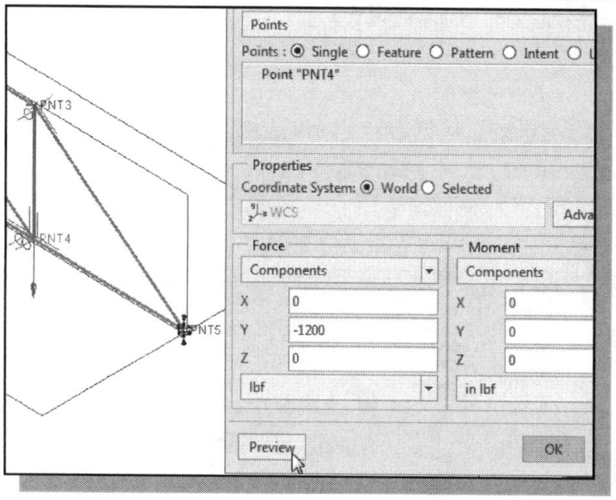

11. Enter **-1200** in the Y component Force value box as shown.

12. Click on the **Preview** button to confirm the applied load is pointing in the correct direction.

13. Click on the **OK** button to accept the second *Load* settings.

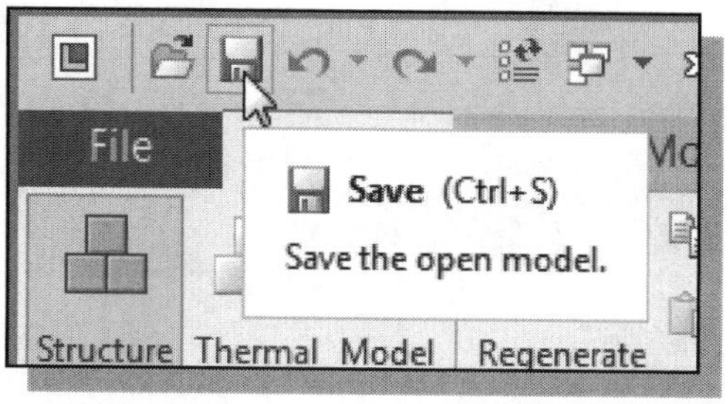

14. On your own, click **Save** to save a copy of the current FEA model before continuing to the next section.

Run the Solver

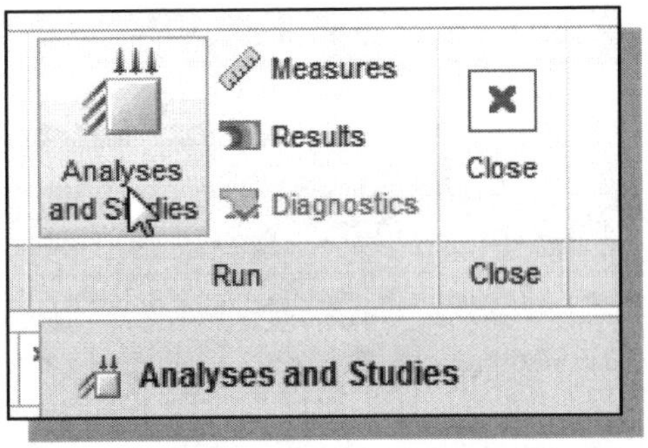

1. Activate the *Simulate* **Analyses and Studies** command by selecting the icon in the Run toolbar as shown.

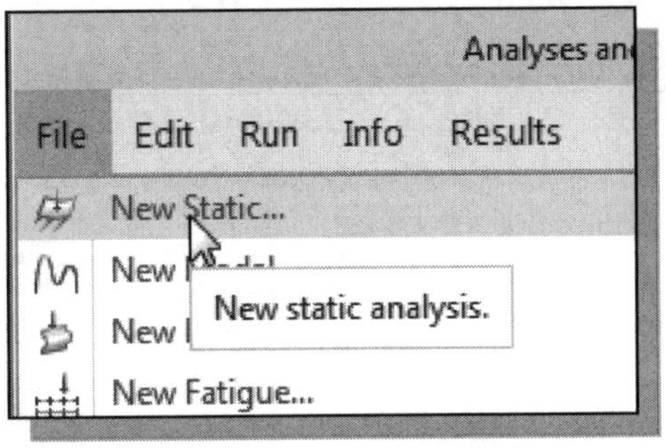

2. In the *Analyses and Design Studies* window, choose **File → New Static** in the pull-down menu. This will start the setup of a basic linear static analysis.

3. Note that the ConstraintSet1 and LoadSet1 are automatically accepted as part of the analysis parameters.

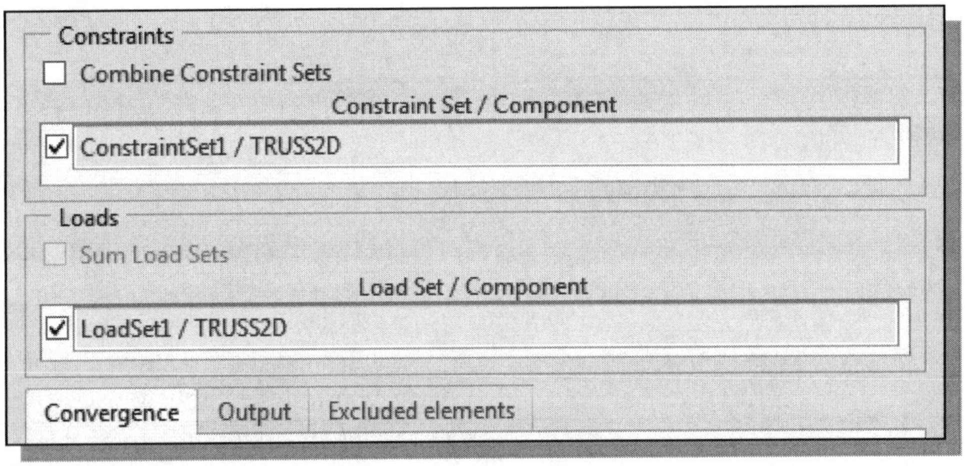

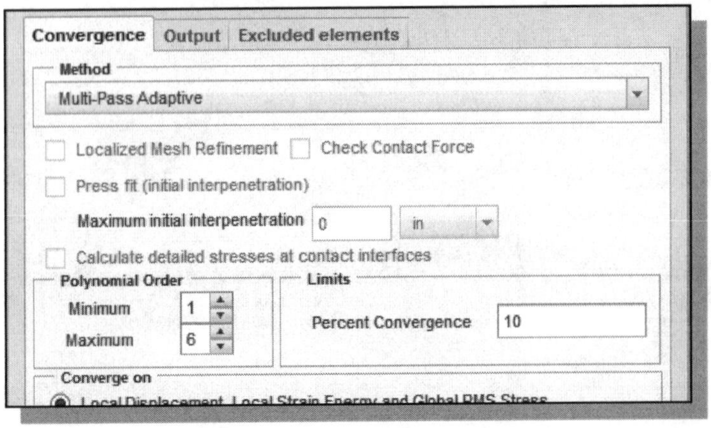

4. Set the Method option to **Multi-Pass Adaptive** and Limits to **10** Percent Convergence as shown.

5. Click **OK** to accept the settings.

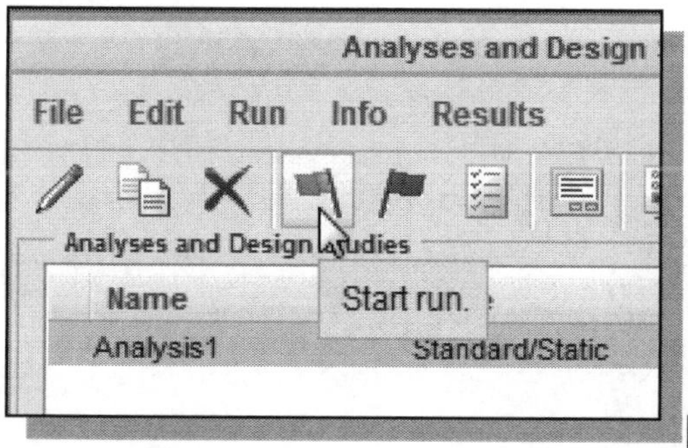

6. Click **Start Run** to begin the solving process.

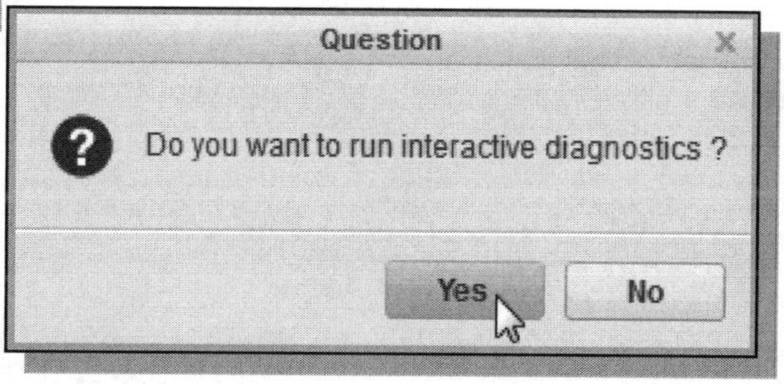

7. Click **Yes** to run the *Simulate* interactive diagnostics.

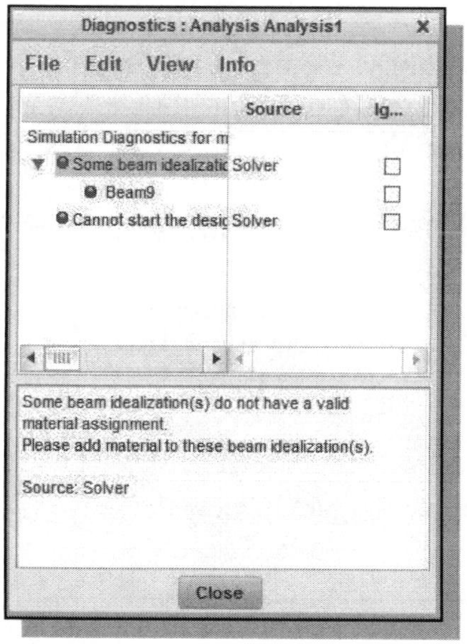

➢ The *Simulate* interactive diagnostics option can be very helpful in identifying and correcting problems associated with our FEA model. *Information*, *warnings* and *fatal errors* regarding the *Solver* are shown in the *Diagnostics* window. Pay extra attention to any **FATAL ERROR** messages, as those must be corrected.

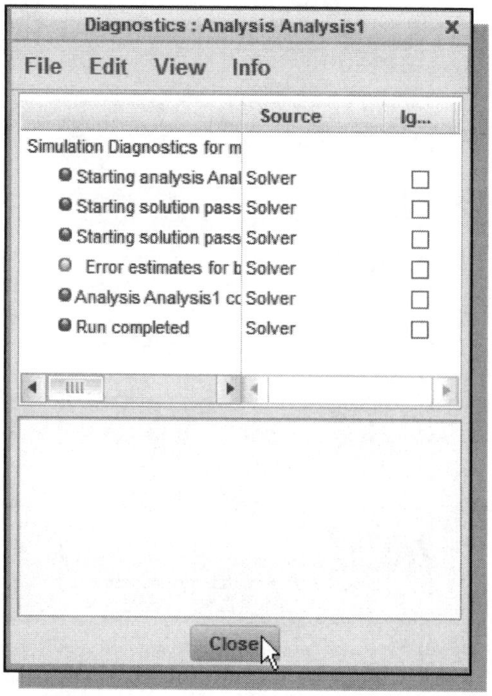

➢ For our analysis, in the *Diagnostics* window, the last message, **Run Completed**, indicates the FEA analysis is a success.

8. Click on the **Close** button to exit the *Diagnostics* window.

View the Results

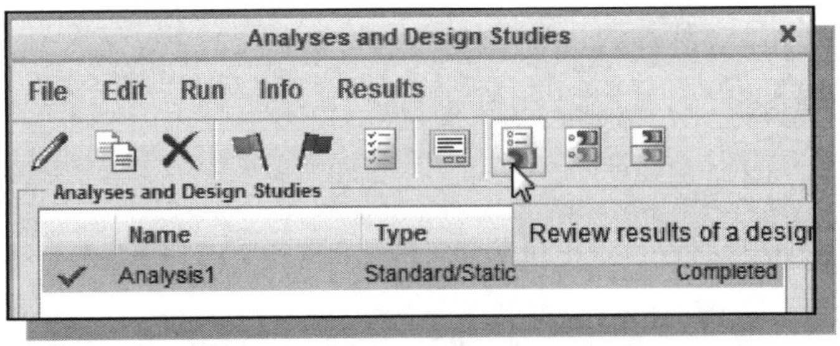

1. In the *Analyses and Design Studies* window, choose **Review results** as shown.

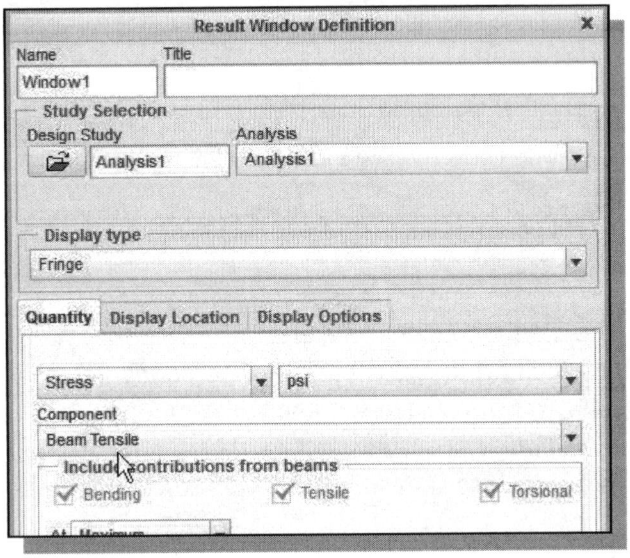

2. Confirm the Display type is set to **Fringe**.

3. Choose **Beam Tensile** as the Stress Component to be displayed.

4. Click **OK and Show** to view the results.

❖ The *Creo Simulate* **Beam Tensile** option allows us to quickly view the tensile stress in beam and truss elements.

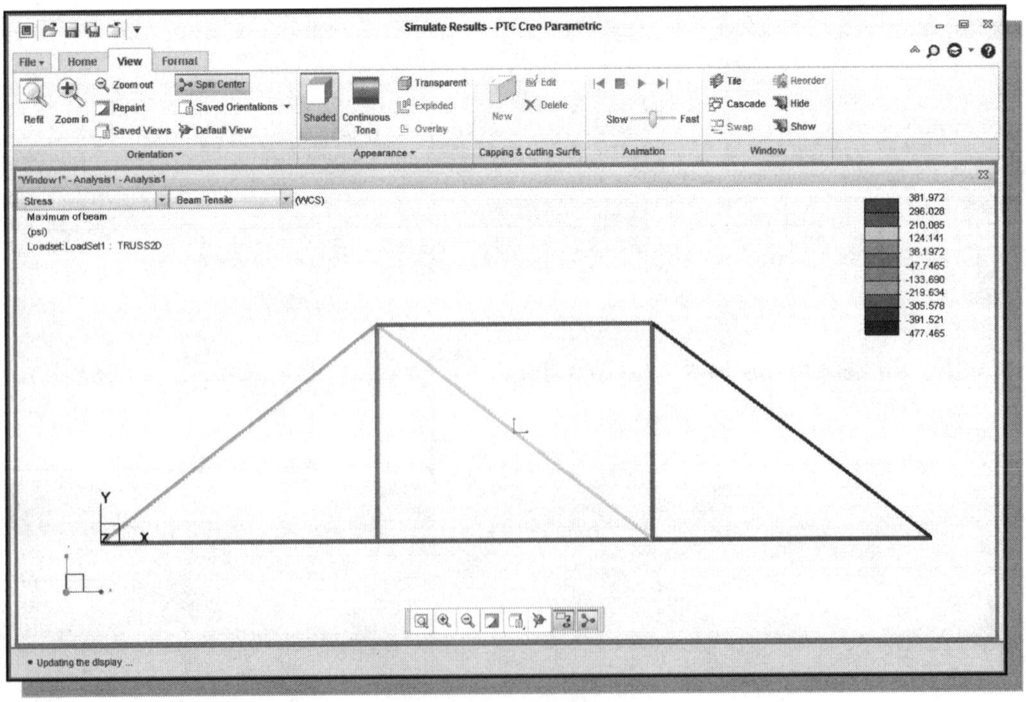

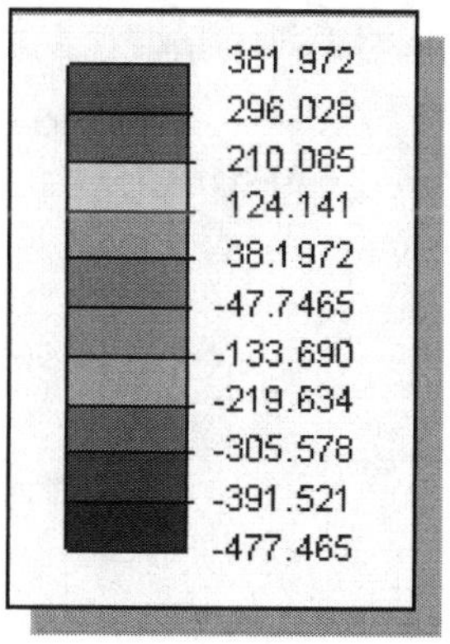

❖ Note the maximum tensile stress is **382 psi** and the minimum tensile stress is **-477.5 psi**. Are these values matching the preliminary analysis performed?

➢ Note that it is a little tough to read the stress values by looking at the color codes. We will next use the **Dynamic Query** command to get a better reading of the stress of each member.

Dynamic Query

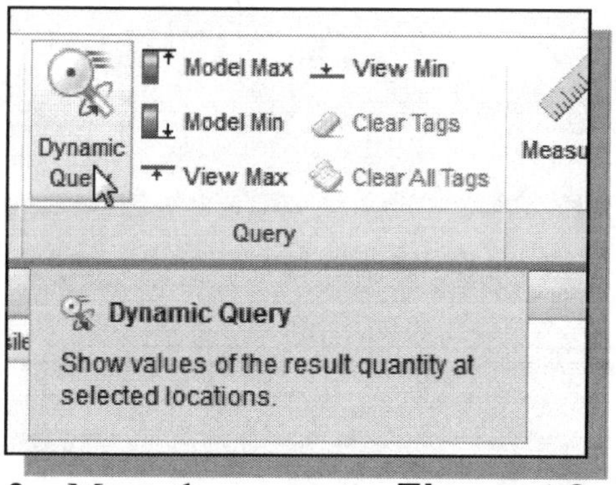

1. Choose **Dynamic Query** in the Query toolbar. The *Query* window appears on the screen.

2. Move the cursor on **Element 3** and read the stress value inside the *Query* window. The negative value implies the member is in compression.

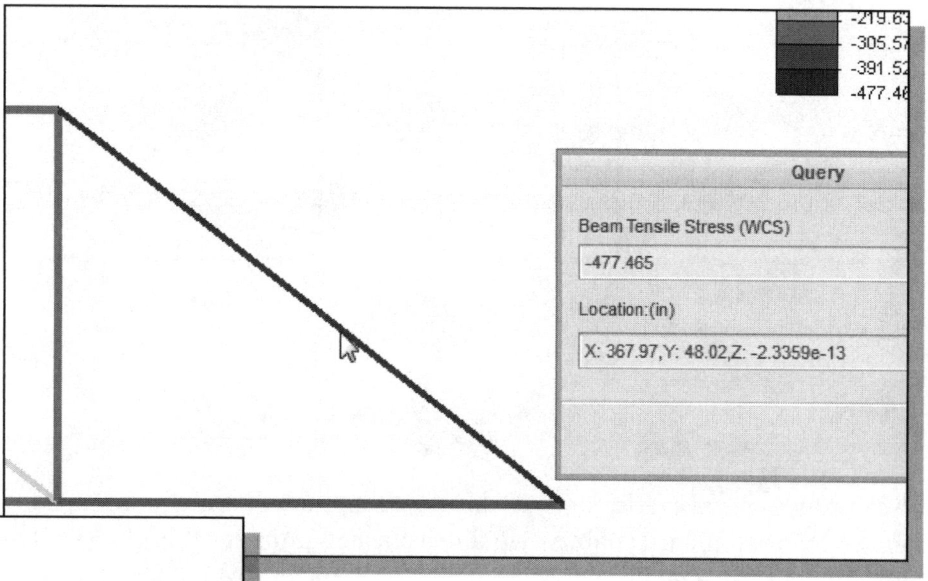

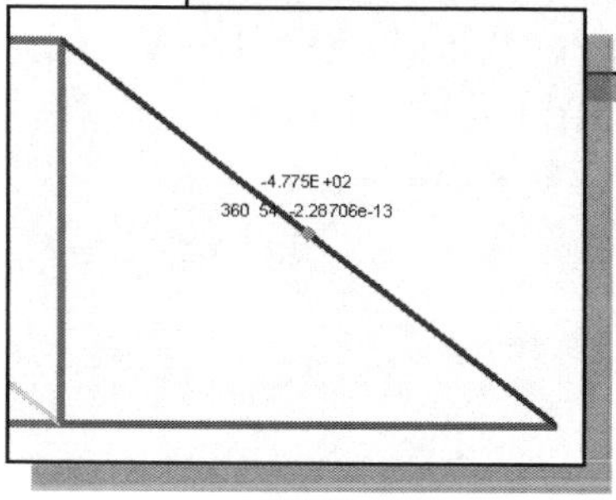

3. Click on the member to display the stress value of the selected location on the member.

4. Move the cursor on **Element 6** and read the stress value inside the *Query* window. This member has the highest stress in the system.

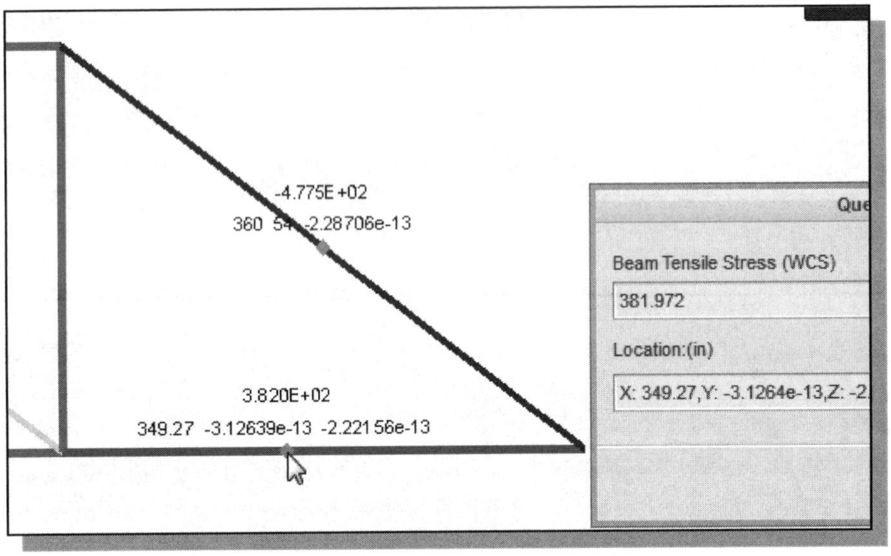

5. Move the cursor on **Element 8** and read the stress value inside the *Query* window. What does the number tell us about this member?

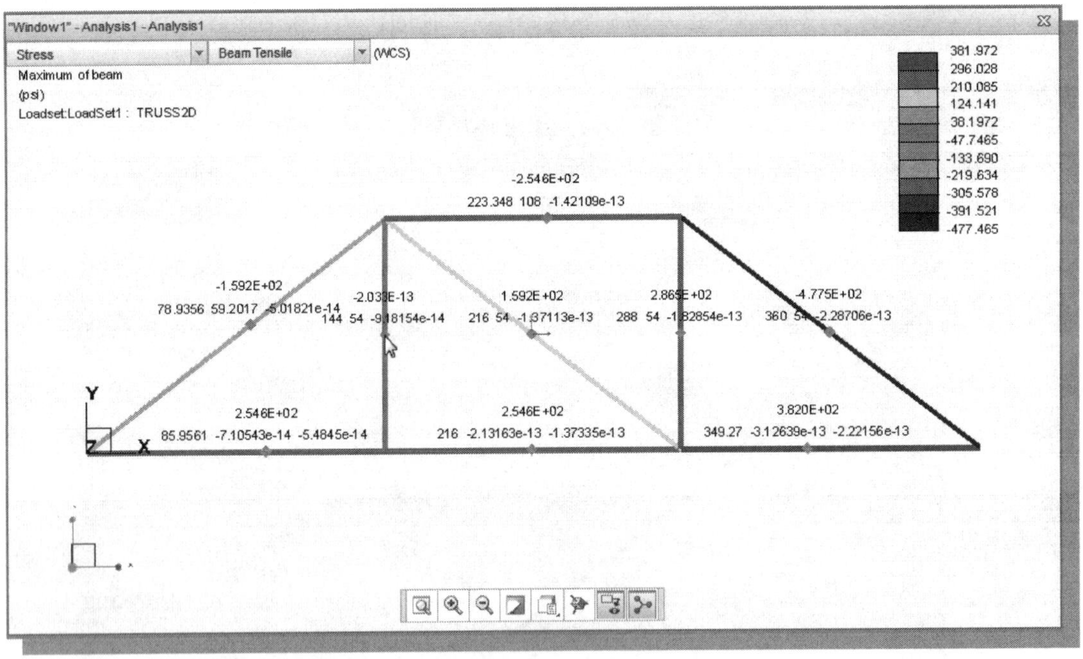

➢ On your own, examine the stresses developed in the other elements and compare them to the preliminary analysis we did at the beginning of this chapter. Does *Creo Simulate* display the stress value for the zero-force member correctly? Which member has the highest stress and what is the stress value?

Review Questions

1. Describe the typical finite element analysis steps.

2. In the tutorial, how were the nodes and elements created in *CREO SIMULATE*?

3. Can loads be applied to the midpoint of a truss element?

4. How do you identify the ZERO-FORCE members in a truss structure?

5. What are the basic assumptions, as defined in Statics, for truss members?

6. What does **BACS** stand for?

7. How is the X axis of the **BACS** determined?

8. What is the purpose of a beam release?

9. How do we assign the material properties for truss element in *Creo Simulate*?

10. How do we define the cross sections for truss elements in *Creo Simulate*? What standard cross sections are available in *Creo Simulate*?

11. How do we determine the X and Y directions of an existing beam element in *Creo Simulate*?

12. How do we determine the applied constraints on an existing system in *Creo Simulate*?

13. How do we obtain the stress values of truss members in *Creo Simulate*?

14. What is the type of FEA analysis performed in the tutorial?

Exercises

Determine the normal stress in each member of the truss structures shown.

1. Material: Steel,
   Diameter: 2.0 in.

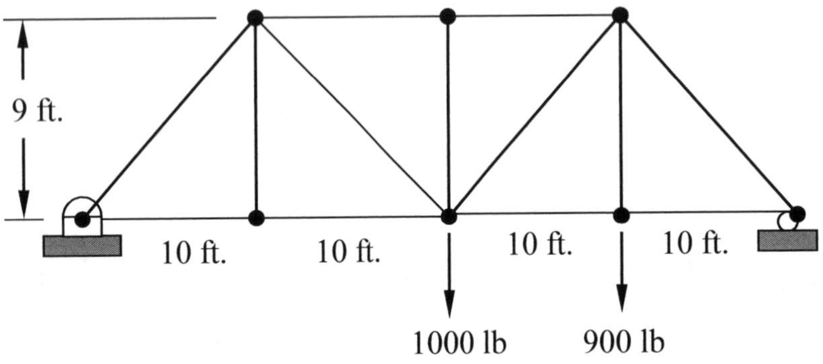

2. Material: Steel,
   Diameter: 2.5 in.

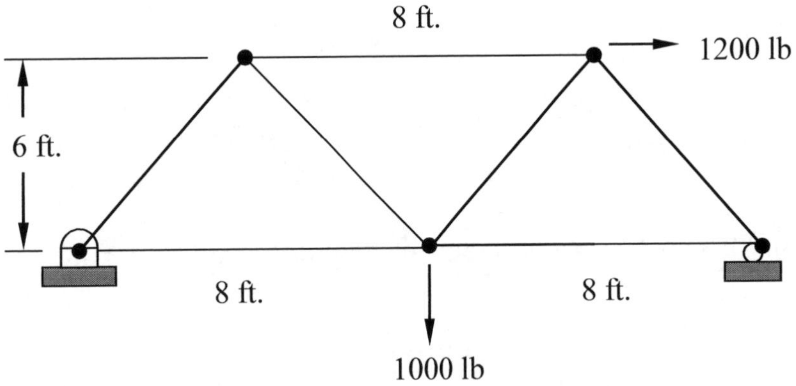

3. Material: Steel,
   Diameter: 3 cm.

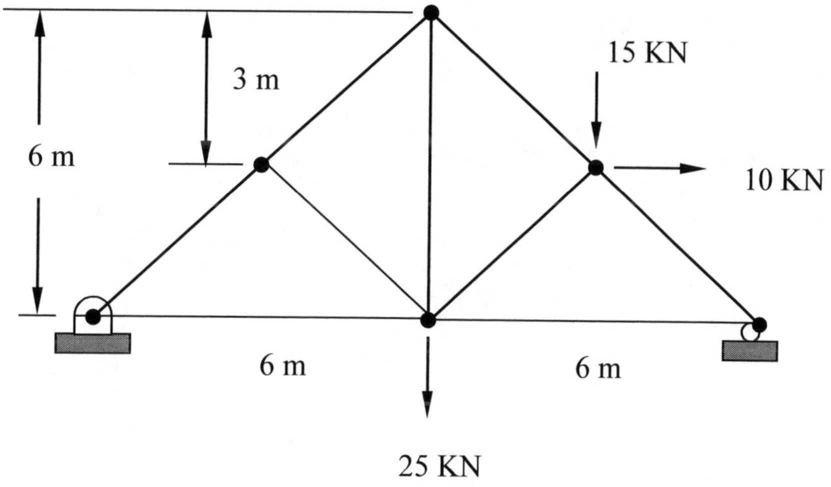

Notes:

Chapter 5
Three-Dimensional Truss Analysis

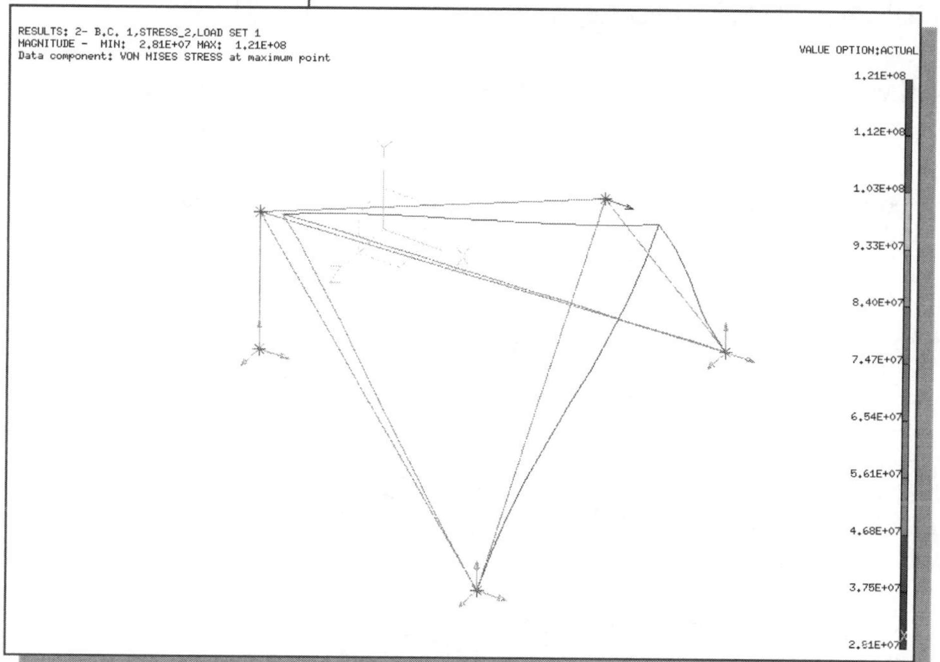

Learning Objectives

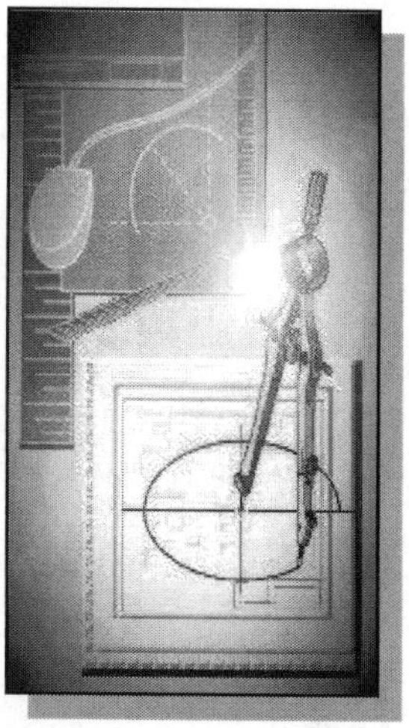

- **Determine the Number of Degrees of Freedom in Elements.**
- **Create 3D FEA Truss Models.**
- **Apply proper boundary conditions to FEA Models.**
- **Use Creo Simulate Solver for 3D Trusses.**
- **Use Creo Simulate to determine Axial Loads.**
- **Use the Multiple Results Display.**

Three-Dimensional Coordinate Transformation Matrix

For truss members positioned in a three-dimensional space, the coordinate transformation equations are more complex than the transformation equations for truss members positioned in two-dimensional space. The coordinate transformation matrix is necessary to obtain the global stiffness matrix of a truss element.

1. The global coordinate system (X, Y and Z axes), chosen for representation of the entire structure.

2. The local coordinate system (*X, Y* and *Z* axes), with the *X* axis aligned along the length of the element.

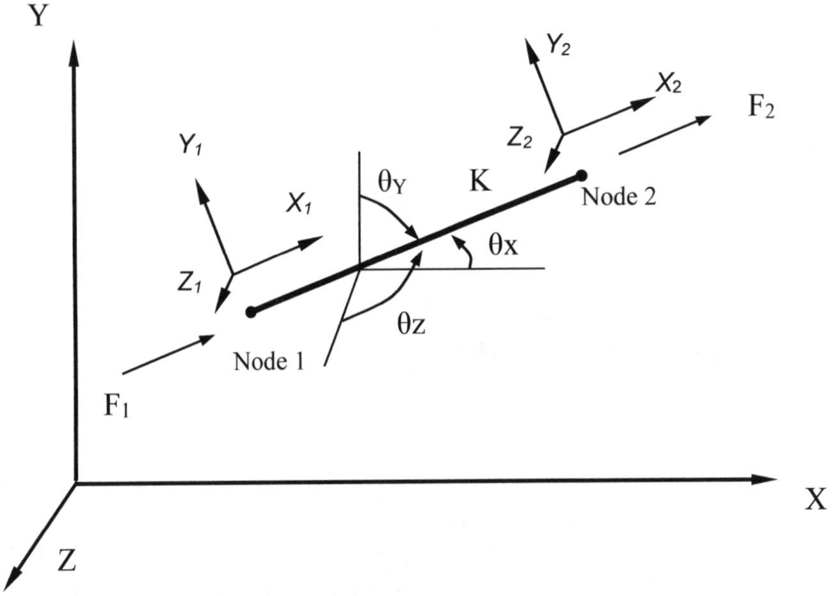

Since truss elements are two-force members, the displacements occur only along the local *X*-axis. The GLOBAL to LOCAL transformation matrix can be written as:

$$\begin{Bmatrix} X_1 \\ X_2 \end{Bmatrix} = \begin{bmatrix} \cos(\theta_X) & \cos(\theta_Y) & \cos(\theta_Z) & 0 & 0 & 0 \\ 0 & 0 & 0 & \cos(\theta_X) & \cos(\theta_Y) & \cos(\theta_Z) \end{bmatrix} \begin{Bmatrix} X_1 \\ Y_1 \\ Z_1 \\ X_2 \\ Y_2 \\ Z_2 \end{Bmatrix}$$

Local

Global

Stiffness Matrix

The displacement and force transformations can be expressed as:

$\{X\} = [l]\{\mathrm{X}\}$ ------- *Displacement transformation equation*

$\{F\} = [l]\{\mathrm{F}\}$ ------- *Force transformation equation*

Combined with the local stiffness matrix, ***{ F } = [K] { X },*** we can then derive the global stiffness matrix for an element:

$$[\mathbf{K}] = \frac{\mathbf{EA}}{\mathbf{L}} \begin{pmatrix} C_X^2 & C_X C_Y & C_X C_Z & -C_X^2 & C_X C_Y & C_X C_Z \\ & C_Y^2 & C_Y C_Z & -C_X C_Y & -C_Y^2 & -C_Y C_Z \\ & & C_Z^2 & -C_X C_Z & -C_Y C_Z & -C_Z^2 \\ & & & C_X^2 & C_X C_Y & C_X C_Z \\ & \text{Symmetry} & & & C_Y^2 & C_Y C_Z \\ & & & & & C_Z^2 \end{pmatrix}$$

where $C_X = cos(\theta_X)$, $C_Y = cos(\theta_Y)$, $C_Z = cos(\theta_Z)$.

The resulting matrix is a 6×6 matrix. The size of the stiffness matrix is related to the number of nodal displacements. The nodal displacements are also used to determine the number of **degrees of freedom** at each node.

Degrees of Freedom

- For a truss element in a one-dimensional space:

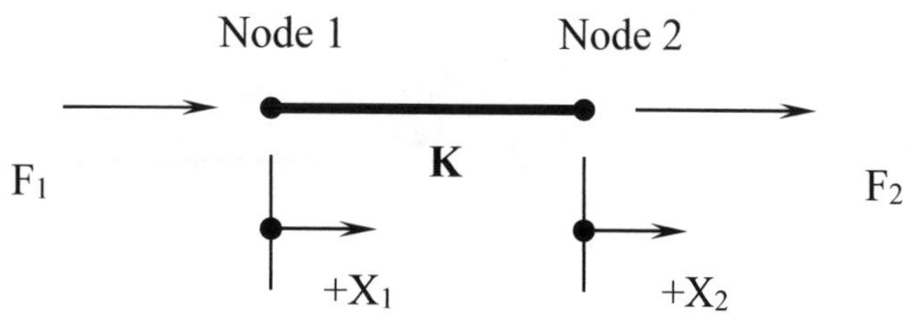

One nodal displacement at each node: one degree of freedom at each node. Each element possesses two degrees of freedom, which forms a 2×2 stiffness matrix for the element. The global coordinate system coincides with the local coordinate system.

♦ For a truss element in a two-dimensional space:

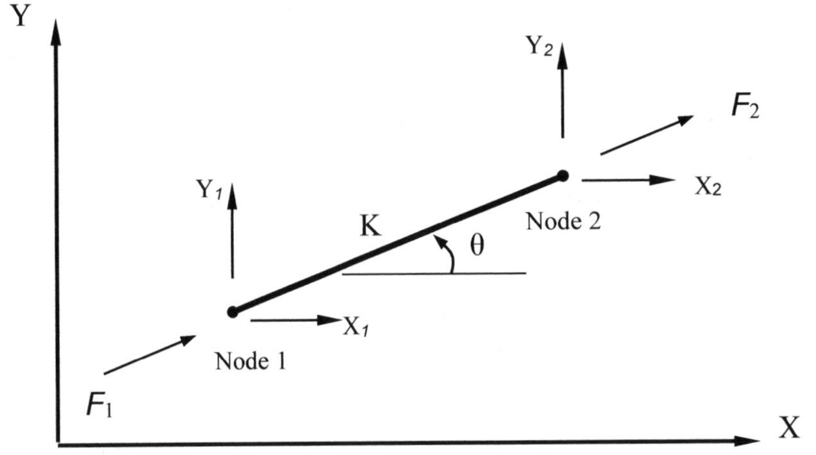

Two nodal displacements at each node: two degrees of freedom at each node. Each element possesses four degrees of freedom, which forms a 4 × 4 stiffness matrix for the element.

♦ For a truss element in a three-dimensional space:

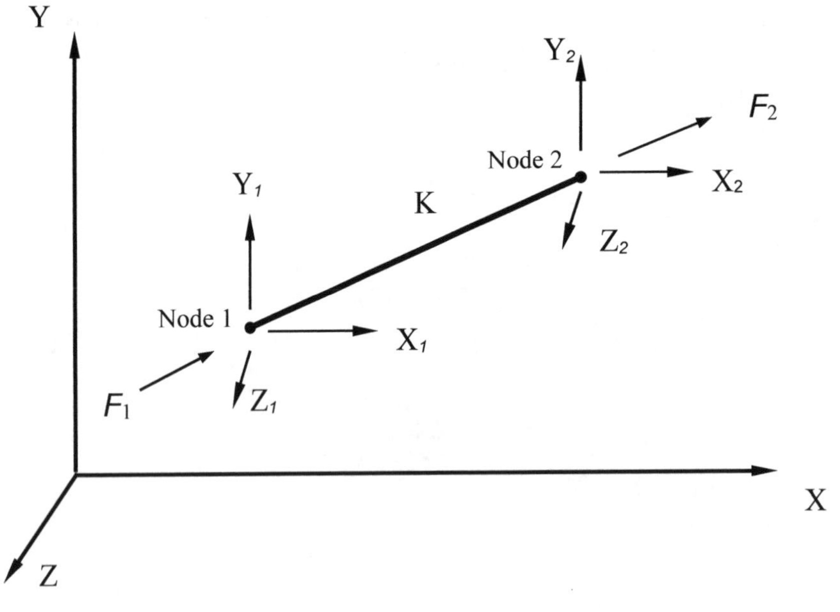

Three nodal displacements at each node: three degrees of freedom at each node. Each element possesses six degrees of freedom, which forms a 6 × 6 stiffness matrix for the element.

Problem Statement

Determine the normal stress in each member of the truss structure shown. All joints are ball-joints.

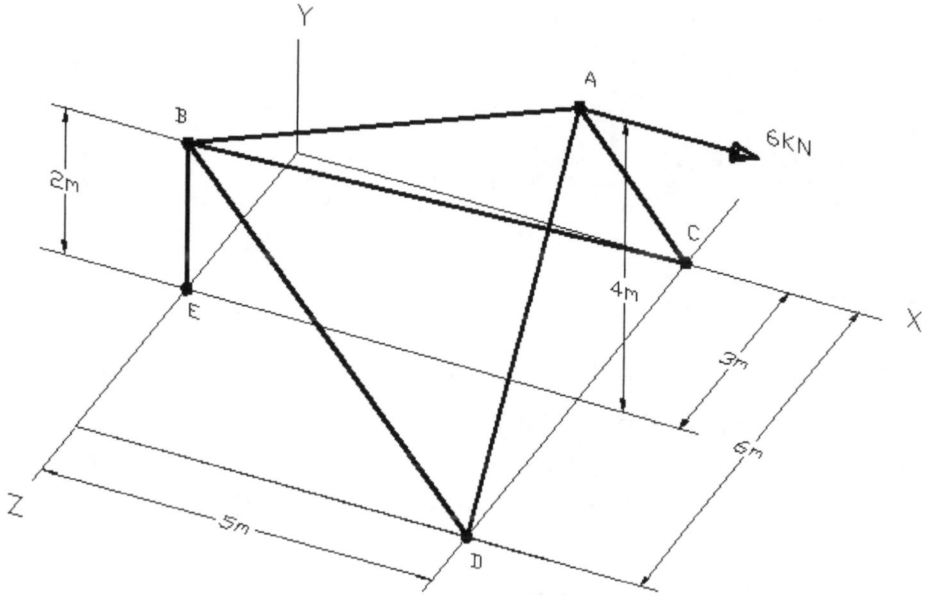

Material: Steel   Cross Section: a b   a = b = 7.5mm

Preliminary Analysis

This section demonstrates using conventional *vector algebra* to solve the three-dimensional truss problem.

The coordinates of the nodes:

A(5,4,3), B(0,2,3), C(5,0,0), D(5,0,6), E(0,0,3)

Free Body Diagram of Node A:

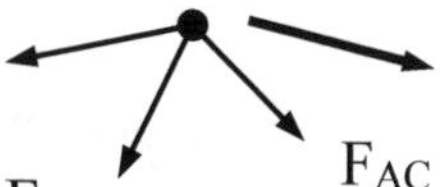

Position vectors AB, AC and AD:

AB = (0-5) i + (2-4) j + (3-3) k = (-5) i + (-2) j

AC = (5-5) i + (0-4) j + (0-3) k = (-4) j + (-3) k

AD = (5-5) i + (0-4) j + (6-3) k = (-4) j + (3) k

Unit vectors along AB, AC and AD:

$$u_{AB} = \frac{(-5)\,i + (-2)\,j}{\sqrt{(-5)^2+(-2)^2}} = -\,0.9285\,i - 0.371\,j$$

$$u_{AC} = \frac{(-4)\,j + (-3)\,k}{\sqrt{(-4)^2+(-3)^2}} = -\,0.8\,j - 0.6\,k$$

$$u_{AD} = \frac{(-4)\,j + (3)\,k}{\sqrt{(-4)^2+(3)^2}} = -\,0.8\,j + 0.6\,k$$

Forces in each member:

$F_{AB} = F_{AB}\,u_{AB} = F_{AB}\,(-\,0.9285\,i - 0.371\,j)$
$F_{AC} = F_{AC}\,u_{AC} = F_{AC}\,(-\,0.8\,j - 0.6\,k)$
$F_{AD} = F_{AD}\,u_{AD} = F_{AD}\,(-\,0.8\,j + 0.6\,k)$
(F_{AB}, F_{AC} and F_{AD} are magnitudes of vectors F_{AB}, F_{AC} and F_{AD})

Applying the equation of equilibrium at node A:

$$\begin{aligned}\sum F_{@A} &= 0 = 6000\,i + F_{AB} + F_{AC} + F_{AD}\\ &= 6000\,i - 0.9285\,F_{AB}\,i - 0.371\,F_{AB}\,j - 0.8\,F_{AC}\,j - 0.6\,F_{AC}\,k\\ &\quad - 0.8\,F_{AD}\,j + 0.6\,F_{AD}\,k\\ &= (6000 - 0.9285\,F_{AB})i + (-\,0.371\,F_{AB} - 0.8\,F_{AC} - 0.8\,F_{AD})j\\ &\quad + (-\,0.6\,F_{AC} + 0.6\,F_{AD})k\end{aligned}$$

Also, since the structure is symmetrical, $F_{AC} = F_{AD}$

Therefore,

6000 – 0.9285 $F_{AB} = 0$, $F_{AB} =$ **6462 N**

-0.371 F_{AB} **– 0.8** F_{AC} **– 0.8** $F_{AD} = 0$,

$F_{AC} = F_{AD} =$ **– 1500 N**

The stresses:

$\sigma_{AB} = 6460 / (5.63 \times 10^{-5}) =$ **115 MPa**

$\sigma_{AC} = \sigma_{AD} = -\,1500 / (5.63 \times 10^{-5}) =$ **-26.7 MPa**

Starting Creo Parametric

1. Select the **Creo Parametric** option on the *Start* menu or select the **Creo Parametric** icon on the desktop to start *Creo Parametric*. The *Creo Parametric* main window will appear on the screen.

2. Click on the **New** icon, located in the *Ribbon toolbar* as shown.

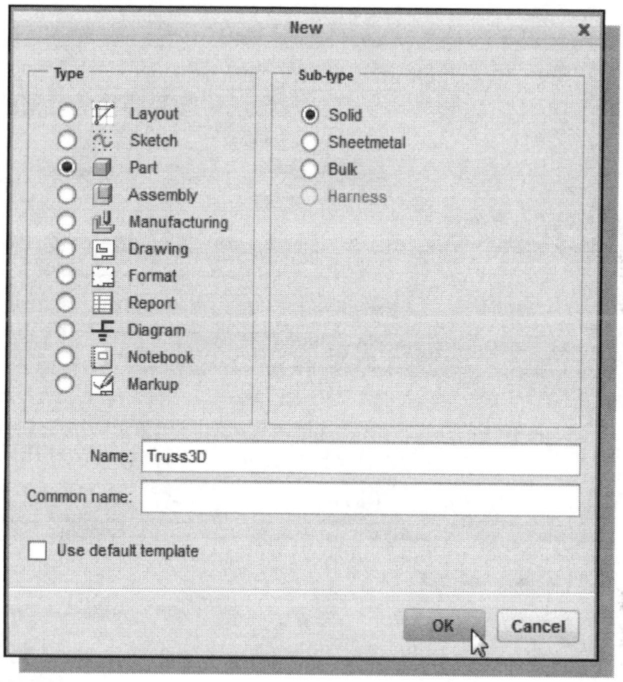

3. In the *New* dialog box, confirm the model's Type is set to **Part** (**Solid** Sub-type).

4. Enter **Truss3D** as the part Name as shown in the figure.

5. Turn ***off*** the **Use default template** option.

6. Click on the **OK** button to accept the settings.

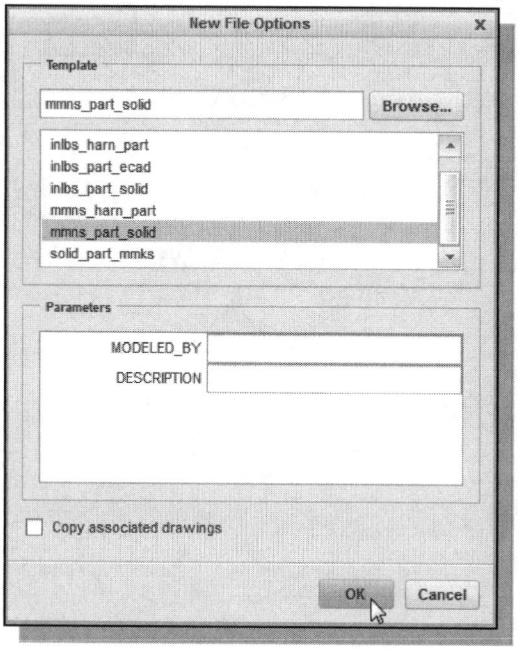

7. In the *New File Options* dialog box, select **mmns_part_solid** in the option list as shown.

8. Click on the **OK** button to accept the settings and enter the *Creo Parametric Part Modeling* mode.

❖ Note that the **mmns_part_solid** template contains several pre-defined settings, such as the units setup, as well as datum planes and views setup.

Create a New Template

The selected template, **mmns_part_solid**, is a template to using the *Millimeter-Newton-Second* units set. Although we can use this setup for our 3D truss example, it is just as easy to adjust the settings and create a new template file.

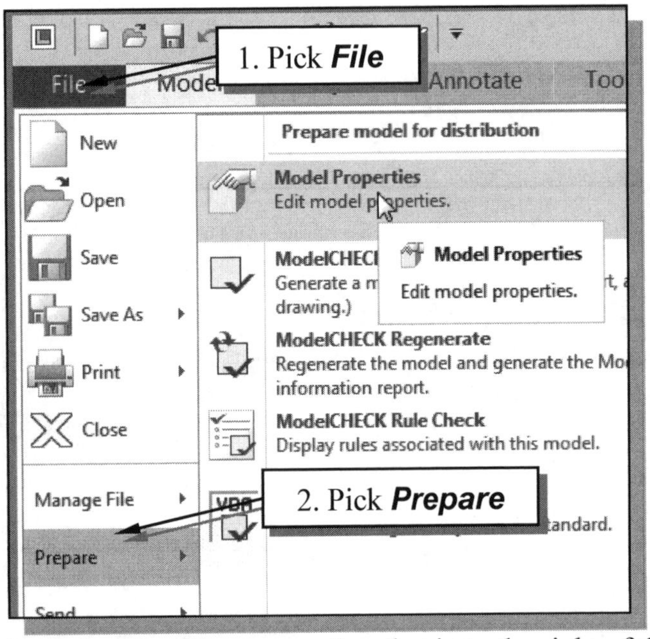

1. Use the left-mouse-button and select the **File** pull-down menu.

2. Use the left-mouse-button and select **Prepare** in the pull-down list as shown.

3. Select **Model Properties** in the expanded list as shown.

4. Select the **Change** option that is to the right of the **Units** option in the ***Model Properties*** window.

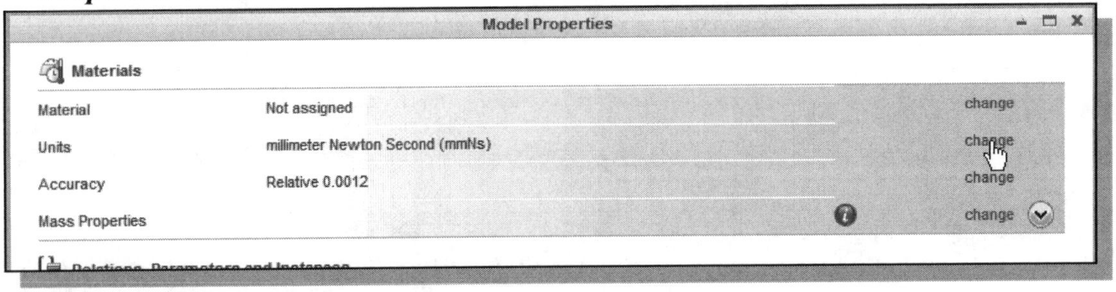

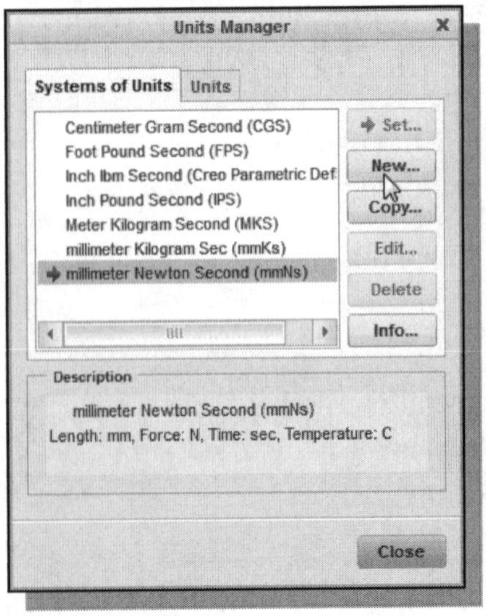

5. In the *Units Manager* window, notice the different units sets that are available. Click **New** to create a new units set.

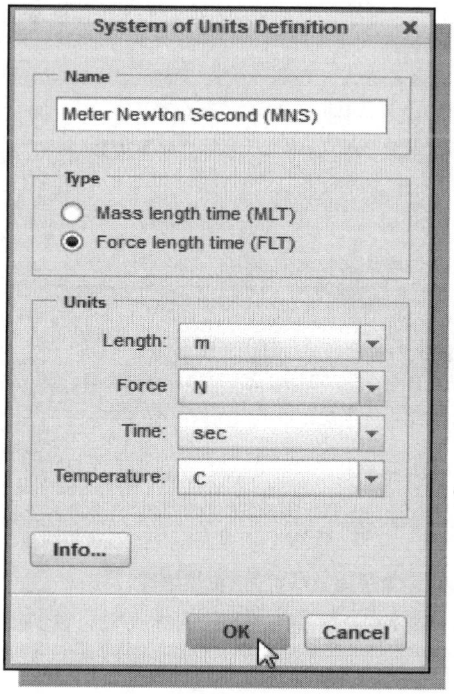

6. Enter **Meter Newton Second (MNS)** as the name of the new *System of Units*.

7. In the Type option, choose **Force Length Time** as shown.

8. Set the Length unit to *meter* (**m**), the Force unit to *Newton* (**N**) and Time unit to *second* (**sec**) as shown.

9. Click **OK** to accept the settings.

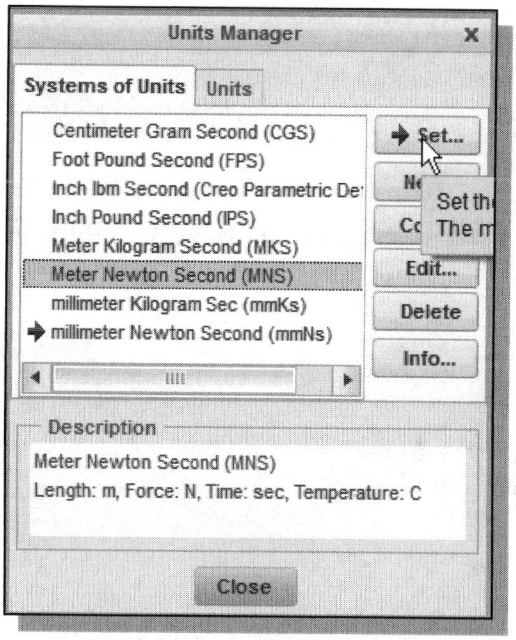

10. Select the new units set we just created.

11. Click on the **Set** button to accept the selection.

➢ Note that the *model units* defined in *Creo Parametric* can also be used in *Creo Simulate*.

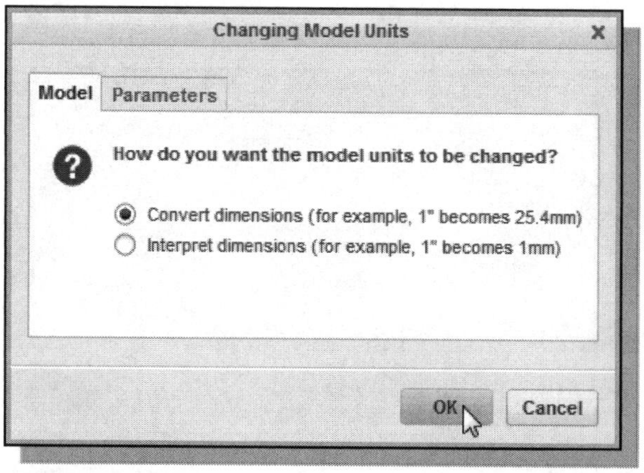

12. In the *Changing Model Units* dialog box, click on the **OK** button to accept the change of the units.

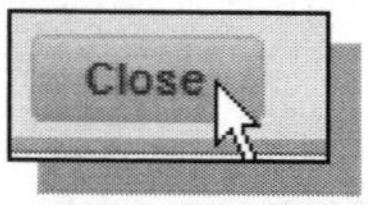

13. Click on the **Close** button to exit the *Units Manager* dialog box.

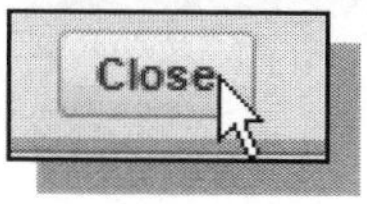

14. Pick **Close** to exit the *Model Properties* window.

Save the Current Setup as a New Template

Before creating elements, we will set up a *Material Property* table. The *Material Property* table contains general material information, such as *Modulus of Elasticity*, *Poisson's Ratio*, etc.

1. Choose **File→ Save a Copy** in the *File pull-down menu* as shown.

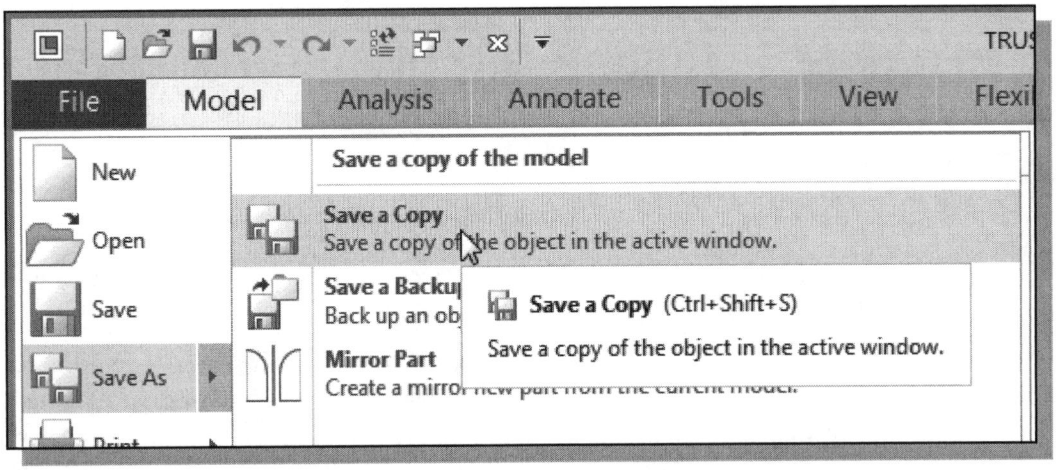

2. Switch to the ***templates*** folder that is under the installation folder (the default installation folder name is *Creo 4.0*). Placing the file in the *templates* folder will make this template available to use when creating a new file.

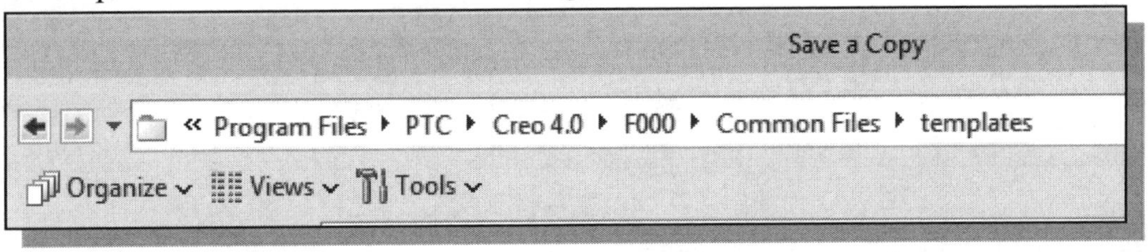

3. Enter **MNS_part_solid** as the new filename and click **OK** to save the current file as a new template.

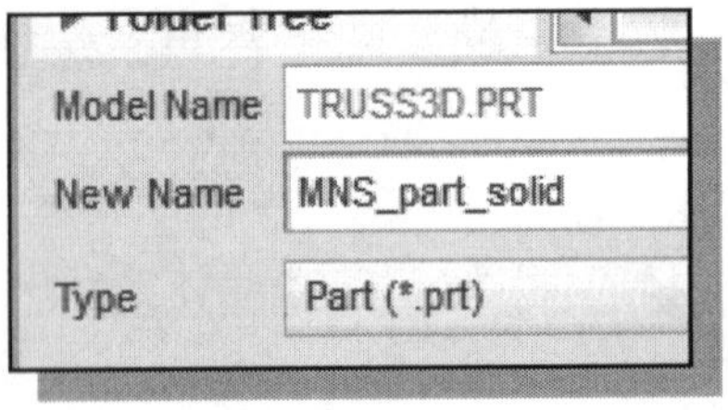

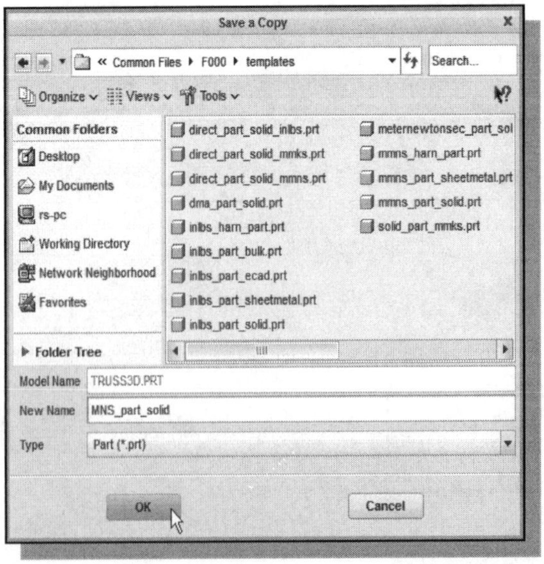

➢ Note that with some OS, you might need to save the template file in another location before moving it to the template folder.

Create 3D Datum Points

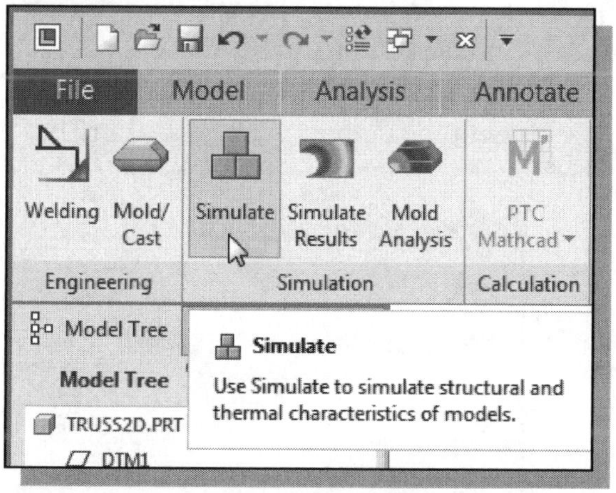

1. Start the *Integrated* mode of *Creo Simulate* by selecting the **Applications → Simulate** option from the *Creo Ribbon* toolbar as shown.

2. Click on the **Refine Model** tab to view the available commands.

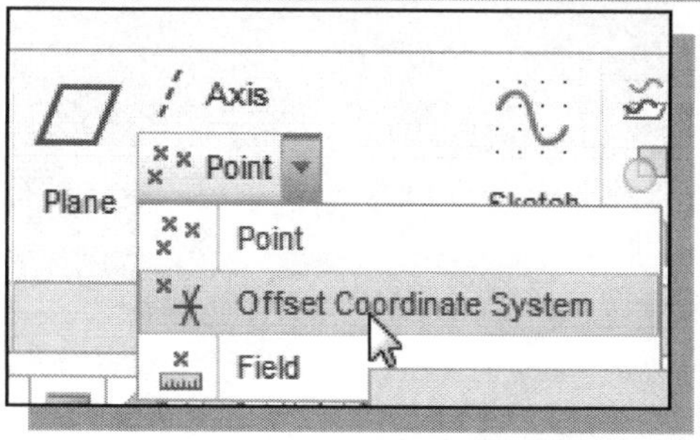

3. Choose **Offset Coordinate System** tool in the icon.

❖ Using the **Offset Coordinate System** tool, we can create datum points by entering X, Y, Z coordinates or importing the coordinates from a text file.

4. The message "*Select a coordinate system to place point.*" is displayed in the message area. Select the **part coordinate system** (**PRT_CSYS**) as shown.

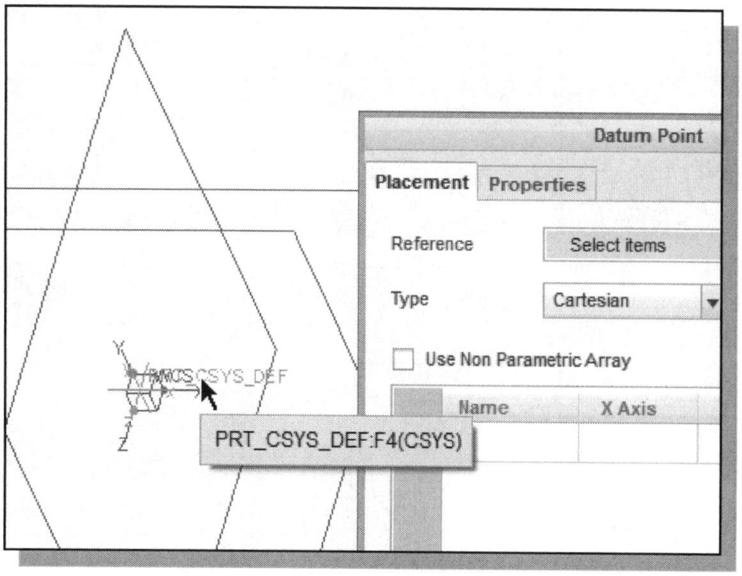

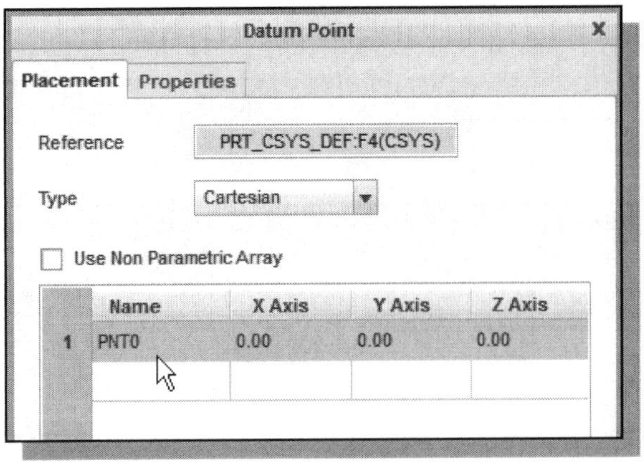

5. Click in the empty row, with the left-mouse-button, and *Creo* will automatically add a new datum point with the default value of **0.0** in all cells.

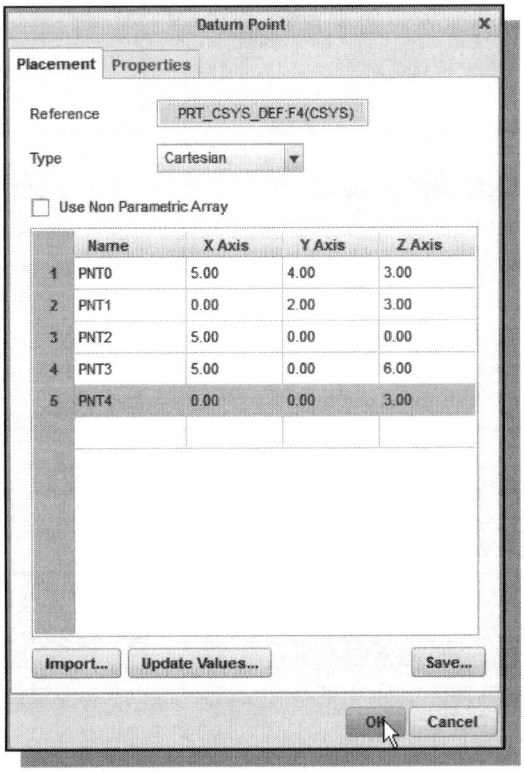

6. On your own, adjust the coordinates of **PTN0** to **5**, **4**, **3**.

7. Repeat the above steps and create the other **4 datum points** as shown.

8. Click **OK** to accept the settings and create the datum points.

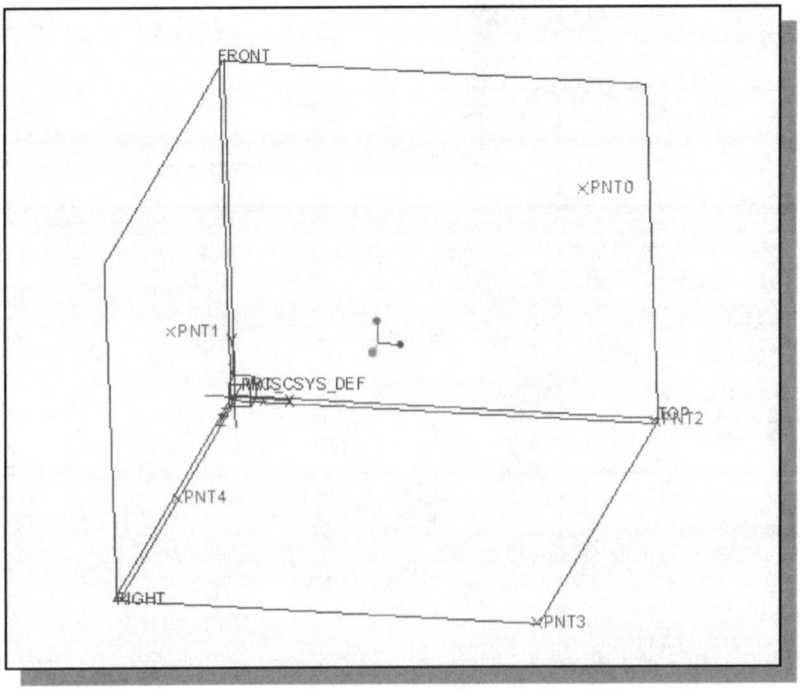

Setting Up an Element Cross Section

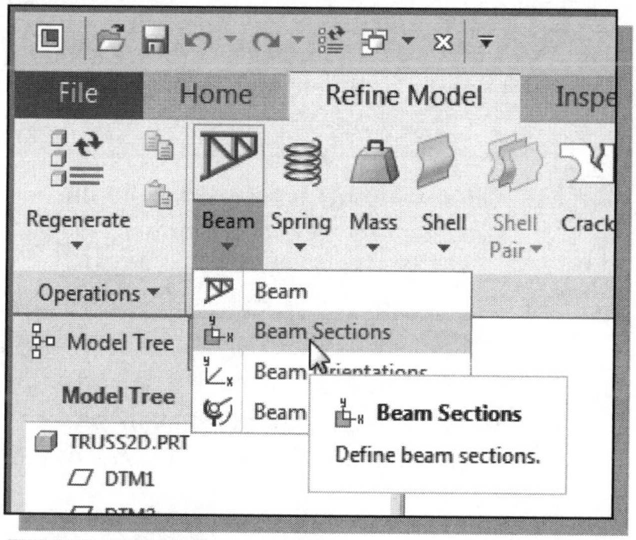

1. Choose **Beam Sections** in the *Idealizations* toolbar as shown.

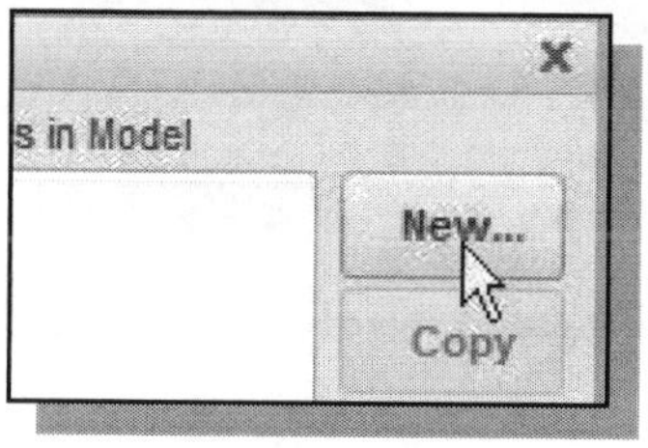

2. Click **New** to create a new beam cross-section.

3. Enter **RectangularRod** as the Name of the beam section.

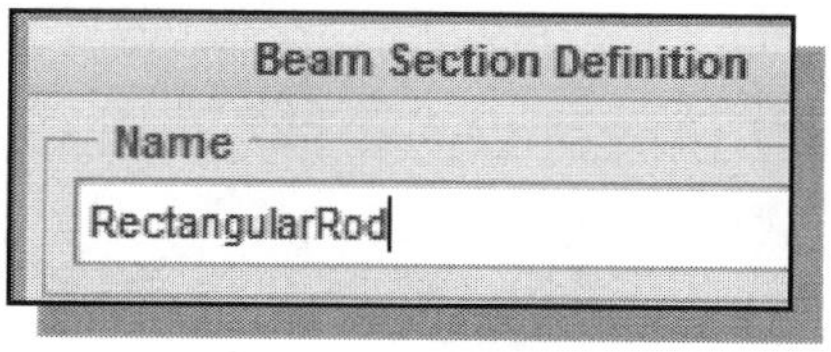

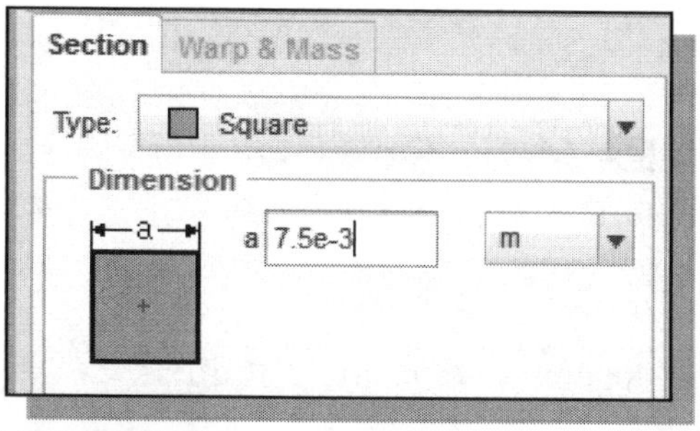

4. Choose **Square** as the cross-section Type as shown.

5. Enter **7.5e-3 m** as the side length of the rectangular beam cross section.

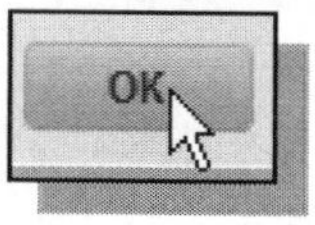

6. Click **OK** to accept the creation of the new beam section.

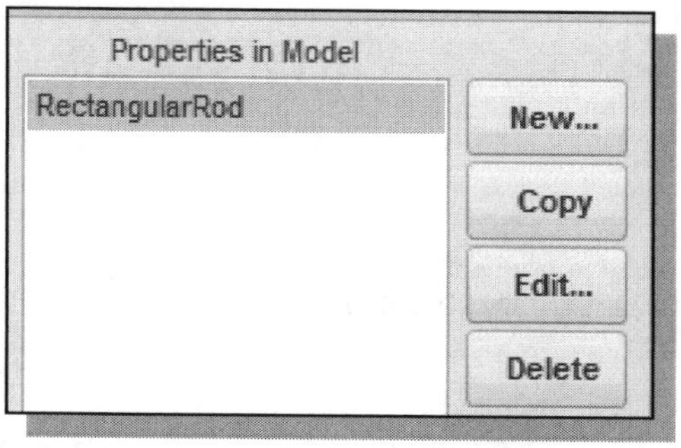

7. Click **Close** to accept the creation of the new beam section.

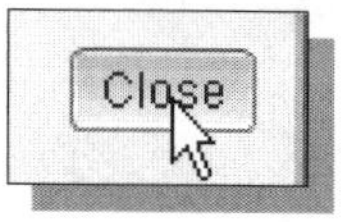

Setting Up Beam Element Releases

In *Creo Simulate*, the *truss element* is considered a special type of *beam element*. This is done by releasing the rotational degrees of freedom of the beam element.

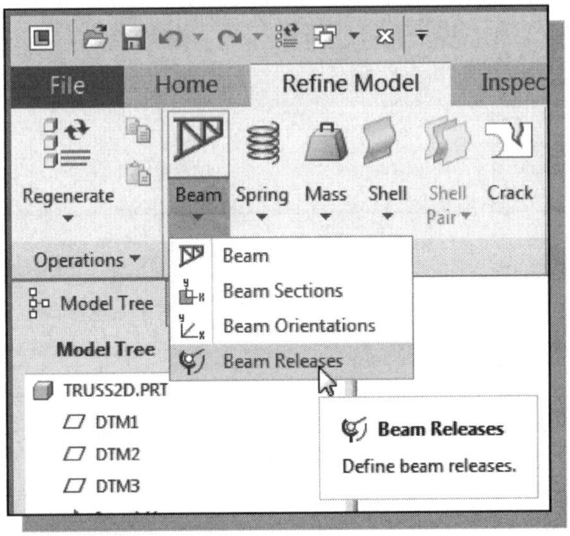

1. Choose **Beam Releases** from the *Idealizations* toolbar as shown.

❖ The **Beam Releases** command can be used to specify the degrees of freedom for a beam end or ends. By default, *Creo Simulate* assumes that all degrees of freedom are locked at the ends of the beam.

2. Click **New** to create a new beam cross-section.

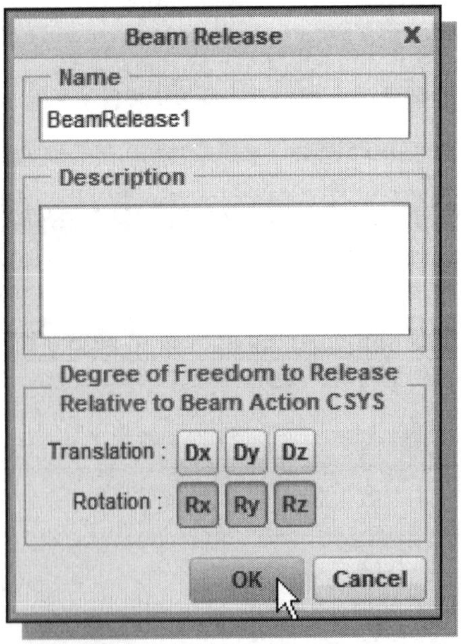

3. Click on the **Rx**, **Ry** and **Rz** icons to allow free rotations about the X, Y and Z axes.

4. Click **OK** to accept the definition of the new beam release.

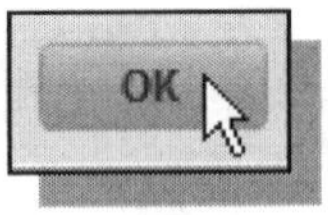

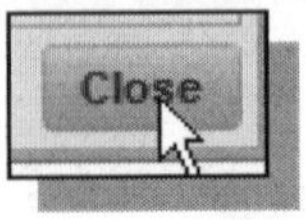

5. Click **Close** to accept the creation of the new beam release.

Select and Examine the Element Material Property

Before creating the FEA elements, we will first set up the *Material Property* for the elements. The *Material Property* contains the general material information, such as *Modulus of Elasticity*, *Poisson's Ratio*, etc., that is necessary for the FEA analysis.

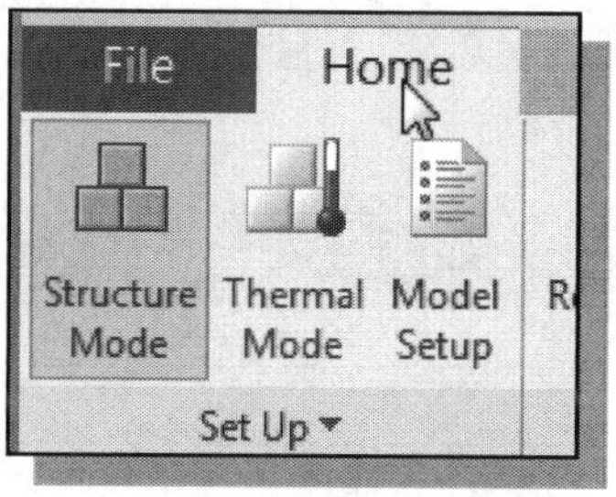

1. Click on the **Home** tab to return to the main *Creo Simulate* commands toolbar.

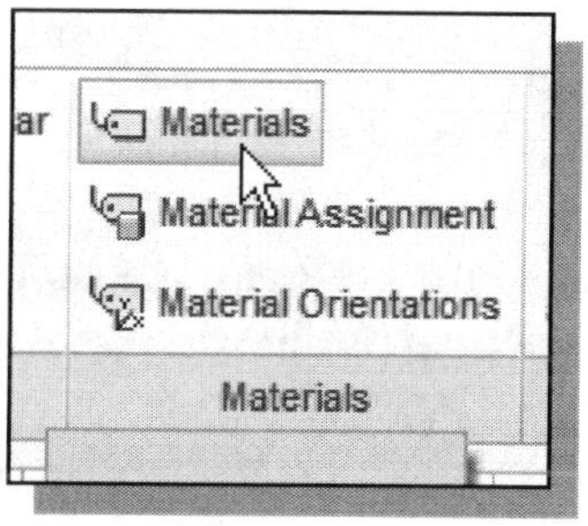

2. Choose **Materials** in the *Materials* group from the *Ribbon* toolbar as shown.

❖ Note the default list of materials libraries, which are available in the pre-defined *Creo Simulate*, is displayed.

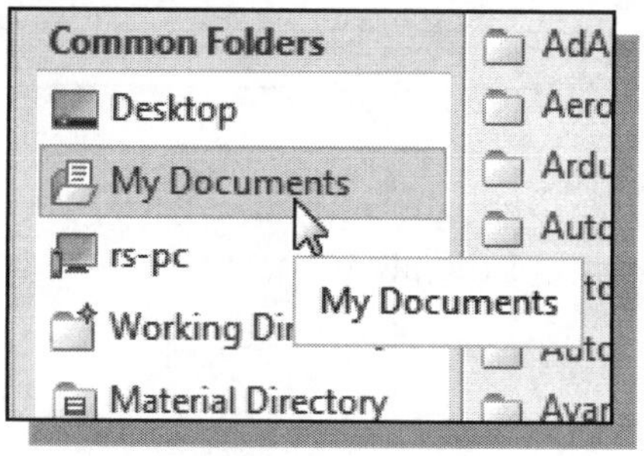

3. On your own, locate the folder where we saved the modified material definition, filename **Steel_low_Carbon_Modified**.

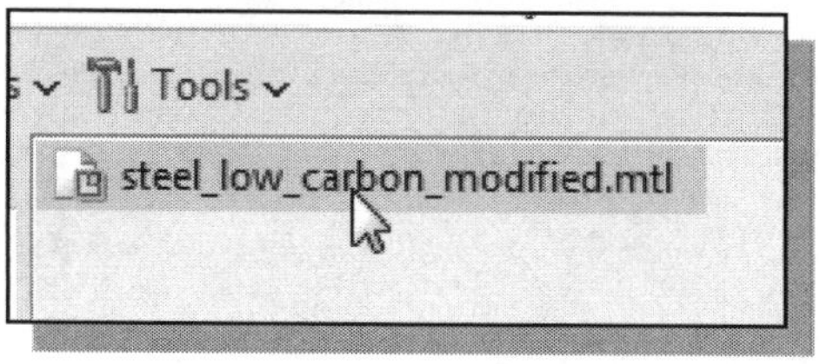

4. **Double-click** on the modified material definition for use with the current model.

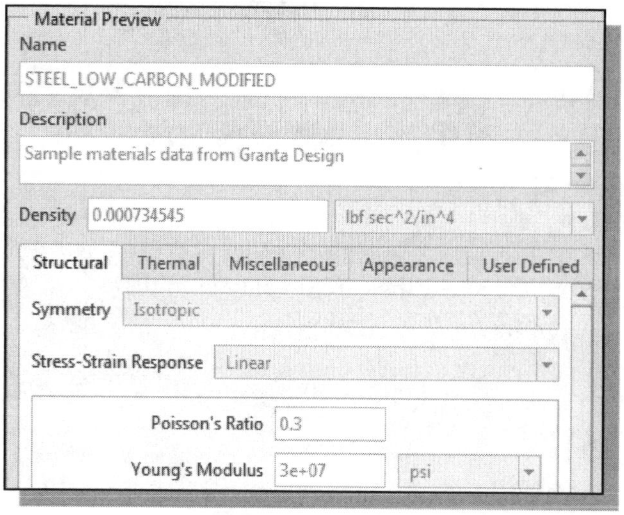

5. Click **Yes** to convert the units to match the model units.

6. Examine the material information in the **Material Preview** area and click **OK** to accept the settings.

Create 2D Truss Elements

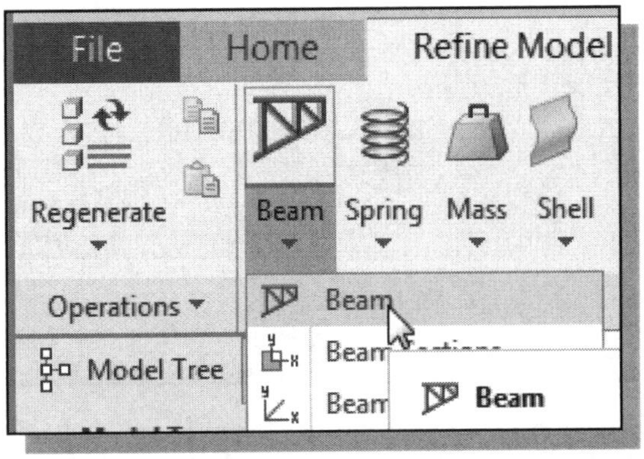

1. Choose **Beam** in the icon panel. The *Beam Definition* window appears.

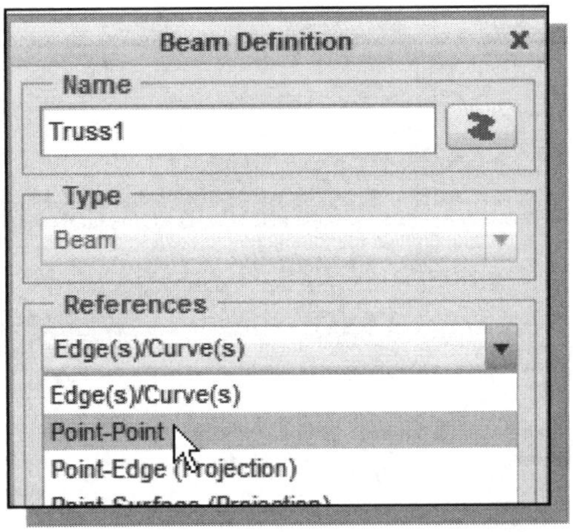

2. Enter **Truss1** as the **Name** of the first element to be created.

3. In the **References** option list, select **Point-Point** as shown.

4. Set the two **Point** options to **Single**.

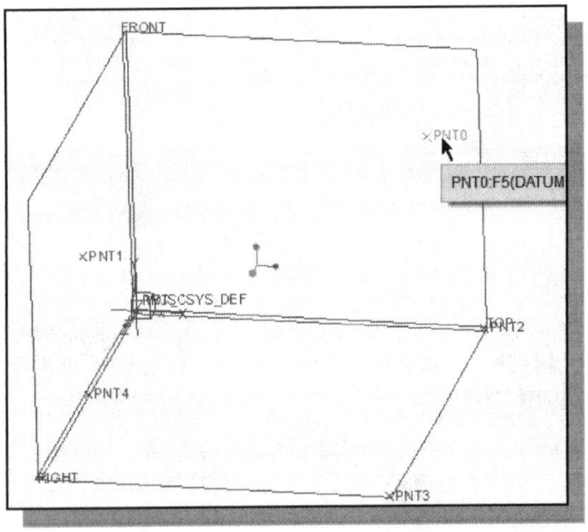

5. Click on the first datum point, **PNT0**, as the first point of the element.

6. Choose **PTN1** as the second point of the element.

❖ Note the selections are displayed in the *Beam Definition* window.

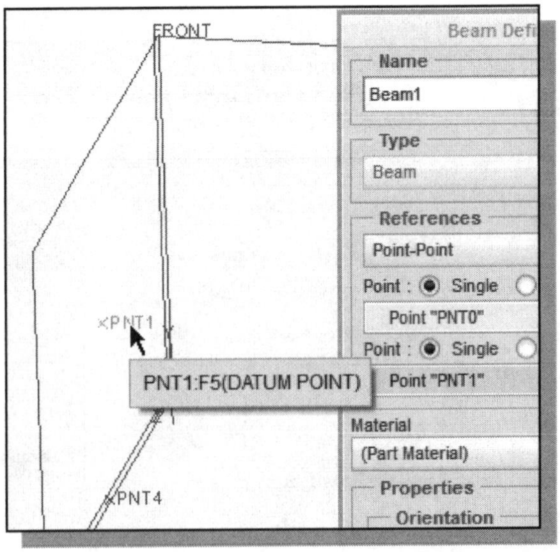

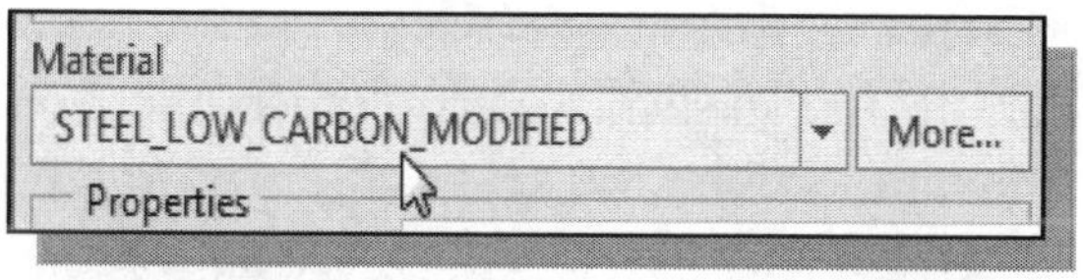

7. Select **Steel_low_Carbon_Modified** as the material of the element.

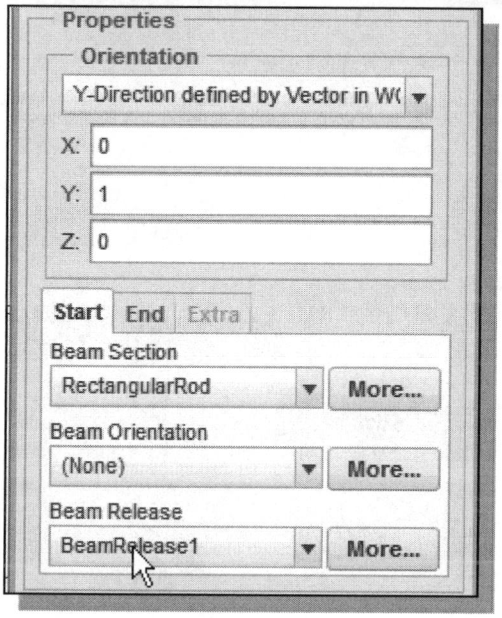

8. Confirm the *Y-Direction* is set to WCS with 0,1,0 as the X, Y and Z directions.

9. Set the Beam Release to **BeamRelease1** as shown.

10. Confirm the other two **Start** options are set: Beam Section to **RectangularRod** and Beam Orientation set to **(None)** as shown.

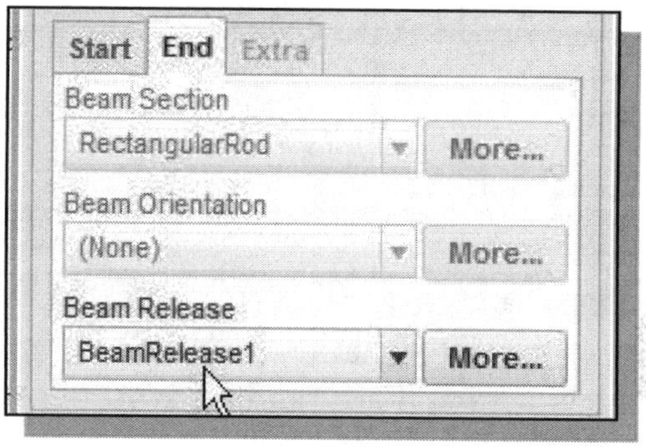

11. Click on the **End** tab and set the Beam Release to **BeamRelease1** as shown.

12. Click on the **OK** button to exit the *Fore Cross Section* window.

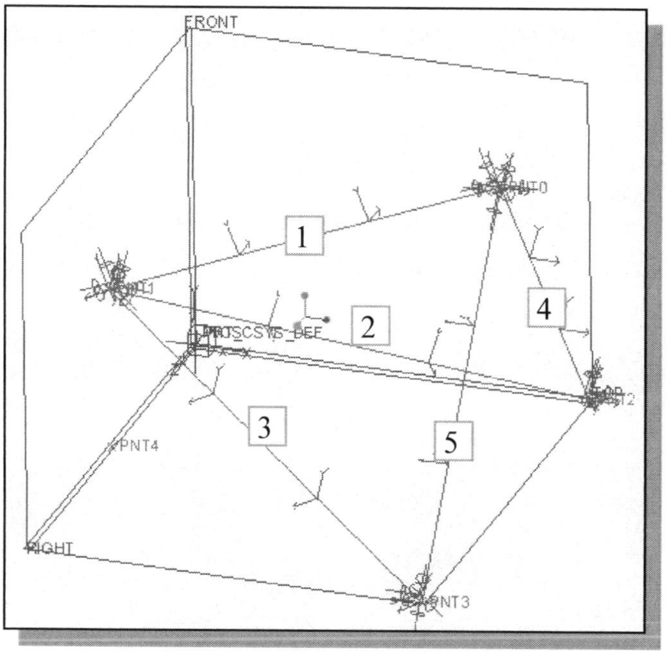

13. On your own, repeat the above procedure, step 1 through step 12, and create the five truss elements as shown.

❖ Be sure to set the **Material** and **Beam Release** options on all the elements correctly. Missing any one of these settings will prevent the *Solver* from finding the correct FEA results.

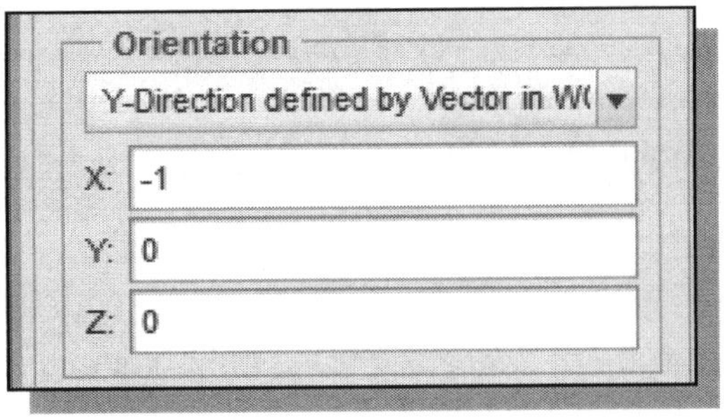

14. For the one vertical element, repeat the above procedure, but set the *Y-Direction* to use WCS with **-1,0,0** as the X, Y and Z components as shown.

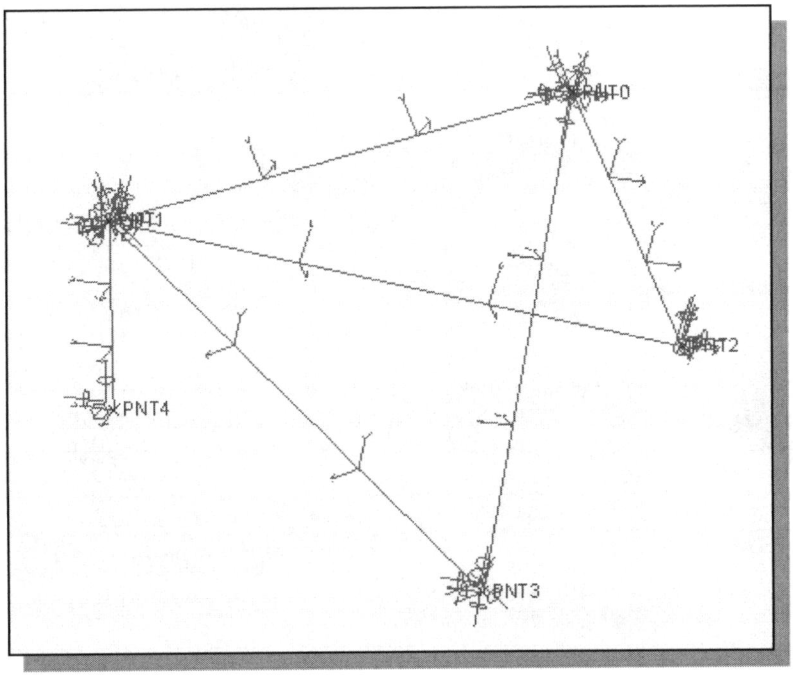

➢ Note that the local X direction of a truss member is always aligned to the member direction, and the Y direction is set at 90° from the X direction.

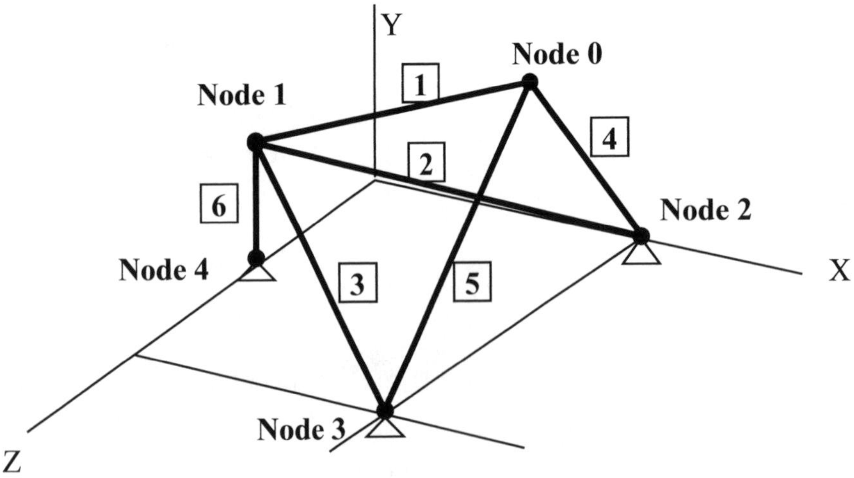

Apply Boundary Conditions – Constraints and Loads

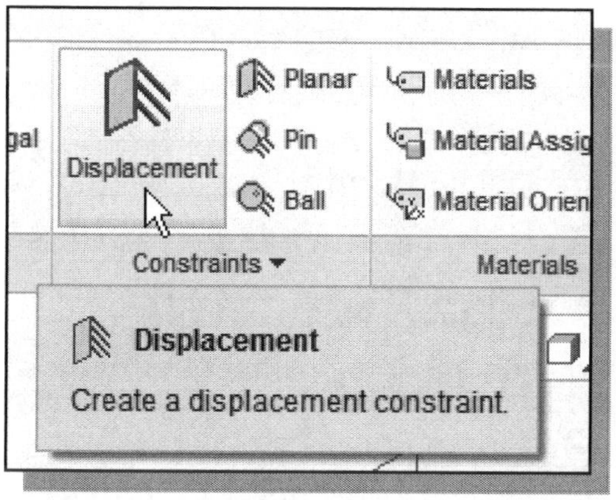

1. Choose **Displacement Constraint** by clicking the icon in the toolbar as shown.

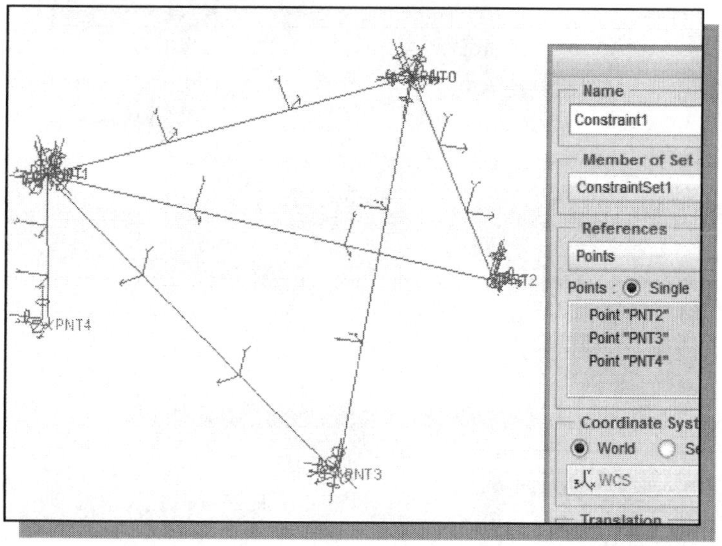

2. Set the References to **Points** and select **PNT2**, **PNT3** and **PNT4** by holding down the [Ctrl] key.

❖ PNT2, PNT3 and PNT4 are fixed support points which cannot move in the X, Y or Z directions. For 3D truss systems, all joints are *ball-joints*; therefore, all rotational movements are free.

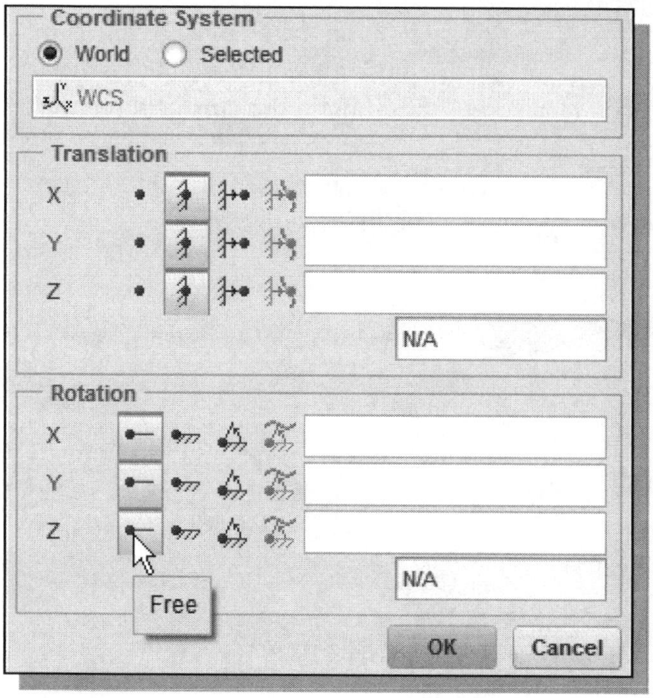

3. Set all Translations to **Fixed** as shown.

4. Set all Rotations to **Free** as shown.

5. Click on the **OK** button to accept the first Displacement constraint settings.

Apply External Loads

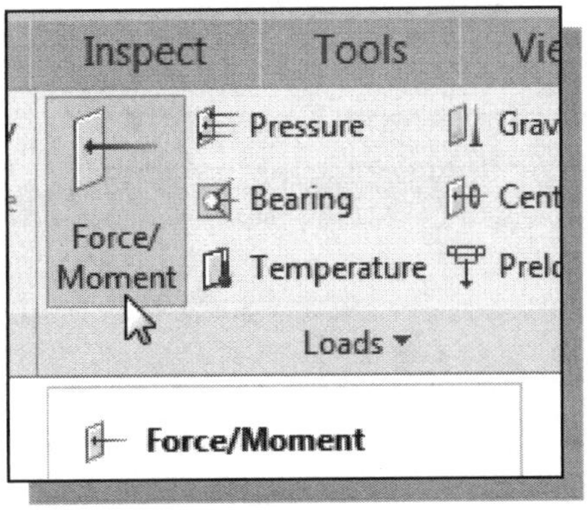

1. Choose **Force/Moment Load** by clicking the icon in the *Loads toolbar* as shown.

2. Set the References to **Points**.

3. Pick the upper right node, **PNT0**, of the truss system as shown in the figure below.

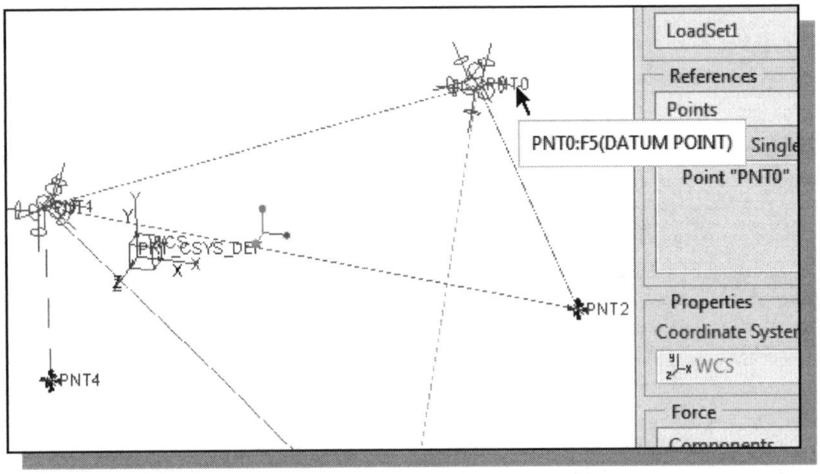

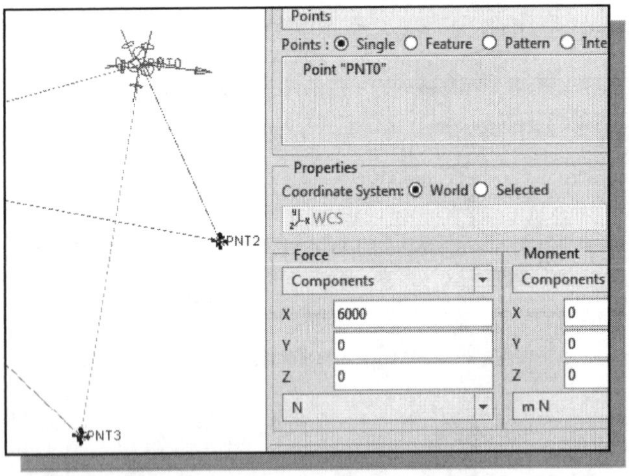

4. Enter **6000** in the X component Force value box as shown.

5. Click on the **Preview** button to confirm the applied load is pointing in the correct direction.

6. Click on the **OK** button to accept the first load settings.

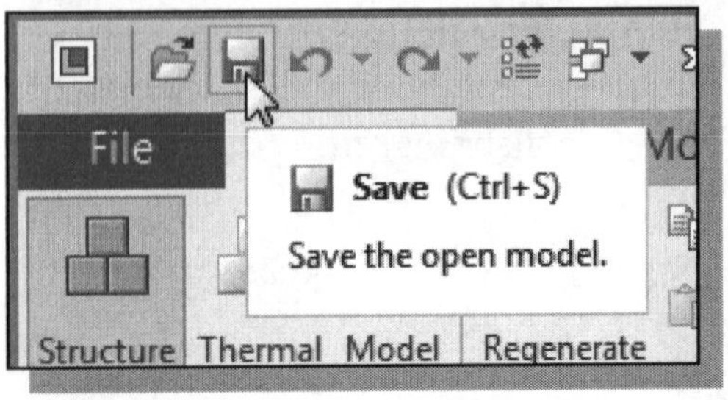

7. On your own, click **Save** to save a copy of the current FEA model.

Run the FEA Solver

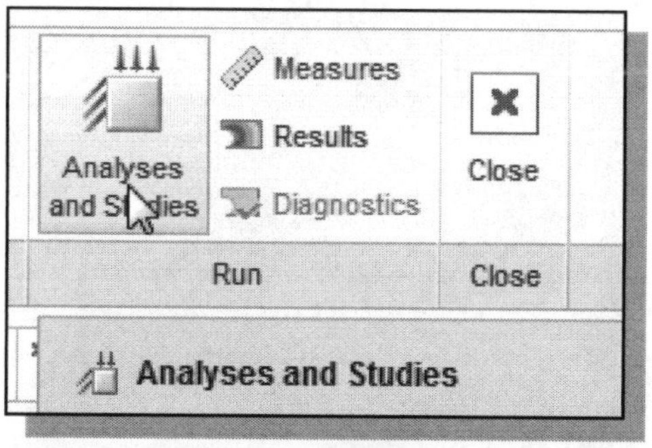

1. Activate the **Analyses and Studies** command by selecting the icon in the toolbar area as shown.

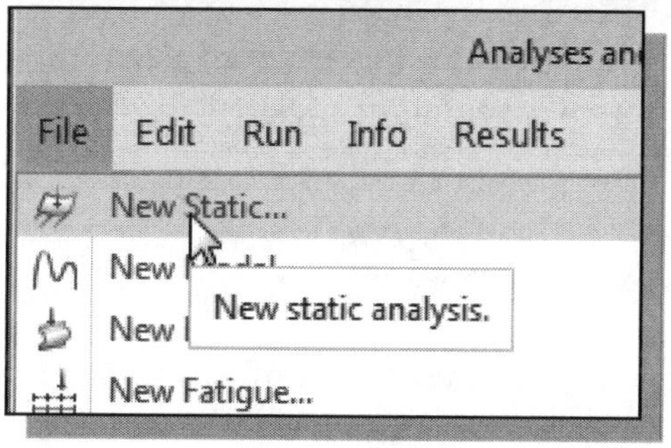

2. In the *Analyses and Design Studies* window, choose **File → New Static** in the pull-down menu. This will start the setup of a basic linear static analysis.

3. Note that ConstraintSet1 and LoadSet1 are automatically accepted as part of the analysis parameters.

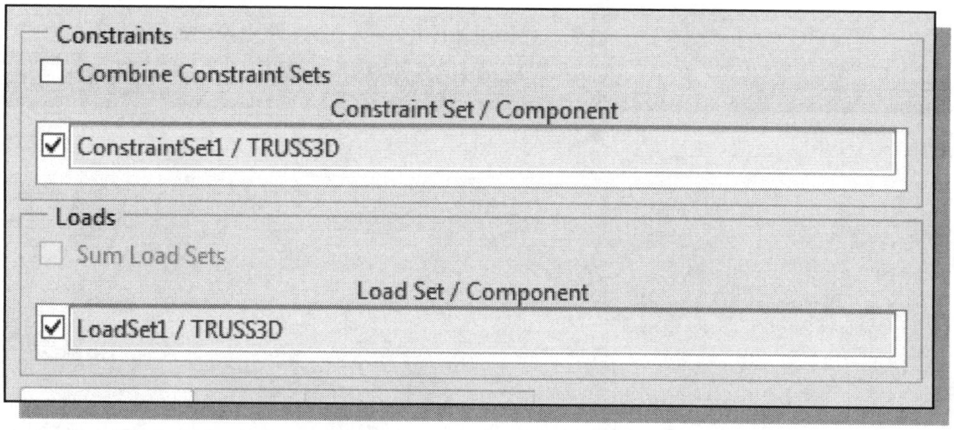

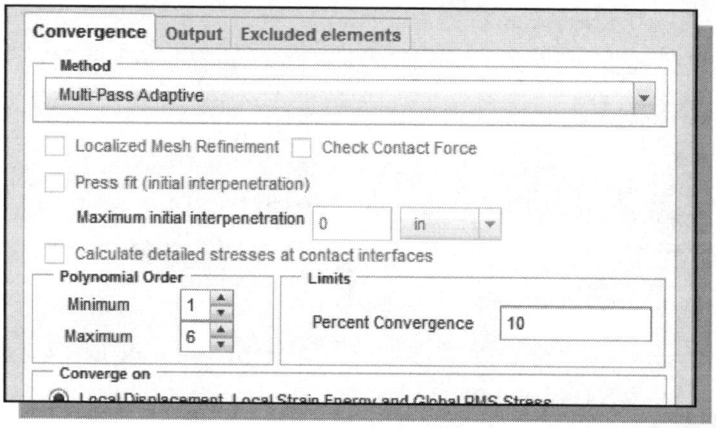

4. Set the Method option to **Multi-Pass Adaptive** and Limits to **10** Percent Convergence as shown.

5. Click **OK** to accept the settings.

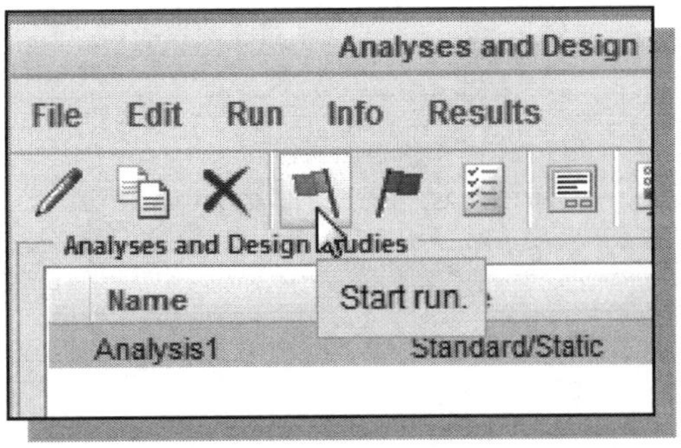

6. Click **Start Run** to begin the *Solving* process.

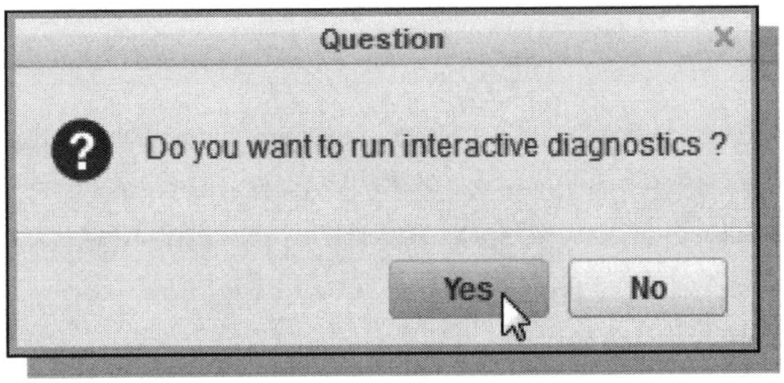

7. Click **Yes** to run the *Creo Simulate* interactive diagnostics.

8. In the *Diagnostics* window, *Creo Simulate* will list any problems or conflicts in our FEA model.

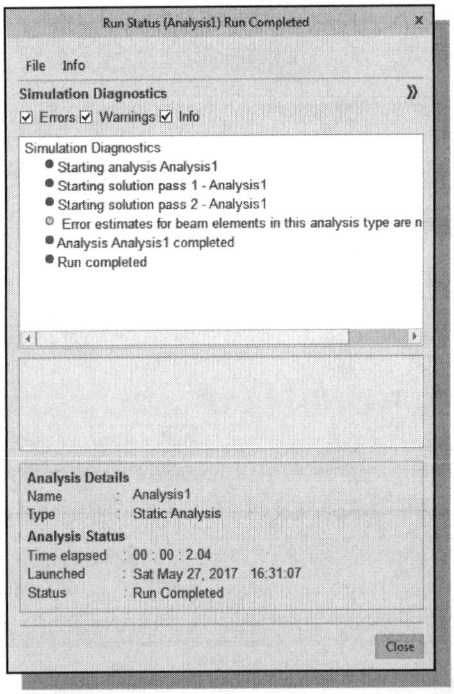

9. Click on the **Close** button to exit the *Diagnostics* window.

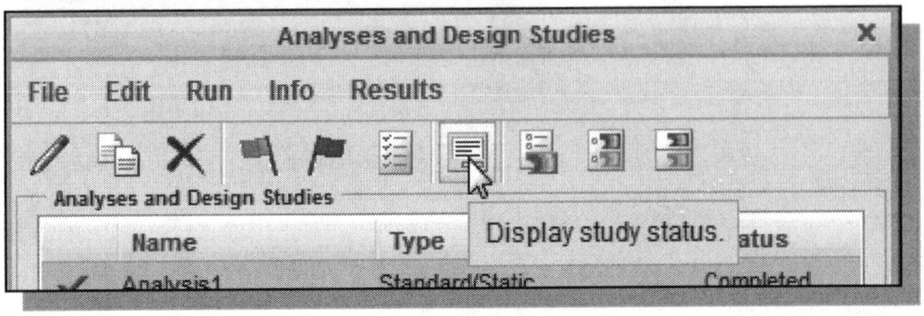

10. In the *Analyses and Design Studies* window, choose **Display study status** as shown.

11. Information related to the *Finite Element Analysis* is displayed in the *Run Status* window.

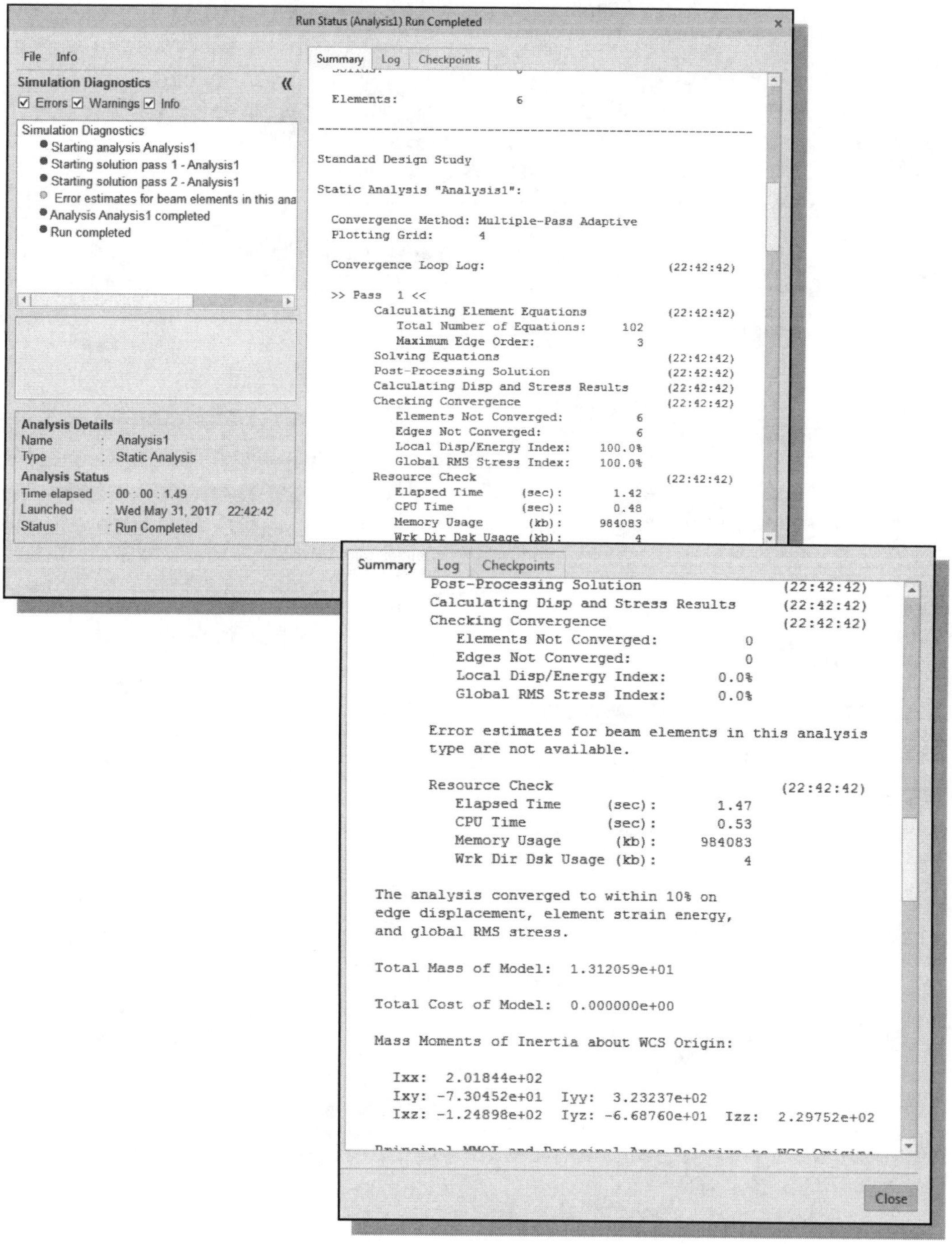

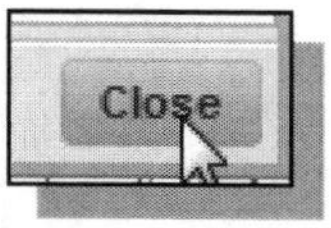

12. Click **Close** to exit the *Run Status* window.

View the FEA Results

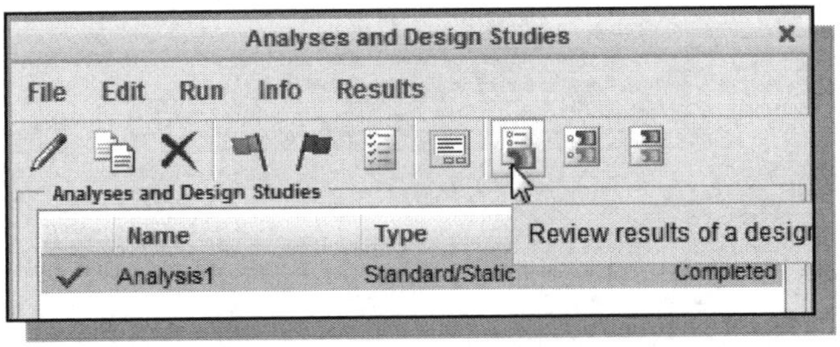

1. In the *Analyses and Design Studies* window, choose **Review results** as shown.

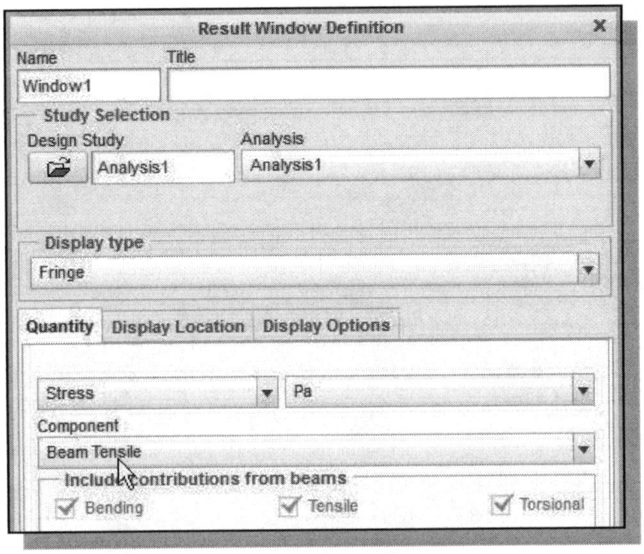

2. Confirm the Display type is set to **Fringe**.
3. Choose **Beam Tensile** as the Stress Component to be displayed.
4. Click **OK and Show** to view the results.

❖ The *Creo Simulate* **Beam Tensile** option allows us to quickly view the tensile stress in beam and truss elements.

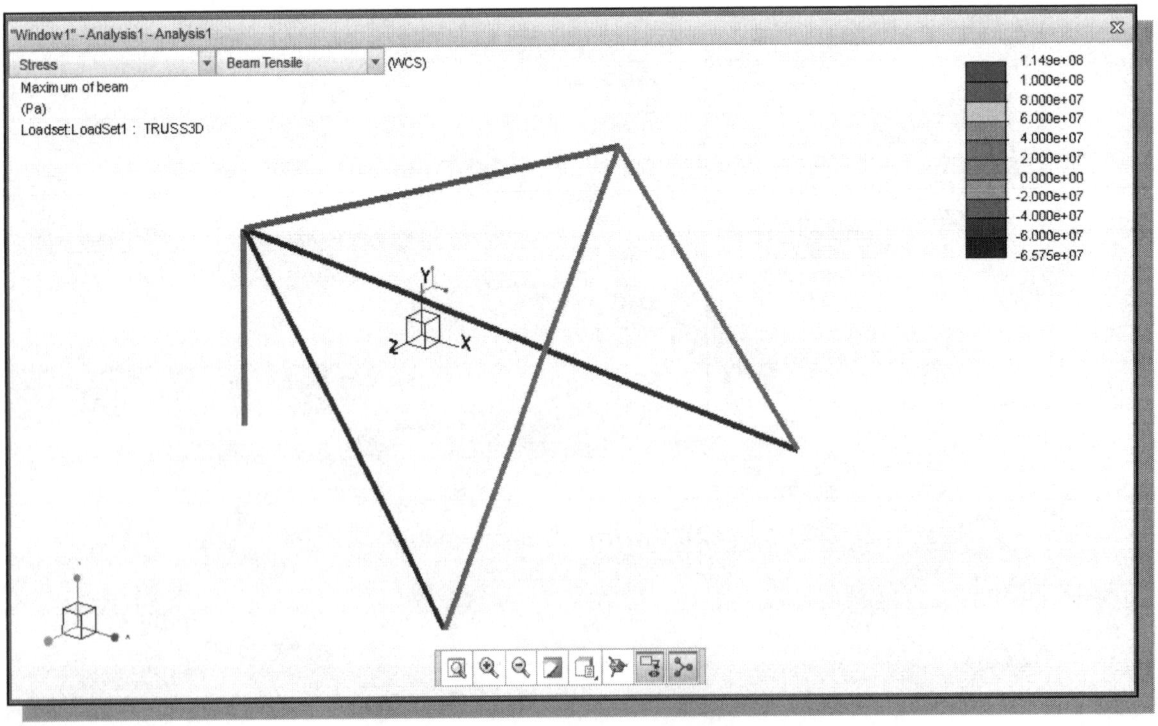

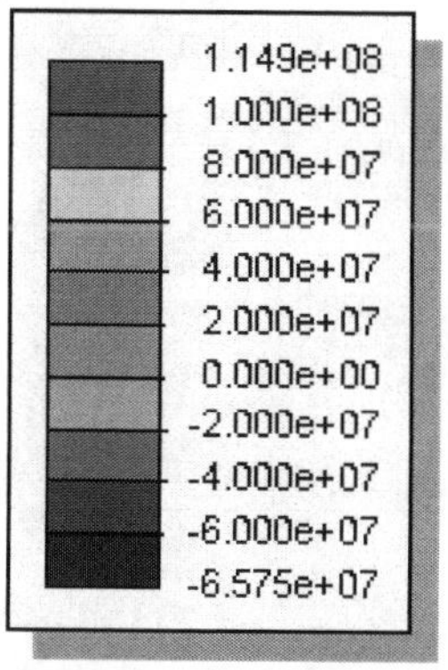

❖ Note the maximum tensile stress is **114.9 MPa** and the minimum tensile stress is **-65.7 MPa**. Are these values matching the preliminary analysis performed?

➢ Note that it is a little tough to read the stress values by looking at the color codes. We will next use the **Dynamic Query** command to get a better reading of the stress of each member.

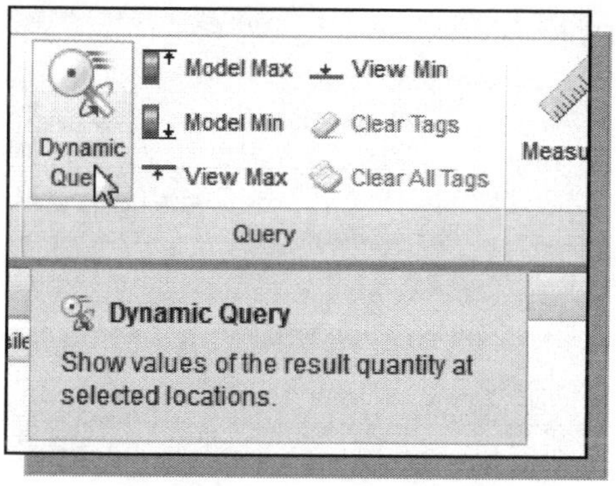

5. Choose **Dynamic Query** in the Query toolbar. The *Query* window appears on the screen.

6. Move the cursor on **Element 1** and read the stress value inside the *Query* window. The negative value implies the member is in compression.

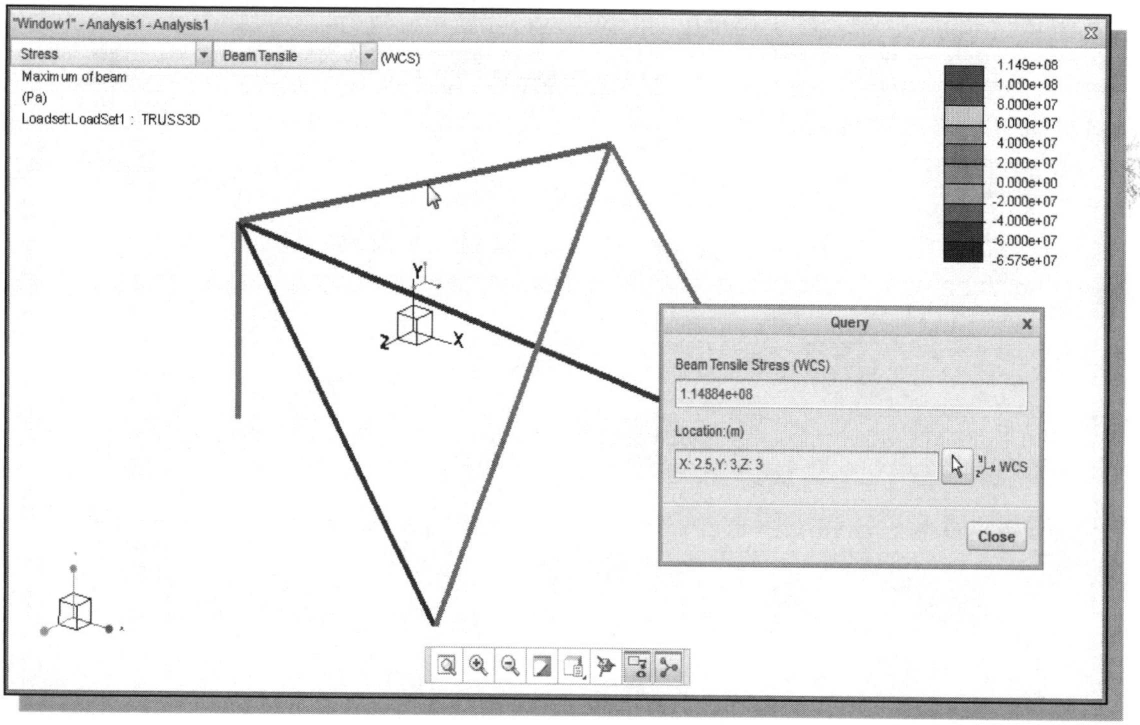

7. Click on the member, **Element 1**, to display the stress value of the selected location on the member.

8. Move the cursor on **Element 3** and read the stress value inside the *Query* window. This member has the highest stress in the system.

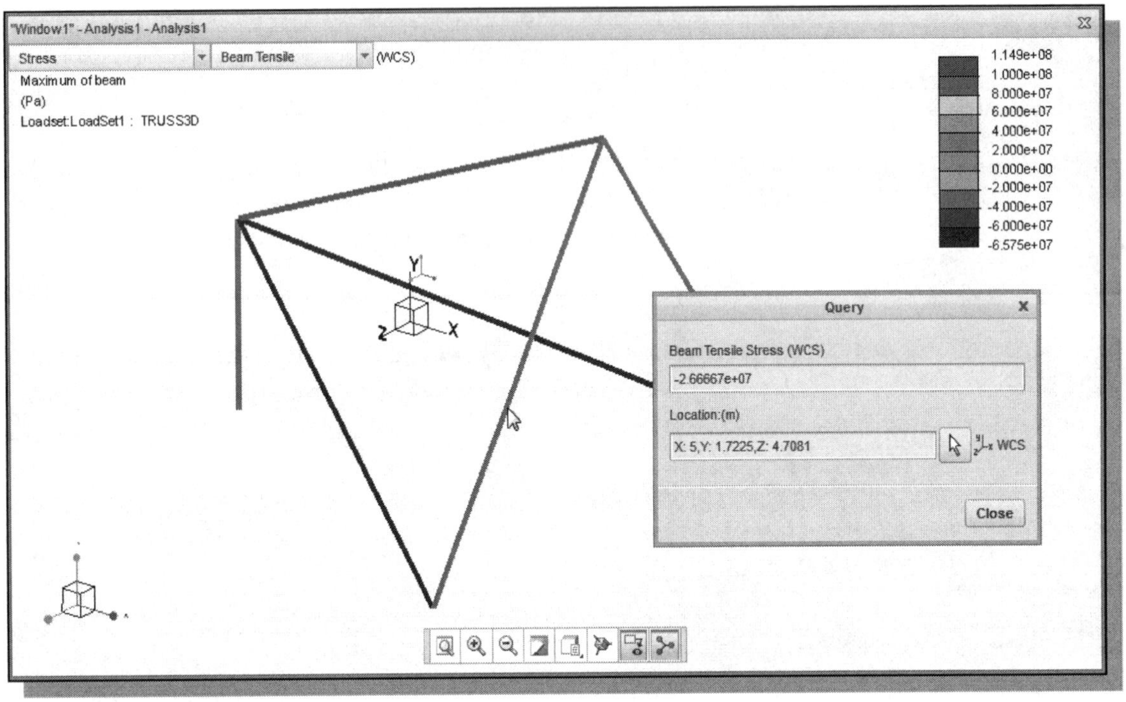

➢ On your own, examine the stresses developed in the other elements and compare them to the preliminary analysis we did at the beginning of this chapter.

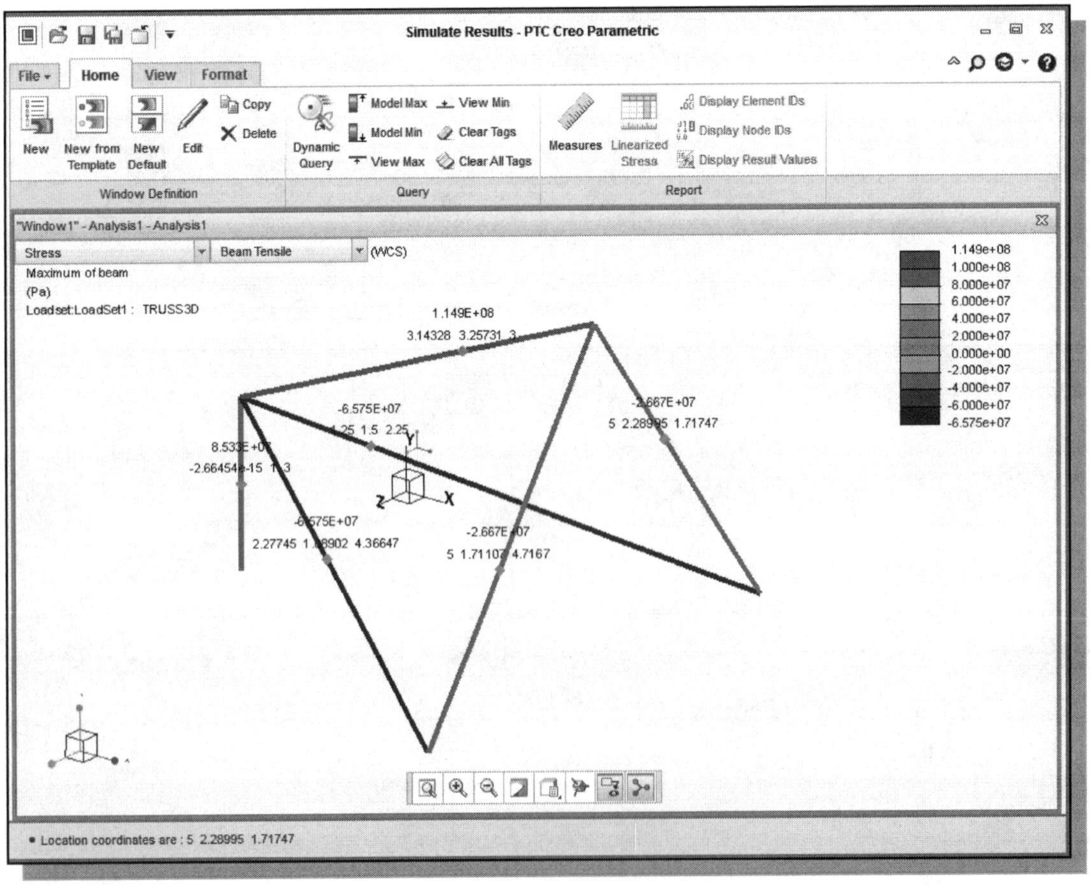

Review Questions

1. For a truss element in three-dimensional space, what is the number of degrees of freedom for the element?

2. How does the number of degrees of freedom affect the size of the stiffness matrix?

3. How do we create 3D points in *Creo Simulate*?

4. In *Creo Simulate*, can we import a set of 3D point data that are stored in a regular text file?

5. Will the *Creo Simulate* FEA software calculate the internal forces of the members?

6. Describe the procedure in setting up the vector analysis of a 3D truss problem.

7. Besides using the vector analysis procedure, can we solve the example problem by the 2D approach described in Chapter 4?

8. Will *Creo Simulate* display the orientation of cross sections of truss elements correctly on the screen?

9. Truss members are also known as two-force members. What does that mean?

10. Can we use several different material properties for truss members in a truss system?

Exercises

Determine the normal stress in each member of the truss structures.

1. All joints are ball-joints.
   Material: Steel

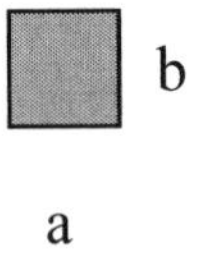

a = b = 2 cm

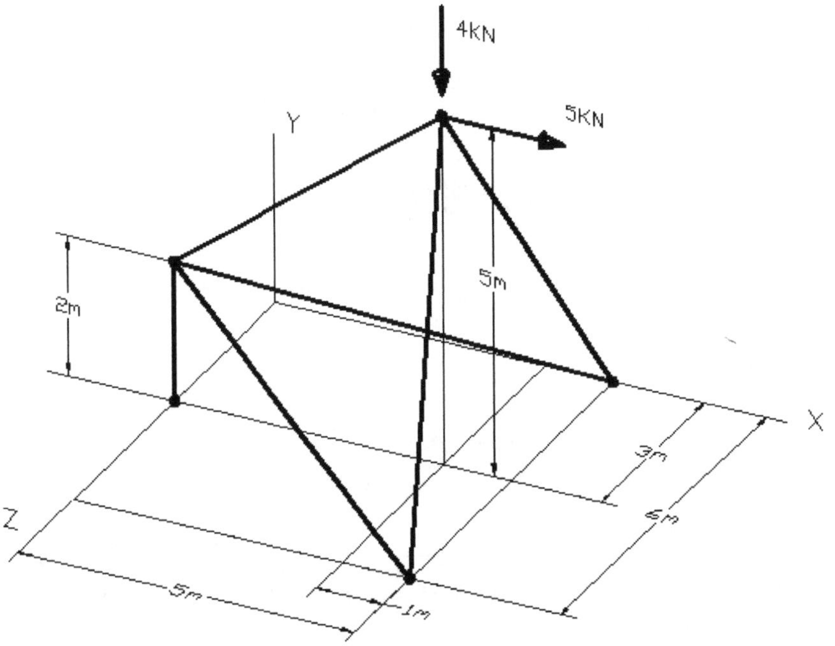

2. All joints are ball-joints, and joints D, F, F are fixed to the floor.
   Material: Steel
   Diameter: 1 in.

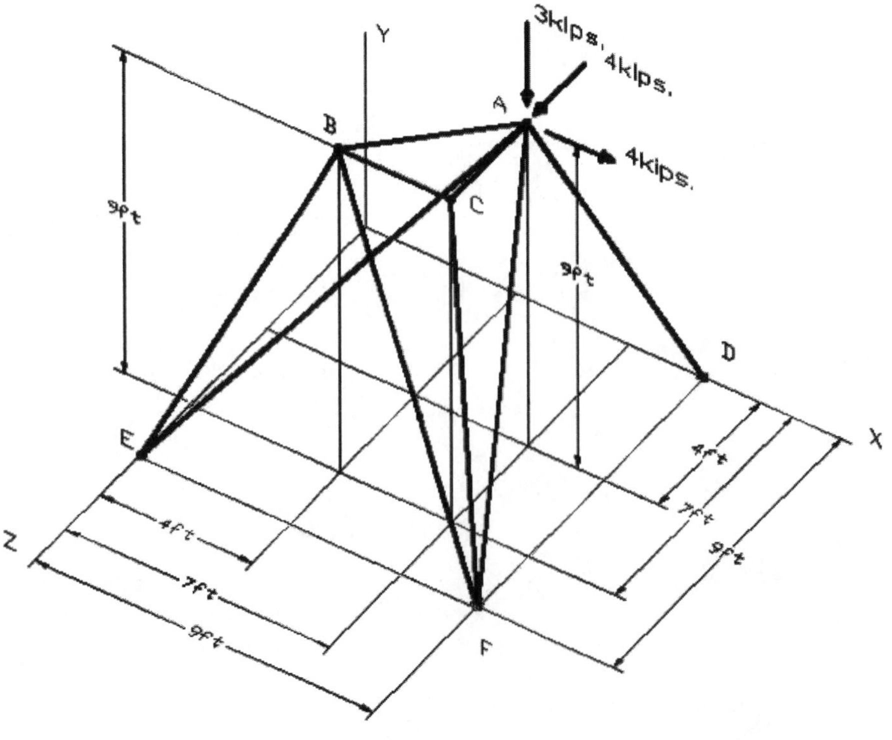

Chapter 6
Basic Beam Analysis

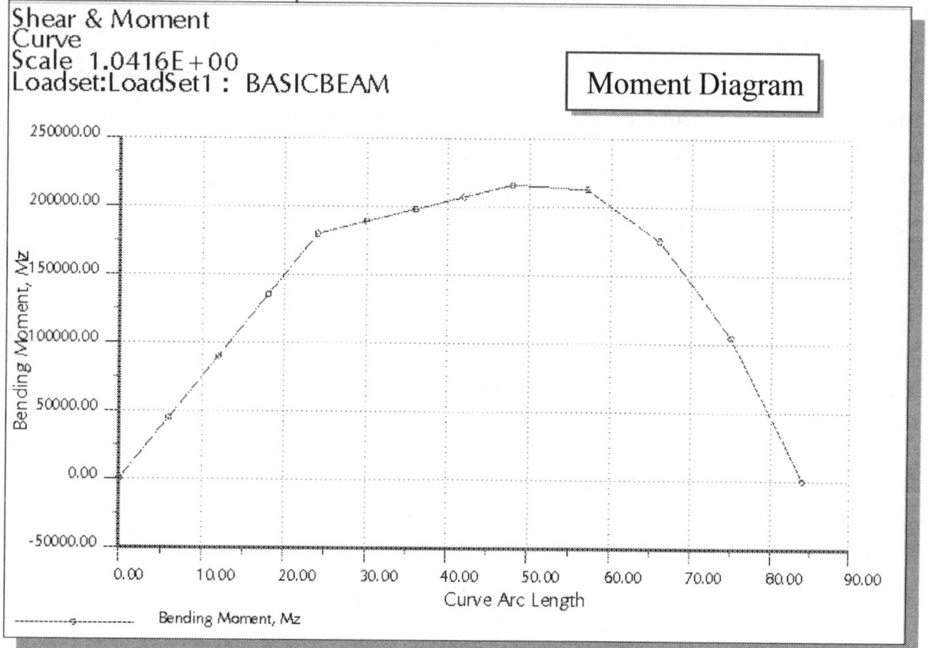

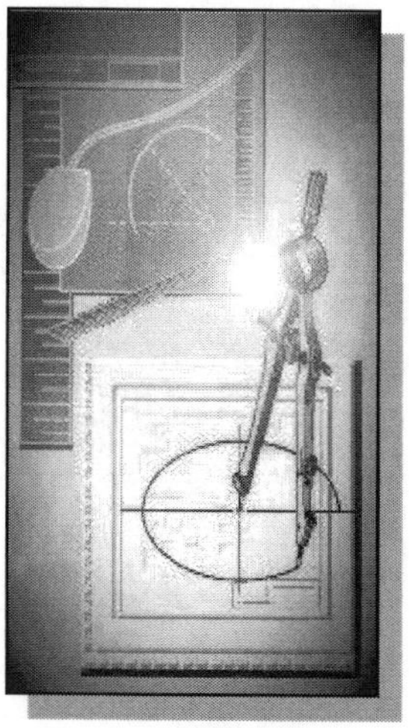

Learning Objectives

- Understand the basic assumptions for Beam elements.
- Apply Constraints and Forces on FE Beam Models.
- Apply Distributed Loads on Beams.
- Create and Display a Shear Diagram.
- Create and Display a Moment Diagram.
- Perform Basic Beam Analysis using Creo Simulate.

Introduction

The truss element discussed in the previous chapters does have a practical application in structural analysis, but it is a very limiting element since it can only transmit axial loads. The second type of finite element to study in this text is the *beam element.* Beams are used extensively in engineering. As the members are rigidly connected, the members at a joint transmit not only axial loads but also bending and shear. This contrasts with truss elements where all loads are transmitted by axial force only. Furthermore, beams are often designed to carry loads both at the joints and along the lengths of the members, whereas truss elements can only carry loads that are applied at the joints. A beam element is a long, slender member generally subjected to transverse loading that produces significant bending effects as opposed to axial or twisting effects.

Modeling Considerations

Consider a simple beam element positioned in a two-dimensional space:

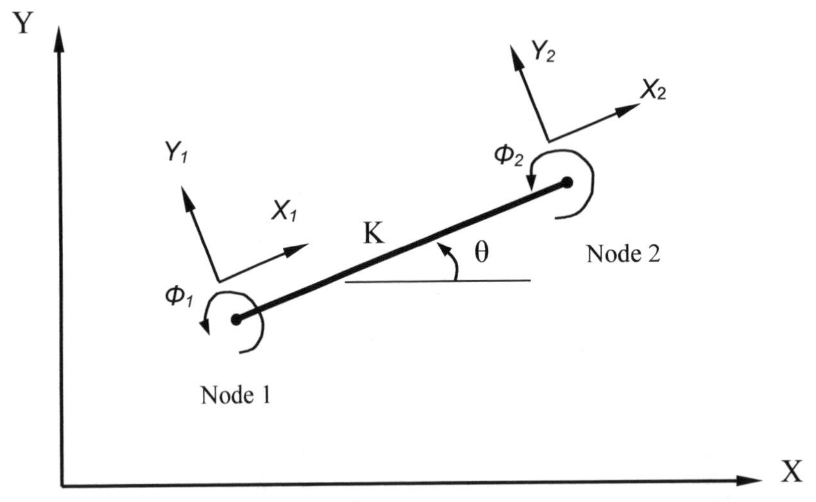

A beam element positioned in two-dimensional space is typically modeled to possess three nodal displacements (two translational and one rotational) at each node: three degrees of freedom at each node. Each element therefore possesses six degrees of freedom. A finite element analysis using beam elements typically provides a solution to the displacements, reaction forces, and moments at each node. The formulation of the beam element is based on the elastic beam theory, which implies the beam element is initially straight, linearly elastic, and loads (forces and moments) are applied at the ends. Therefore, in modeling considerations, place nodes at all locations that concentrated forces and moments are applied. For a distributed load, most finite element procedures replace the distributed load with an equivalent load set, which is applied to the nodes available along the beam span. Accordingly, in modeling considerations, place more nodes along the beam spans with distributed loads to lessen the errors.

Problem Statement

Determine the maximum normal stress that loading produces in the steel member.

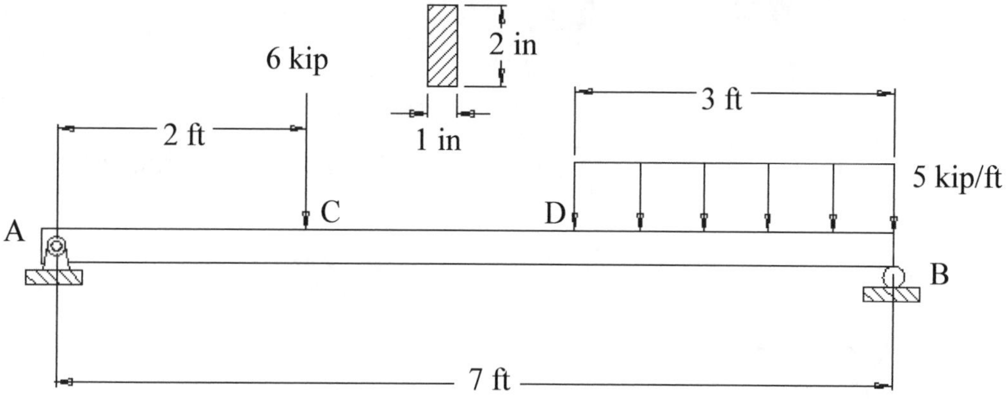

Preliminary Analysis

Free Body Diagram of the member:

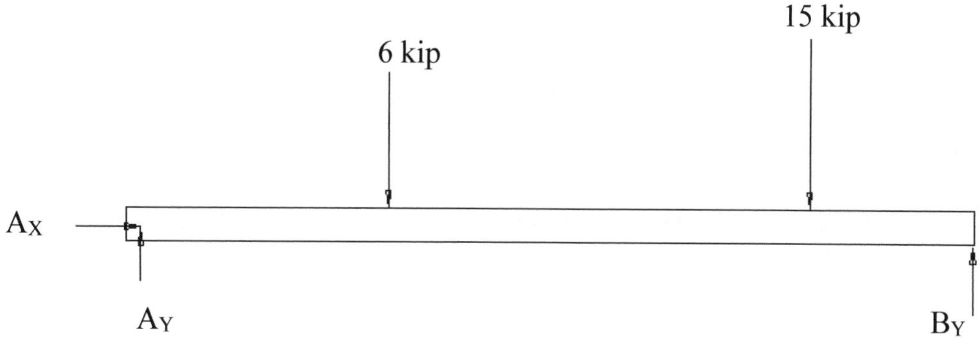

Applying the equations of equilibrium:

$$\sum M_{@A} = 0 = B_Y \times 7 - 6 \times 2 - 15 \times 5.5$$
$$\sum F_X = 0 = A_X$$
$$\sum F_Y = 0 = A_Y + B_Y - 6 - 15$$

Therefore,

$$B_Y = 13.5 \text{ kip}, A_X = 0 \text{ and } A_Y = 7.5 \text{ kip}$$

Next, construct the shear and moment diagrams for the beam. Although this is not necessary for most FEA analyses, we will use this example to build up our confidence with *Creo Simulate's* analysis results.

- Between A and C, 0 ft. $< X_1 <$ 2 ft.:

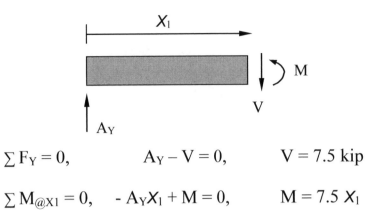

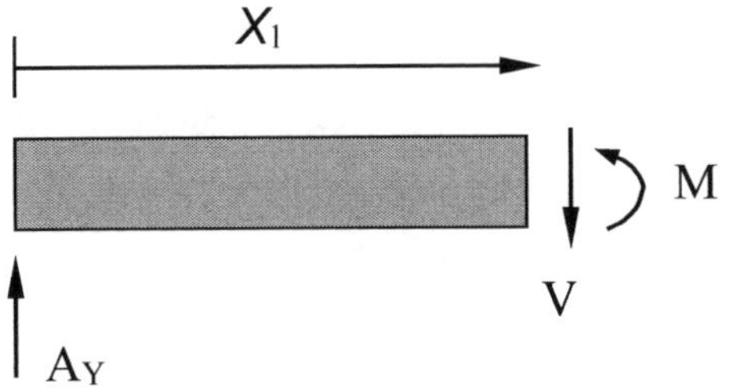

$\sum F_Y = 0, \quad A_Y - V = 0, \quad V = 7.5$ kip

$\sum M_{@X1} = 0, \quad -A_Y X_1 + M = 0, \quad M = 7.5\,X_1$

- Between A and D, 2 ft. $< X_2 <$ 4 ft.:

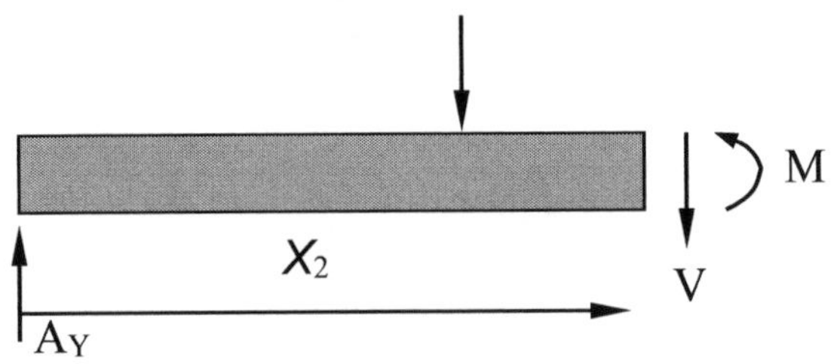

$\sum F_Y = 0, \quad A_Y - V - 6 = 0, \quad V = 1.5$ kip

$\sum M_{@X2} = 0, \quad -A_Y X_2 + M + 6\,(X_2 - 2) = 0, \quad M = 1.5\,X_2 + 12$

- Between A and B, 4 ft. $< X_3 <$ 7 ft.:

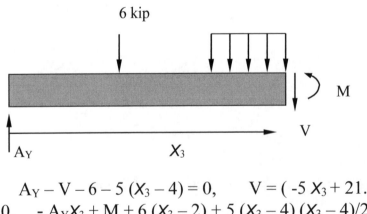

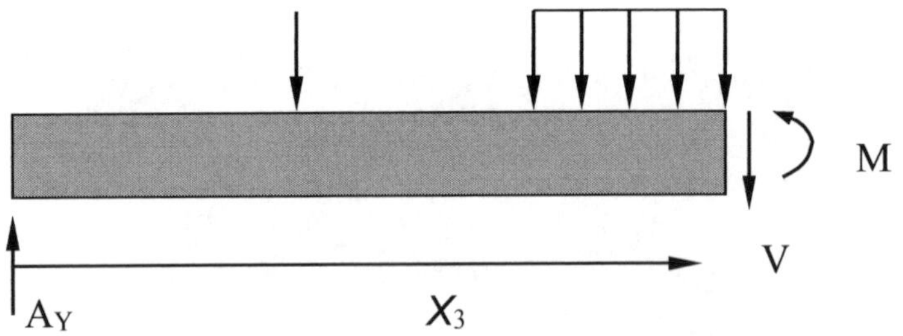

$\sum F_Y = 0, \quad A_Y - V - 6 - 5\,(X_3 - 4) = 0, \quad V = (-5\,X_3 + 21.5)$ kip

$\sum M_{@X3} = 0, \quad -A_Y X_3 + M + 6\,(X_3 - 2) + 5\,(X_3 - 4)\,(X_3 - 4)/2 = 0$

$$M = -2.5\,X_3^2 + 21.5\,X_3 - 28$$

Shear and Moment diagrams:

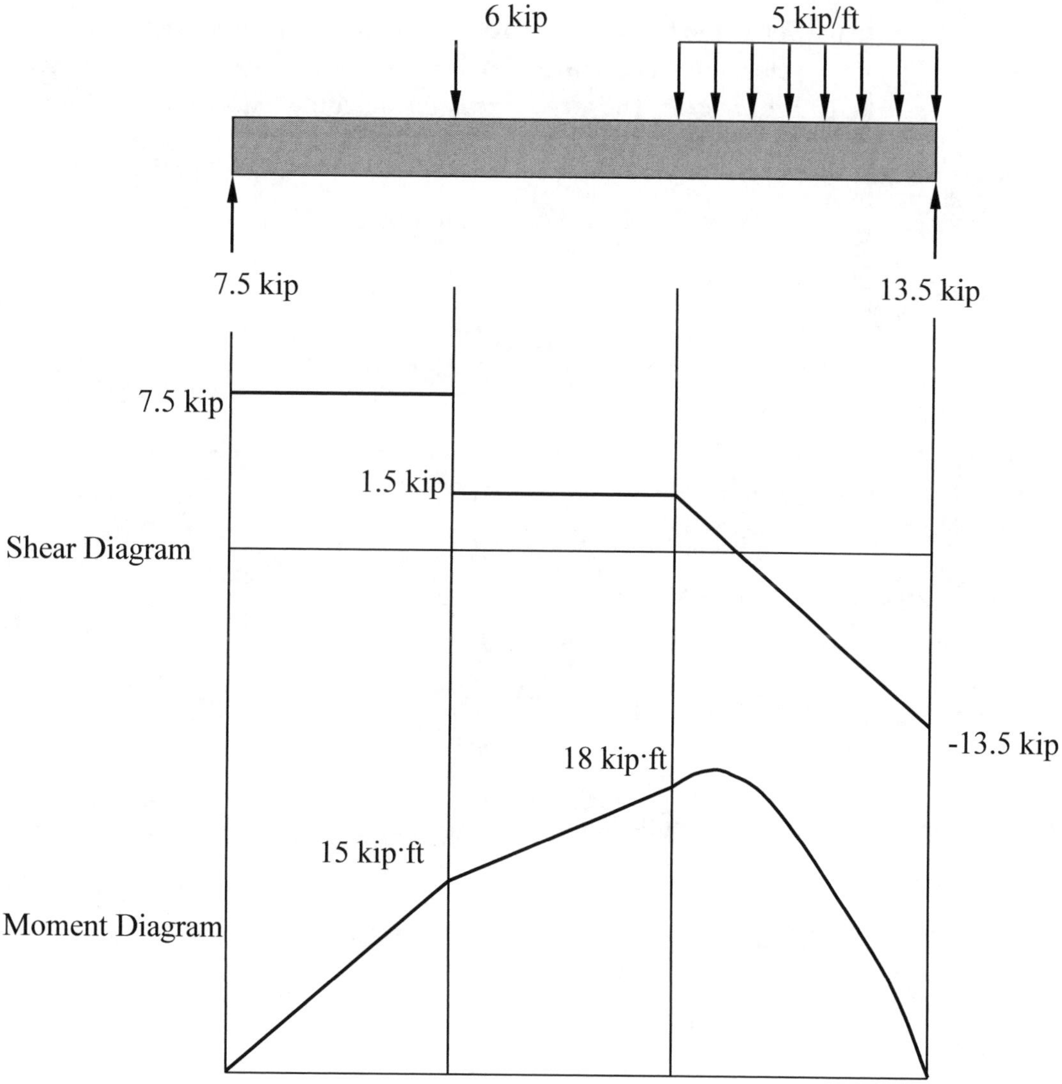

Determine the maximum normal stress developed in the beam:

$$V = -5X_3 + 21.5 = 0, \quad X_3 = 4.3 \text{ ft.}$$

$$M = -2.5X_3^2 + 21.5X_3 - 28 = 18.225 \text{ kip-ft} = 218700 \text{ lb-in}$$

Therefore,

$$\sigma_{max} = \frac{MC}{I} = 18.255\,(h/2)/(bh^3/12) = 47239 \text{ kip/ft}^2 = 3.28E{+}5 \text{ psi}$$

Starting Creo Parametric

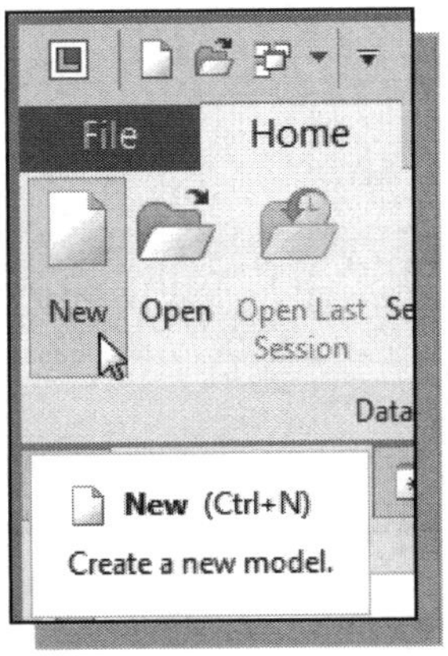

1. Select the **Creo Parametric** option on the *Start* menu or select the **Creo Parametric** icon on the desktop to start *Creo Parametric*. The *Creo Parametric* main window will appear on the screen.

2. Click on the **New** icon, located in the *Ribbon toolbar* as shown.

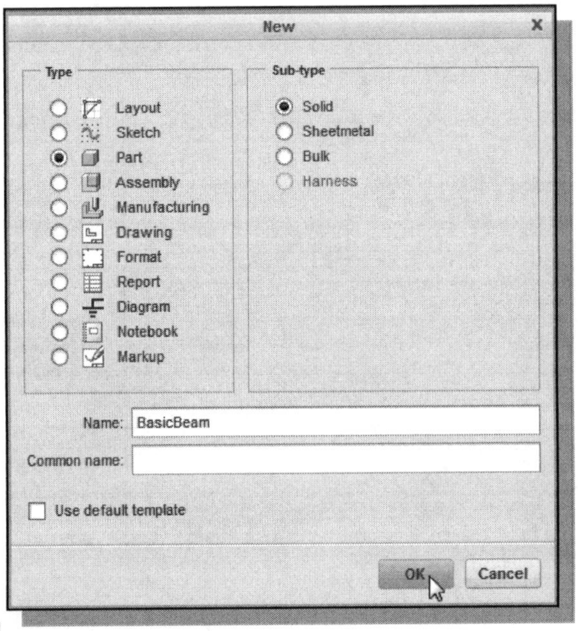

3. In the *New* dialog box, confirm the model's Type is set to **Part** (**Solid** Sub-type).

4. Enter **BasicBeam** as the part Name as shown in the figure.

5. Turn ***off*** the **Use default template** option.

6. Click on the **OK** button to accept the settings.

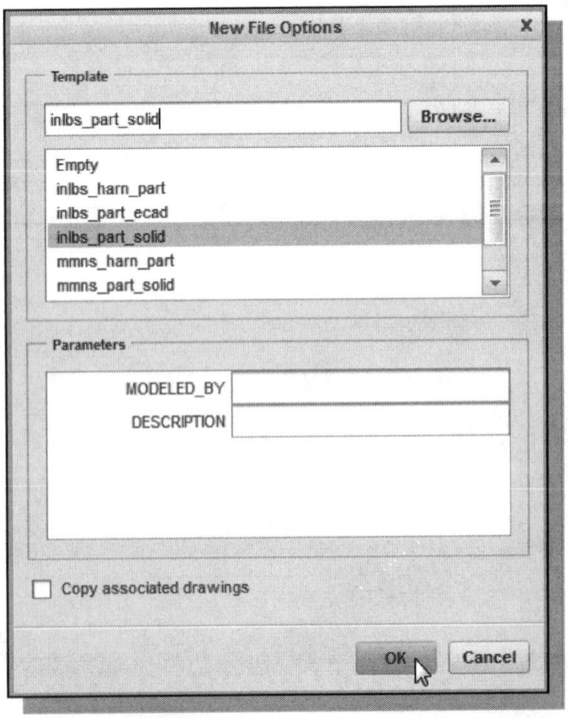

7. In the *New File Options* dialog box, select **inlbs_part_solid** in the option list as shown.

8. Click on the **OK** button to accept the settings and enter the *Creo Parametric Part Modeling* mode.

❖ Note that the **inlbs_part_solid** template contains several pre-defined settings, such as the units setup, as well as datum planes and views setup.

New Template Setup

The selected template, **inlbs_part_solid**, is a template using the *Inch-lbm-Second* units set. Although we can use this setup for our basic beam example, it is just as easy to adjust the settings and create a new template file.

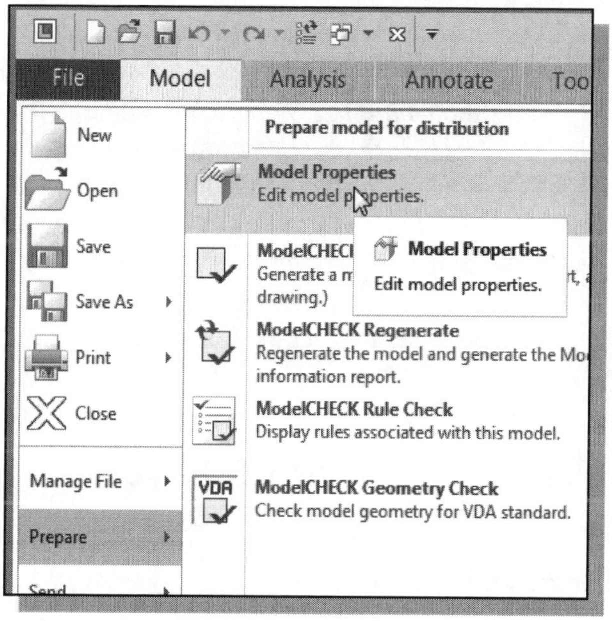

1. Use the left-mouse-button and select **File** in the pull-down menu area.

2. Use the left-mouse-button and select **Prepare** in the pull-down list as shown.

3. Select **Model Properties** in the expanded list as shown.

➢ Note that the *Creo Parametric* menu system is context-sensitive, which means that the menu items and icons of the non-applicable options are grayed out (temporarily disabled).

4. Select the **Change** option that is to the right of the **Units** option in the *Model Properties* window.

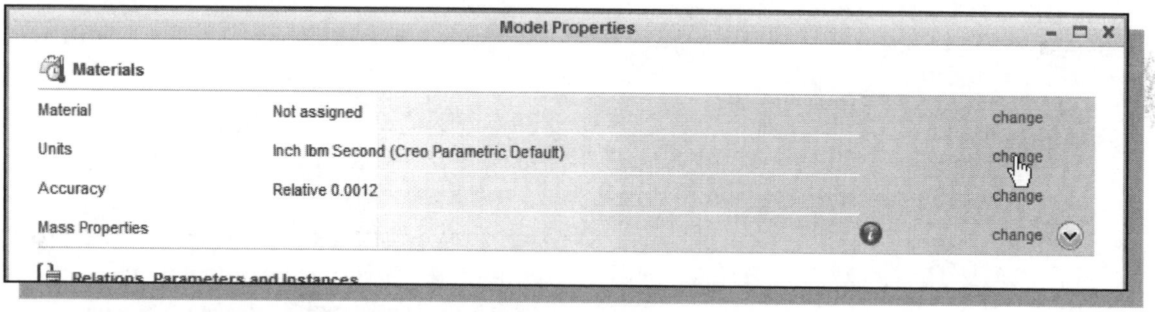

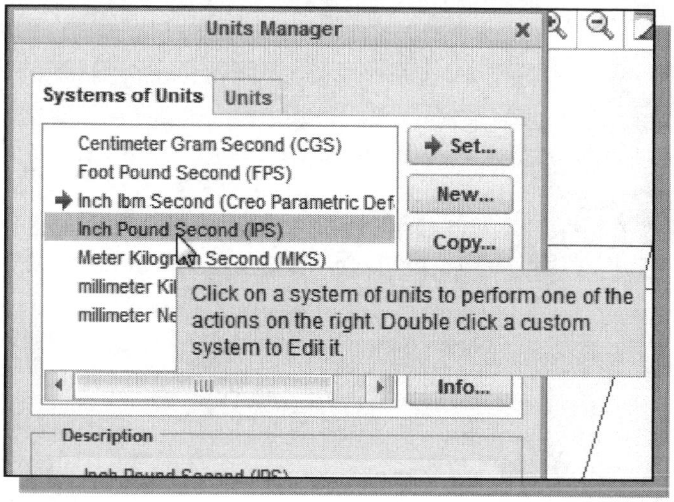

5. In the *Units Manager* window, note the different units sets that are available. Click **Inch Pound Second** in the **Systems of Units** to create a new units list as shown.

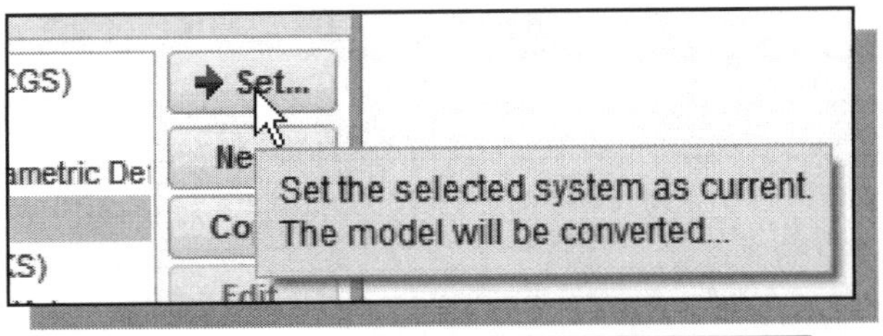

6. Click **Set** to set the **Inch Pound Second** units as the current System of Units set.

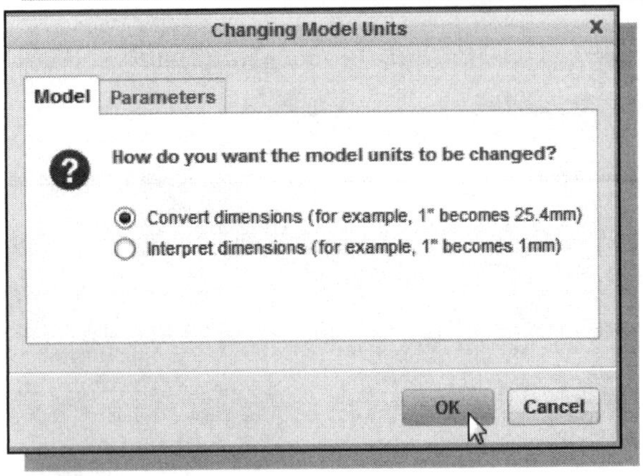

7. In the *Changing Model Units* dialog box, click on the **OK** button to accept the default option to change the units.

➢ Note that *Creo Parametric* allows us to change model units even after the model has been constructed.

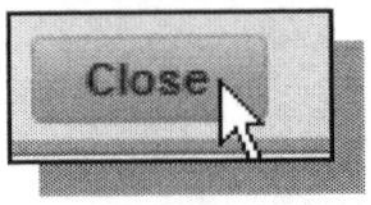

8. Click on the **Close** button to exit the *Units Manager* dialog box.

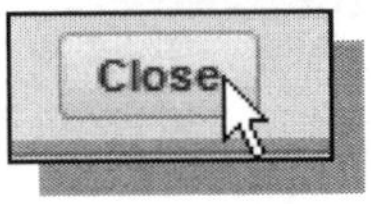

9. Pick **Close** to exit the *Model Properties* window.

Set up an Isometric View

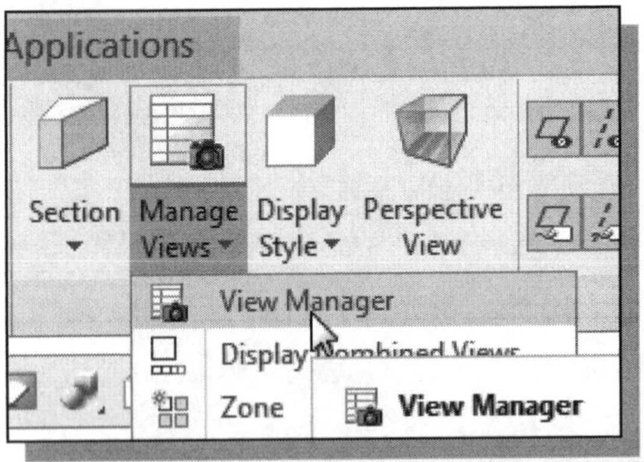

1. Select the **View** tab in the *Ribbon* toolbar. Pick the **View Manager** option as shown.

❖ The View Manager option can be used to set up and adjust various display and viewing settings.

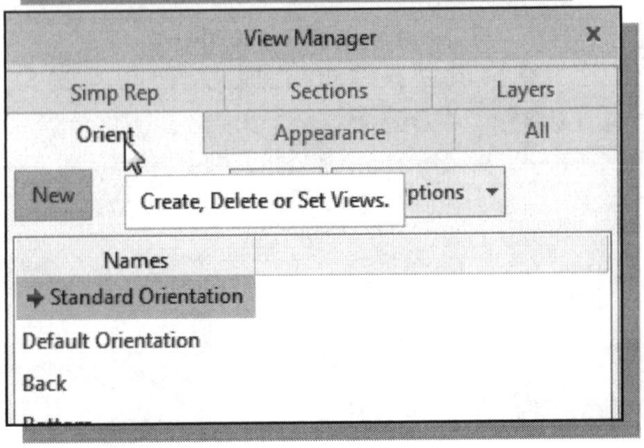

2. Click on the **Orient** tab as shown.

❖ The **Orient** option is used to control the ***saved view*** list, which can be accessed through the *View* toolbar. Note the six standard 2D orthographic views have been set up.

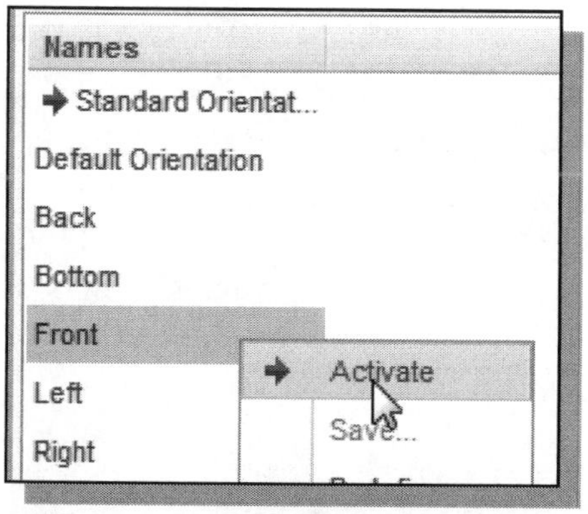

3. Choose **Front** in the saved **Names** list.

4. Press and hold the right-mouse-click to bring up the option menu and choose **Activate** as shown.

❖ We will first adjust the display to the **Front** view angle then use the *Dynamic Rotate* option to set the **Isometric** view.

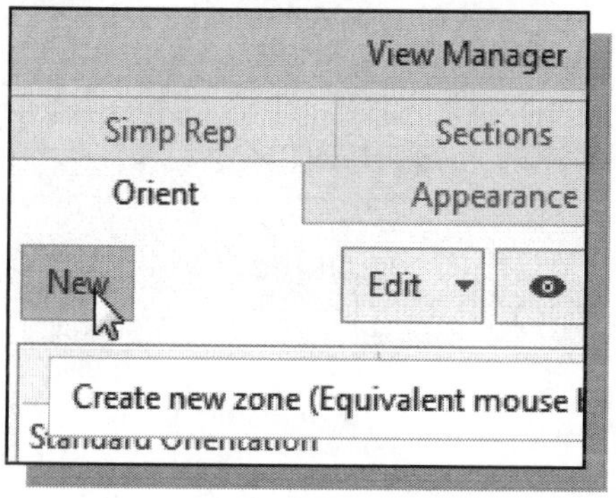

5. Click **New** to create a new ***saved view***.

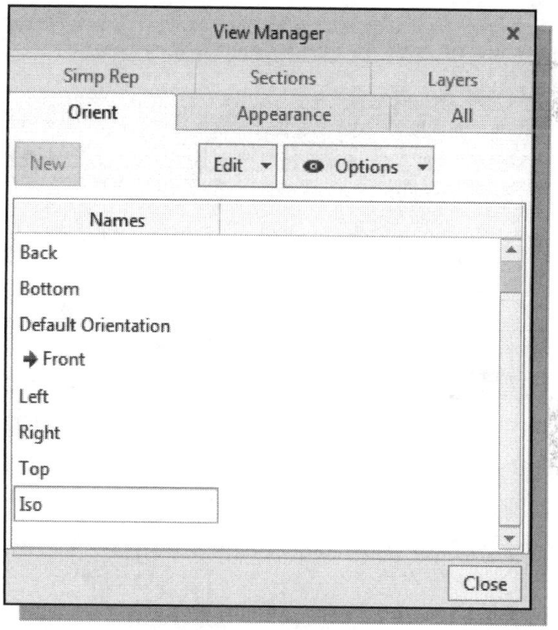

6. Enter **Iso** as the new saved view **Name**. Press the [**ENTER**] key once to accept the input.

7. Choose **Edit → Redefine** to edit the definition.

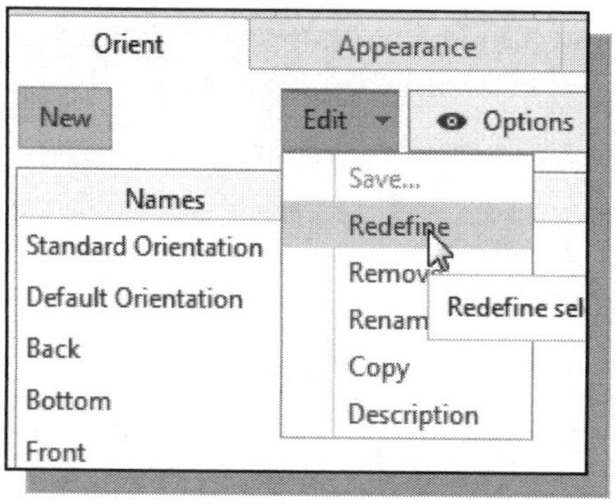

8. Choose the **Dynamic orient** option in the **Type** list as shown.

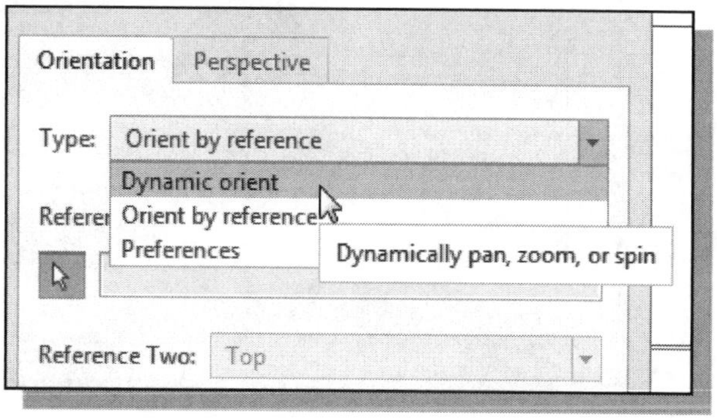

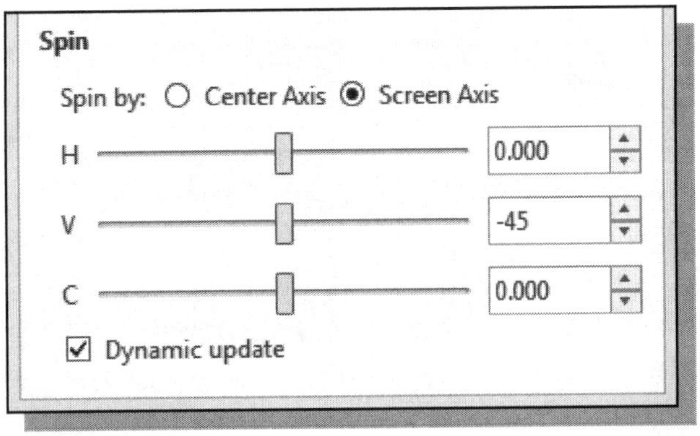

9. Enter **-45.0** as the angle to spin about the vertical axis as shown.

❖ Note the coordinate system, near the center of the screen, rotates to the new angle as we enter the new spin angle.

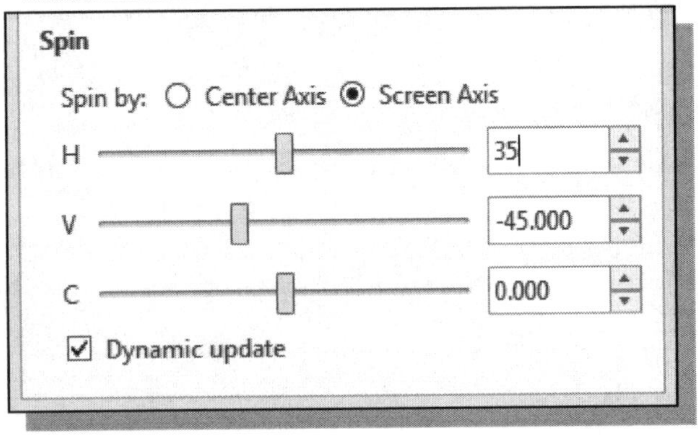

10. Enter **35.0** as the angle to spin about the horizontal axis. The coordinate system is rotated further.

11. Click **OK** to accept the settings for the **ISO** view.

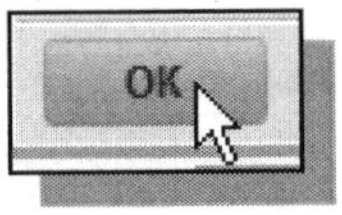

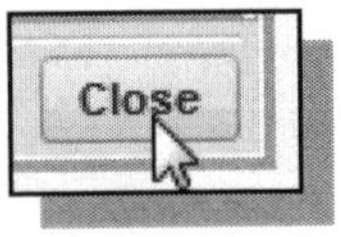

12. Click **Close** to end the View Manager command.

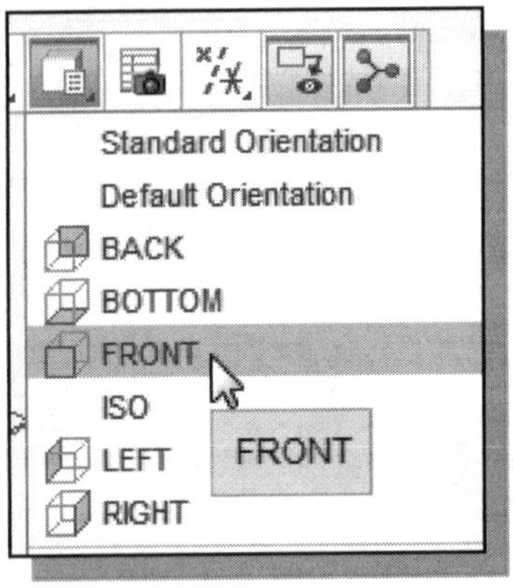

13. Select **FRONT** in the named view list and notice the model display is now switched to the standard Front view.

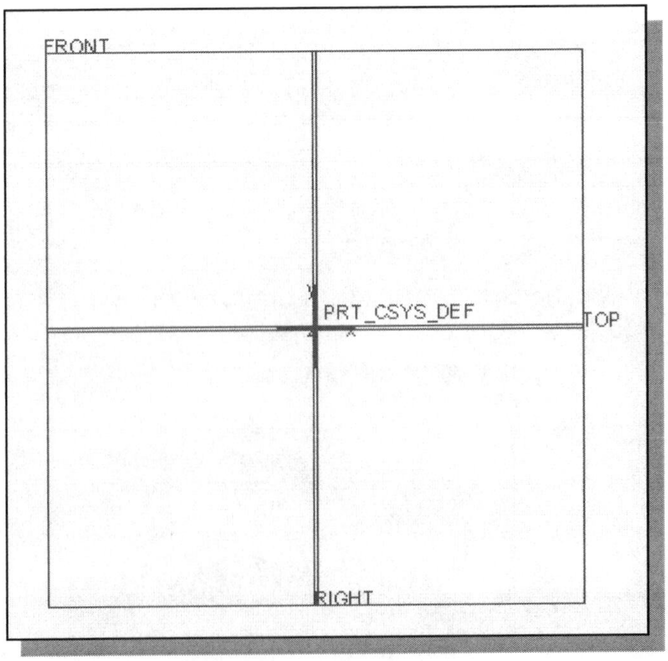

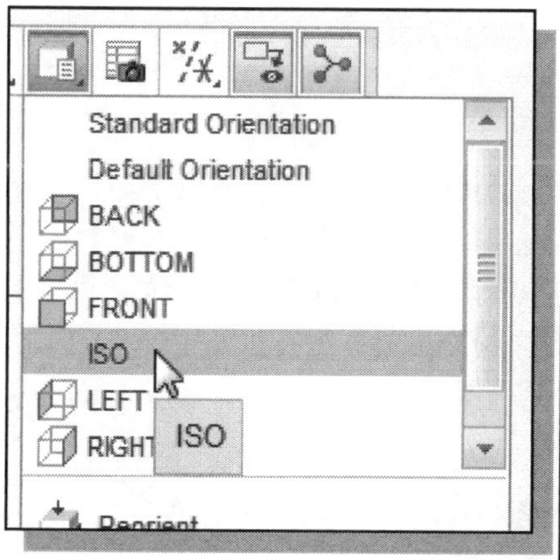

14. Select **ISO** in the named view list and notice the model display is now switched to the newly set up Isometric view.

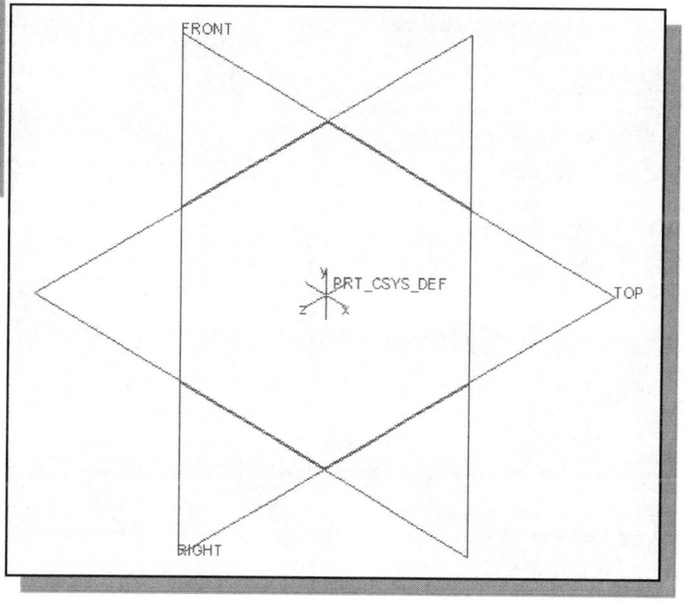

Save Current Setup as a New Template

1. Choose **File→ Save a Copy** in the *File pull-down menu* as shown.

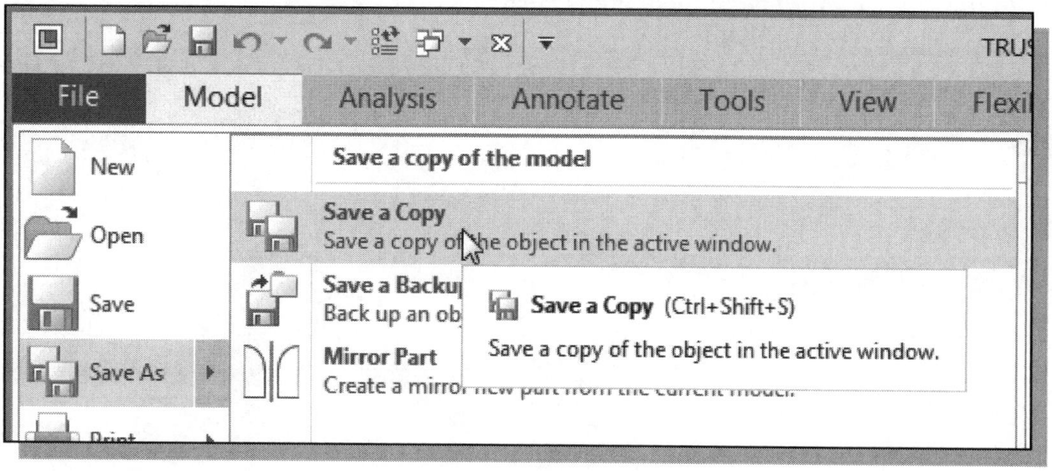

2. Switch to the ***templates*** folder that is under the installation folder (the default installation folder name is *PTC/Creo 4.0*).

3. On your own, use **lnlbfS_part_solid.PRT** as the new template Name and click **OK** to save the current file as a new template.

The Integrated Mode of Creo Simulate

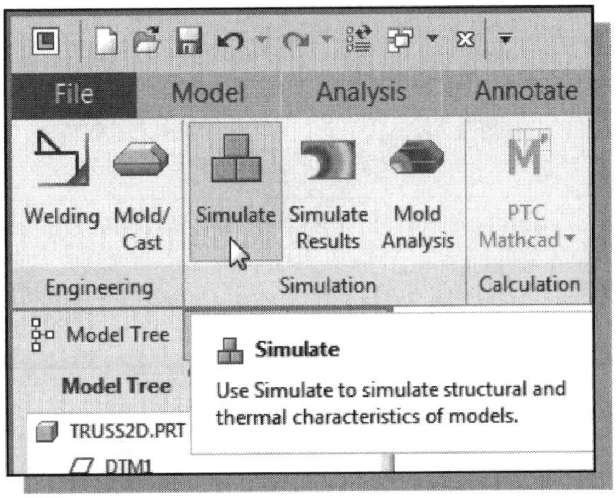

1. Start the *Integrated* mode of *Creo Simulate* by selecting the **Applications → Simulate** option from the *Creo Parametric* pull-down menu as shown.

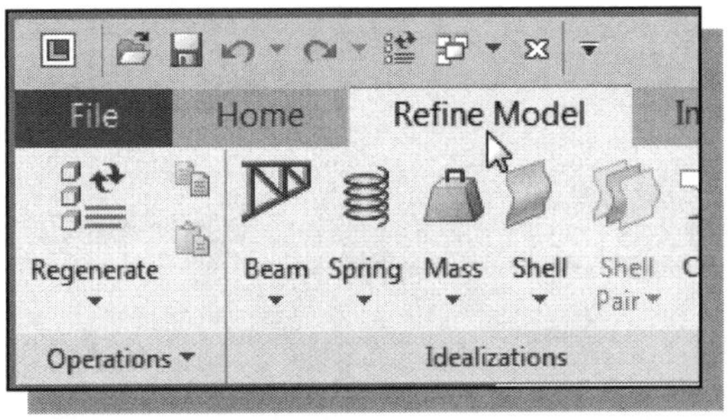

2. Click on the **Refine Model** tab to view the available commands.

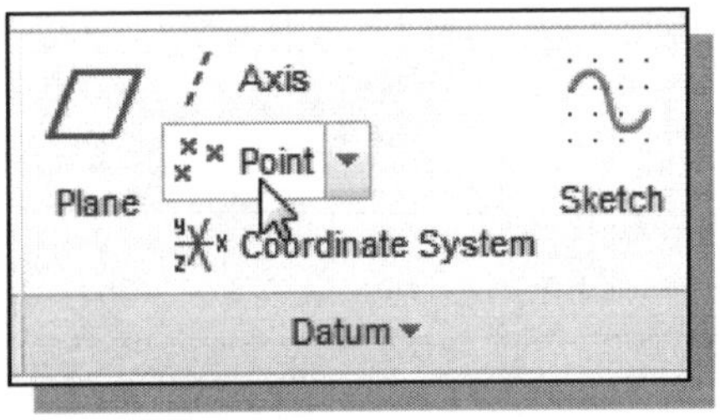

3. Choose **Datum Point** tool in the icon panel displayed in the *Ribbon.*

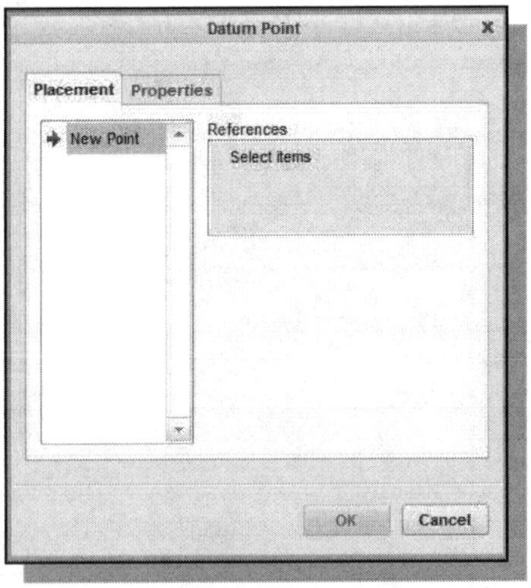

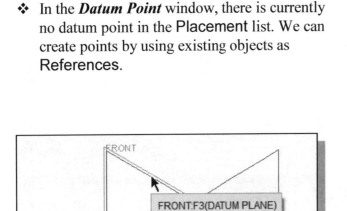

❖ In the ***Datum Point*** window, there is currently no datum point in the Placement list. We can create points by using existing objects as References.

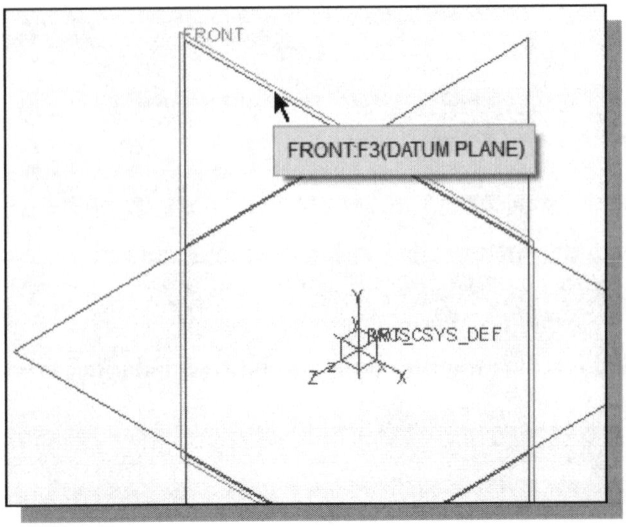

4. Inside the *graphics area,* select the **FRONT** plane by clicking on one of the edges as shown.

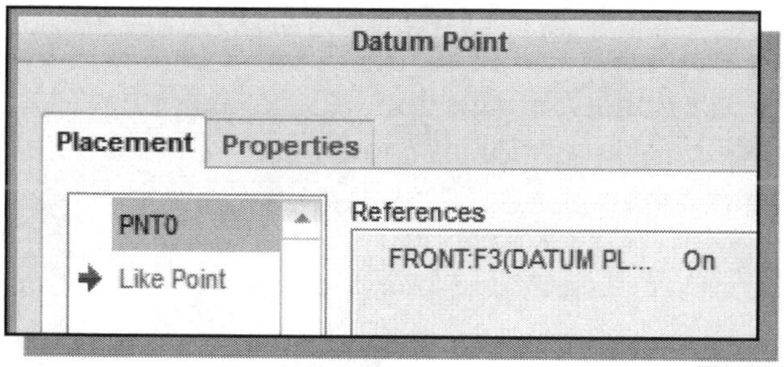

❖ The new point is placed on the FRONT plane.

5. Grab one of the handle points and place it on one of the datum planes, **RIGHT**, as shown. RIGHT is now being used as an Offset reference to position the point.

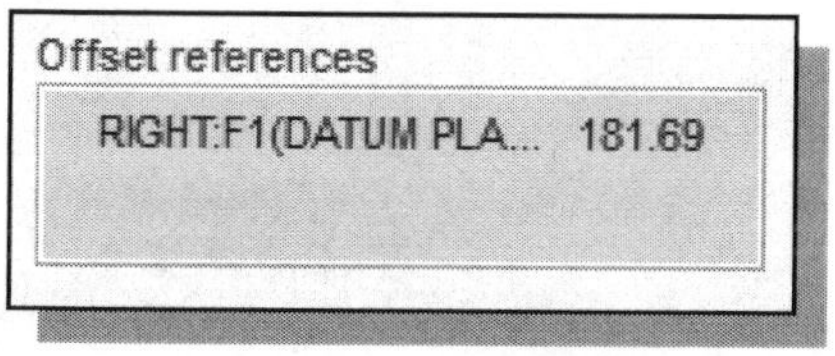

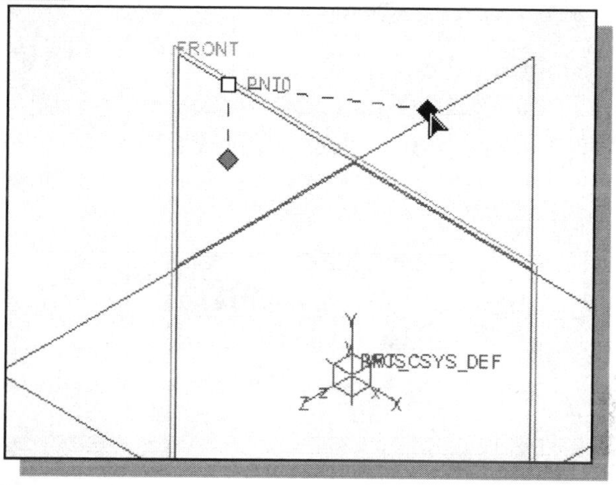

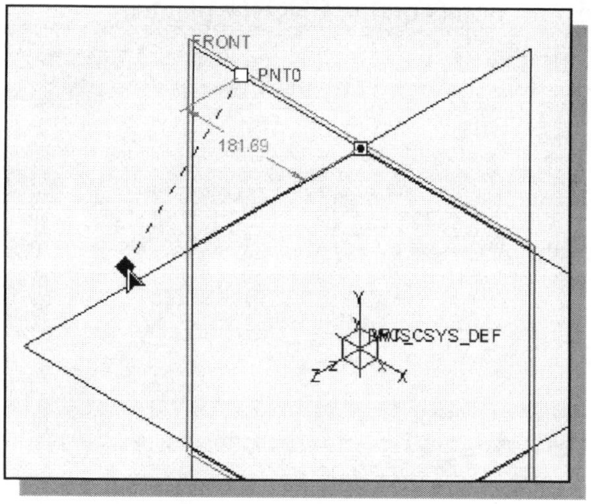

6. Repeat the above step and place the other handle point on **TOP**.

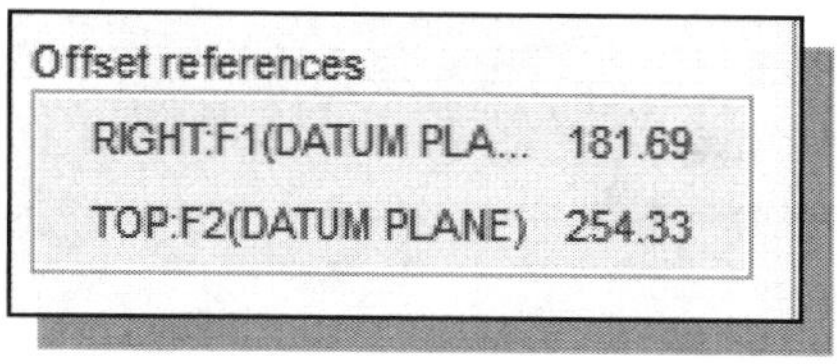

7. Click on one of the Offset values inside the References option box, in the *Datum Point* window, and change the value to **0.0** as shown.

8. Click on the second value and also adjust it to **0.0** as shown.

❖ We have created a datum point located exactly at the origin of the coordinate system.

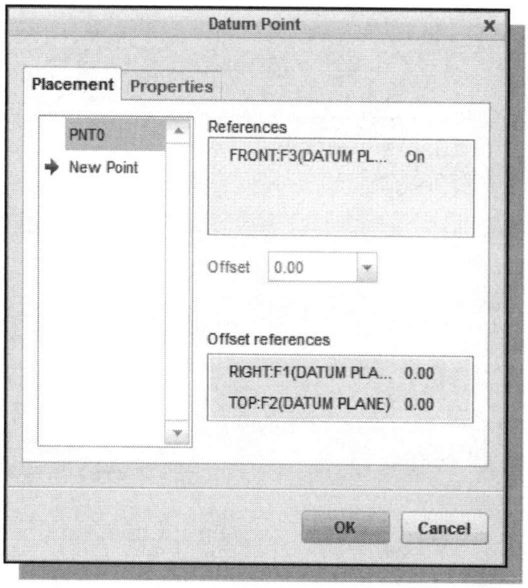

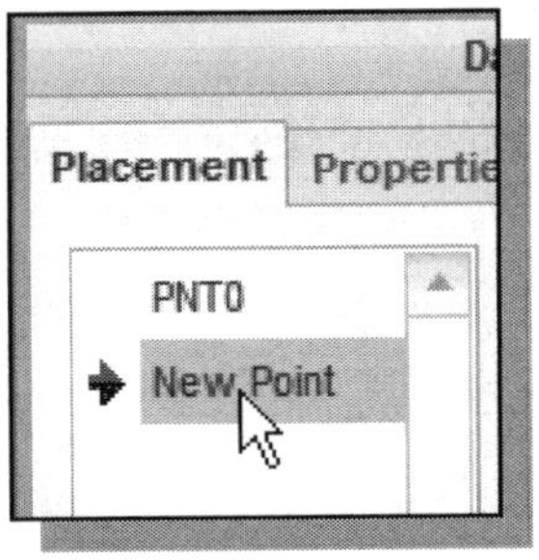

9. In the *Datum Point* window, create a new datum point by clicking on the **New Point** option as shown.

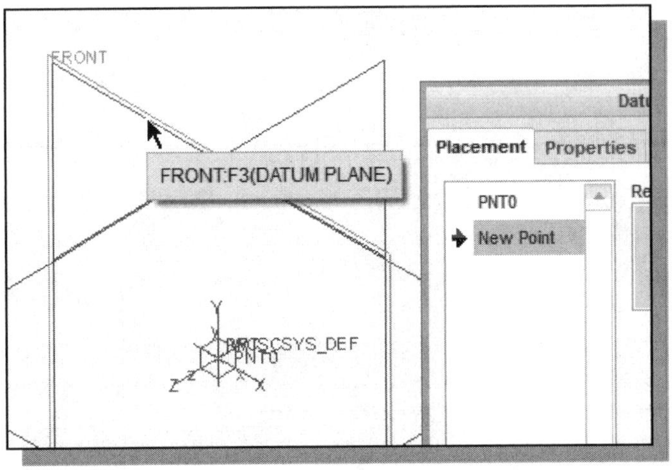

10. Inside the graphics area, select the **FRONT** plane by clicking on one of the edges as shown.

11. Attach the handles to the other two datum planes, **RIGHT** and **TOP**. Adjust the distances as shown in the figure below.

12. On your own, create the additional three datum points on the **FRONT** plane as shown. Note the four points are aligned horizontally.

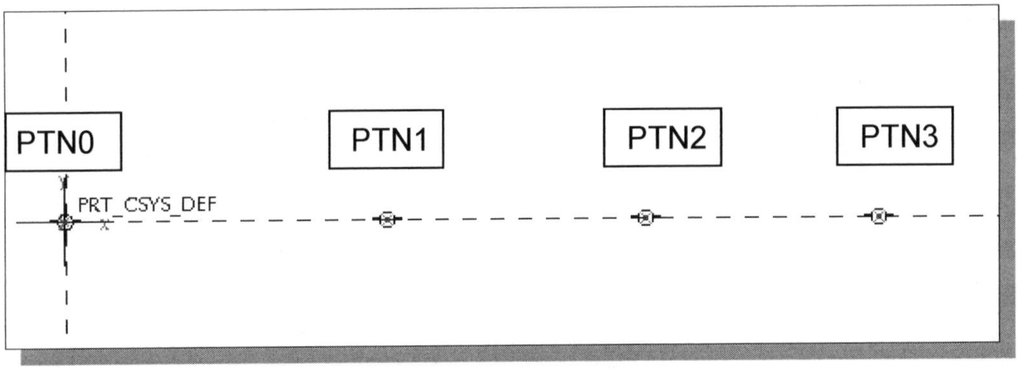

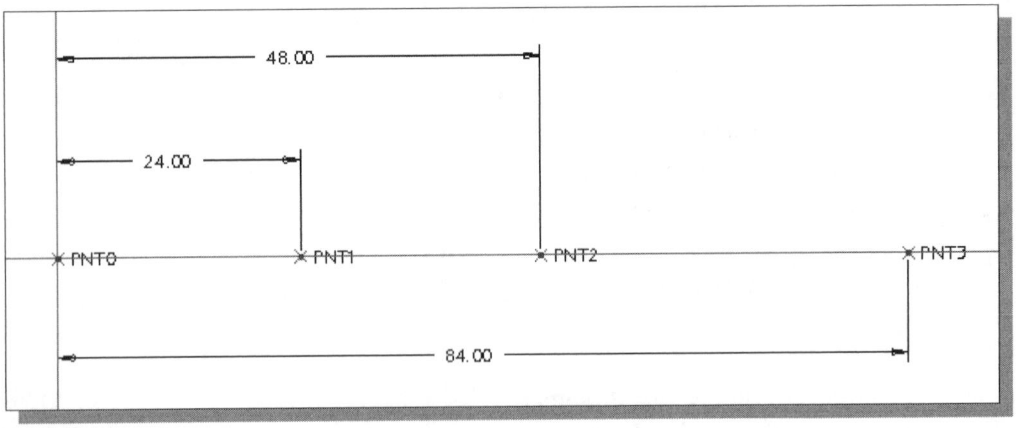

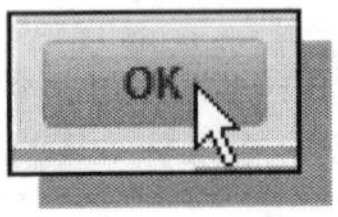

13. Click on the **OK** button to accept the creation of the datum points.

Create a Datum Curve for the Distributed Load

In *Creo Simulate*, beam elements can also be established with curves. One advantage of using curves is we can apply distributed loads on curves for beam analysis.

1. Choose **Datum → Curve → Curve through Points** tool in the *Datum toolbar*.

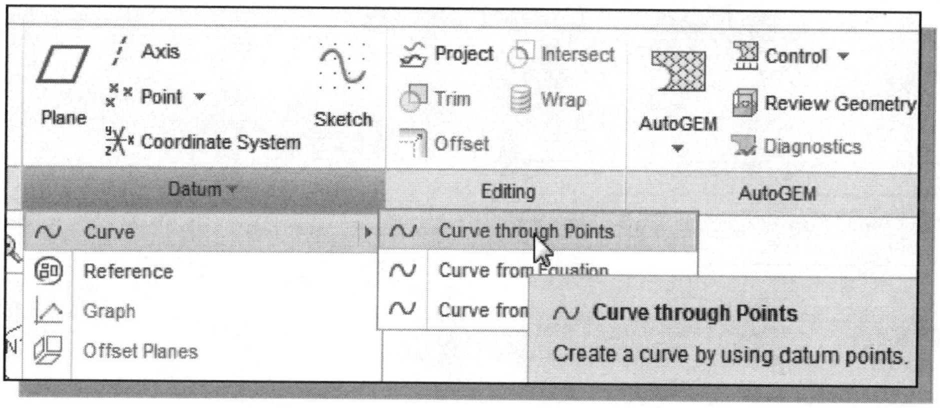

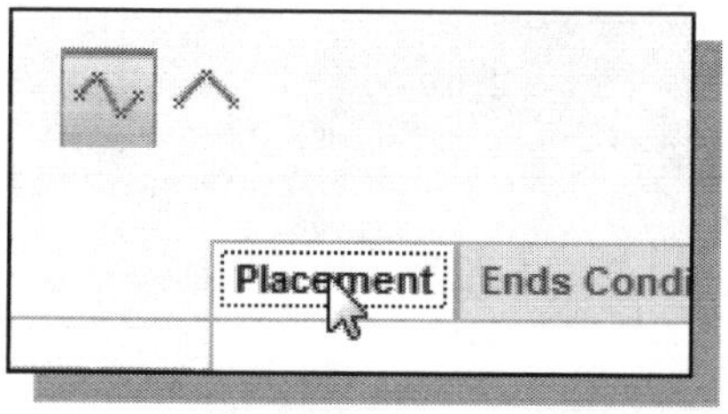

2. In the *Dashboard* area, click the **Placement** tab to view the available options.

- Note that Creo Simulate automatically created a datum curve.

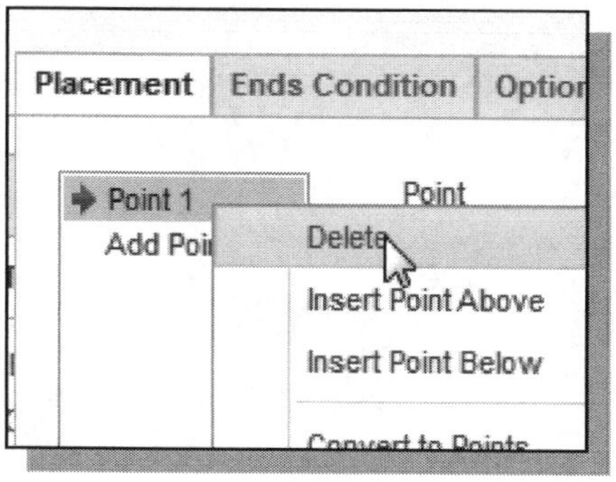

3. **Right-click** on **Point 1** to bring up the option menu and select **Delete** to remove the pre-defined curve as shown.

4. In the *graphics area*, select **PNT0** as the first point for the curve definition.

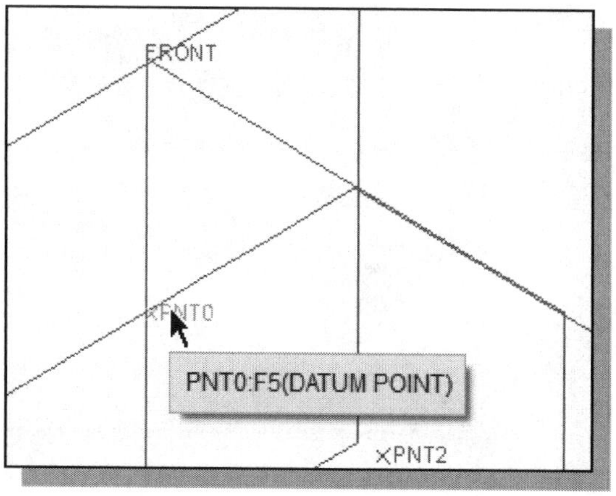

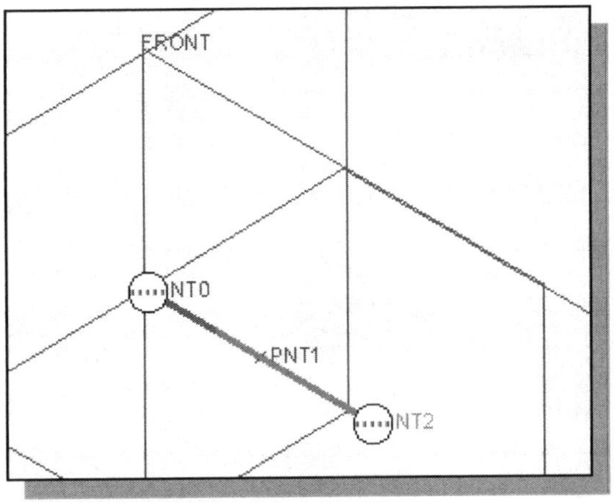

5. In the *graphics area*, select **PNT2** as the second point for the first curve definition. Note that PTN1 is the location for the applied point-load.

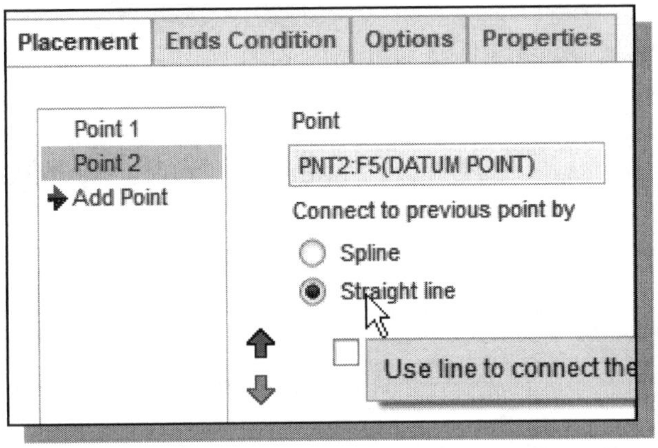

6. Select **Straight line** to define the connection option.

7. Click **OK** to create the first datum curve.

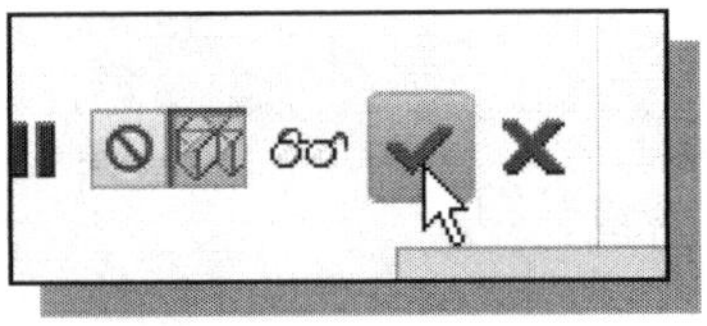

8. On your own, repeat the above procedure to create a second beam, connecting between **PNT2** and **PNT3**.

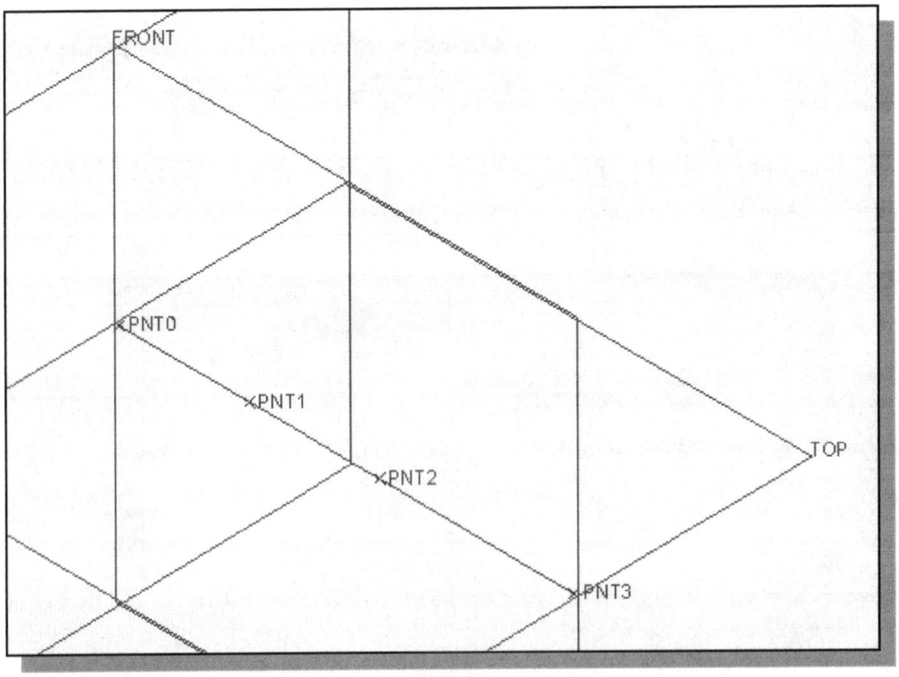

❖ Two curves were created to represent the Beam system. By establishing a curve between **PNT2** and **PNT3**, a distributed load can now be applied to the system.

Set up an Element Cross Section

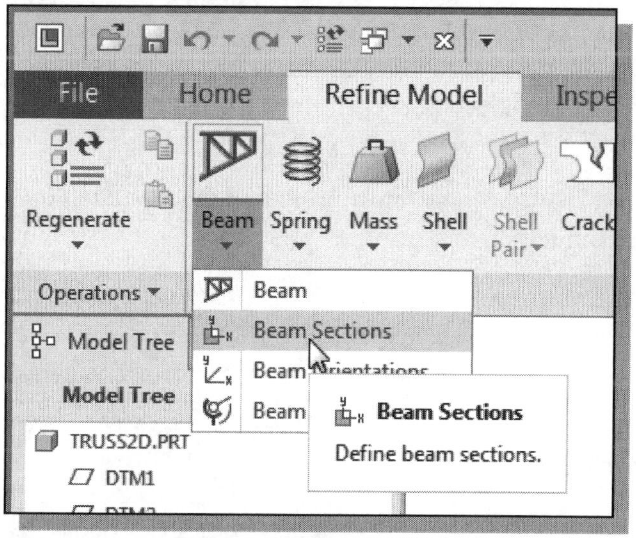

1. Choose **Beam Sections** in the **Refine Model** tab as shown.

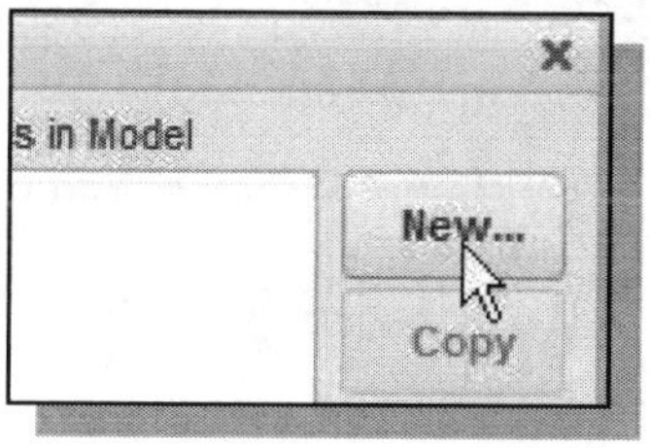

2. Click **New** to create a new beam cross-section.

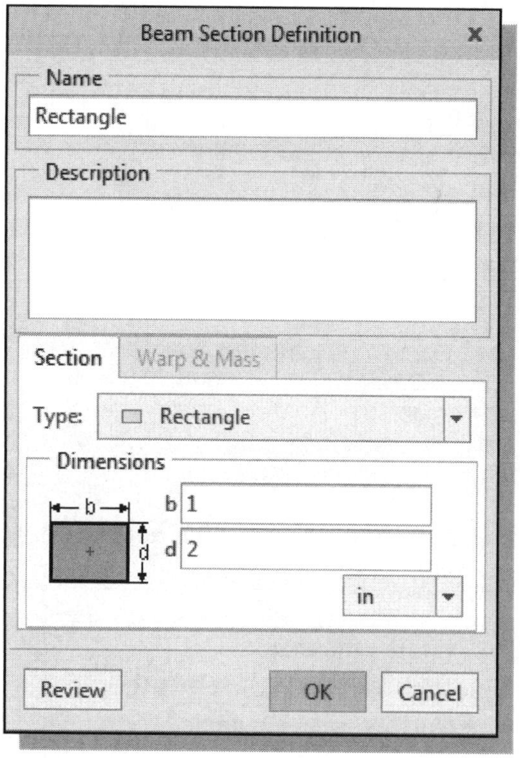

3. Choose **Rectangle** as the cross-section **Type** as shown.

4. Enter **1.0** and **2.0** as the width and height of the rectangular beam cross section.

5. Click **OK** to accept the creation of the new beam section.

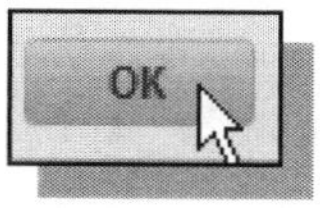

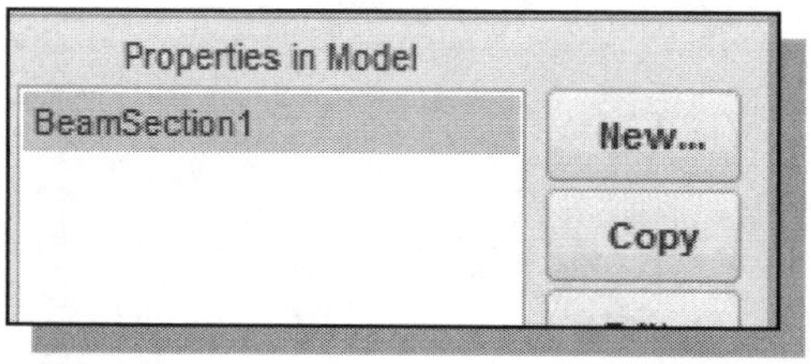

6. Click **Close** to accept the creation of the new beam section.

Select and Examine the Element Material Property

Before creating the FEA elements, we will first set up the *Material Property* for the elements. The *Material Property* contains the general material information, such as *Modulus of Elasticity*, *Poisson's Ratio*, etc., that is necessary for the FEA analysis.

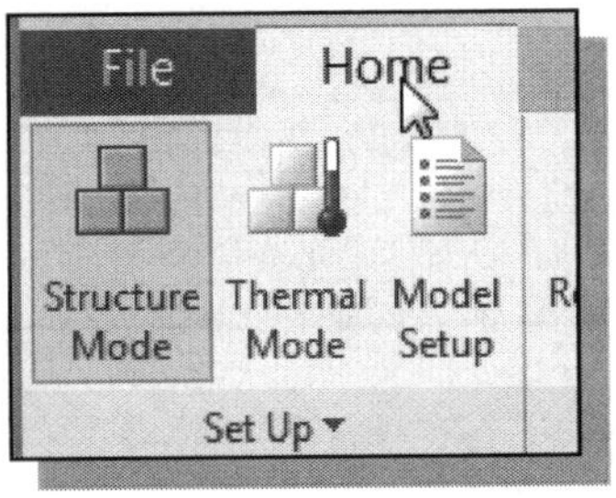

1. Click on the **Home** tab to return to the main *Creo Simulate* commands toolbar.

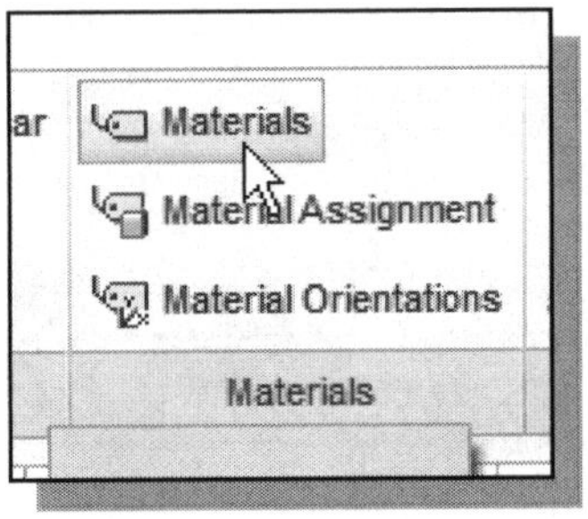

2. Choose **Materials** in the *Materials* group from the *Ribbon* toolbar as shown.

❖ Note the default list of materials libraries, which are available in the pre-defined *Creo Simulate*, is displayed.

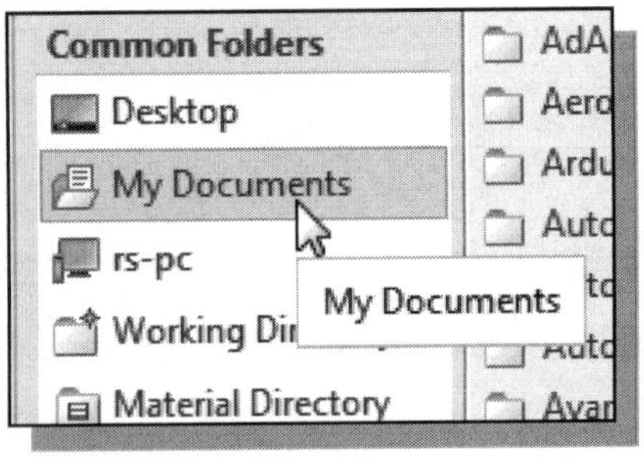

3. On your own, locate the folder where we saved the modified material definition, filename **Steel_low_Carbon_Modified**.

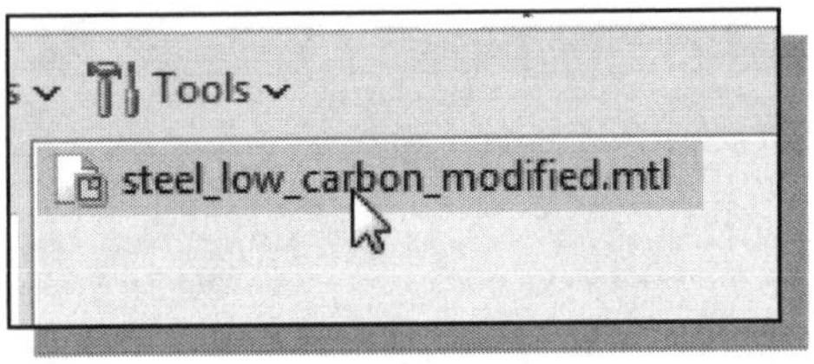

4. **Double-click** on the modified material definition for use with the current model.

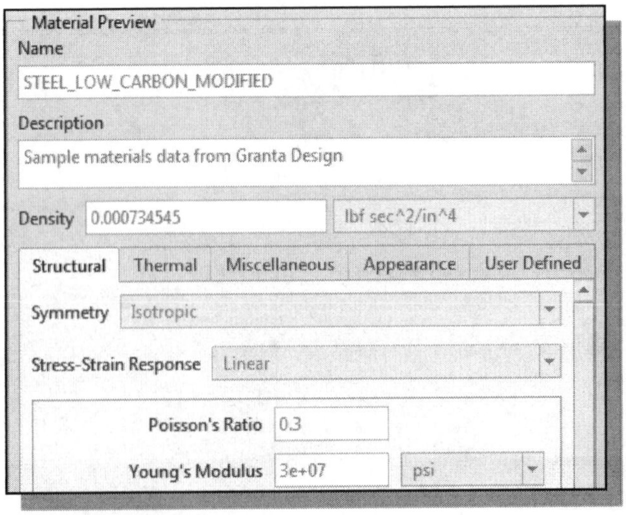

5. Click **Yes** to convert the units to match the model units.

6. Examine the material information in the **Material Preview** area and click **OK** to accept the settings.

Create 3D Elements

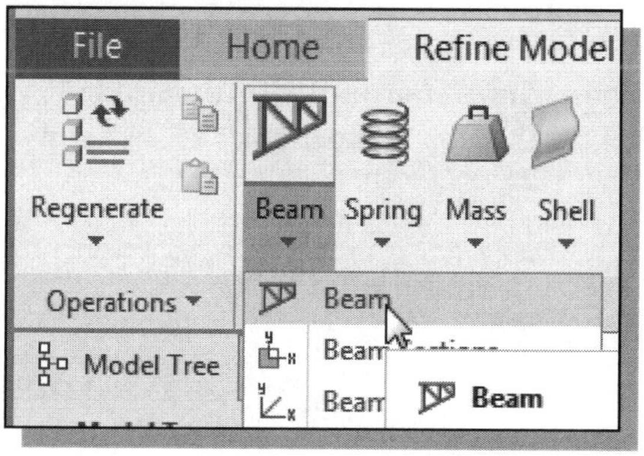

1. Choose **Beam** in the icon panel. The *Beam Definition* window appears.

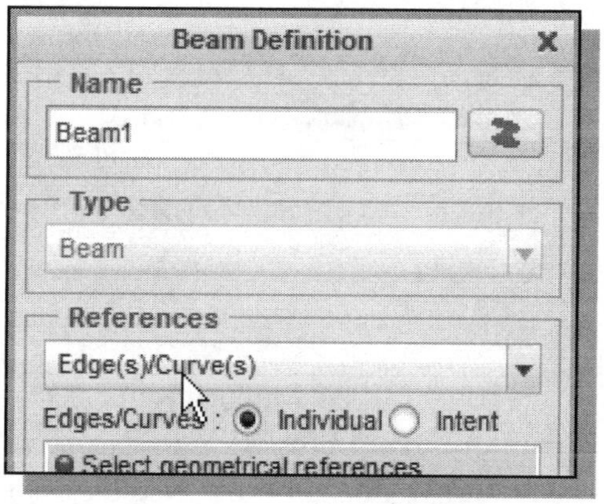

2. Confirm the References option is set to **Edge(s)/Curve(s)**.

3. Press the [**CTRL**] key and pick from left to right the datum curves we just created.

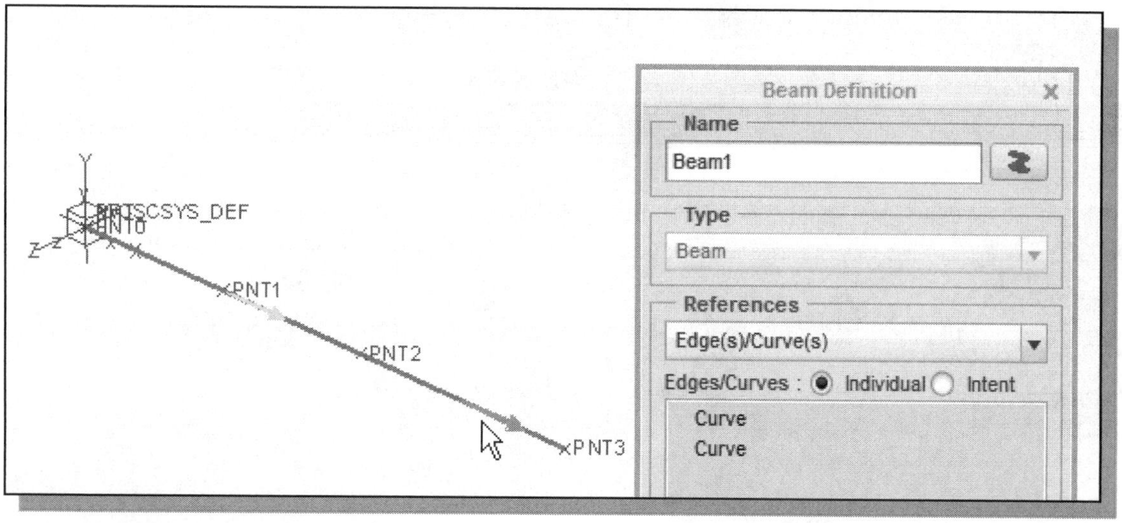

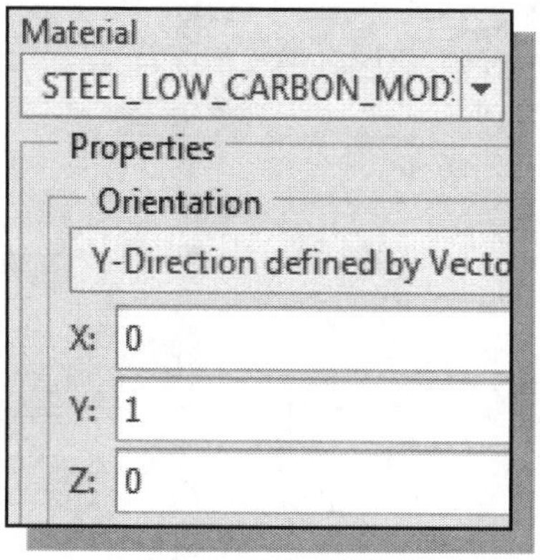

4. Select **Steel_low_Carbon_Modified** as the material of the element.

5. Confirm the *Y-Direction* is set to WCS with **0**,**1**,**0** as the X, Y and Z directions.

6. Click on the **OK** button to create the beam.

Apply Boundary Conditions – Constraints

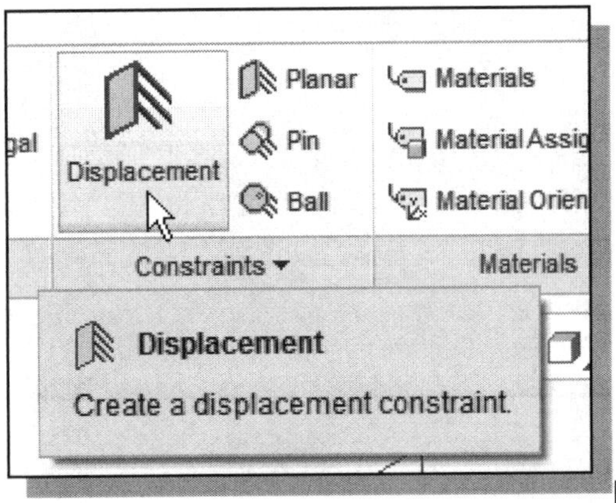

1. Choose **Displacement Constraint** by clicking the icon in the toolbar as shown.

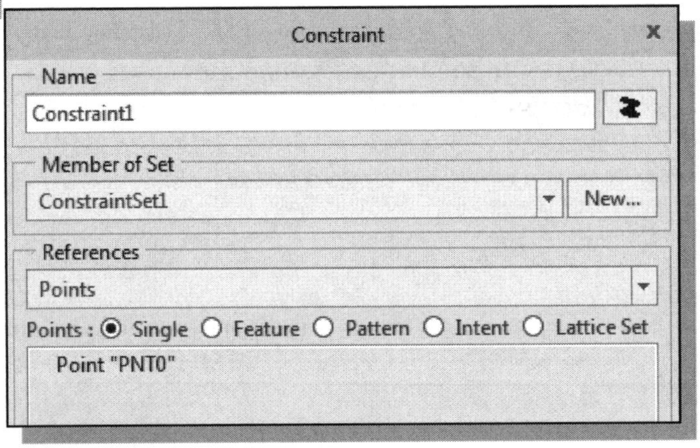

2. Set the References to **Points** and select **PNT0** as shown.

❖ PNT0 is a fixed support point; it cannot move in the X, Y or Z directions. For 2D truss systems, all joints are *pin-joints*; therefore, the Z rotational movement is free at PNT0.

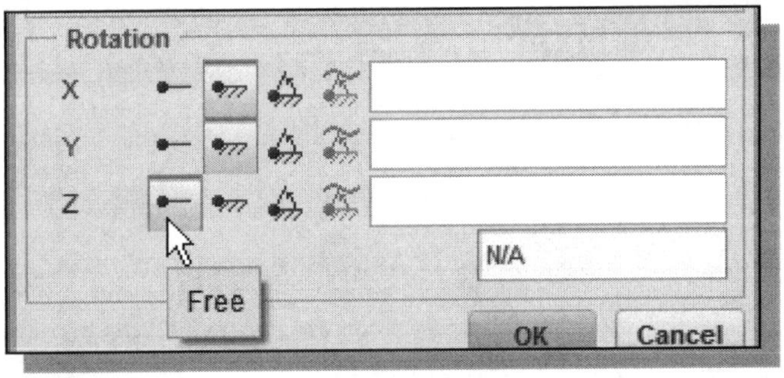

3. Set Z Rotation to **Free** as shown.

4. Click on the **OK** button to accept the first displacement constraint settings.

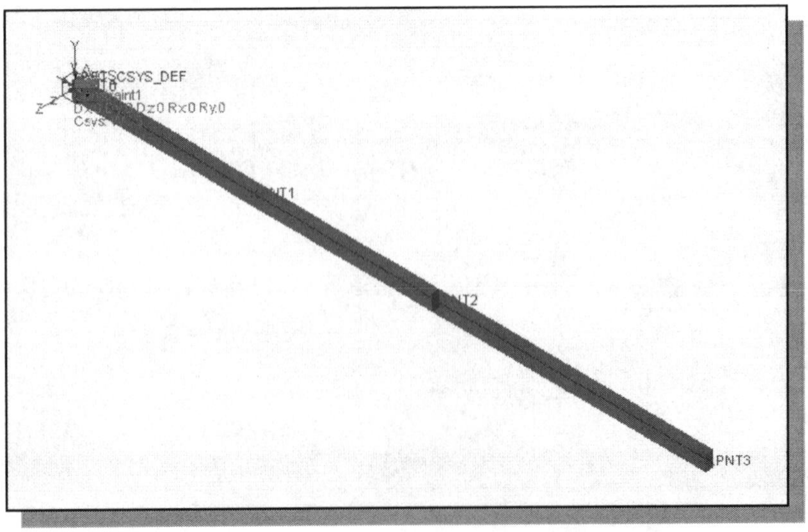

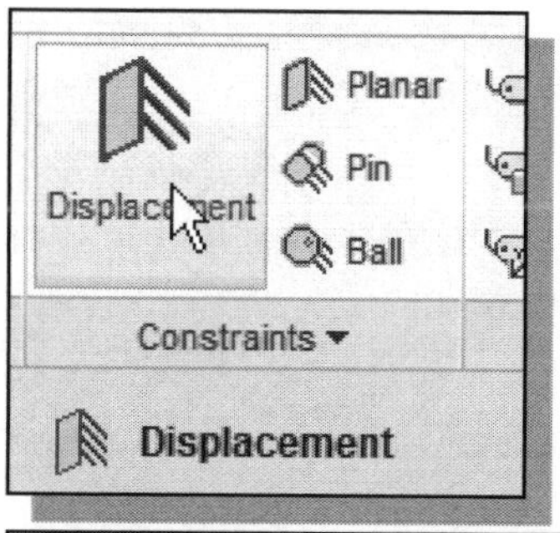

5. Choose **Displacement Constraint** by clicking the icon in the toolbar as shown.

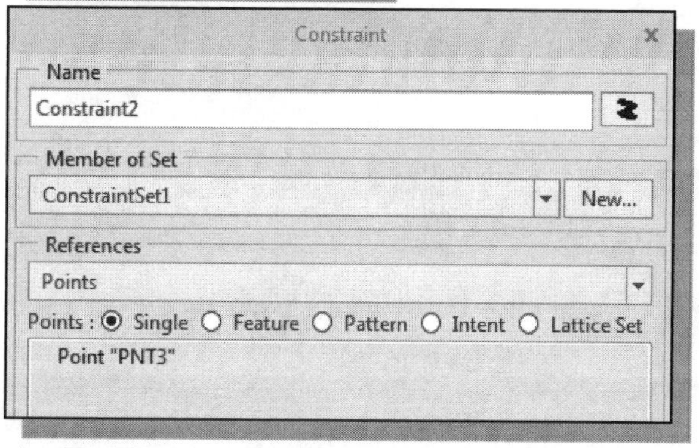

6. Confirm the *Constraint* set is still set to **ConstraintSet1**.

❖ Note that a constraint set can contain multiple constraints.

7. Set the References to **Points** and select **PNT3** as shown.

❖ PNT3 has a *roller* support; it can move in the X direction, but not in the Y or Z directions. The Z rotational movement should also be set to **Free**.

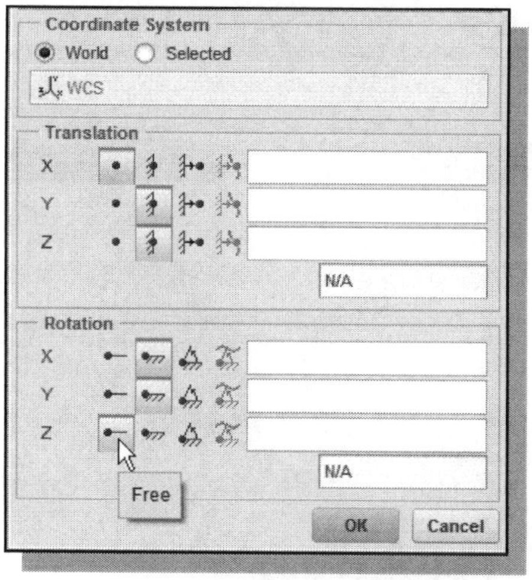

8. Set X Translation to **Free** as shown.

9. Set Z Rotation to **Free** as shown.

10. Click on the **OK** button to accept the second displacement constraint settings.

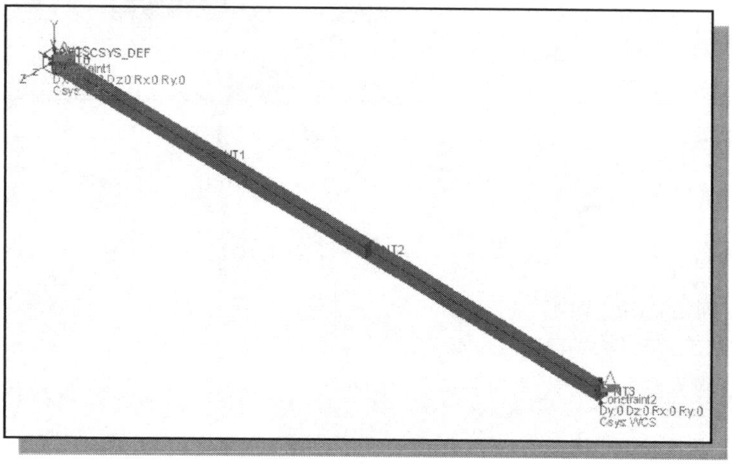

Apply External Loads

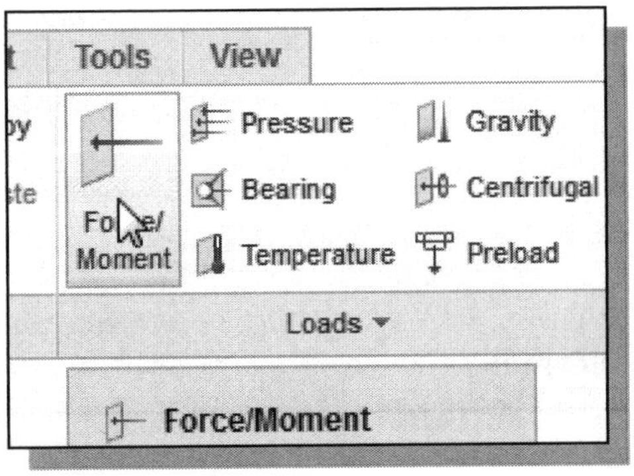

1. Choose **Force/Moment Load** by clicking the icon in the toolbar as shown.

2. Set the References to **Points**.

3. Pick the second node, **PNT1**, of the beam system as shown.

4. Enter **-6000** in the Y component Force value box as shown in the figure below.

5. Click on the **Preview** button to confirm the applied load is pointing in the correct direction.

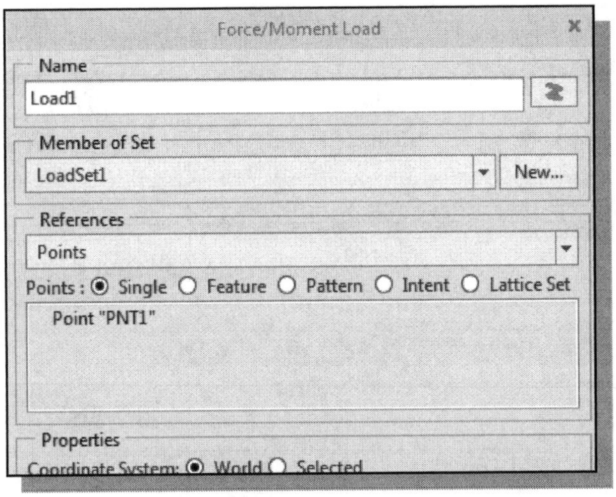

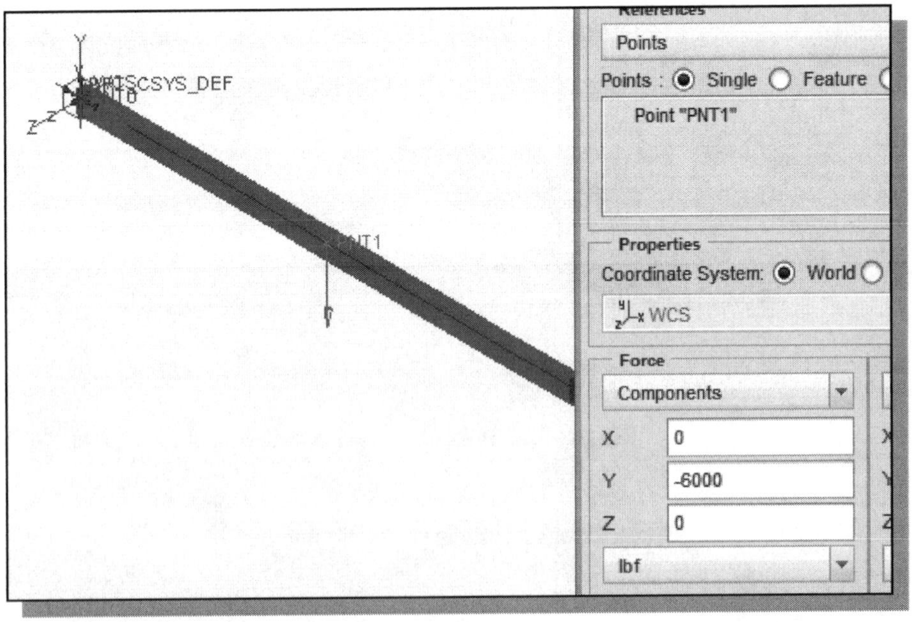

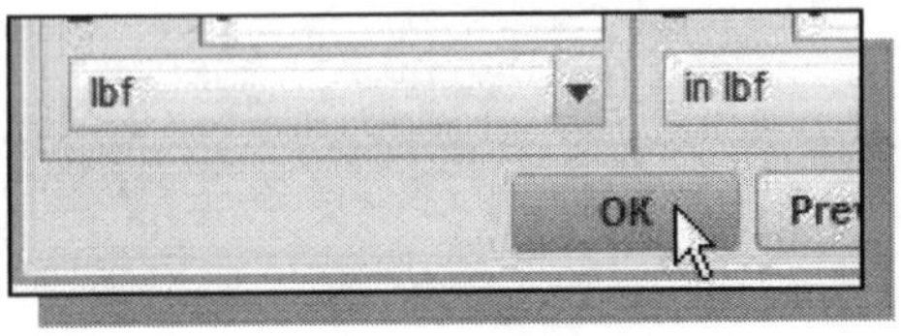

6. Click on the **OK** button to accept the first load settings.

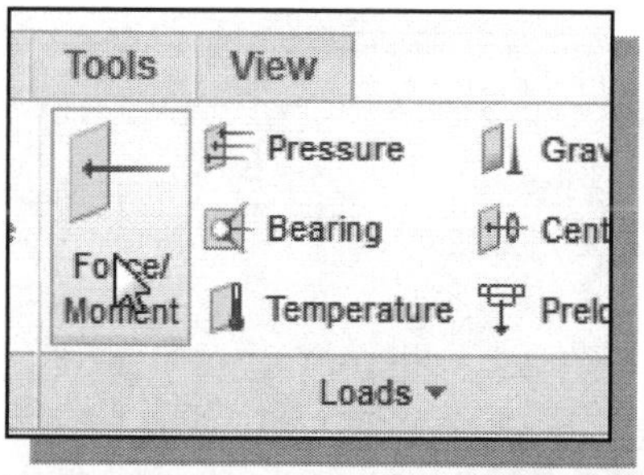

7. Choose **Force/Moment Load** by clicking the icon in the *Loads toolbar* as shown.

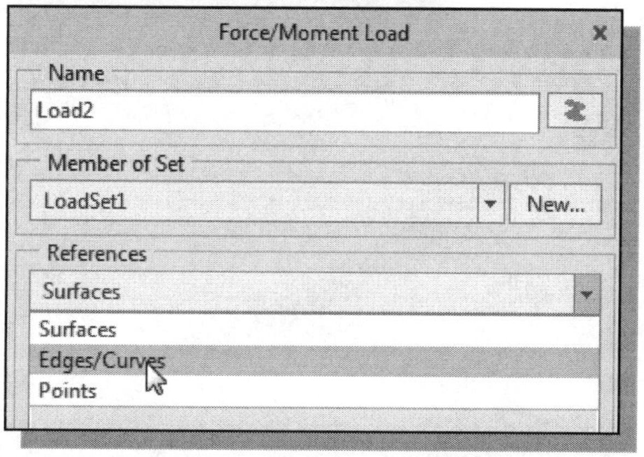

8. Confirm the *Load* set is still set to **LoadSet1**.

❖ Note that a load set can contain multiple loads.

9. Set the References to **Edges/Curves** as shown.

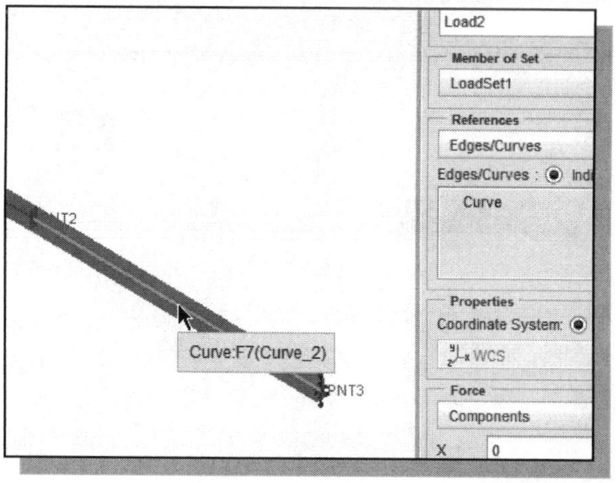

10. Pick the datum curve, **Curve**, as shown.

11. Click **Advanced** to display additional options for the settings of the load.

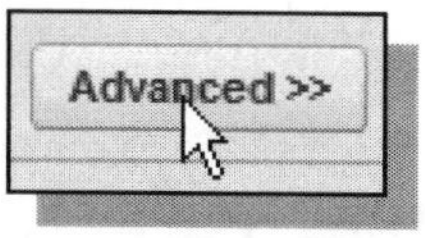

12. Set the Distribution option to **Force Per Unit Length**.

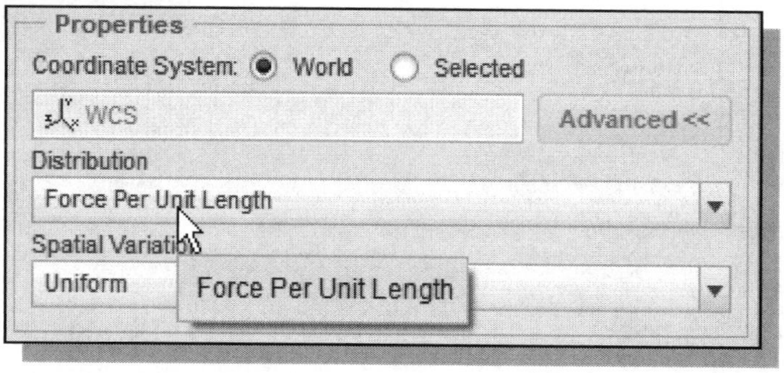

13. Set the Spatial Validation option to **Uniform** as shown.

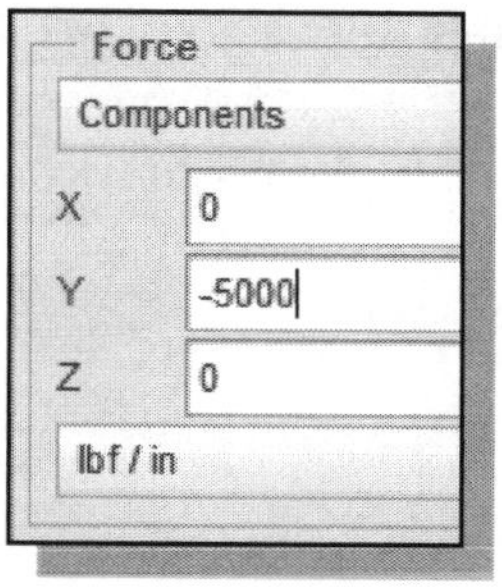

14. Enter **-5000** in the **Y** component **Force** value box as shown.

15. Click on the **OK** button to accept the second load settings.

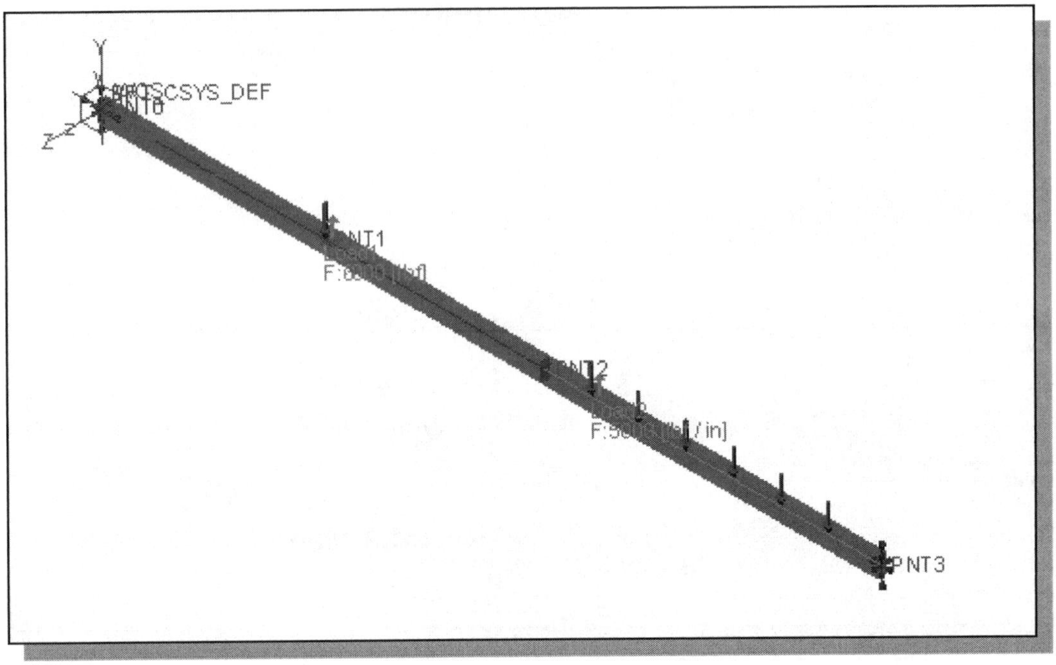

❖ Note that we can also use functions to describe any other type of distributed loads, such as a triangular distributed load as shown.

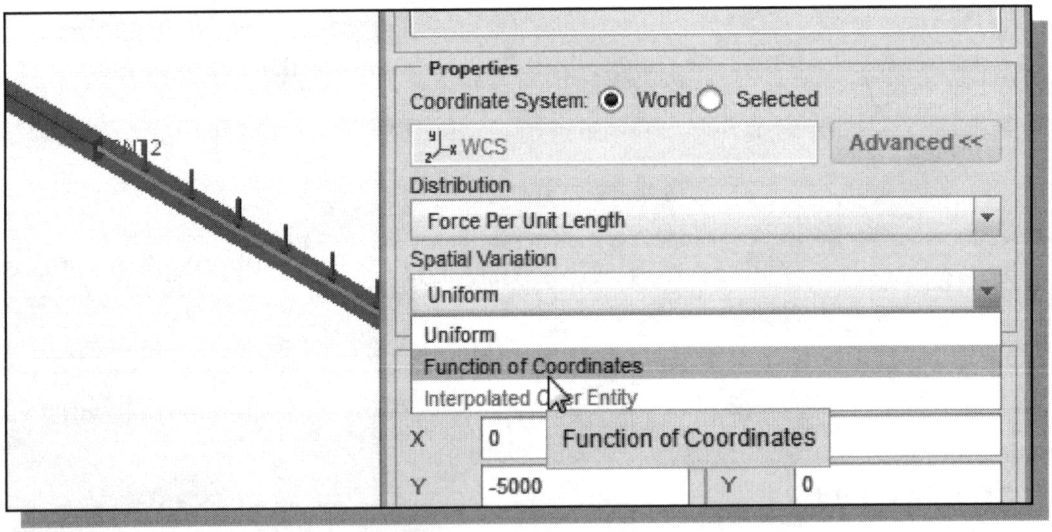

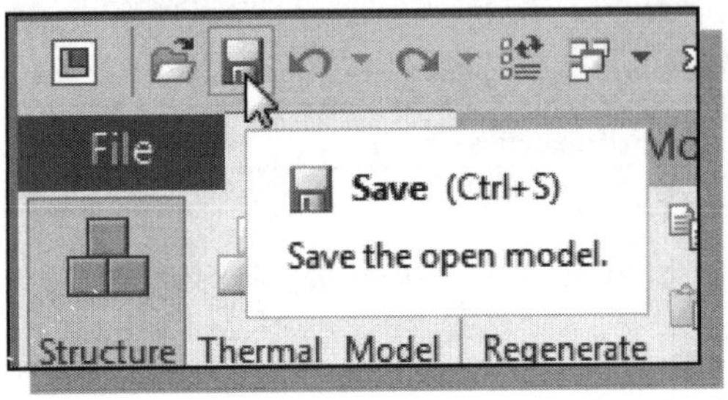

16. On your own, click **Save** and save a copy of the current FEA model.

Run the FEA Solver

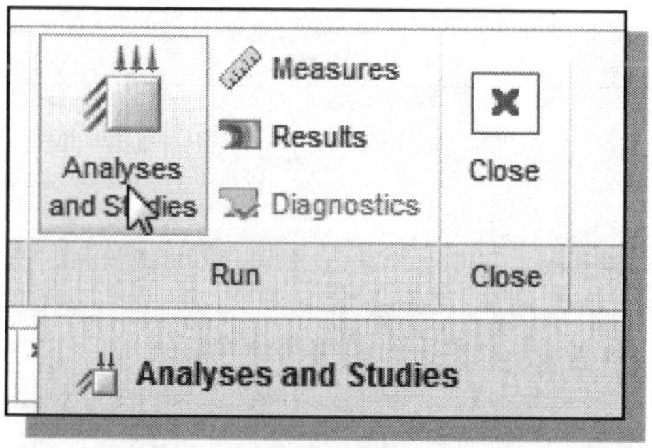

1. Activate the **Analyses and Studies** command by selecting the icon in the *Ribbon* toolbar area as shown.

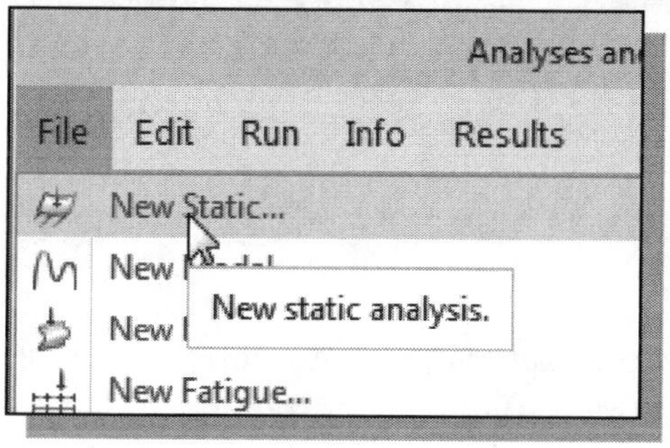

2. In the *Analyses and Design Studies* window, choose **File → New Static** in the pull-down menu. This will start the setup of a basic linear static analysis.

3. Note that ConstraintSet1 and LoadSet1 are automatically accepted as parts of the analysis parameters.

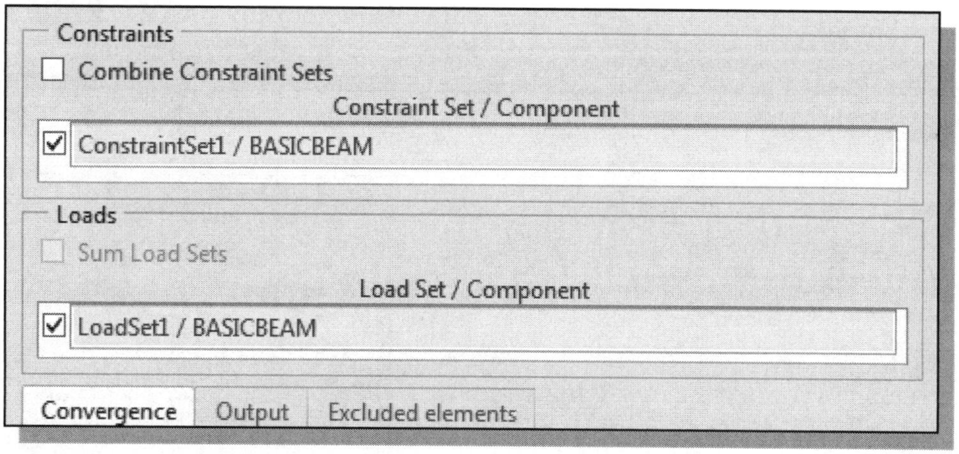

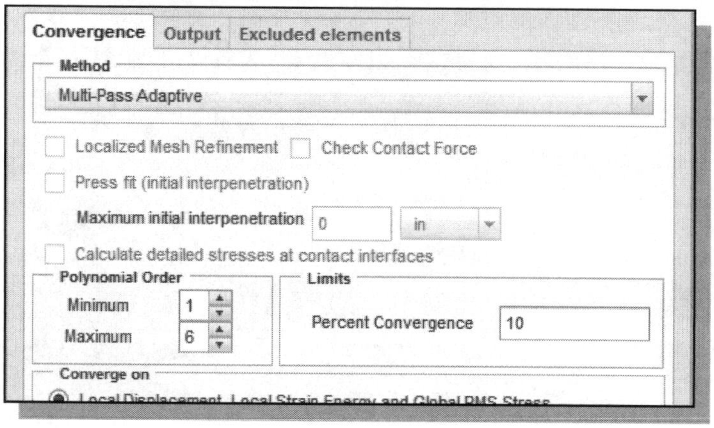

4. Set the Method option to **Multi-Pass Adaptive** and Limits to **10** Percent Convergence as shown.

5. Click **OK** to accept the settings.

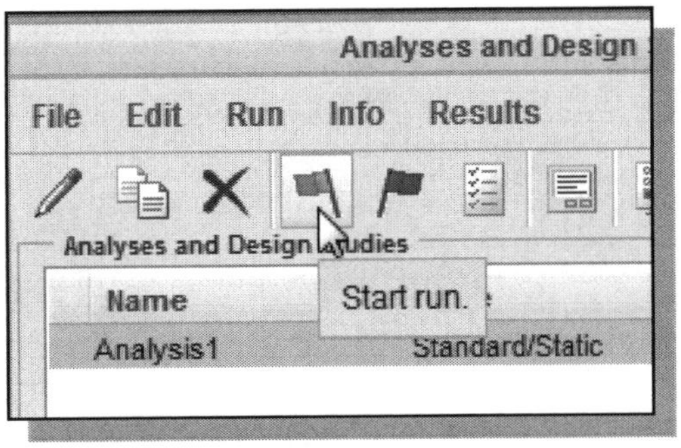

6. Click **Start Run** to begin the solving process.

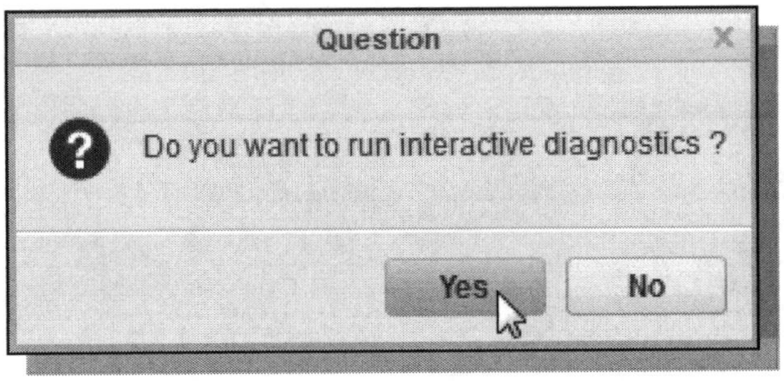

7. Click **Yes** to run the *Simulate* interactive diagnostics.

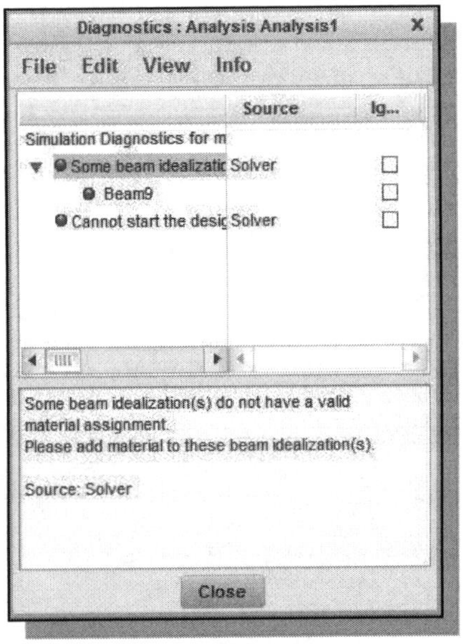

➢ The *Simulate* interactive diagnostics option can be very helpful in identifying and correcting problems associated with our FEA model. *Information*, *warnings* and *fatal errors* regarding the *Solver* are shown in the *Diagnostics* window. Pay extra attention to any **FATAL ERROR** messages, as those must be corrected.

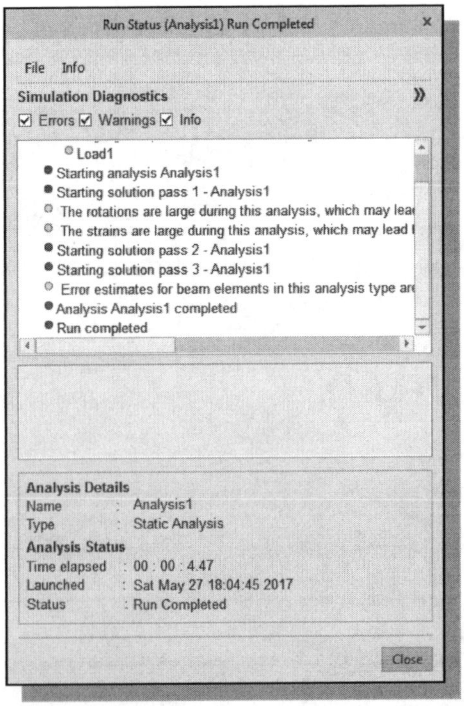

➢ For our analysis, in the *Diagnostics* window, the last message, **Run Completed**, indicates the FEA analysis is a success.

8. Click on the **Close** button to exit the *Diagnostics* window.

View the FEA Results

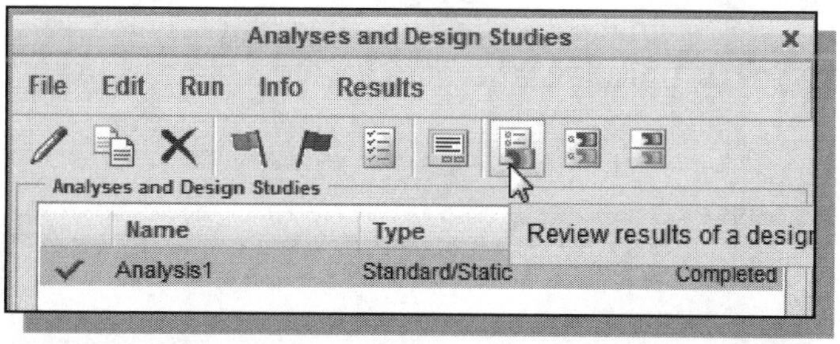

1. In the *Analyses and Design Studies* window, choose **Review results** as shown.

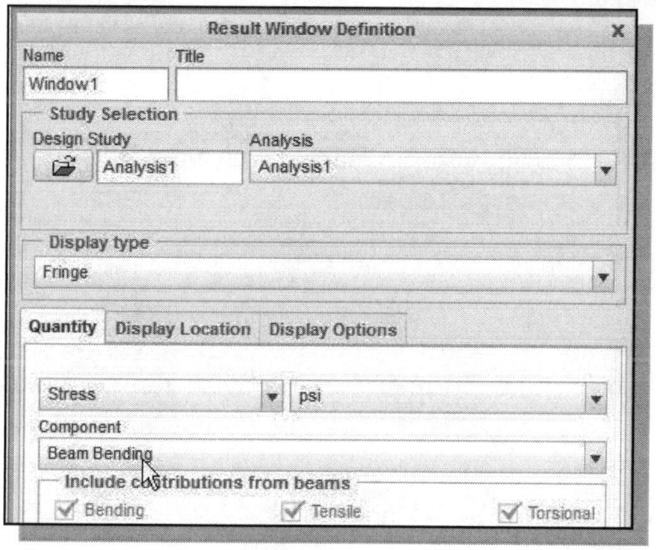

2. Confirm the **Display type** is set to **Fringe**.

3. Choose **Beam Bending** as the **Stress Component** to be displayed.

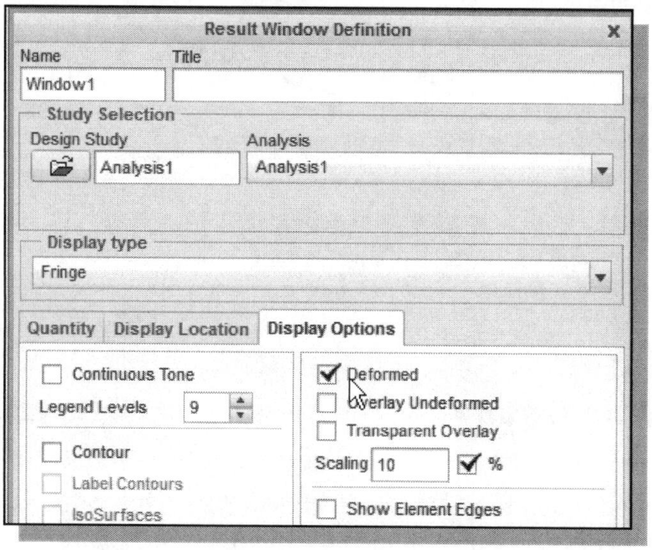

4. Click on the **Display Options** tab and switch ***on*** the **Deformed** option as shown.

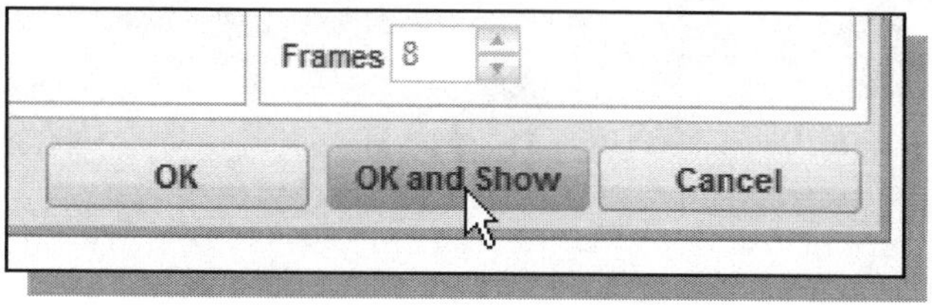

5. Click **OK and Show** to display the results.

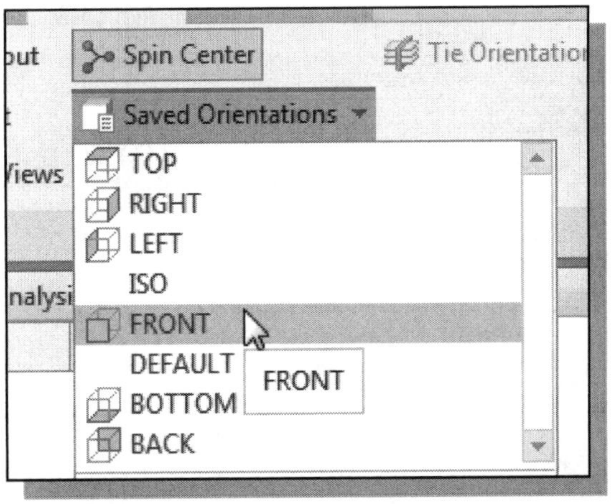

6. Under the *View* tab, select **FRONT** view in the saved name list to adjust the display to front view.

❖ The *Creo Simulate* **Beam Bending** option allows us to quickly view the *bending stress* in beam elements.

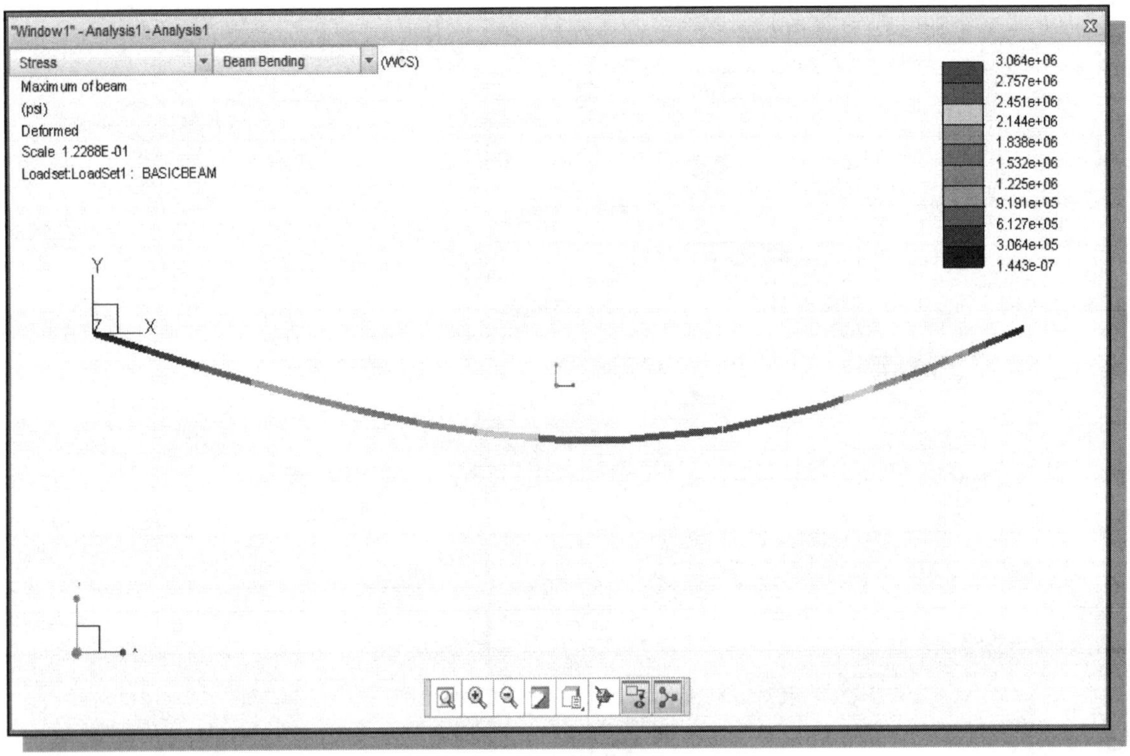

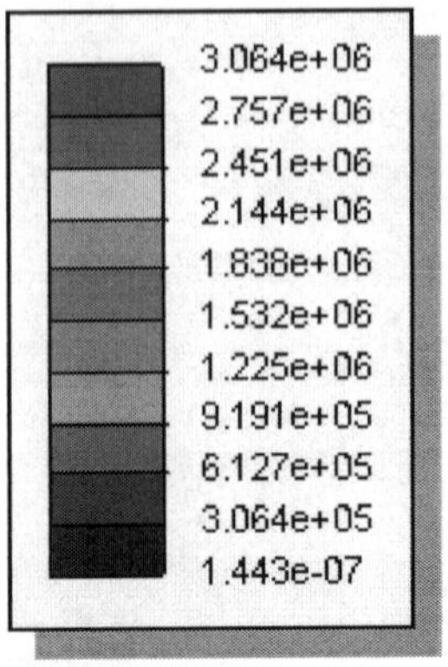

❖ Note the maximum bending stress is **3.064e6 psi**. This value is almost ten times of the value obtained during the preliminary analysis (**3.28e5 psi** on page 6-5).

What Went Wrong?

With the discrepancy between the FEA result and the preliminary analysis, let's examine the FEA *shear diagram*, which is also available in *Creo Simulate.*

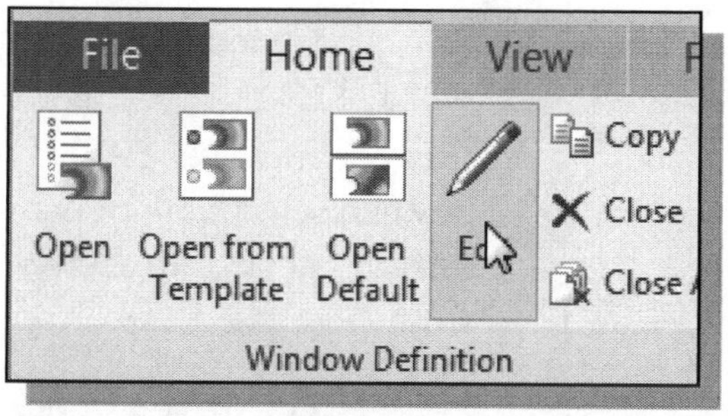

1. In the *Standard* toolbar area, click the **Edit the selected definition** command as shown.

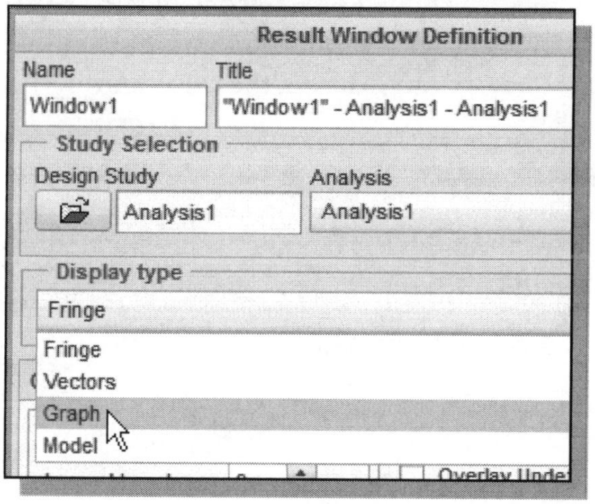

2. Choose **Graph** under the Display type option list as shown.

3. In the Graph Ordinate (Vertical) Axis option list, select **Shear & Moment** as shown.

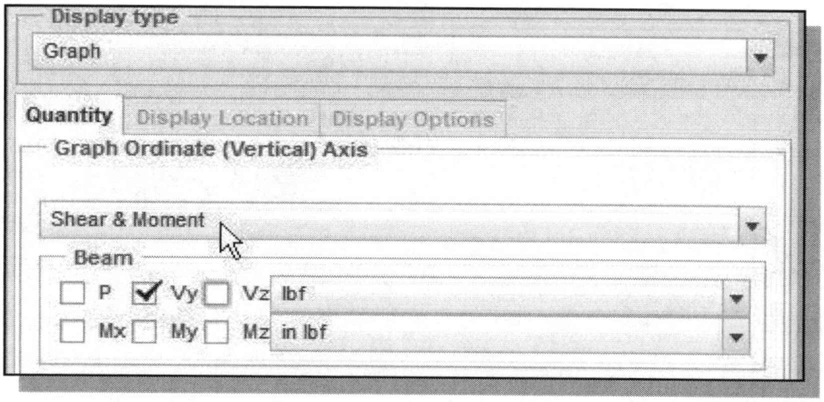

4. Leaving only the **Y shear force** switched ***on***, turn ***off*** all of the other components as shown in the figure above.

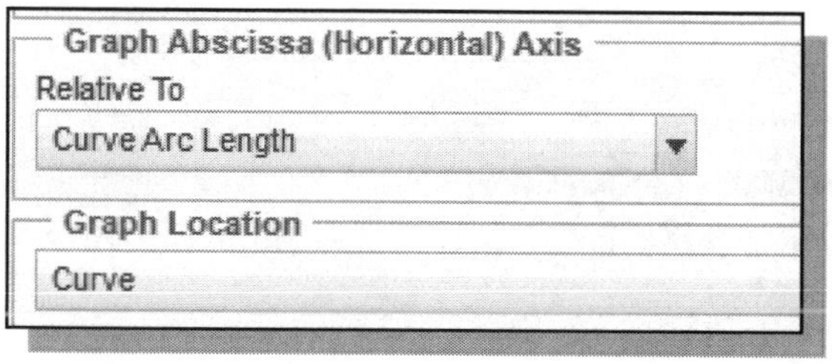

5. In the Graph Abscissa (Horizontal) Axis section, confirm the options are set to **Curve Arc Length** and **Curve** as shown.

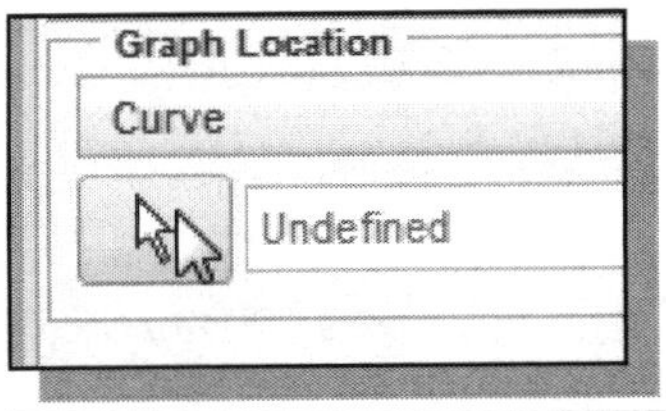

6. Click on the **arrow** icon to select curves to display the *shear diagram*.

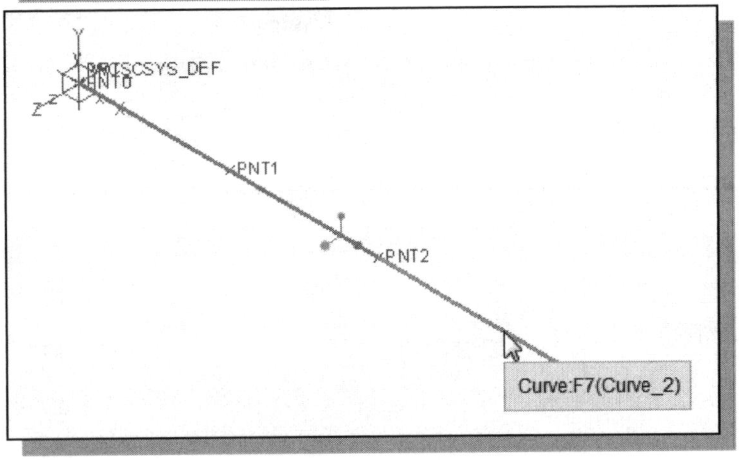

7. Press down the [**CTRL**] key and select both of the **datum curves, left curve first**.

8. Click once with the **middle-mouse-button** to accept the selection.

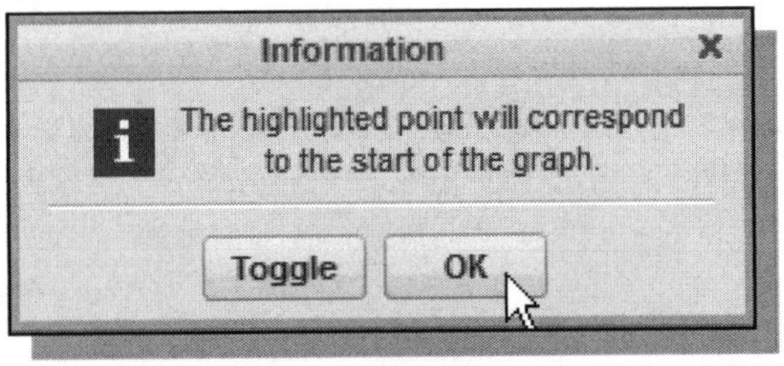

9. Confirm the **PNT0** point is highlighted and will be used as the start point of the graph.

10. Click **OK** to continue.

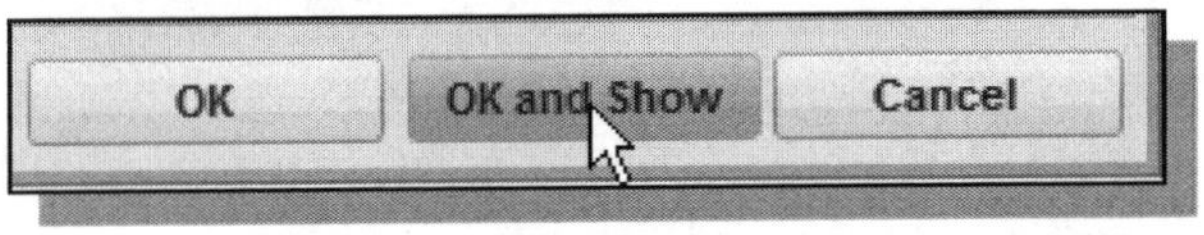

11. Click **OK and Show** to display the results.

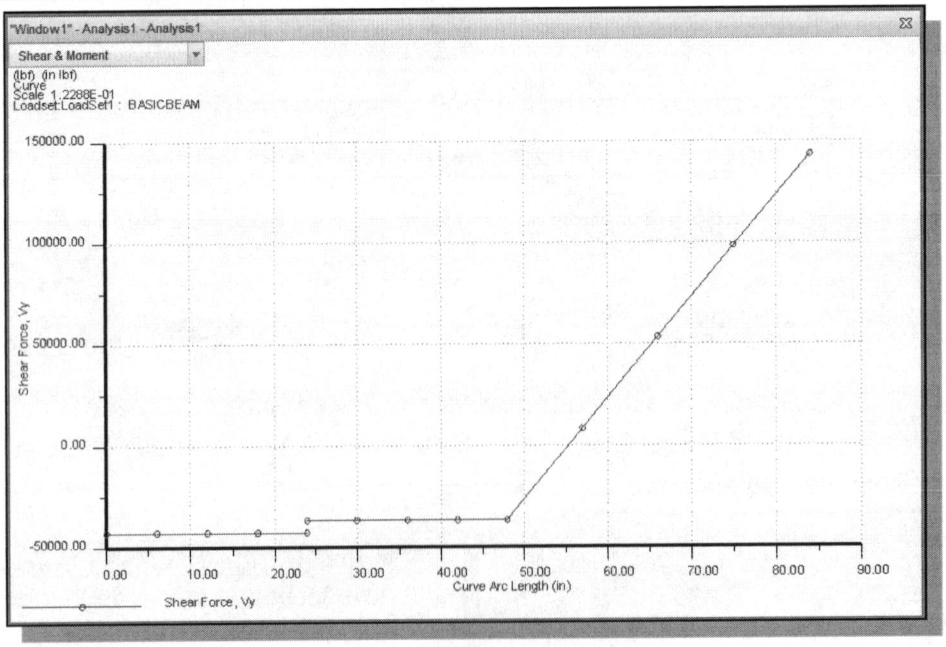

❖ The shear diagram shows the FEA model has a fairly large load on the right side of the beam.

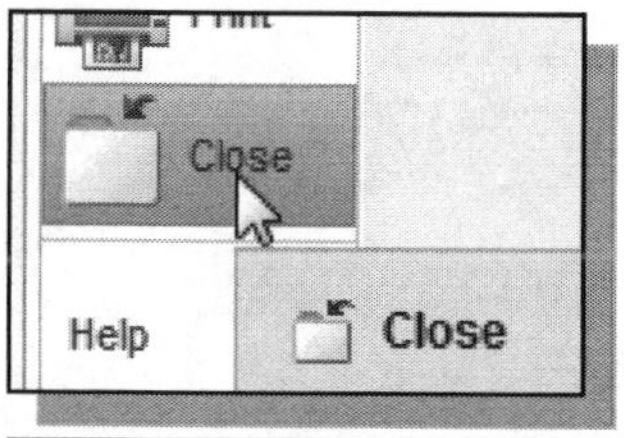

12. Select **File → Close** to close the *Results* window. Also close the *Analysis* window to return to the *Creo Simulate* window.

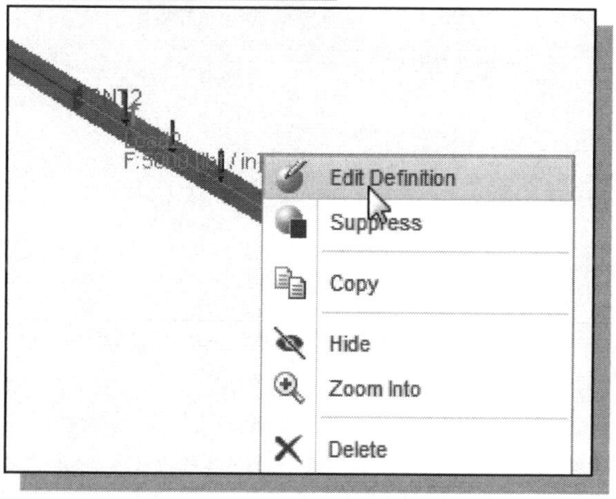

13. Select the **distributed load**, Load2, in the graphics window.

14. Press down the right-mouse-button to display the option list and select **Edit Definition** as shown.

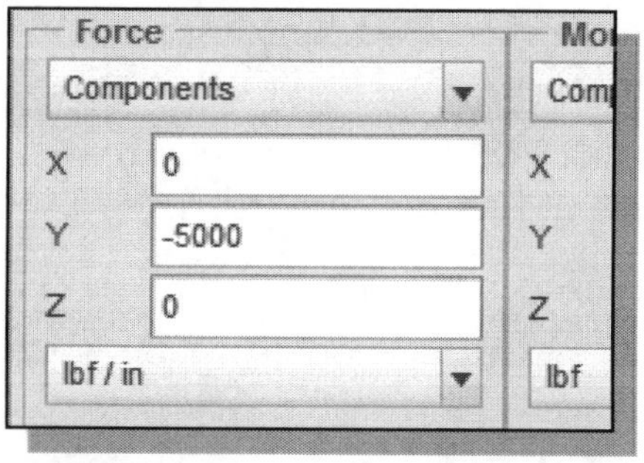

15. The distributed load we entered was **-5000 lb/in**; and it should be **5kip/ft**. The System of Units we are using is **inches-lbf**. **Note that the units used <u>MUST</u> be consistent throughout the analysis.**

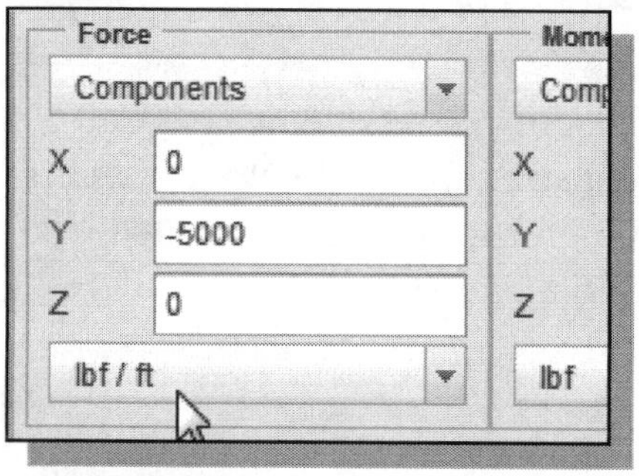

16. Choose the correct unit to reflect the distributed load as shown.

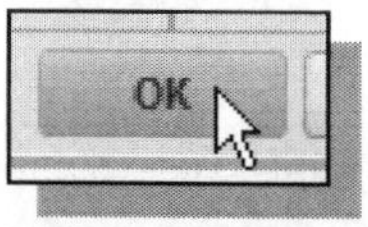

17. Click **OK** to accept the settings and exit the *Edit Definition* window.

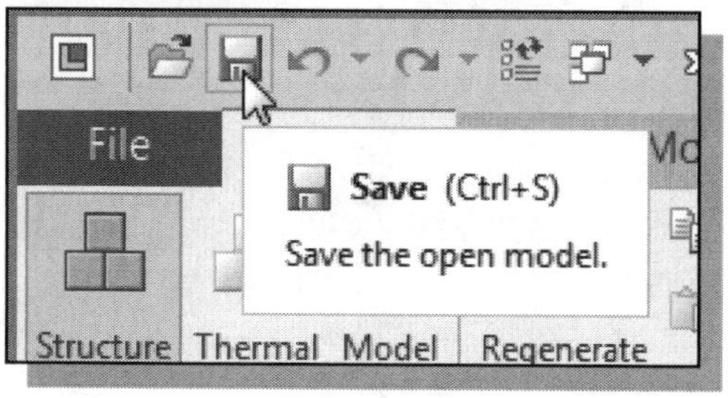

18. On your own, click **Save** and save a copy of the current FEA model.

Run the Solver

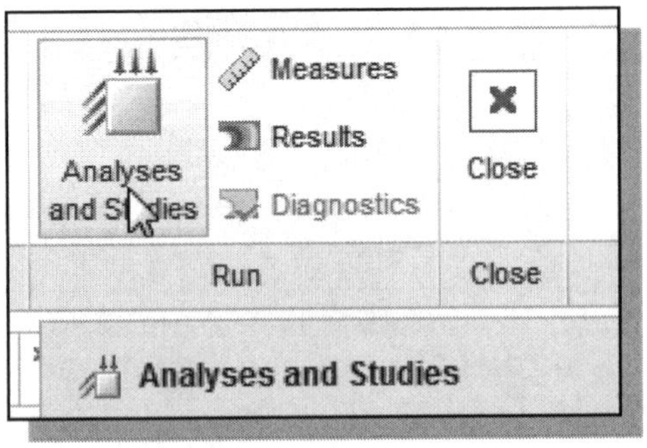

1. Activate the **Analyses and Studies** command by selecting the icon in the *Run* toolbar as shown.

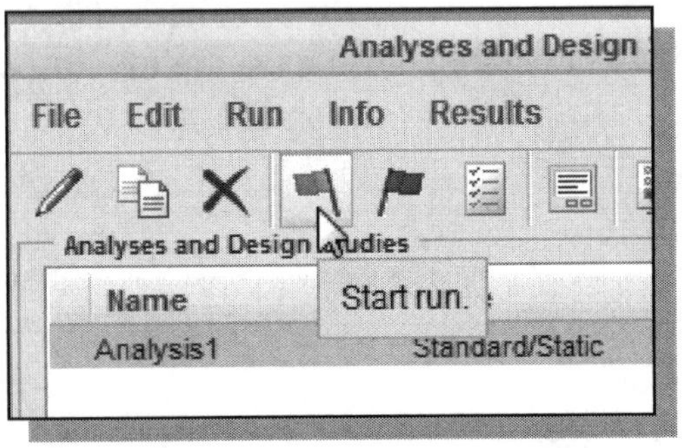

2. Click **Start run** to begin the *Solving* process.

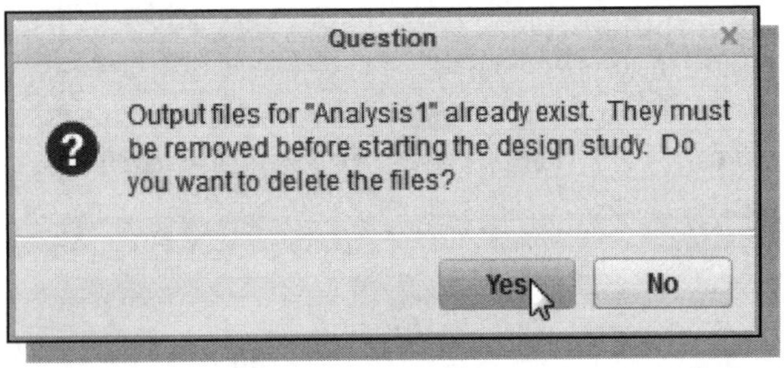

3. Click **Yes** to remove the old output files of "Analysis1."

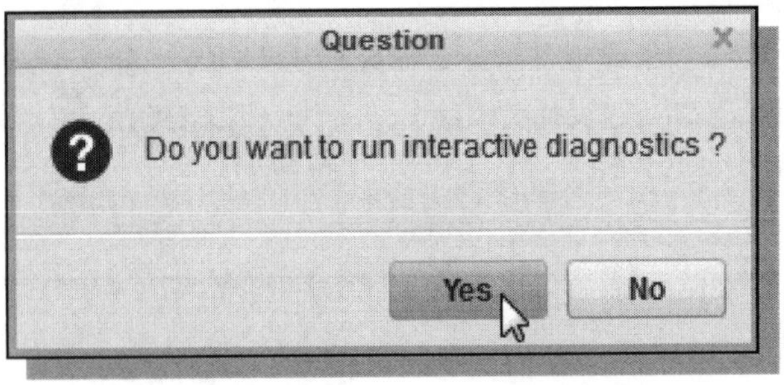

4. Click **Yes** to run interactive diagnostics; this will help us identify and correct problems associated with our FEA model.

5. Click on the **Close** button to exit the *Diagnostics* window.

6. In the *Analyses and Design Studies* window, choose **Review results** as shown.

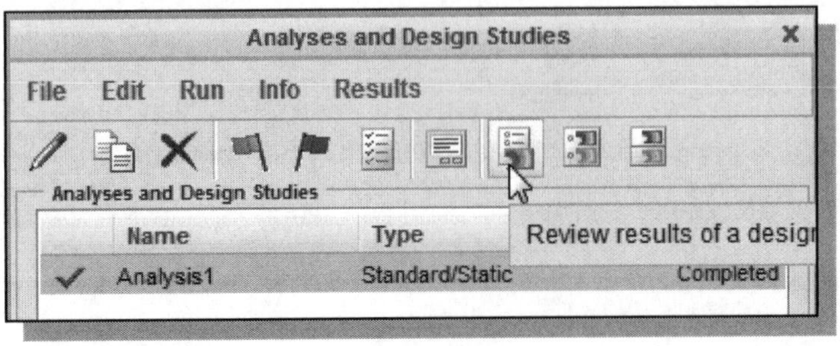

View the FEA results

➢ Reactions at supports

1. Choose **Model** as the Display type as shown.

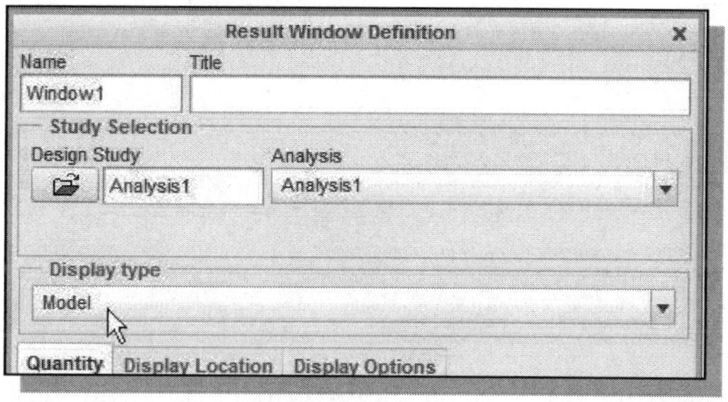

2. Choose **Reactions at Point Constraints** as the Quantity option.

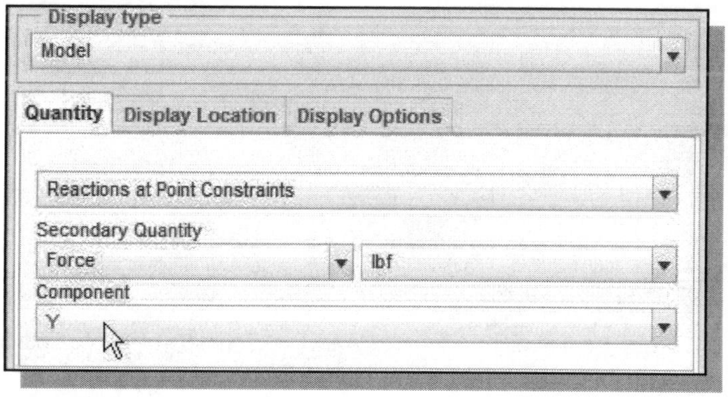

3. Choose **Force** as the Second Quantity option.

4. Choose **Y** as the force Component to be displayed.

5. Click on the **Display Options** tab and switch ***on*** the **Deformed** and **Overlay Undeformed** options as shown.

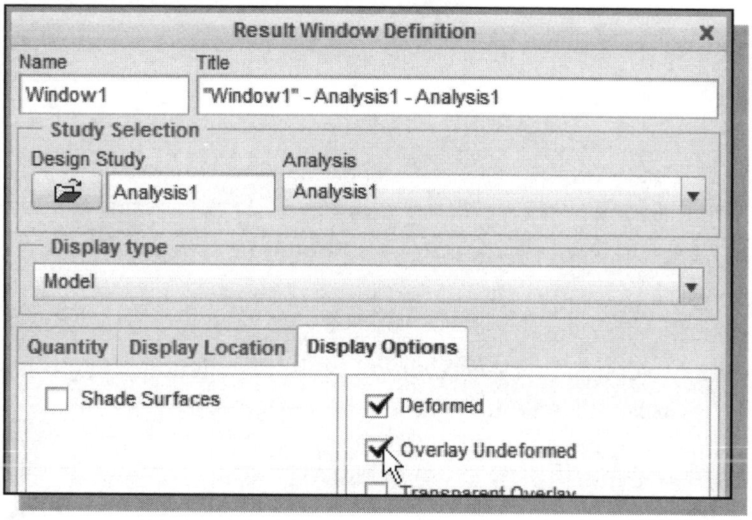

6. Click **OK and Show** to display the results.

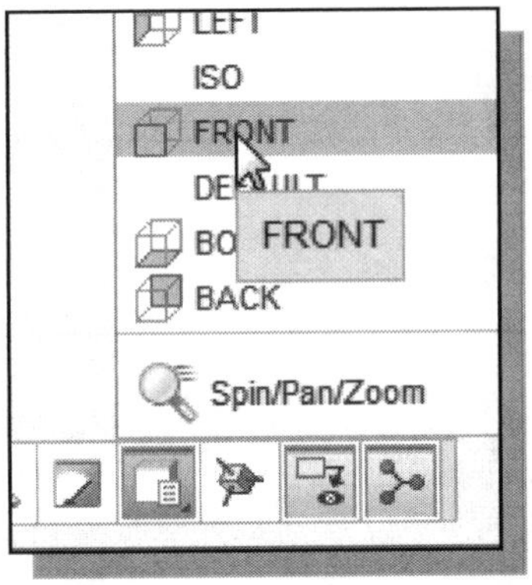

7. Select the **FRONT** view in the saved name list to adjust the display to the *Front* view.

❖ The *Creo Simulate* **Reaction Force** option allows us to quickly view the *Reactions at Point Constraints* in beam elements.

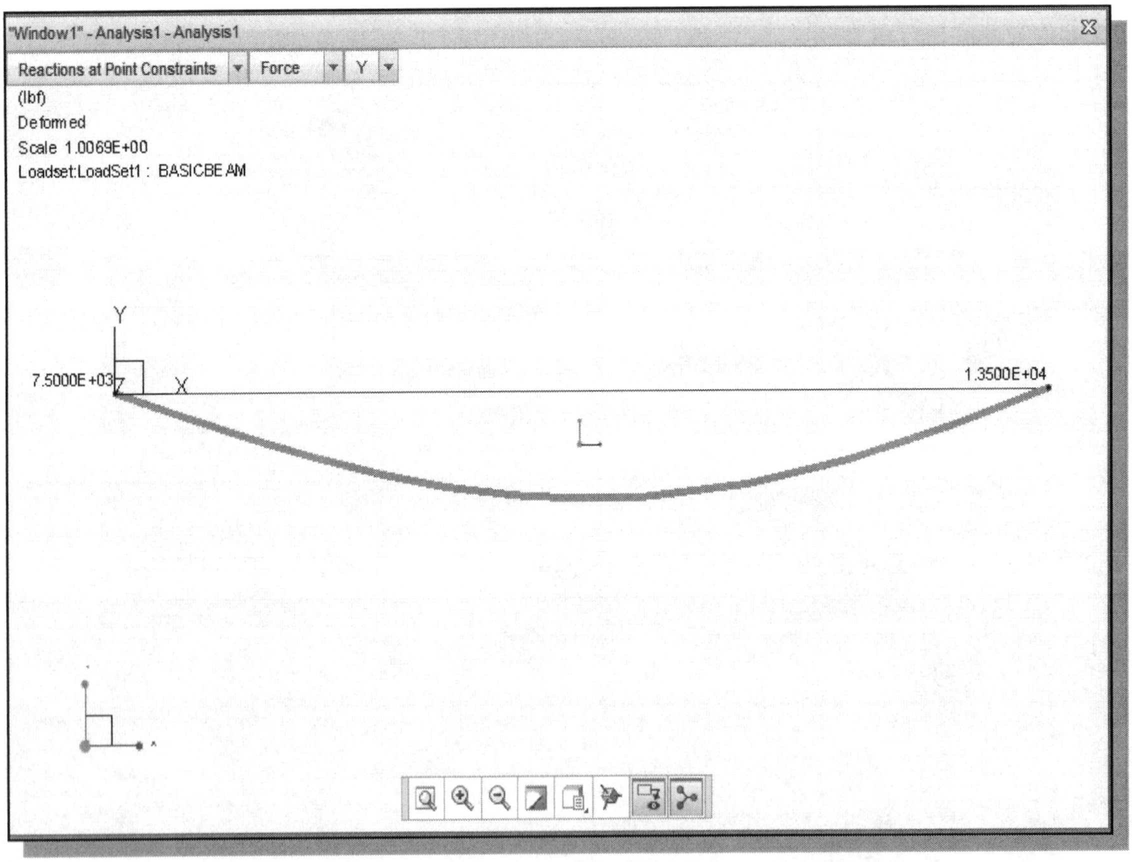

❖ Do values of the reactions at the supports match with the values obtained in our preliminary analysis?

❖ Note that the current version of *Creo Simulate* will not calculate reactions at points that have beam releases or points using a local coordinate system.

➢ Bending Stress

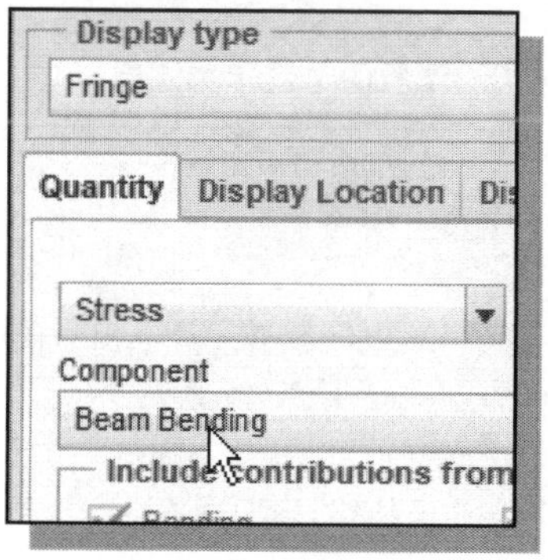

1. Use the **Edit** option and set the Display type to **Fringe**.

2. Choose **Beam Bending** as the Stress Component to be displayed.

3. Click on the **Display Options** tab and switch ***on*** the **Deformed** option as shown.

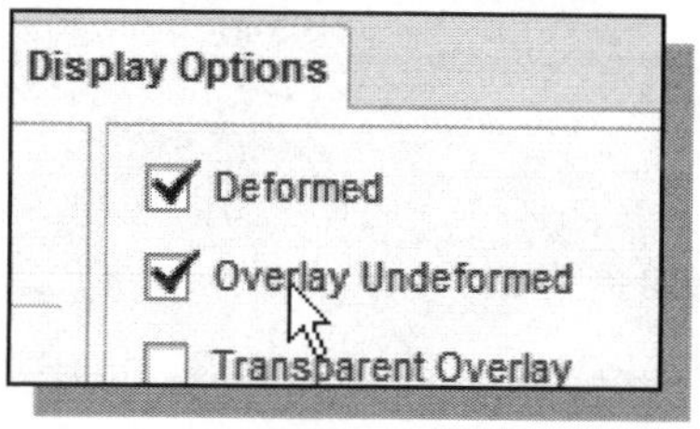

4. In the Display location tab, activate the **Use All** option as shown.

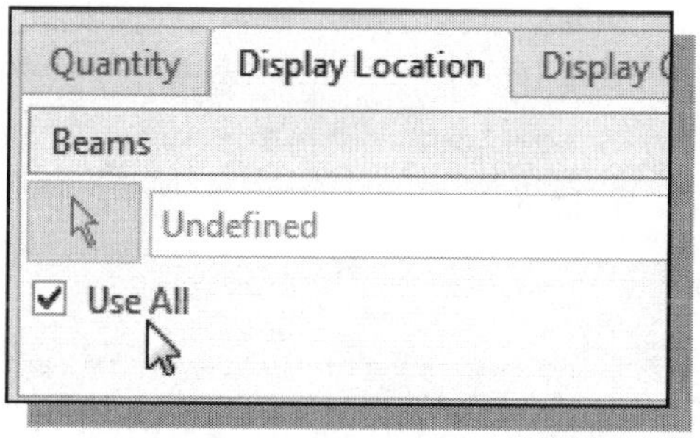

5. Click **OK and Show** to display the results.

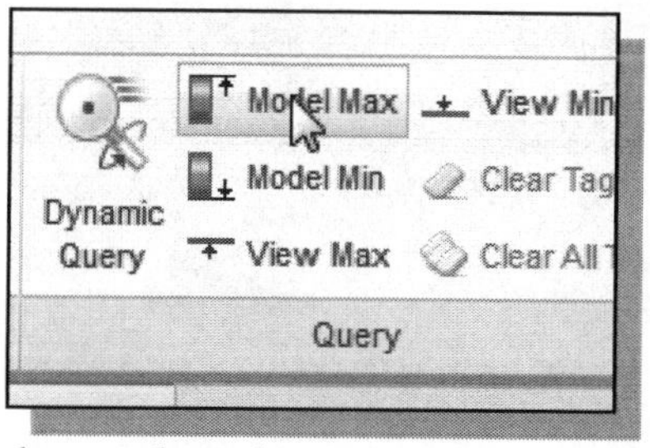

6. Click **Model Max** to display the max stress value and location.

➢ The max. bending stress is **3.24e5**, which is a bit lower than the max. stress value obtained during the preliminary analysis.

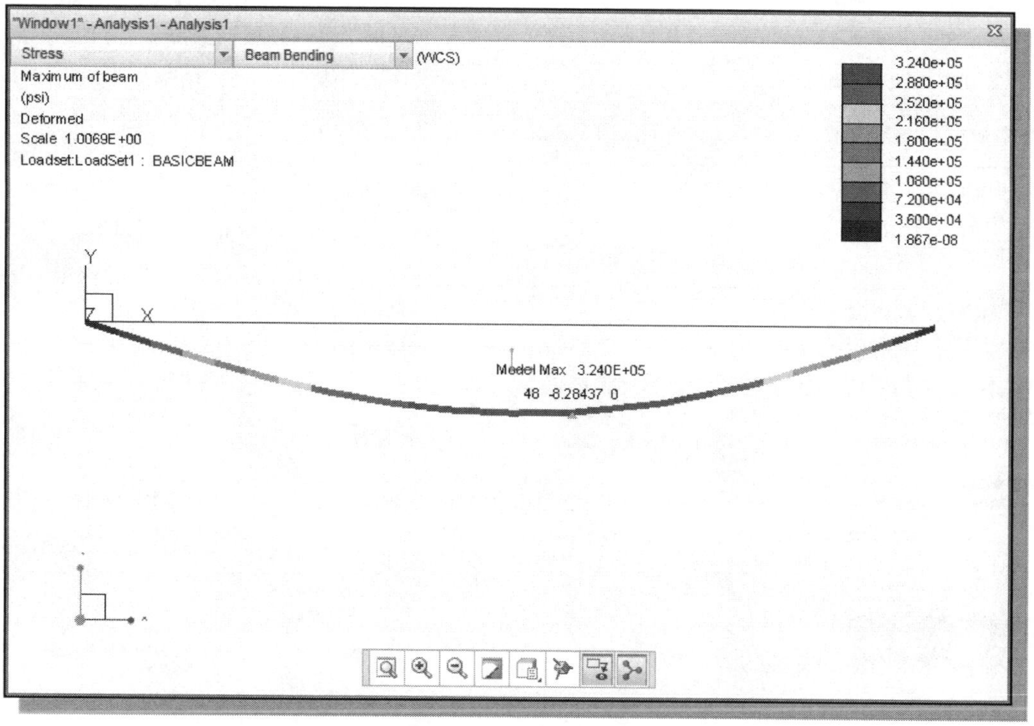

➢ Shear Diagram

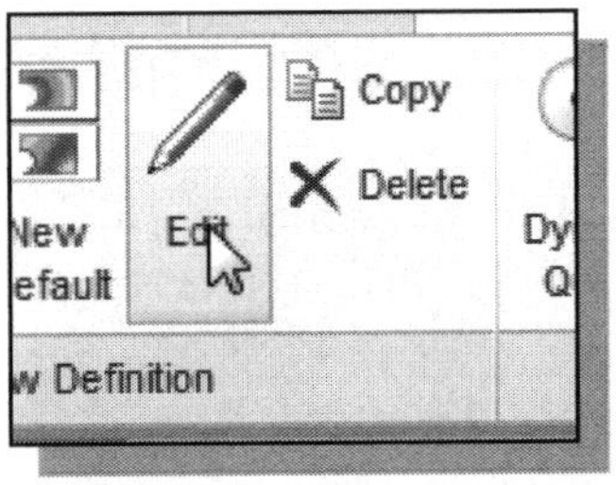

1. In the *Standard* toolbar area, click **Edit the current definition** command as shown.

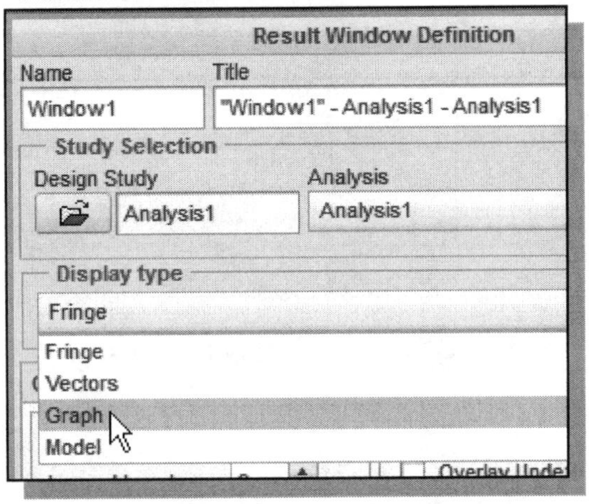

2. Choose **Graph** under the Display type option list as shown.

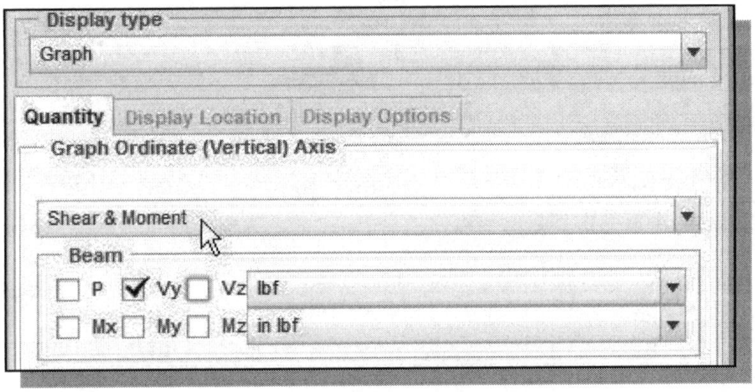

3. In the Graph Ordinate (Vertical) Axis option list, select **Shear & Moment** as shown.

4. Leaving only the **Y shear force** switched ***on***, turn ***off*** all of the other components as shown in the figure.

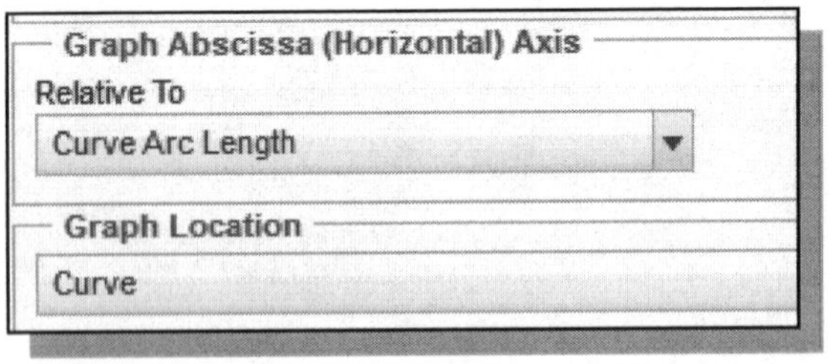

5. In the Graph Abscissa (Horizontal) Axis section, confirm the options are set to **Curve Arc Length** and **Curve** as shown.

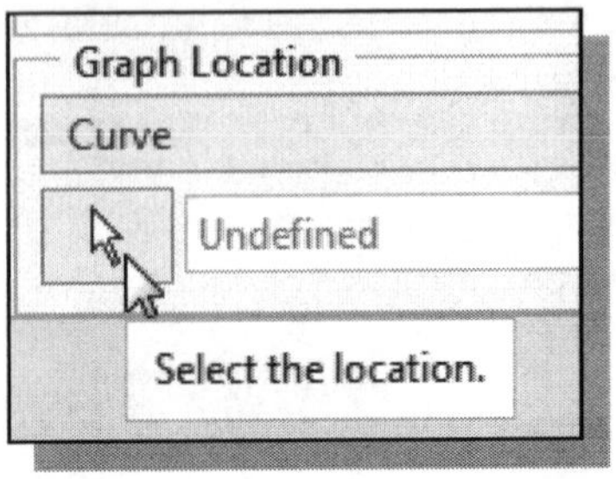

6. Click on the **arrow** icon to select curves to display the shear diagram.

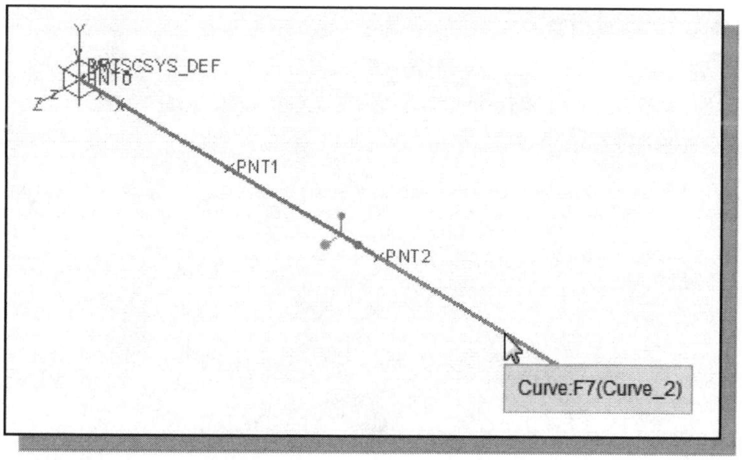

7. Press down the [**CTRL**] key and select both of the **datum curves**, **left curve first**.

8. Click once with the middle-mouse-button to accept the selection.

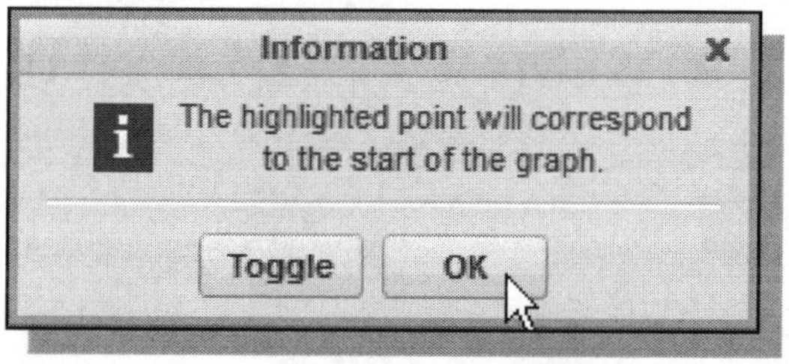

9. Confirm the **PNT0** point is highlighted and click **OK** to continue.

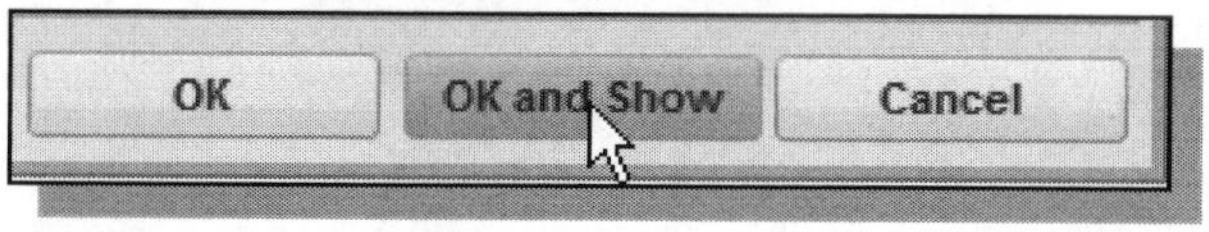

10. Click **OK and Show** to display the results.

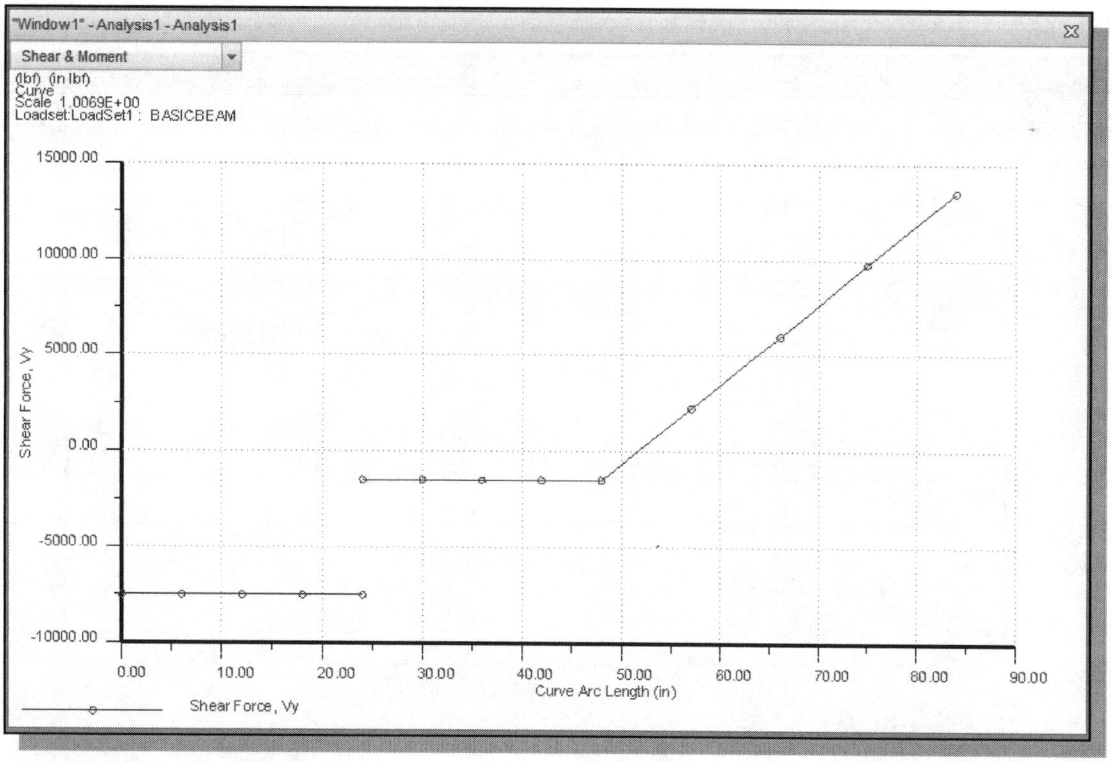

❖ The shear diagram is displaying a graph similar to the graph we constructed in the preliminary analysis.

➤ Moment Diagram

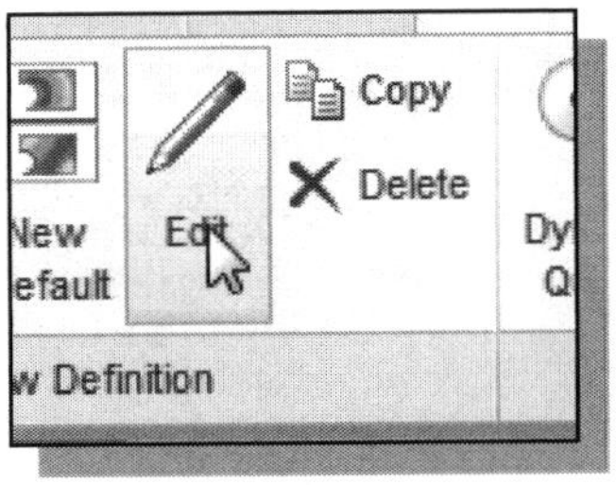

1. In the *Standard* toolbar area, click **Edit the current definition** command as shown.

2. Leaving only the **moment about Z** switched ***on***, turn ***off*** all of the other components as shown in the figure.

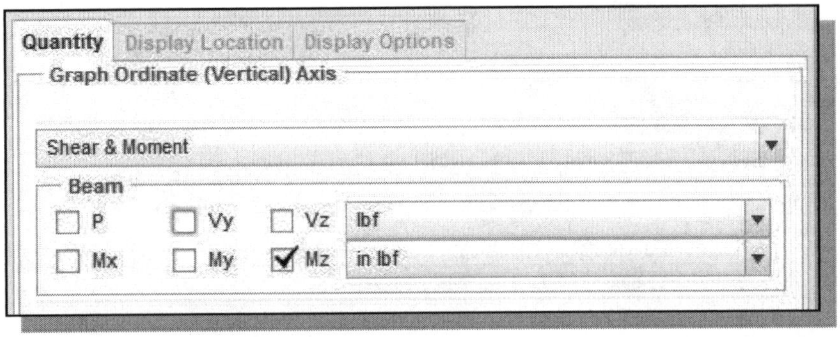

3. Click **OK and Show** to display the results.

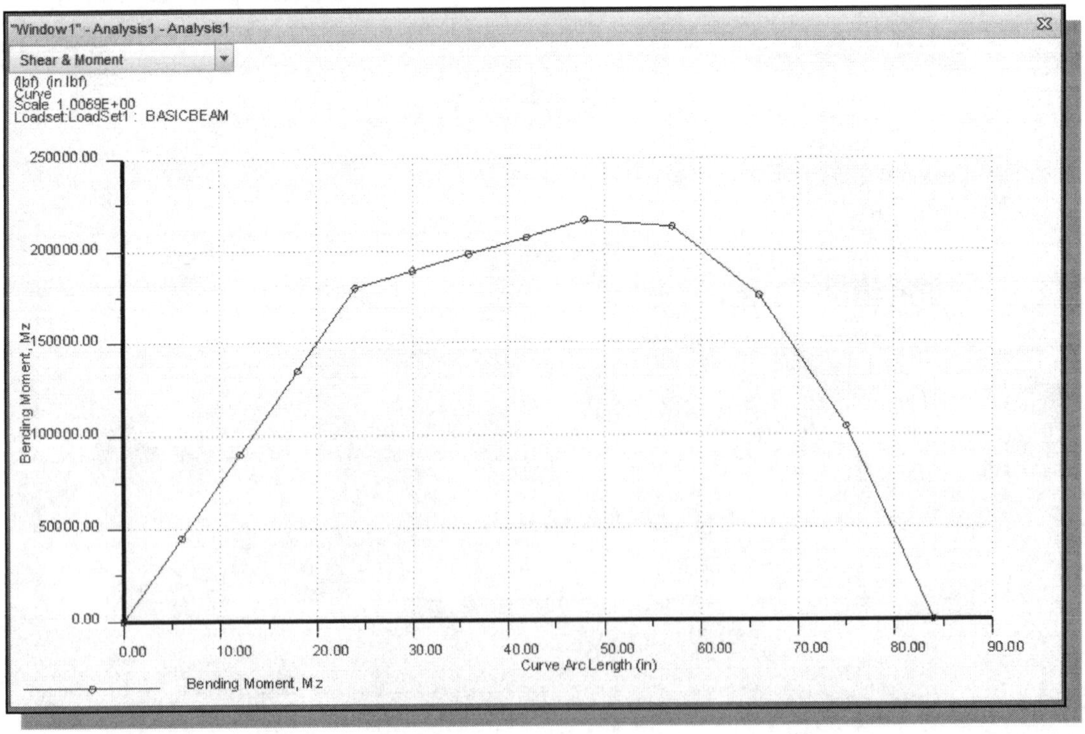

❖ The moment diagram is displaying a graph similar to the graph we constructed in the preliminary analysis, but the right side of the graph shows straight segments rather than a smooth curve.

Refine the FE Model

By default, *Creo Simulate* performs five calculations between user defined nodes in the FE model. To increase the number of calculations, we can simply increase the number of nodes.

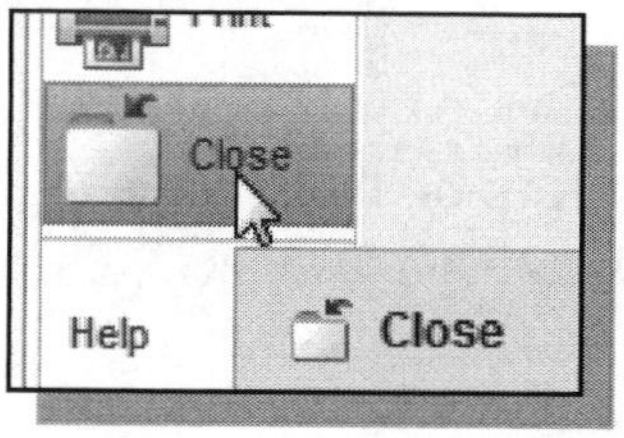

1. Select **File → Close** to close the *Results* window and return to the *Creo Simulate* window.

2. Select **Don't Save** to exit the save option.

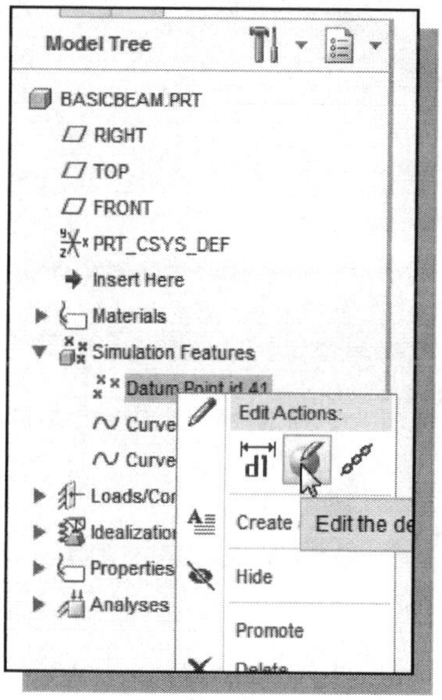

3. In the browser window, expand the **Simulation Features** by clicking on the [+] sign.

4. Press down the **right-mouse-button** on **Datum Point** to display the option list and select **Edit Definition** as shown.

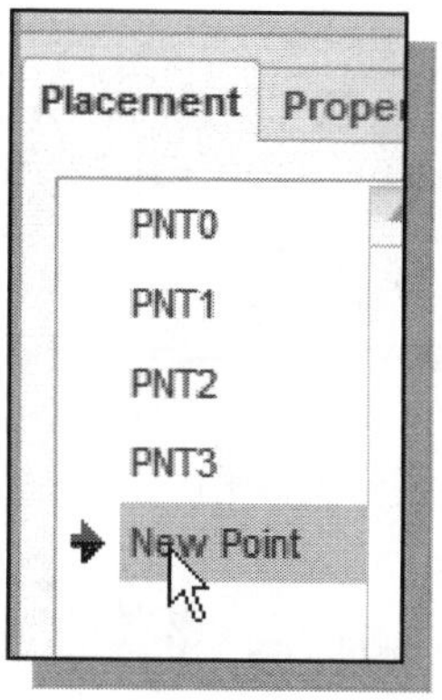

5. Click on the **Point** button to activate the **Create Points** command.

6. Create 3 additional datum points, PNT4, PNT5, and PNT6, as shown in the figure.

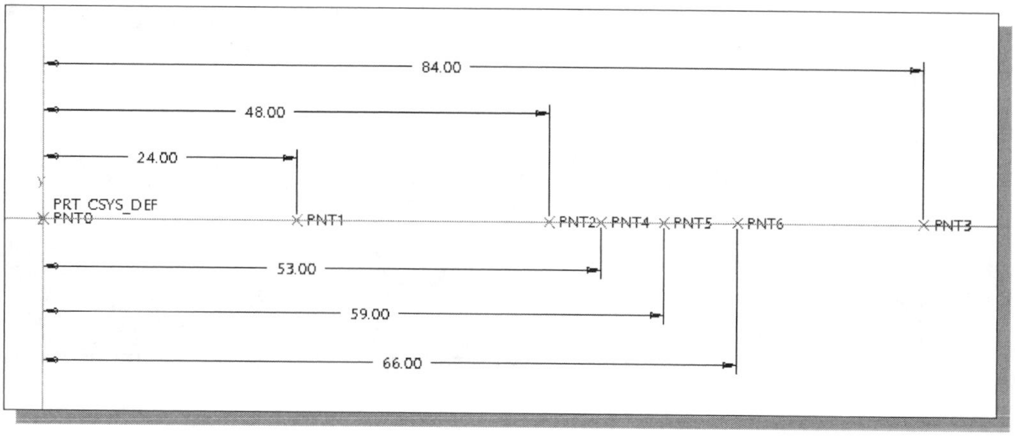

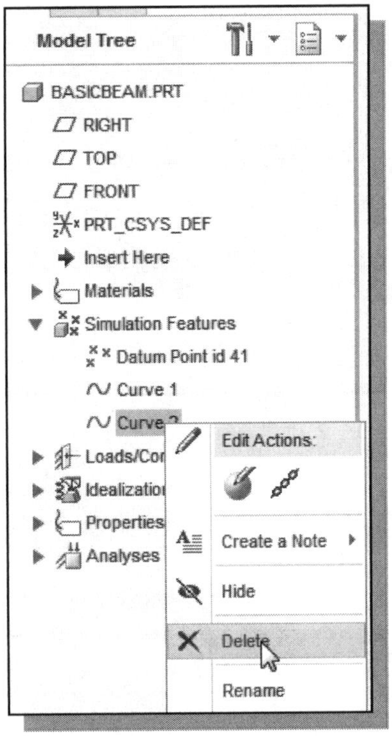

❖ Now we have created more nodes; we will also need to define new datum curves.

7. In the browser window, expand the **Simulation Features** by clicking on the [**+**] sign.

8. Press down the **right-mouse-button** on the **second datum curve** to display the option list and select **Delete** as shown.

9. Click **Yes** to delete associated features.

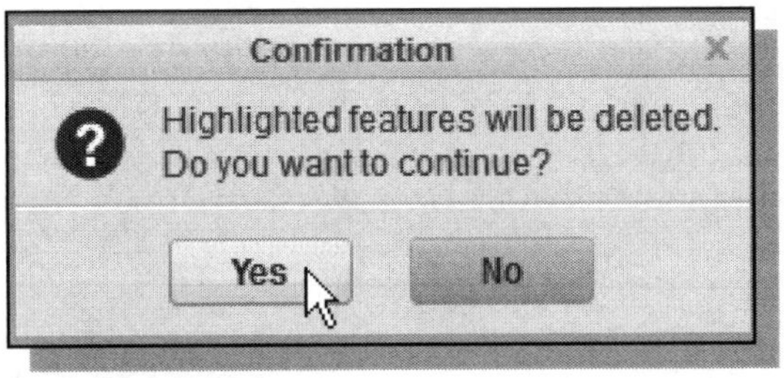

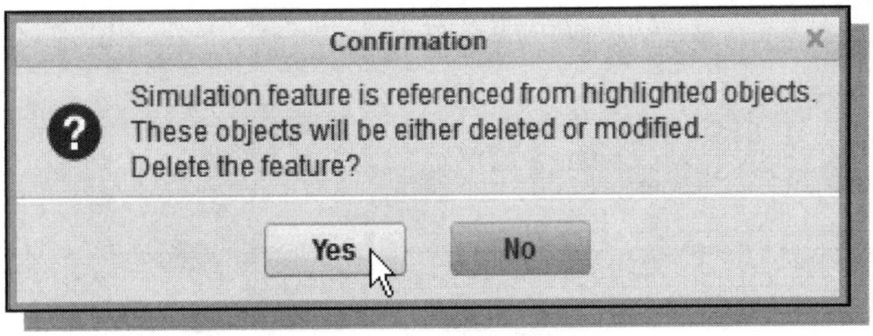

10. Click **Yes** to delete associated features.

11. Switch to the **Refine model** tab.

12. Choose **Datum → Curve→ Curve through Points** tool in the *Datum* toolbar displayed in the *Creo Ribbon* toolbar.

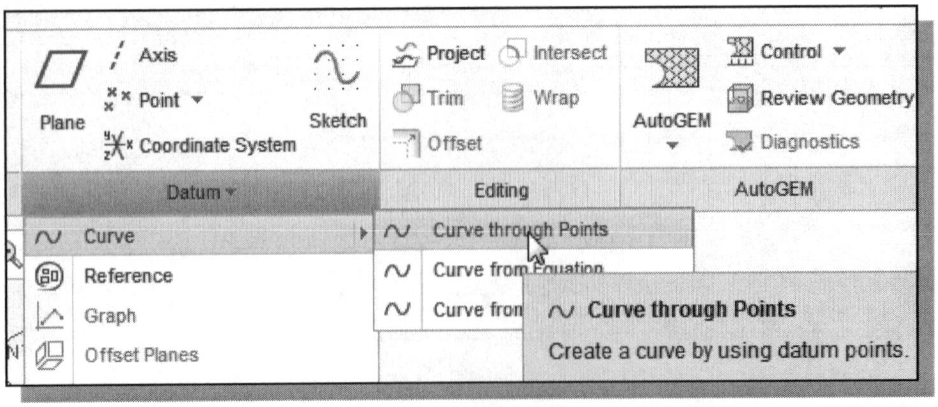

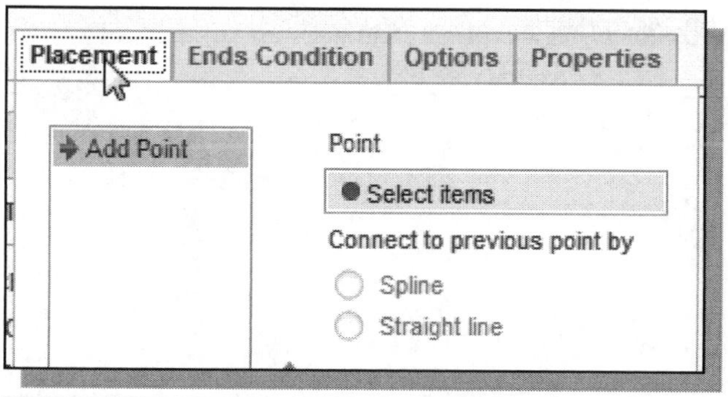

13. In the *Dashboard* area, click the **Placement** tab to view the available options.

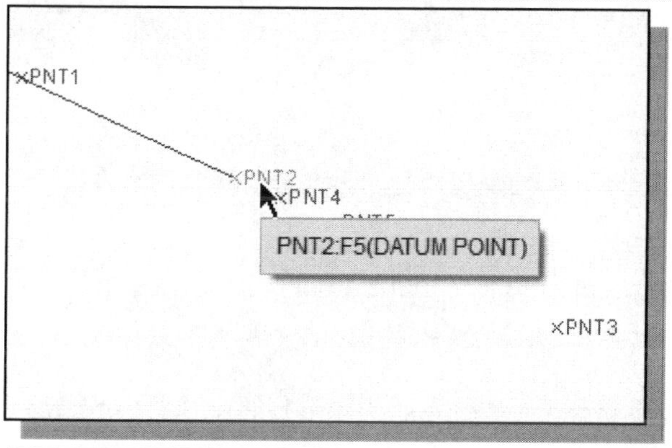

14. In the *graphics area*, select **PNT2** as the first point for the curve definition.

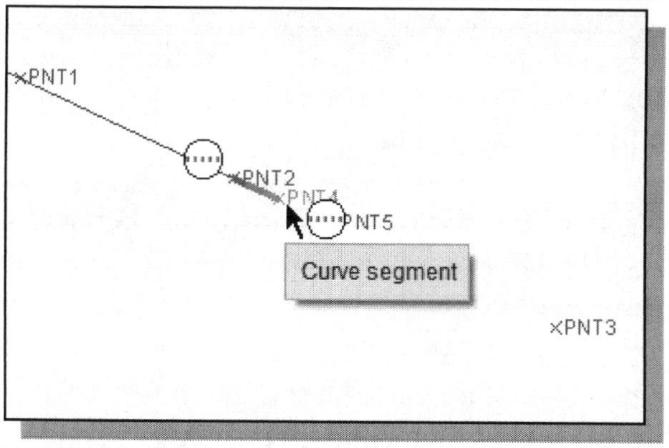

15. In the graphics area, select **PNT4** as the second point for the curve definition.

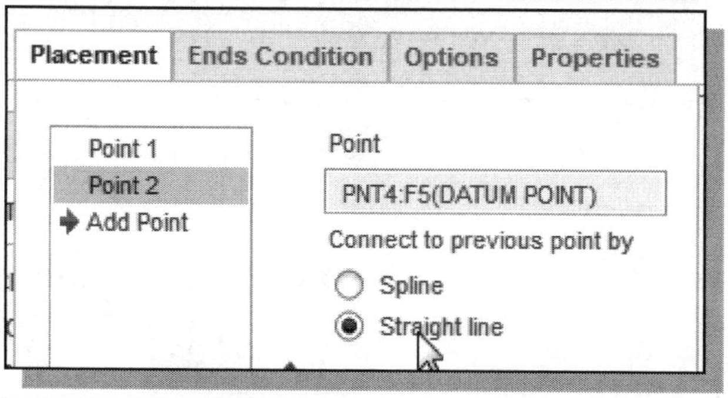

16. Select **Straight line** to define the connection option.

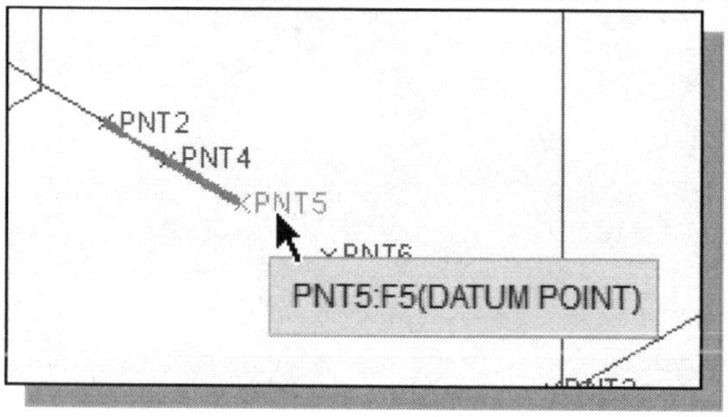

17. In the graphics area, select **PNT5** as the third point for the curve definition.

18. On your own, continue to select **PNT6** and **PNT3** to form the second datum curve.

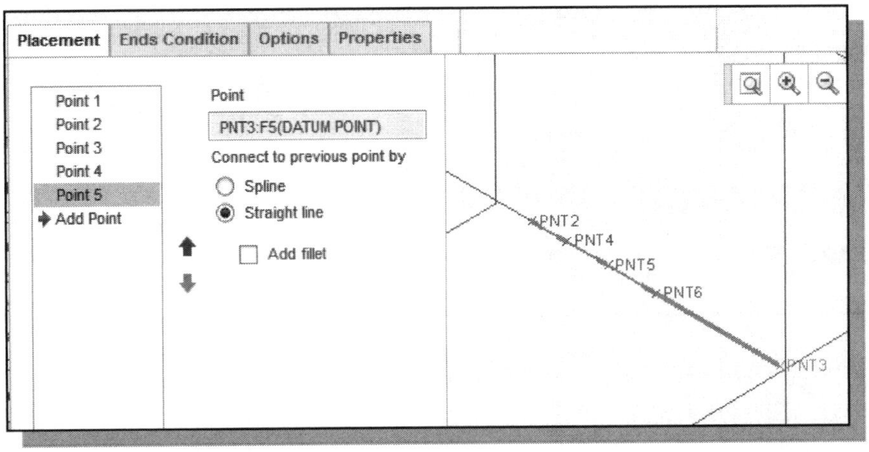

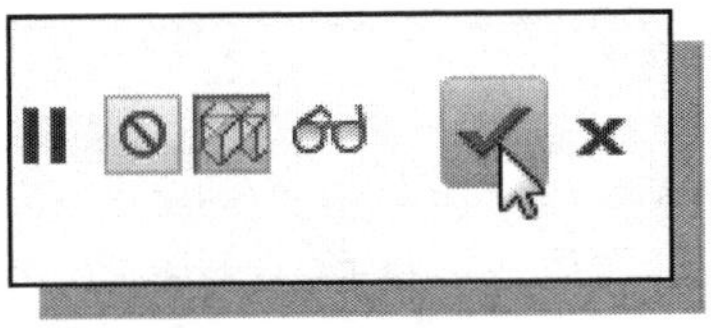

19. Click **OK** to create the new datum curve.

20. With the new set of datum curves, we will also need to update the beam elements for the system.

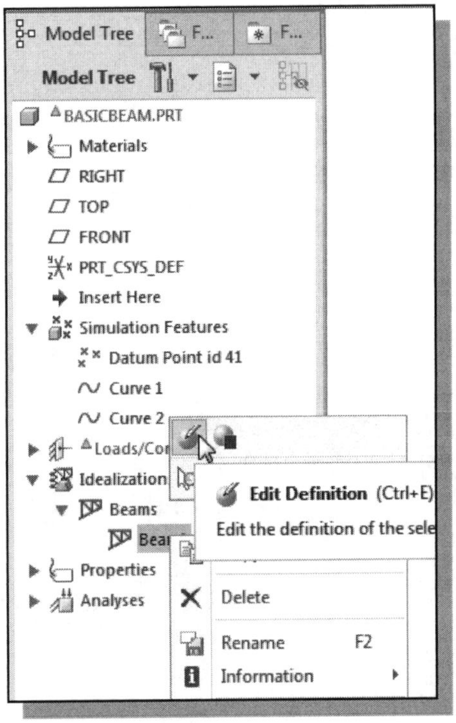

21. In the *browser window*, expand the **Idealizations** by clicking on the [**+**] sign.

22. Press down the **right-mouse-button** on **Beam1** to display the option list and select **Edit Definition** as shown.

23. Right click on the **second curve**, in the References list, and choose **Remove**. This is the datum curve we have deleted.

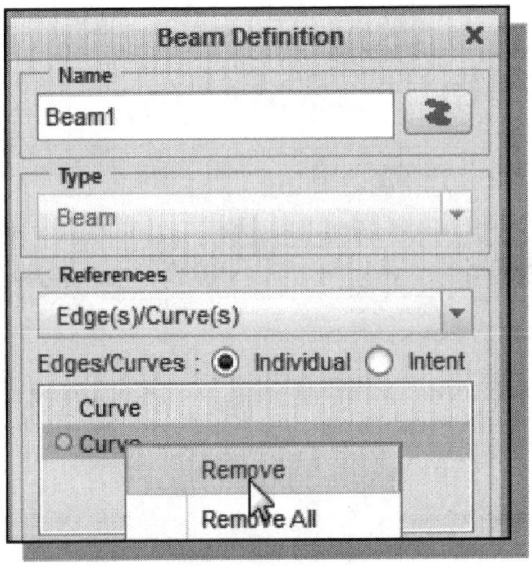

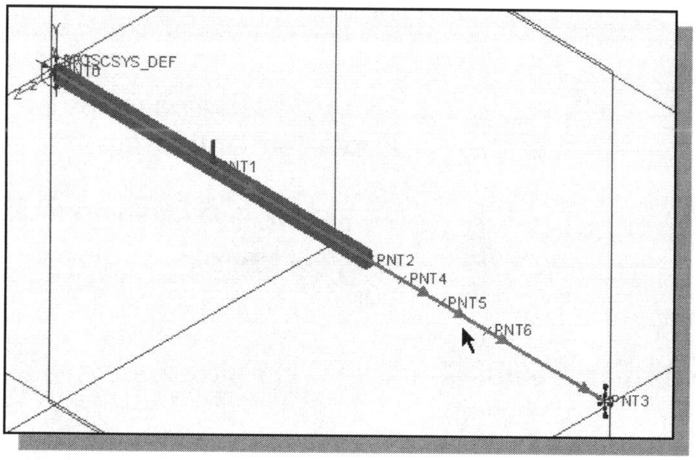

24. Hold down the [**CTRL**] key and select the new four datum curve we just created.

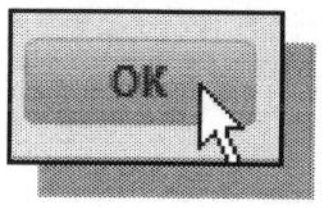

25. Click **OK** to accept the new settings.

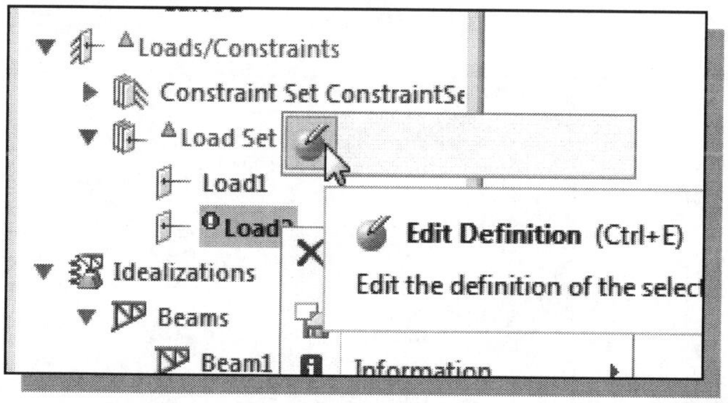

26. Press down the **right-mouse-button** on **Load2** to display the option list and select **Edit Definition** as shown.

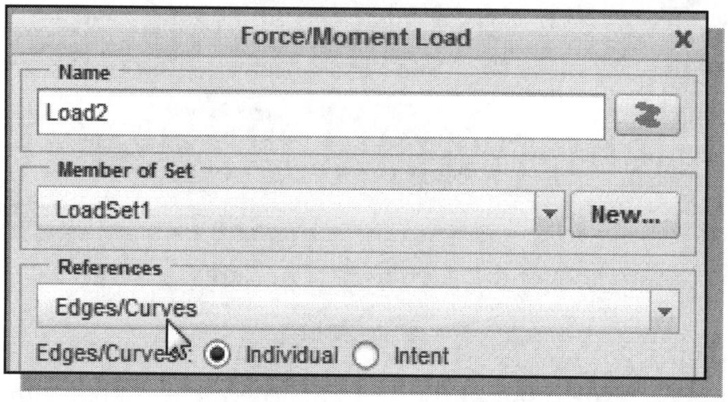

27. Confirm the *Load* set is still set to **LoadSet1**.

28. Set the References to **Edges/ Curves** as shown.

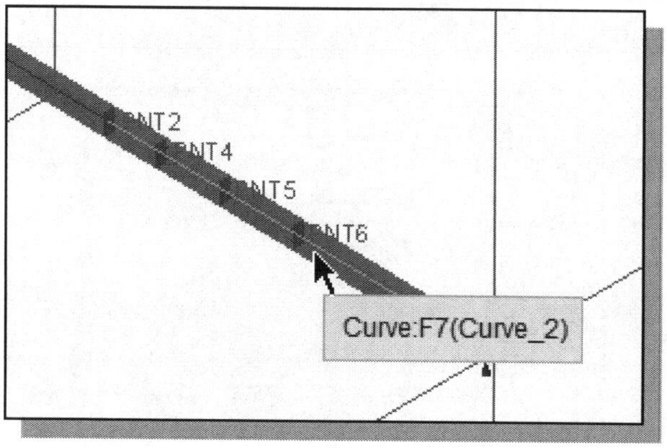

29. Pick the **second datum curve** we just created as shown.

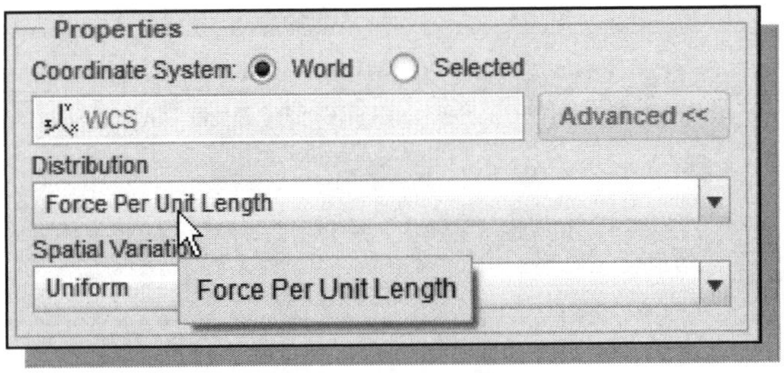

30. Set the Distribution option to **Force Per Unit Length**.

31. Set the Spatial Validation option to **Uniform** as shown.

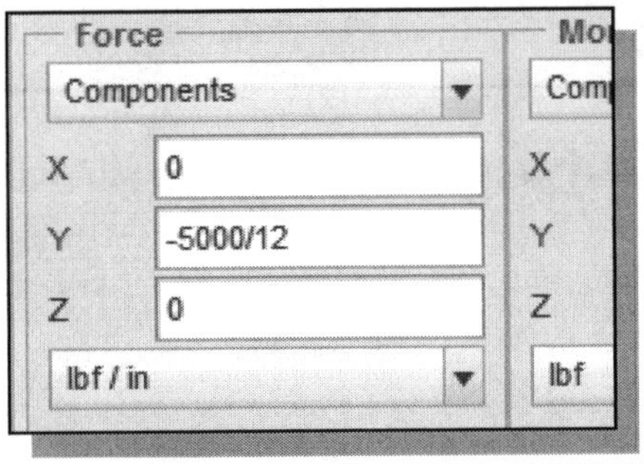

32. Enter **-5000/12** in the Y component Force value box as shown.

    (Load: **5000** lbf/ft **=5000/12** lbf/in)

33. Click on the **Preview** button to confirm the load is applied correctly to the system.

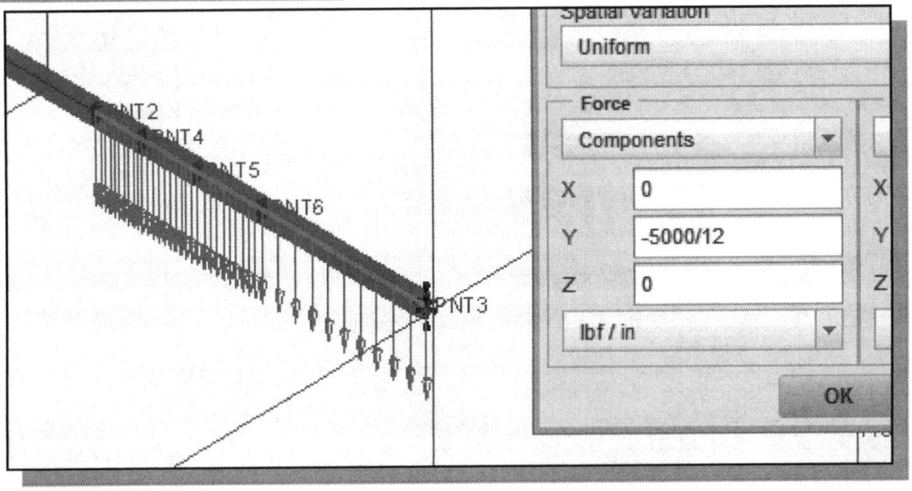

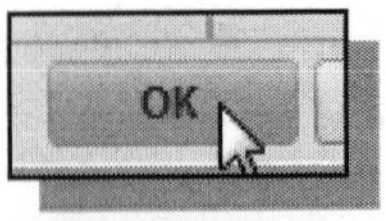

34. Click on the **OK** button to accept the second load settings.

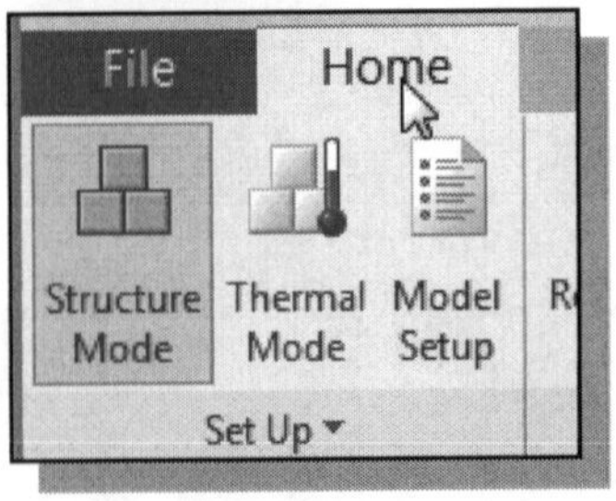

35. Click on the **Home** tab to return to the main *Creo Simulate* commands toolbar.

36. Activate the **Analyses and Studies** command by selecting the icon in the *Run* toolbar as shown.

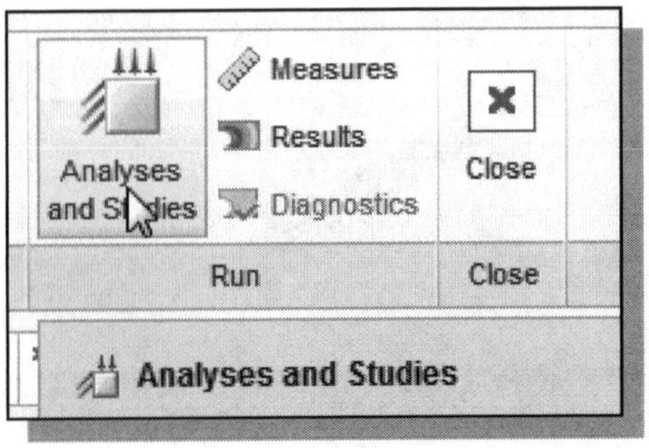

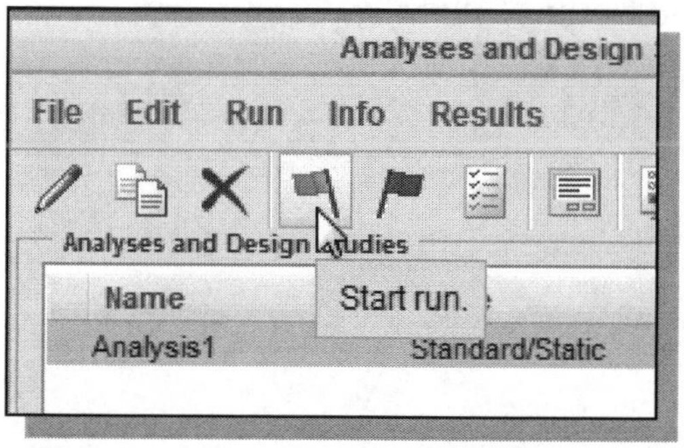

37. Click **Start run** to begin the *Solving* process.

38. On your own, perform the necessary FEA analysis and examine the resulting bending stress, shear and moment diagrams.

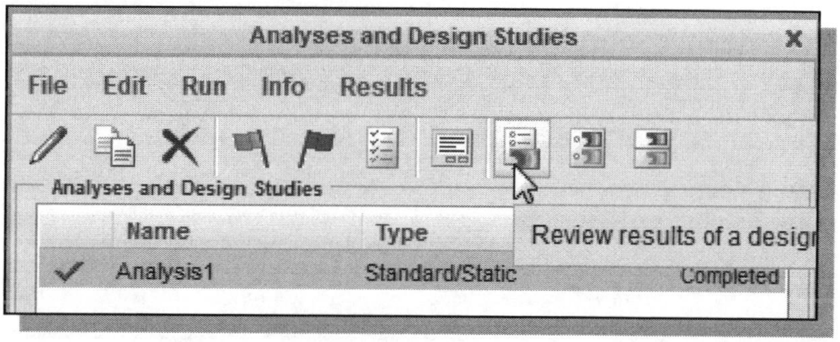

➢ Note that *Creo Simulate* has added additional node points and performed more calculations to produce a much better curve near the high stress area of concern.

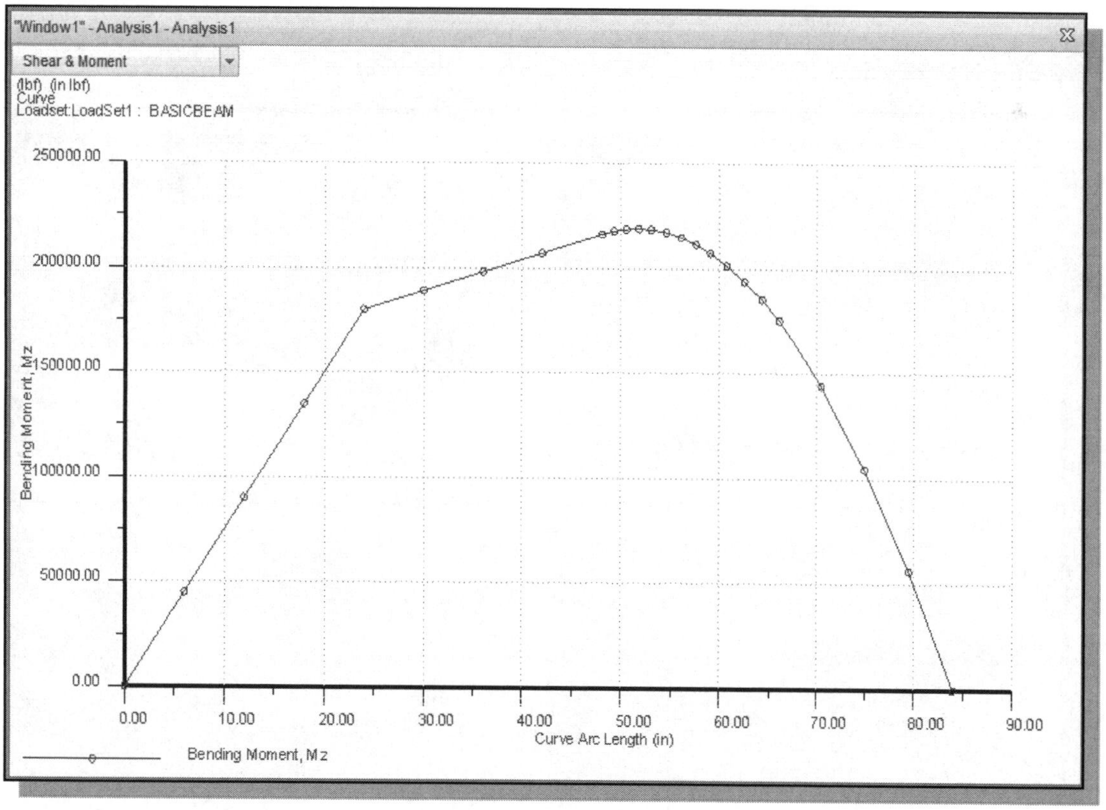

➤ On your own, examine the Y-Shear diagram and compare it to the preliminary analysis done at the beginning of the chapter. Is there any difference?

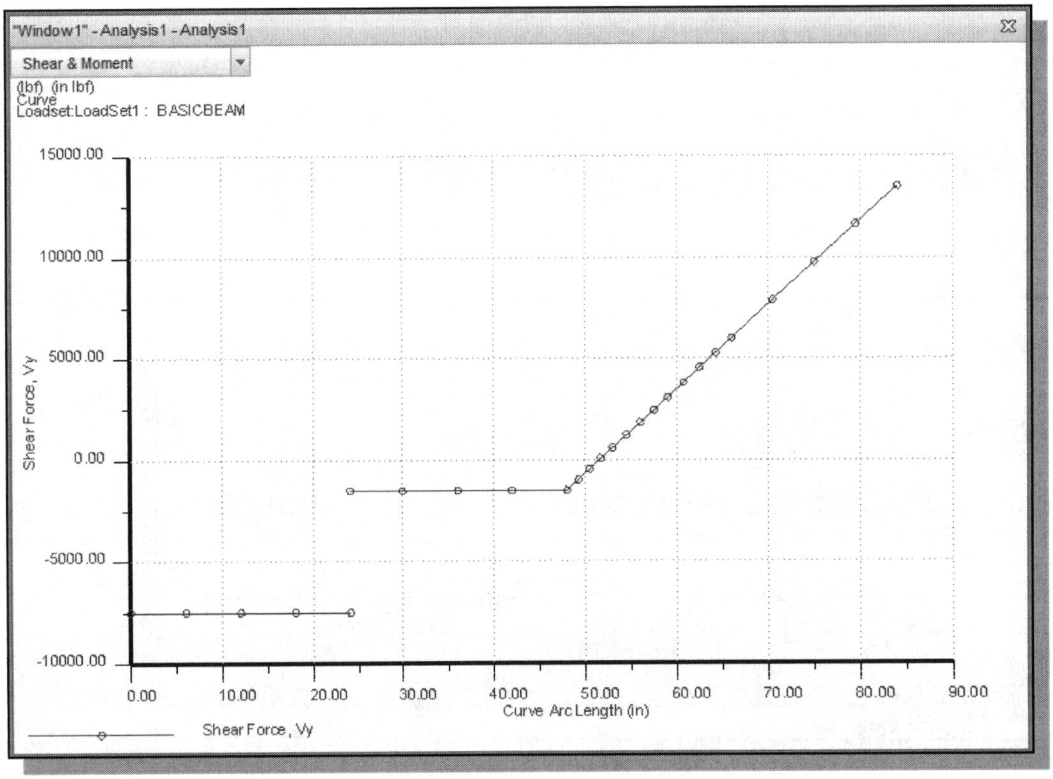

➤ Note that *Creo Simulate's* calculated max bending stress also matches the result of the preliminary analysis.

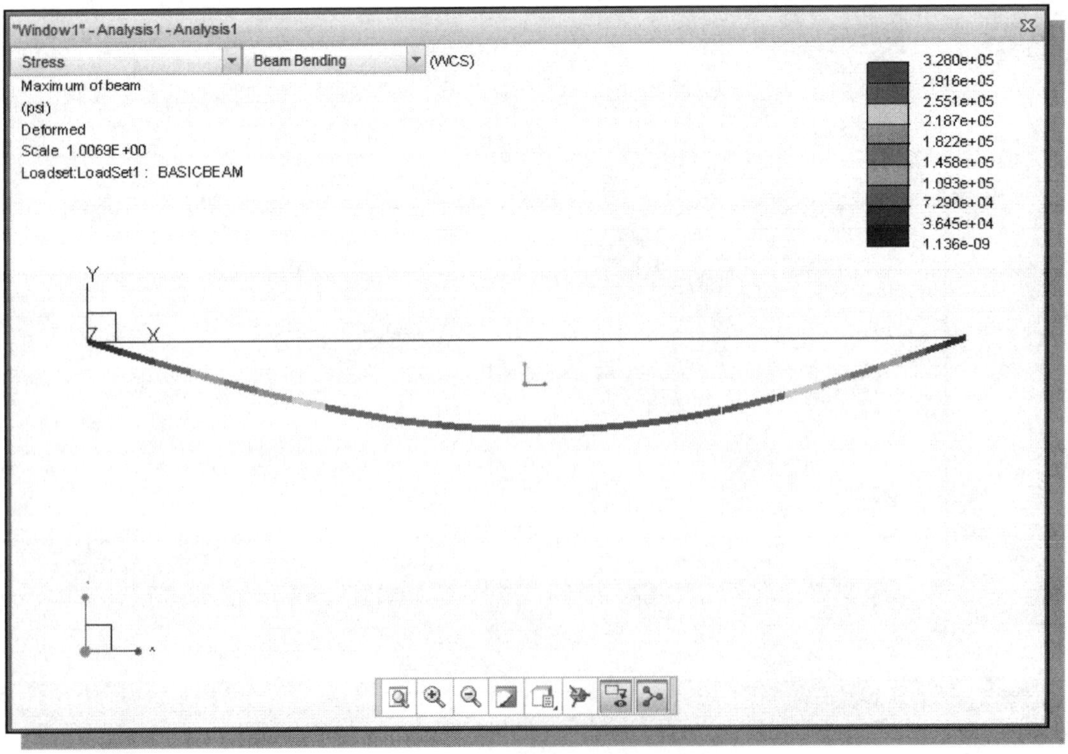

Review Questions

1. For a *beam element* in three-dimensional space, what is the number of degrees of freedom it possesses?

2. What are the assumptions for the beam element?

3. What are the differences between *truss element* and *beam element*?

4. What are the relationships between the shear diagram and the moment diagram?

5. For beam elements, will the *Creo Simulate* FEA software calculate the reaction forces at the supports?

6. Can we apply both a point load and a distributed load to the same beam element?

7. How do we improve the quality of the curves shown in the shear and moment diagrams?

8. For a 2D roller support in a beam system, how should we set the constraints in *Creo Simulate*?

9. What are the main differences between a *Creo Parametric* template file and a regular *Creo Parametric* part file?

10. List and describe the general procedure to set up a display view in *Creo Parametric*.

Exercises

Determine the maximum stress produced by the loads and create the shear and moment diagram.

1. Material: Steel; Diameter 2.5 in.

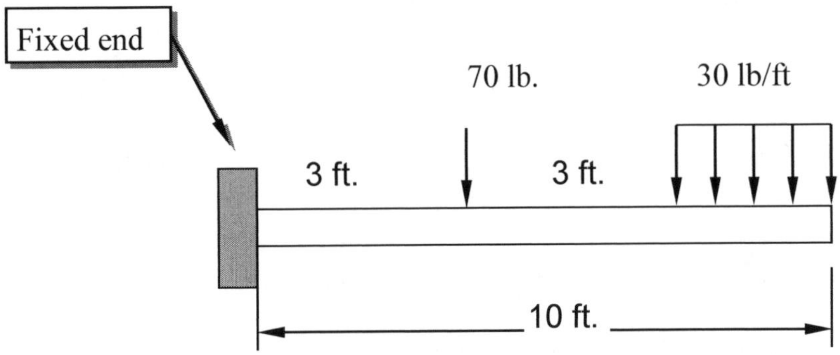

2. Material: Aluminum

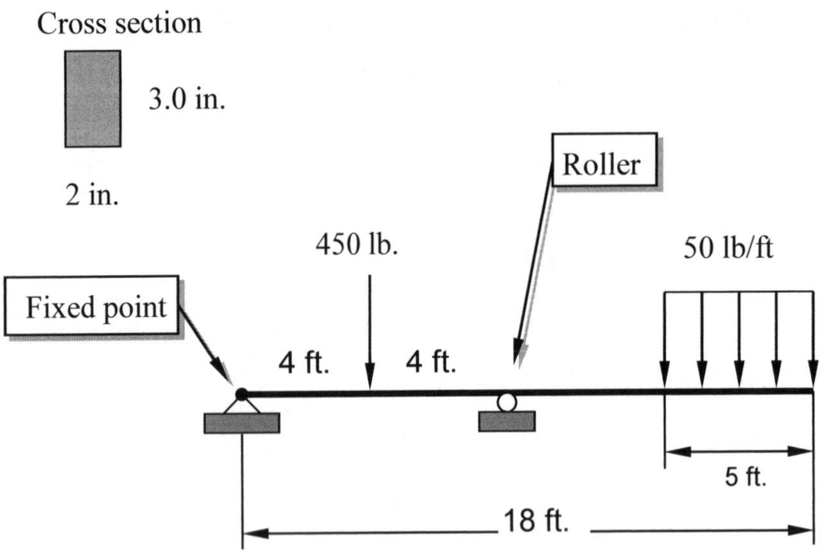

Chapter 7
Beam Analysis Tools

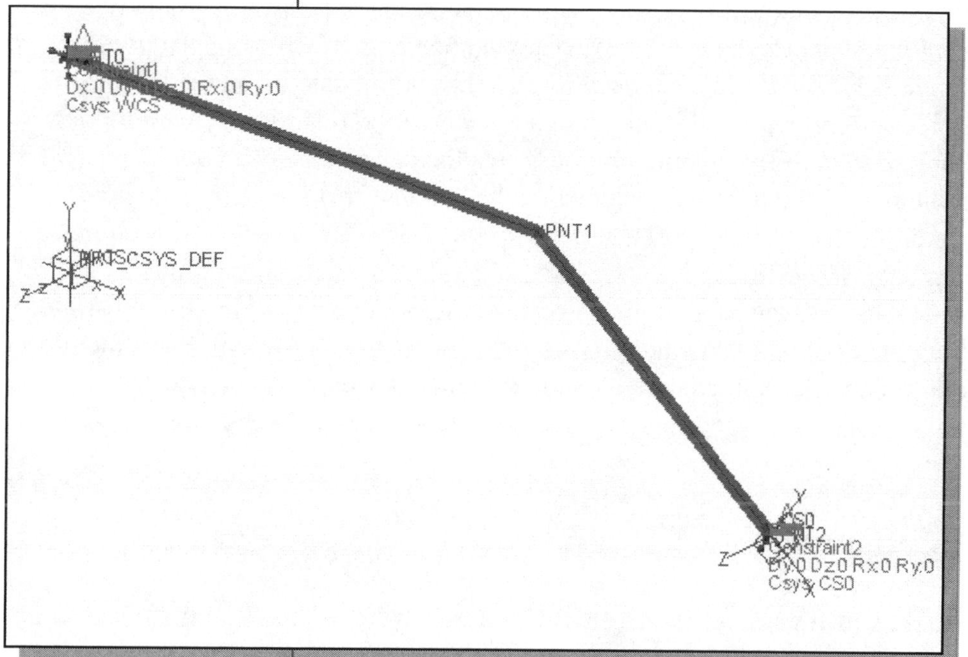

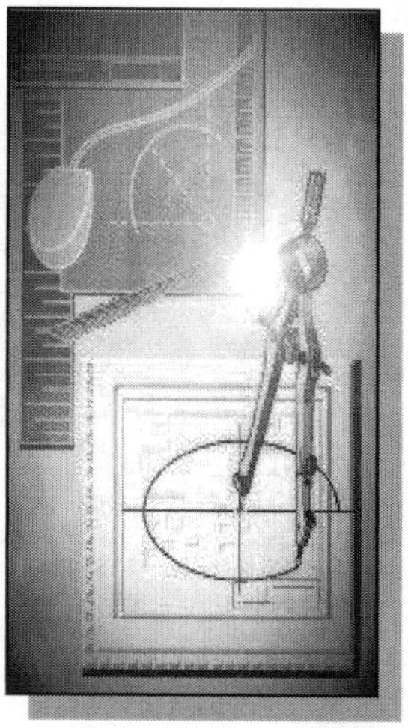

Learning Objectives

- **Create FE models using the AutoGEM command.**
- **Create User Coordinate Systems.**
- **Create Inclined Supports.**
- **Analyze Structures with Combined Stresses.**
- **View the Results with the Animation options.**

Introduction

The beam element is the most common of all structural elements as it is commonly used in buildings, towers, bridges and many other structures. The use of beam elements in finite element models provides the engineer with the capability to solve rather complex structures, which could not be easily done with conventional approaches. In many cases, the use of beam elements also provides a very good approximation of the actual three-dimensional members, and there is no need to do a three-dimensional analysis. *Creo Simulate* provides an assortment of tools to make the creation of finite element models easier and more efficient. This chapter demonstrates the use of *automatic mesh-generation* to help generate nodes and elements, the use of a *displacement coordinate system* to account for inclined supports, and the application of *elemental loads* along individual elements. The effects of several internal loads that occur simultaneously on a member's cross-section, such as axial load and bending, are considered in the illustrated example. Although the illustrated example is a two-dimensional problem, the principle can also be applied to three-dimensional beam problems as well as other types of elements.

Problem Statement

Determine the state of stress at *point C* (measured 1.5 m from *point A*), and the maximum normal stress that the load produces in the member (Structural Steel A36).

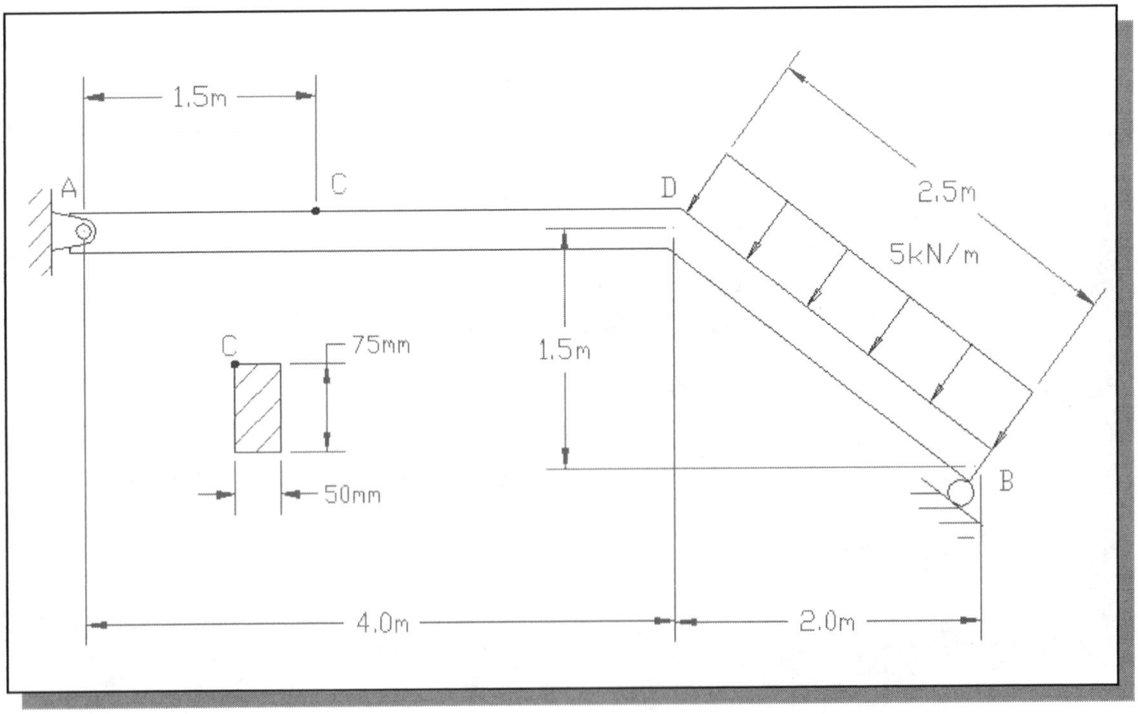

Preliminary Analysis

Free Body Diagram of the member:

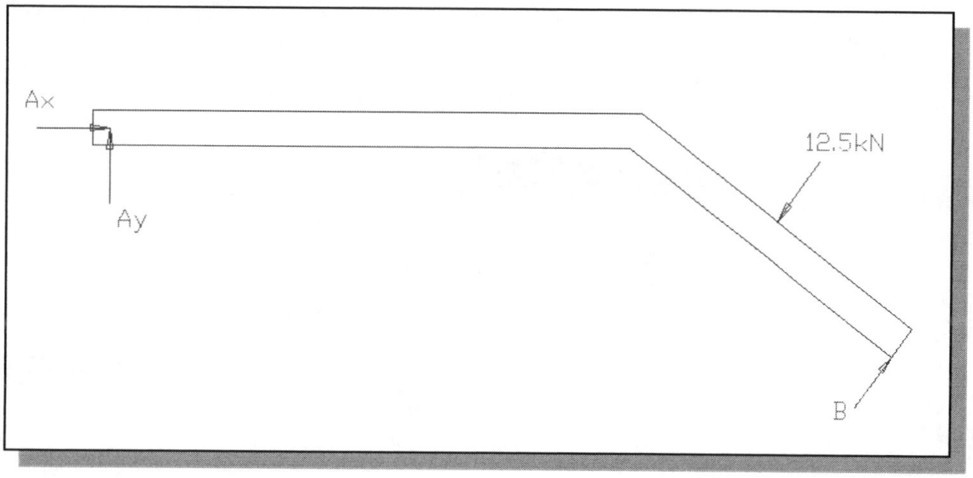

Applying the equations of equilibrium:

$$\sum M_{@A} = 0, \sum F_X = 0, \sum F_Y = 0,$$

Therefore $B = 9.76$ kN, $A_X = 1.64$ kN and $A_Y = 2.19$ kN

Consider a segment of the horizontal portion of the member:

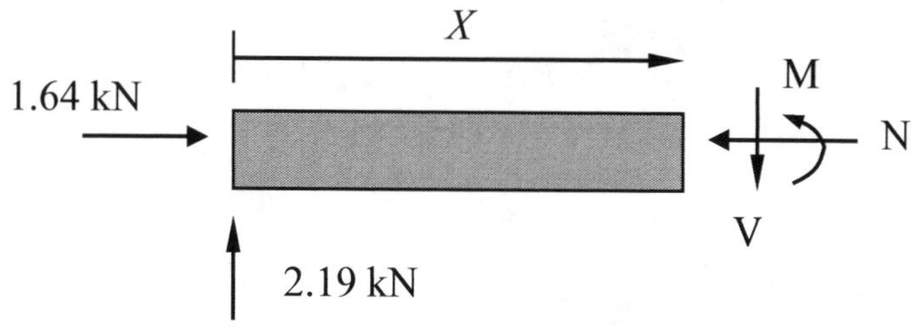

$$\sum F_X = 0, \quad 1.64 - N = 0, N = 1.64 \text{ kN}$$
$$\sum F_Y = 0, \quad 2.19 - V = 0, V = 2.19 \text{ kN}$$
$$\sum M_{@X} = 0, \quad - A_Y X + M = 0, \quad M = 2.19 \text{ x kN-m}$$

At *point* ***C***, X = 1.5 m and

$$N = 1.64 \text{ kN}, V = 2.19 \text{ kN}, M = 3.29 \text{ kN-m}$$

- The state of stress at *point C* can be determined by using the principle of superposition. The stress distribution due to each loading is first determined, and then the distributions are superimposed to determine the resultant stress distribution. The principle of superposition can be used for this purpose provided a linear relationship exists between the stress and the loads.

Stress Components

Normal Force:

The normal stress at *C* is a compressive uniform stress.

σ_{normal_force} = 1.64kN/(0.075 x 0.05)m^2 = 0.437 MPa

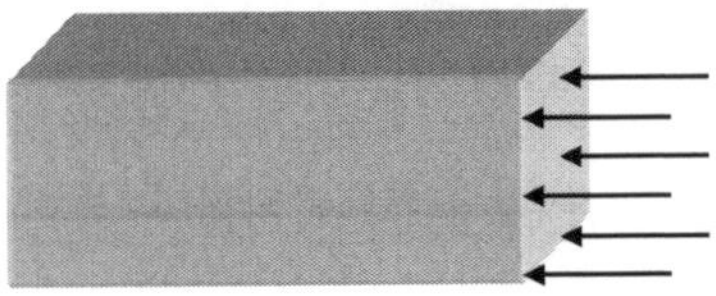

Shear Force:

Point C is located at the top of the member. No shear stress existed at *point C*.

$\tau_{shear\_force} = 0$

Bending Moment:

Point C is located at 37.5 mm from the neutral axis. The normal stress at *C* is a compressive uniform stress.

$\sigma_{bending\_moment}$ = (3.29kN-m x 0.0375m)/(1/12 x 0.05 x (0.075)^3)m^4
= 70.16 MPa

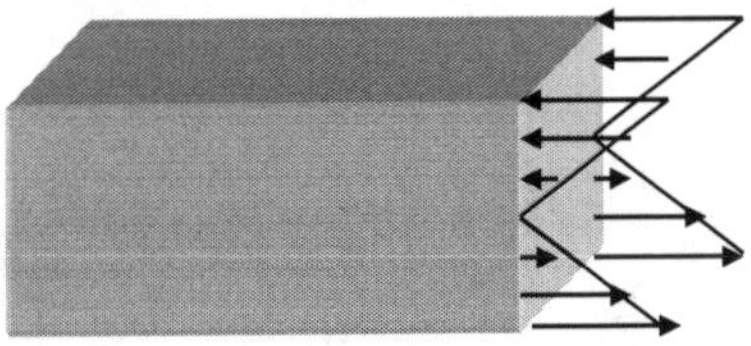

Superposition:

The shear stress is zero and combining the normal stresses gives a compressive stress at *point C*:

σ_C = 0.437 MPa + 70.16 MPa = 70.6 MPa

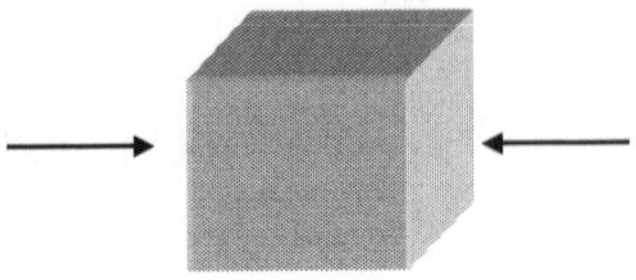

Examine the horizontal segment of the member:

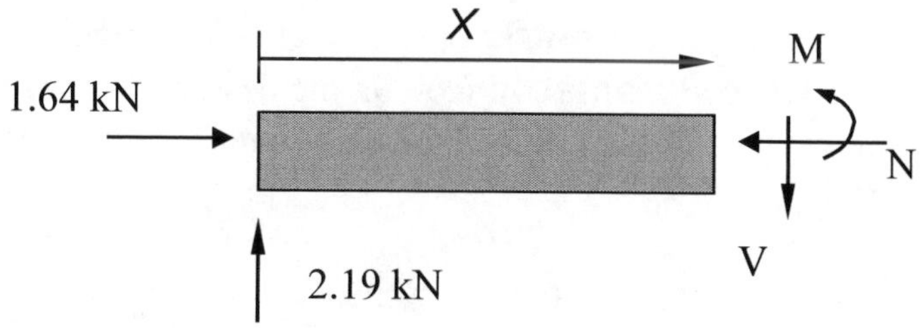

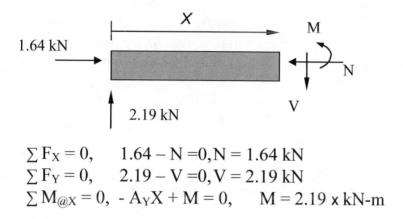

$$\sum F_X = 0, \quad 1.64 - N = 0, N = 1.64 \text{ kN}$$
$$\sum F_Y = 0, \quad 2.19 - V = 0, V = 2.19 \text{ kN}$$
$$\sum M_{@X} = 0, \quad -A_Y X + M = 0, \quad M = 2.19 \text{ x kN-m}$$

The maximum normal stress for the horizontal segment will occur at *point D*, where X = 4m.

$$\sigma_{normal\_force} = 1.64\text{kN}/(0.05 \times 0.075)\text{m}^2 = 0.437 \text{ MPa}$$

$$\sigma_{bending\_moment} = (8.76\text{kN-m} \times 0.0375\text{m})/(1/12 \times 0.05 \times (0.075)^3)\text{m}^4$$
$$= 186.9\text{MPa}$$

$$\sigma_{max@D} = 0.437 \text{ MPa} + 186.9 \text{ MPa} = 187.34 \text{ MPa}$$

➢ Does the above calculation provide us the maximum normal stress developed in the structure? To be sure, it would be necessary to check the stress distribution along the inclined segment of the structure. We will rely on the *Creo Simulate* solutions to find the maximum stress developed. The above calculation (the state of stress at *point C* and *point D*) will serve as a check of the *Creo Simulate* FEA solutions.

Starting Creo Parametric

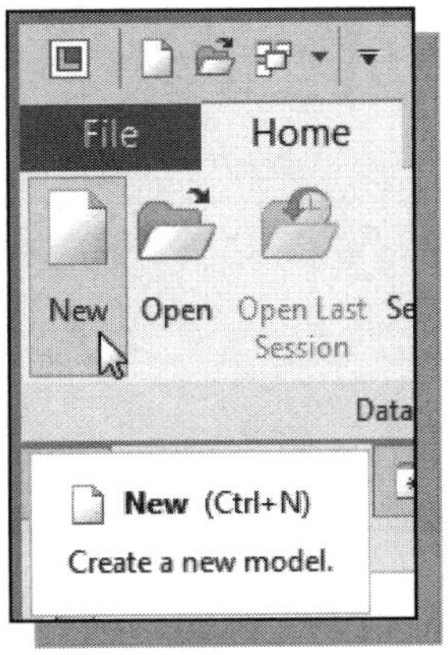

1. Select the **Creo Parametric** option on the *Start* menu or select the **Creo Parametric** icon on the desktop to start *Creo Parametric*. The *Creo Parametric* main window will appear on the screen.

2. Click on the **New** icon, located in the *Ribbon toolbar* as shown.

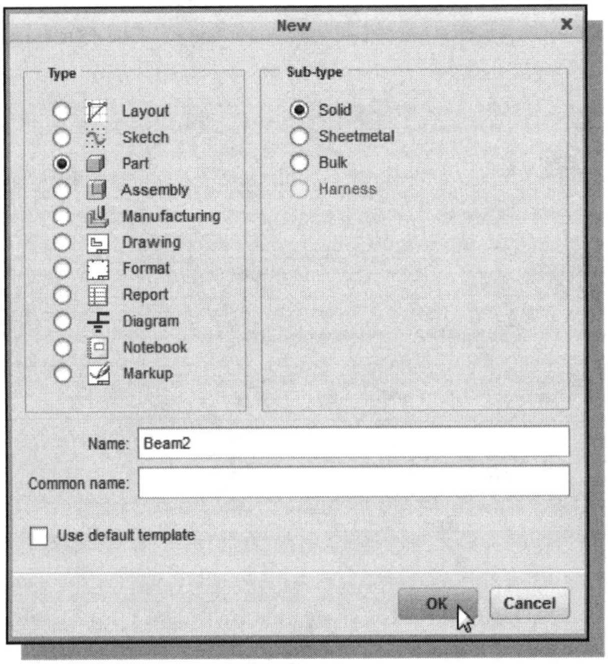

3. In the *New* dialog box, confirm the model's Type is set to **Part** (**Solid** Sub-type).

4. Enter **Beam2** as the part Name as shown in the figure.

5. Turn *off* the **Use default template** option.

6. Click on the **OK** button to accept the settings.

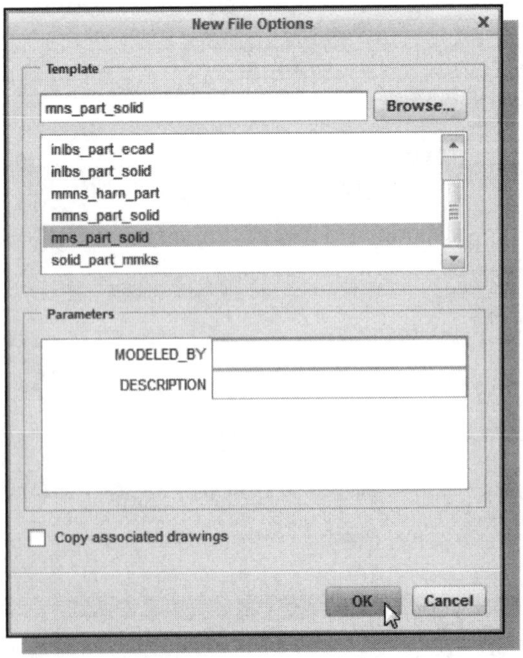

7. In the *New File Options* dialog box, select **mns_part_solid** in the option list as shown. (Use the *Browse* option to locate the file if it is not shown in the list.)

8. Click on the **OK** button to accept the settings and enter the *Creo Parametric Part Modeling* mode.

The Integrated Mode of Creo Simulate

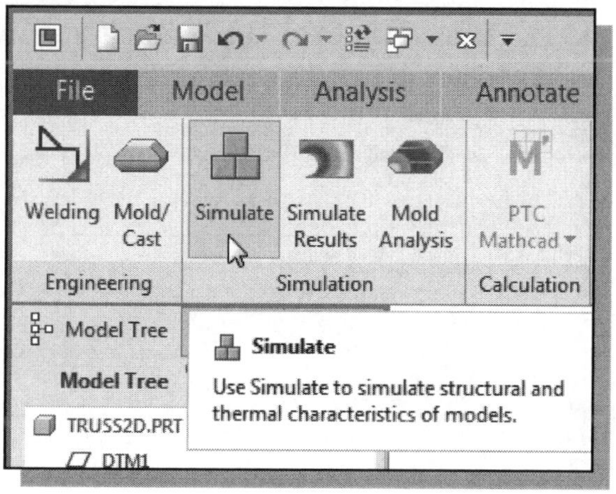

1. Start the *Integrated* mode of *Creo Simulate* by selecting the **Applications → Creo Simulate** option in the *Creo Simulate Ribbon* as shown.

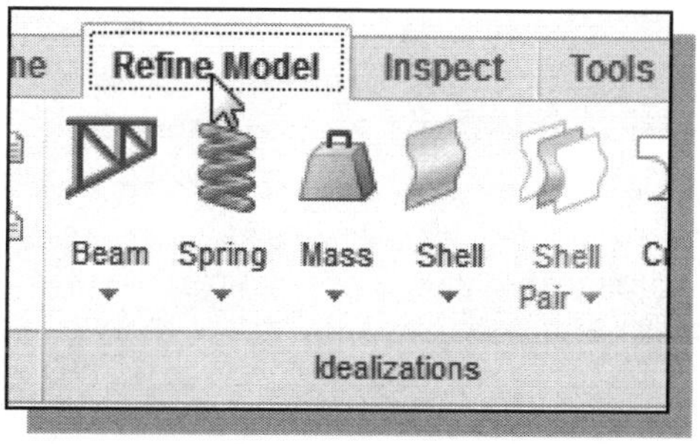

2. Click on the **Refine Model** tab to view the available commands.

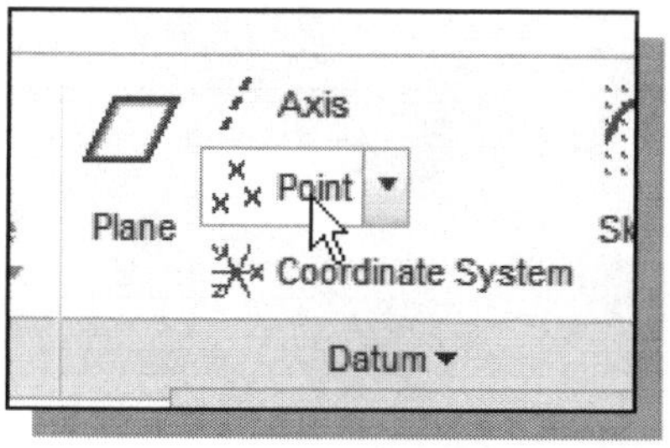

3. Choose **Datum Point** tool in the icon panel displayed on the *Creo Parametric Ribbon* toolbar.

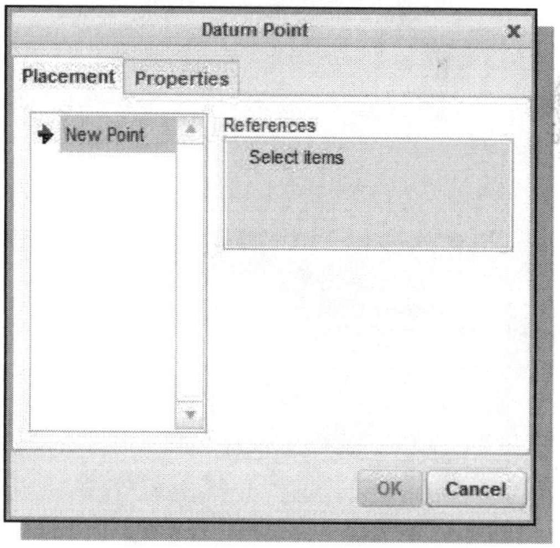

❖ In the ***Datum Point*** window, there is currently no datum point in the Placement list. We can create points by using existing objects as References.

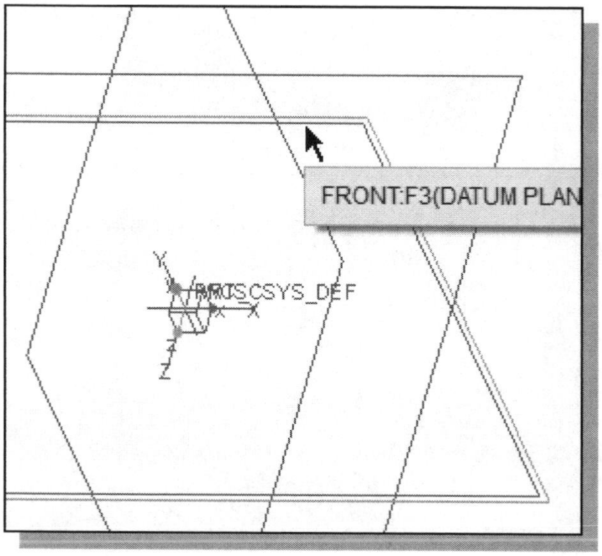

4. Inside the graphics area, select the **FRONT** plane, by clicking on one of the edges, as shown.

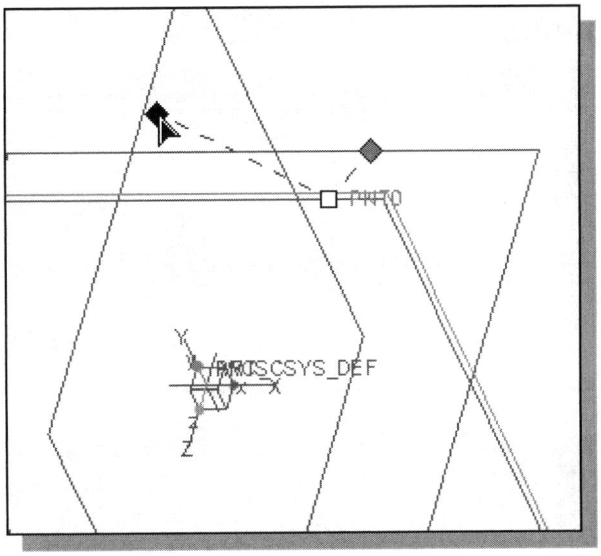

5. Grab one of the handle points and place it on datum plane **RIGHT**, as shown. RIGHT is now being used as an Offset reference to position the point.

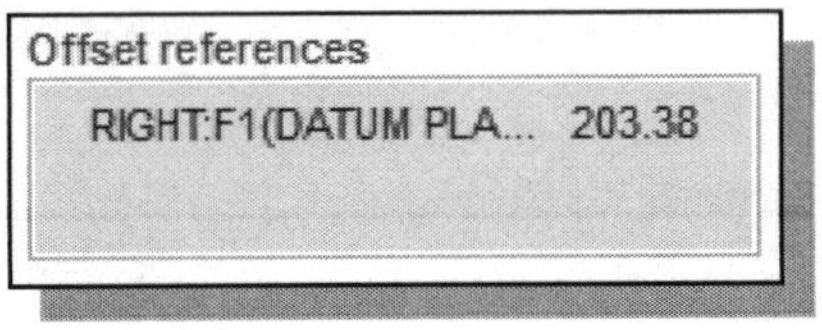

6. Repeat the above step and place the other handle point on **TOP**.

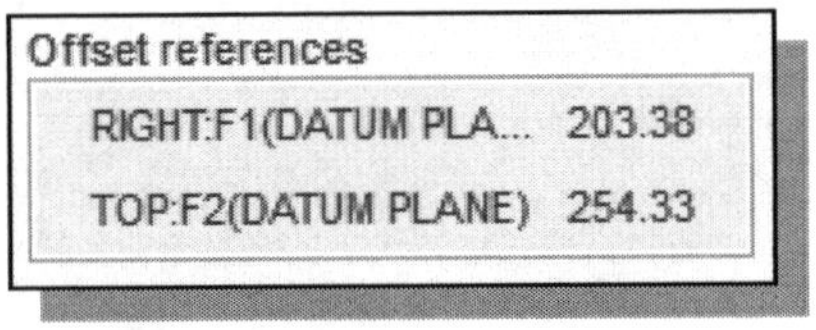

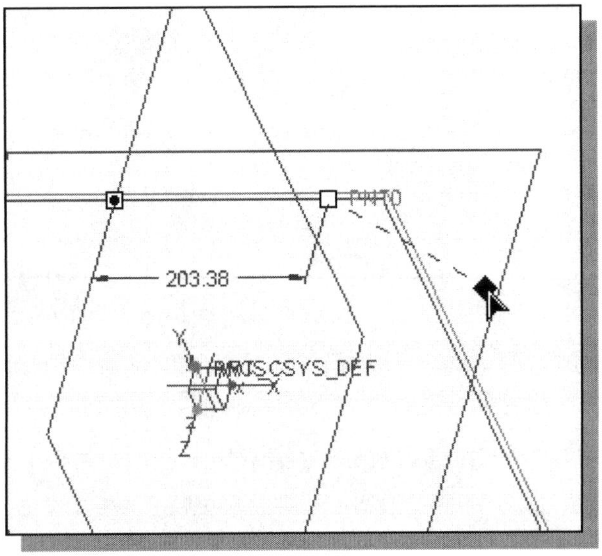

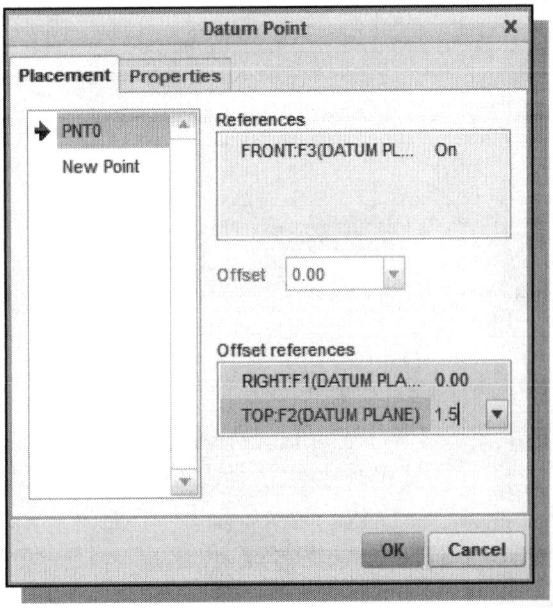

7. In the *Datum Point* window, click on the Offset value for the RIGHT plane, which is inside the References option box, and change the value to **0.0** as shown.

8. Click on the second value and adjust the Offset value to **1.5** for the TOP plane, as shown.

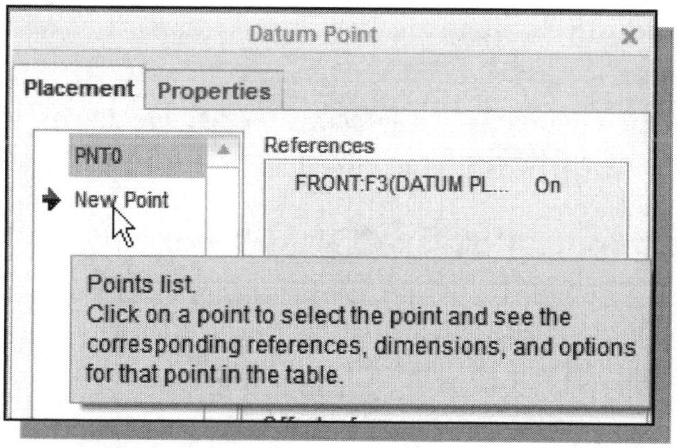

9. In the *Datum Point* window, create a new datum point by clicking on the **New Point** option as shown.

10. Repeat the above steps and create PNT1, located at (**4,1.5**).

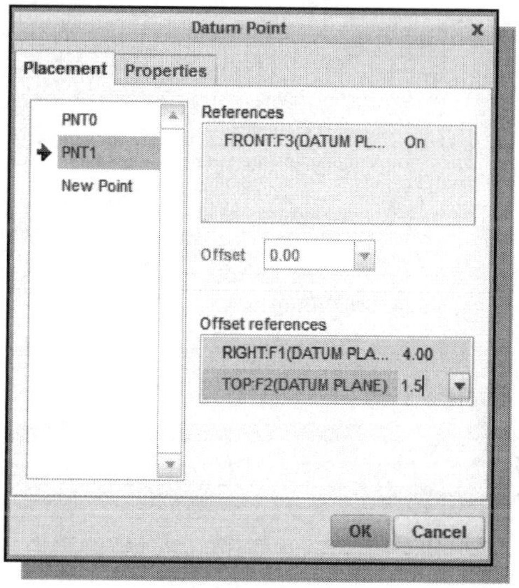

11. Repeat the above steps and create PNT2, located at (**6,0.0**).

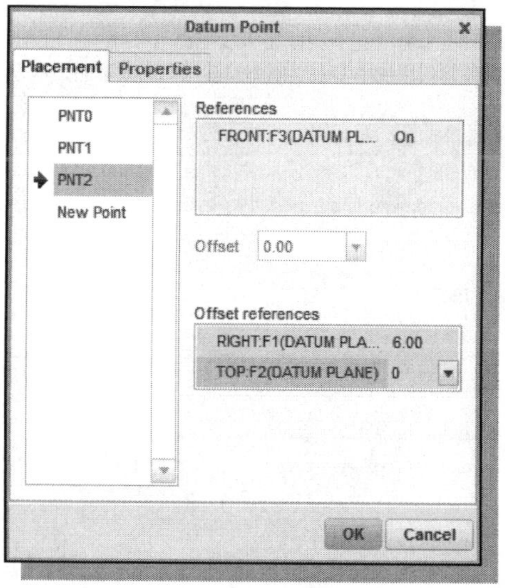

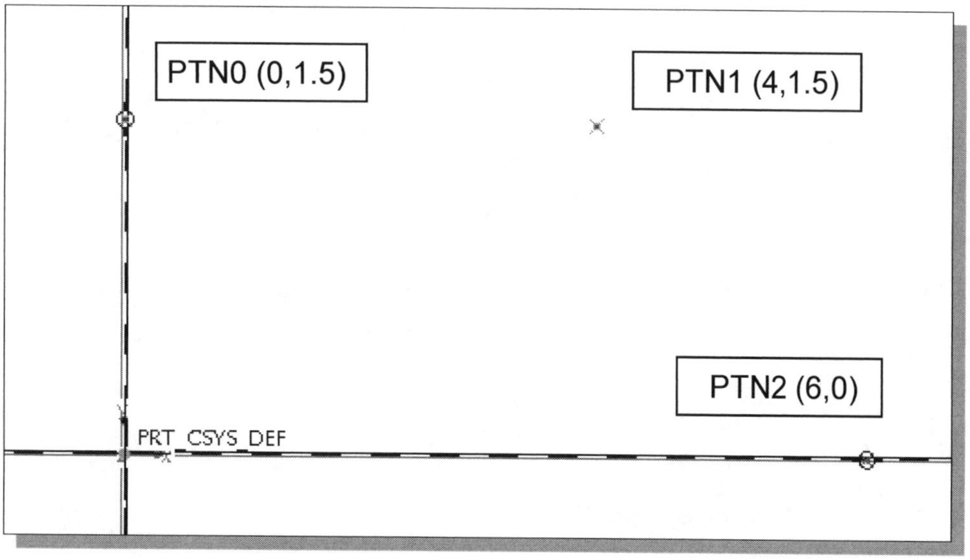

Create Datum Curves

In *Creo Simulate*, beam elements can also be established with curves. One advantage of using curves is we can apply distributed loads on curves for beam analysis.

1. Choose **Datum → Curve → Curve through Points** tool in the *Datum toolbar*.

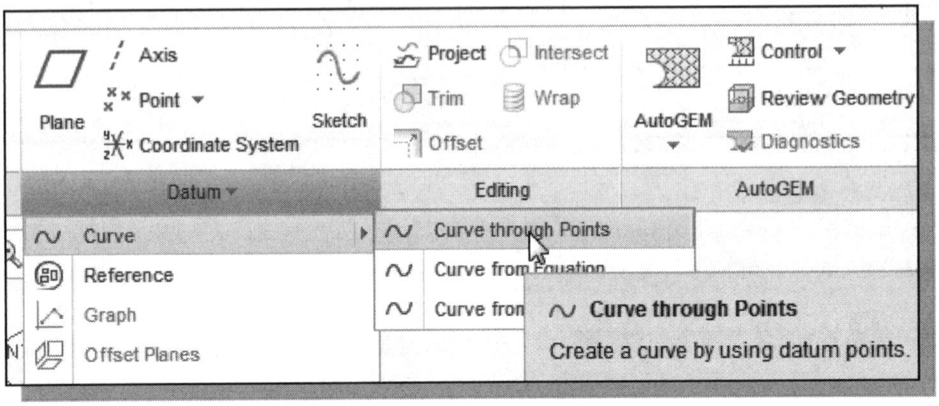

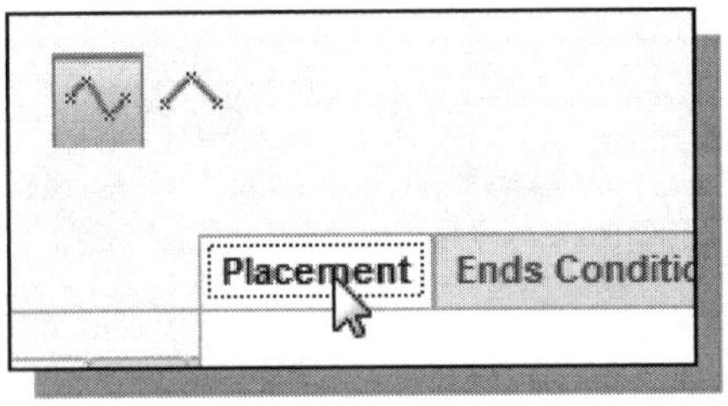

2. In the *Dashboard* area, click the **Placement** tab to view the available options.

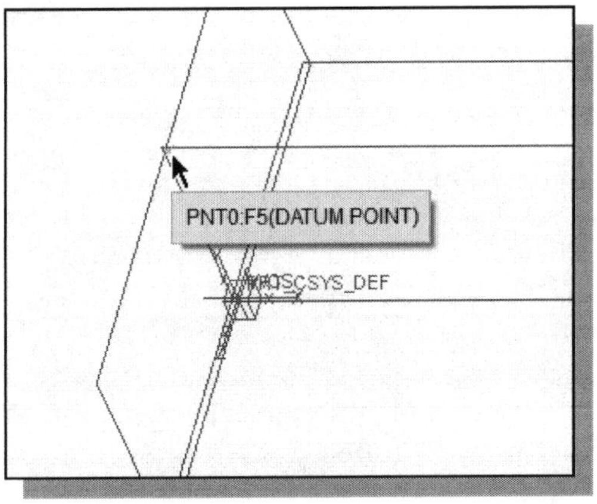

3. On your own, delete the pre-defined point in the placement list.

4. In the graphics area, select **PNT0** as the first point for the curve definition.

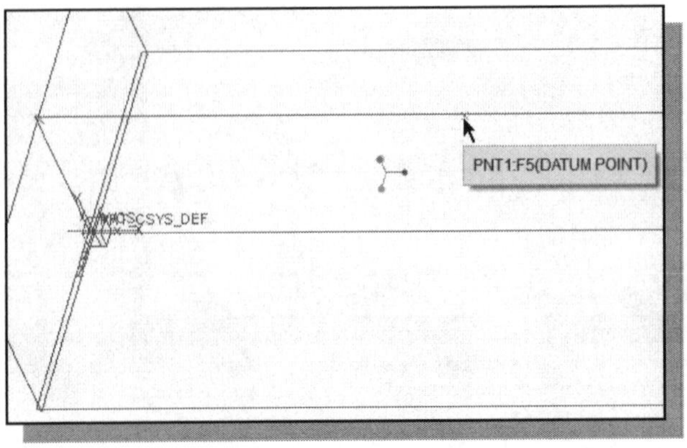

5. In the *graphics area*, select **PNT1** as the second point for the curve definition.

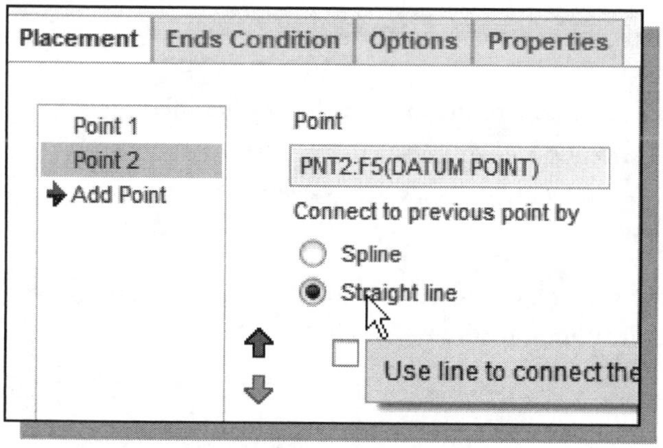

6. Select **Straight line** to define the connection option.

7. Click **OK** to create the first datum curve.

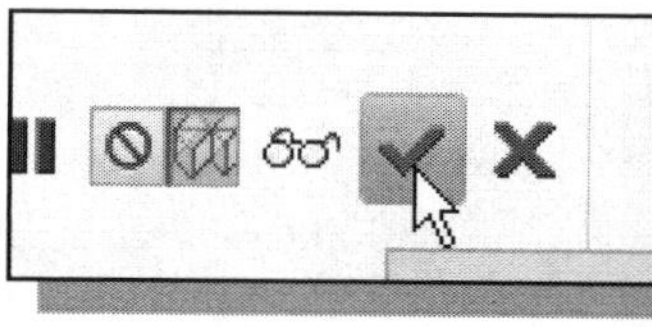

8. On your own, repeat the above procedure to create a second beam, connecting **PNT1** and **PNT2**.

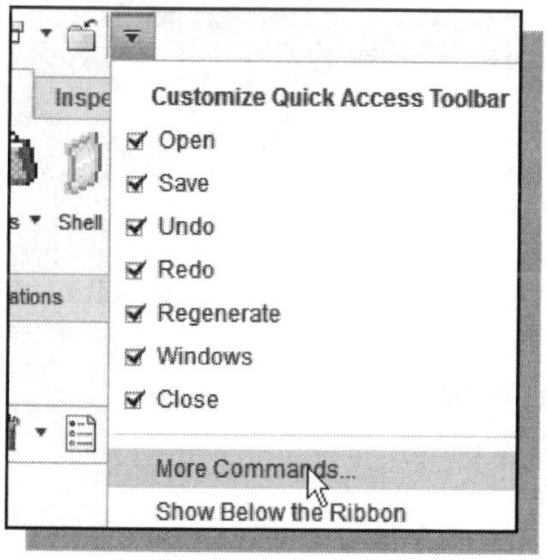

9. In the *Quick Access* toolbar, select **More Commands** to modify the default *Creo Parametric* options.

10. Set the default Model Orientation to **Isometric** as shown.

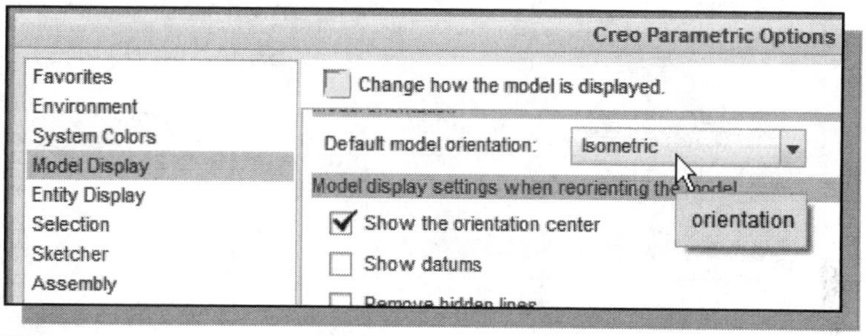

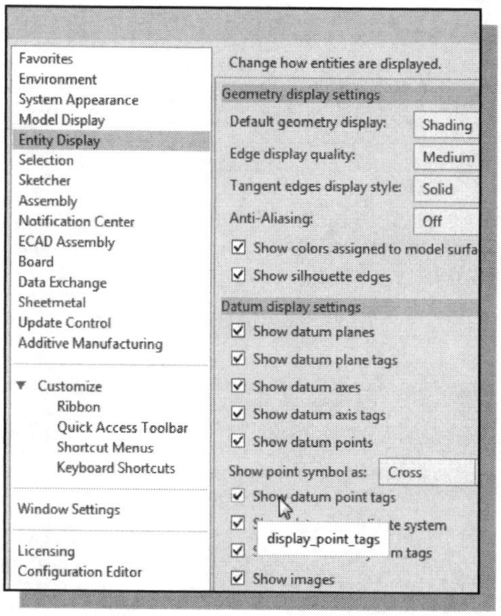

11. Switch ***on*** the **Show datum point tags** option as shown.

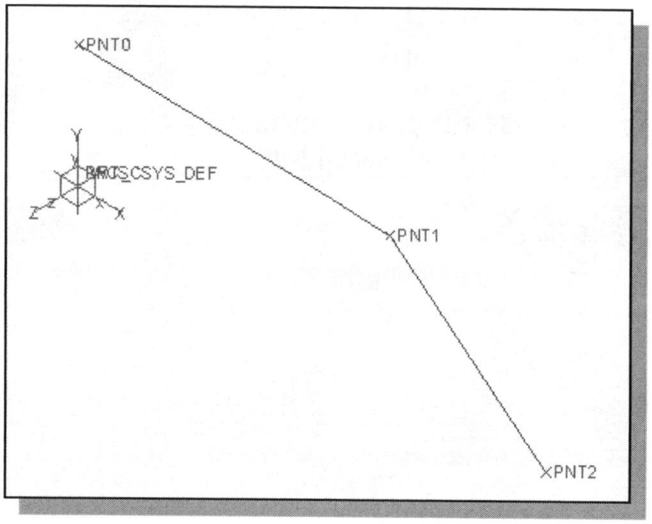

Select and Examine the Element Material Property

Before creating the FEA elements, we will first set up the *Material Property* for the elements. The *Material Property* contains the general material information, such as *Modulus of Elasticity*, *Poisson's Ratio*, etc., that is necessary for the FEA analysis.

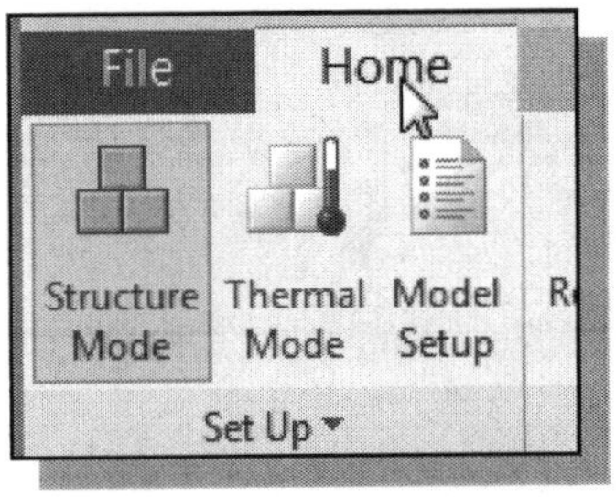

1. Click on the **Home** tab to return to the main *Creo Simulate* commands toolbar.

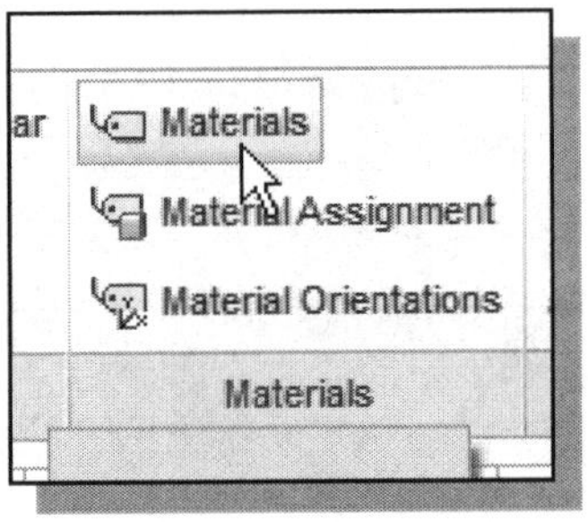

2. Choose **Materials** in the *Materials* group from the *Ribbon* toolbar as shown.

❖ Note the default list of materials libraries, which are available in the pre-defined *Creo Simulate*, is displayed.

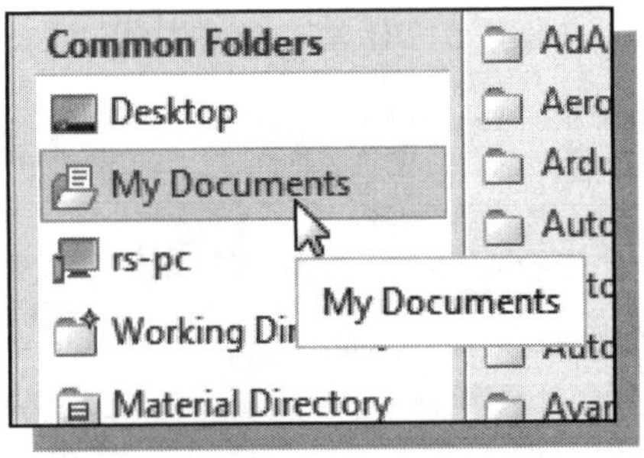

3. On your own, locate the folder where we saved the modified material definition, filename **Steel_low_Carbon_Modified**.

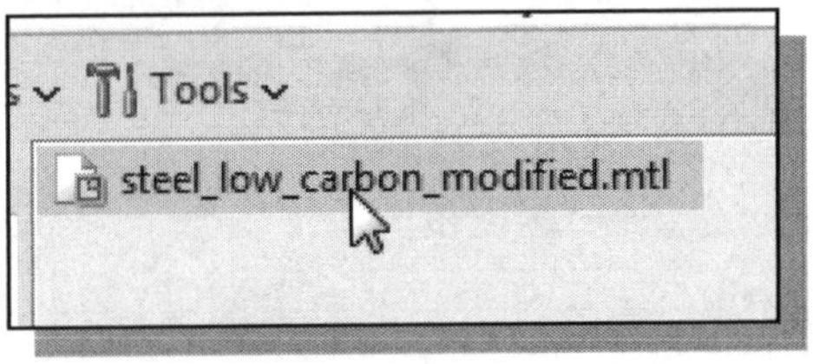

4. **Double-clicking** on the modified material definition for use with the current model.

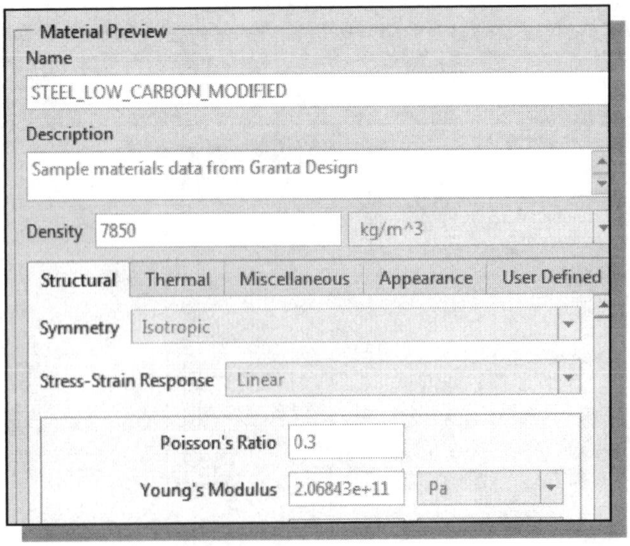

5. Click **Yes** to convert the units to match the model units.

6. Examine the material information in the **Material Preview** area and click **OK** to accept the settings.

Set up the Element Cross Section

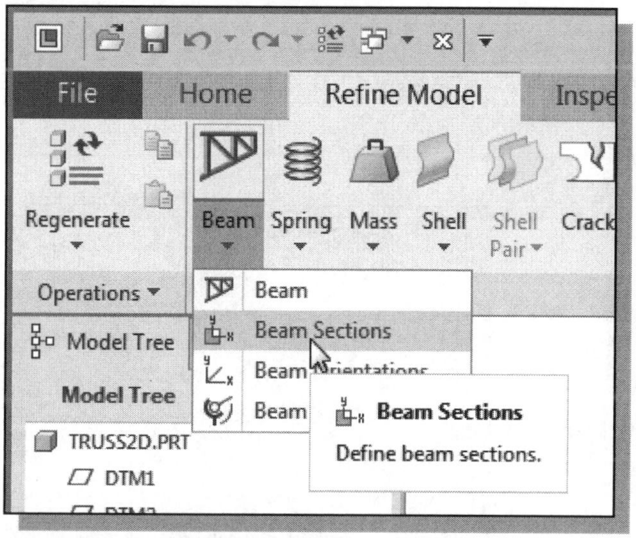

1. Switch to the **Refine Model** tab.

2. Choose **Beam Sections** in the *Idealizations* toolbar as shown.

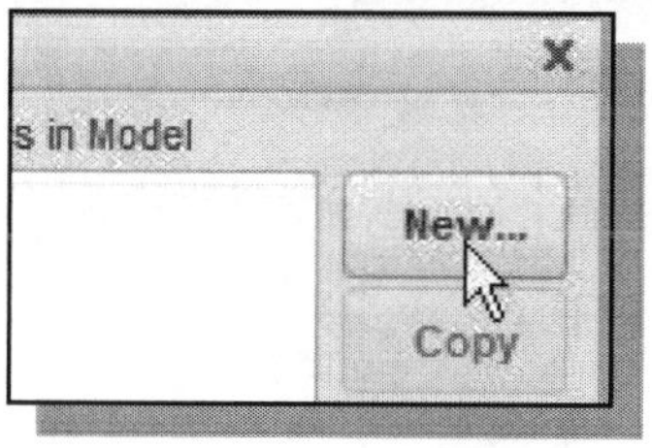

3. Click **New** to create a new beam cross-section.

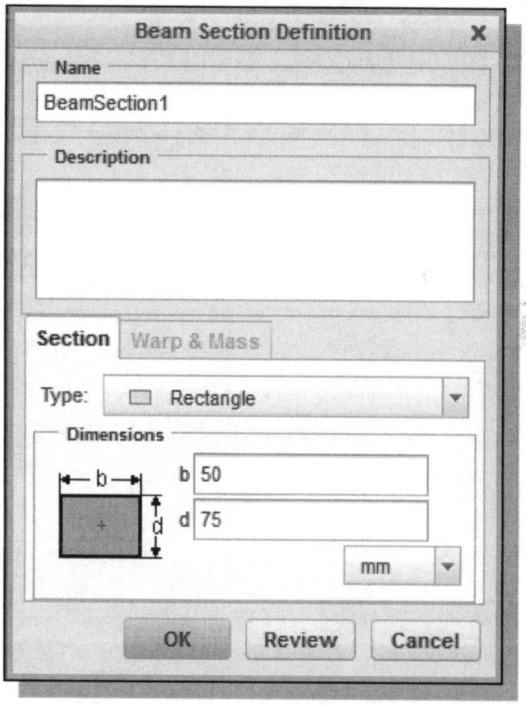

4. Choose **Rectangle** as the cross-section Type as shown.

5. Enter **50 mm** and **75 mm** as the width and height of the rectangular beam cross section.

6. Click **OK** to accept the creation of the new beam section.

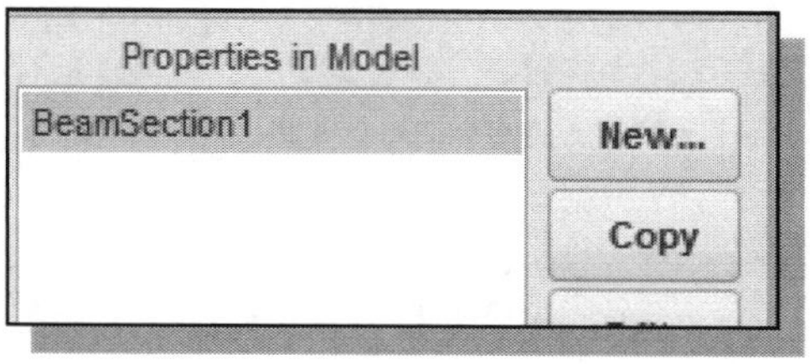

7. Click **Close** to accept the creation of the new beam section.

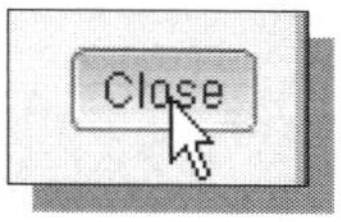

Create Beam Elements

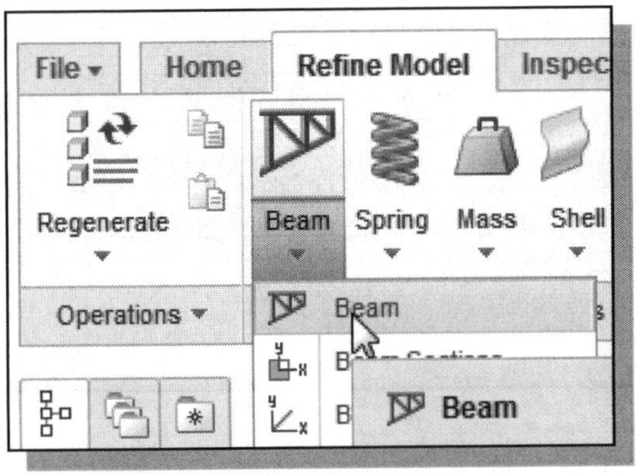

1. Choose **Beam** in the icon panel. The *Beam Definition* window appears.

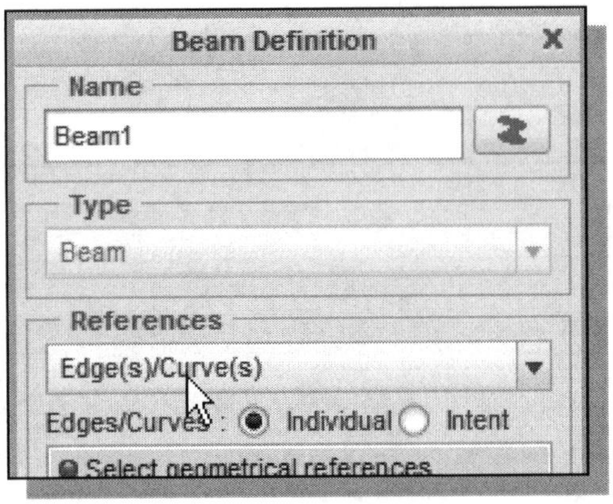

2. Confirm the References option is set to **Edge(s)/Curve(s)**.

3. Press the [**CTRL**] key and click on both of the datum curves we just created. Note the arrows identify the X direction of BACS.

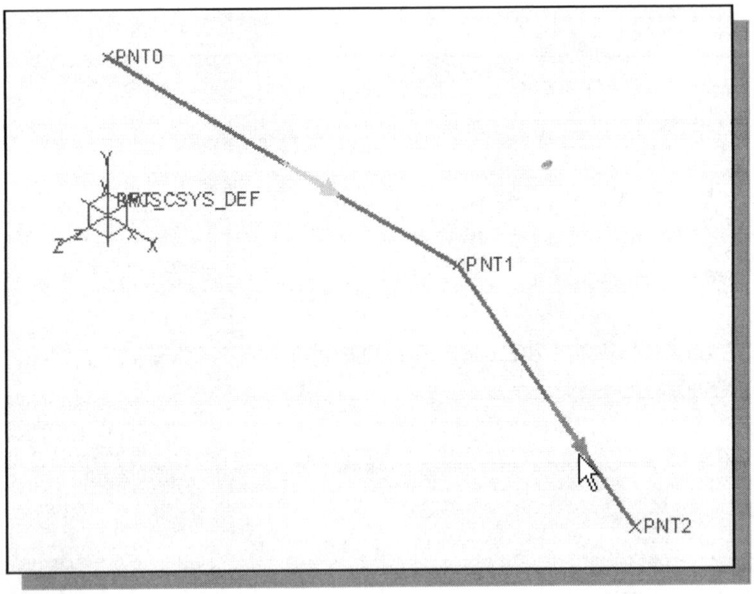

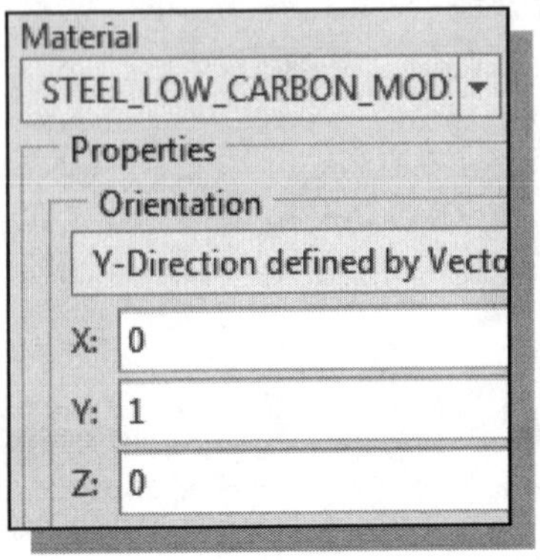

4. Select **Steel_low_Carbon_Modified** as the material of the element.

5. Confirm the *Y Direction* is set to WCS with **0,1,0** as the X, Y and Z directions.

6. Click on the **OK** button to create the beam.

Apply the First Displacement Constraint

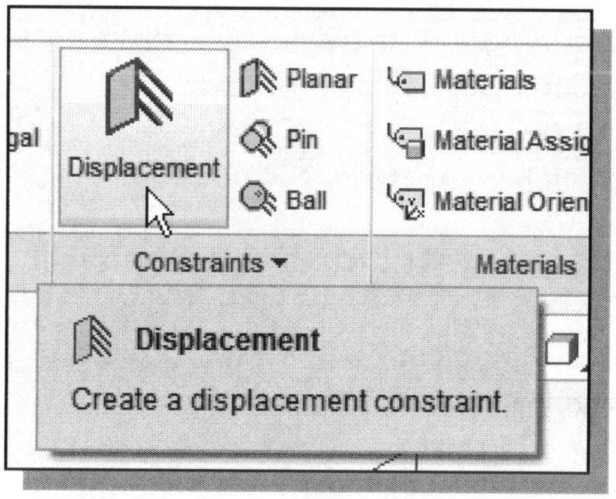

1. Switch to the **Home** tab as shown.

2. Choose **Displacement Constraint** by clicking the icon in the *Constraints* toolbar as shown.

3. Set the References to **Points** and select **PNT0** as shown.

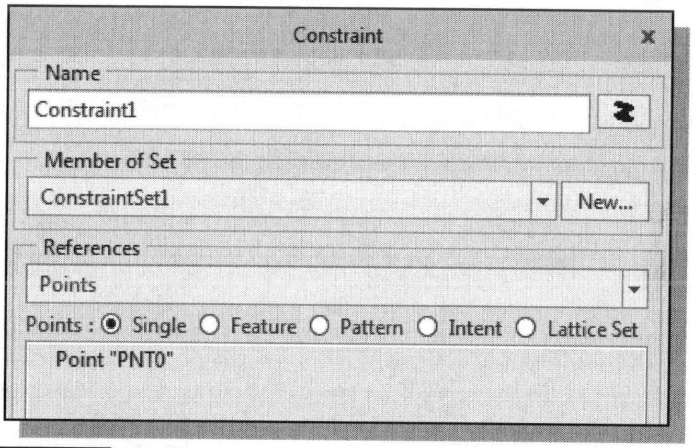

❖ PNT0 is a fixed support point; it cannot move in the X, Y or Z directions. For the 2D simply supported beam, the joint at PNT0 is a pin-joint; the Z rotational movement should be set to **Free**.

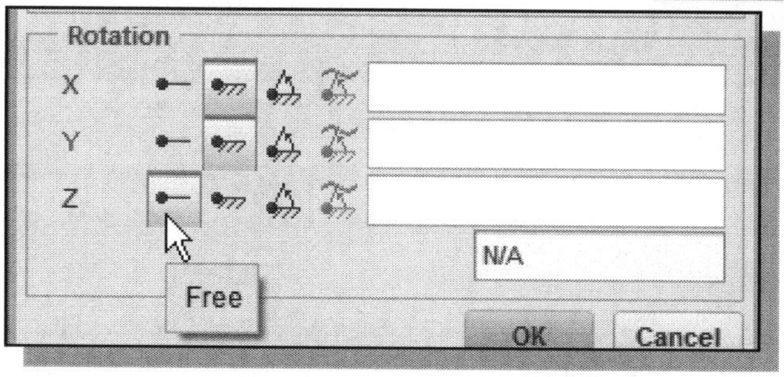

4. Set Z Rotation to **Free** as shown.

5. Click on the **OK** button to accept the first displacement constraint settings.

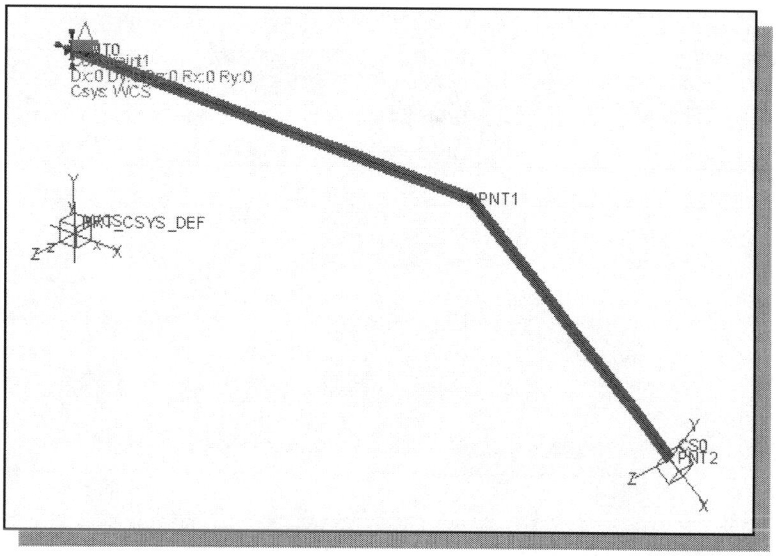

Create a New Coordinate System for the Second Support

For the support at the end of the inclined beam, we will need to create a new coordinate system since the direction of the support is not parallel to any of the WCS axes.

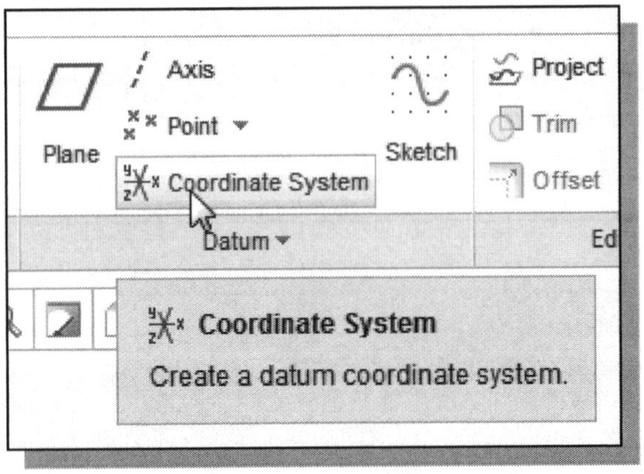

1. Switch to the **Refine Model** tab.

2. Choose the **Coordinate System** tool in the *Datum* icon panel from the *Ribbon* toolbar of the *Creo Parametric* main screen.

❖ In *Creo Simulate*, additional coordinate systems can be created to allow constraints and loadings to be set at any specific angle.

3. Select **PNT2** as the location to align the origin of the new coordinate system.

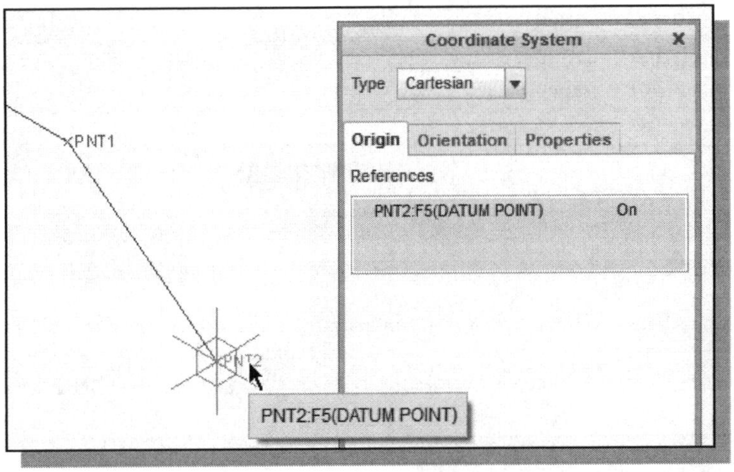

4. Switch to the **Orientation** tab and click in the first **Use** box, then select the second **datum curve** to orient the X direction of the new coordinate system.

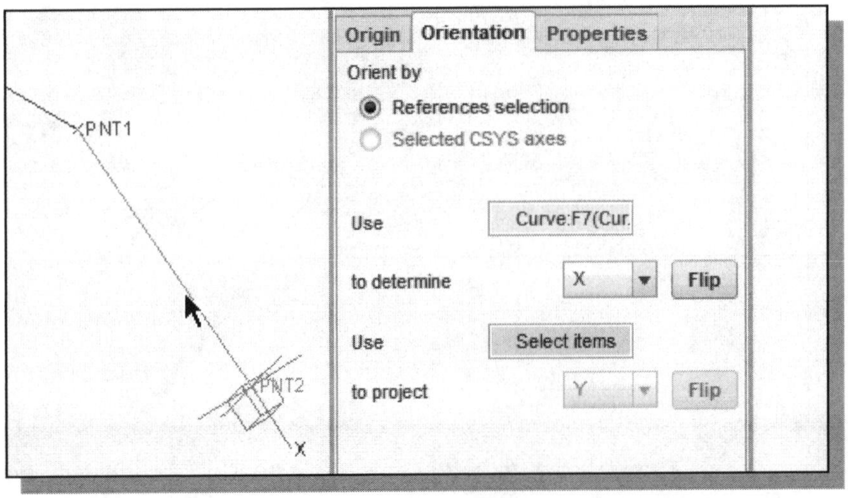

5. Select the **TOP** datum plane to align the Y direction of the new coordinate system as shown.

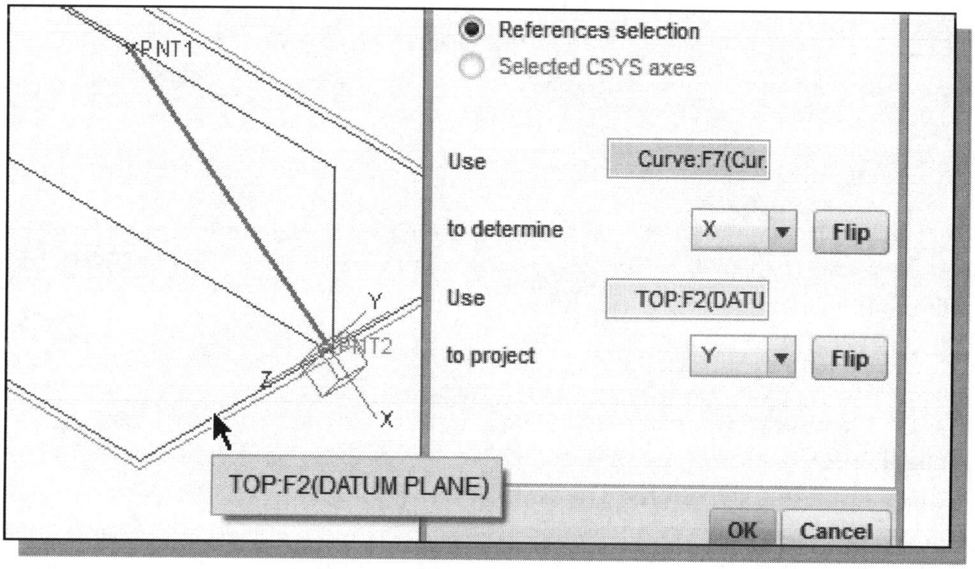

6. Click **OK** to complete the creation of the new coordinate system, which is labeled as **CS0**.

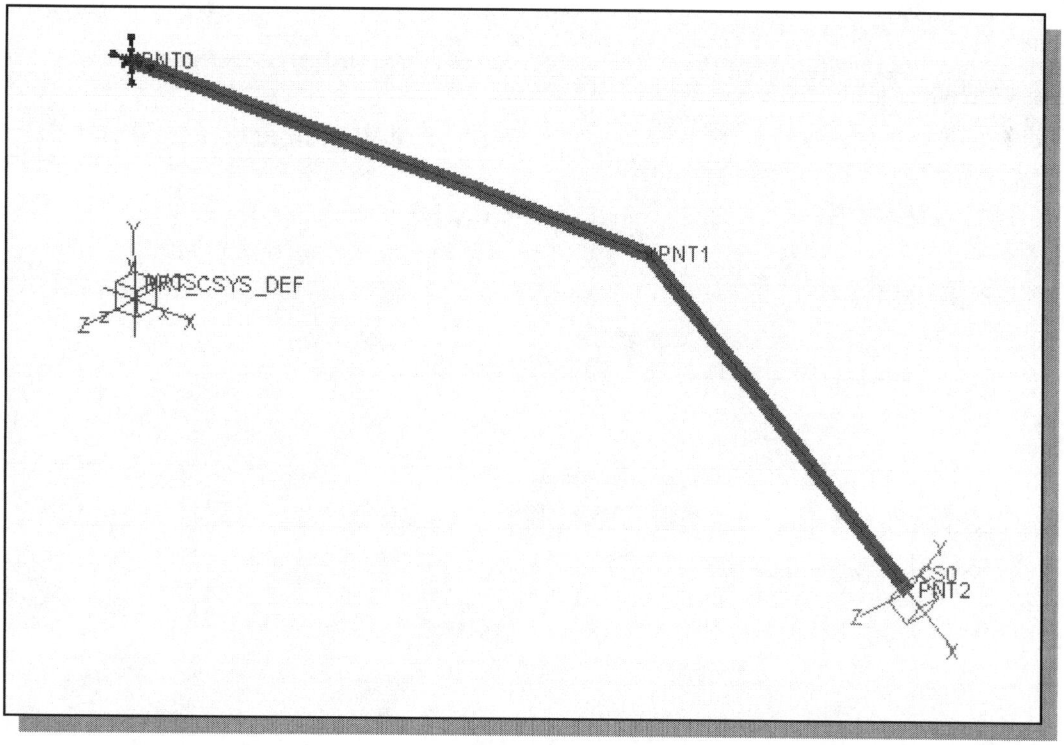

➢ The new coordinate system can now be used to set up the support at the end of the inclined beam.

Apply the Second Displacement Constraint

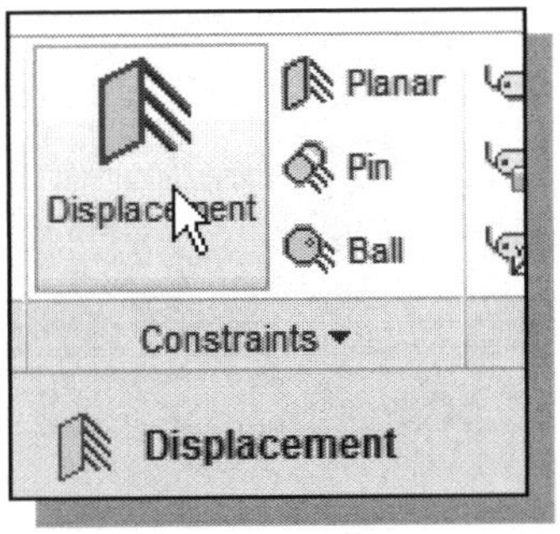

1. Choose **Displacement Constraint** by clicking the icon in the toolbar as shown.

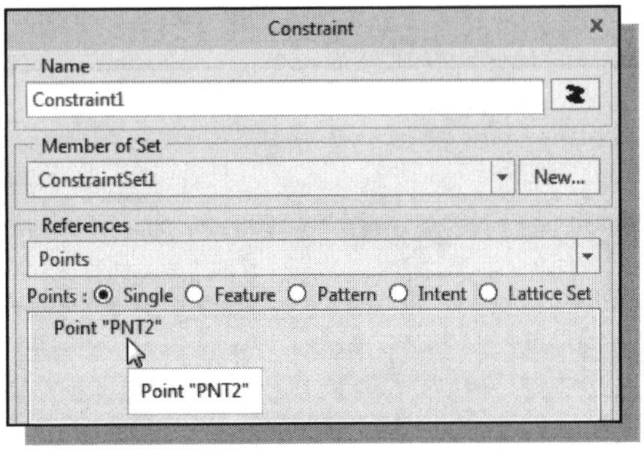

2. Set the References to **Points** and select **PNT2** as shown.

❖ PNT2 has a roller support; it can move in the CS0-X direction, but not in the CS0-Y or CS0-Z directions. The CS0-Z rotational movement should also be set to **Free**.

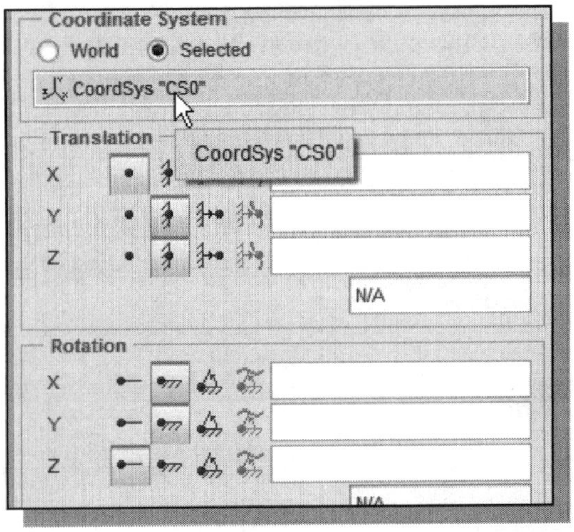

3. Set the Coordinate System to **CS0** as shown.

4. Set X Translation to **Free** as shown.

5. Set Z Rotation to **Free** as shown.

6. Click on the **OK** button to accept the first displacement constraint settings.

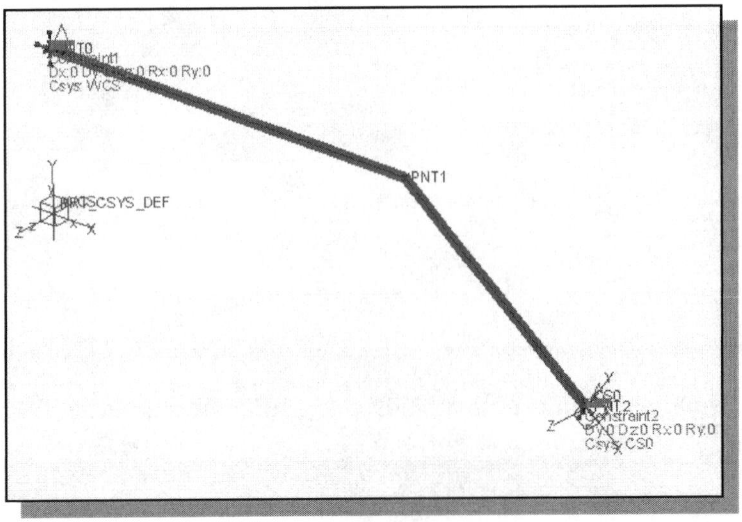

Apply the External Loads

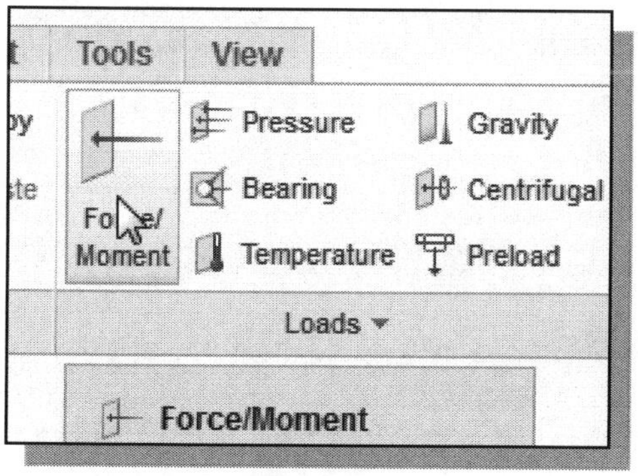

1. Choose **Force/Moment Load** by clicking the icon in the *Loads toolbar* as shown.

2. Confirm the *Load* set is still set to **LoadSet1**.

   ❖ Note that a load set can contain multiple loads.

3. Set the References to **Edges/ Curves** as shown.

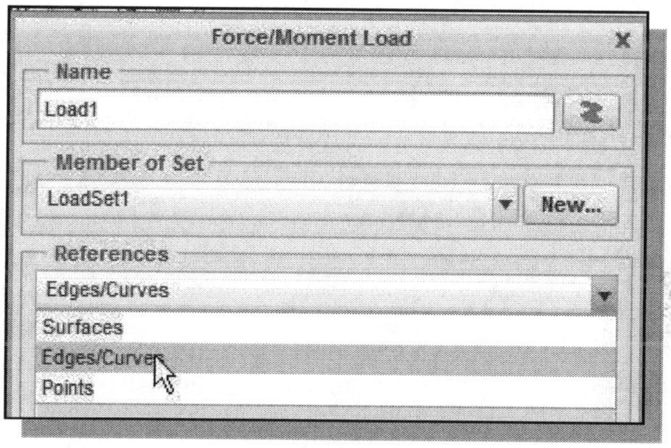

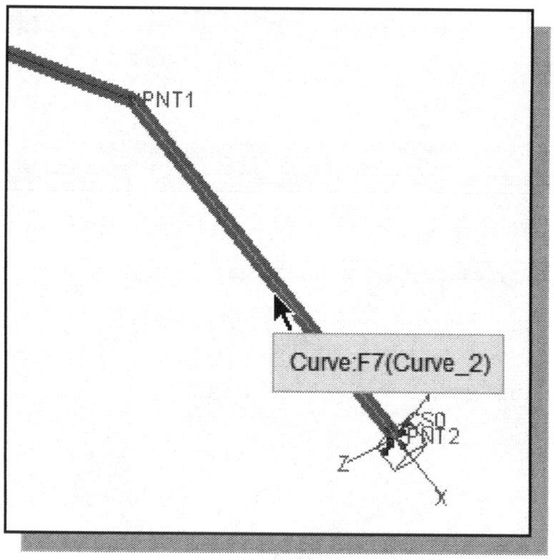

4. Pick the inclined datum curve, **Curve**, as shown.

5. Choose **CS0** as the coordinate system for the distributed load.

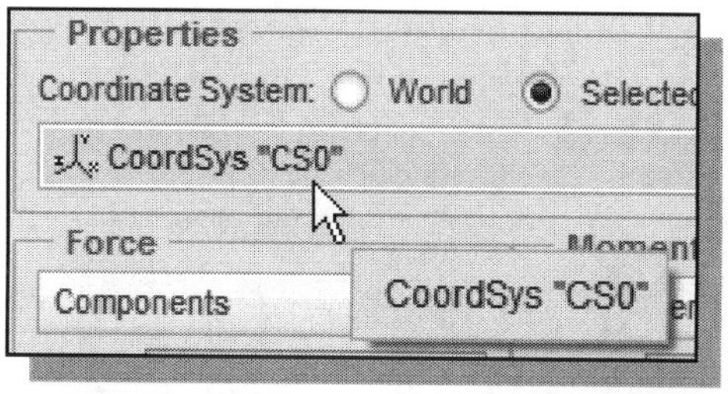

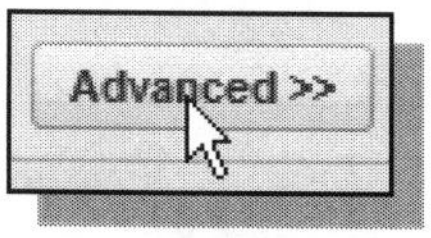

6. Click **Advanced** to display additional options for the settings of the load.

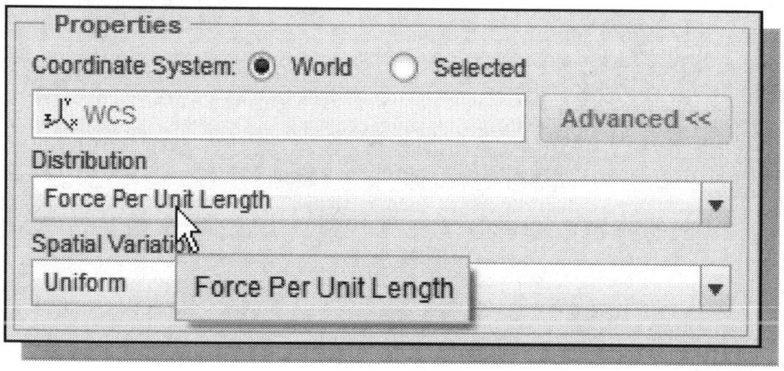

7. Set the Distribution option to **Force Per Unit Length**.

8. Set the Spatial Validation option to **Uniform**.

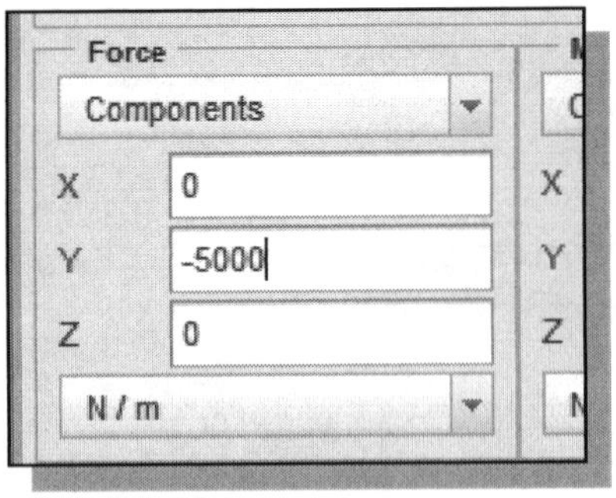

9. Enter **-5000 N/m** in the Y component Force value box as shown.

10. Click on the **Preview** button to confirm the applied load is pointing in the correct direction.

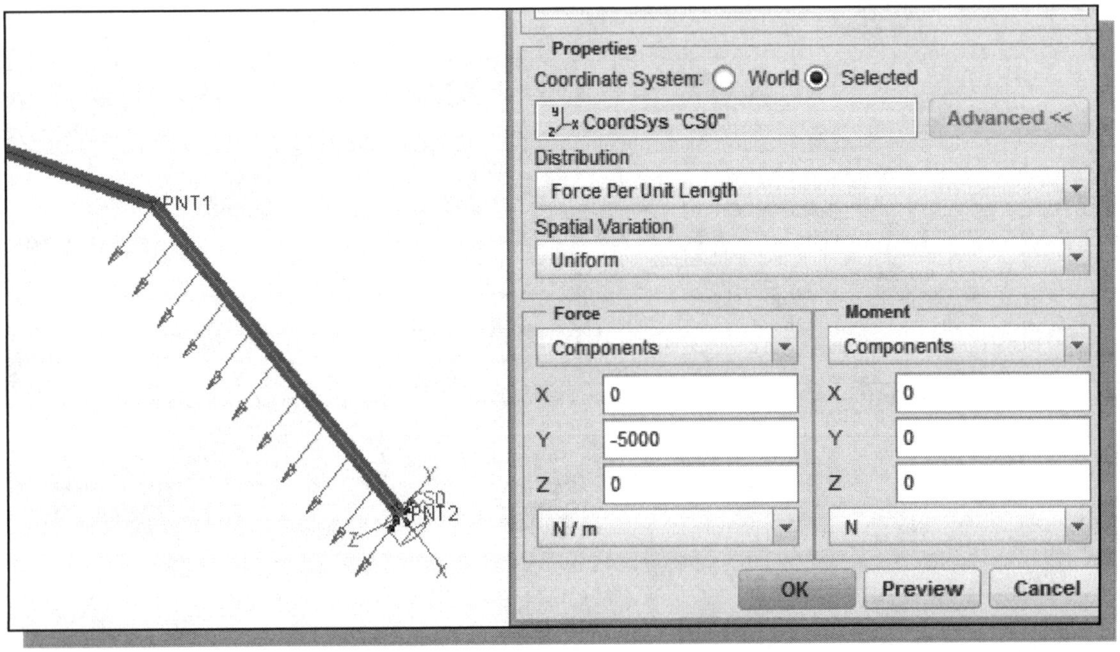

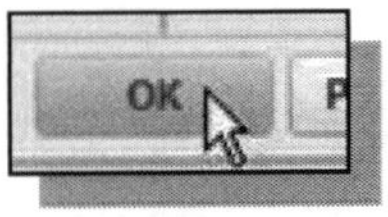

11. Click on the **OK** button to accept the second *Load* settings.

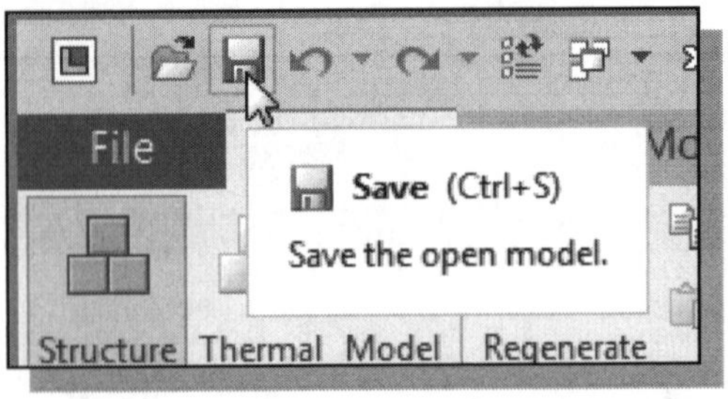

12. On your own, click **Save** and save a copy of the current FEA model.

❖ Note that the CS0 coordinate system is used to set up both the second constraint and the distributed load on the inclined beam. With *Creo*, there is no limit to the number of coordinate systems that we can create in a *Creo Simulate* FE model.

Run the FEA Solver

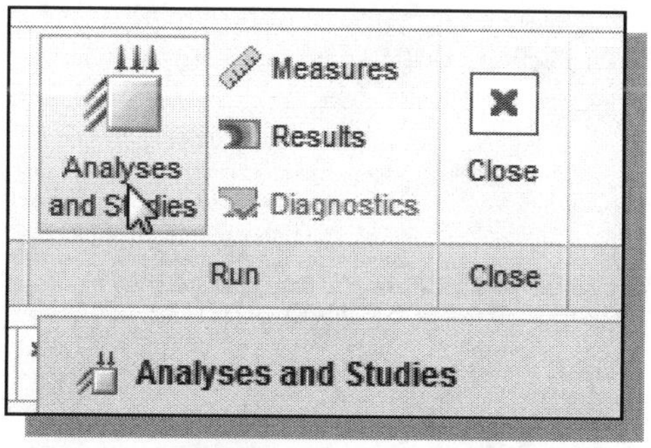

1. Activate the *Creo* **Analyses and Studies** command by selecting the icon in the toolbar area as shown.

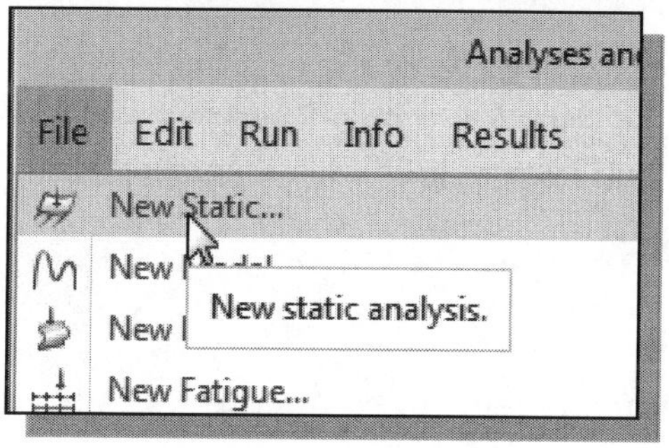

2. In the *Analyses and Design Studies* window, choose **File → New Static** in the *File pull-down menu*. This will start the setup of a basic linear static analysis.

3. Note that **ConstraintSet1** and **LoadSet1** are automatically accepted as part of the analysis parameters.

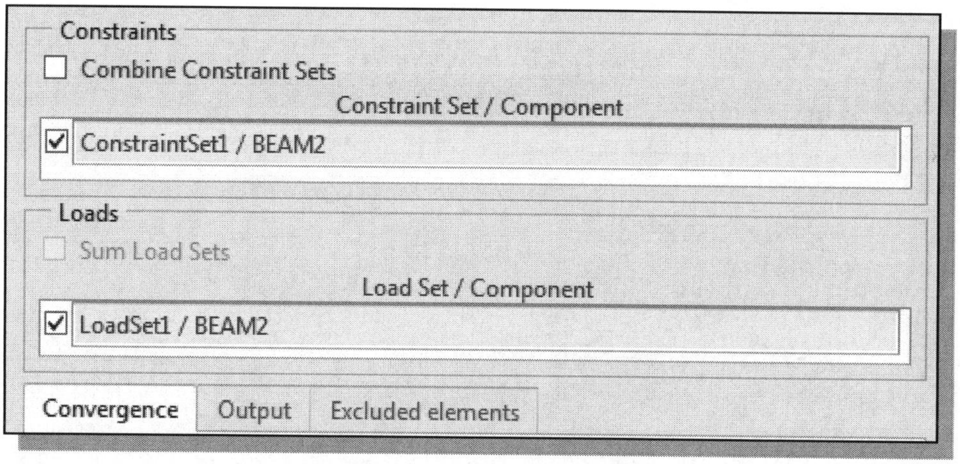

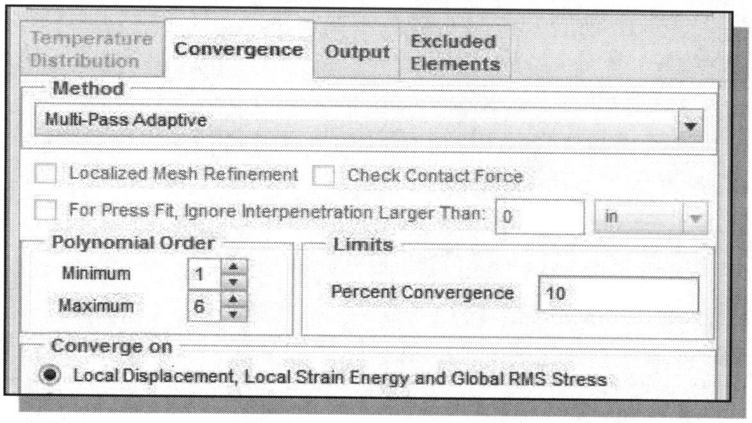

4. Set the Method option to **Multi-Pass Adaptive** and Limits to **10** Percent Convergence as shown.

5. Click **OK** to accept the settings.

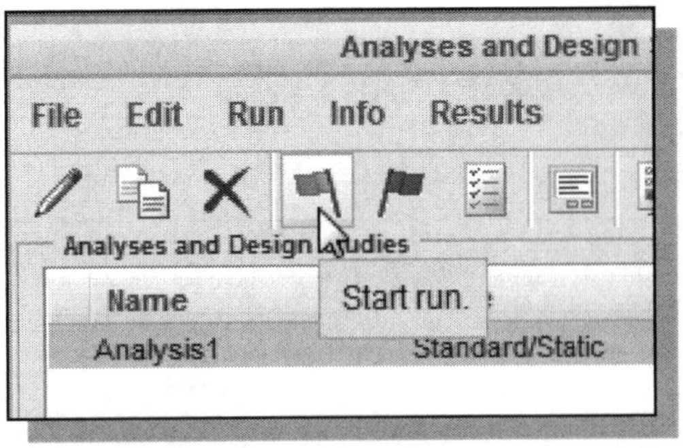

6. Click **Start run** to begin the *Solving* process.

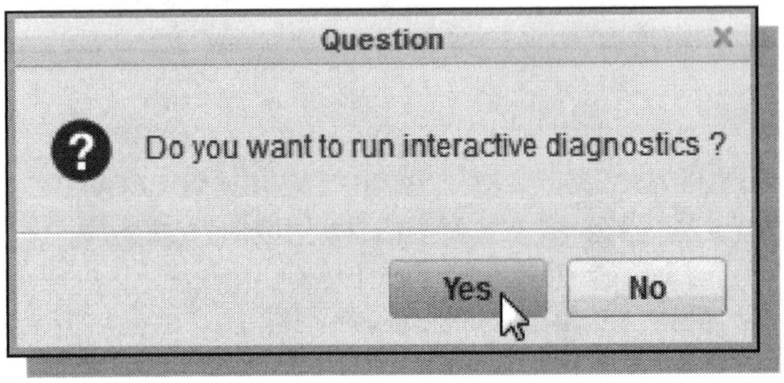

7. Click **Yes** to run interactive diagnostics; this will help us identify and correct problems associated with our FEA model.

8. In the *Diagnostics* window, *Creo Simulate* lists any problems or conflicts of our FEA model.

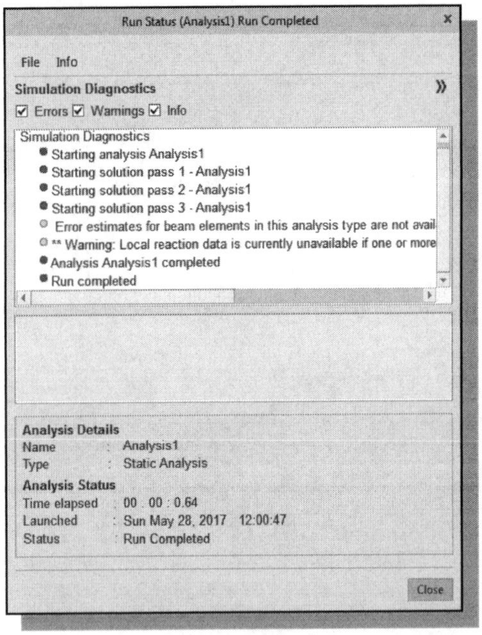

9. Click on the **Close** button to exit the *Diagnostics* window.

10. In the *Analyses and Design Studies* window, choose **Review results** as shown.

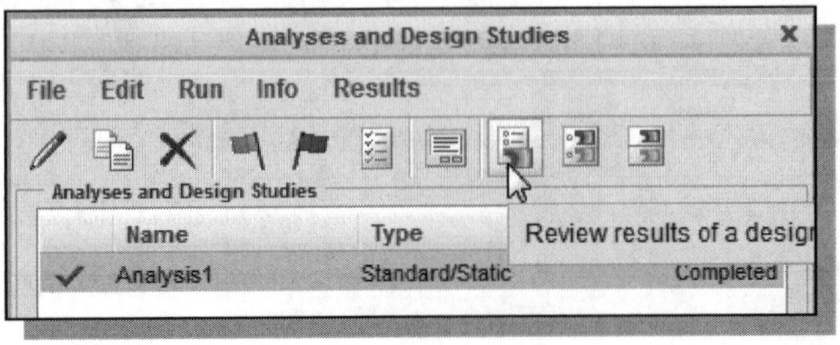

View the FEA Results

➢ Maximum Principal Stress

1. Confirm the Display type is set to **Fringe**.

2. Choose **Maximum Principal** as the Stress Component to be displayed.

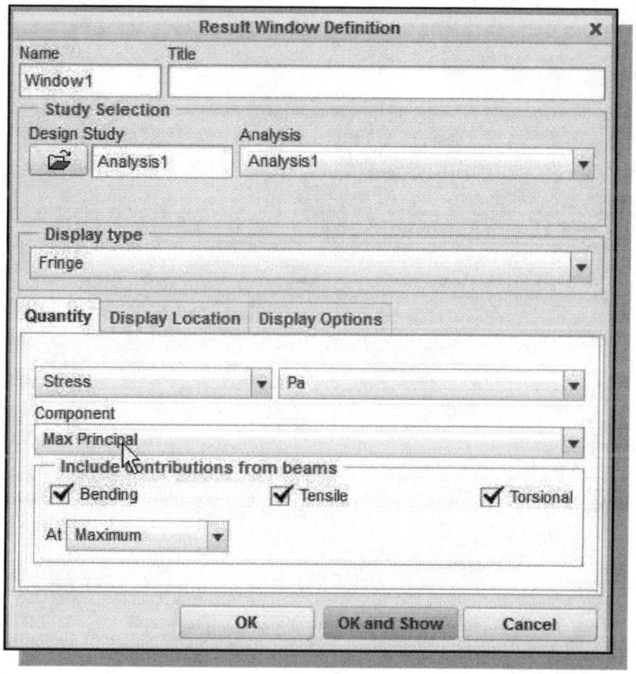

3. Click on the **Display Options** tab and switch ***on*** the **Deformed** and **Animate** options as shown.

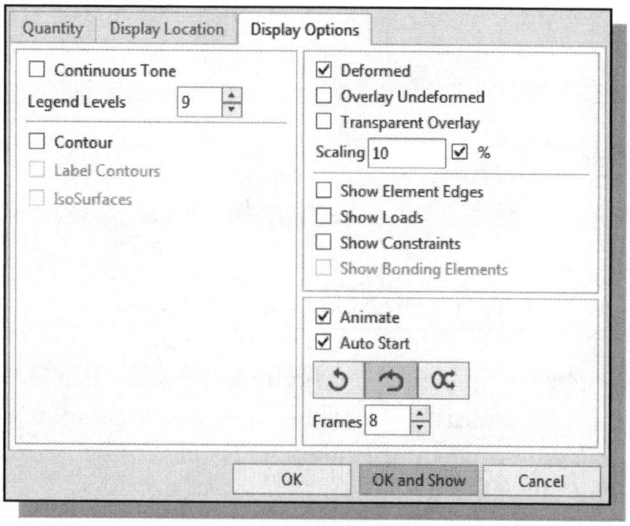

4. Click **OK and Show** to display the results.

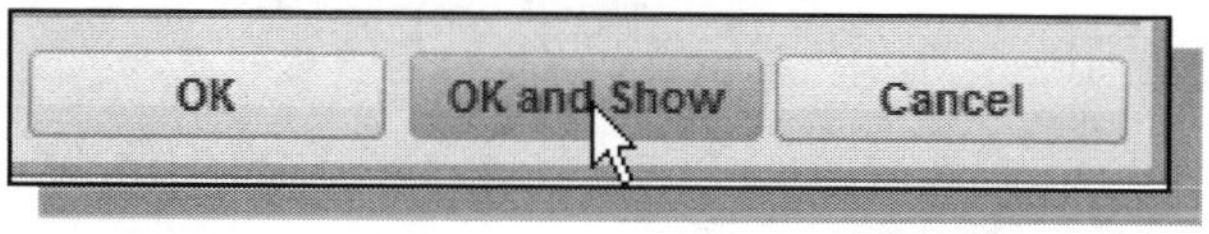

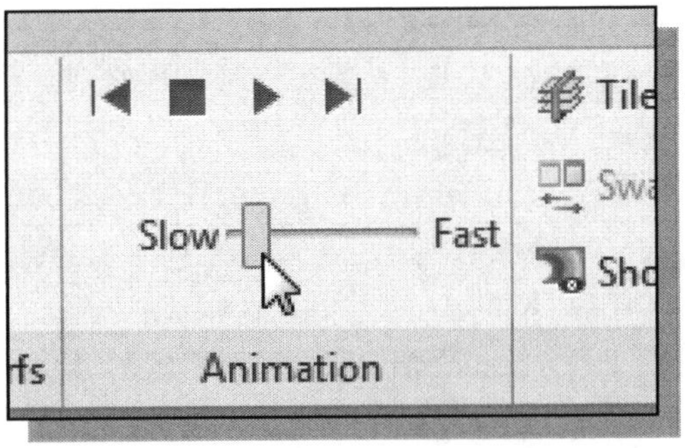

5. On your own, use the **Animation control panel** in the *View* tab to adjust the speed of the animation.

❖ The *Creo Simulate* **Maximum Principal** option allows us to quickly view the *combined normal stresses* in beam elements.

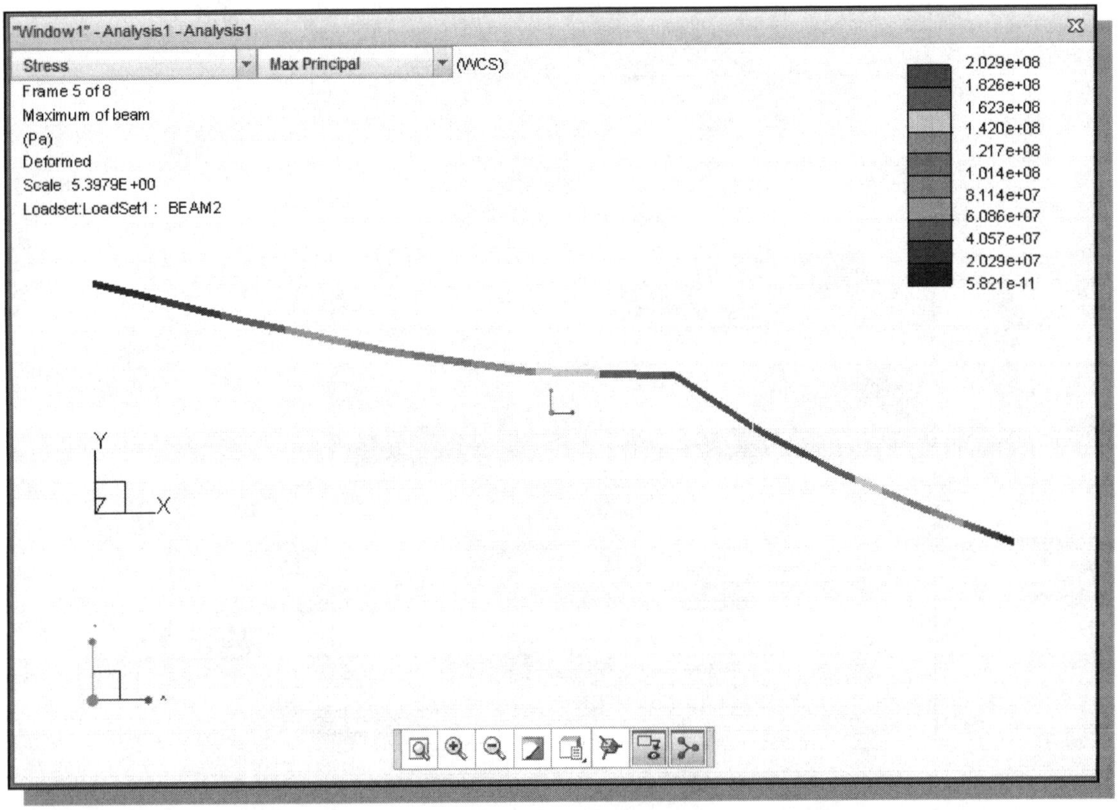

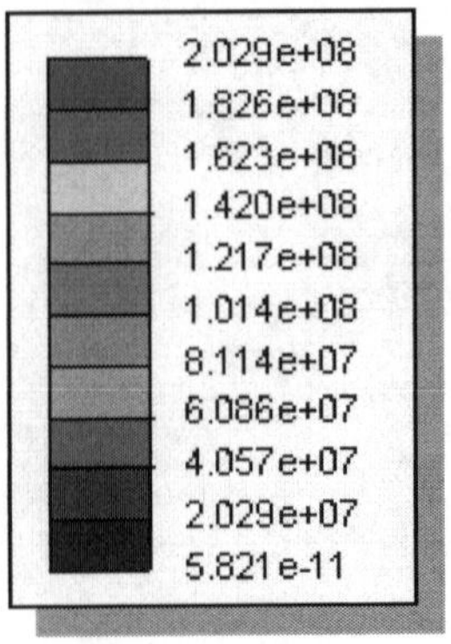

➢ The Max. Principal stress of the FEA analysis is **2.029e5**, which is a bit higher than the max. stress value at point D. This result matches the result of our preliminary analysis. The FEA result indicates the max. stress occurs on the inclined beam. To verify the location where this max. stress occurs, let's examine the shear and moment diagrams.

➢ Shear and Moment Diagrams

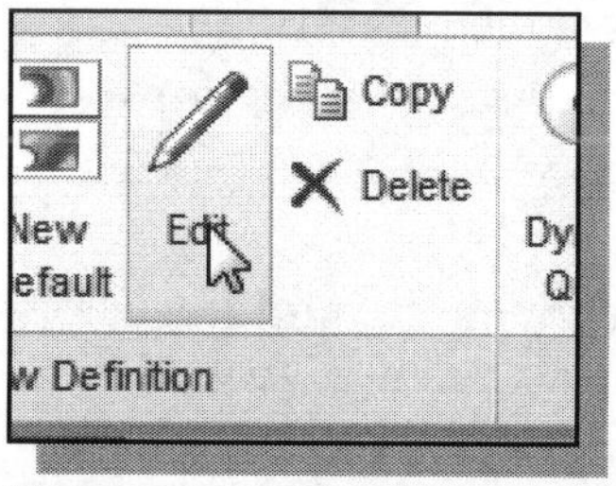

1. In the *Standard* toolbar area, click **Edit the current definition** command as shown.

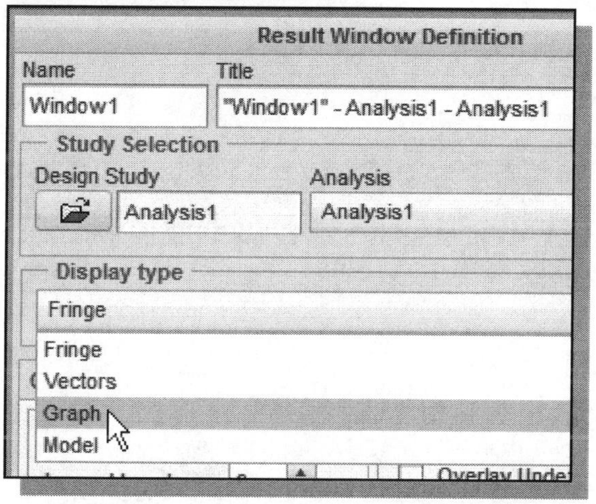

2. Choose **Graph** under the Display type option list as shown.

3. In the **Graph Ordinate (Vertical) Axis** option list, select **Shear & Moment** as shown.

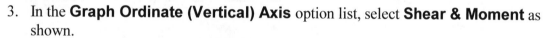

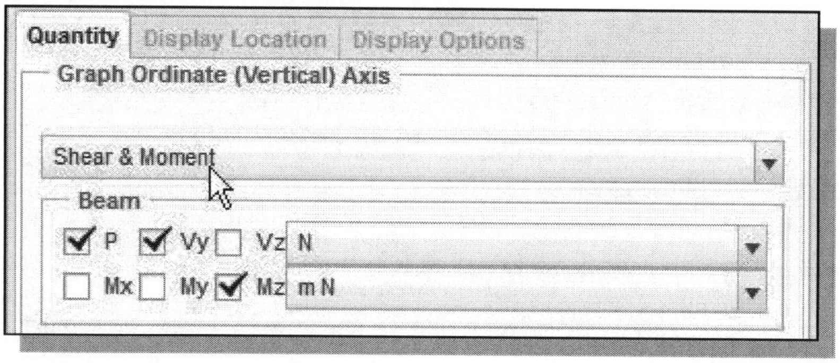

4. Switch ***on*** the **P** (axial load), **Y** (shear force) and **Z** (bending moment) components as shown in the above figure.

5. In the Graph Abscissa (Horizontal) Axis section, confirm the options are set to **Curve Arc Length** and **Curve** as shown.

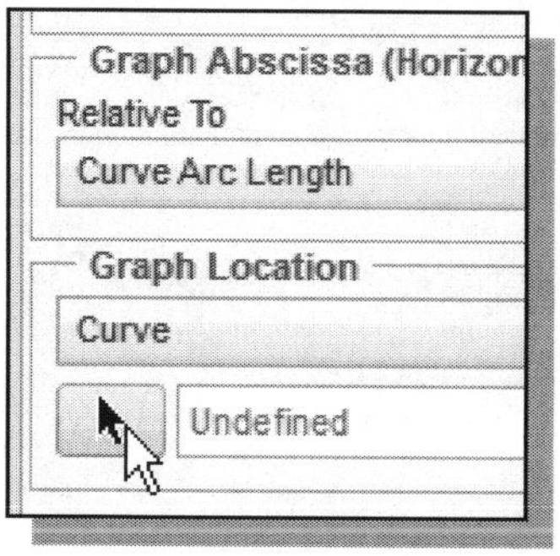

6. Click on the **arrow button** to select curves to display the shear diagram.

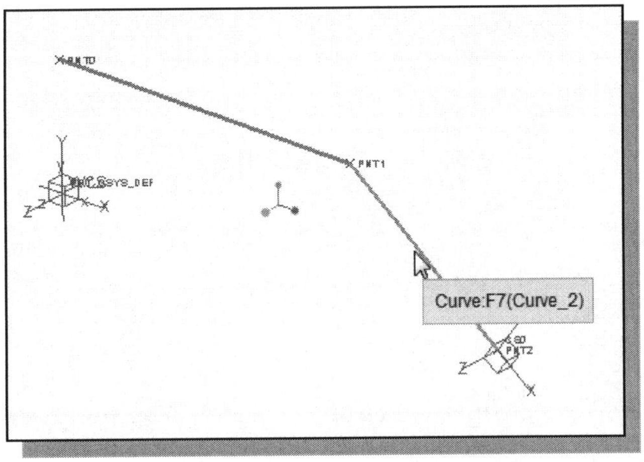

7. Press down the [**CTRL**] key and select both of the **datum curves**, left curve first.

8. Click once with the **middle-mouse-button** to accept the selection.

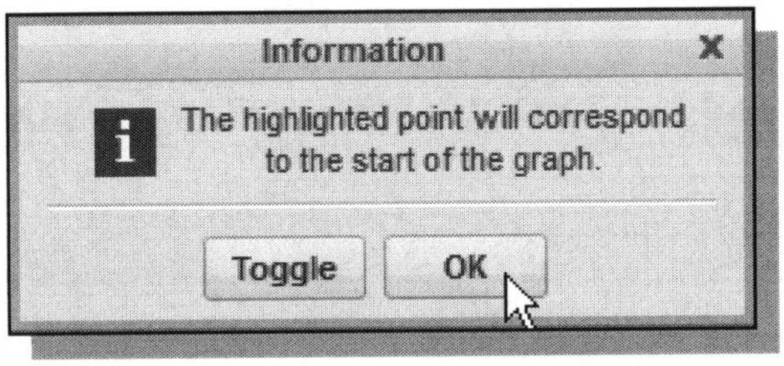

9. Set the **PNT0** point is highlighted and click **OK** to continue.

10. Click **OK and Show** to display the results.

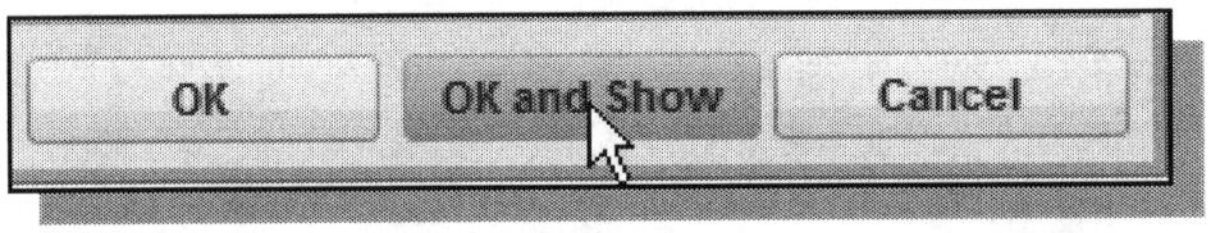

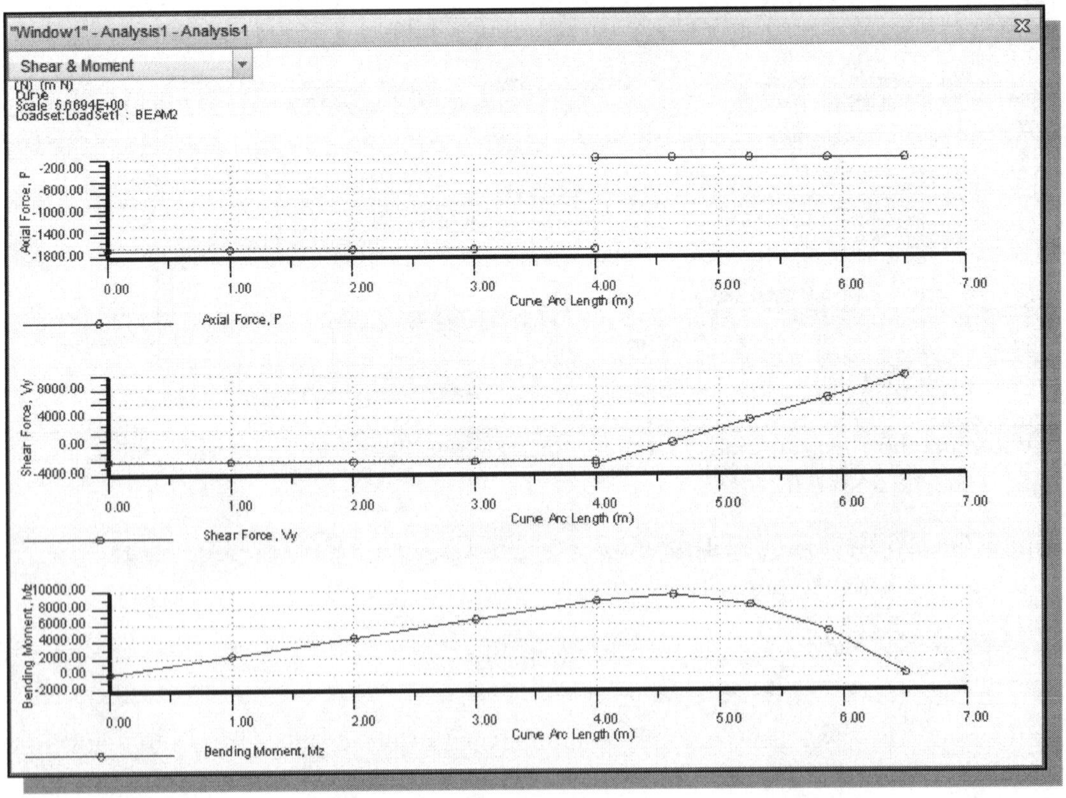

❖ The displayed graph identifies the max. bending stress is located about 0.6 meter along the inclined beam, but more datum points are needed on the inclined beam for better curves.

Refine the FE Model

By default, *Creo Simulate* performs five calculations between user defined nodes in the FE model. To increase the number of calculations, we can simply increase the number of nodes.

1. Select **File → Exit Results** to close the *Result* window and return to the *Creo Simulate* window.

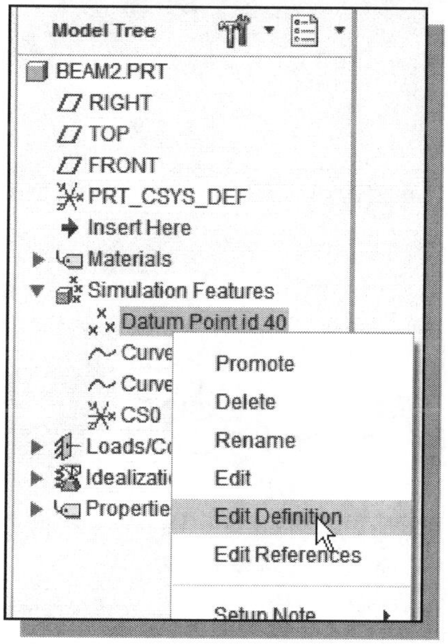

2. In the browser window, expand the **Simulation Features** by clicking on the [+] sign.

3. Select the **Datum Points** option.

4. Press down the right-mouse-button on the highlighted item to display the option list and select **Edit Definition** as shown.

5. On your own, create 3 additional datum points, PNT3, PNT4, PNT5, as shown in the figure below.

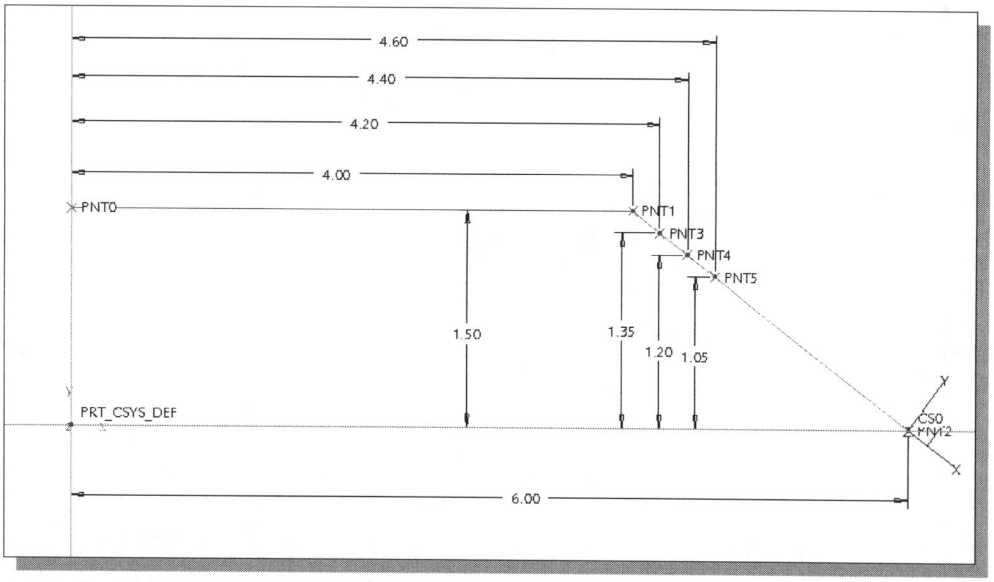

6. On your own, redefine the second **datum curve** and adjust the ***beams***, ***constraints*** and ***load*** sets if necessary.

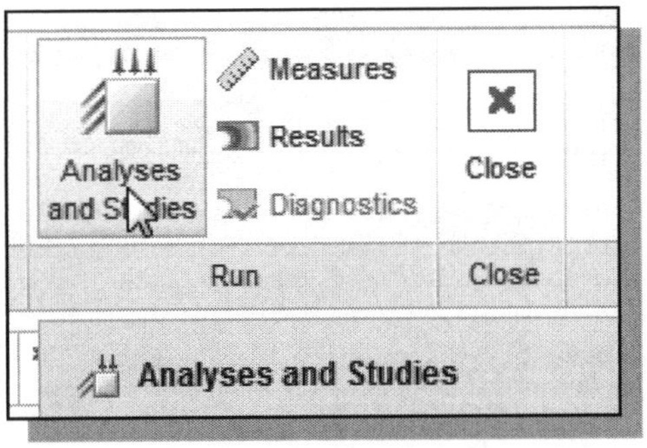

7. Activate the *Creo Simulate* **Analyses and Studies** command by selecting the icon in the toolbar area as shown.

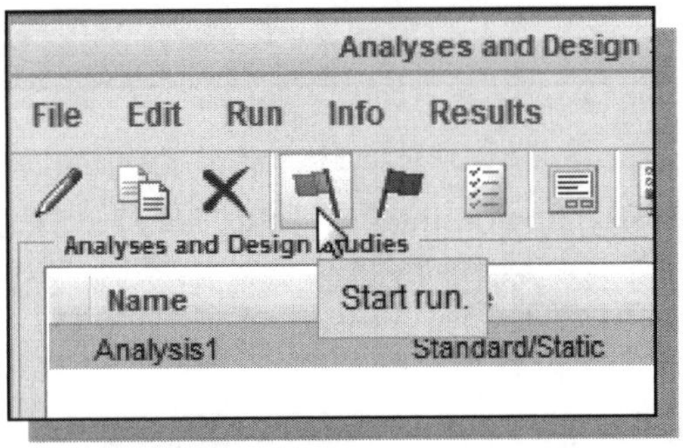

8. Click **Start run** to begin the *Solving* process.

9. On your own, perform the necessary FEA analysis.

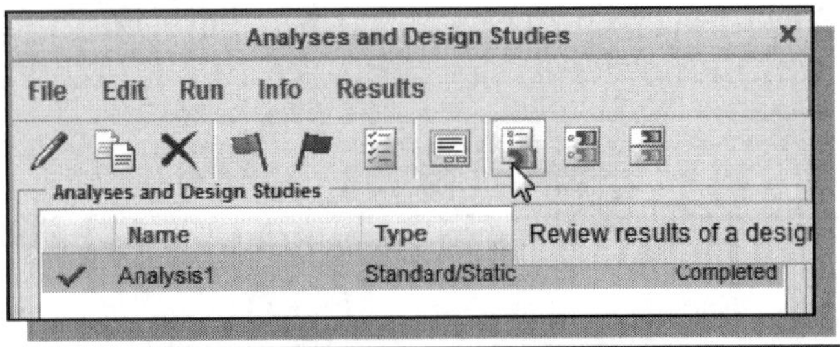

10. In the *Analyses and Design Studies* window, choose **Review results** as shown.

11. Examine the resulting bending stress, shear and moment diagrams.

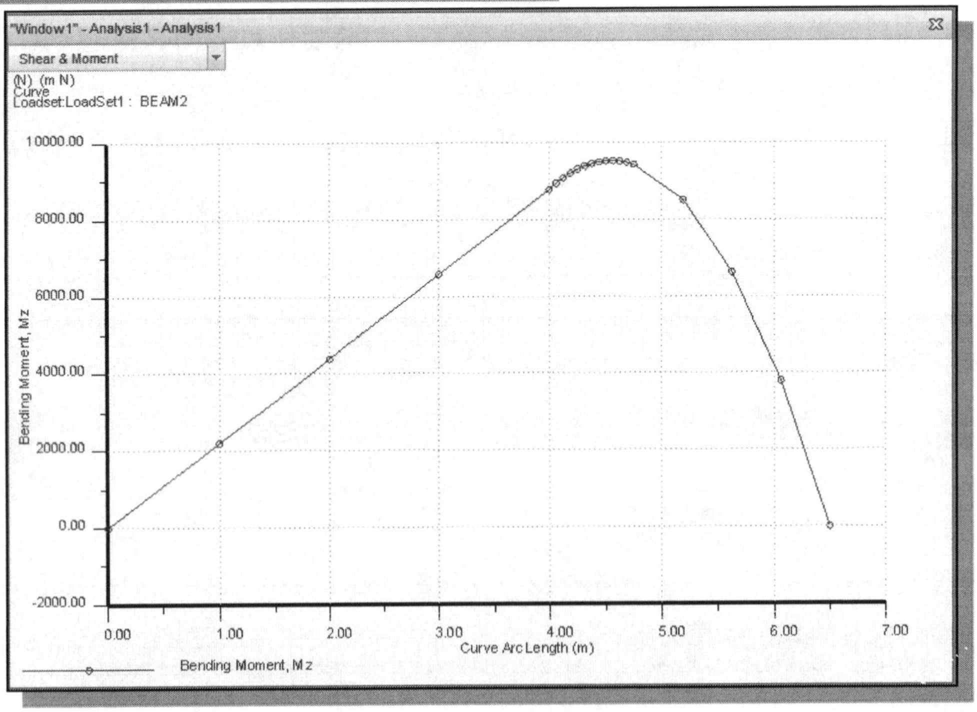

➢ Note *Creo Simulate* has added more nodes and performed more calculations near the location where the max. bending moment occurs.

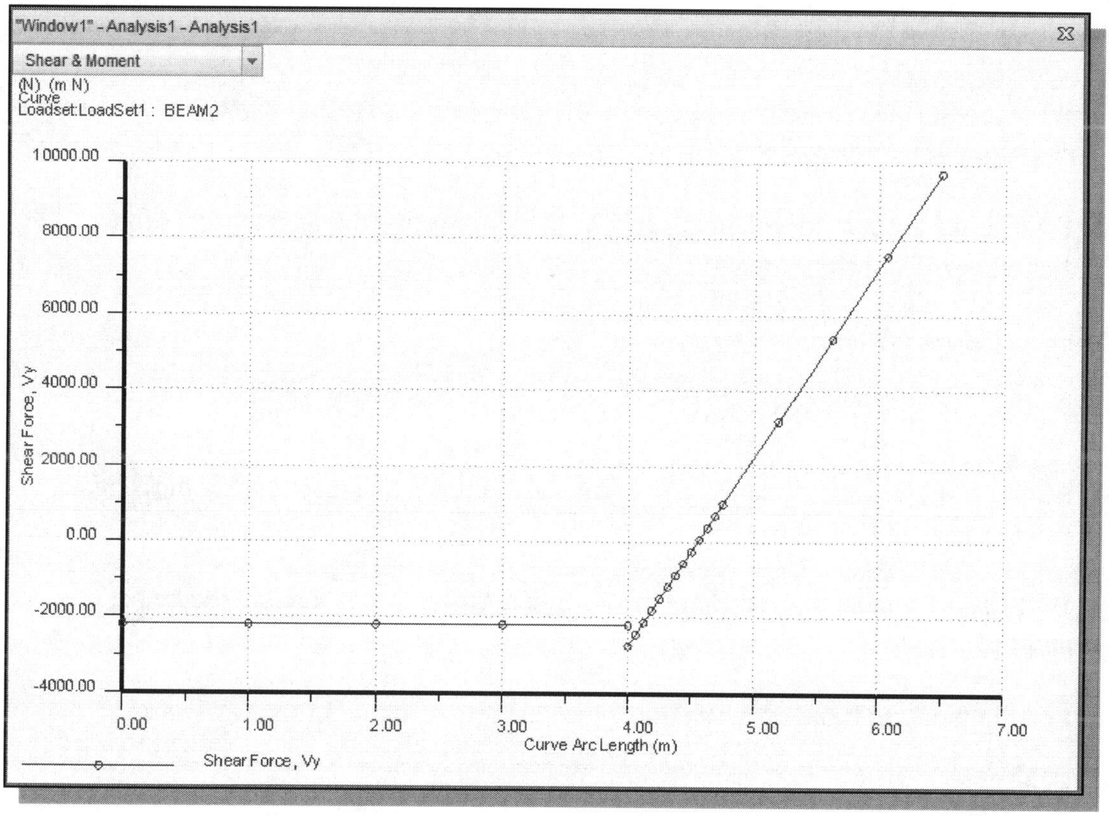

➢ Note the modified FEA model shows the normal stress at point D, matching the results of the preliminary analysis.

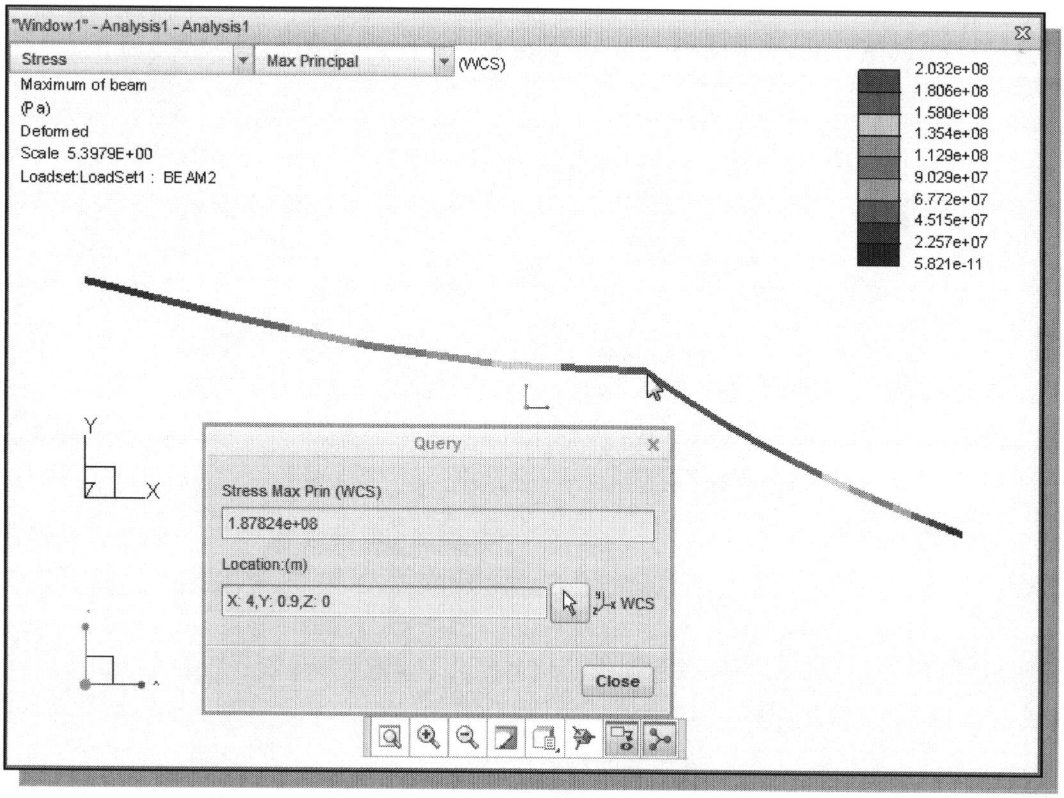

Review Questions

1. What are the main differences between the *Integrated* mode and the *Independent* mode of *Creo Simulate*?

2. How do we account for inclined supports and apply angled loads on individual elements in *Creo Simulate*?

3. Describe the method of *Superposition.*

4. Can we create additional coordinate systems in *Creo Simulate*?

5. Which command option do we use to display multiple result screens in the *Review results* window?

6. Will *Creo Simulate* calculate and display the reaction forces that are located at inclined supports?

7. List the standard beam cross section types that are available in *Creo Simulate.*

8. In *Creo Simulate*, list and describe the two options to establish beam elements you have learned so far.

9. Are the *Creo Simulate* analysis results stored inside the *Creo Parametric* part files?

10. How do we control the speed of animation of the *Creo Simulate* analysis results?

Exercises

Determine the maximum stress produced by the loads and create the shear and moment diagrams for the structures.

1. Fixed-end support.
   Material: Steel
   Diameter 1.0 in.

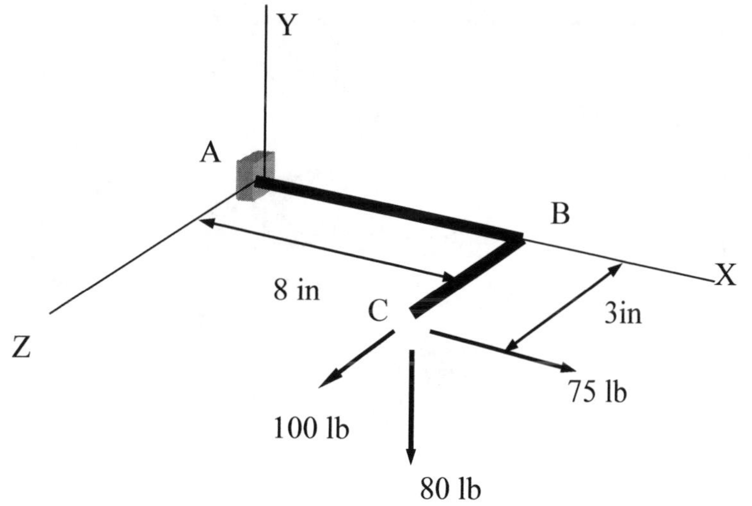

2. A Pin-Joint and a roller support.
   Material: Aluminum Alloy 6061 T6

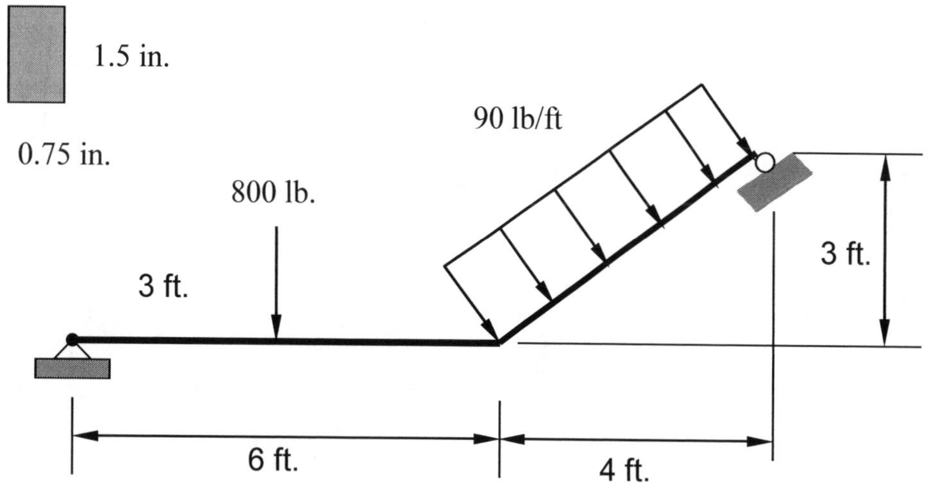

3. A Pin-Joint support and a roller support.
   Material: Steel

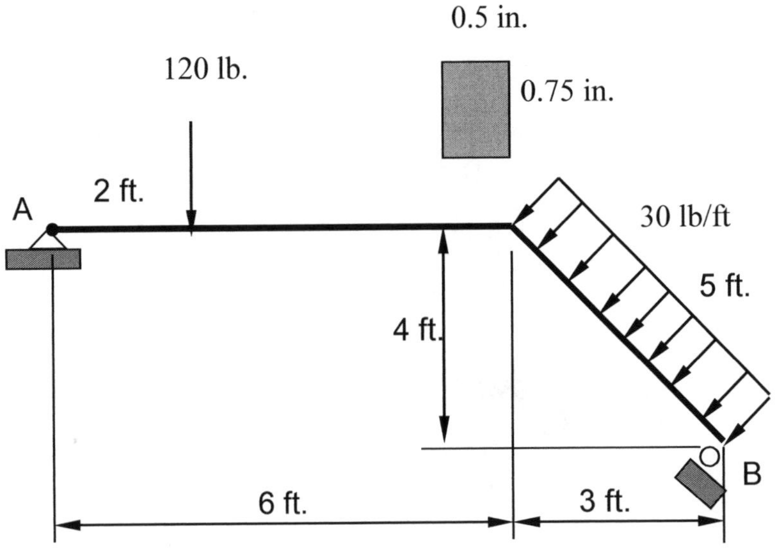

Chapter 8
Statically Indeterminate Structures

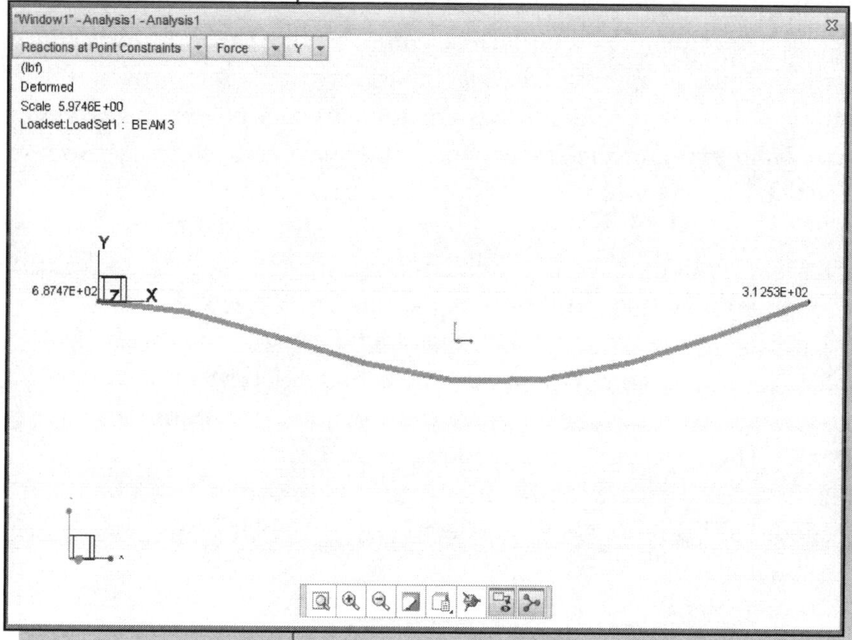

Learning Objectives

- ♦ Perform Statically Indeterminate Beam Analysis.
- ♦ Understand and Apply the Principle of Superposition.
- ♦ Identify Statically Indeterminate Structures.
- ♦ Apply and Modify Boundary Conditions on Beams.
- ♦ Generate Shear and Moment diagrams.

Introduction

Up to this point, we have dealt with a very convenient type of problem in which the external reactions can be solved by using the *equations of Statics* (Equations of Equilibrium). This type of problem is called *statically determinate*, where the *equations of Statics* are directly applicable to *determine* the loads in the system. Fortunately, many real problems are statically determinate, and our effort is not in vain. An additional large class of problems may be reasonably approximated as though they are *statically determinate*. However, there are those problems that are not statically determinate, and these are called *statically indeterminate structures*. Static indeterminacy can arise from redundant constraints, which means more supports than are necessary to maintain equilibrium. The reactions that can be removed leaving a stable system statically determinate are called *redundant*. For statically indeterminate structures, the number of unknowns exceeds the number of static equilibrium equations available. Additional equations, beyond those available from *Statics*, are obtained by considering the geometric constraints of the problem and the force-deflection relations that exist. Most of the derivations of the force-deflection equations are fairly complex; the equations can be found in design handbooks. There are various procedures for solving statically indeterminate problems. In recent years, with the improvements in computer technology, the finite element procedure has become the most effective and popular way to solve statically indeterminate structures.

Equations of equilibrium:

$$\sum M = 0$$
$$\sum F = 0$$

Statically determinate structure:

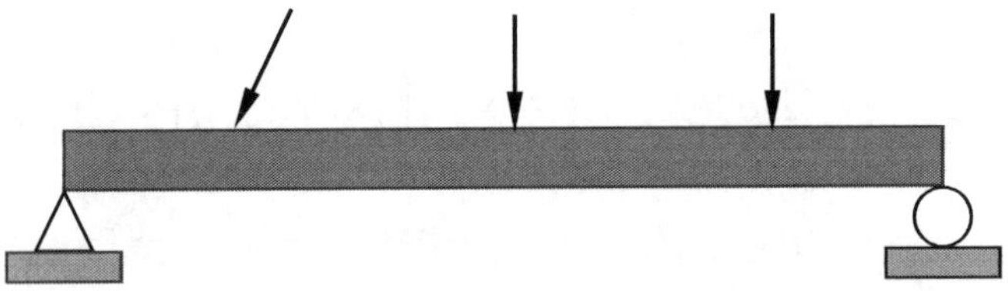

Statically indeterminate structure:

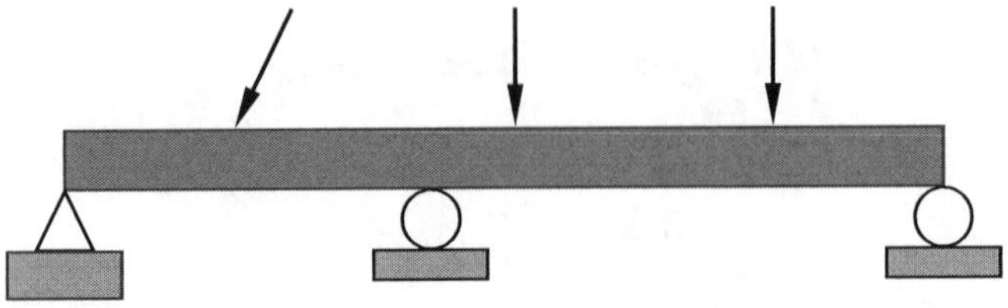

Problem Statement

Determine the reactions at point A and point B. Also determine the maximum normal stress that the loading produces in the Aluminum Alloy member (Diameter 2″). Al-6061 is an Aluminum hardened alloy, with magnesium and silicon as the major elements.

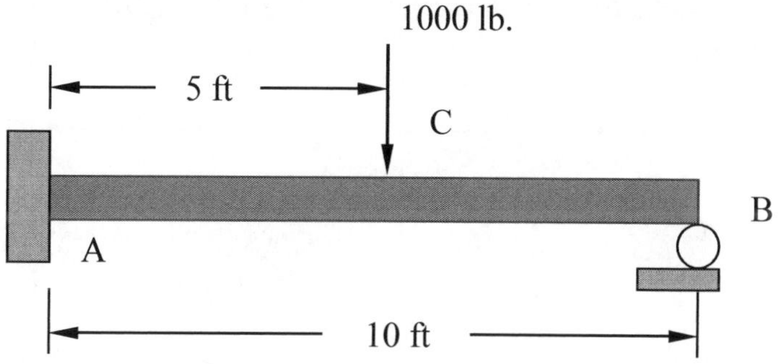

Preliminary Analysis

The reactions at point A and B can be determined by using the principle of superposition. By removing the roller support at point B, the displacement at point B due to the loading is first determined. Since the displacement at point B should be zero, the reaction at point B must cause an upward displacement of the same amount. The displacements are superimposed to determine the reaction. The *principle of superposition* can be used for this purpose, and once the reaction at B is determined, the other reactions are determined from the equations of equilibrium.

1

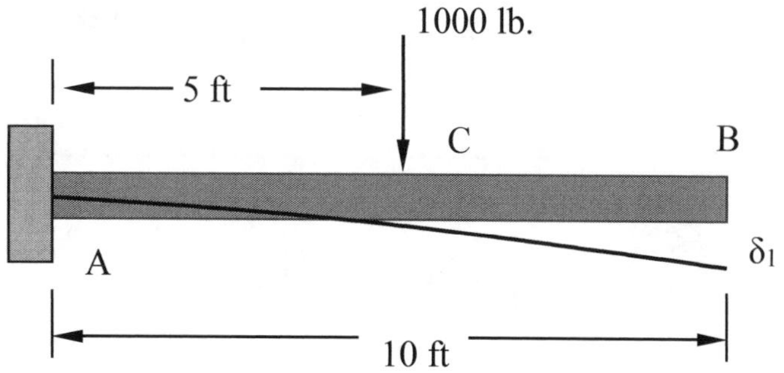

The displacement δ_1 can be obtained from most strength of materials textbooks and design handbooks:

$$\delta_1 = \frac{5\ PL^3}{48\ EI}$$

2

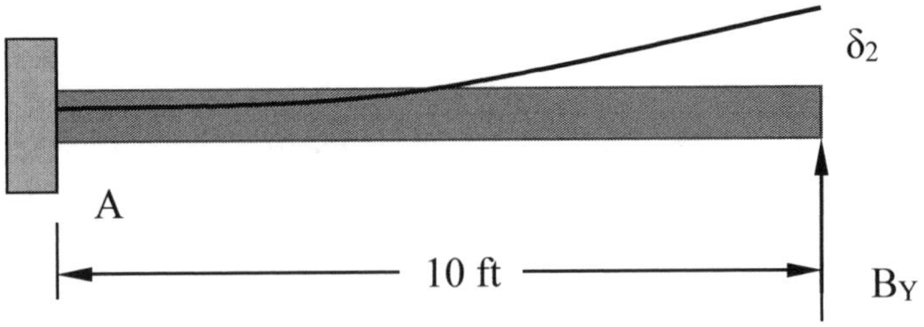

The displacement δ_2 due to the load can also be obtained from most strength of materials textbooks and design handbooks:

$$\delta_2 = \frac{B_Y L^3}{3\,E\,I}$$

3 (Superposition of 1 and 2)

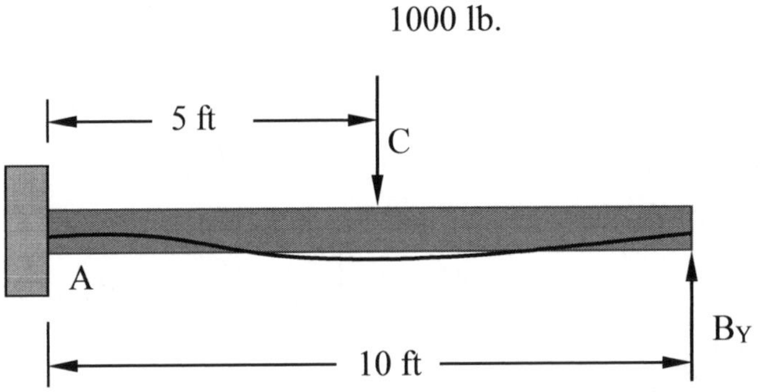

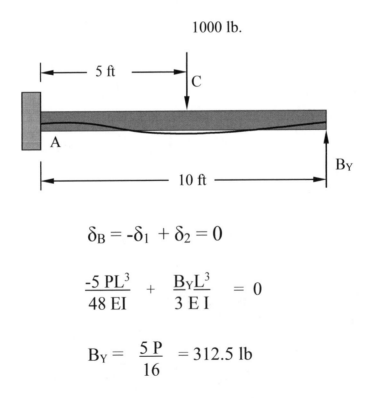

$$\delta_B = -\delta_1 + \delta_2 = 0$$

$$\frac{-5\,PL^3}{48\,EI} + \frac{B_Y L^3}{3\,E\,I} = 0$$

$$B_Y = \frac{5\,P}{16} = 312.5 \text{ lb}$$

Free Body Diagram of the system:

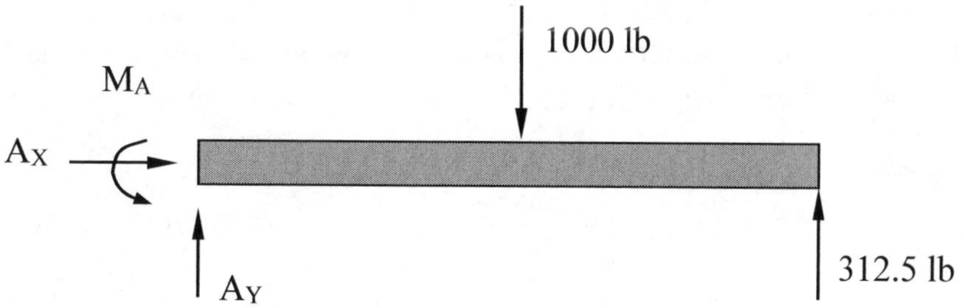

The reactions at A can now be solved:

$$\sum F_X = 0, \quad A_X = 0$$

$$\sum F_Y = 0, \quad 312.5 - 1000 + A_Y = 0, \quad A_Y = 687.5 \text{ lb.}$$

$$\sum M_{@A} = 0, \; 312.5 \text{ x } 10 - 1000 \text{ x } 5 + M_A = 0, \; M_A = 1875 \text{ ft-lb}$$

The maximum normal stress at point A is from the bending stress:

$$\sigma = \frac{MC}{I}$$

for the circular cross section:

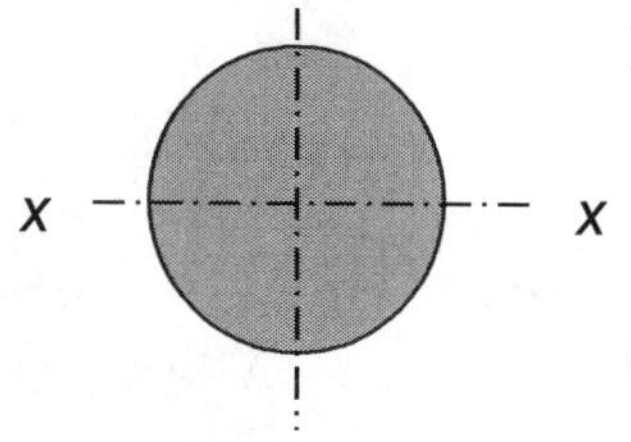

$$I_{X\text{-}X} = \frac{\pi r^4}{4}$$

$$C = r$$

Therefore,

$$\sigma_A = \frac{MC}{I} = \frac{4 M r}{\pi r^4} = 4.125 \text{ x } 10^6 \text{ lb/ft}^2$$

Starting Creo Parametric

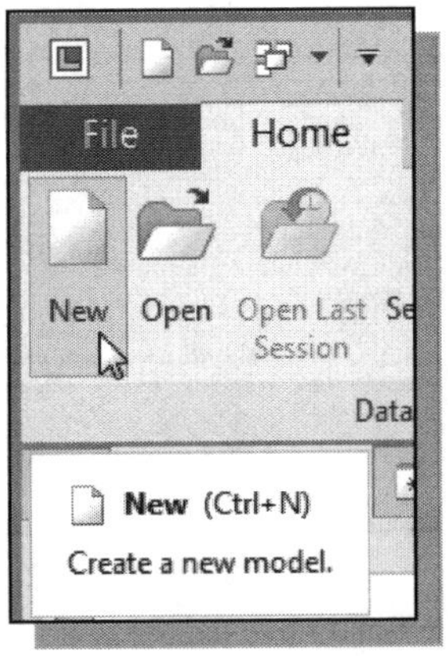

1. Select the **Creo Parametric** option on the *Start* menu or select the **Creo Parametric** icon on the desktop to start *Creo Parametric*. The *Creo Parametric* main window will appear on the screen.

2. Click on the **New** icon, located in the *Ribbon toolbar* as shown.

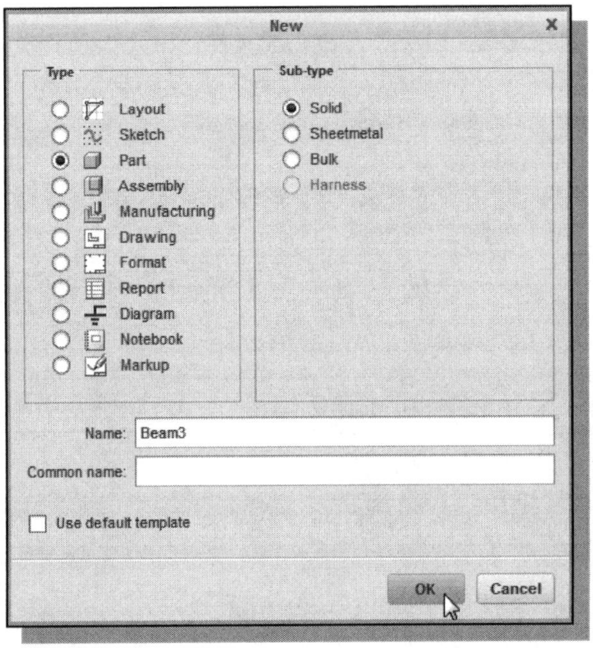

3. In the *New* dialog box, confirm the model's Type is set to **Part** (**Solid** Sub-type).

4. Enter **Beam3** as the part Name as shown in the figure.

5. Turn ***off*** the **Use default template** option.

6. Click on the **OK** button to accept the settings.

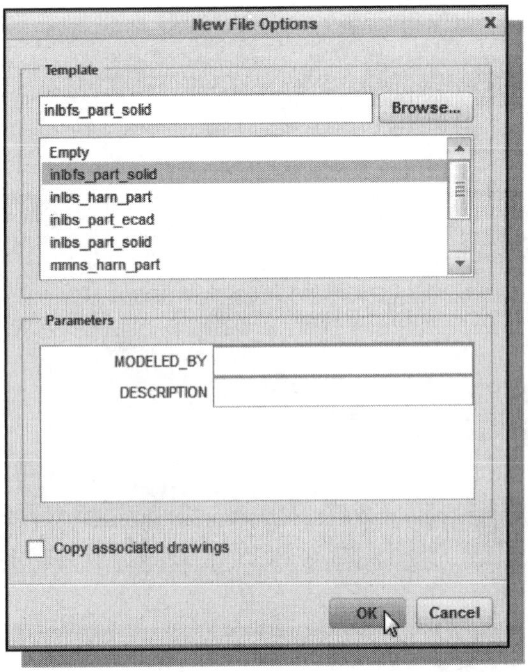

7. In the *New File Options* dialog box, select **inlbfs_part_solid** in the option list as shown. (Use the *Browse* option to locate the file if it is not shown in the list.)

8. Click on the **OK** button to accept the settings and enter the *Creo Parametric Part Modeling* mode.

Create a Wireframe Model in Creo Parametric

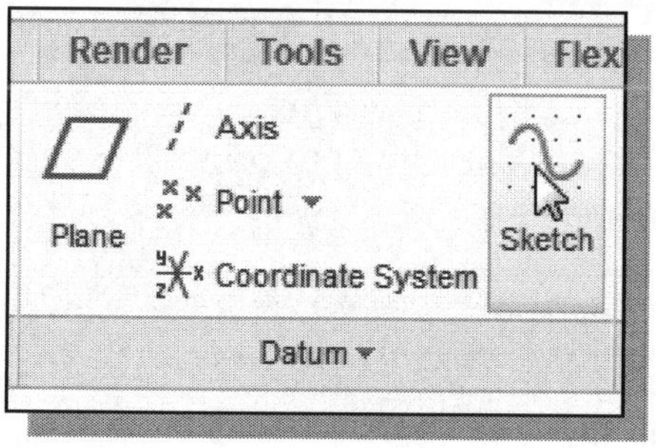

1. Click **Sketch** tool in the icon panel on the *Ribbon* toolbar in the *Creo Parametric* main screen as shown.

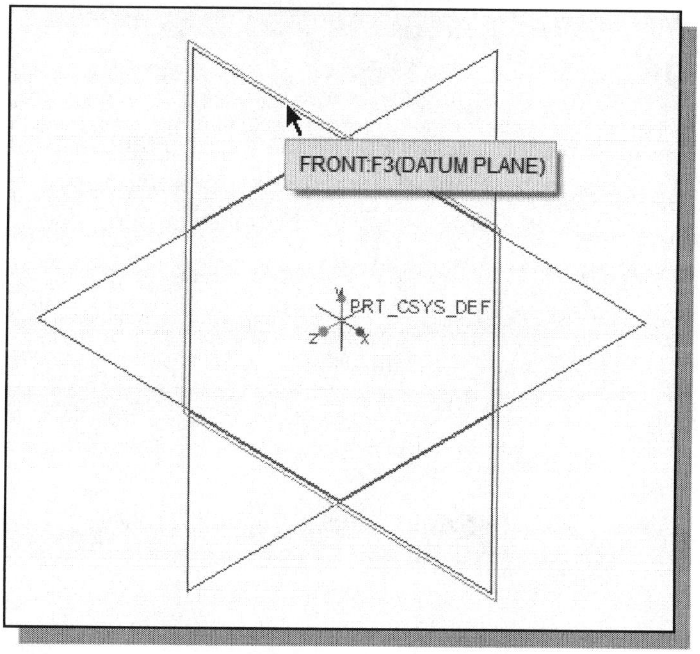

2. In the graphics area, select **FRONT** by clicking on the text FRONT as shown.

❖ Notice an arrow appears on the edge of FRONT. The arrow direction indicates the viewing direction of the sketch plane.

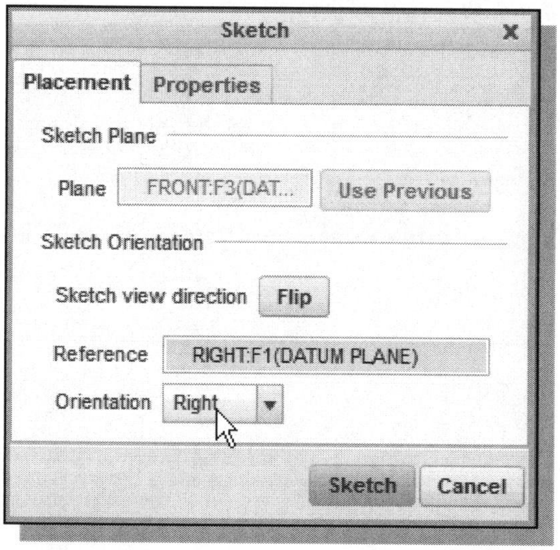

3. Click inside the Reference option box in the Sketch Orientation window as shown. The message "*Select a reference, such as surface, plane or edge to define view orientation.*" is displayed in the message area.

4. In the *graphics area*, select **RIGHT** by clicking on the text.

5. Pick **Sketch** to exit the *Sketch* window and proceed to enter the *Creo Parametric Sketcher* mode.

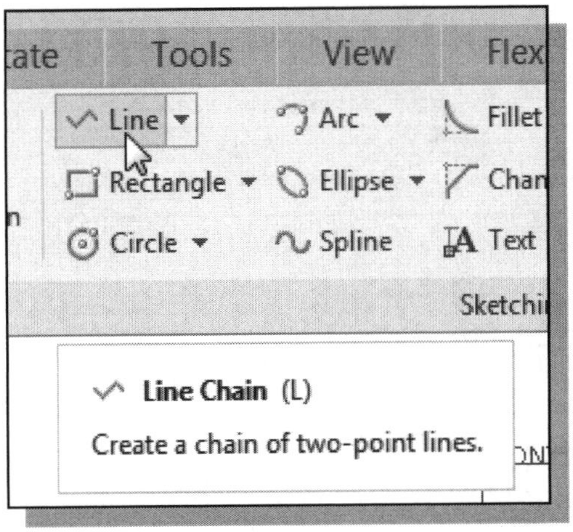

6. Click on the **Line** button to activate the Create 2 point lines command.

7. Click the **Sketch View** icon to orient the sketching plane parallel to screen.

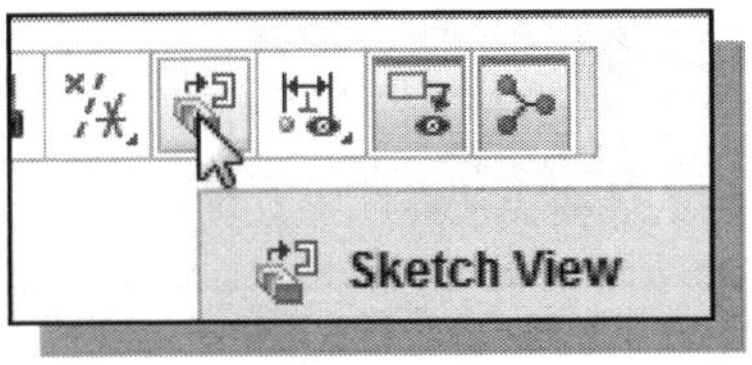

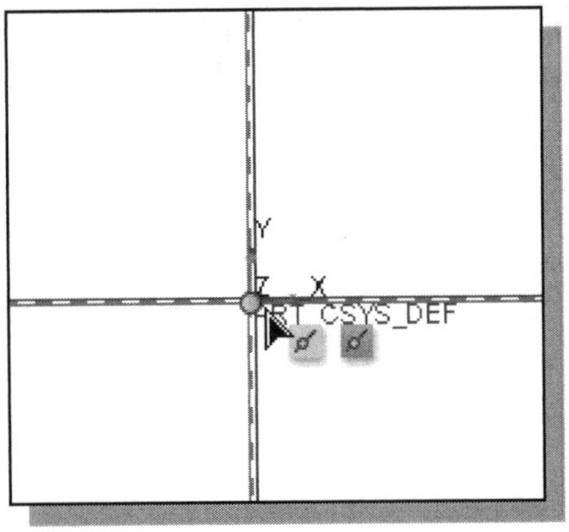

8. Click at the intersection of the two references as the starting point of the line; this location can be viewed as the origin of the XY plane where the beam system will be placed.

9. Move the cursor toward the right and create a horizontal line segment as shown.

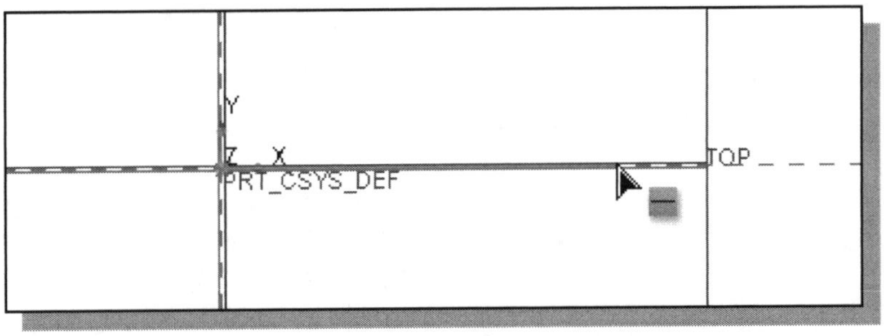

10. Move the cursor toward the right and create another horizontal line segment, with equal length constraint, as shown.

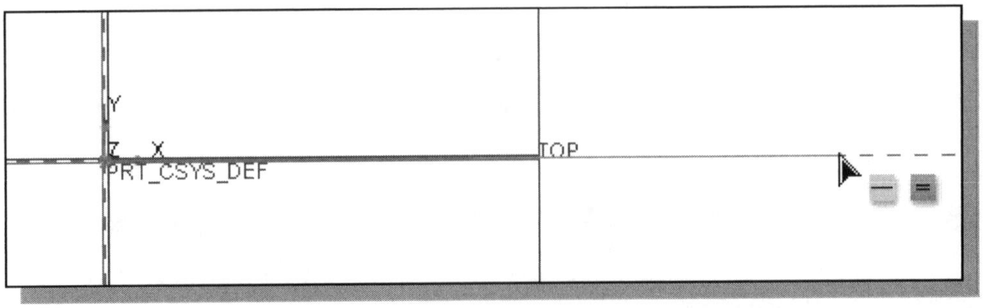

11. Click once with the **middle-mouse-button** to accept the selections and create the two line segments.

12. On your own, create and modify the length dimension (60) so that the sketch appears as shown in the figure.

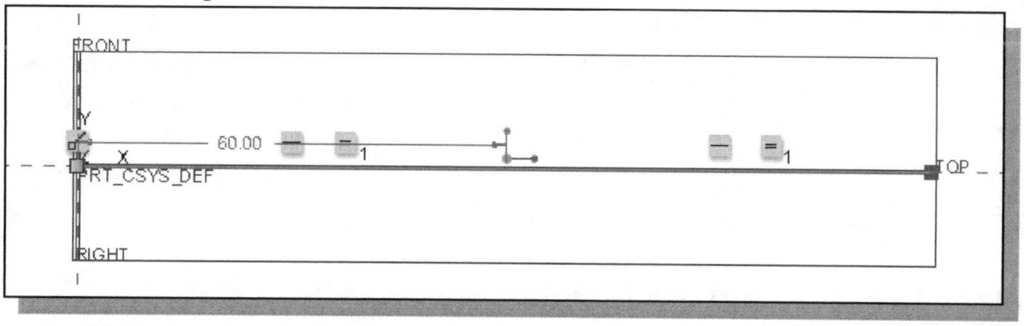

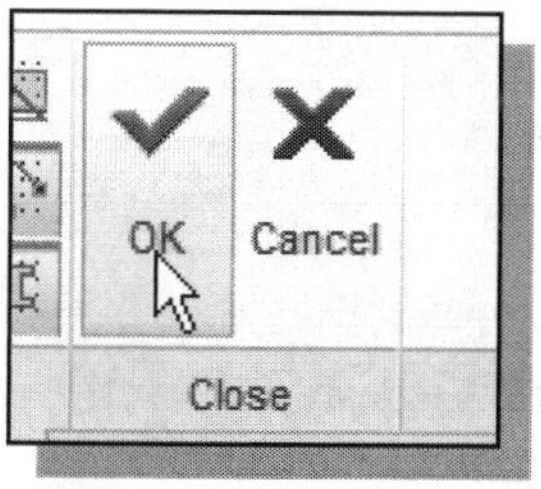

13. Click on the **Done** button to accept the creation of the datum points.

A CAD Model is NOT an FEA Model

The two lines we just created represent the system that we will be analyzing. This model is created under *Creo Parametric*, and it is a CAD model that contains geometric information about the system. The CAD model can be used to provide geometric information needed in the finite element model. However, the CAD model is not an FEA model. In the previous chapters, we created datum points and datum curves to provide the geometric information, which we could just as easily have done in *Creo Parametric*. The CAD model is typically developed to provide geometric information necessary for manufacturing. All details must be specified and all dimensions are required. The manufactured part and the CAD model are identical in terms of geometric information. The FEA model uses the geometric information of the CAD model as the starting point, but the FEA model usually will adjust some of the basic geometric information of the CAD model. The FEA model usually will contain additional nodes and elements. Idealized boundary conditions and external loads are also required in the model. The goal of finite element analysis is to gain sufficient reliable insights into the behaviors of the real-life system. Many assumptions are made in the finite element analysis procedure to simplify the analysis, since it is not possible or practical to simulate all factors involved in real-life systems. The finite element analysis procedure provides an idealized approximation of the real-life system. It is therefore not practical to include all details of the system in the FEA model; the associated computational cost cannot be justified in doing so. It is a common practice to begin with a more simplified FEA model. Once the model has been solved accurately and the result has been interpreted, it is feasible to consider a more refined model in order to increase the accuracy of the prediction of the actual system. In *Creo Simulate*, creating an FEA model by using the geometric information of the CAD model is known as ***Idealization***.

The Integrated Mode of Creo Simulate

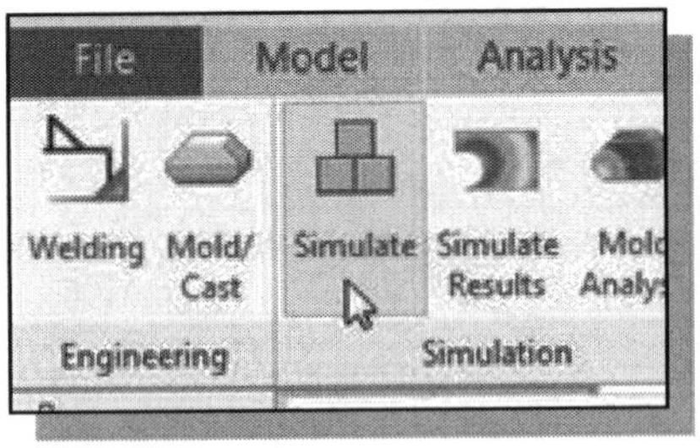

1. Start the *Integrated* mode of *Creo Simulate* by selecting the **Applications → Simulate** option in the *Ribbon* area as shown.

Select and Examine the Element Material Property

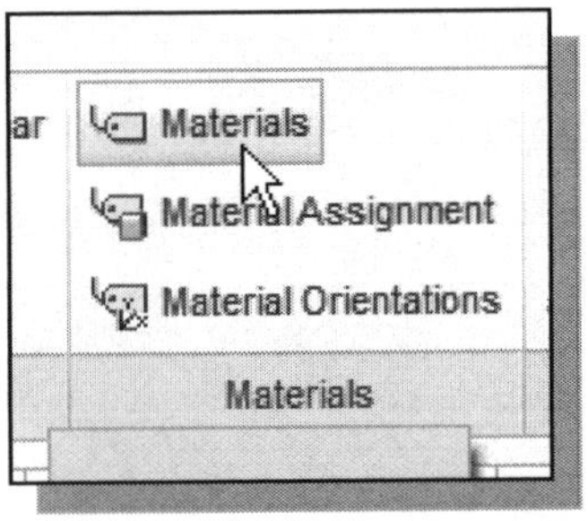

1. Choose **Materials** in the *Materials* group from the *Ribbon* toolbar as shown.

❖ Note the default list of materials libraries, which are available in the pre-defined *Creo Simulate*, is displayed.

2. Select **Al-Mg-Si_alloy_Wrought** in the *Materials* list as shown.

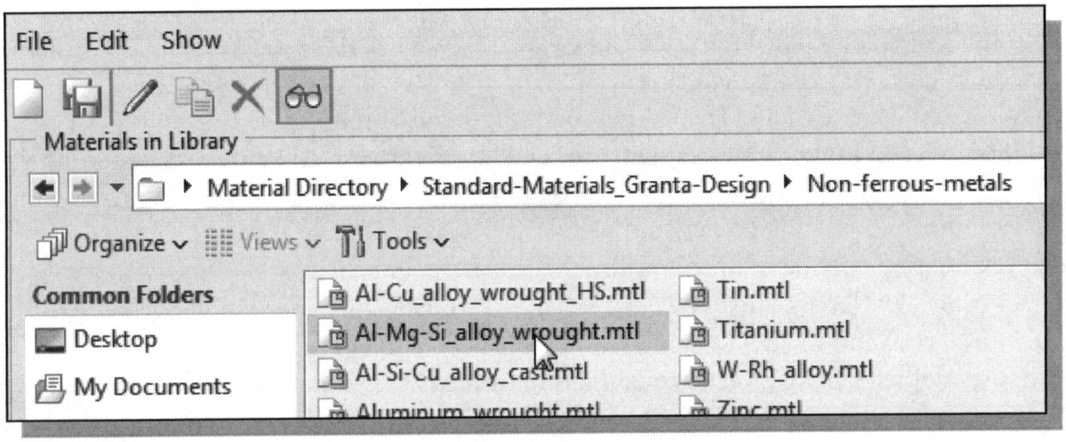

3. **Double-clicking** on the selected material definition to make the selected material available for use in the current FEA model.

4. On your own, use the *Edit* option to examine the material information and click **OK** to accept the settings.

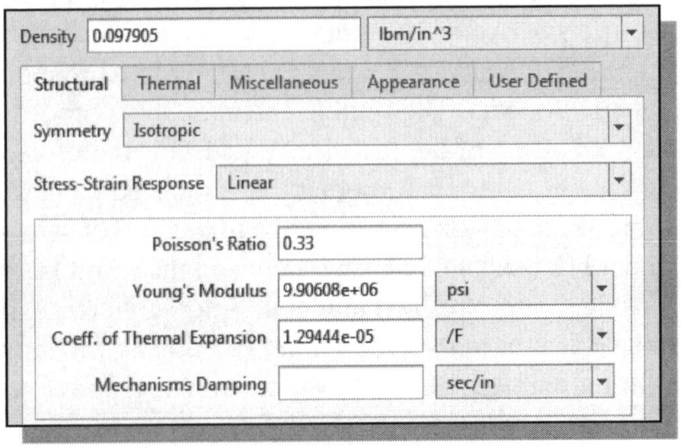

Set up the Element Cross Section

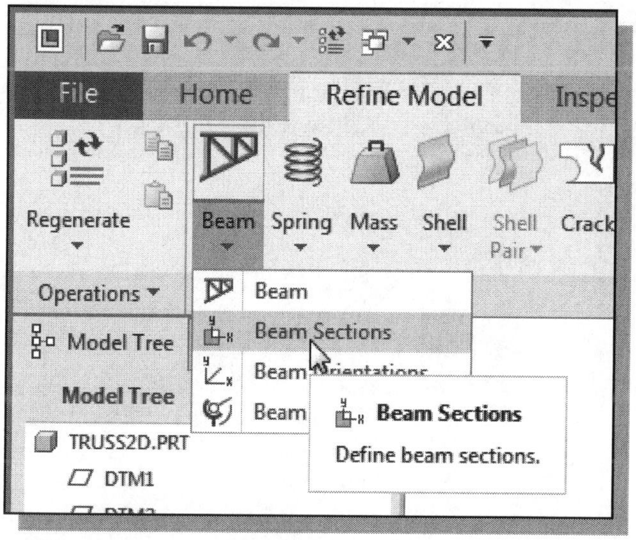

1. Choose **Beam Sections** in the **Refine Model** tab as shown.

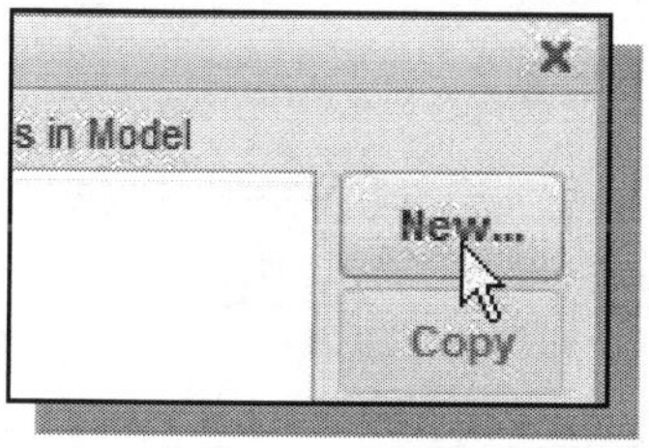

2. Click **New** to create a new beam cross-section.

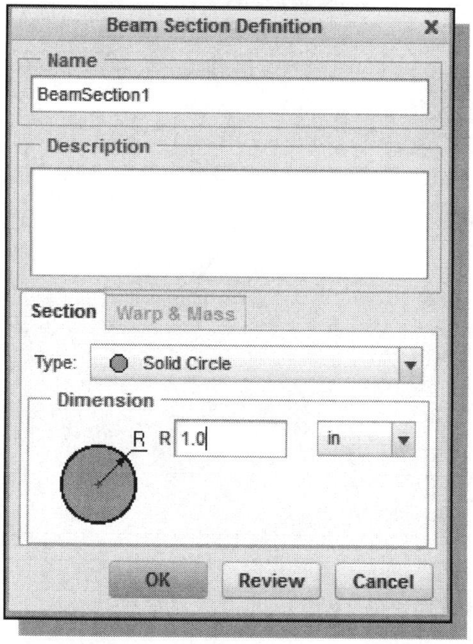

3. Choose **Solid Circle** as the cross-section **Type** as shown.

4. Enter **1.0** as the radius of the circular beam cross section.

5. Click **OK** to accept the creation of the new beam section.

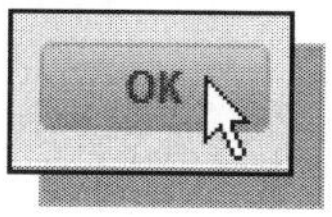

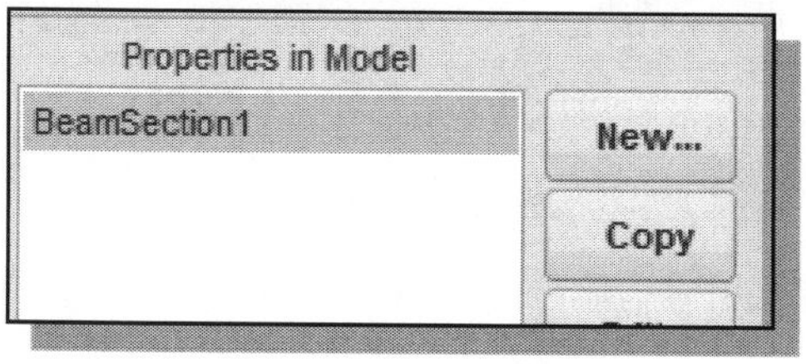

6. Click **Close** to accept the creation of the new beam section.

Create the Beam Elements

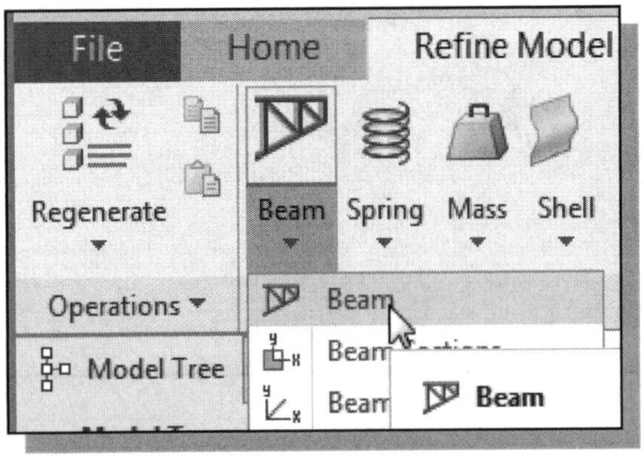

1. Choose **Beam** in the icon panel. The *Beam Definition* window appears.

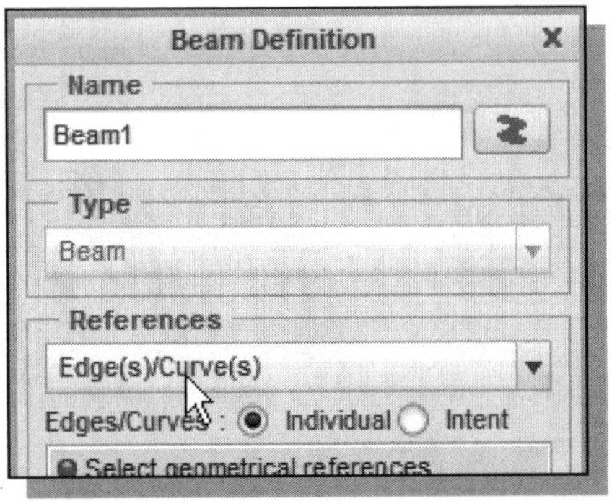

2. Confirm the References option is set to **Edge(s)/Curve(s)**.

3. Click on the CAD wireframe model that we created in *Creo Parametric*. If necessary, click on the arrow to set the BACS X direction toward the right, as shown.

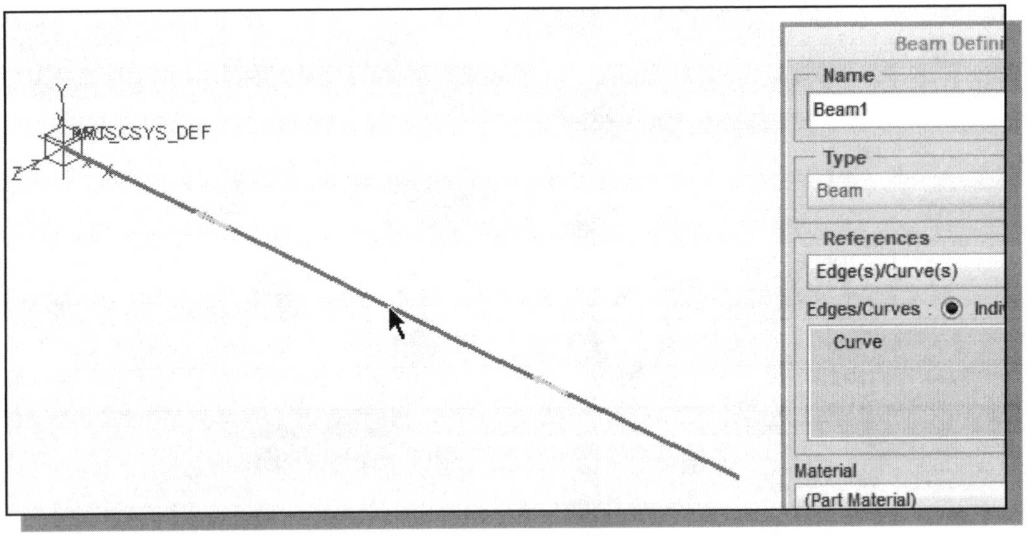

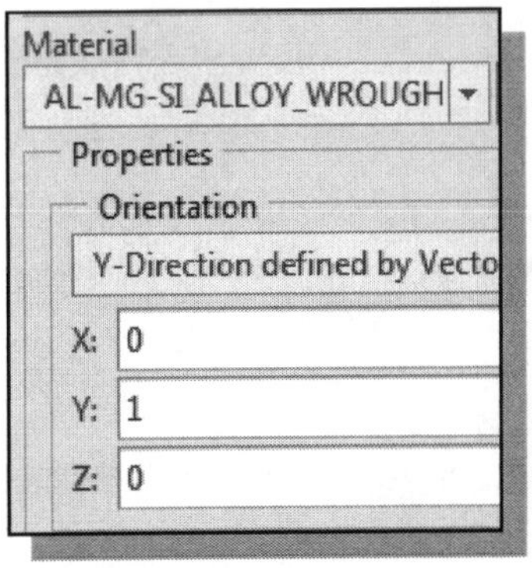

4. Choose **Al-Mg-Si_alloy_Wrought** as the Material of the element.

5. Confirm the *Y-Direction* is set to WCS with **0,1,0** as the X, Y and Z directions.

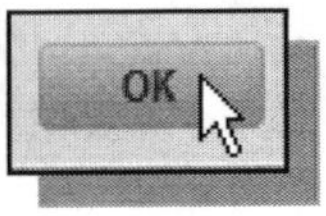

6. Click on the **OK** button to create the beam.

Apply the Displacement Constraints

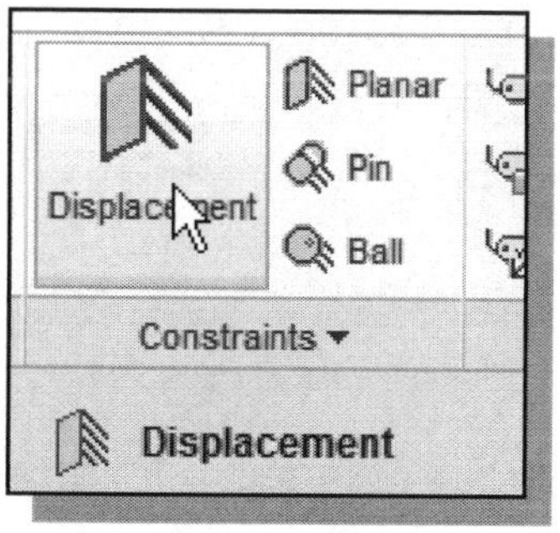

1. Choose **Displacement Constraint** by clicking the icon in the **Home** tab as shown.

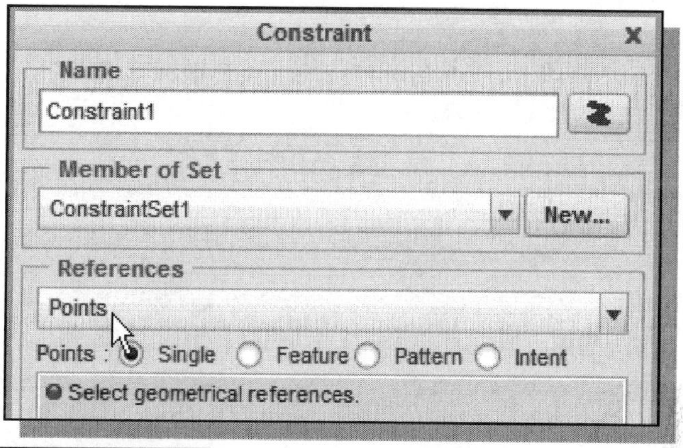

2. Set the References option to **Points**.

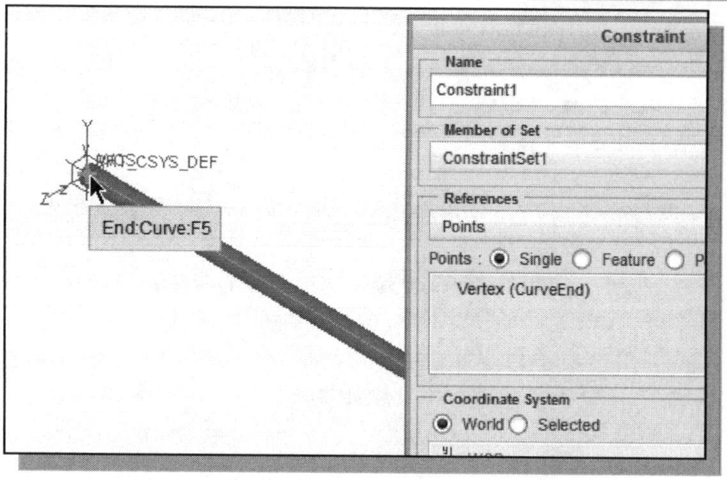

3. Select the first **endpoint** of the first line segment, which is located at the origin, as shown.

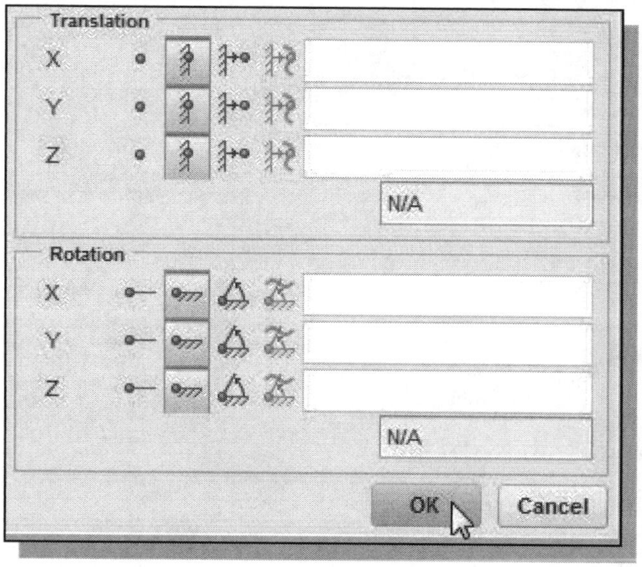

4. Set all Translations and Rotations to **Fixed**.

5. Click on the **OK** button to accept the first displacement constraint settings.

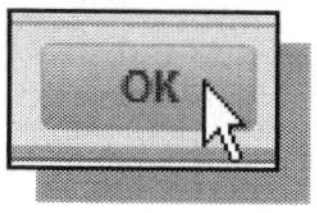

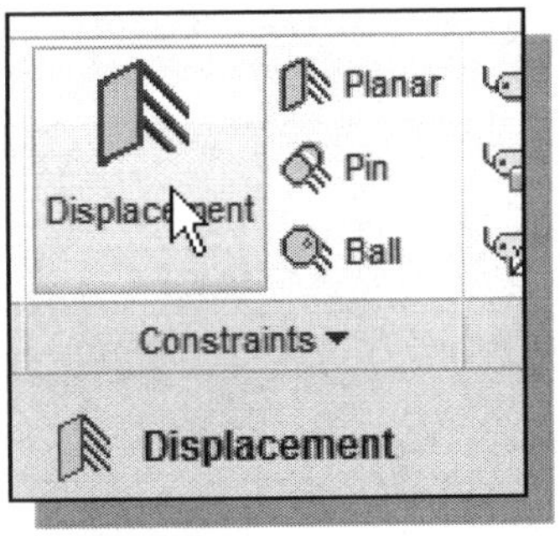

6. Choose **Displacement Constraint** by clicking the icon in the toolbar as shown.

7. Confirm the *Constraint* set is still set to **ConstraintSet1**.

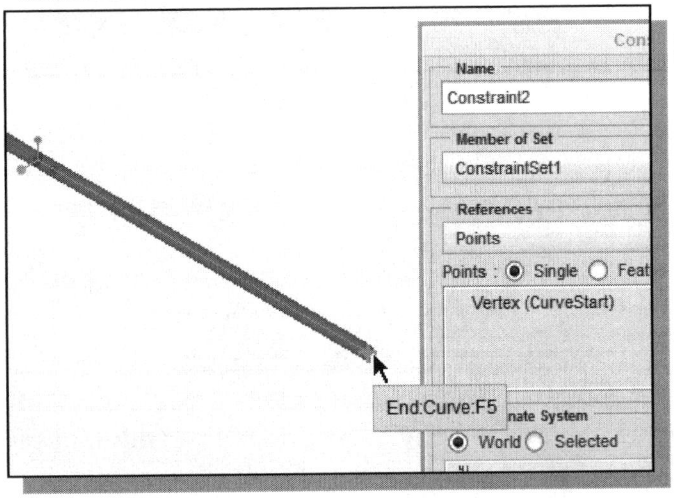

8. Set the References to **Points** and select the **last endpoint** of the second line segment as shown.

❖ The end of the beam structure has a roller support; it can move in the X direction, but not in the Y or Z directions. The *Z Rotational* movement should also be set to **Free**.

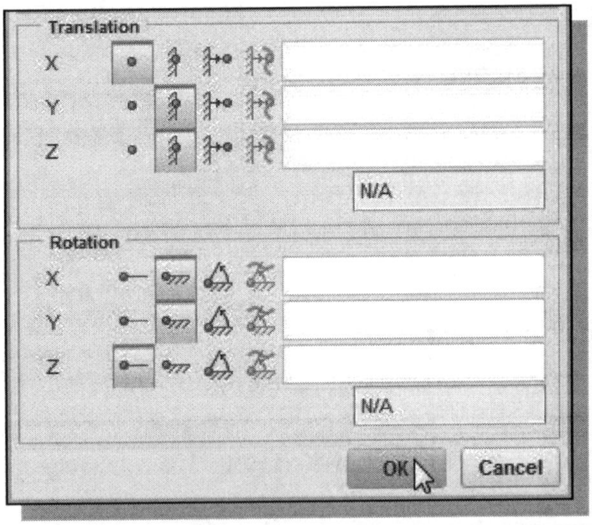

9. Set X Translation to **Free** as shown.

10. Set Z Rotation to **Free** as shown.

11. Click on the **OK** button to accept the second displacement constraint settings.

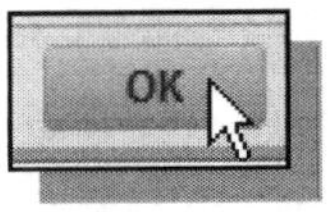

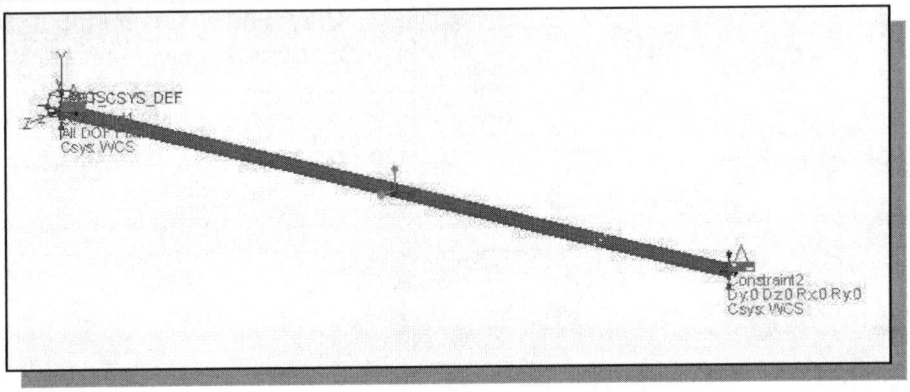

Apply the External Loads

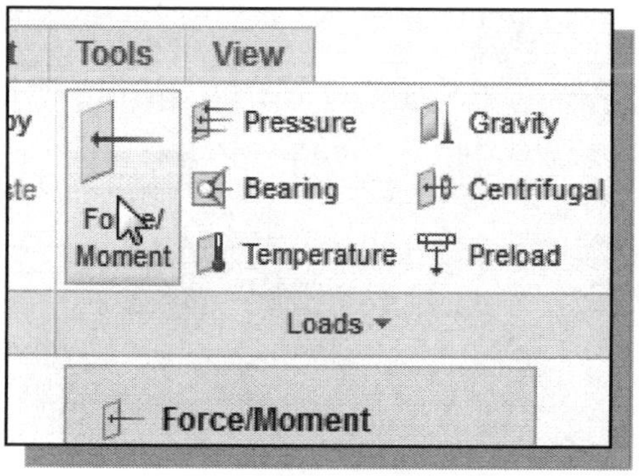

1. Choose **Force/Moment Load** by clicking the icon in the toolbar as shown.

2. Set the References option to **Points**.

3. Pick the **end point** of one of the curves, located at the center of the beam system, as shown.

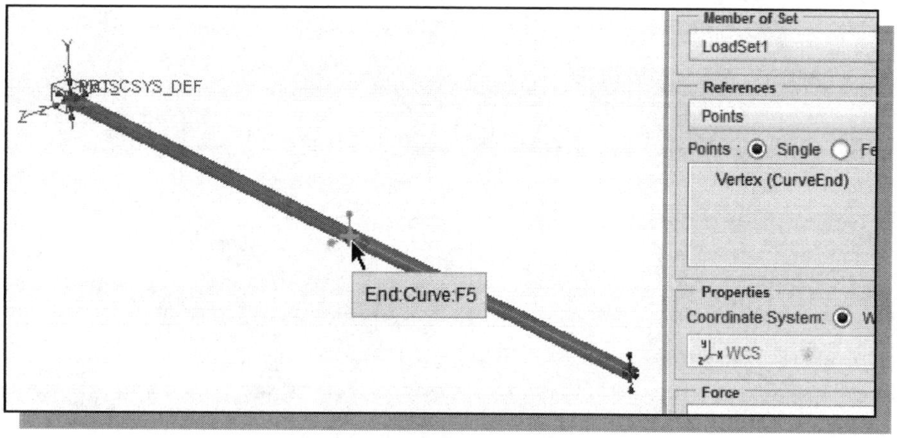

4. Enter **-1000** in the Y component Force value box as shown.

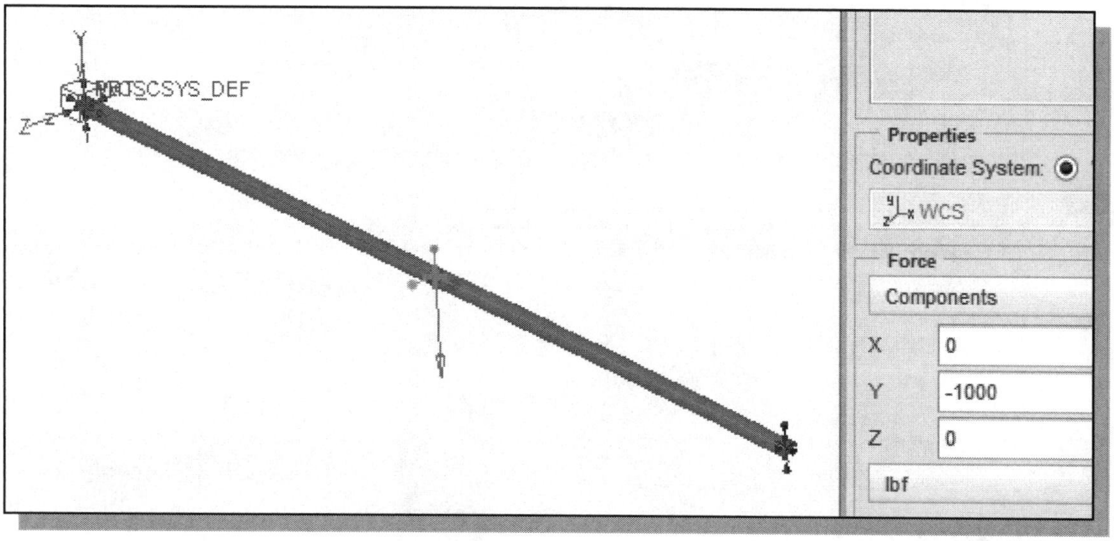

5. Click on the **OK** button to accept the first load settings.

Run the FEA Solver

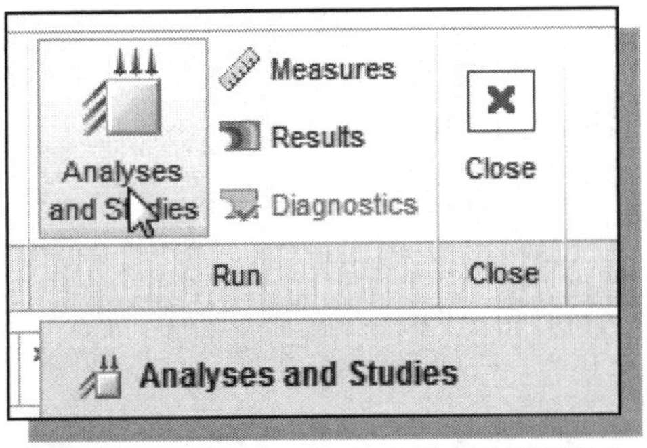

1. Activate the *Creo* **Analyses and Studies** command by selecting the icon in the toolbar area as shown.

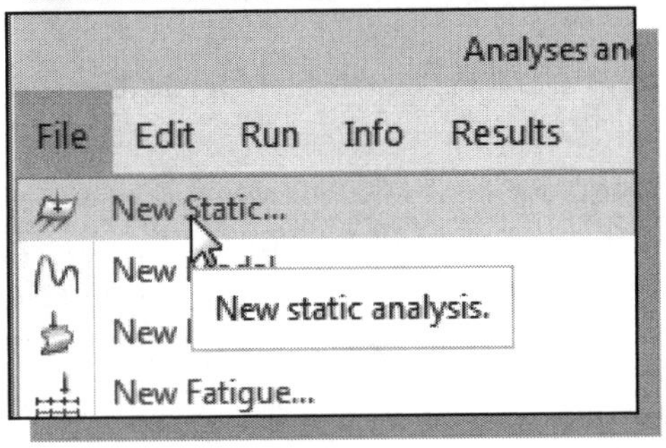

2. In the *Analyses and Design Studies* window, choose **File → New Static** in the *File pull-down menu*. This will start the setup of a basic linear static analysis.

3. Note that **ConstraintSet1** and **LoadSet1** are automatically accepted as part of the analysis parameters.

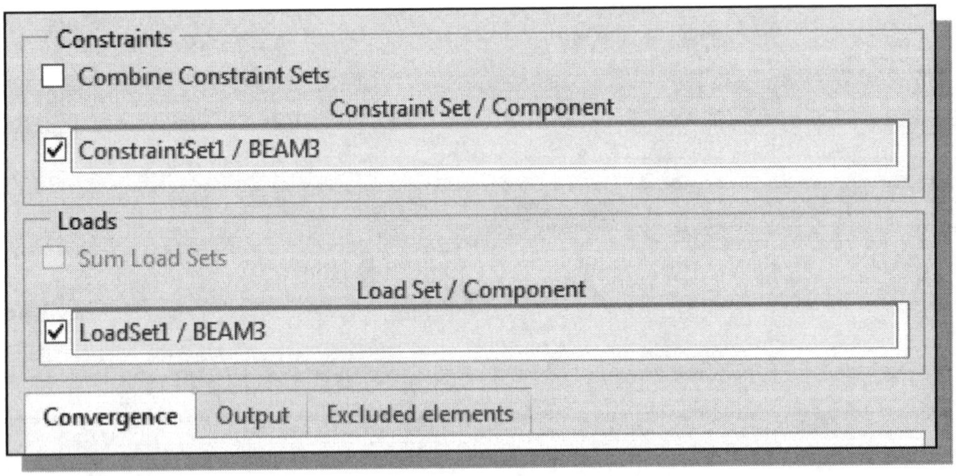

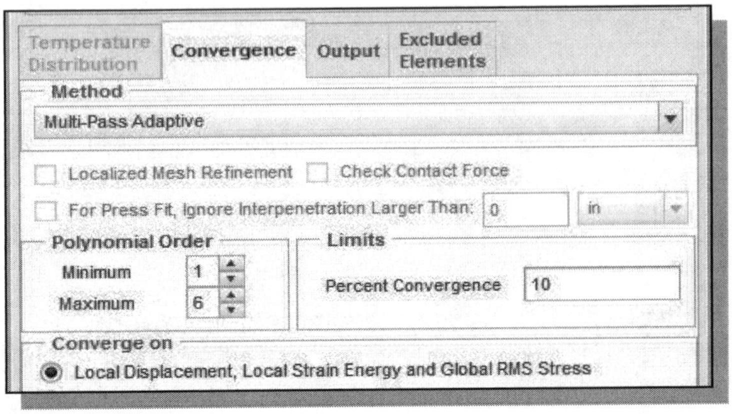

4. Set the Method option to **Multi-Pass Adaptive** and Limits to **10** Percent Convergence as shown.

5. Click **OK** to accept the settings.

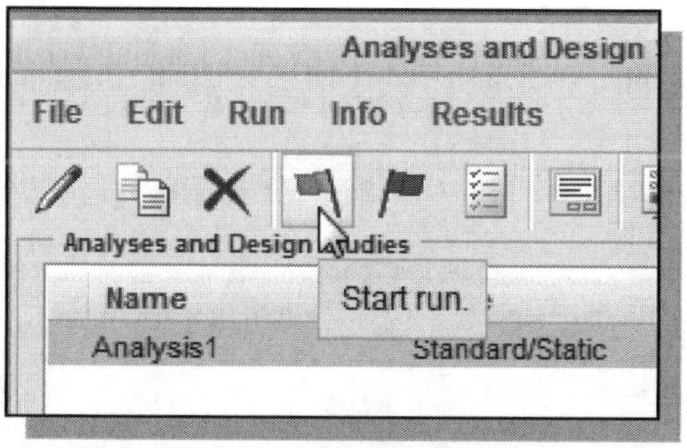

6. Click **Start run** to begin the *Solving* process.

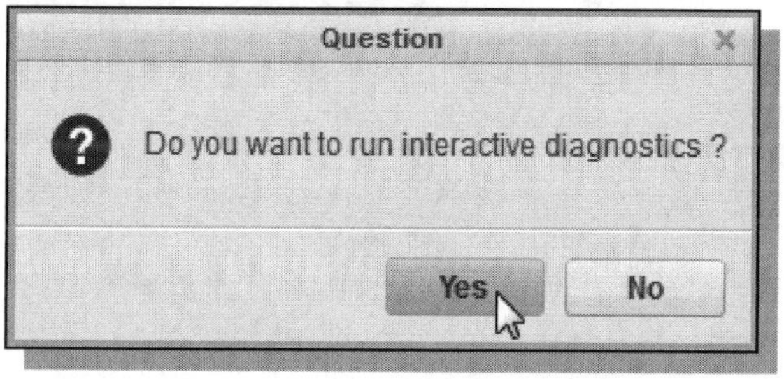

7. Click **Yes** to run interactive diagnostics; this will help us identify and correct problems associated with our FEA model.

8. In the *Diagnostics* window, *Creo Simulate* lists any problems or conflicts of our FEA model.

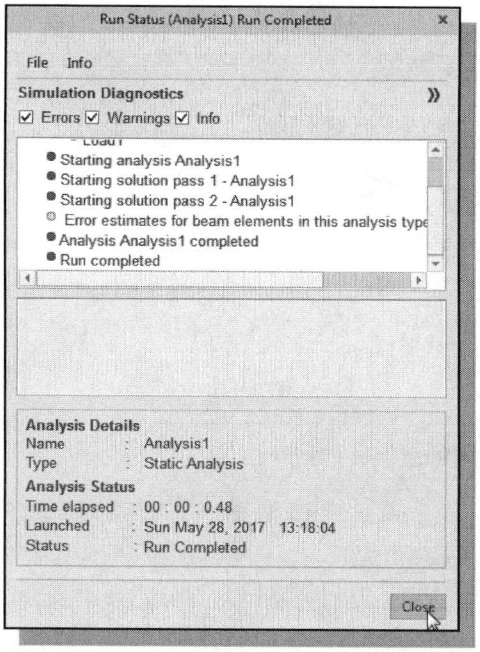

9. Click on the **Close** button to exit the *Diagnostics* window.

10. In the *Analyses and Design Studies* window, choose **Review results** as shown.

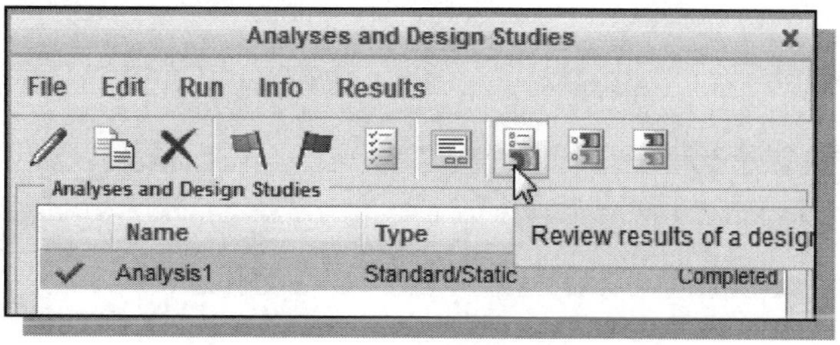

View the FEA Results

➢ **Reactions at supports**

1. Choose **Model** as the Display type as shown.

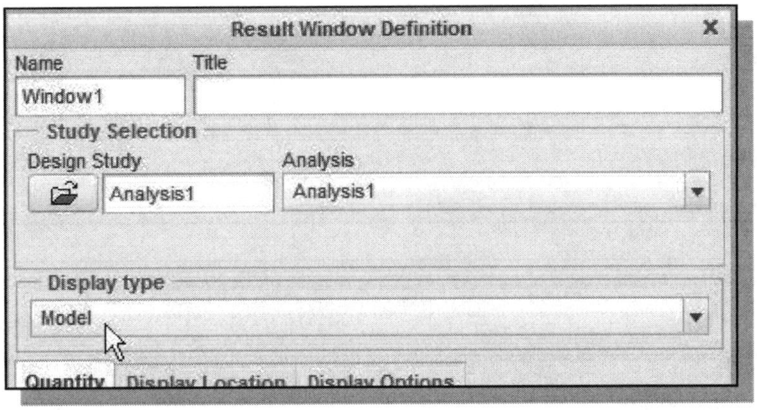

2. Choose **Reaction at Point Constraints** as the Quantity option.

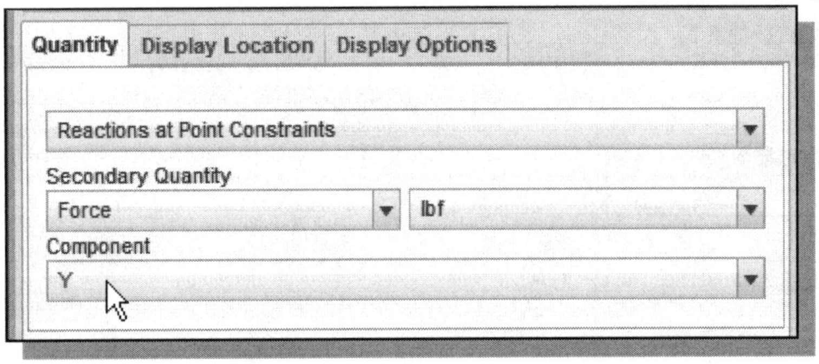

3. Choose **Force** and **lbf** as the Secondary Quantity options.

4. Choose **Y** as the Force Component to be displayed.

5. Click on the **Display Options** tab and switch ***on*** the **Deformed** option as shown.

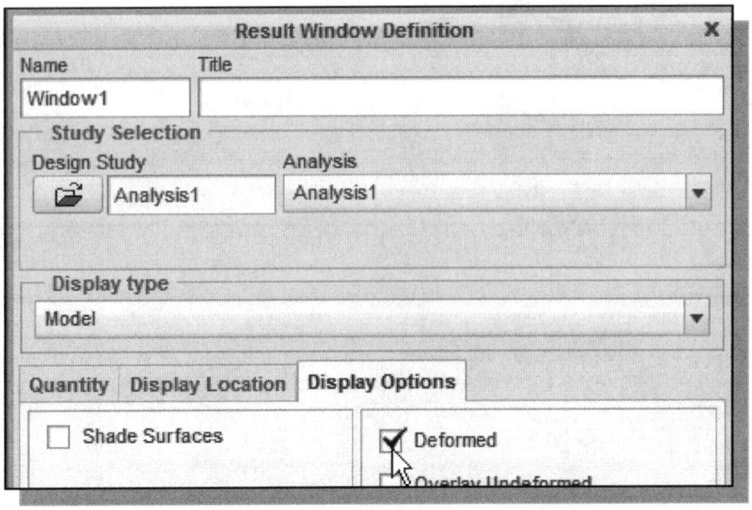

6. Click **OK and Show** to display the results.

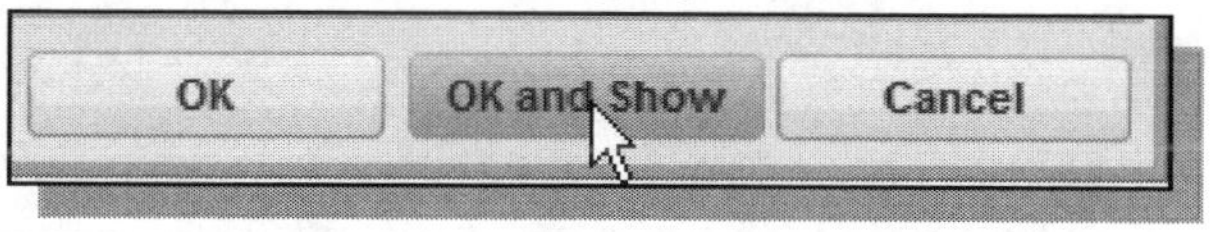

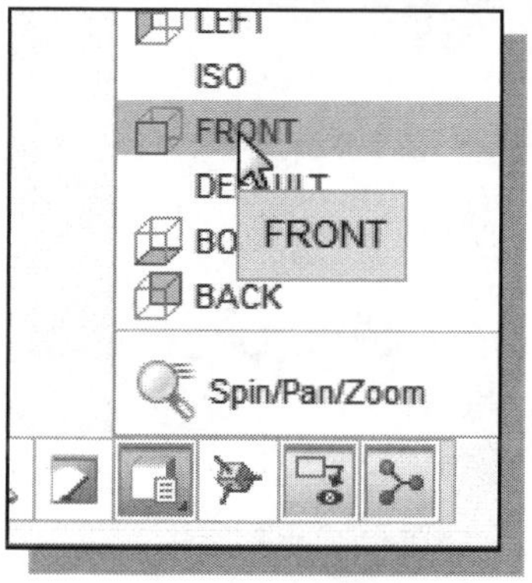

7. Select **FRONT** view in the saved name list to adjust the display to the Front view.

❖ The *Creo Simulate* **Reaction Force** option allows us to quickly view the *Reactions at Point Constraints* in beam elements.

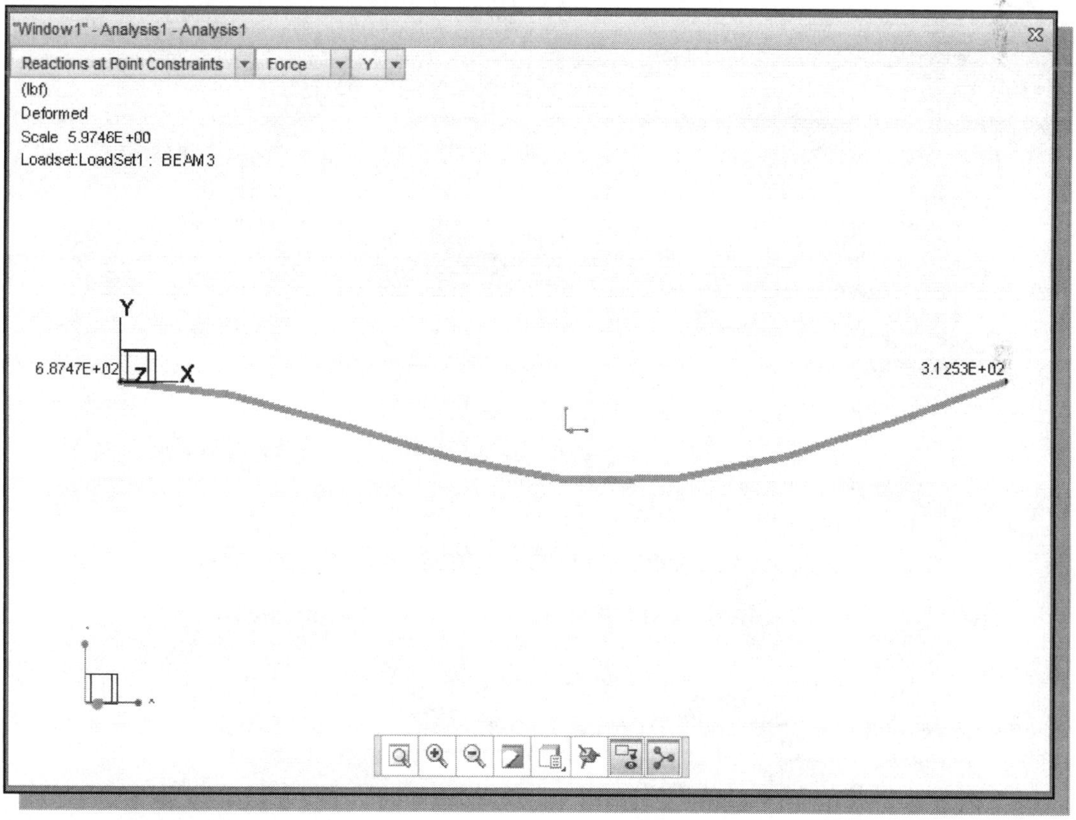

➢ Do values of the reactions at the supports match with the values obtained in our preliminary analysis?

❖ Note that the current version of *Creo Simulate* will not calculate reactions at points that have beam releases or points using local coordinate system.

➢ Shear Diagram

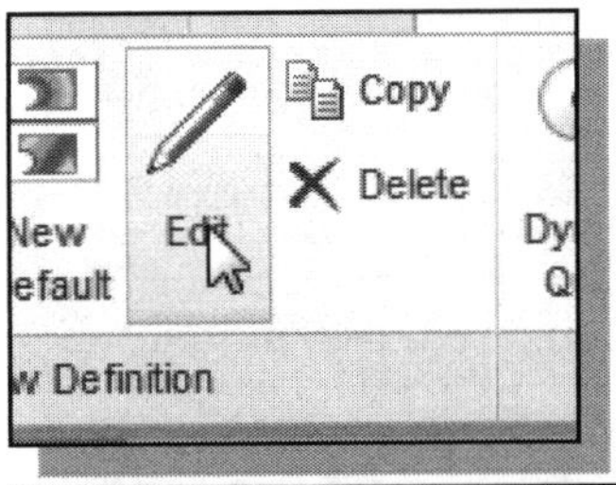

1. In the *Standard* toolbar area, click **Edit the current definition** command as shown.

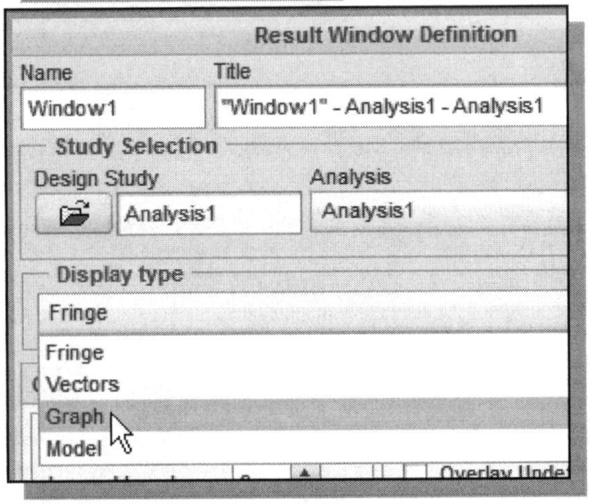

2. Choose **Graph** under the Display type option list as shown.

3. In the Graph Ordinate (Vertical) Axis option list, select **Shear & Moment** as shown.

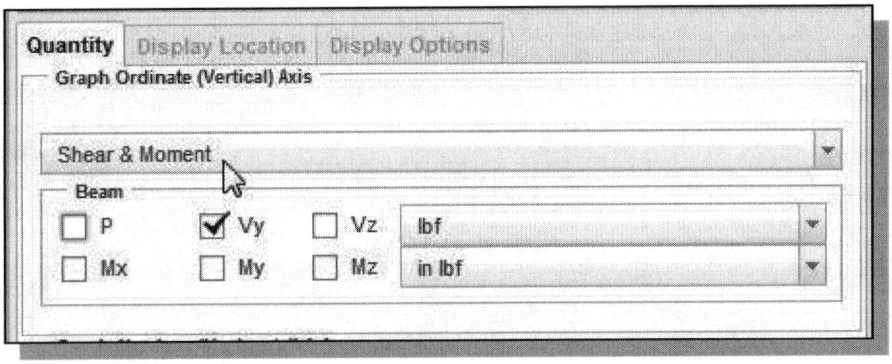

4. Switch ***on*** the **Y** (shear force) component as shown in the above figure.

5. In the Graph Abscissa (Horizontal) Axis section, confirm the options are set to **Curve Arc Length** and **Curve** as shown.

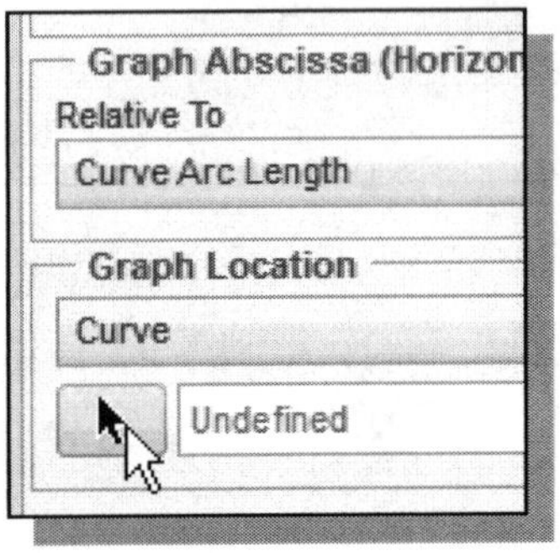

6. Click on the **arrow button** to select curves to display the shear diagram.

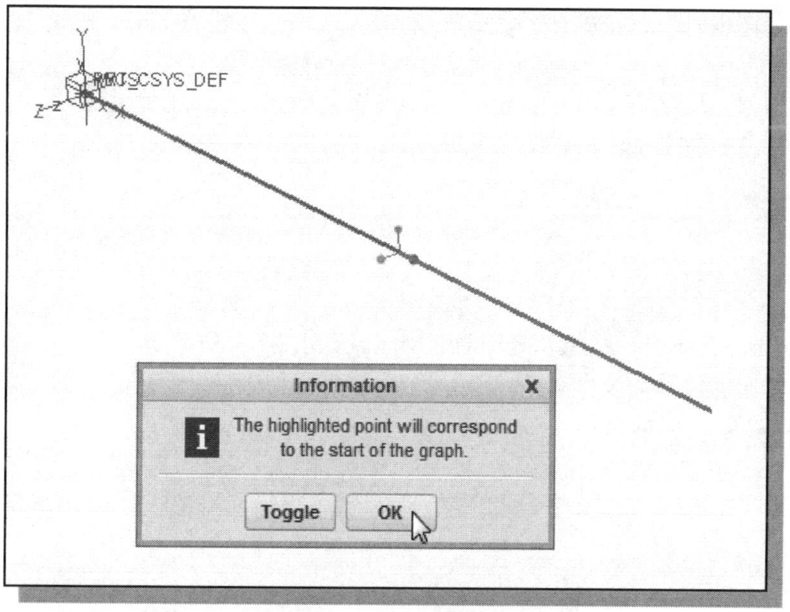

7. Select the two **Beam curves**; select the left curve first.

8. Click once with the **middle-mouse-button** to accept the selection.

9. Set the **left endpoint** as the start of the graph and click **OK** to continue.

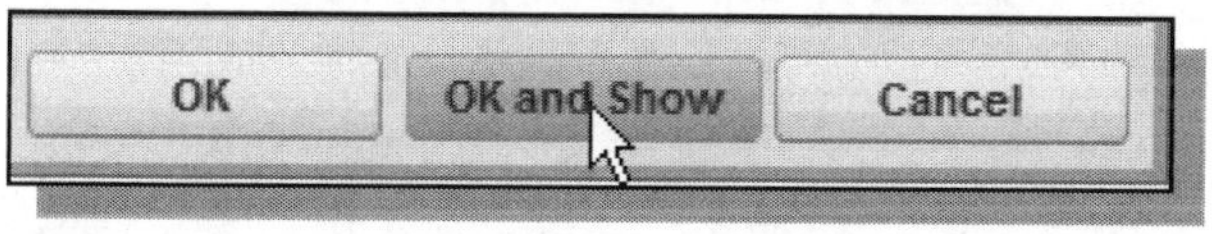

10. Click **OK and Show** to display the results.

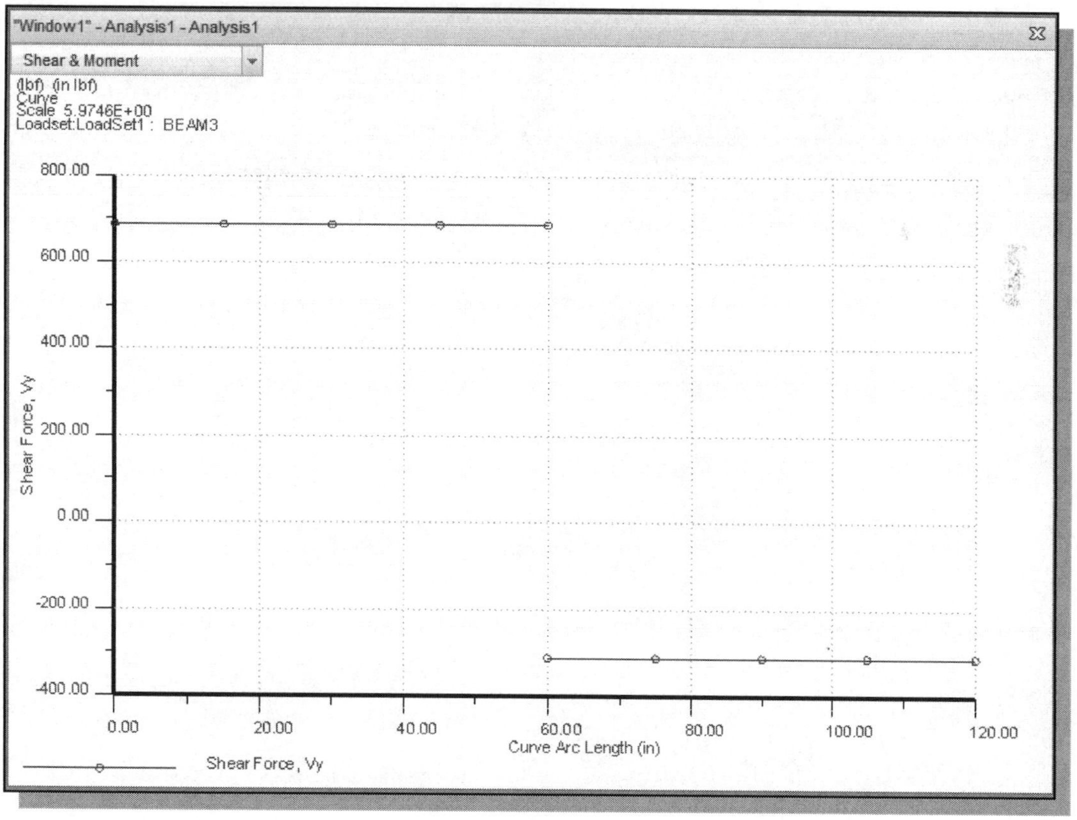

❖ Does the shear diagram match the calculations we performed during the preliminary analysis?

➢ Moment Diagram

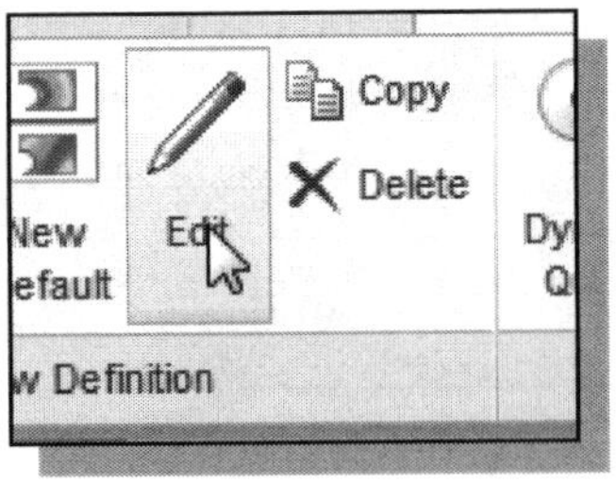

1. In the *Standard* toolbar area, click **Edit the current definition** command as shown.

2. Leaving only the **moment about Z** switched ***on***, turn ***off*** all of the other components as shown in the figure.

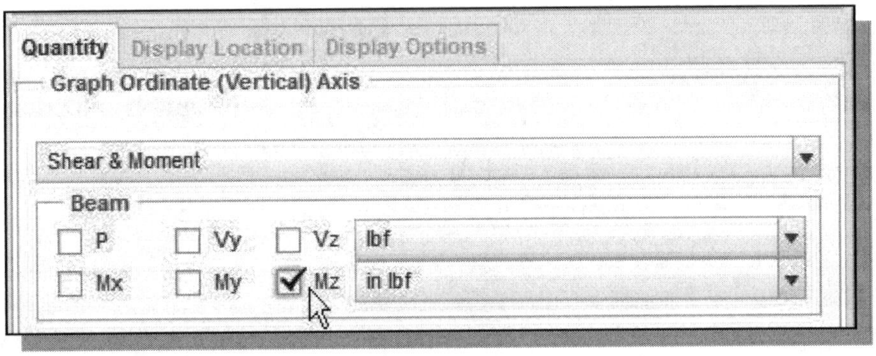

3. Click **OK and Show** to display the results.

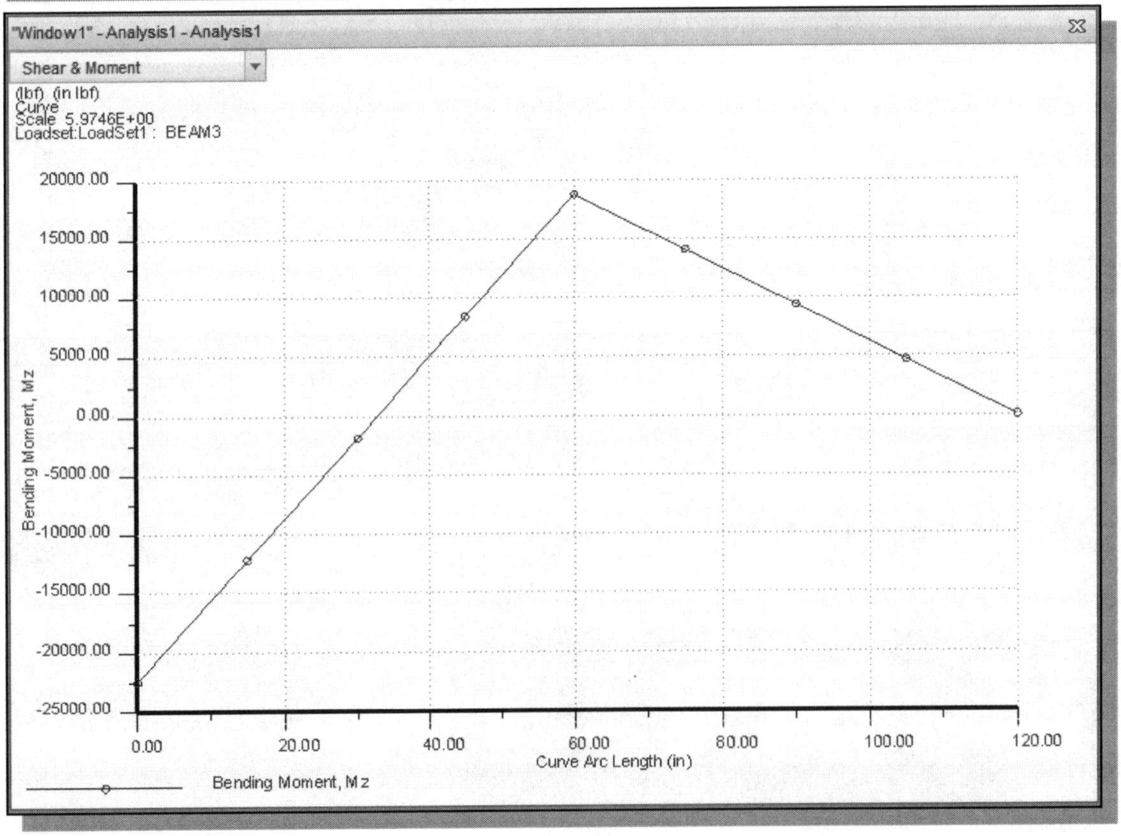

➢ Bending Stress

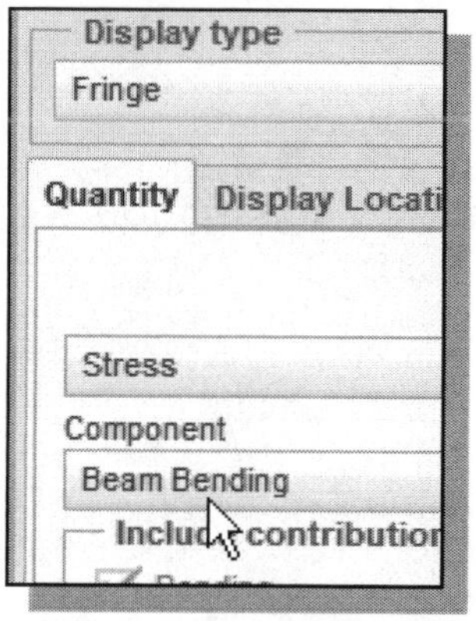

1. On your own, set the Display type to **Fringe**.

2. Choose **Beam Bending** as the Stress Component to be displayed.

3. Also set the *Stress Units* to **psi**.

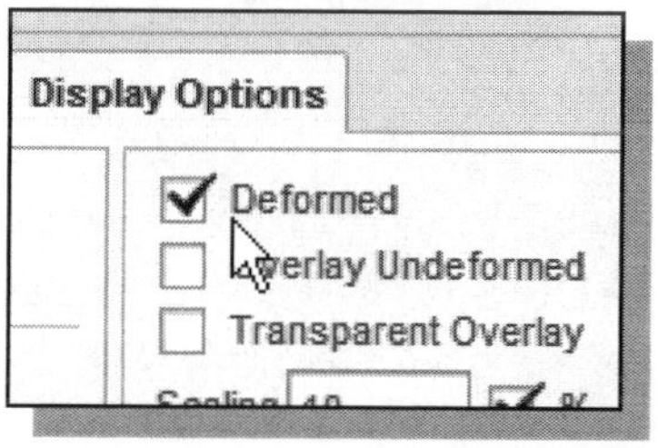

4. Click on the **Display Options** tab and switch ***on*** the **Deformed** option as shown.

5. Click **OK and Show** to display the results.

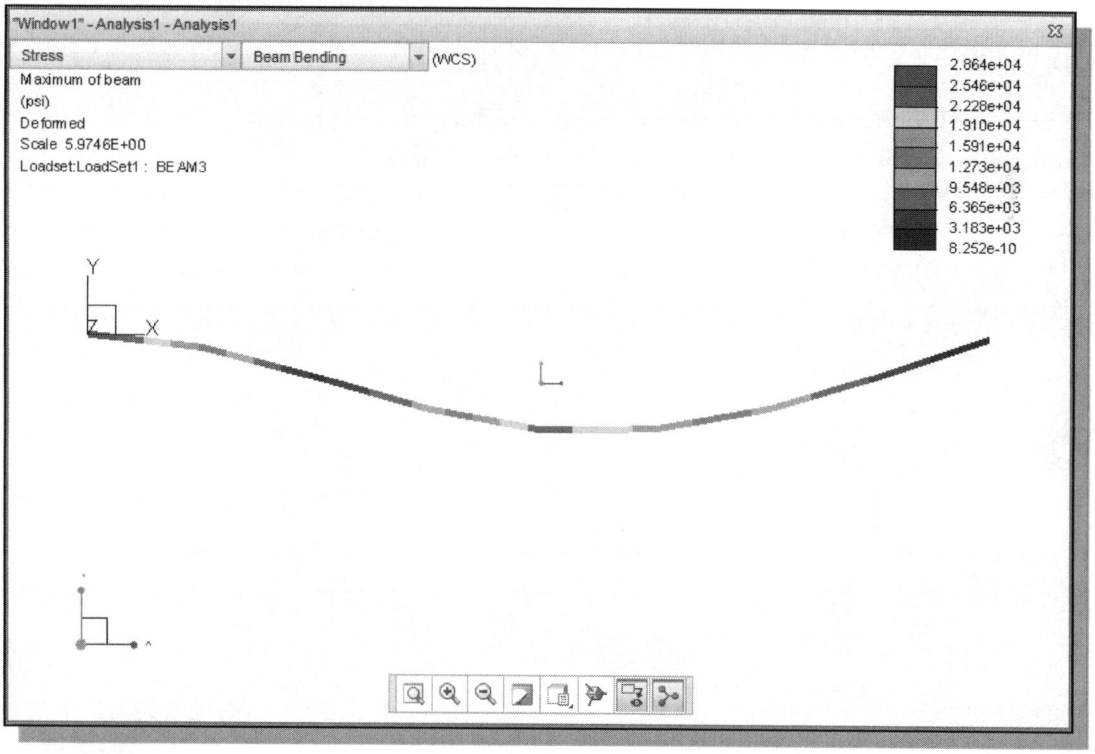

➢ The computer solution for the maximum stress is **2.864E+4 lb/in²**, which occurs at the fixed-end of the member. Our preliminary calculation at the fixed end gives the value of **4.125E+6 lb/ft²**.

Review Questions

1. What are the equations of *Statics* (Equations of Equilibrium)?

2. What is the type of structure where there are more supports than are necessary to maintain equilibrium?

3. In the example problem, how did we solve the *statically indeterminate* structure?

4. What is the quick-key combination to dynamically Pan the 3D model in *Creo Simulate*?

5. Is there any special procedure in setting up statically indeterminate structures in *Creo Simulate*?

6. List and describe the differences between a CAD model and an FEA model.

7. In *Creo Simulate*, what does ***Idealization*** mean?

8. What are the differences in the way the FEA model was created in the tutorial vs. the approach done in the last chapter?

9. Can the beam cross sections be changed after beam elements have been created?

10. What is the percent error of the maximum bending stress obtained by the FEA analysis and the preliminary calculation in the tutorial?

Exercises

Determine the maximum stress produced by the loads and create the shear and moment diagrams.

1. A cantilever beam with a roller support.
   Material: Steel

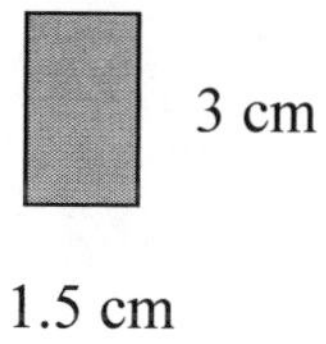

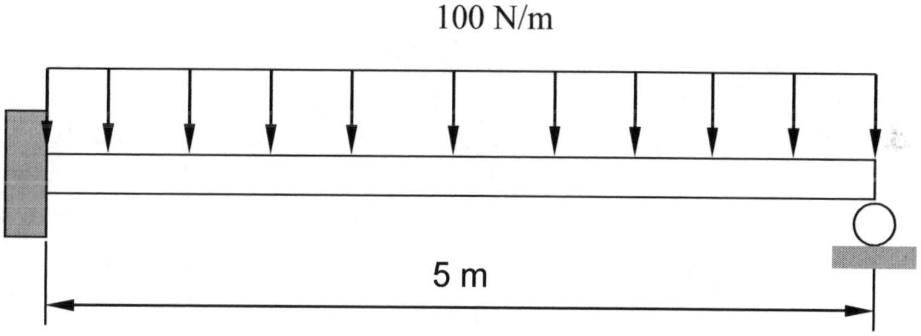

2. A cantilever beam with a roller support.
   Material: Steel
   Diameter 2.0 in.

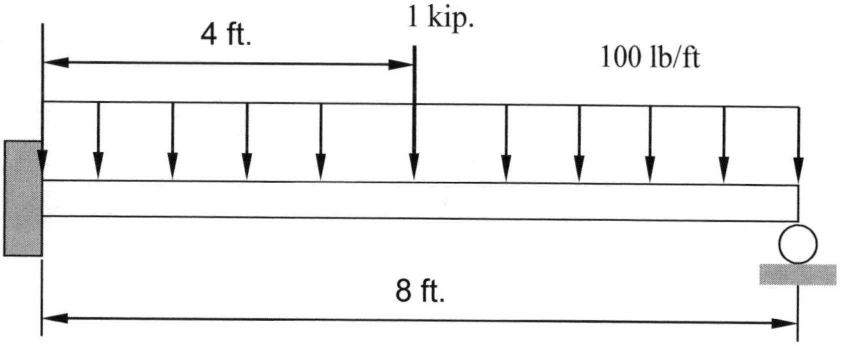

3. Simply Supported beam with an extra roller support.
   Material: Aluminum Alloy 6061 T6

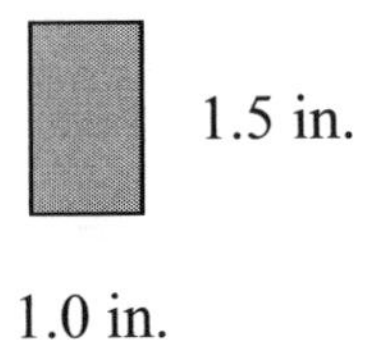

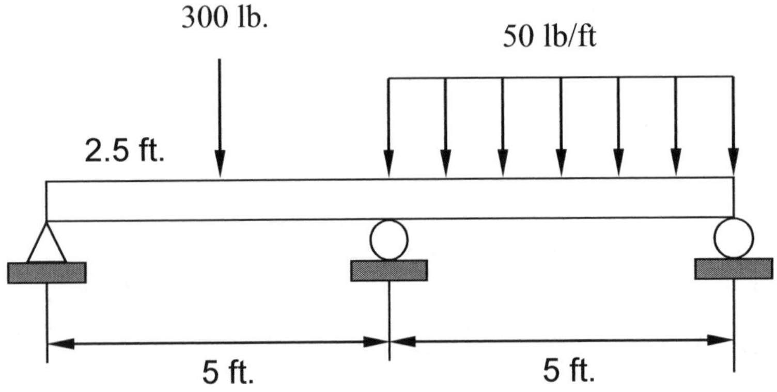

Chapter 9
Two Dimensional Solid Elements

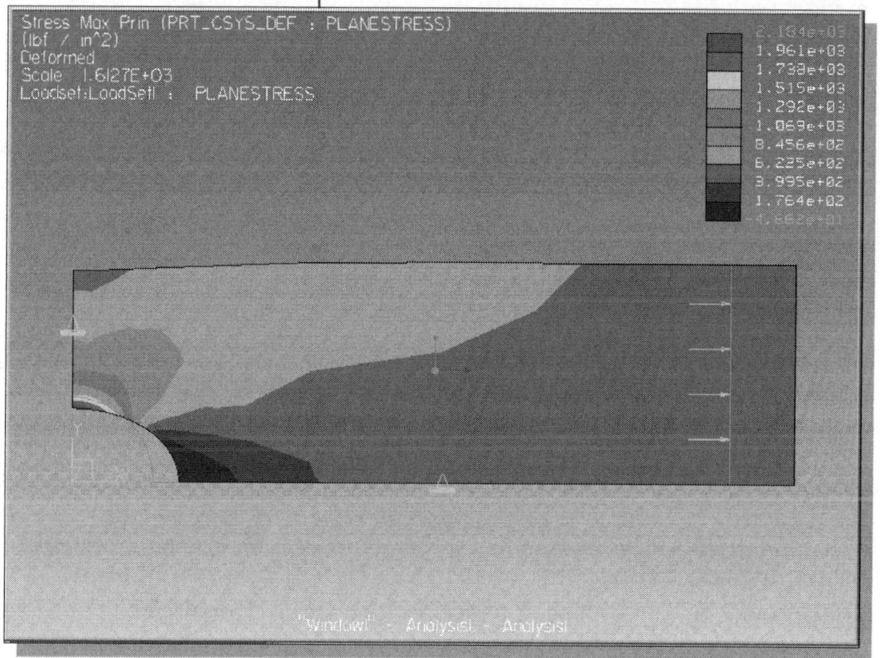

Learning Objectives

- ♦ Understand the basic assumptions for 2D elements.
- ♦ Understand the basic differences between H-Element and P-Element.
- ♦ Create and Refine 2D solid elements using the AutoGEM.
- ♦ Understand the Stress concentration effects on Shapes.
- ♦ Perform Basic Plane Stress Analysis using Creo Simulate.

Introduction

In this chapter, we will examine the use of two-dimensional solid finite elements. Three-dimensional problems can be reduced to two dimensions if they satisfy the assumptions associated with two-dimensional solid elements. There are four basic types of two-dimensional solid elements available in *Creo Simulate*:

1. **Plane Stress Elements**: Plane stress is defined to be a state of stress that the *normal stress* and the *shear stresses* directed perpendicular to the plane are assumed to be zero.

2. **Plane Strain Elements**: Plane strain is defined to be a state of strain that the *normal strain* and the *shear strains* directed perpendicular to the plane are assumed to be zero.

3. **Axisymmetric Elements**: Axisymmetric structures, such as storage tanks, nozzles and nuclear containment vessels, subjected to uniform internal pressures, can be analyzed as two-dimensional systems using *Axisymmetric Elements*.

4. **Thin-Shell Elements**: Thin-shell elements are flat planar surface elements with a thickness that is small compared to their other dimensions. Unlike the plane stress elements, thin-shell elements can have loads that are not parallel to the object surfaces.

The recent developments in computer technology have triggered tremendous advancements in the development and use of 2D and 3D solid elements in FEA. Many problems that once required sophisticated analytical procedures and the use of empirical equations can now be analyzed through the use of FEA.

At the same time, users of FEA software must be cautioned that it is very easy to fall into the trap of blind acceptance of the answers produced by the FEA software. Unlike the line elements (*Truss* and *Beam* elements), where the analytical solutions and the FEA solutions usually matched perfectly, more care must be taken with 2D and 3D solid elements since only very few analytical solutions can be easily obtained. On the other hand, the steps required to perform finite element analysis using 2D and 3D solid elements are in general less complicated than that of line elements.

This chapter demonstrates the use of 2D solid elements to solve the classical stress concentration problem – a plate with a hole in it. This project is simple enough to demonstrate the necessary steps for the finite element process, and the FEA results can be compared to the analytical solutions to assure the accuracy of the FEA procedure. We will demonstrate the use of *plane stress elements* for this illustration. Generally, members that are thin and whose loads act only in the plane can be considered to be under plane stress. For *plane stress/strain* analysis, *Creo Simulate* requires the surface to be aligned with the **XY plane** of either the WCS or any coordinate system.

Problem Statement

Determine the maximum normal stress that loading produces in the **AL6061** plate.

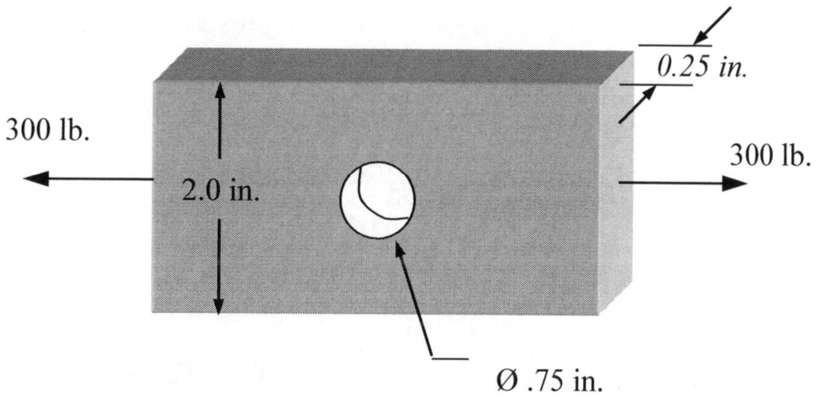

Preliminary Analysis

- **Maximum Normal Stress**
  The nominal normal stress developed at the smallest cross section (through the center of the hole) in the plate is

$$\sigma_{nominal} = \frac{P}{A} = \frac{300}{(2 - 0.75) \times .25} = 960 \text{ psi.}$$

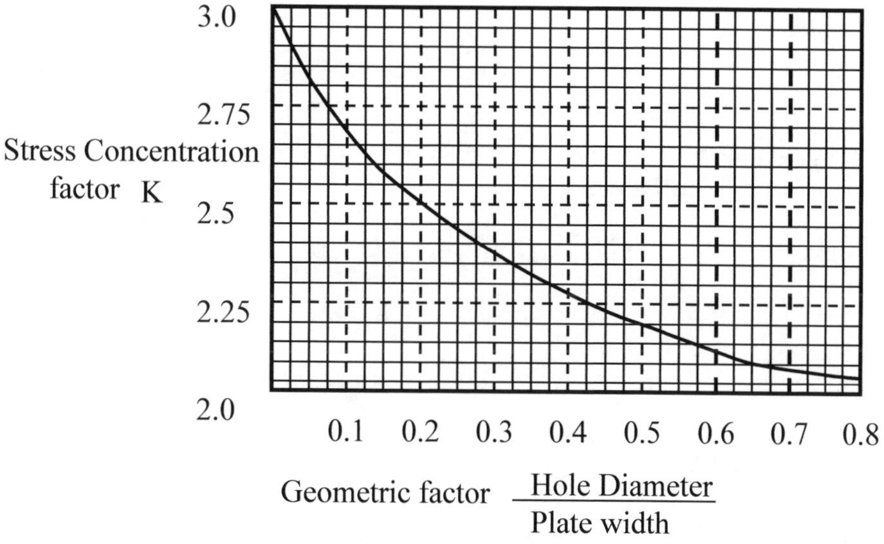

Geometric factor = .75/2 = 0.375
Stress concentration factor K is obtained from the graph, **K = 2.27**

$$\sigma_{MAX} = K\,\sigma_{nominal} = 2.27 \times 960 = 2180 \text{ psi.}$$

- **Maximum Displacement**

We will also estimate the displacement under the loading condition. For a statically determinant system the stress results depend mainly on the geometry. The material properties can be in error and still the FEA analysis comes up with the same stresses. However, the displacements always depend on the material properties. Thus, it is necessary to always estimate both the stress and displacement prior to a computer FEA analysis.

The classic one-dimensional displacement can be used to estimate the displacement of the problem:

$$\delta = \frac{PL}{EA}$$

Where P=force, L=length, A=area, E=elastic modulus, and δ=deflection.

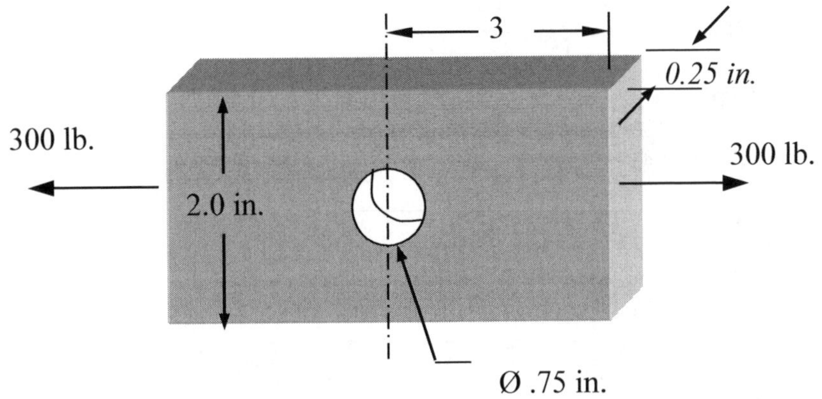

A lower bound of the displacement of the right-edge, measured from the center of the plate, is obtained by using the full area:

$$\delta_{lower} = \frac{PL}{EA} = \frac{300 \times 3}{10E6 \times (2 \times 0.25)} = 1.8E\text{-}4 \text{ in.}$$

and an upper bound of the displacement would come from the reduced section:

$$\delta_{upper} = \frac{PL}{EA} = \frac{300 \times 3}{10E6 \times (1.25 \times 0.25)} = 2.88E\text{-}4 \text{ in.}$$

but the best estimate is a sum from the two regions:

$$\delta_{average} = \frac{PL}{EA} = \frac{300 \times 0.375}{10E6 \times (1.25 \times 0.25)} + \frac{300 \times 2.625}{10E6 \times (2.0 \times 0.25)}$$

$$= 3.6E\text{-}5 + 1.58E\text{-}4 = 1.94E\text{-}4 \text{ in.}$$

Geometric Considerations of Finite Elements

For *linear statics analysis*, designs with symmetrical features can often be reduced to expedite the analysis.

For our plate problem, there are two planes of symmetry. Thus, we only need to create an FE model that is one-fourth of the actual system. By taking advantage of symmetry, we can use a finer subdivision of elements that can provide more accurate and faster results.

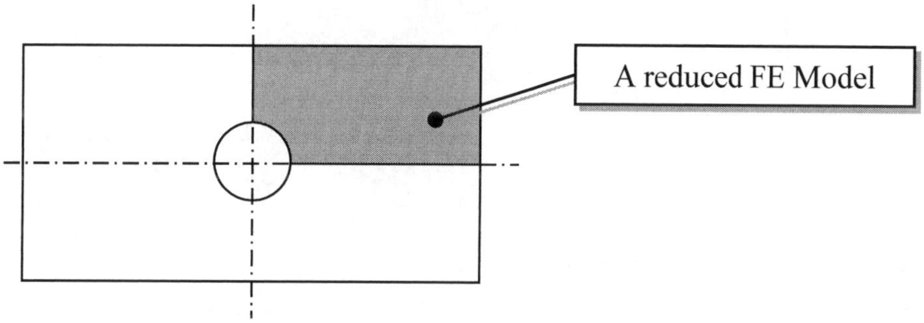

In plane stress analysis, only the x and y displacements are active. The constraints in all other directions can be set to *free* as a reminder of which ones are the true unknowns. For our model, deformations will occur along the axes of symmetry; we will therefore place roller constraints along the two center lines as shown in the figure below.

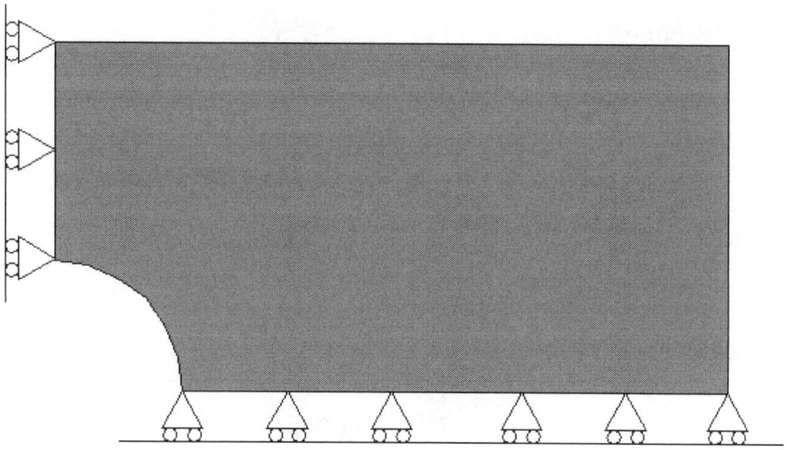

❖ One should also be cautious of using symmetrical characteristics in FEA. The symmetry characteristics of boundary conditions and loads should be considered. Also note the symmetry characteristic that is used in the *linear statics analysis* does not imply similar symmetrical results in vibration or buckling modes.

Starting Creo Parametric

1. Select the **Creo Parametric** option on the *Start* menu or select the **Creo Parametric** icon on the desktop to start *Creo Parametric*. The *Creo Parametric* main window will appear on the screen.

2. Click on the **New** icon, located in the *Ribbon toolbar* as shown.

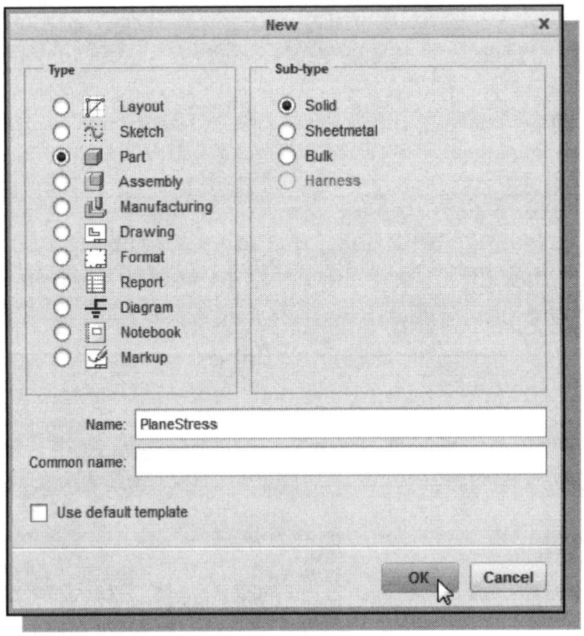

3. In the *New* dialog box, confirm the model's Type is set to **Part** (**Solid** Sub-type).

4. Enter **PlaneStress** as the part Name as shown in the figure.

5. Turn ***off*** the **Use default template** option.

6. Click on the **OK** button to accept the settings.

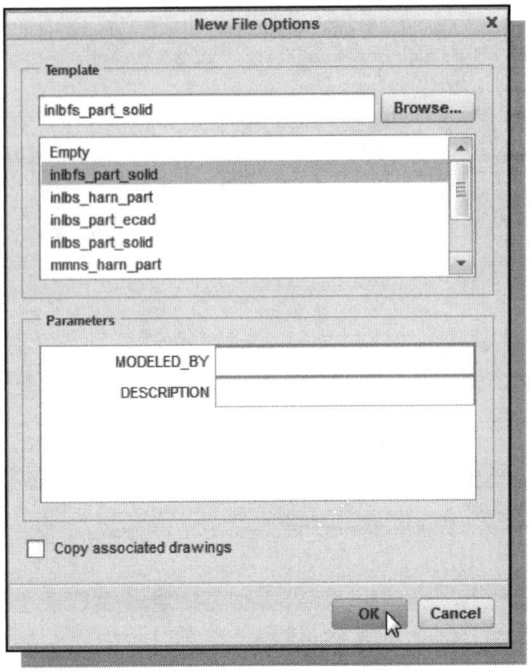

7. In the *New File Options* dialog box, select **inlbfs_part_solid** in the option list as shown.

8. Click on the **OK** button to accept the settings and enter the *Creo Parametric Part Modeling* mode.

Create a CAD Model in Creo Parametric

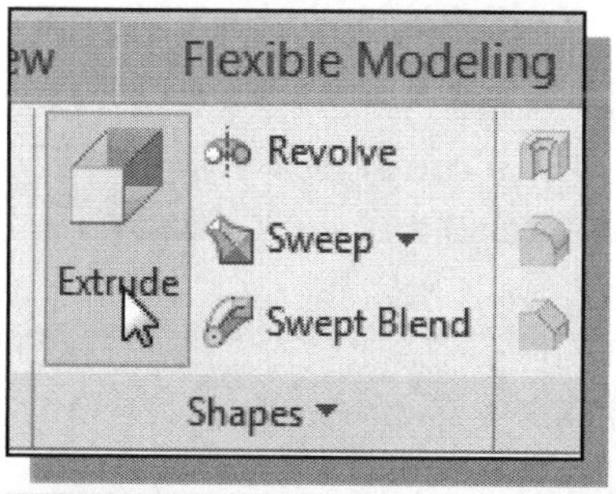

1. In the *Shapes* toolbar, select the **Extrude** tool option as shown.

- The *Feature Option Dashboard*, which contains applicable construction options, is displayed in the *Creo Parametric Ribbon* toolbar area.

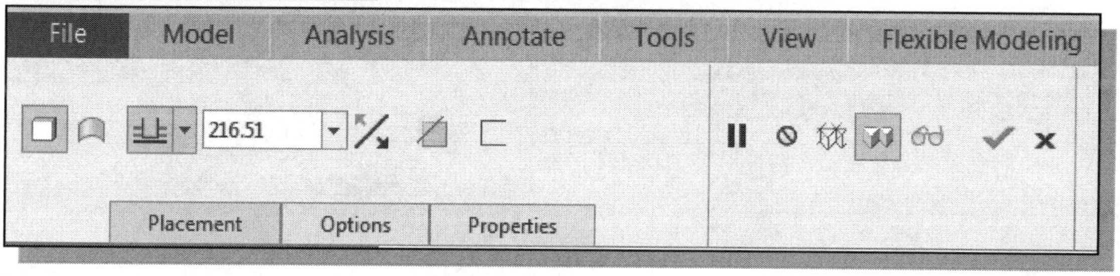

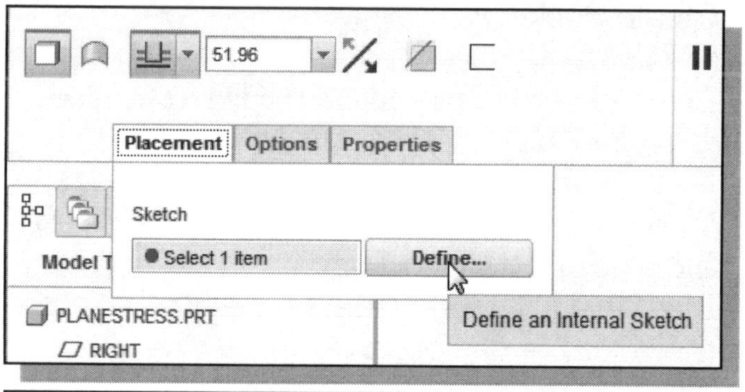

2. Click the **Placement** option and choose **Define** to begin creating a new *internal sketch.*

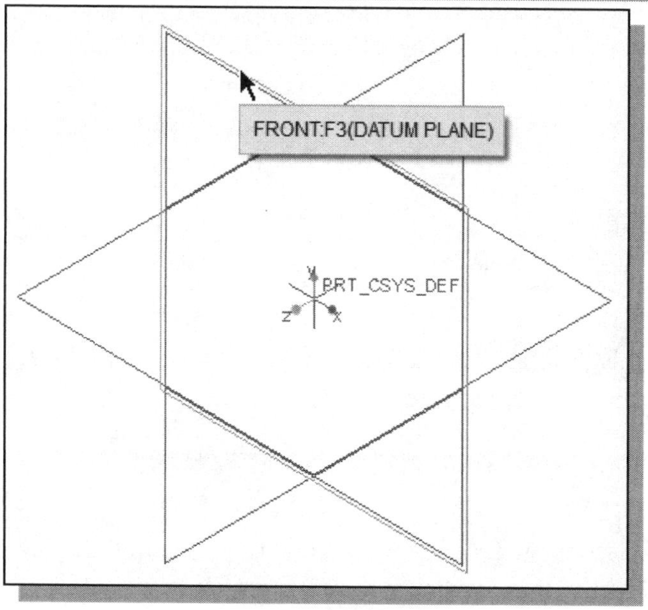

3. In the graphics area, select **FRONT** by clicking on the text FRONT as shown.

❖ Notice an arrow appears on the edge of FRONT. The arrow direction indicates the viewing direction of the sketch plane.

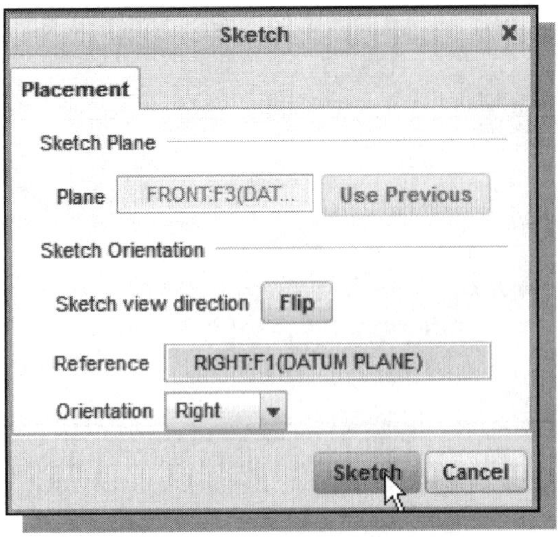

4. In the graphics area, select **RIGHT** by clicking on the text.

5. Pick **Sketch** to exit the *Sketch* window and proceed to enter the *Creo Parametric Sketcher* mode.

6. Click **Sketch View** to orient the sketching plane parallel to the screen.

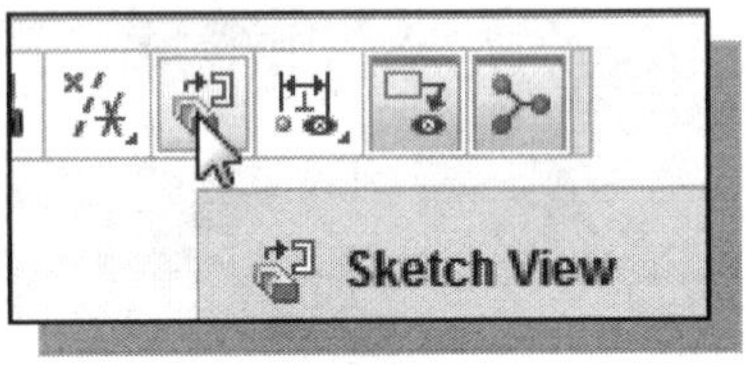

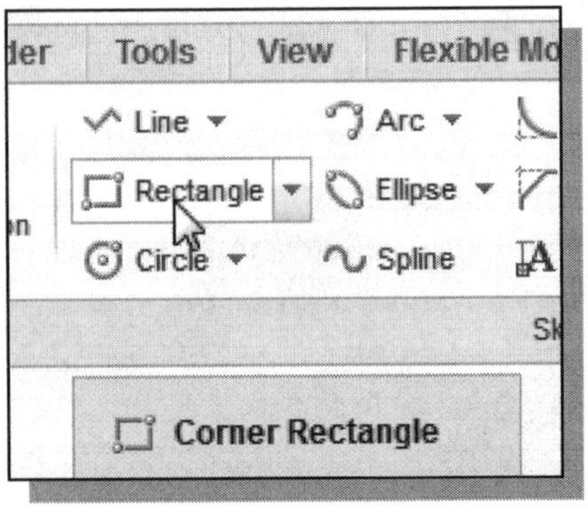

7. Click on the **Rectangle** button to activate the Create Rectangle command.

8. Click near the **intersection of the two references** as the first corner of the rectangle.

9. Move the cursor toward the **right side** of the screen and create a rectangle as shown.

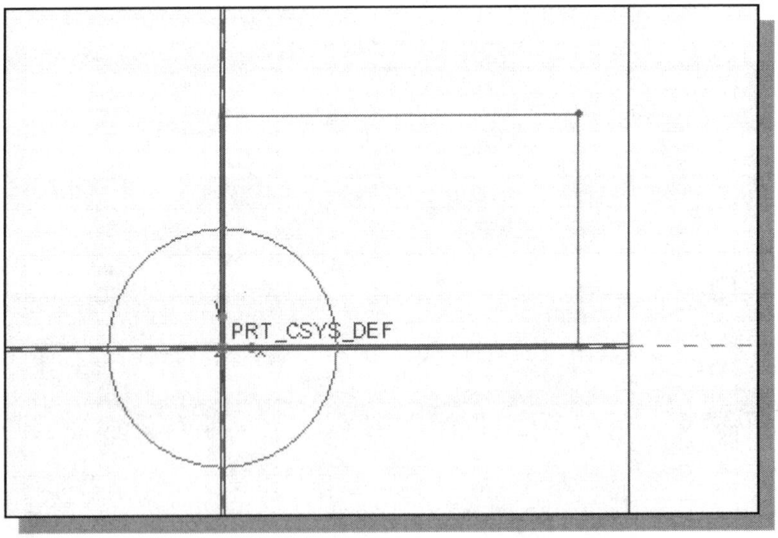

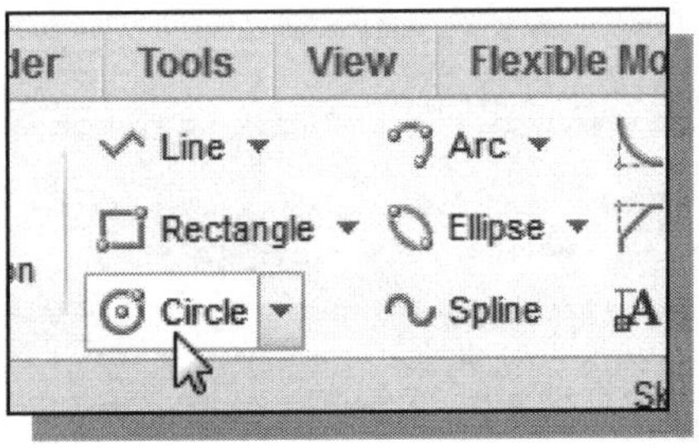

10. On your own, create a **circle** with the center aligned to the origin as shown in the figure above.

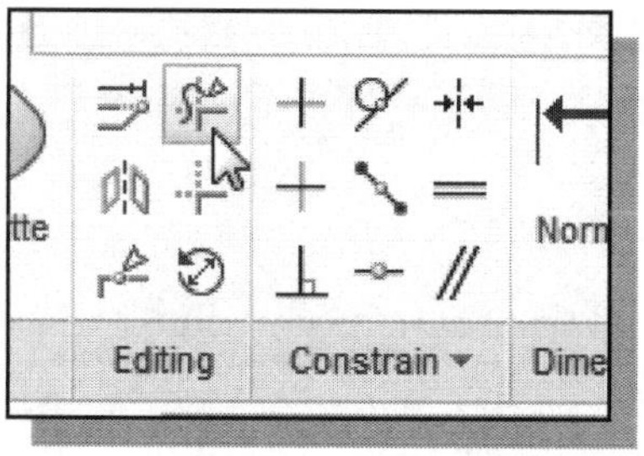

11. Click on the **Delete Segment** button to activate the Quick Trim command.

12. On your own, adjust the sketch so that it appears as shown in the figure below.

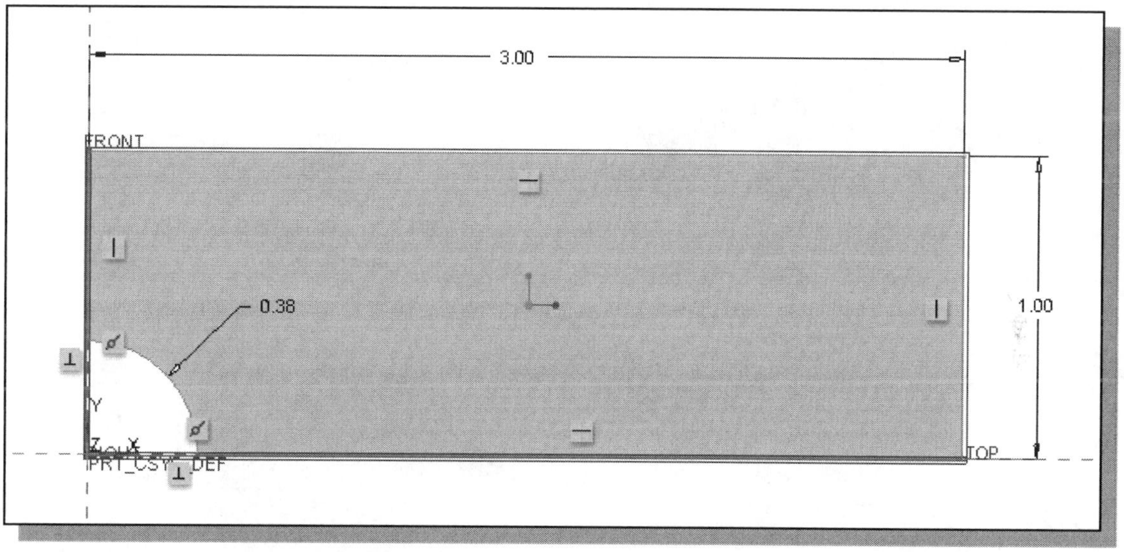

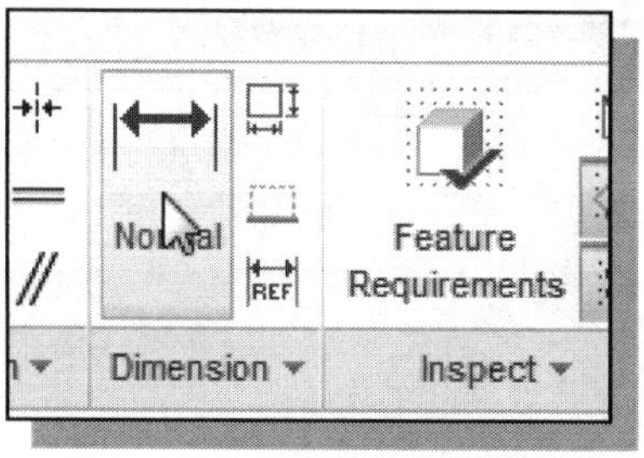

13. Use the **Dimension** command to create the *three dimensions* of the sketch as shown in the figure above.

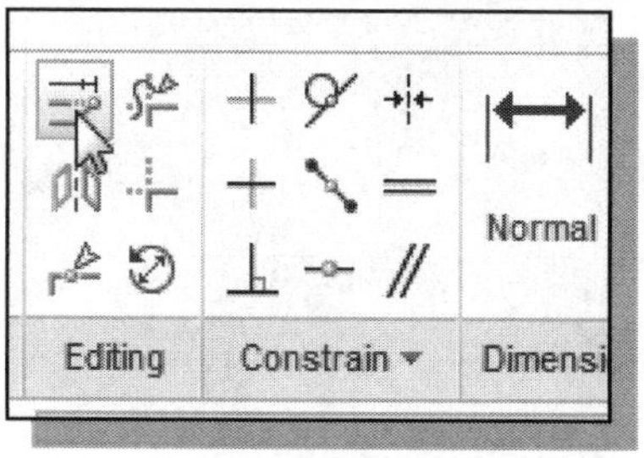

14. On your own, use the **Modify** command to edit the dimensions as shown in the figure above.

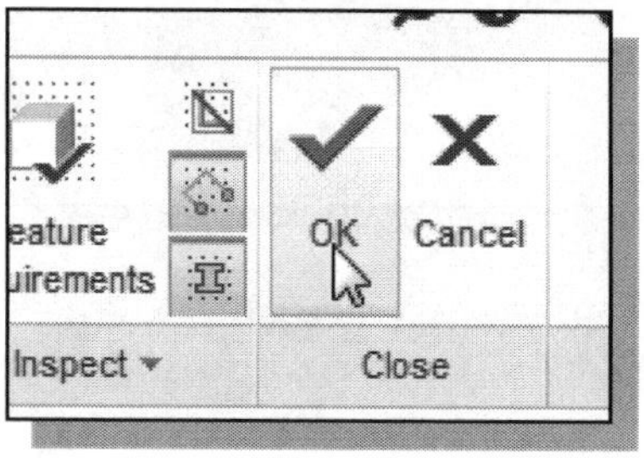

15. Click on the **OK** button to accept the creation of the sketch.

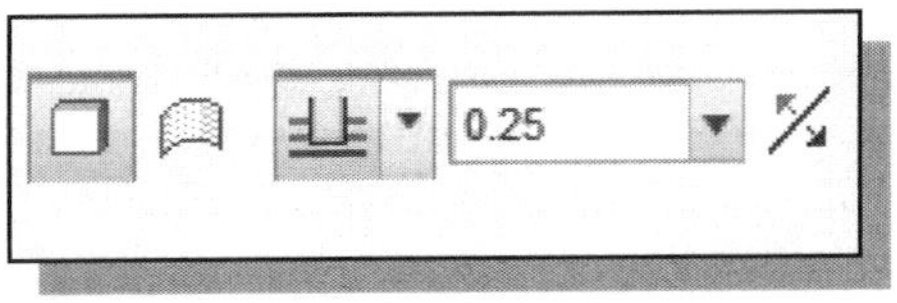

16. In the *extrude distance* value box, enter **0.25** as the thickness of the plate.

17. Click on the arrow to switch the extrusion direction so that the arrow points into the screen. Note that for **plane stress analysis**, the surface must be aligned to the XY plane of any coordinate system. In our case, we will align the front surface to the XY plane of the default part coordinate system.

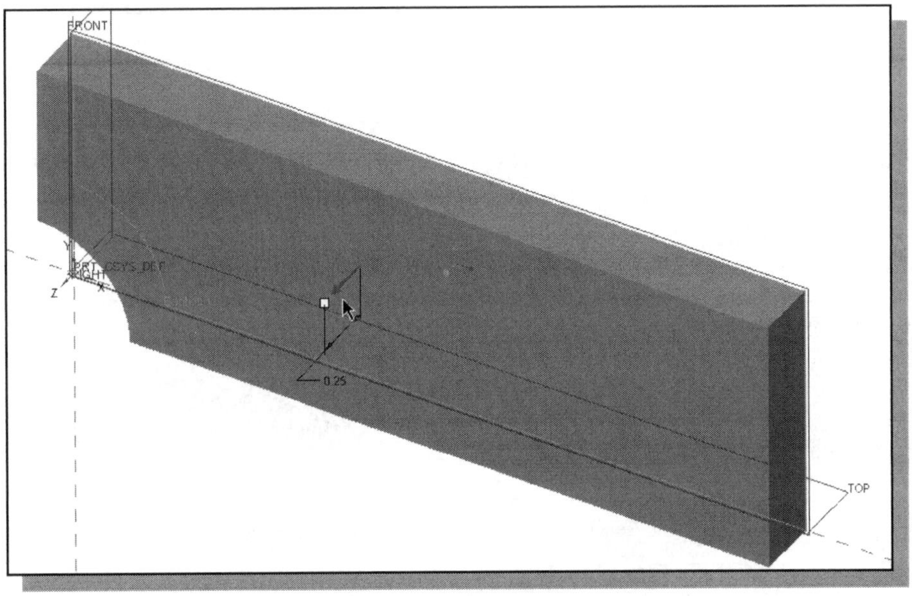

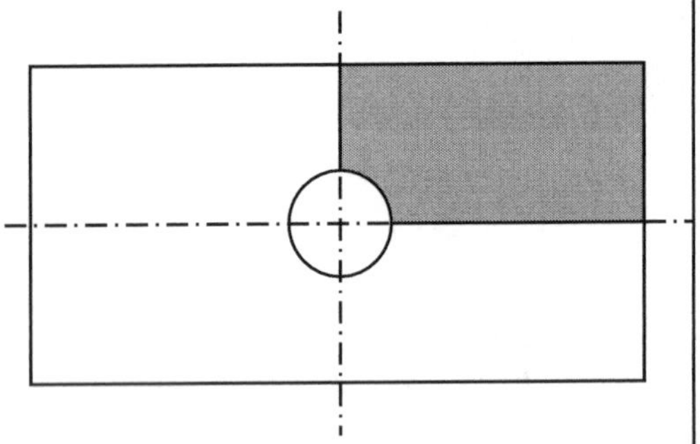

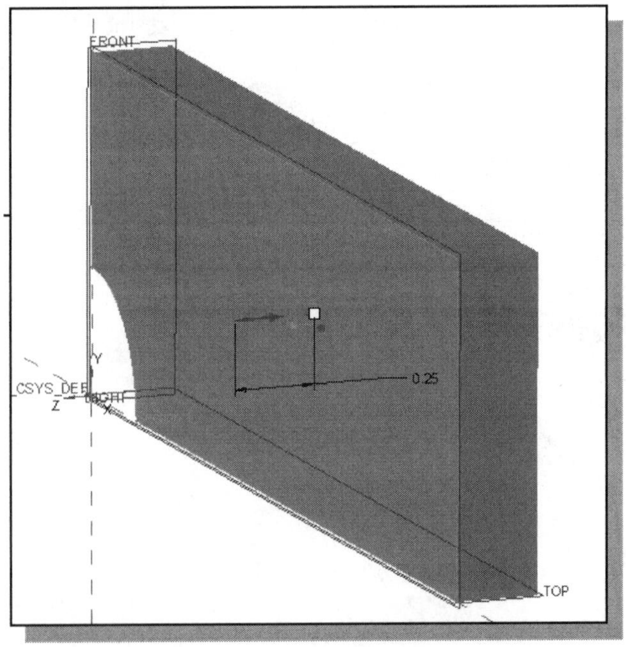

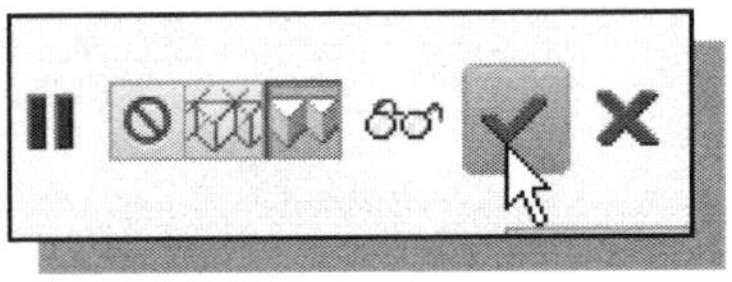

18. Click **Done** to accept the settings and create the solid model.

Select and Examine the Part Material Property

Before switching to *Creo Simulate*, we will first set up the *Material Property* for the part. The *Material Property* contains the general material information, such as *Modulus of Elasticity*, *Poisson's Ratio*, etc., that is necessary for the FEA analysis.

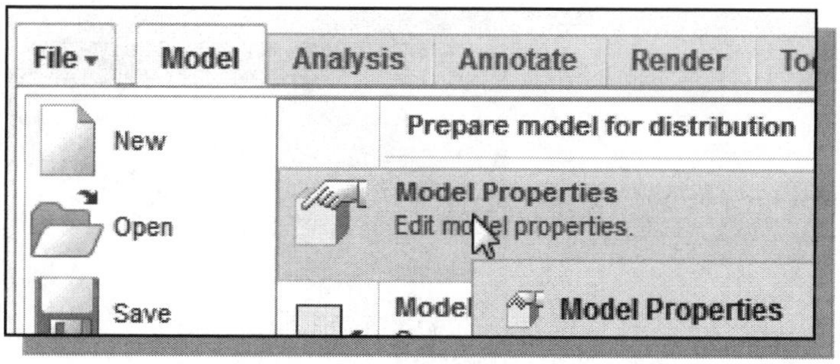

1. Choose **File →Prepare →Model Properties** from the *File pull-down menu.*

2. Select the **Change** option that is to the right of the **Material** option in the ***Model Properties*** window.

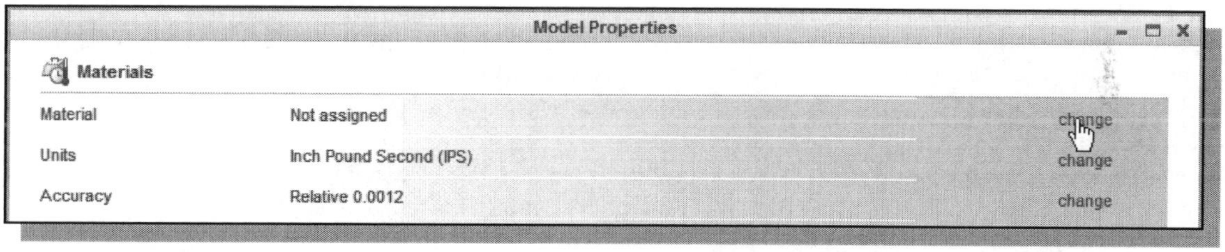

3. Select **AL6061** in the **Materials** list as shown.

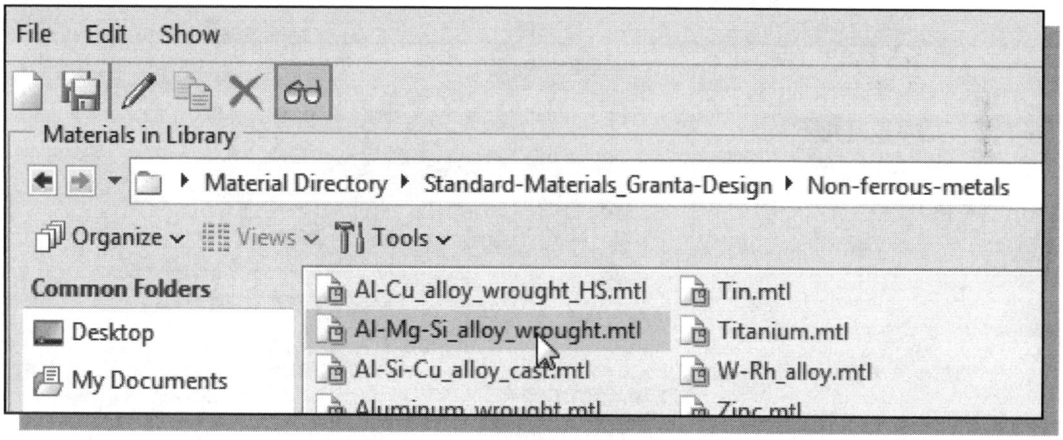

4. **Double-click** on the selected material definition to make the selected material available for use in the current FEA model.

5. On your own, use the Edit option to examine the material information and click **OK** to accept the settings.

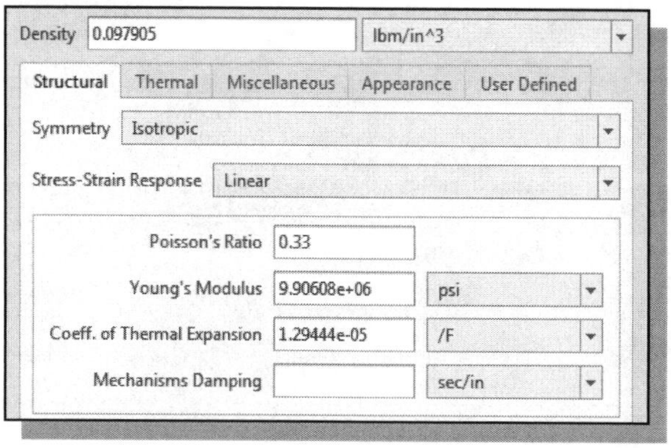

Use the Model setup of Creo Simulate

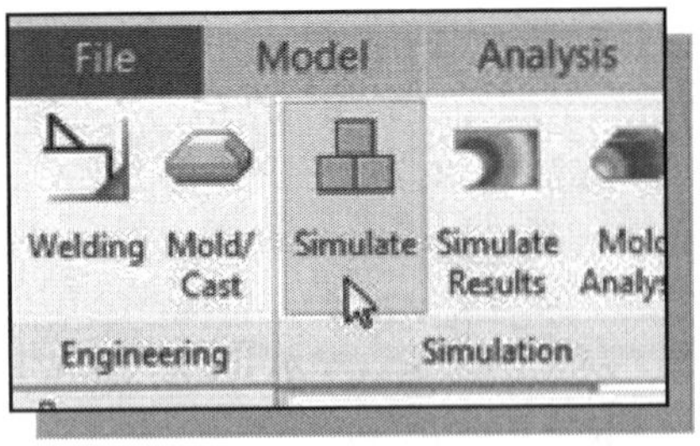

1. Start the *Integrated* mode of *Creo Simulate* by selecting the **Applications → Simulate** option in the *Ribbon* area as shown.

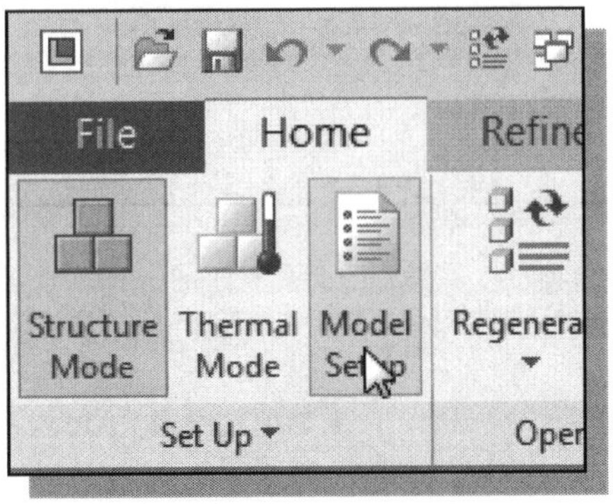

2. In the *Creo Simulate Ribbon* toolbar, choose **Structure** as the analysis type.

3. Click **Model Setup** to examine the *Simulate* model settings.

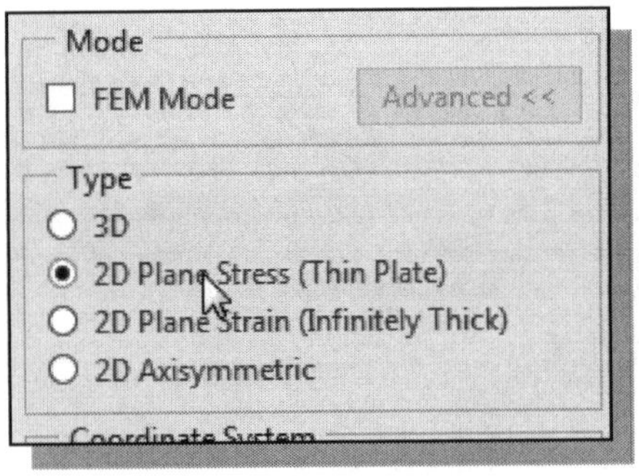

4. Click **Advanced** to show the additional settings that are available.

5. Choose **2D Plane Stress (Thin Plate)** as the model Type as shown in the figure.

6. Select the default part coordinate system, **PRT_CSYS_DEF**, as shown in the figure below.

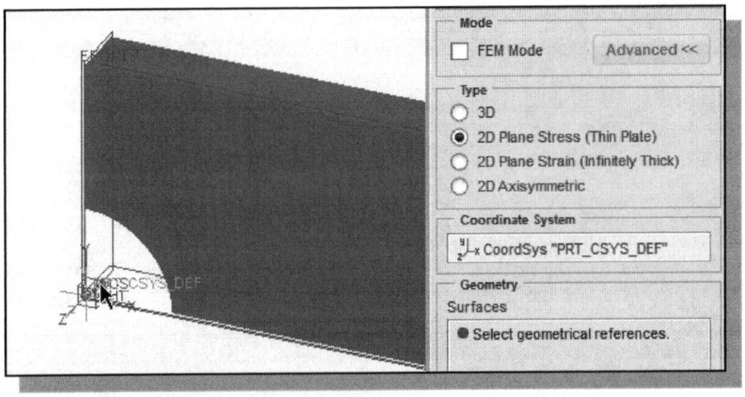

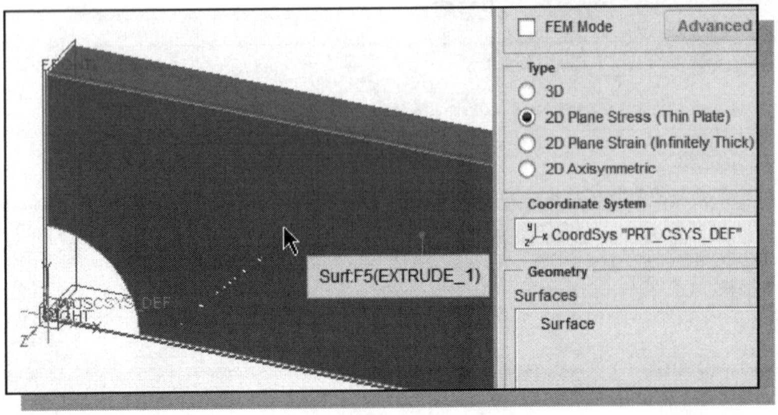

7. Select the Front face of the model as the surface to perform the analysis.

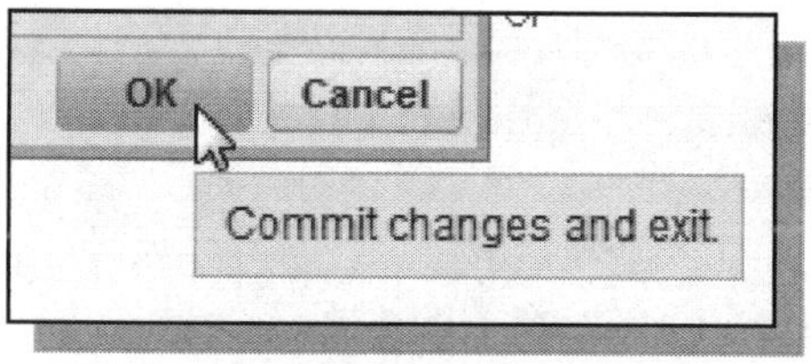

8. In the *Creo Simulate Model Setup* window, click on the **OK** button to accept the settings and proceed to create a 2D Plane Stress FEA model.

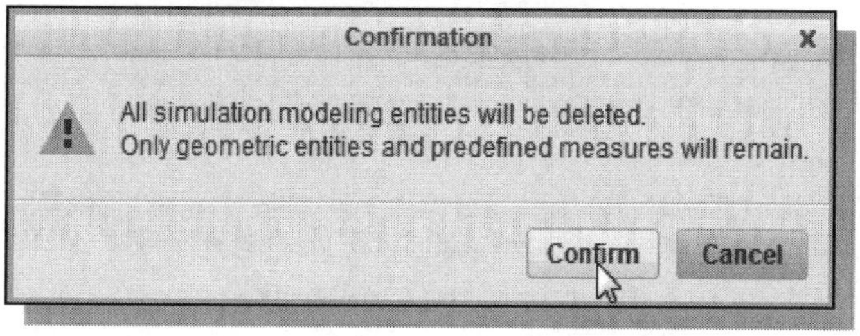

9. Click **Confirm** to proceed with the creation of the FEA model.

Apply the Boundary Conditions - Constraints

In the previous chapters, we have created the elements before applying the constraints and loads. With *Creo Simulate*, we can also apply the constraints and loads onto the CAD model. This is known as the *Geometry Based Analysis* approach.

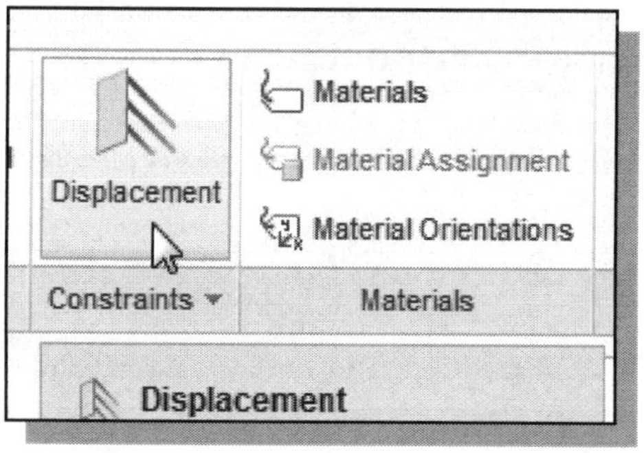

1. Choose **Displacement Constraint** by clicking the icon in the *constraints toolbar* as shown.

2. Set the References to **Edge/Curves** and pick the **left vertical edge** of the front surface as shown.

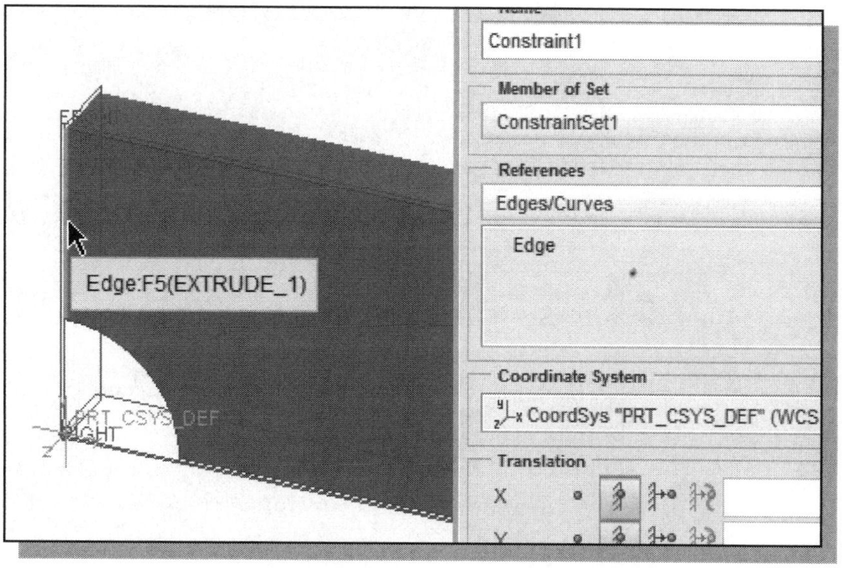

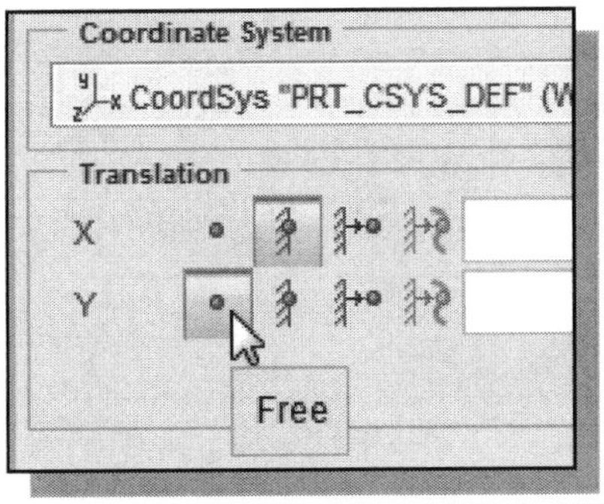

3. Confirm the X Translation option to **Fixed** as shown.

4. Set the Y Translation option to **Free** as shown.

5. Click on the **OK** button to accept the first displacement constraint settings.

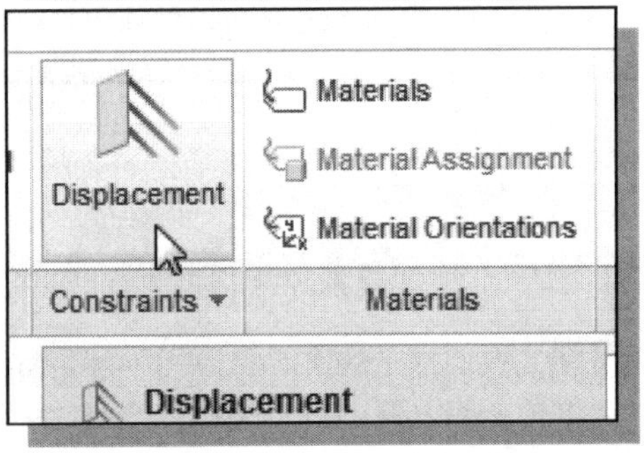

6. Choose **Displacement Constraint** by clicking the icon in the toolbar as shown.

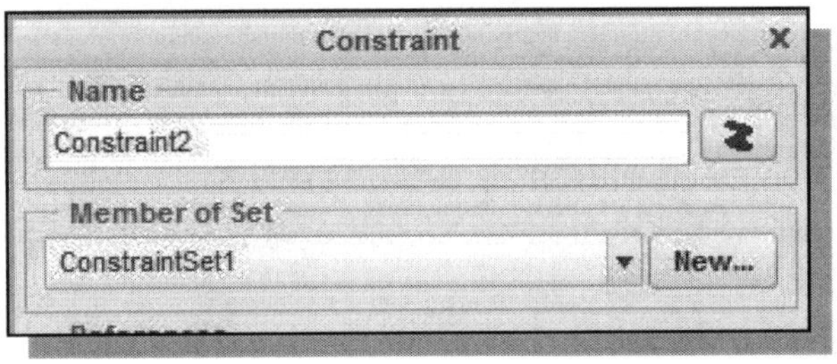

7. Confirm the *Constraint* set is still set to **ConstraintSet1**.

❖ Note that a constraint set can contain multiple constraints.

8. Set the References to **Edge/Curves** and select the **bottom edge** of the front surface as shown.

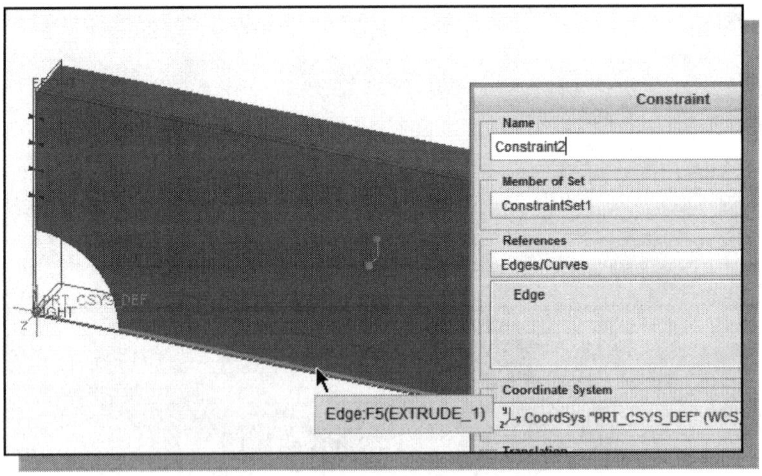

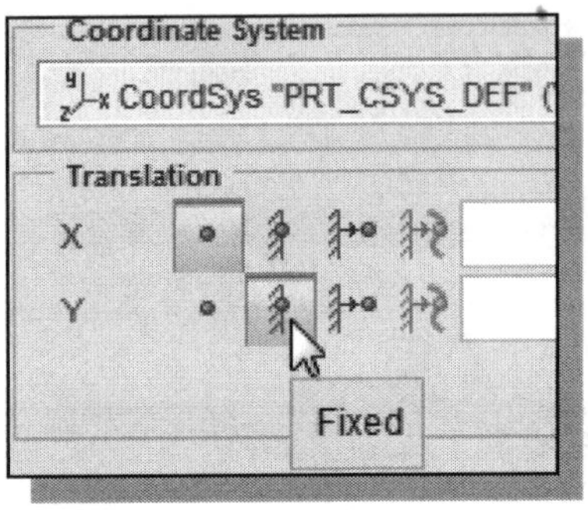

9. Set X Translation to **Free** as shown.

10. Set Y Tanslation to **Fixed** as shown.

11. Click on the **OK** button to accept the second displacement constraint settings.

Apply the External Loads

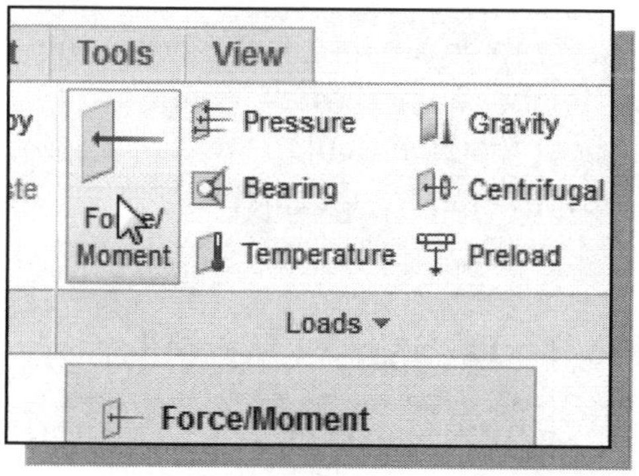

1. Choose **Force/Moment Load** by clicking the icon in the toolbar as shown.

2. Set the References to **Edges/Curves**.

3. Pick the **right vertical edge** of the front surface as shown.

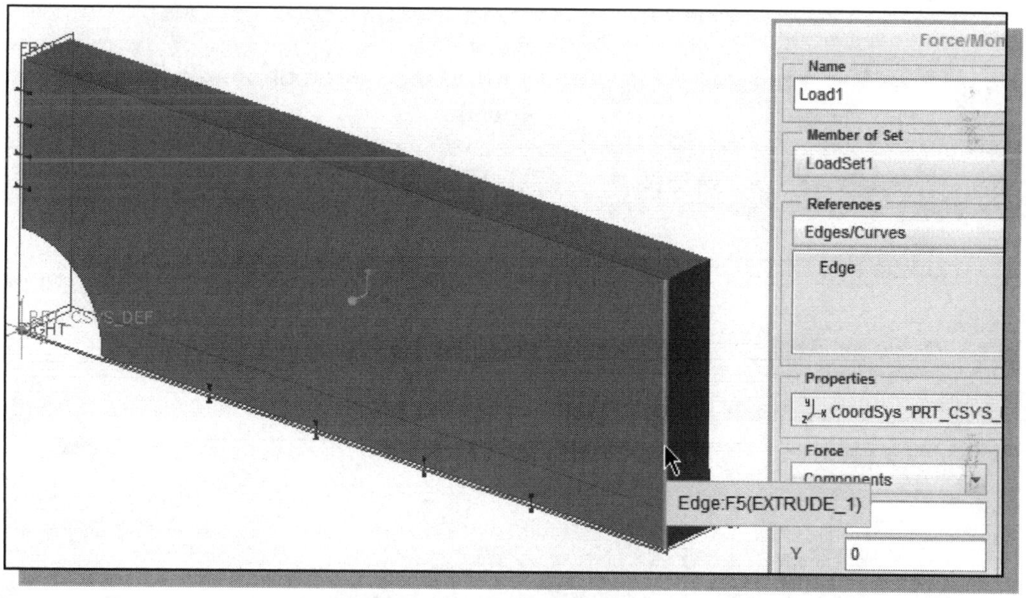

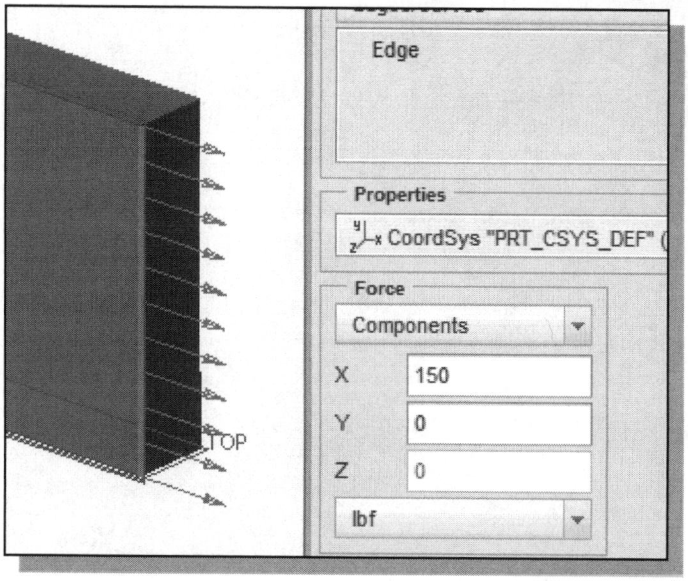

4. Enter **150** in the X component Force value box as shown.

5. Click on the **Preview** button to confirm the applied load is pointing in the correct direction.

6. Click on the **OK** button to accept the first load settings.

FEA Surface Idealization

To perform a 2D analysis in *Creo Simulate*, it is necessary to create a surface FEA model. Although we have specified the use of 2D plane stress analysis when we entered *Creo Simulate*, it is still necessary to identify the actual surfaces that are involved in the analysis as well as to provide some of the additional information for the analysis.

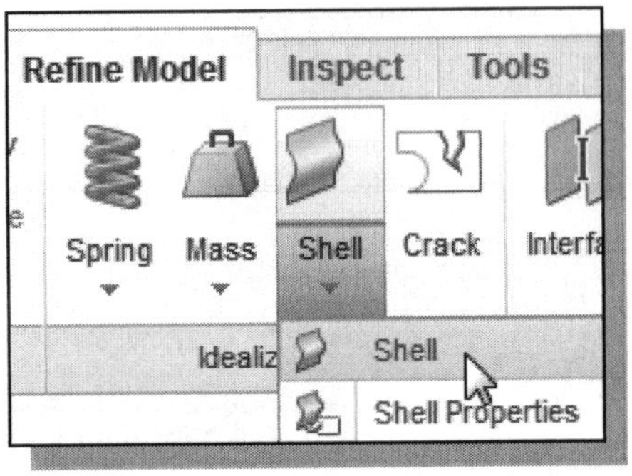

1. Choose **Refine Model** in the *Ribbon* toolbar.

2. Activate the **Shell** command by selecting the icon in the toolbar area as shown.

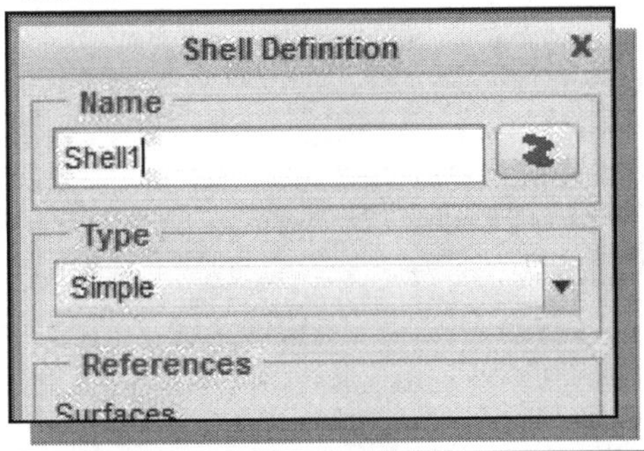

3. Confirm the Type option is set to **Simple** as shown.

4. Select the **front surface** as the reference surface to create the shell model.

5. Set the Thickness property to **0.25** as shown.

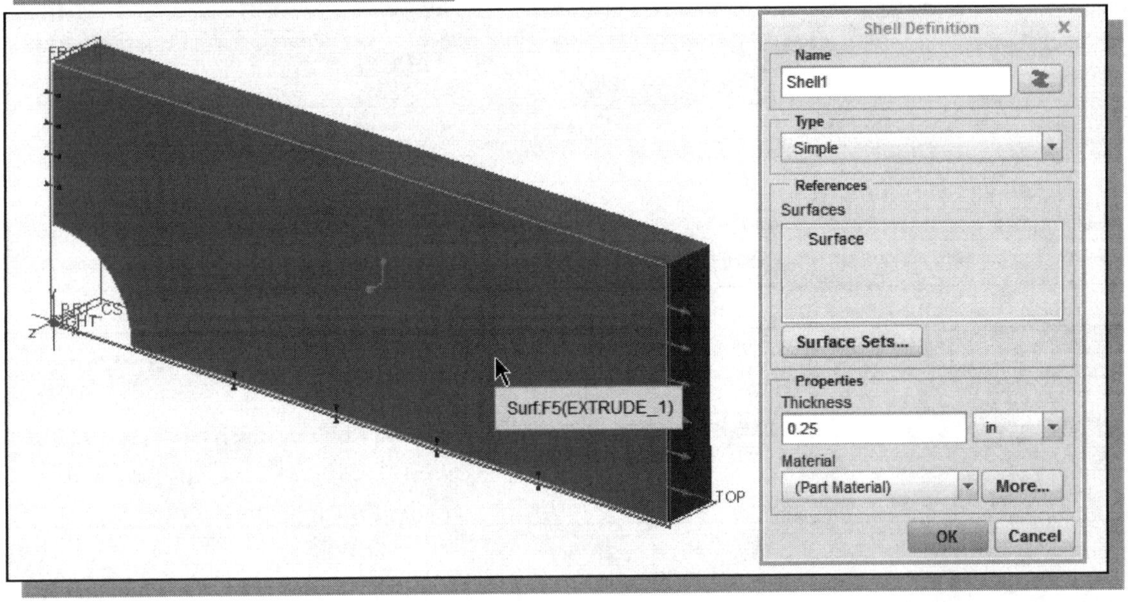

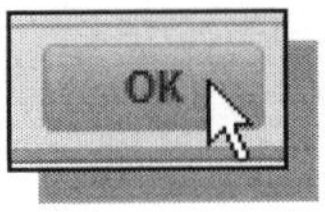

6. Click **OK** to accept the selection and create the surface FEA model.

H-Element versus P-Element

There are two different approaches to FEA: H-element method and P-element method. Finite elements used by commercial programs in the 1970s and 80s were all H-elements. The stress analysis of the H-element concentrates primary at the nodes. The H-element uses a low order interpolating polynomial, which usually is linear or quadratic. Strain is obtained by taking the derivatives of the displacement and the stress computed from the strain. For a first order interpolating polynomial within each element, this would cause the strain and stress components to be constant within the element; this caused a discontinuity in the stress field between elements, and will lead to inaccurate values for the maximum local and global stresses. This is why the H-element method requires the refinement of mesh around the high stress areas. The process of mesh refinement is called convergence analysis, on H-elements; this style of convergence is called H-convergence. Oftentimes, this convergence analysis leads to a fairly large set of differential equations, and then simultaneous linear algebraic equations. Another drawback of the H-element is its inability to adapt to shape extremes in terms of skew and large size variation.

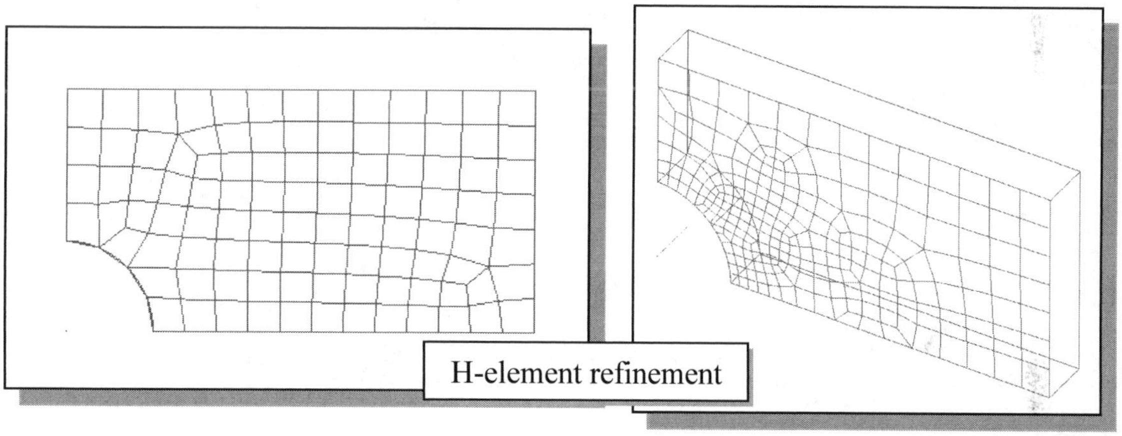

H-element refinement

With the improvements in computer power and efficiency, a new type of element, the P-element, was developed in late 1980s. The P-element is unique because convergence is obtained by increasing the order of the interpolating polynomials in each individual element. Instead of changing the mesh, higher order polynomials are used in the element. Generally speaking, the P-method uses a constant mesh, which is usually coarser than an H-element mesh. The FEA solver will detect the areas where high gradients occur and those elements will have their order of the interpolating polynomials automatically increased. This allows for the monitoring of expected error in the solution and then automatically increases the polynomial order as needed. This is the main benefit of using P-elements. We can use a mesh that is relatively coarse, thus computational time will be low, and still get reasonable results. The accepted level of error in a solution using the highest settings is generally less then ten percent; often times, an analysis with less than a two percent error can be accomplished. Also, the mesh restrictions for this element size are not nearly as stringent for the P-element. One should also realize that despite the automatic process of the P-elements analysis, areas of the model that are of particular interest or with more complex geometry or loading could still benefit from the user specifying an increase in mesh density.

Create the 2D Mesh

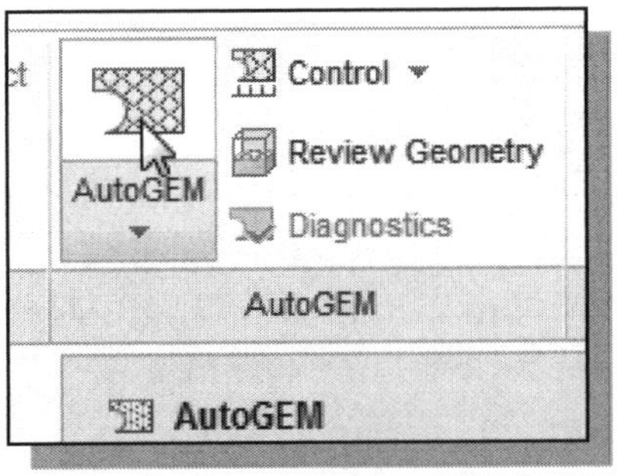

1. Activate the **AutoGEM** command by selecting the icon in the toolbar area as shown.

2. Click **No** to close the retrieve mesh message.

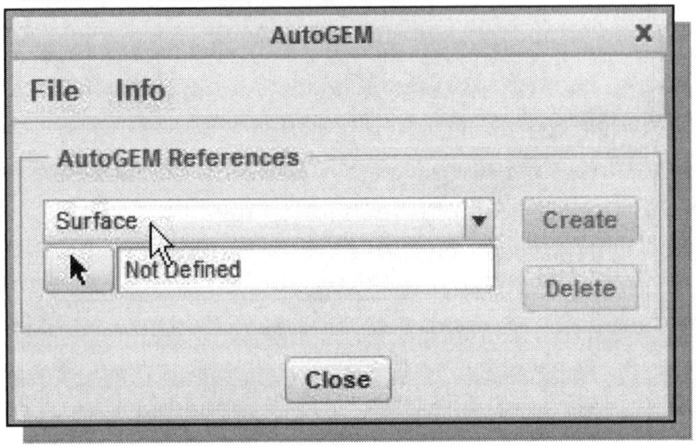

3. Switch the AutoGEM References option to **Surface** as shown.

4. Activate the **Select geometry** option by clicking on the arrow button as shown.

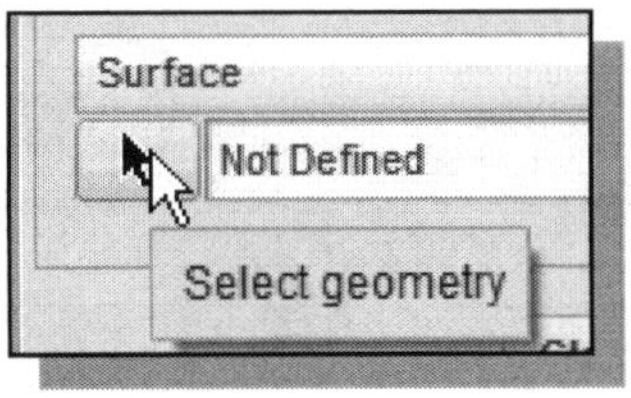

5. On your own, select the **front surface** of the model.

6. Click once with the **middle-mouse-button** to accept the selection.

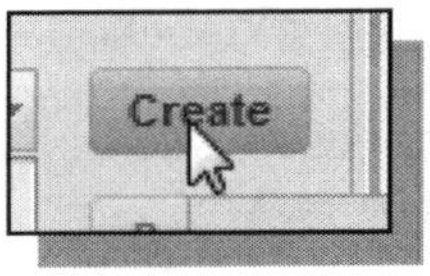

7. Inside the *AutoGEM* dialog box, click **Create** to generate the P-mesh.

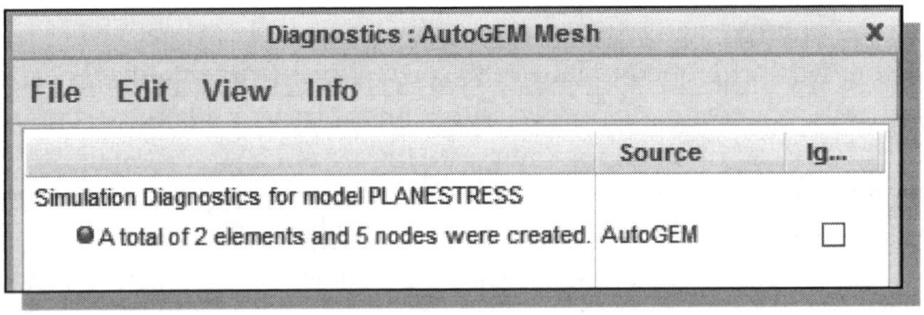

➢ Note that *Creo Simulate* generated only **two elements** and **five nodes**. This is the main characteristic of the P-element method, using very coarse mesh for the analysis.

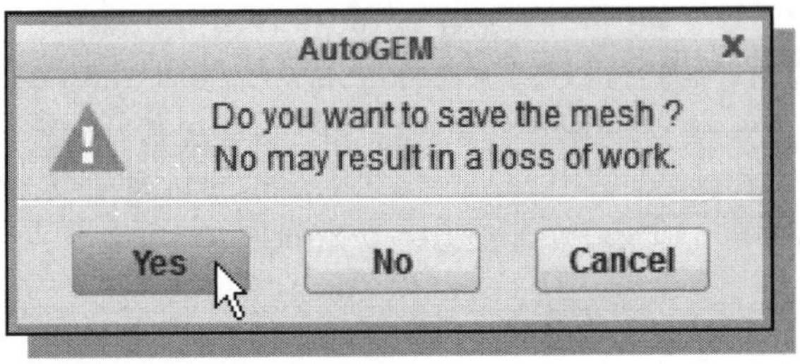

8. On your own, exit the *AutoGEM* dialog box.

9. Click **Yes** to save the mesh as shown.

Run the FEA Solver

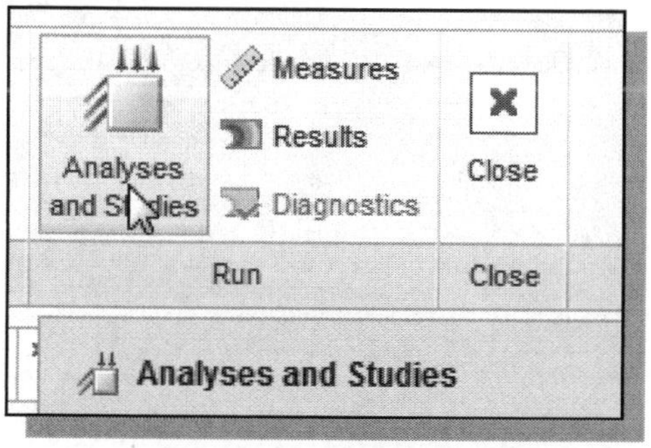

1. Activate the *Creo* **Analyses and Studies** command by selecting the icon in the toolbar area as shown.

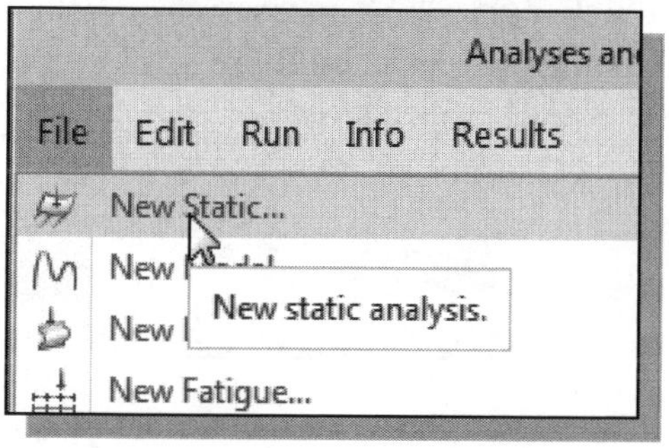

2. In the *Analyses and Design Studies* window, choose **File → New Static** in the pull-down menu. This will start the setup of a basic linear static analysis.

3. Note that the **ConstraintSet1** and **LoadSet1** are automatically accepted as part of the analysis parameters.

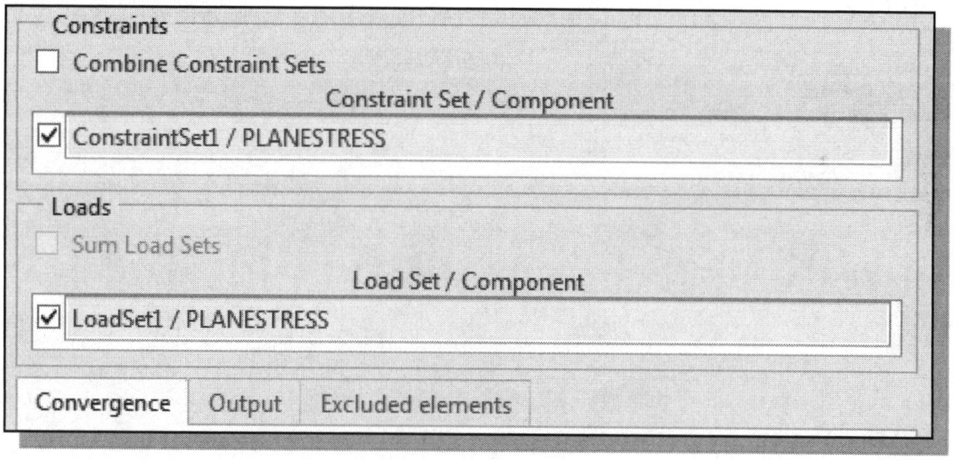

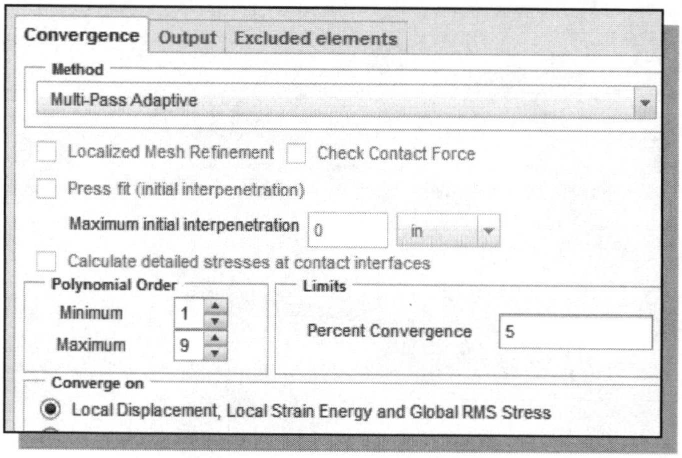

4. Set the Method option to **Multi-Pass Adaptive**, P-Order to **9** and Limits to **5** Percent Convergence as shown.

5. Click **OK** to accept the settings.

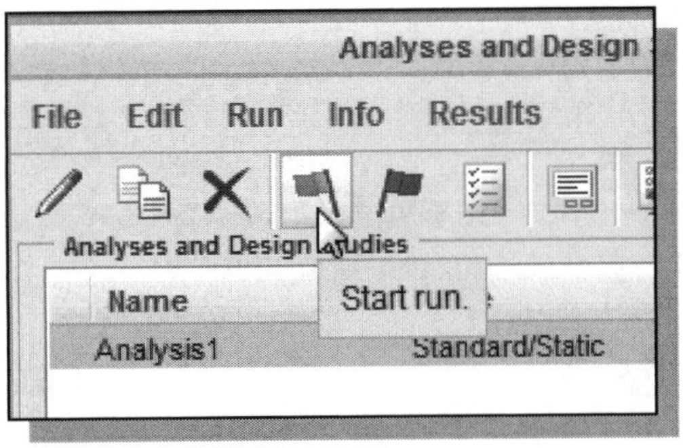

6. Click **Start run** to begin the *Solving* process.

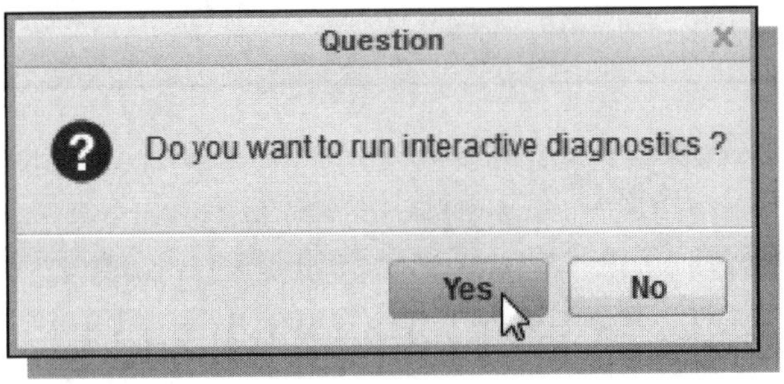

7. Click **Yes** to run interactive diagnostics; this will help us identify and correct problems associated with our FEA model.

8. In the *Diagnostics* window, *Creo Simulate* will list any problems or conflicts in our FEA model.

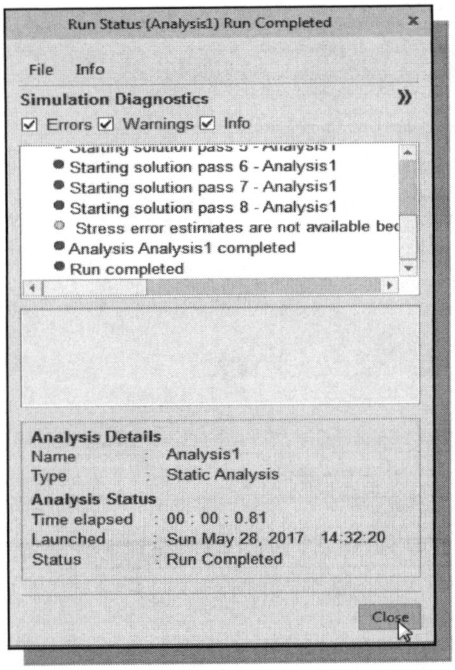

9. Click on the **Close** button to exit the *Diagnostics* window.

10. In the *Analyses and Design Studies* window, choose **Review results** as shown.

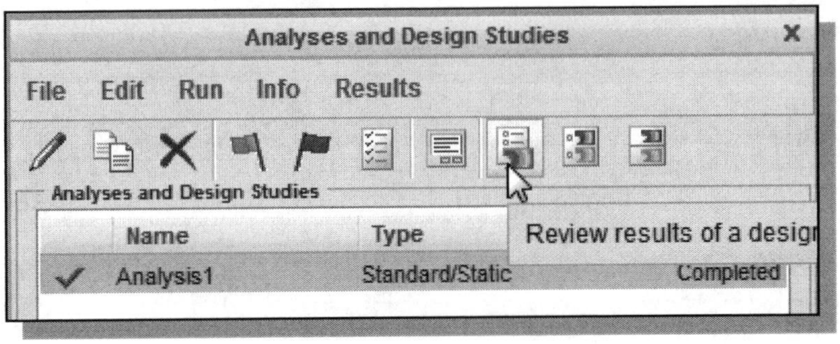

View the FEA Results

➢ Maximum Principal Stress

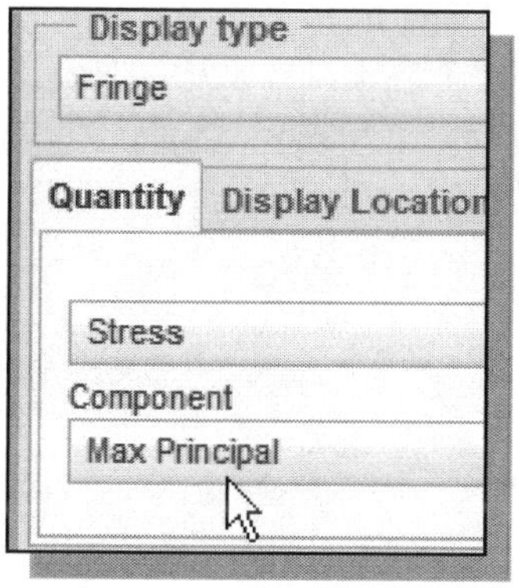

1. Confirm the Display type is set to **Fringe**.

2. Choose **Maximum Principal** as the Stress Component to be displayed.

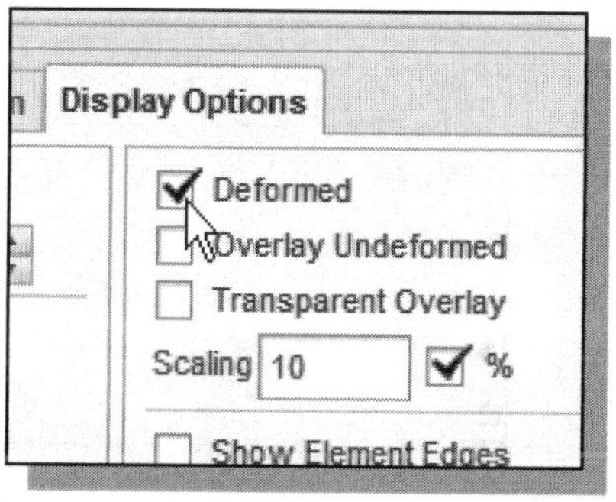

3. Click on the **Display Options** tab and switch ***on*** the **Deformed** option as shown.

4. Click **OK and Show** to display the results.

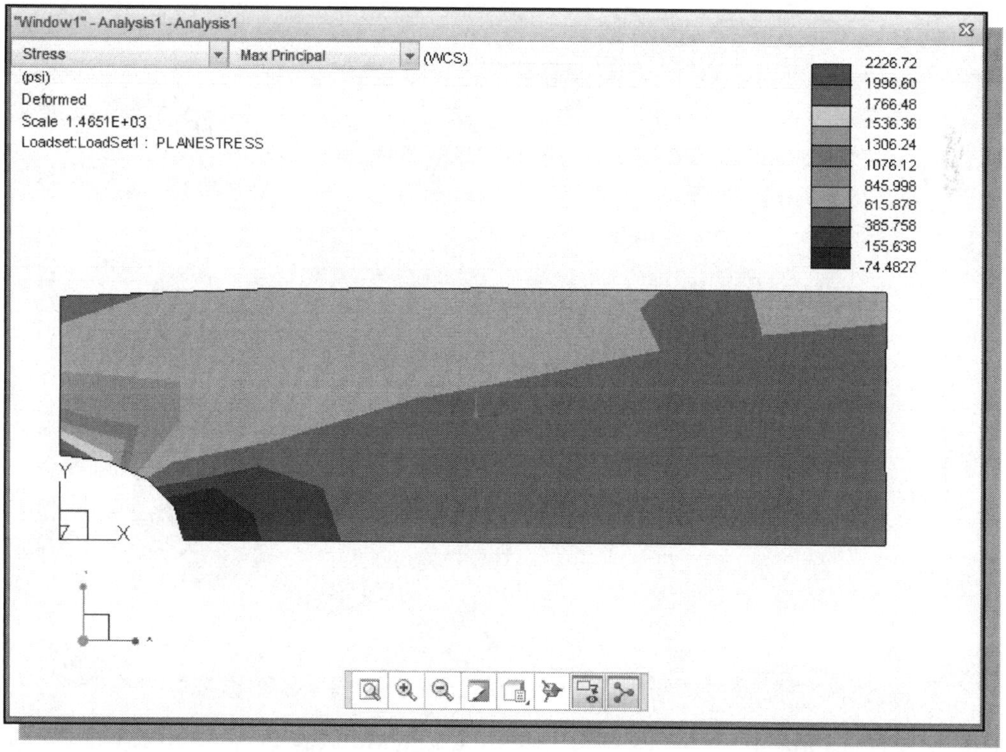

➢ The computer solution for the maximum stress is **2.226E+3 lb/in²**, which occurs at the top of the circular arc. This number is very close to the **2180** we calculated during our preliminary analysis.

- **X Displacement**

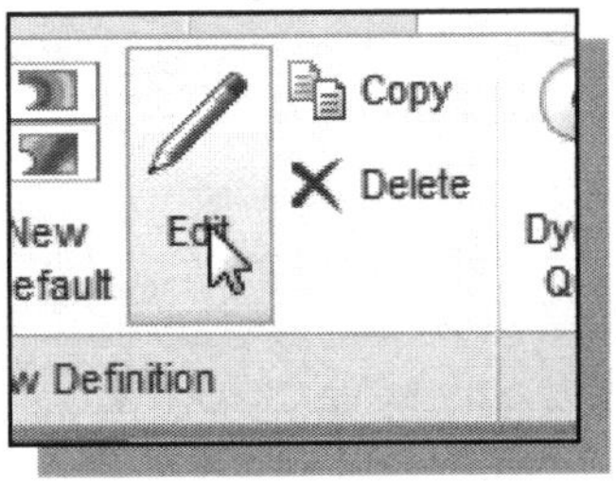

1. In the *Standard* toolbar area, click **Edit the selected definition** command as shown.

2. Confirm **Fringe** is the Display type as shown.

3. Choose **Displacement** as the Quantity option.

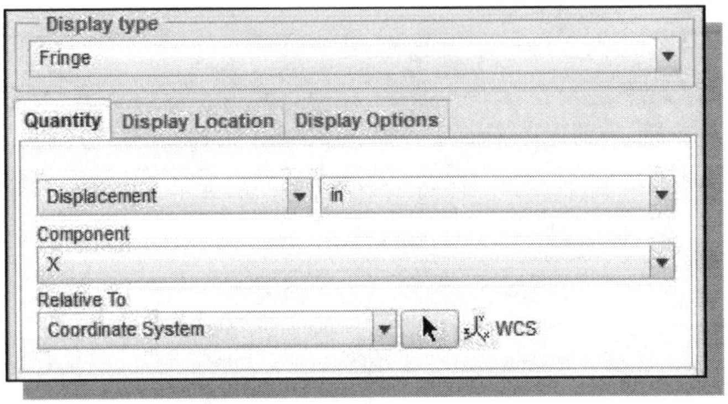

4. Choose **X** as the Displacement Component to be displayed.

5. Click **OK and Show** to display the results.

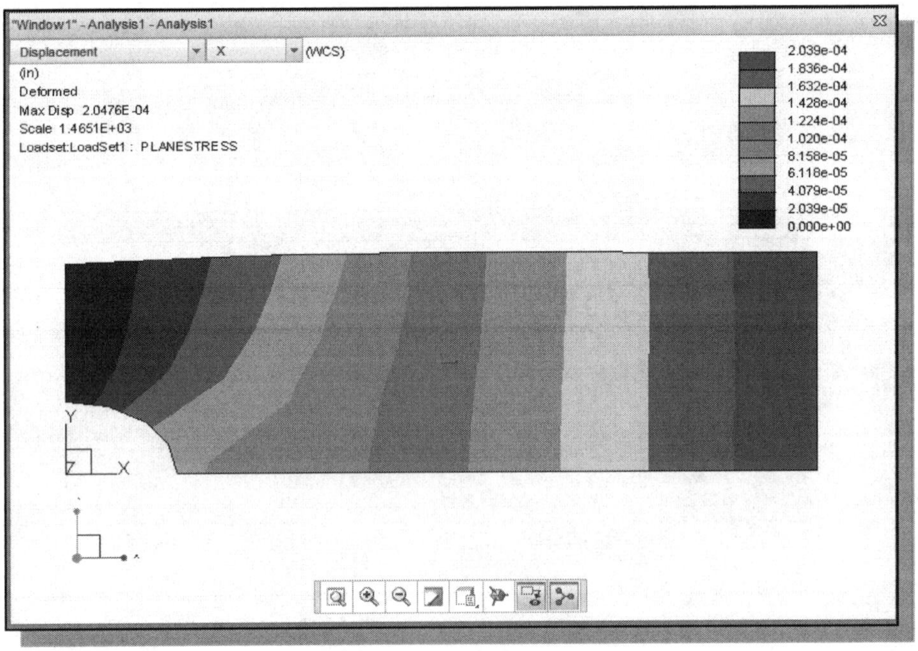

- The *max. X displacement* is **2.068e-4**, which is within the upper and lower bounds of our preliminary analysis.

6. On your own, close the results display window and return to *Creo Simulate*.

Refinement of the P-mesh

In *Creo Simulate*, several options are available to refine the default AutoGEM mesh. Let's examine the effects of refining of the P-mesh.

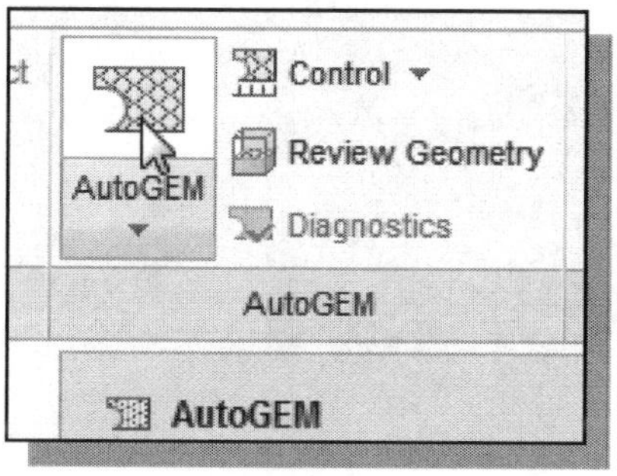

1. Activate the **AutoGEM** command by selecting the icon in the toolbar area as shown.

2. Click **Yes** to retrieve the current elements.

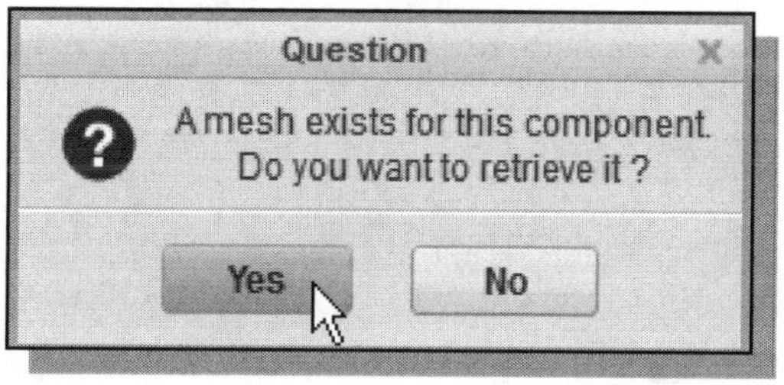

3. Click **Delete** to remove the 2 element mesh that is part of the FEA model.

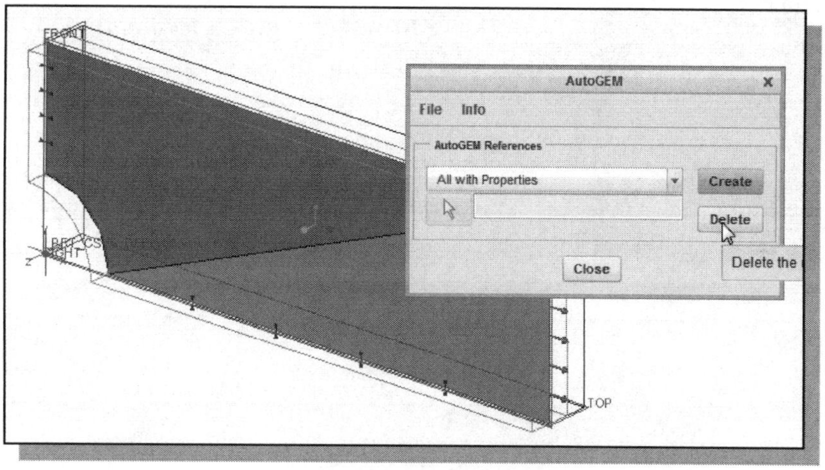

4. In the *AutoGEM* dialog box, click **Close** to exit the AutoGEM command.

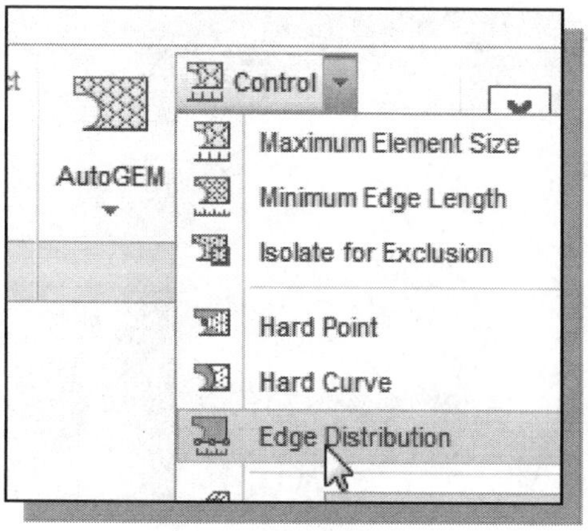

5. Select the **AutoGEM → Control** option from the *Creo Simulate* pull-down menu as shown.

6. Set the mesh control type to **Edge Distribution** as shown.

❖ The **Edge Distribution** option allows us to set the number of nodes on the selected edges.

7. Select both the arc and the left vertical edge by holding down the [**Ctrl**] key and clicking with the left-mouse-button.

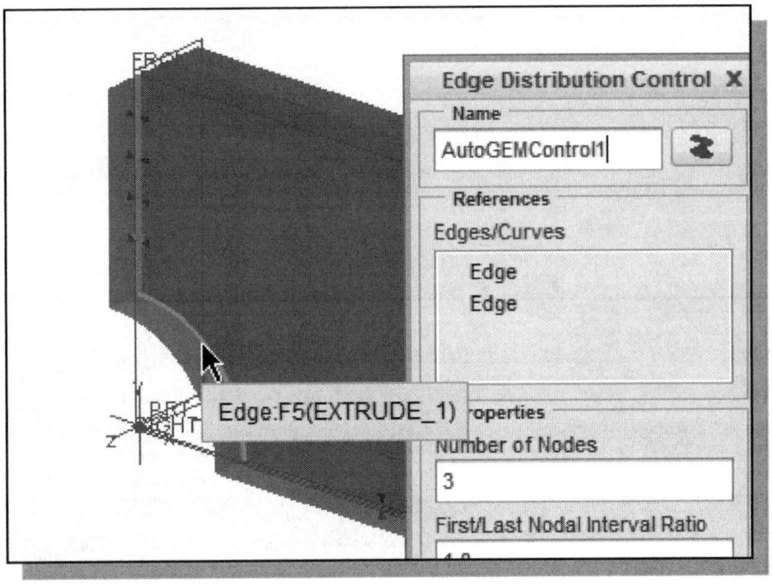

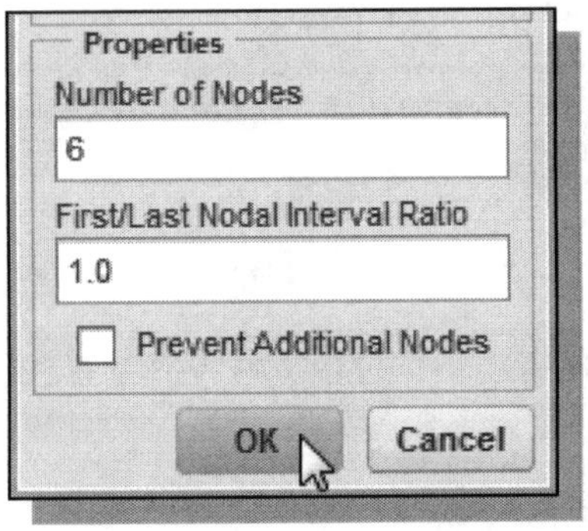

8. Set the Number of Nodes to **6** and the Nodal Interval Ratio to **1.0** as shown.

9. Click **OK** to accept the settings.

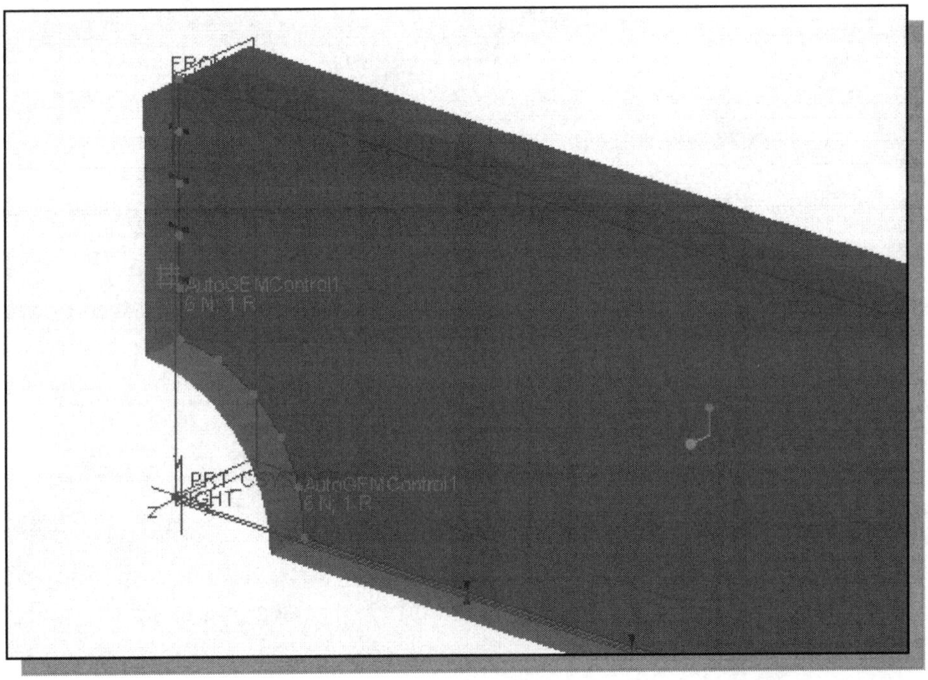

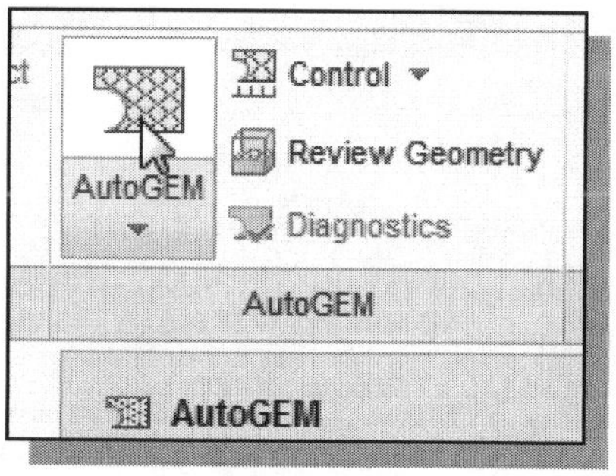

10. Activate the **AutoGEM** command by selecting the icon in the toolbar area as shown.

11. Click **Yes** to retrieve the current elements.

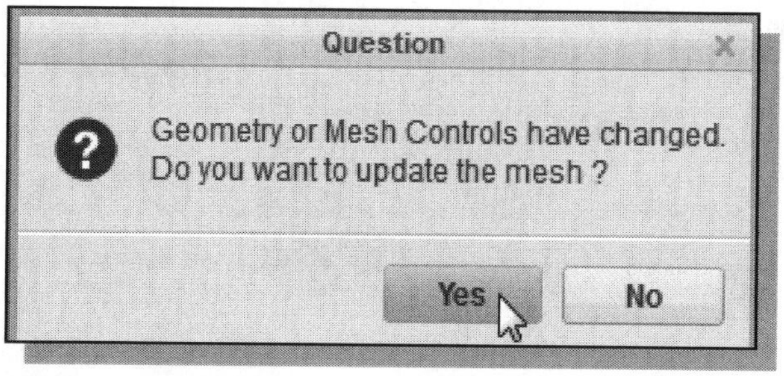

12. Click **Yes** to update the current mesh.

➢ Note that *Creo Simulate* now generated eleven elements and eighteen nodes. We will compare the FEA results of this much-refined mesh to the results of the original two elements mesh.

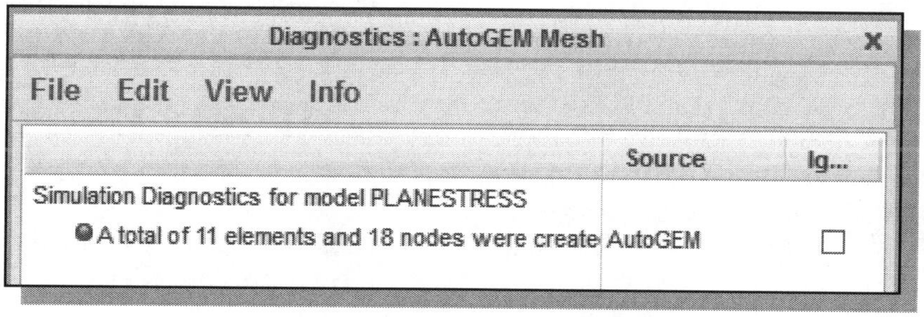

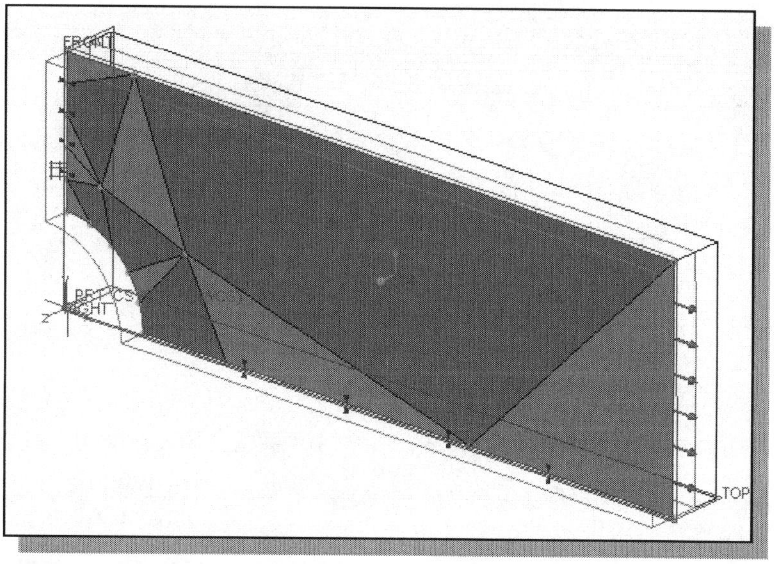

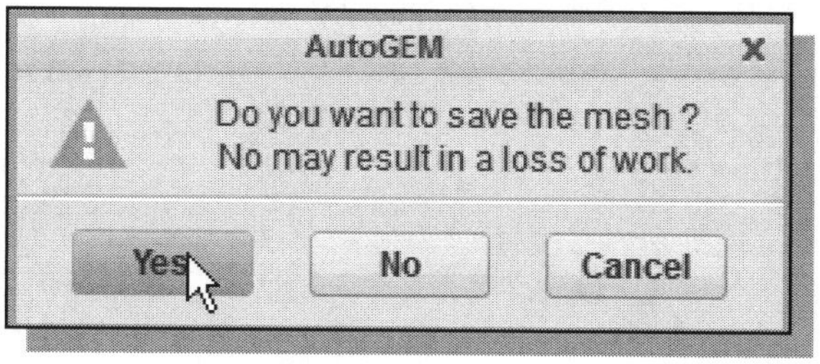

13. On your own, exit the *AutoGEM* dialog box by clicking on the **Close** button.

14. Click **Yes** to save the mesh.

Run the FEA Solver

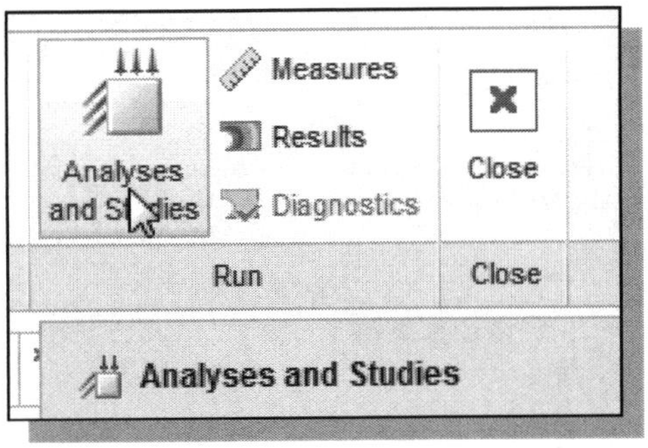

1. Activate the *Creo Simulate* **Analyses and Studies** command by selecting the icon in the toolbar area as shown.

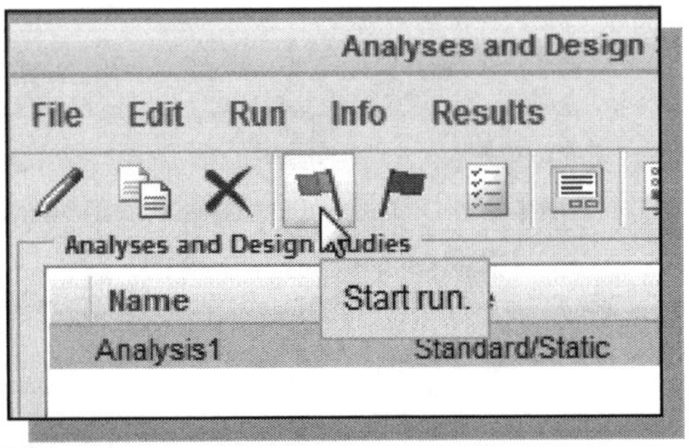

2. Click **Start run** to begin the *Solving* process.

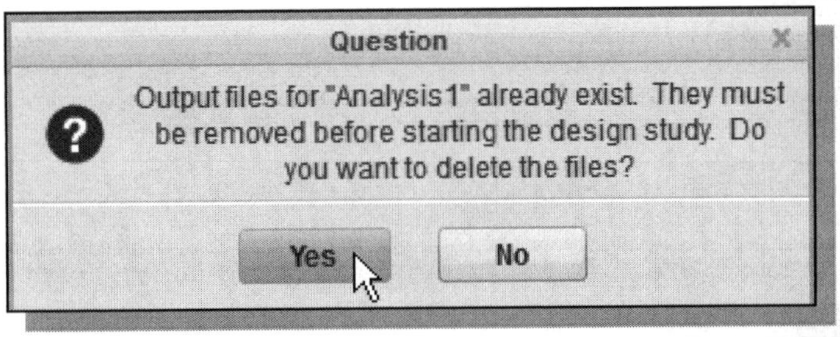

3. Click **Yes** to delete the old output files.

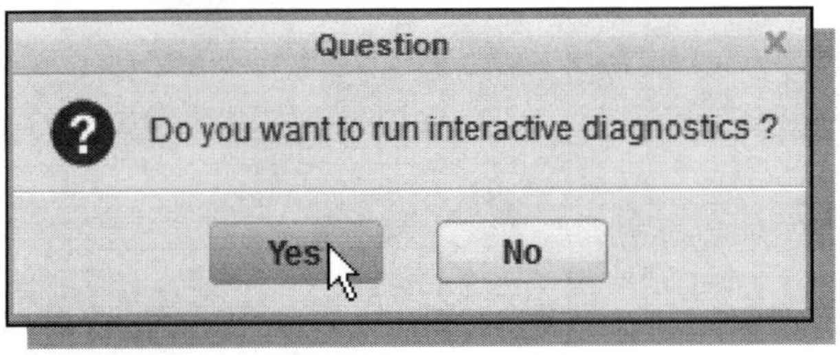

4. Click **Yes** to run interactive diagnostics; this will help us identify and correct problems associated with our FEA model.

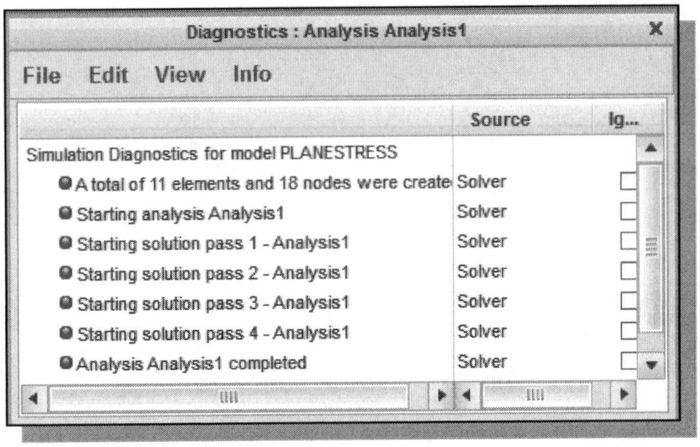

5. With the increased elements, we actually reduced the number of passes that are required to reach convergence.

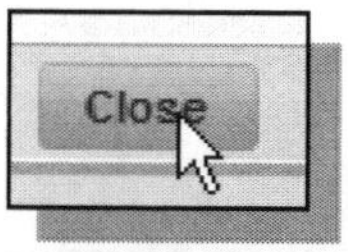

6. Click on the **Close** button to exit the *Diagnostics* window.

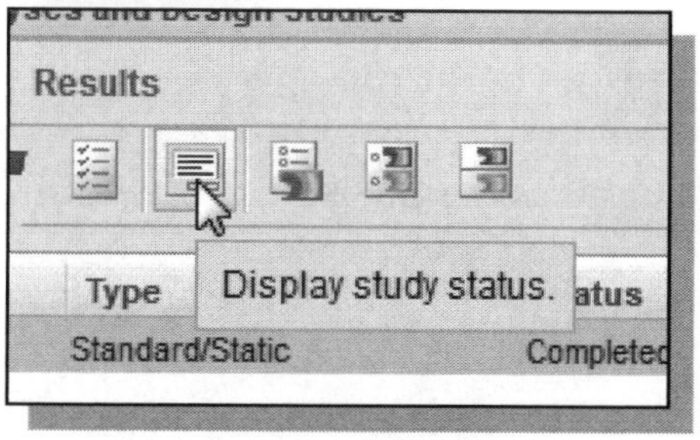

7. In the *Analyses and Design Studies* window, choose **Display study status** as shown.

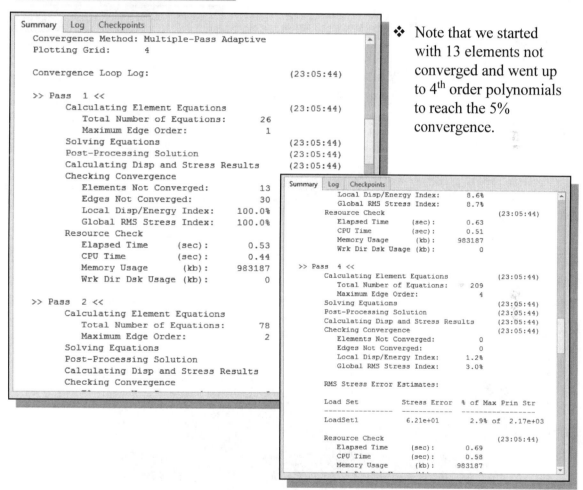

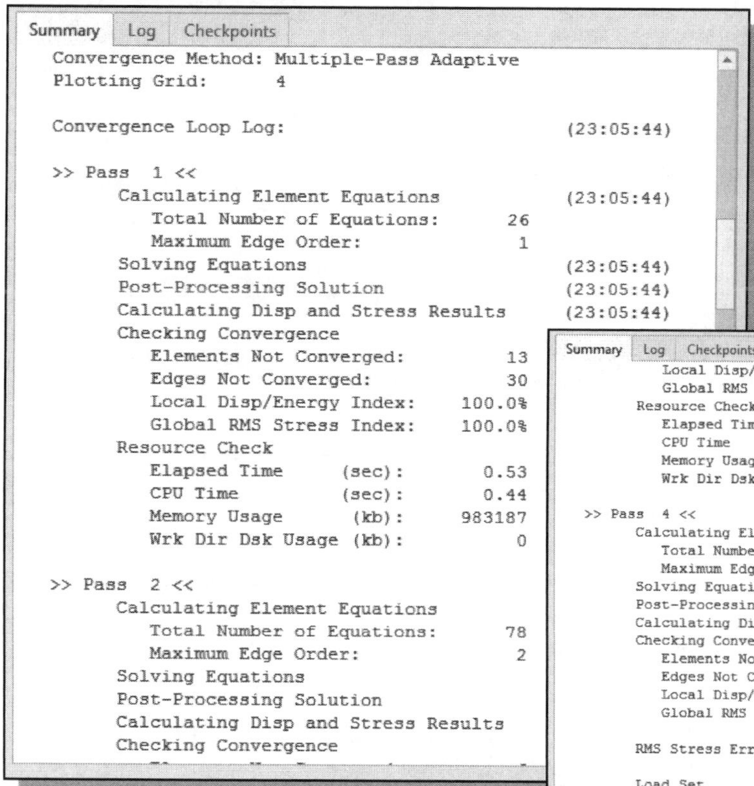

❖ Note that we started with 13 elements not converged and went up to 4th order polynomials to reach the 5% convergence.

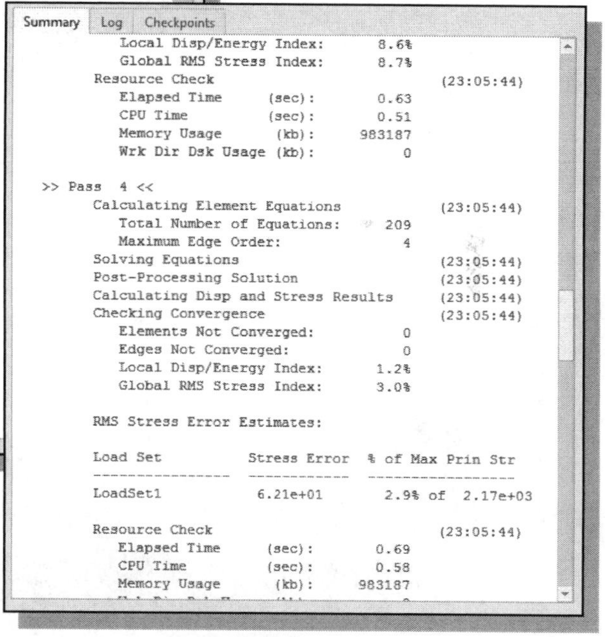

8. In the *Analyses and Design Studies* window, choose **Review results** as shown.

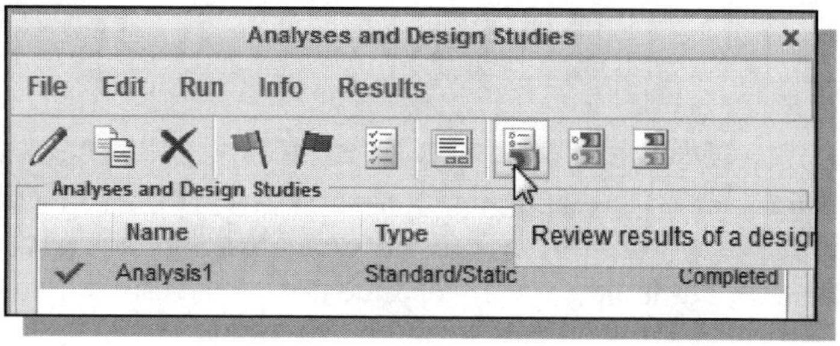

View the FEA Results

➢ Maximum Principal Stress

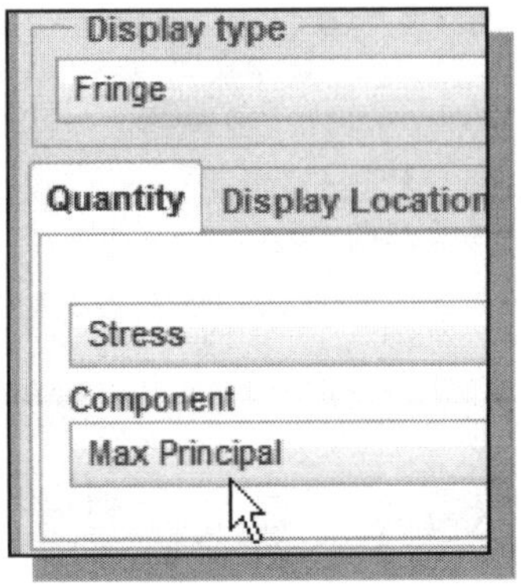

1. Confirm the Display type is set to **Fringe**.
2. Choose **Maximum Principal** as the Stress Component to be displayed.

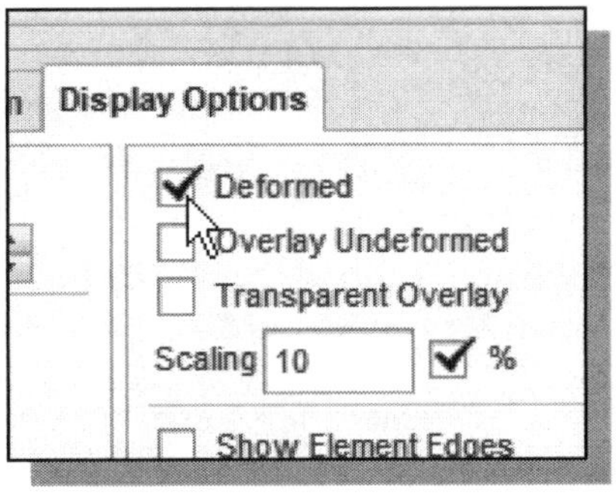

3. Click on the **Display Options** tab and switch ***on*** the **Deformed** option as shown.

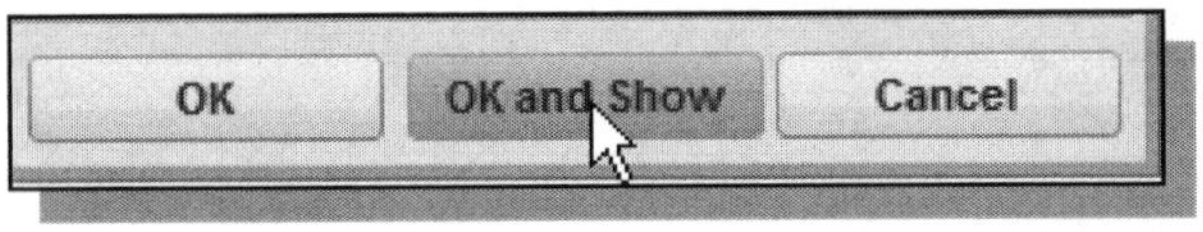

4. Click **OK and Show** to display the results.

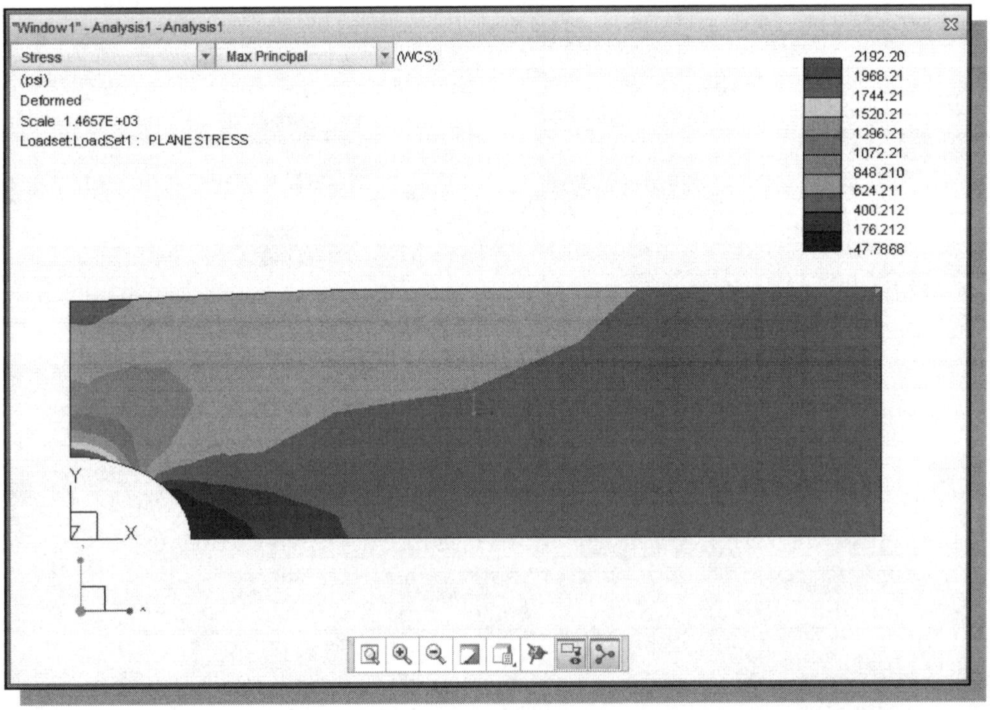

➢ The computer solution for the maximum stress now is **2.192E+3 lb/in²**, which is a little lower than the two elements mesh. Note the added elements have helped the element convergence, but it didn't really change the stress results that much.

➤ **X Displacement**

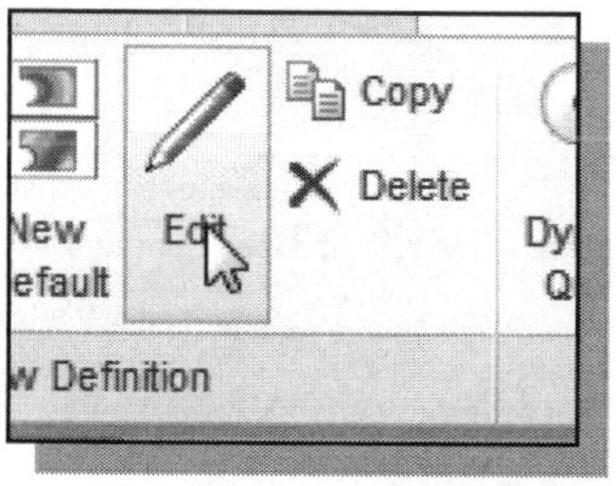

1. In the *Standard* toolbar area, click **Edit the selected definition** command as shown.

2. Confirm **Fringe i**s the Display type as shown.

3. Choose **Displacement** as the Quantity option.

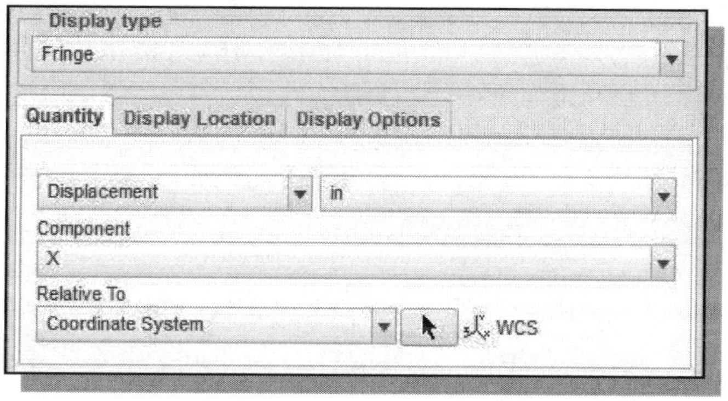

4. Choose **X** as the Displacement Component to be displayed.

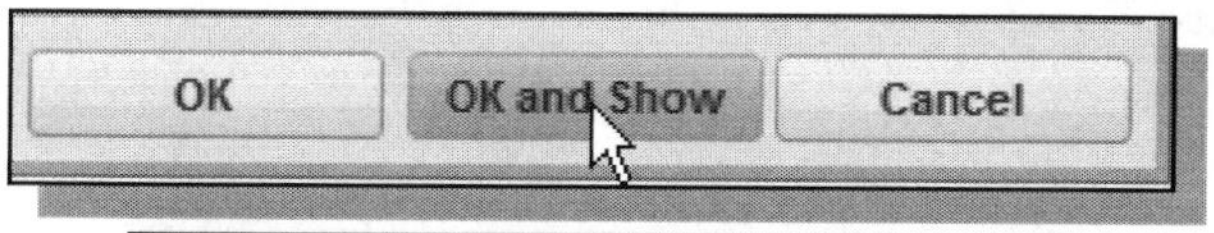

5. Click **OK and Show** to display the results.

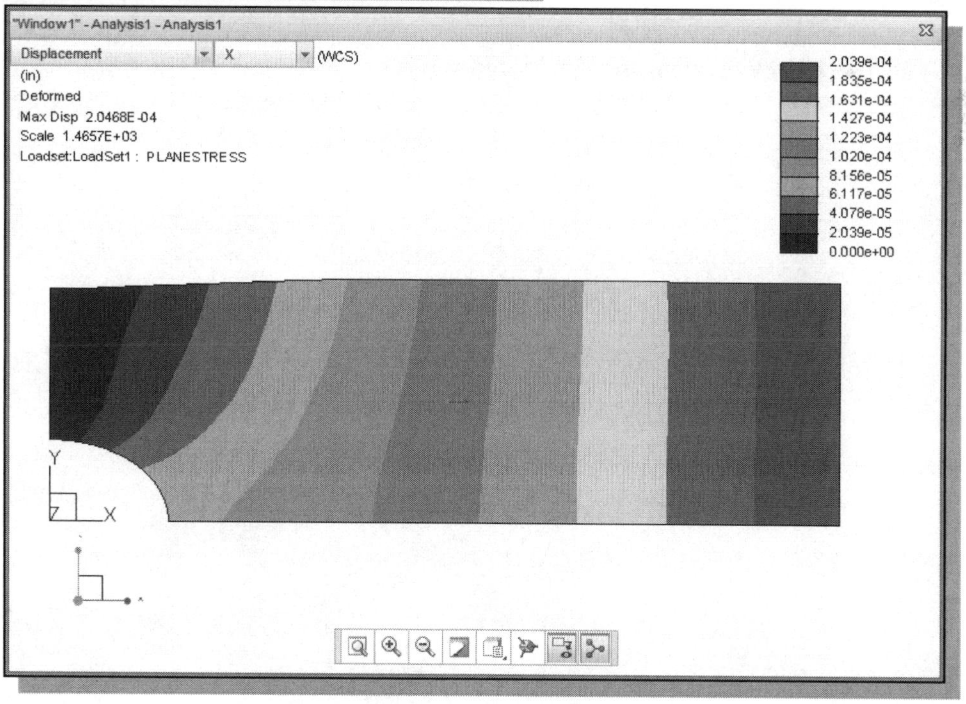

➤ The *max. X displacement* is **2.048e-4**, which is almost the same as our previous result with the two elements mesh.

Review Questions

1. What are the four basic types of two-dimensional solid elements?

2. What type of situation is more suitable to perform a *Plane Stress* analysis than a *Plane Strain* analysis?

3. For plane stress analysis using two-dimensional elements, the surface of the mesh must be on which plane of a coordinate system?

4. Which type of two-dimensional solid element is most suitable for members that are thin and whose loads act only in the plane?

5. Why is it important to recognize the symmetrical nature of design in performing FEA analysis?

6. List and describe the advantages and disadvantages of the P-element method over the H-element method.

7. Why is it important to not just concentrate on the stress results but also examine the displacements of FEA results?

8. In the tutorial, did the change in the refinement of the P-elements help the stress results? By how much?

9. What does the ***Edge Distribution*** mesh option allows us to do?

Exercises

Determine the maximum stress produced by the loads.

1. 1000 lbf of load is applied to the right vertical edge with the left edge being fixed.
   Material: Steel Plate
   Thickness: 0.25 inches

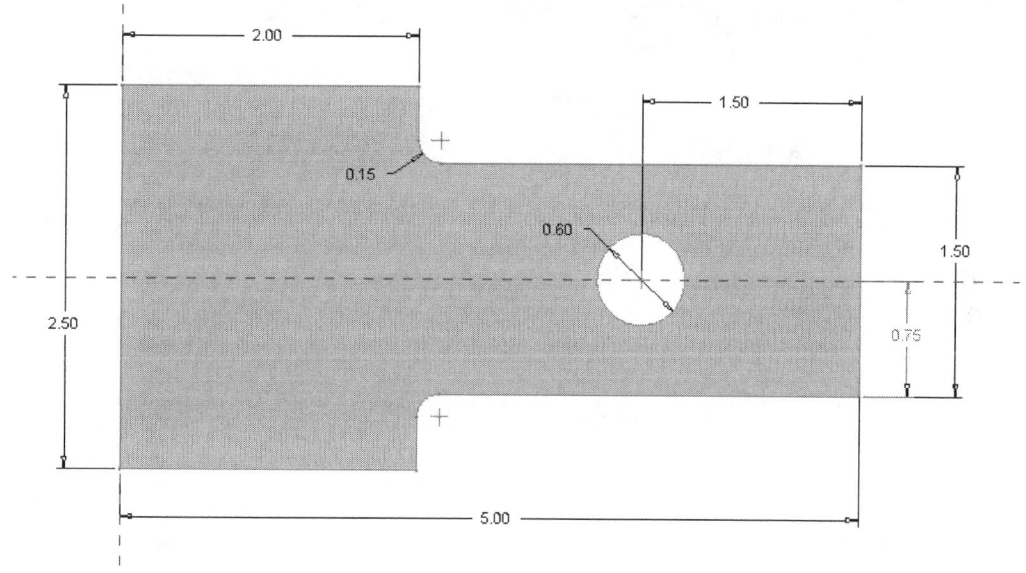

2. Material: Steel Plate
   Thickness: 25 mm

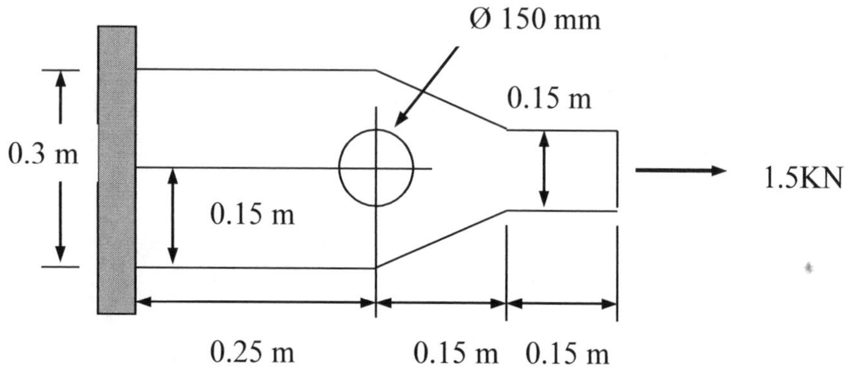

3. A force of 800 N is applied to the handle of the wrench.
   Material: Steel
   Thickness: 7 mm

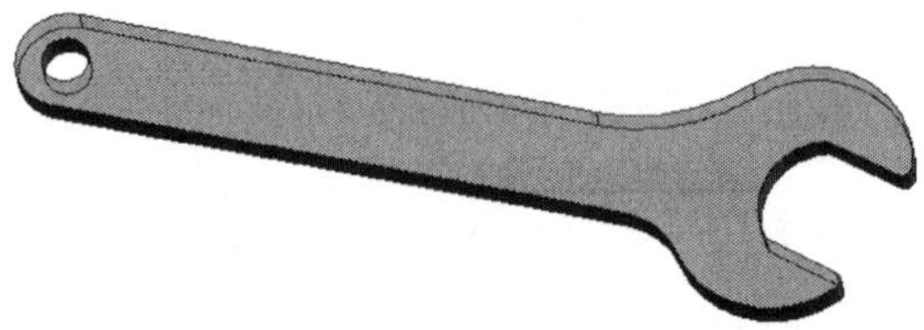

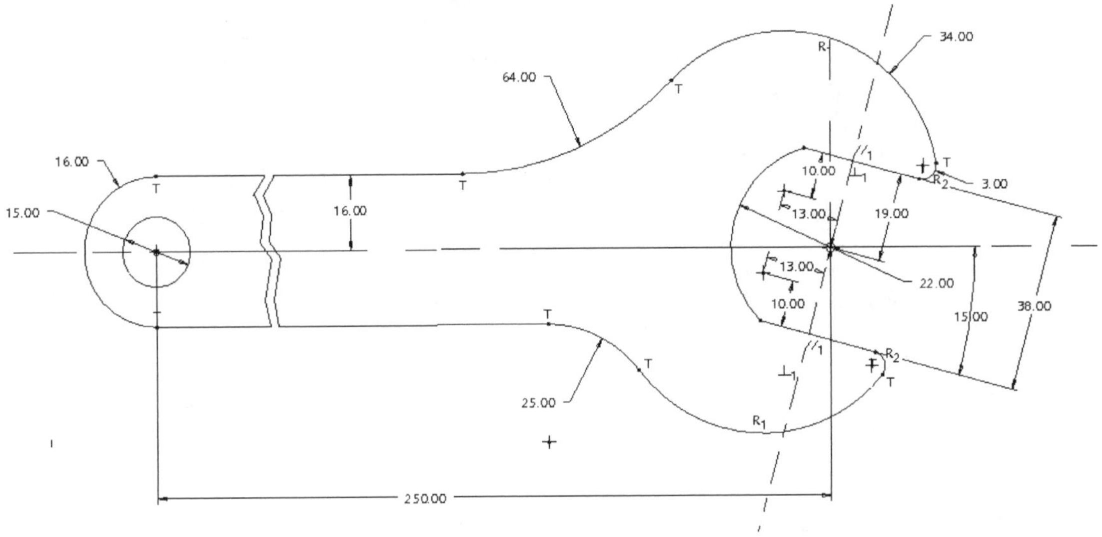

Chapter 10
Three-Dimensional Solid Elements

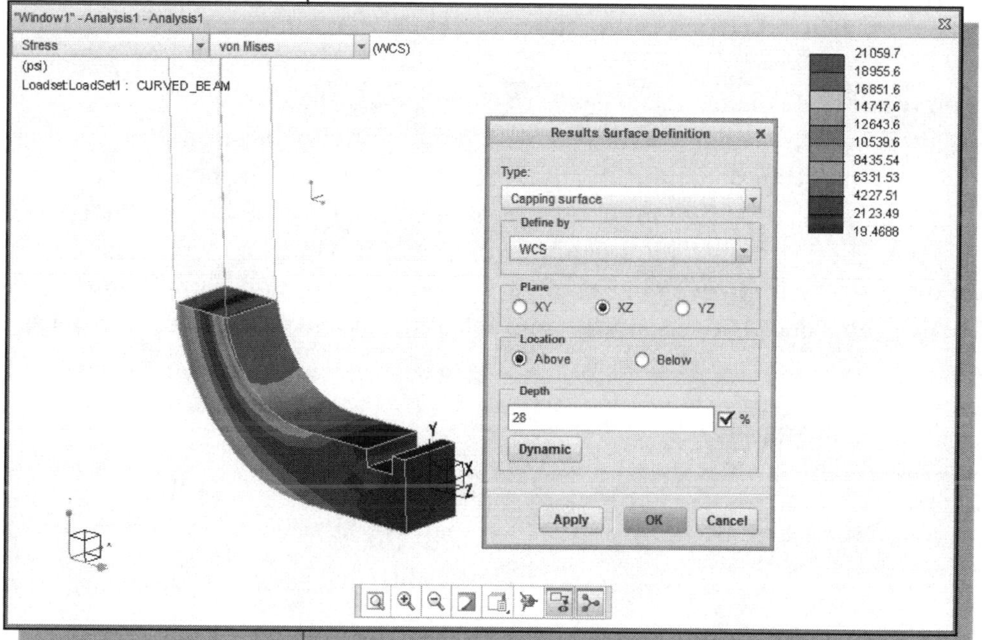

Learning Objectives

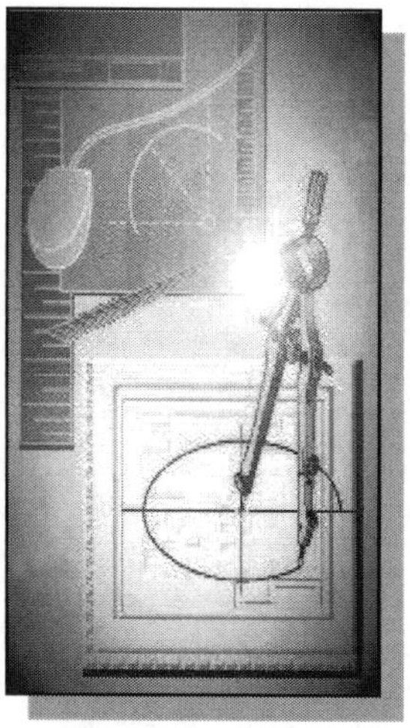

- Perform 3-D Finite Element Analysis.
- Understand the concepts of Failure Criteria.
- Create 3-D Solid models and FE models.
- Use the Creo Simulate Geometry Based Analysis.
- Use the different display options to display Stress Results.

Introduction

In this chapter, the general FEA procedure for using three-dimensional solid elements is illustrated. A finite element model using three-dimensional solid elements may look the most realistic as compared to the other types of FE elements. However, this type of analysis also requires more elements, which implies more mathematical equations and therefore more computational resources and time.

The main objective of finite element analysis is to calculate the stresses and displacements for specified loading conditions. Another important objective is to determine if *failure* will occur under the effect of the applied loading. It should be pointed out that the word *failure* as used for *failure criteria* is somewhat misleading. In the elastic region of the material, the system's deformation is recoverable. Once the system is stressed beyond the elastic limit, even in a small region of the system, deformation is no longer recoverable. This does not necessarily imply that the system has failed and cannot carry any further load.

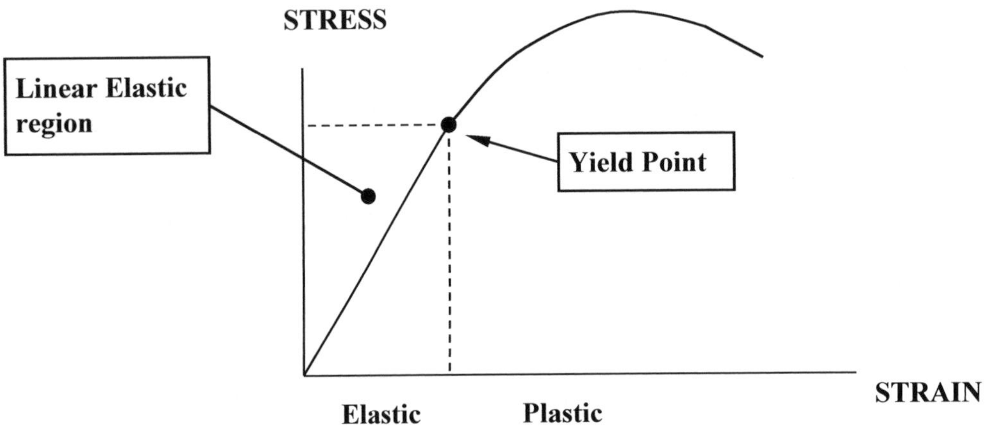

Stress-Strain diagram of typical ductile material

Several theories propose different failure criteria. In general, all these theories provide fairly similar results. The most widely used failure criteria are the **Von Mises yield criterion** and the **Tresca yield criterion**. Both the *Von Mises* and *Tresca* stresses are scalar quantities, and when compared with the yield stress of the material they indicate whether a portion of the system has exceeded the elastic state.

Von Mises Criterion: (Note that σ_1, σ_2, σ_3 are the three principal stresses.)

$$2\,\sigma_{yp}^2 = (\sigma_1 - \sigma_2)^2 + (\sigma_2 - \sigma_3)^2 + (\sigma_3 - \sigma_1)^2$$

Tresca Yield Criterion:

$$\frac{1}{2}\,\sigma_{yp} = \frac{1}{2}\,(\sigma_1 - \sigma_3)$$

This chapter illustrates the general FEA procedure of using three-dimensional solid elements. The creation of a solid model is first illustrated and *solid elements* are generated using the *Creo Simulate* AutoGEM command. In theory, all designs could be modeled with three-dimensional solid elements. The three-dimensional solid element is the most versatile type of element compared to the more restrictive one-dimensional or two-dimensional elements. The procedure involved in performing a three-dimensional solid FEA analysis is very similar to that of a two-dimensional solid FEA analysis, as was demonstrated in Chapter 9. As one might expect, the number of node-points involved in a typical three-dimensional solid FEA analysis is usually much greater than that of a two-dimensional solid FEA analysis.

Problem Statement

Determine the maximum normal stress in the AL6061 member shown; the c-link design is assembled to the frame at the upper hole (Ø 0.25) and a vertical load of 200 lbs. is applied at the top of the notch on the lower arm as shown.

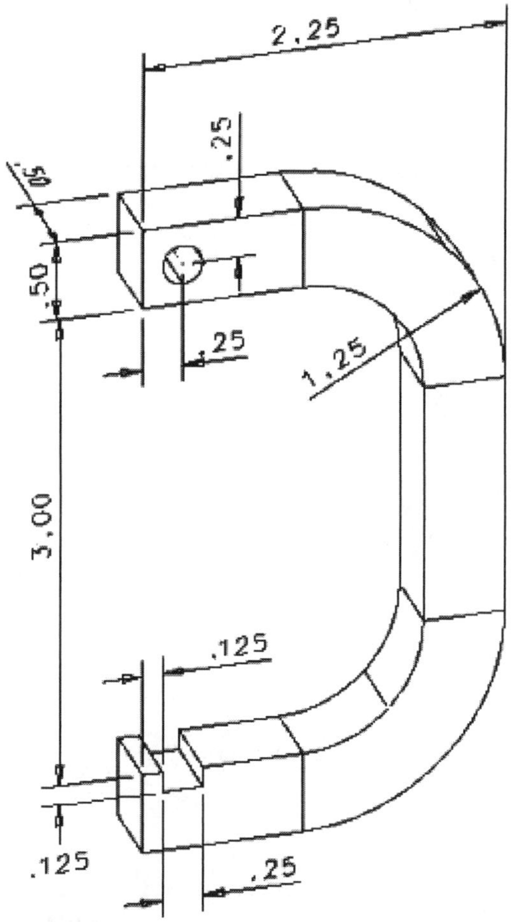

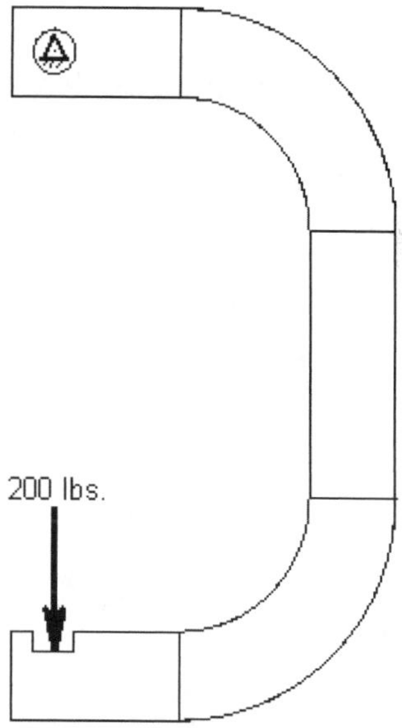

Preliminary Analysis

The analysis of stresses due to bending moment has been restricted to straight members as illustrated in Chapter 6 through Chapter 8. Note that a good approximation may be obtained if the curvature of the member is relatively small in comparison to the cross section of the member. However, when the curvature is large, the stress distribution is no longer linear; it becomes hyperbolic. This is due to the fact that in a curved member, the neutral axis of a transverse section does not pass through the centroid of that section.

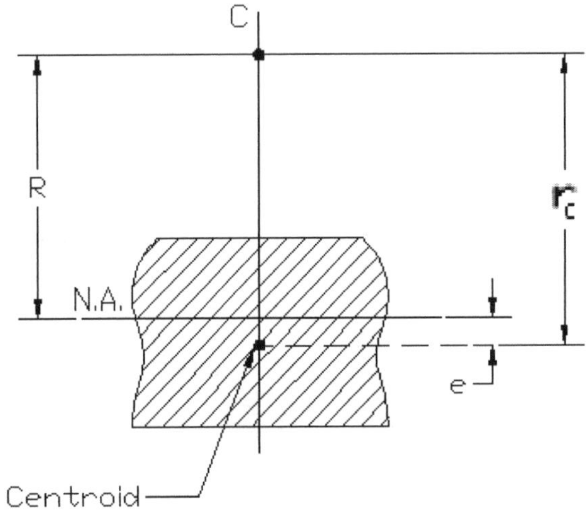

The distance R from the center of curvature C to the neutral axis is then defined by the equation:

$$R = \frac{A}{\int \frac{dA}{r_c}}$$

(A is the cross-sectional area.)

For some of the more commonly used shapes, the R values are as shown:

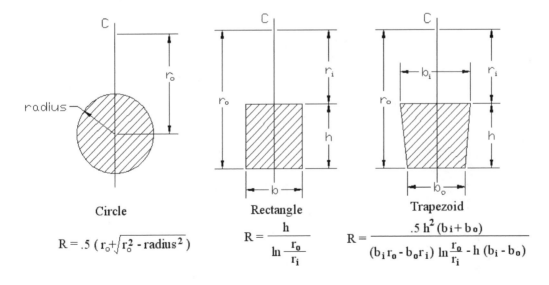

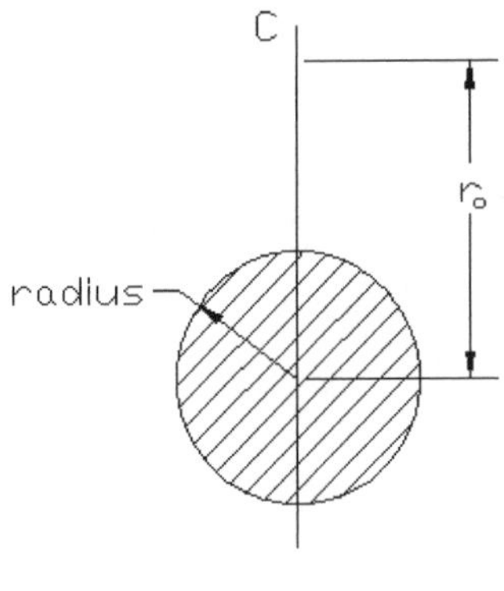

Circle

$$R = .5\,(r_o + \sqrt{r_o^2 - \text{radius}^2})$$

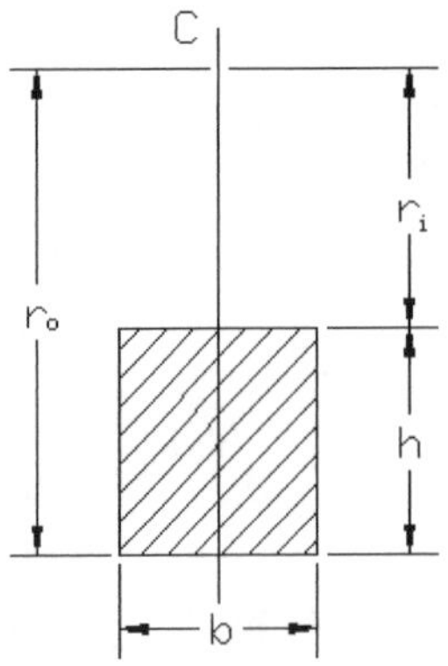

Rectangle

$$R = \frac{h}{\ln \frac{r_o}{r_i}}$$

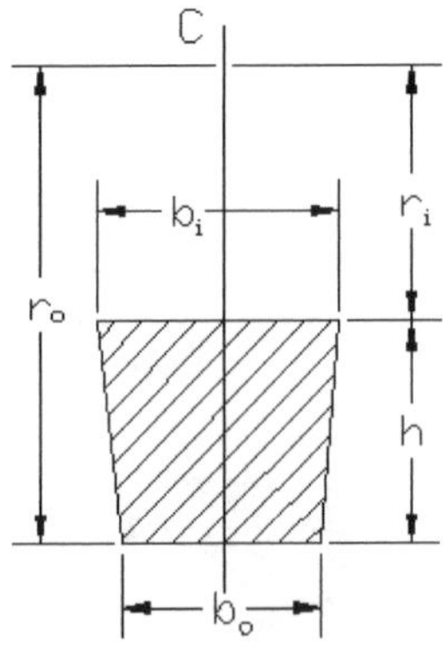

Trapezoid

$$R = \frac{.5\,h^2\,(b_i + b_o)}{(b_i r_o - b_o r_i)\ln \frac{r_o}{r_i} - h\,(b_i - b_o)}$$

The bending stress in a **curved beam**, at location **r** measured from the center of curvature **C**, can be expressed as:

$$\sigma_r = \frac{M(R-r)}{Aer}$$

For our analysis, we can establish an equivalent load system in the upper portion of the member, through the centroid of the cross section, as shown.

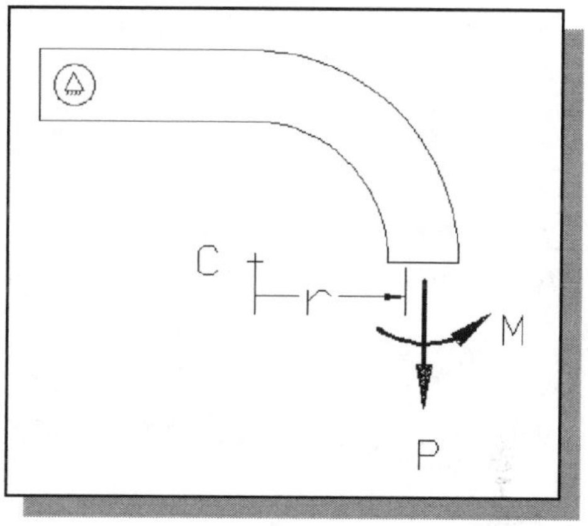

P = 200 lbs.

M = 350 in-lb.

The normal stress at the cross section is a combination of the stresses from the normal force P and the bending moment M. The stress distributions are as shown in the figure below.

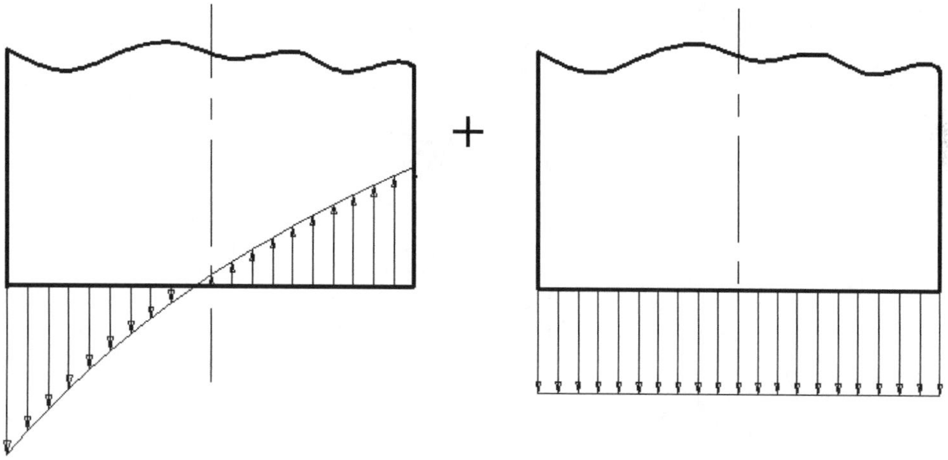

The maximum normal stress occurs on the inside edge, the edge that is closer to the center of the curvature C.

The normal stress from the normal force P:

$$\sigma_{normal\_force} = \frac{P}{A} = \frac{200\ lb}{.25\ in^2} = 800\ psi$$

The bending stress component:

$$\sigma_{\text{bending moment}} = \frac{\mathbf{M(R\text{-}r)}}{\mathbf{Aer}}$$

To calculate the bending stress, we will first calculate the **R** and **e** values:

$$\mathbf{R} = \frac{h}{\ln \frac{r_o}{r_i}} = .5/\,(\ln(1.25/.75)) = 0.9788$$

$$e = 1 - 0.9788 = 0.0212$$

The maximum bending stress, which is at the inside edge (r = 0.75), can now be calculated:

$$\sigma_{\text{bending moment}} = 350\,(0.9788 - .75)\,/\,(0.25 \times .0212 \times .75) = 20146 \text{ psi}$$

Therefore

$$\sigma_{\text{maximum}} = \sigma_{\text{normal force}} + \sigma_{\text{bending moment}} = 800 + 20146 = 20946 \text{ psi}$$

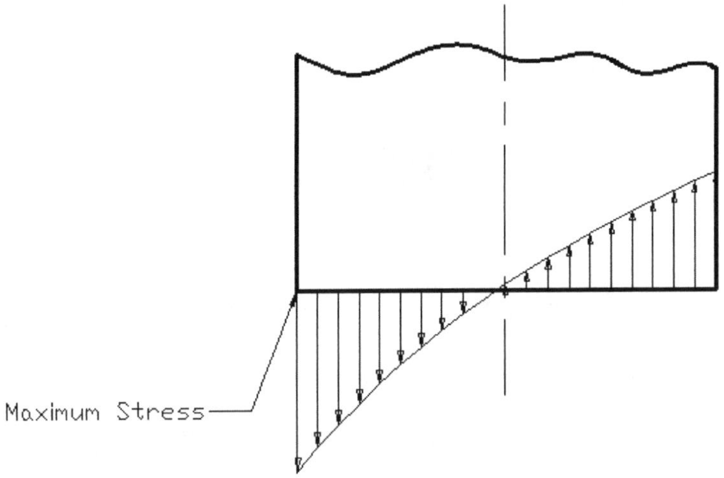

The normal stress at the outer edge can also be calculated:

$$\sigma_{\text{bending @ 1.25}} = 350\,(0.9788 - 1.25)\,/\,(0.25 \times .0212 \times 1.25) = -14327 \text{ psi}$$

Therefore

$$\sigma_{@1.25} = \sigma_{\text{normal force}} + \sigma_{\text{bending moment}} = 800 - 14327 = -13527 \text{ psi}$$

Starting Creo Parametric

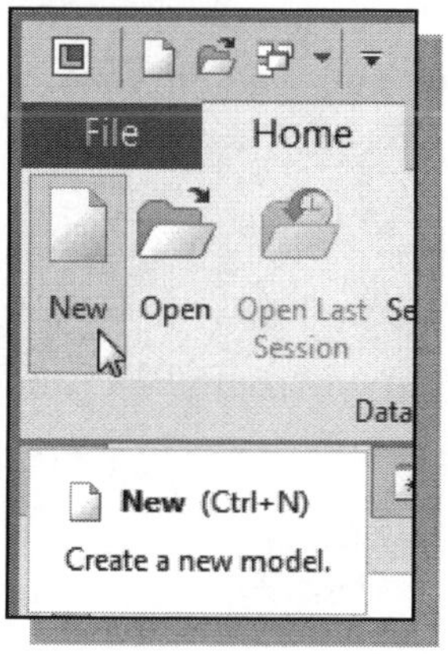

1. Select the **Creo Parametric** option on the *Start* menu or select the **Creo Parametric** icon on the desktop to start *Creo Parametric*. The *Creo Parametric* main window will appear on the screen.

2. Click on the **New** icon, located in the *Ribbon toolbar* as shown.

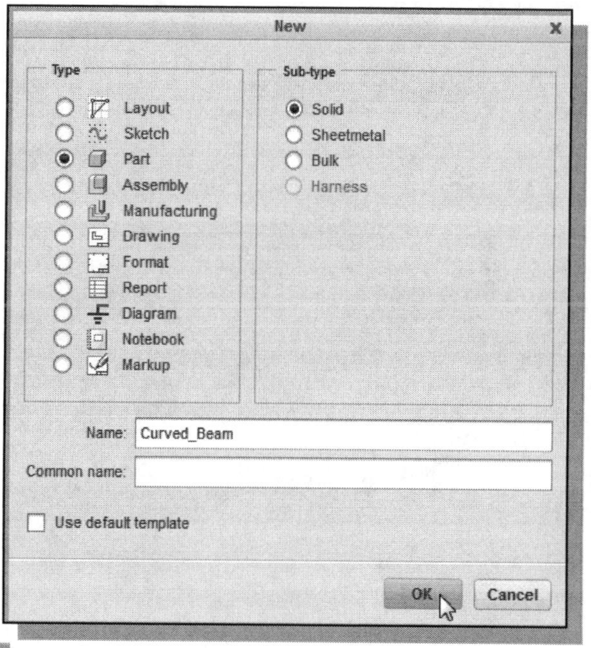

3. In the *New* dialog box, confirm the model's Type is set to **Part** (**Solid** Sub-type).

4. Enter **Curved_Beam** as the part Name as shown in the figure.

5. Turn *off* the **Use default template** option.

6. Click on the **OK** button to accept the settings.

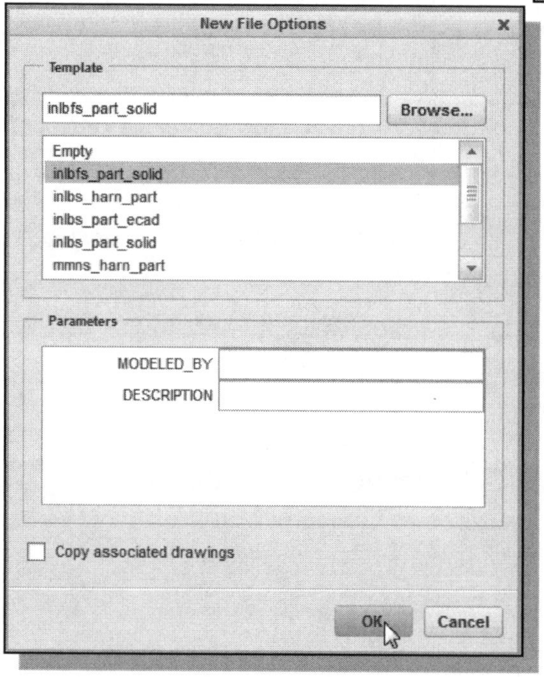

7. In the *New File Options* dialog box, select **inlbfs_part_solid** in the option list as shown.

8. Click on the **OK** button to accept the settings and enter the *Creo Parametric Part Modeling* mode.

Create a CAD Model in Creo Parametric

For our CAD model, we will create a swept feature. A ***swept*** operation allows us to move a planar section through a path in space to form a three-dimensional object. The path of the sweep can be a straight line or a curve.

➢ Define the Sweep Trajectory

The *sweep trajectory* is the *sweep path* that the *sweep section* is moved along. We will create a datum sketch for this.

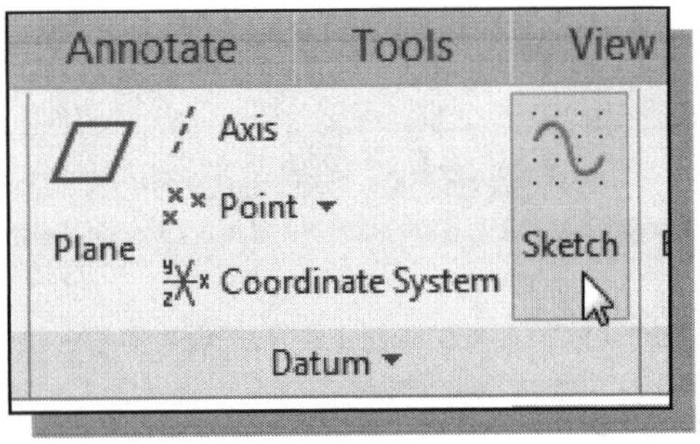

1. Click **Datum Sketch** in the *Ribbon* in the main *Creo* window as shown.

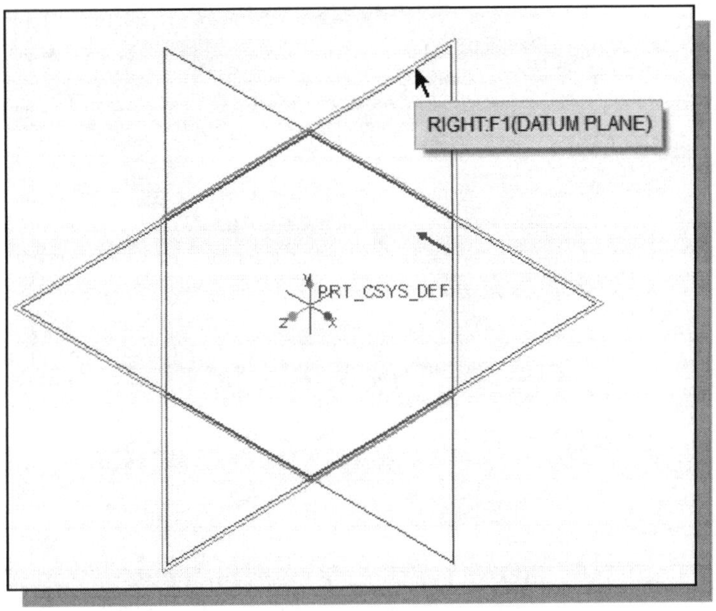

2. In the graphics area, select **RIGHT** by clicking on the plane as shown.

❖ Notice an arrow appears on the edge of RIGHT. The arrow direction indicates the viewing direction of the sketch plane.

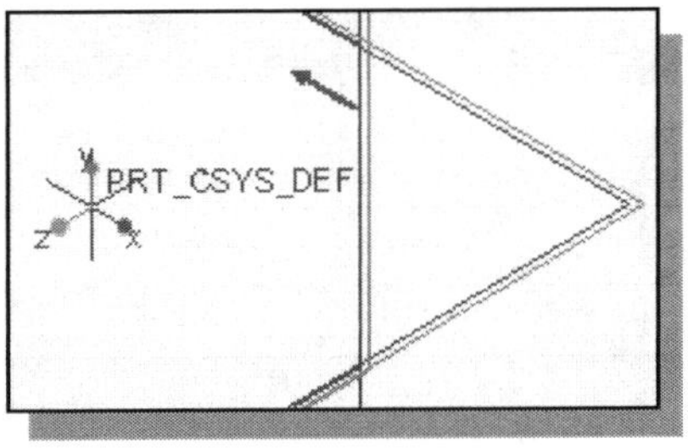

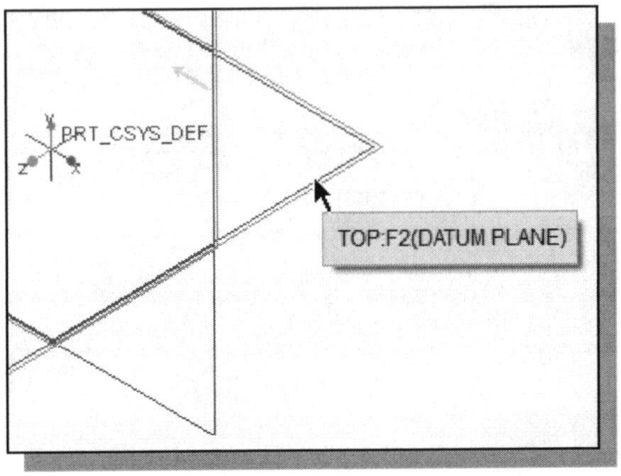

3. In the graphics area, confirm the **TOP** datum plane is set as the Orientation reference.

4. Click **Sketch** to enter the *Creo Parametric Sketcher* mode.

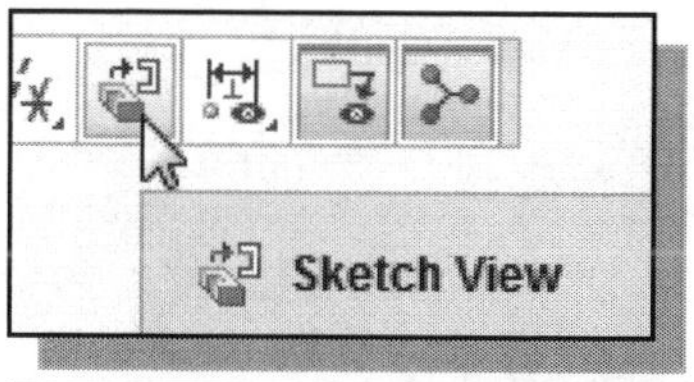

5. Click **Sketch View** to orient the sketching plane parallel to the screen.

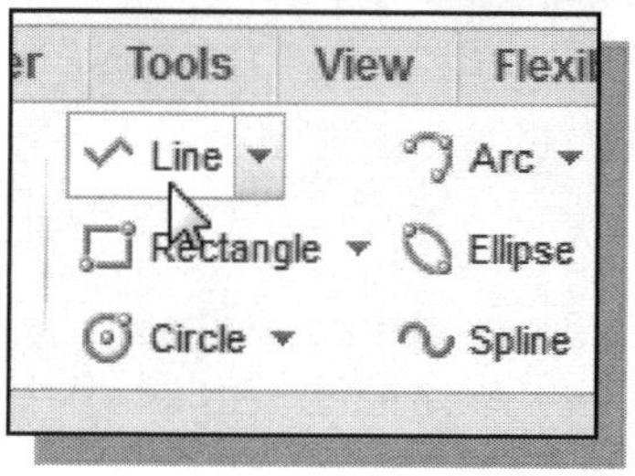

6. Pick **Line** in the *Sketching* toolbar.

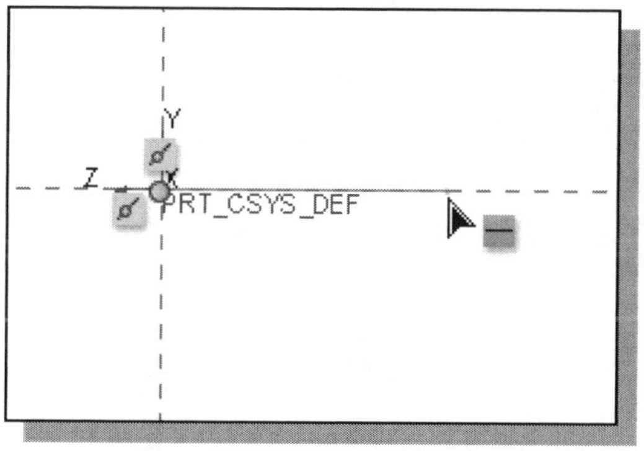

7. Click near the **intersection of the two references** as the first point of the line segment.

8. Move the cursor toward the **right** and create a **horizontal line**, aligned to the horizontal reference as shown.

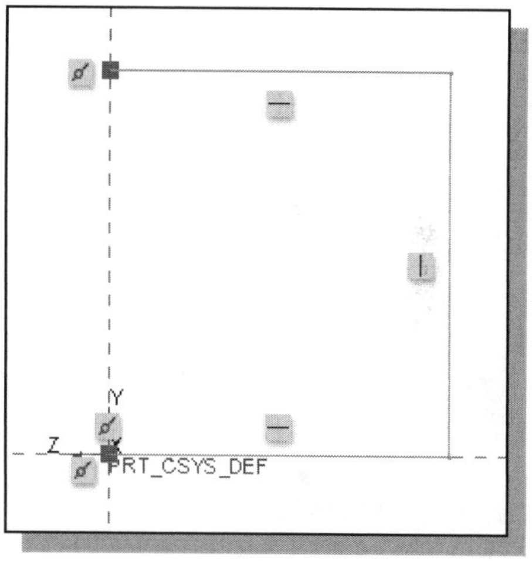

9. On your own, continue with the Line command and create a vertical and another horizontal line segment as shown.

10. Inside the graphics area, click once with the **middle-mouse-button** to end the Line command.

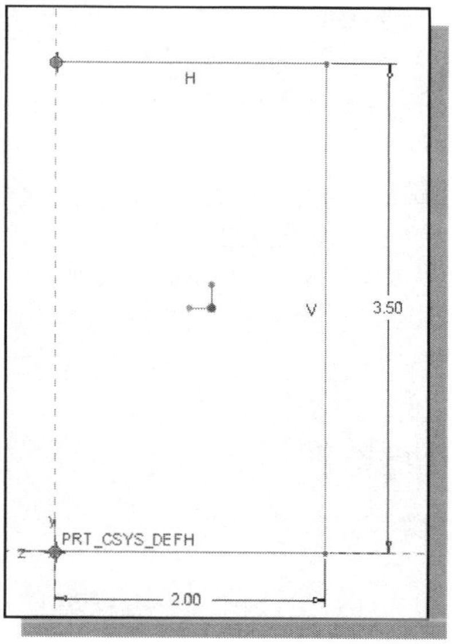

11. On your own, use the **Modify** command to edit the two dimensions as shown in the figure.

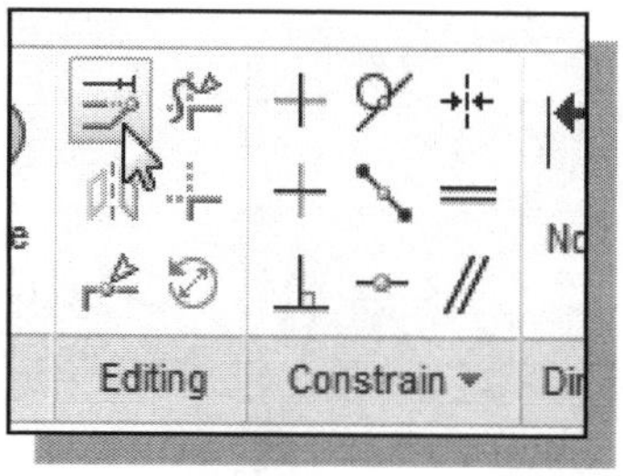

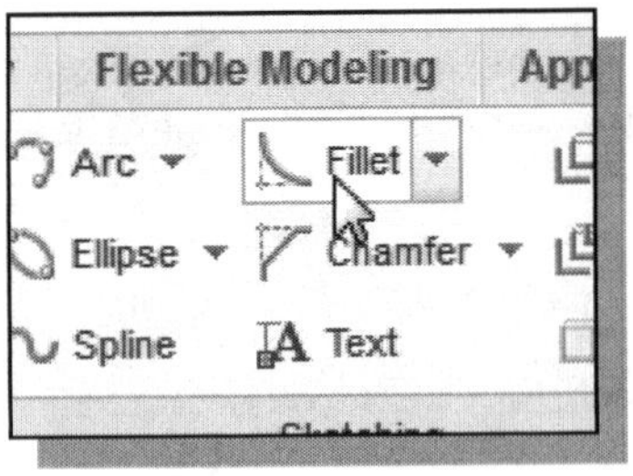

12. Pick **Circular Fillet** in the *Sketching* toolbar to create fillets.

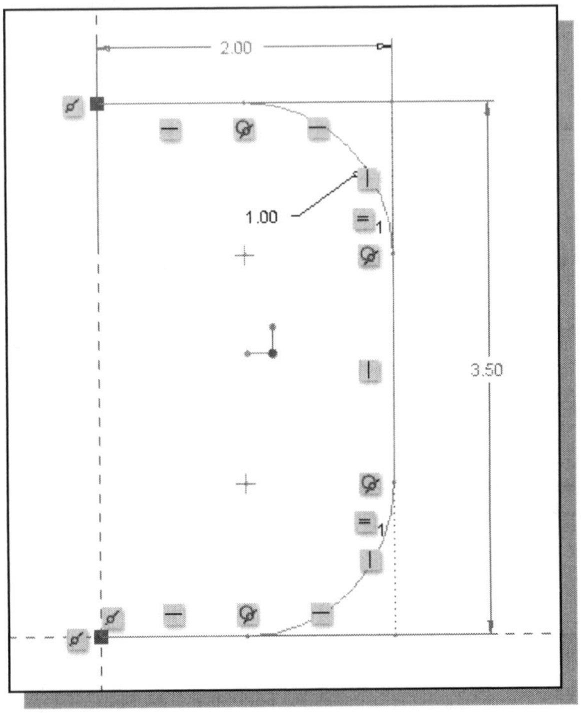

13. On your own, create the two fillets, **Radius 1.0**, at the two corners as shown in the figure.

14. On your own, adjust the three dimensions so that they appear as shown in the figure.

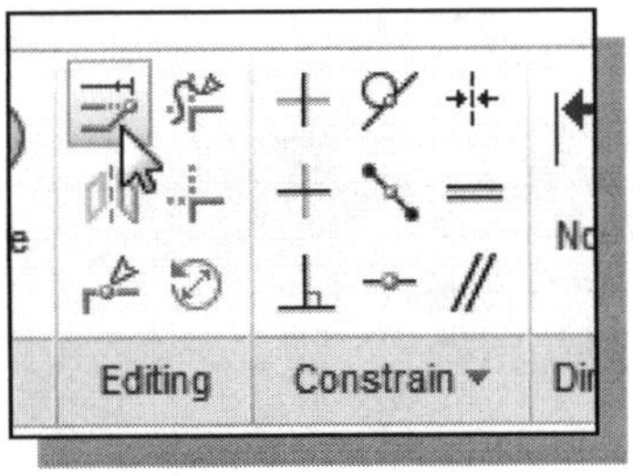

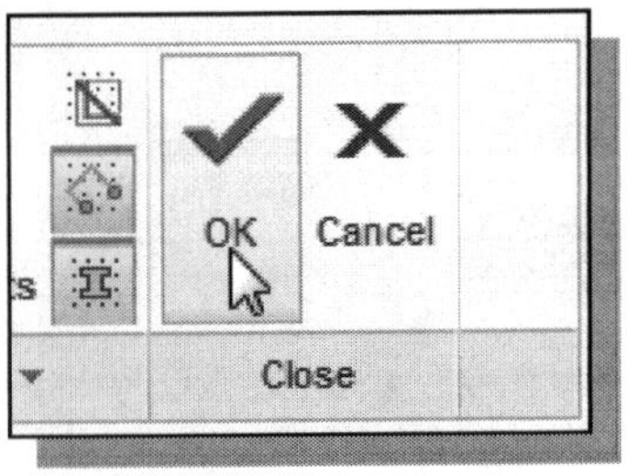

15. Click on the **OK** button to accept the creation of the sketch.

❖ Note that we just completed the *Sketched Trajectory*. We can now begin to create a **Swept** feature.

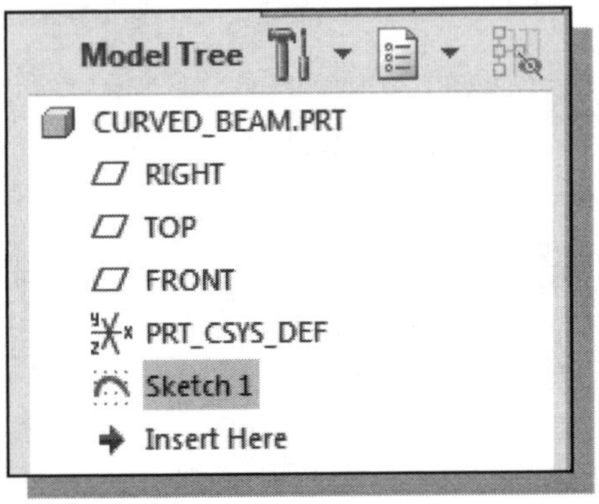

16. Confirm the *datum sketch* is highlighted in the *Model Tree* window as shown.

➢ The preselected sketch will be used as the *sweep trajectory* for the swept feature.

The Sweep Command

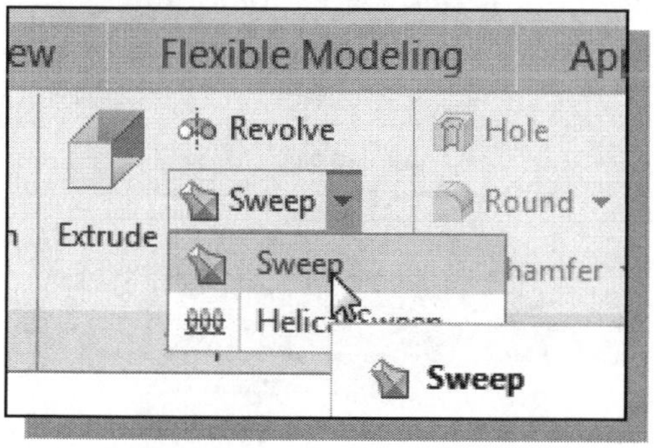

1. In the *Shapes* toolbar, select **Sweep** as shown.

❖ Two elements are required to create the swept feature: Trajectory and Section. The first element to define is the trajectory, which can use a datum sketch or reference existing geometry.

➢ Define the Sweep Section

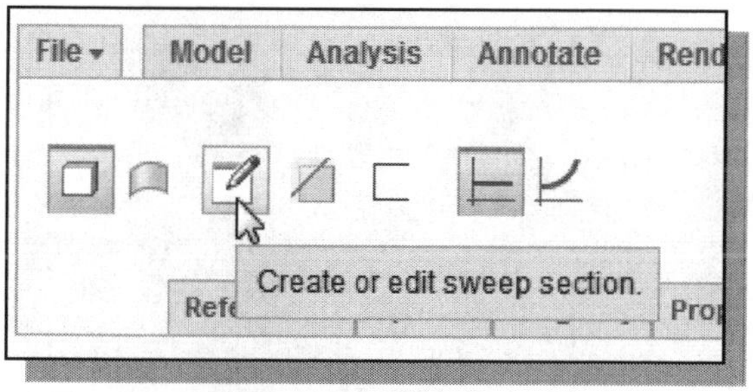

1. Click **Create or edit sweep section** as shown.

2. Click **Sketch View** to orient the sketching plane parallel to the screen.

➢ Note the origin is aligned to the start of the trajectory path.

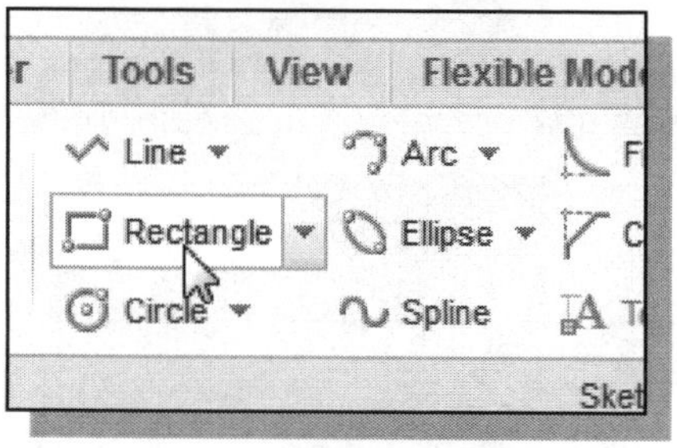

3. Click on the **Rectangle** button to activate the Create Rectangle command.

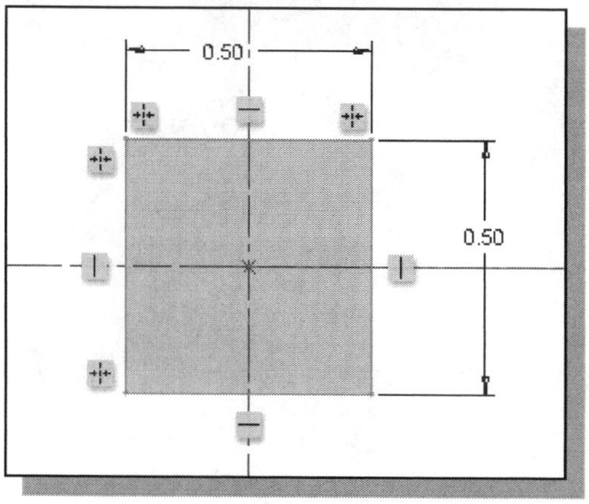

4. On your own, create the rectangle by placing the two corners as shown in the figure.
5. On your own, adjust the two dimensions so that the square is centered as shown in the figure.

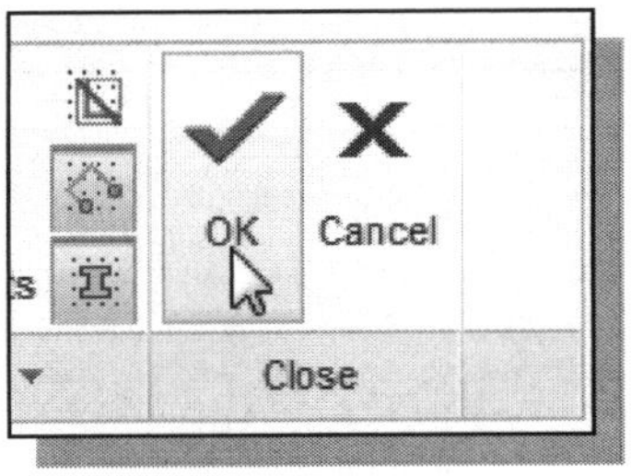

6. Click on the **OK** button to accept the creation of the sketch.

7. Click **Accept** to create the solid feature.

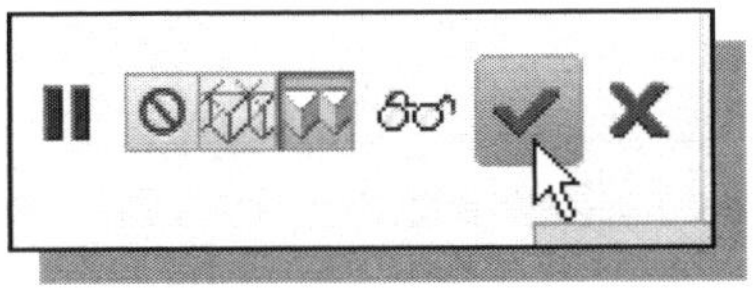

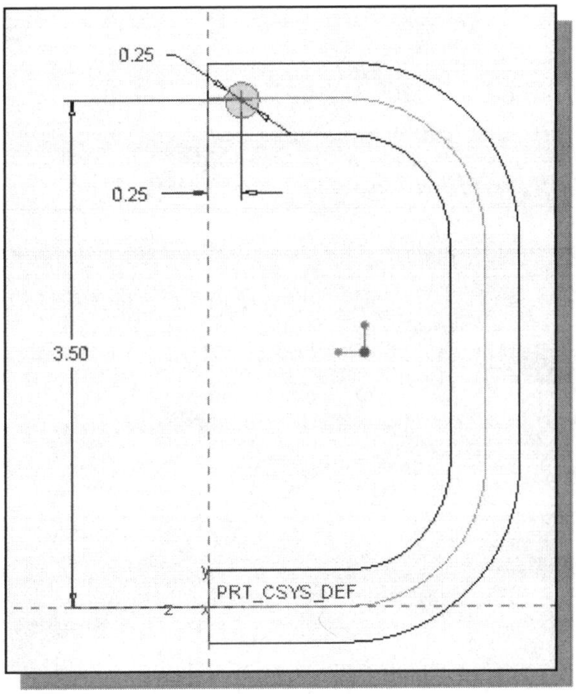

8. On your own, create a cut feature on the top section of the solid model as shown.

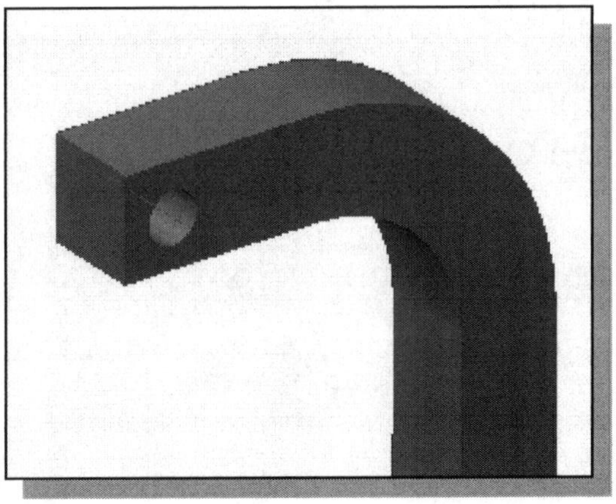

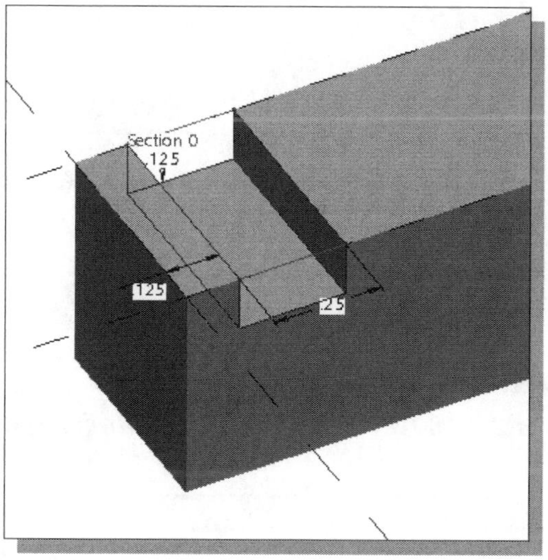

9. On your own, create another cut feature on the bottom section of the solid model as shown.

❖ Your solid model should appear as shown in the figure below.

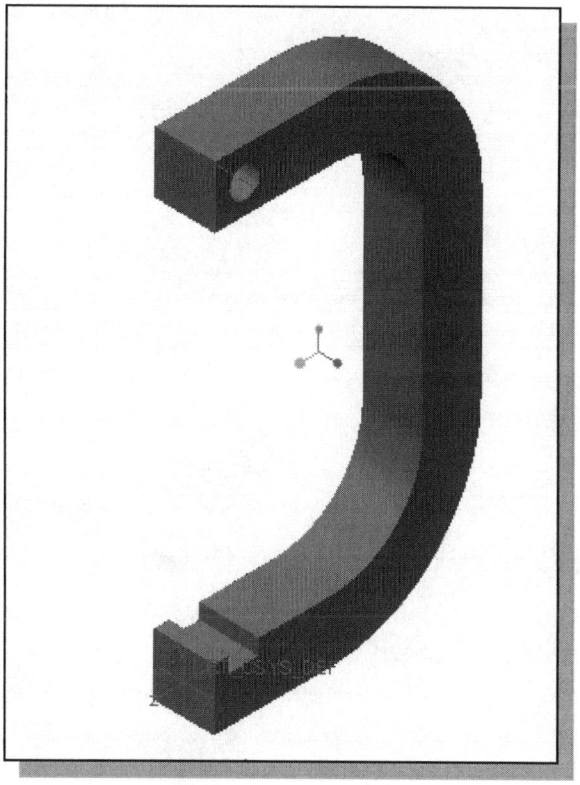

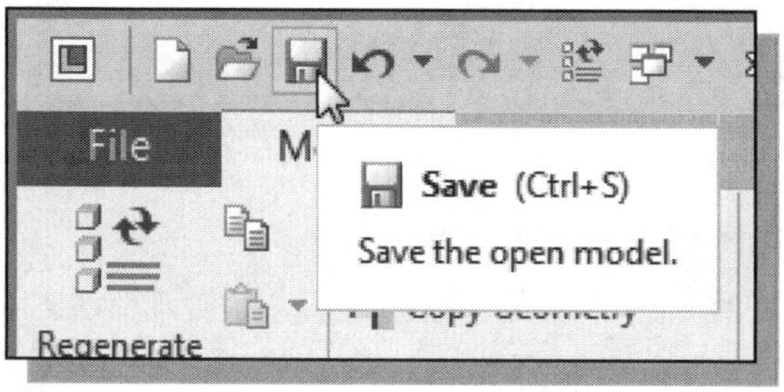

10. Click **Save** and save the CAD model to disk.

Select and Examine the Part Material Property

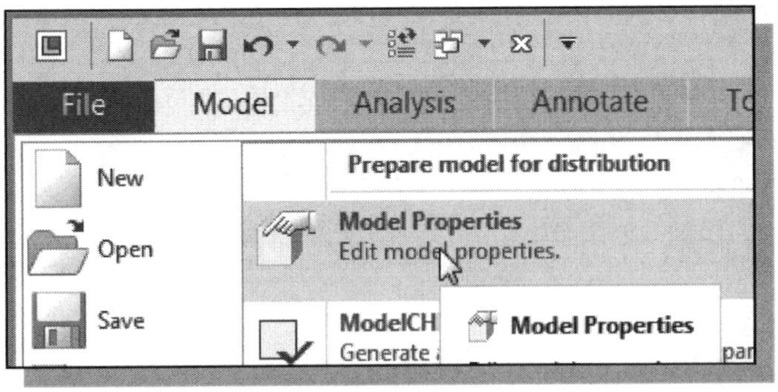

1. Choose **File →Prepare →Model Properties** from the *File pull-down menu.*

2. Select the **Change** option that is to the right of the **Material** option in the ***Model Properties*** window.

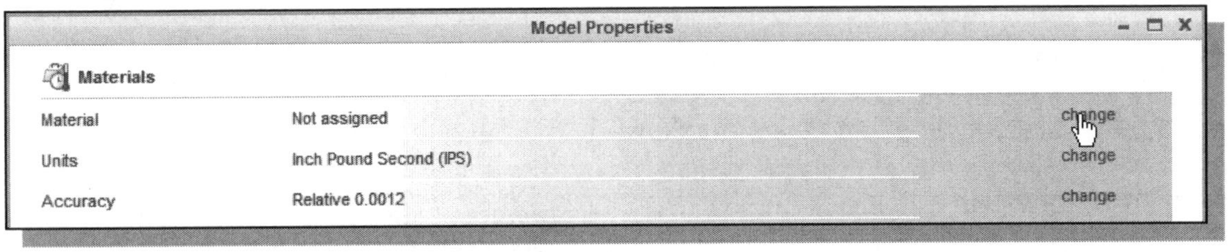

3. Select **Al-Mg-Si_alloy_Wrought** in the *Materials* list as shown.

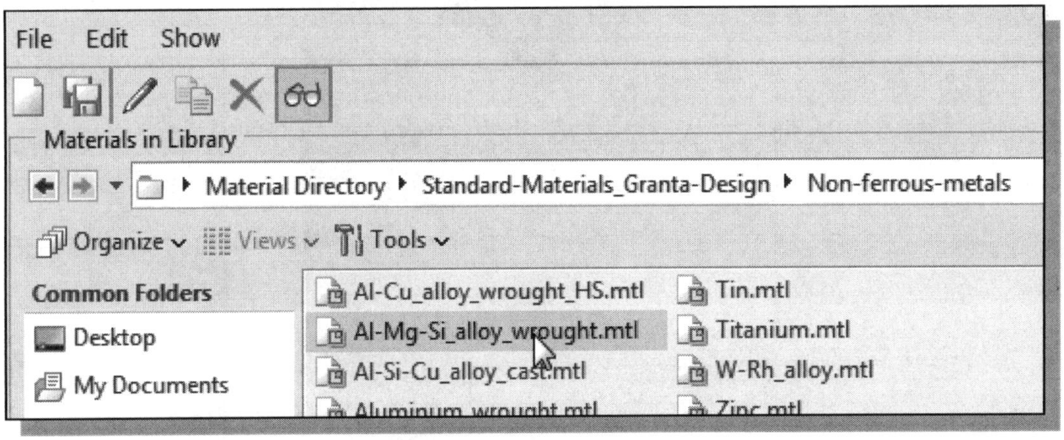

4. **Double-click** on the selected material definition to make the selected material available for use in the current FEA model.

5. On your own, use the *Edit* option to examine the material information and click **OK** to accept the settings.

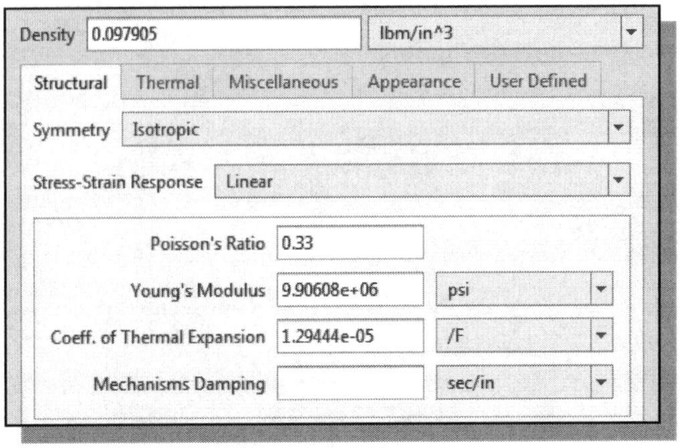

The Integrated Mode of Creo Simulate

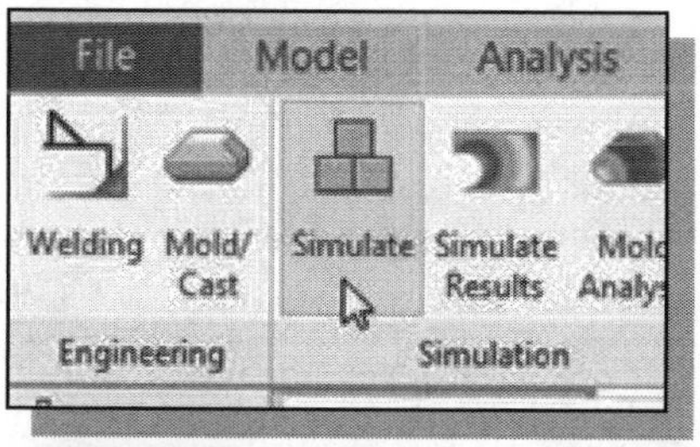

1. Start the *Integrated* mode of *Creo Simulate* by selecting the **Applications → Simulate** option in the *Ribbon* area as shown.

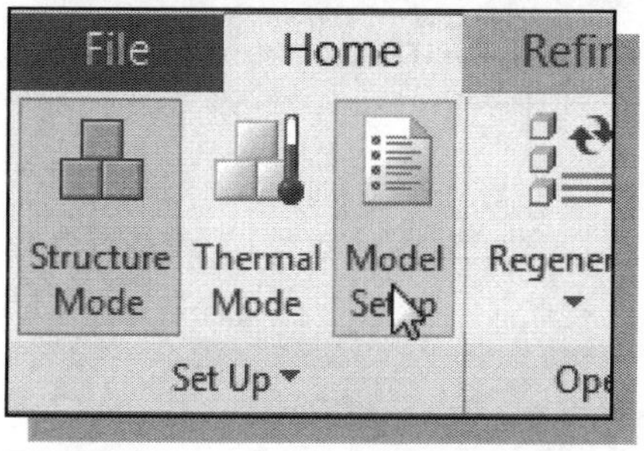

2. In the *Creo Simulate Ribbon* toolbar, confirm **Structure** is set as the analysis type.

3. Click **Model Setup** to examine the *Simulate* model settings.

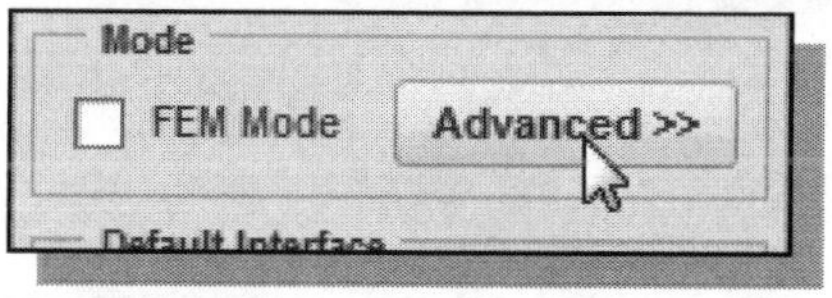

4. Click **Advanced** to show the additional settings that are available.

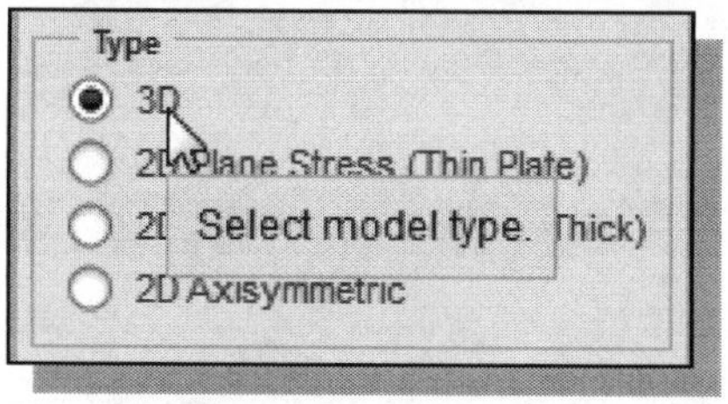

5. Confirm the selection of **3D** as the model Type as shown in the figure.

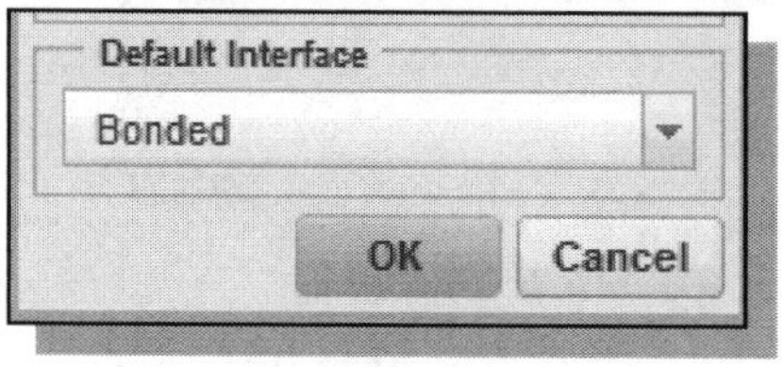

6. Confirm the default interface is set to **Bonded** as shown.

- The **Bonded** interface is used to connect components or surfaces that touch each other and act as if they are bonded and stuck together. Loads between bonded components are transferred through their common interface. Note other options, such as **Free** and **Contact**, are also available.

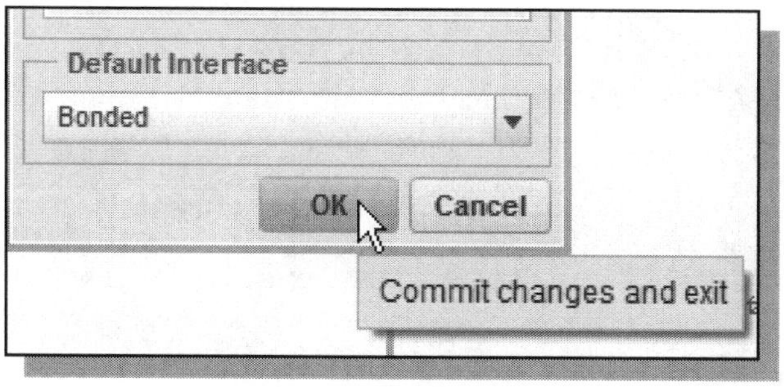

7. In the *Creo Simulate Model Setup* window, click on the **OK** button to accept the settings and proceed to create a 3D FEA model.

Apply the Boundary Conditions – Constraints

In the previous chapters, we created the elements before applying the constraints and loads. With *Creo Simulate*, we can also apply the constraints and loads onto the CAD model. This is known as the *Geometry Based Analysis* approach.

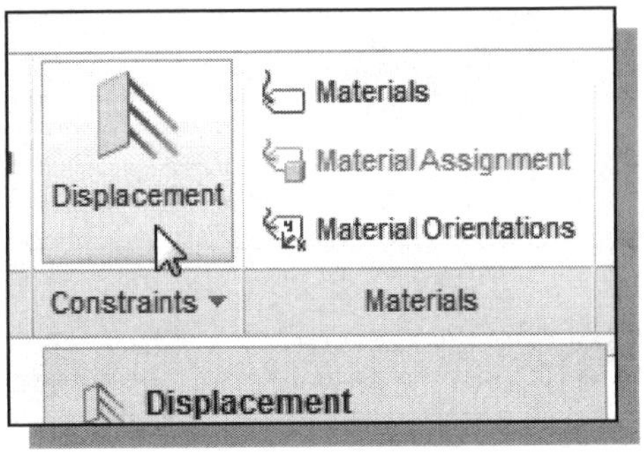

1. Choose **Displacement Constraint** by clicking the icon in the toolbar as shown.

2. Set the References to **Surfaces** and pick the **circular surface** of the top cylindrical section as shown.

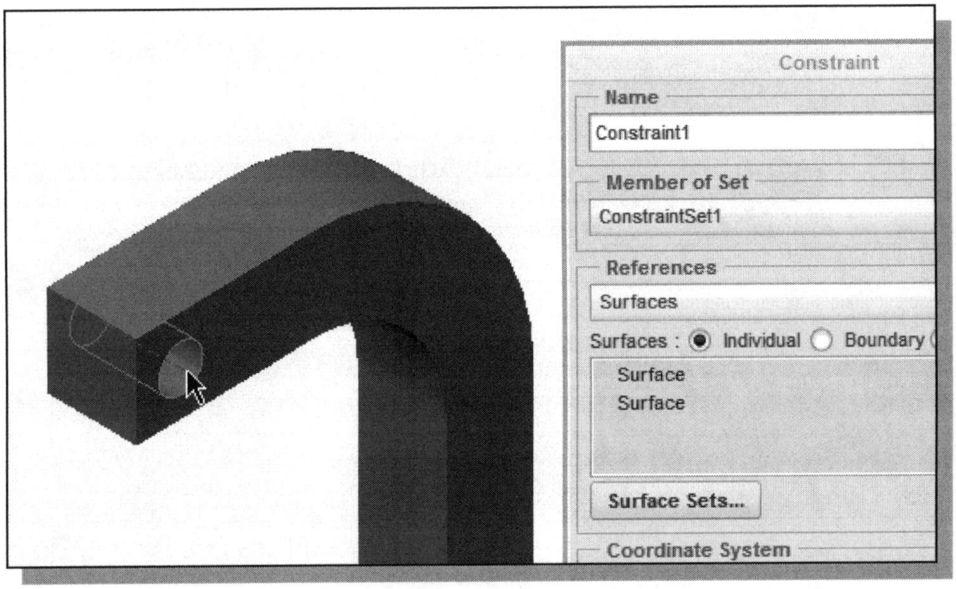

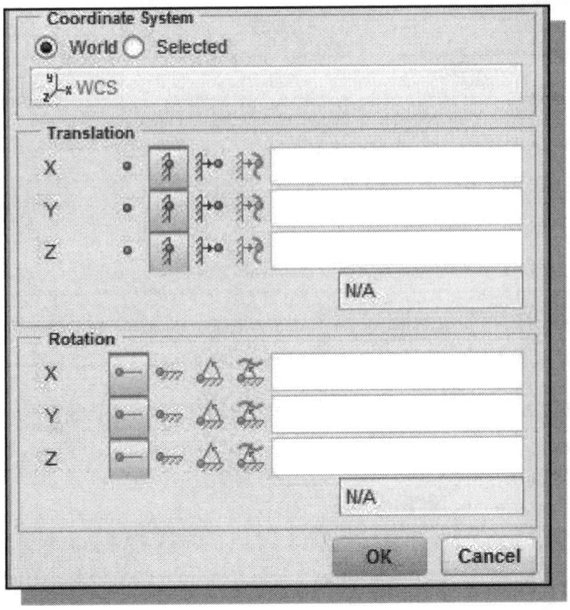

3. Confirm the Translation components are set to **Fixed** as shown. Note the Rotation options are not applicable for the selected surface.

4. Click on the **OK** button to accept the first displacement constraint settings.

Apply the External Loads

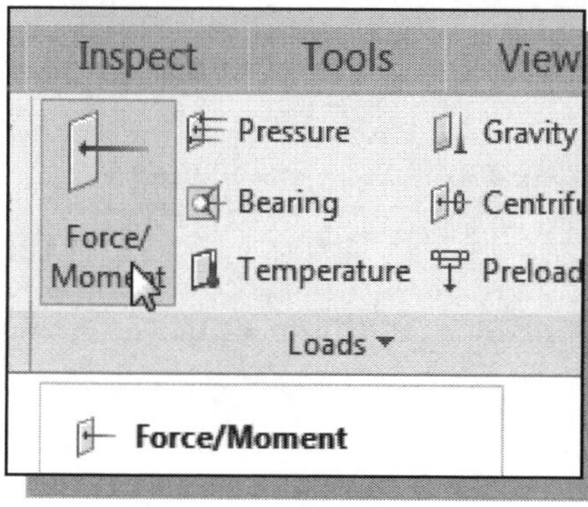

1. Choose **Force/Moment Load** by clicking the icon in the toolbar as shown.

2. Set the References option to **Surfaces**.

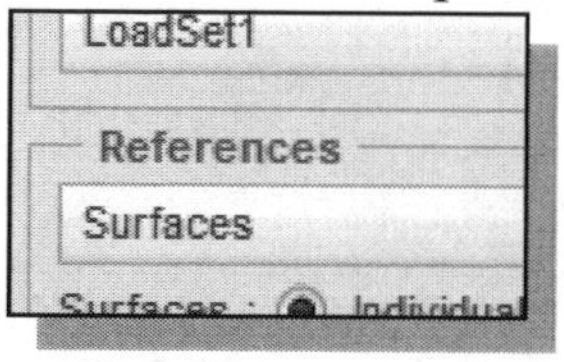

3. Pick the **small horizontal surface** of the bottom portion as shown.

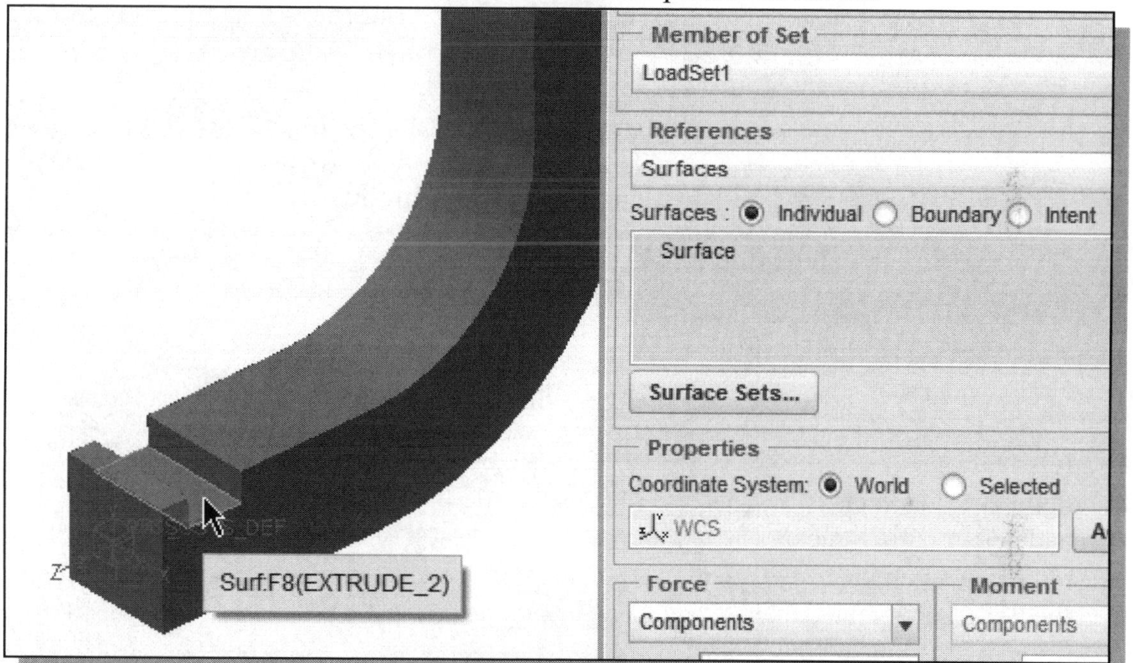

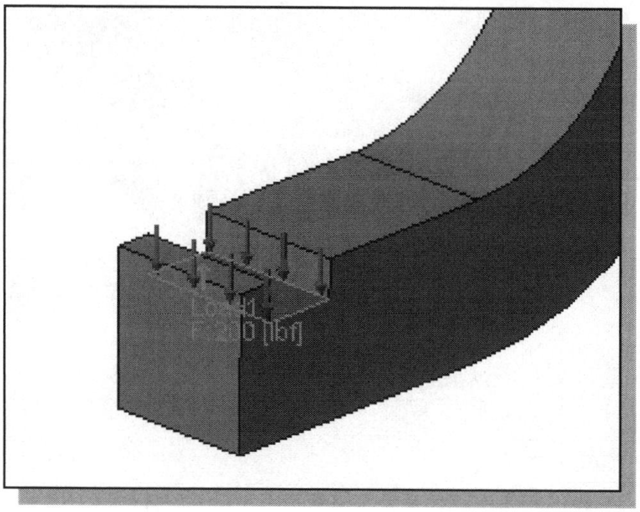

4. Enter **-200 lbf** in the Y component Force value box.

5. Click on the **Preview** button to confirm the applied load is pointing in the correct direction.

6. Click on the **OK** button to accept the first load settings

Create the 3D Mesh

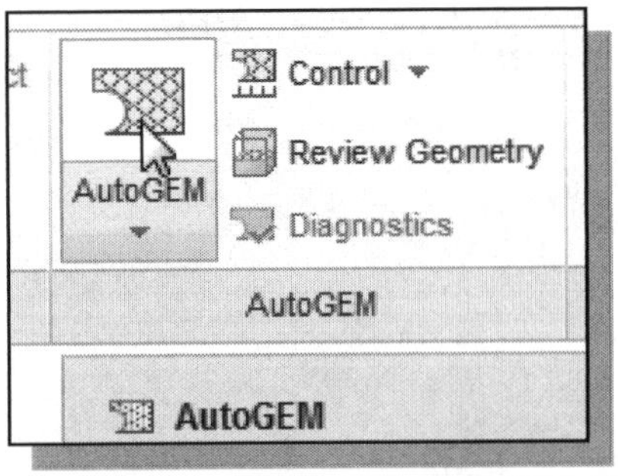

1. Choose **Refine Model** in the *Ribbon* toolbar.

2. Activate the **AutoGEM** command by selecting the icon in the toolbar area as shown.

3. Click **No** to close the retrieve mesh message.

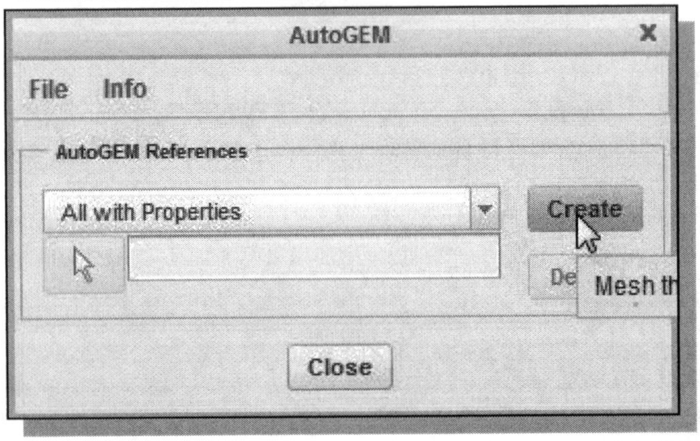

4. Confirm the AutoGEM References option is set to **All with Properties** as shown.

5. Inside the *AutoGEM* dialog box, click **Create** to generate the P-mesh.

➢ Note that *Creo Simulate* generated relatively small numbers of elements and nodes. This is the main characteristic of the P-element method, using a very coarse mesh for the analysis.

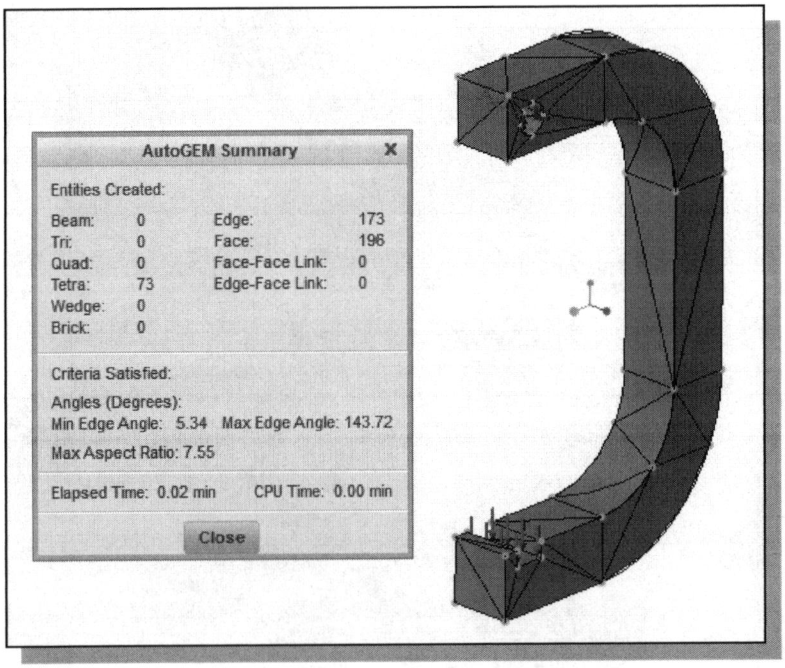

6. Exit the *AutoGEM* dialog box by clicking on the **Close** button.

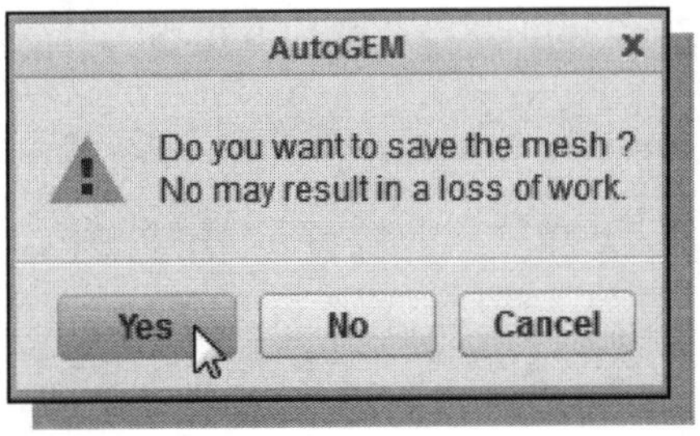

7. Click **Yes** to save the mesh.

Run the FEA Solver

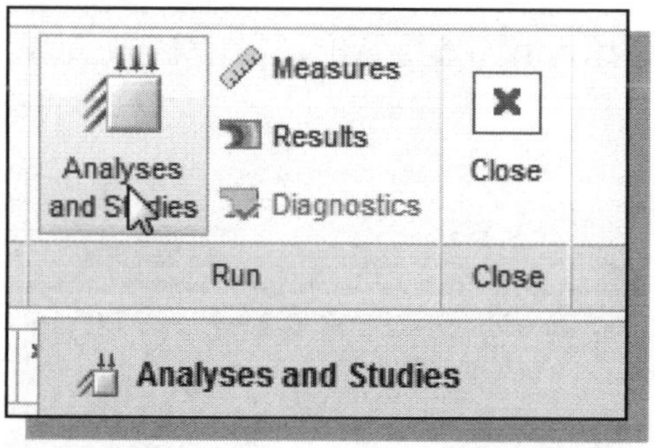

1. Choose **Home** in the *Ribbon* toolbar.

2. Activate the *Creo Simulate* **Analyses and Studies** command by selecting the icon in the toolbar area as shown.

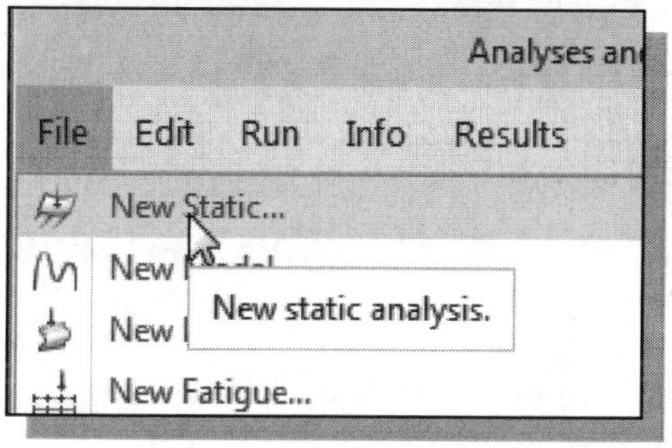

3. In the *Analyses and Design Studies* window, choose **File → New Static** in the pull-down menu. This will start the setup of a basic linear static analysis.

4. Note that **ConstraintSet1** and **LoadSet1** are automatically accepted as part of the analysis parameters.

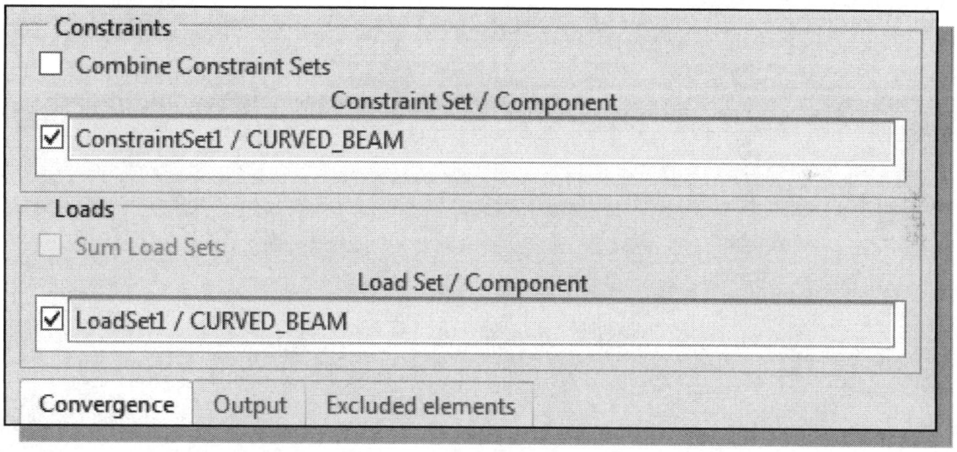

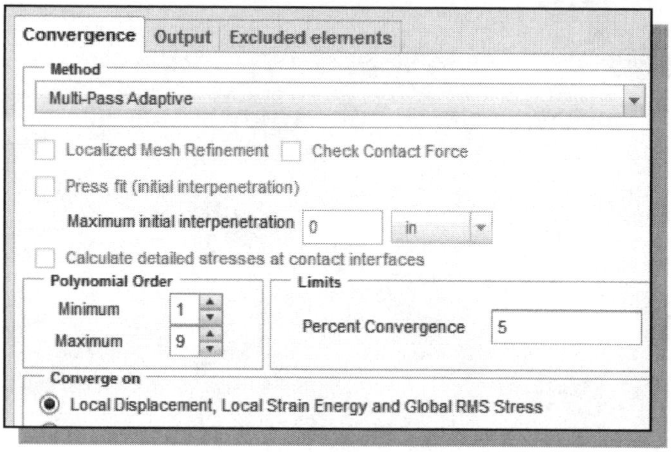

5. Set the Method option to **Multi-Pass Adaptive**, P-order to **9** and Limits to **5** Percent Convergence as shown.

6. Click **OK** to accept the settings.

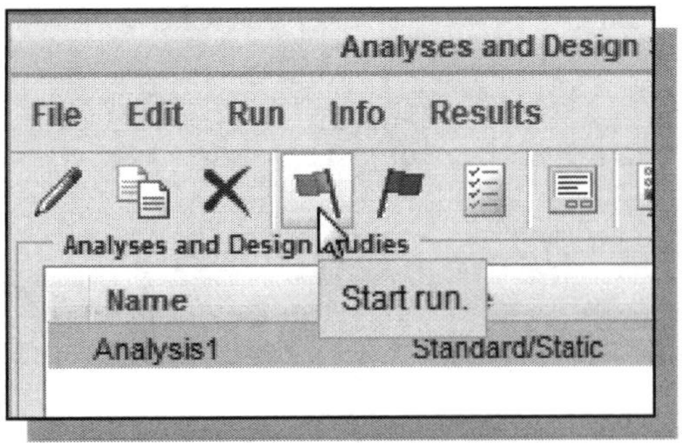

7. Click **Start run** to begin the *Solving* process.

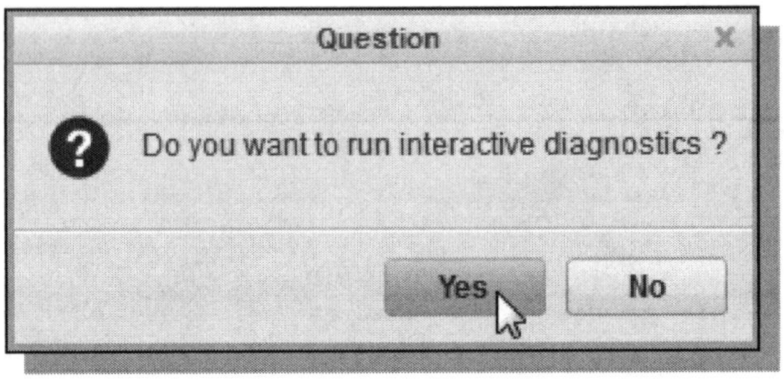

8. Click **Yes** to run interactive diagnostics; this will help us identify and correct problems associated with our FEA model.

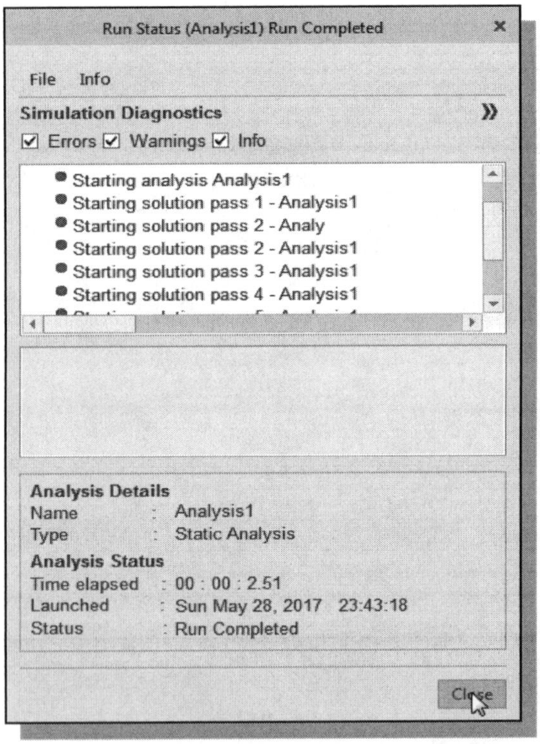

9. In the *Diagnostics* window, *Creo Simulate* will list any problems or conflicts of our FEA model.

10. Click on the **Close** button to exit the *Diagnostics* window.

11. In the *Analyses and Design Studies* window, choose **Display study status** as shown.

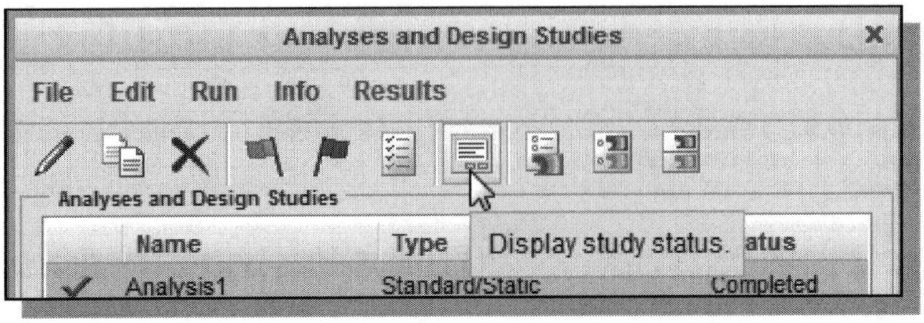

❖ Note that we started with 75 elements not converged and went up to 5th order polynomials to reach the 5% convergence.

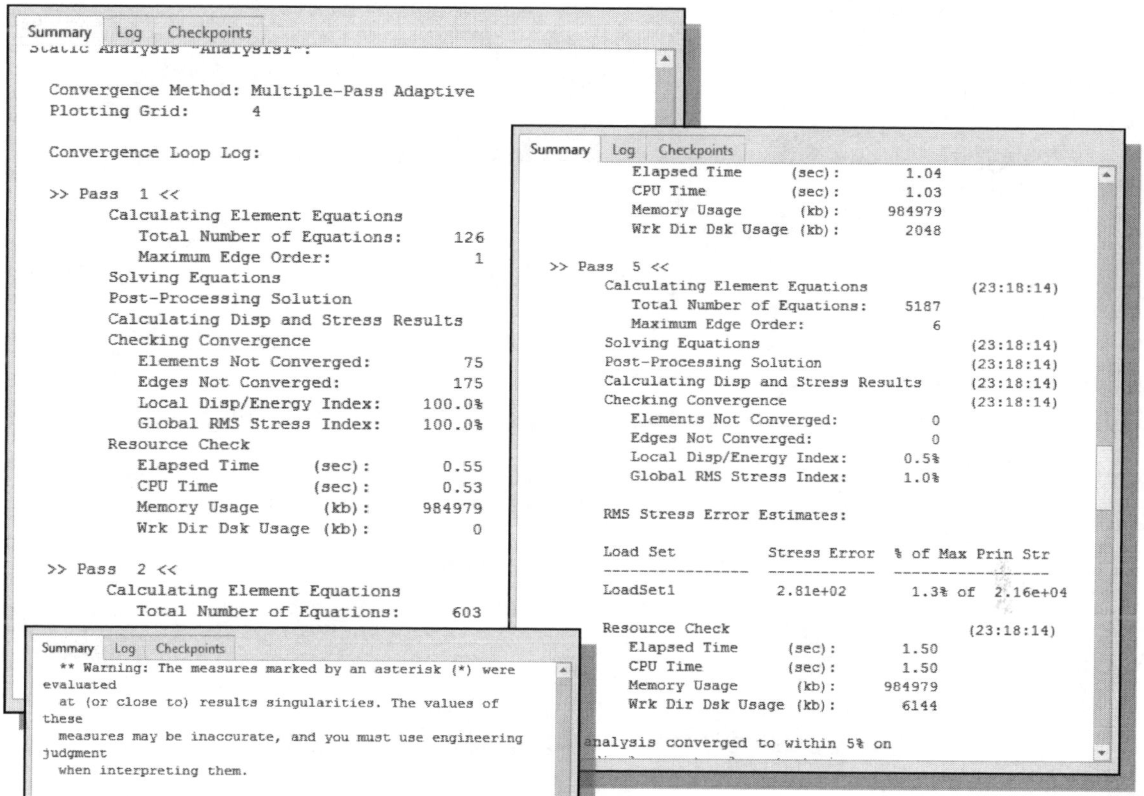

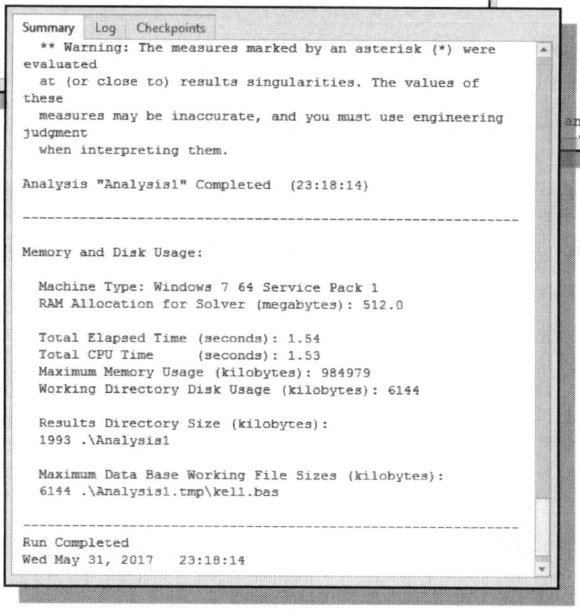

12. On your own, examine the detailed information regarding the analysis and the design.

13. In the *Analyses and Design Studies* window, choose **Review results** as shown.

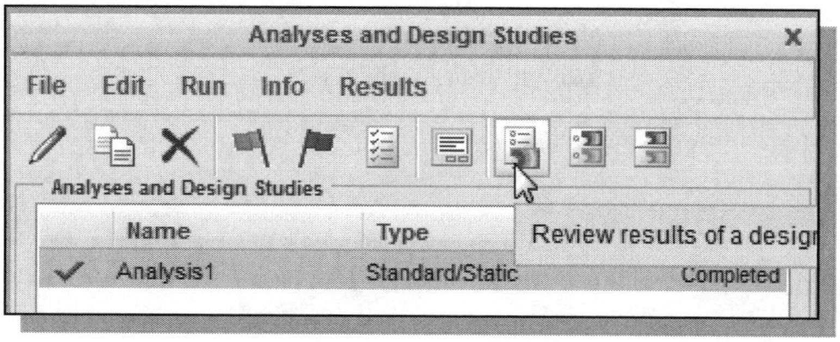

View the FEA Results

➢ Von Mises Stress

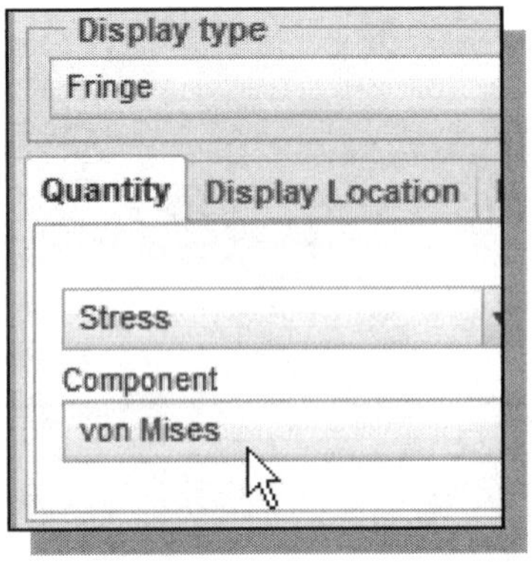

1. On your own, set the Display type to **Fringe**.

2. Choose **Von Mises** as the Stress Component to be displayed.

3. Click on the **Display Options** tab and switch ***on*** the **Deformed** option as shown.

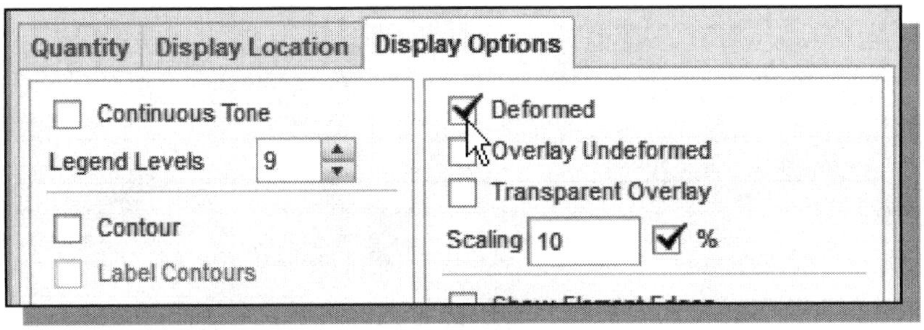

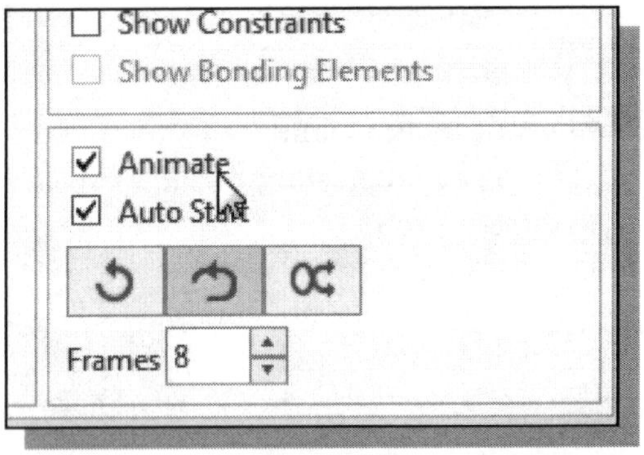

4. Switch ***on*** the **Animate** option as shown.

5. Click **OK and Show** to display the results.

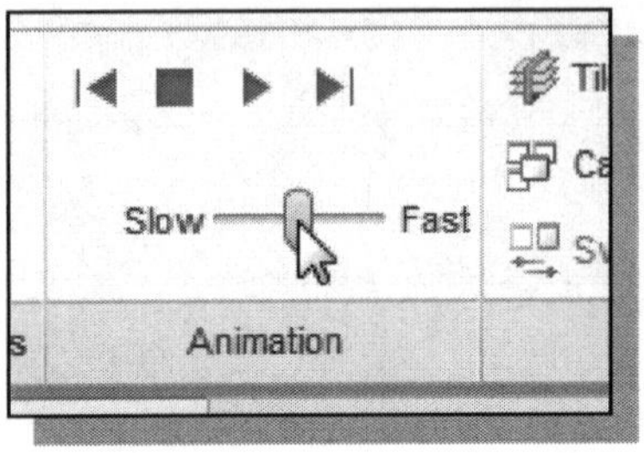

6. Note the ***animation controls*** are available in the *View* toolbar area as shown.

❖ Note that the **Dynamic Rotation** command is also available while the animation is running.

➢ *Creo Simulate* confirms that the highest stress occurs on the inside edge of the curved portion adjacent to the vertical section of the design. The computer solution for the maximum stress is **2.106E+4 lb/in²**, which occurs at the top of the circular arc. This number is very close to the **20946** we calculated during our preliminary analysis.

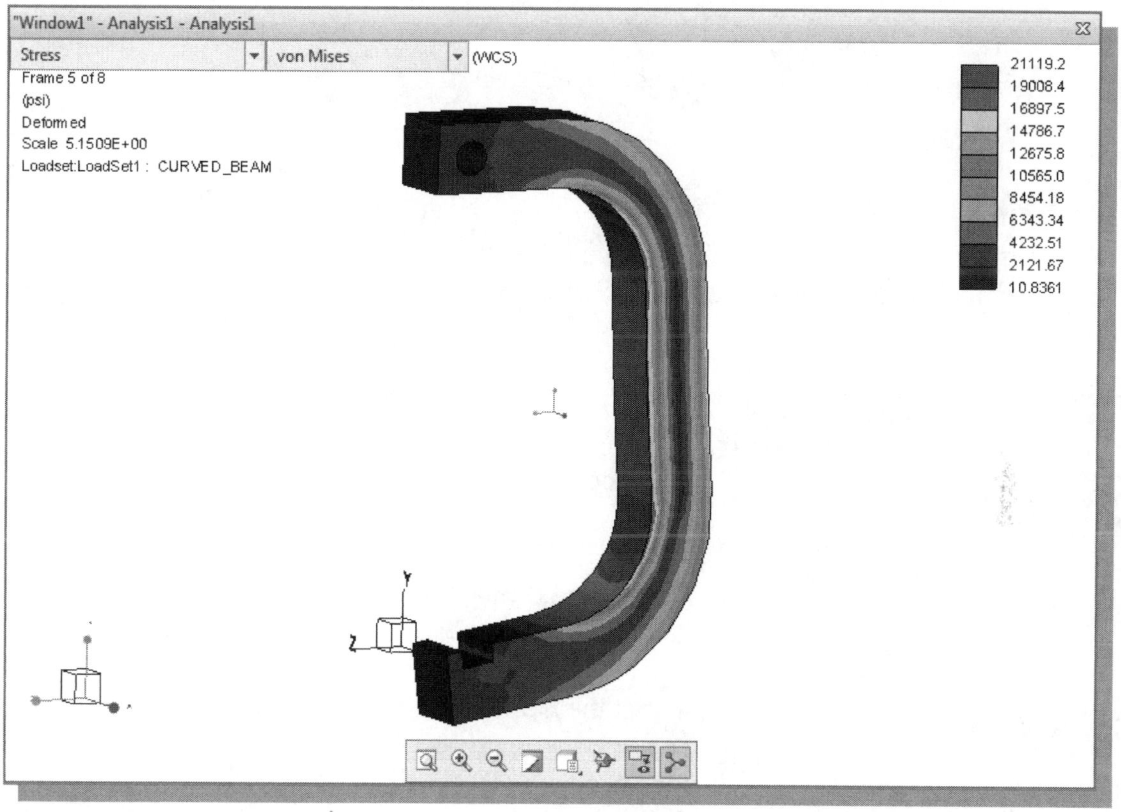

➢ Viewing with the Cutting/Capping Option

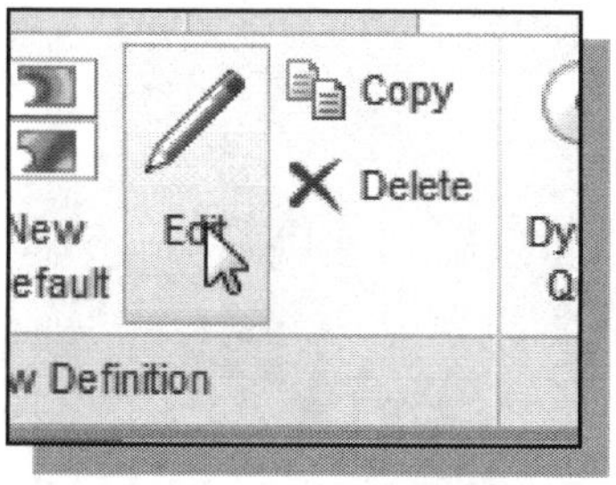

1. Click **Edit the current definition** icon as shown.

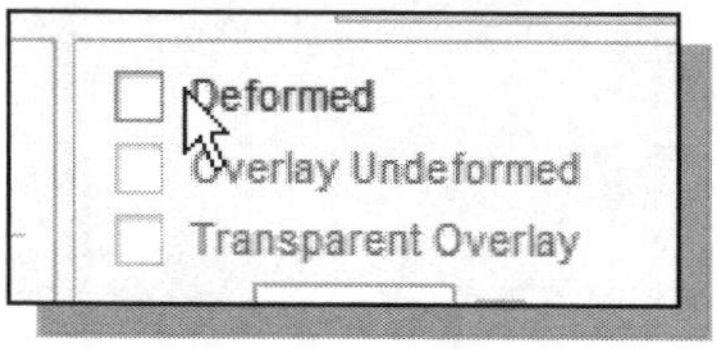

2. Turn ***off*** the display of **Deformed** and the **Animate** options.

3. Click **OK and Show** to display the results.

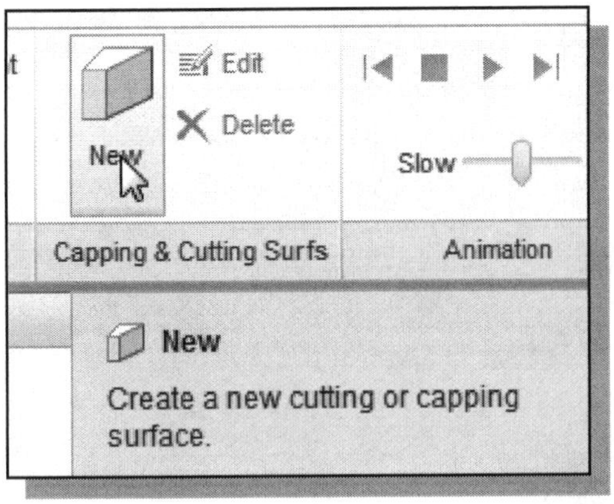

4. Select the **View → New Cutting/Capping Surfs** option from the pull-down menu to perform additional result display options as shown.

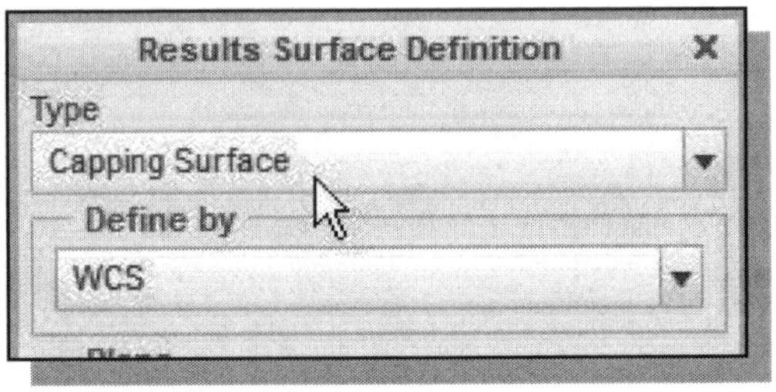

5. In the *Results Surface Definition* dialog box, set the Type option to **Capping Surface** as shown.

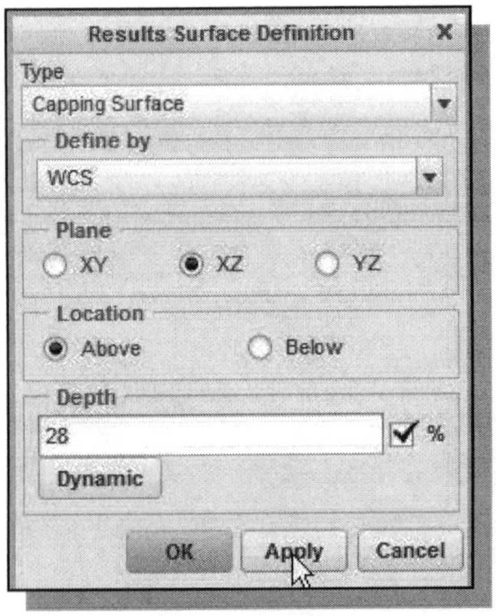

6. Select the **XZ** Plane option.

7. Set the Location option to **Above**.

8. Set the Depth option to **28%** as shown.

9. Click **Apply** to view the results.

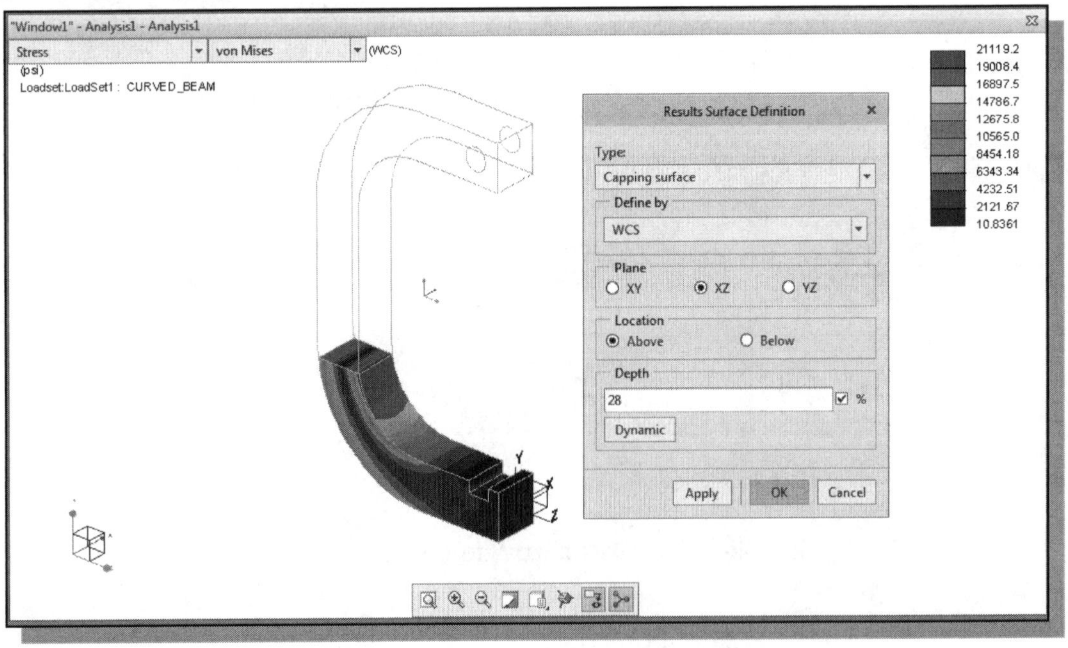

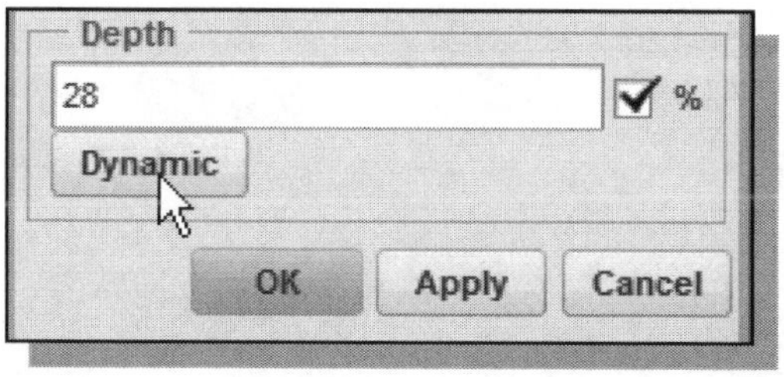

10. Click on the **Dynamic** button to dynamically move the cutting surface by dragging with the **left-mouse-button**.

➢ You are also encouraged to create and compare FEA analyses of the same problem using one-dimensional beam elements and two-dimensional plane-stress elements. What type of maximum stresses would you expect to see using the 1D and 2D elements? What are the advantages/limitations of performing multiple analyses using different types of FEA elements?

Review Questions

1. What are the most widely used failure criteria?

2. What are the main objectives of finite element analysis?

3. What will happen to the system if it is stressed beyond the elastic limit?

4. What are the necessary elements to create a swept feature in *Creo Parametric*?

5. What is the purpose of doing a convergence study?

6. Under what condition is the bending stress developed in a curved beam no longer linear?

7. What does the **Cutting/Capping Surfs** option allow us to do?

8. Can a curved beam problem be analyzed with the 1D beam element? What is the main limitation of performing such an analysis?

9. What are the advantages of using 3D elements over using the 1D or 2D elements?

10. What is the purpose of performing a multi-pass adaptive analysis?

Exercises

1. For the steel U-shape design (diameter: ¾″), determine the maximum stress developed under a loading of P=250 lb. (Hint: Add two square features at the two ends to be used for applying the proper constraint and load.)

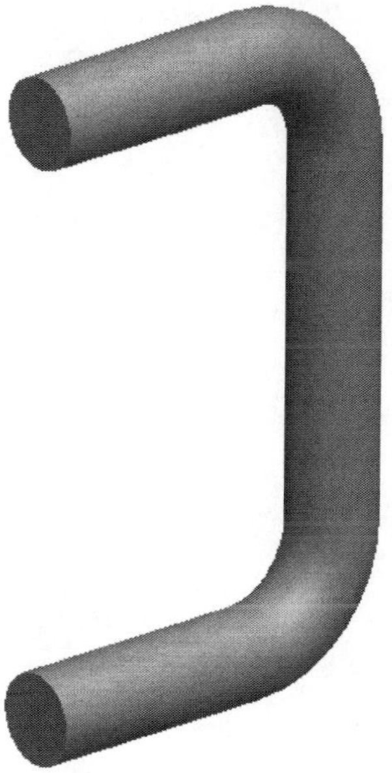

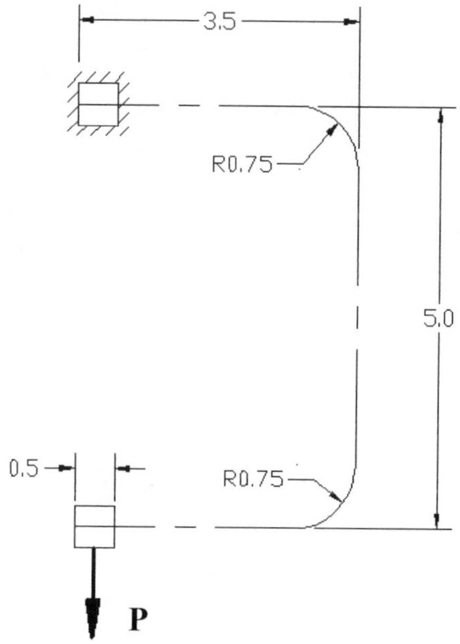

2. For the above problem, perform additional FEA analyses using the two cross-sections shown below and compare the results.

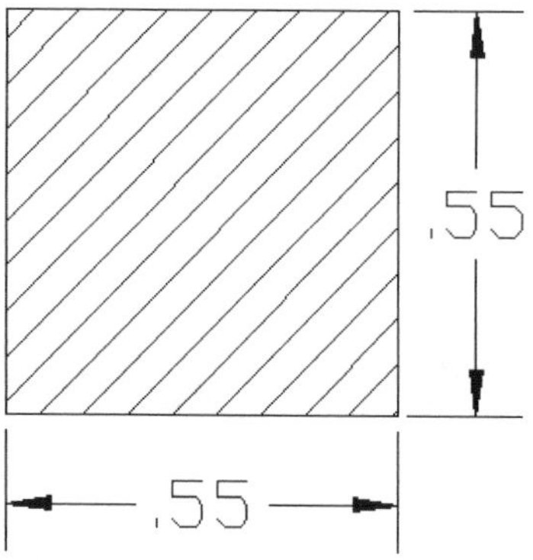

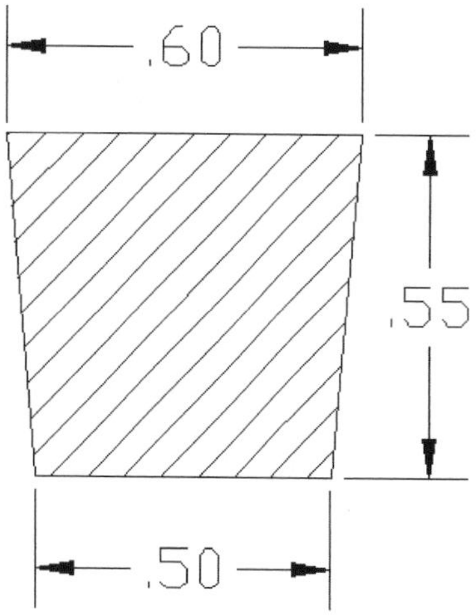

Notes:

Chapter 11
Axisymmetric and Thin Shell Elements

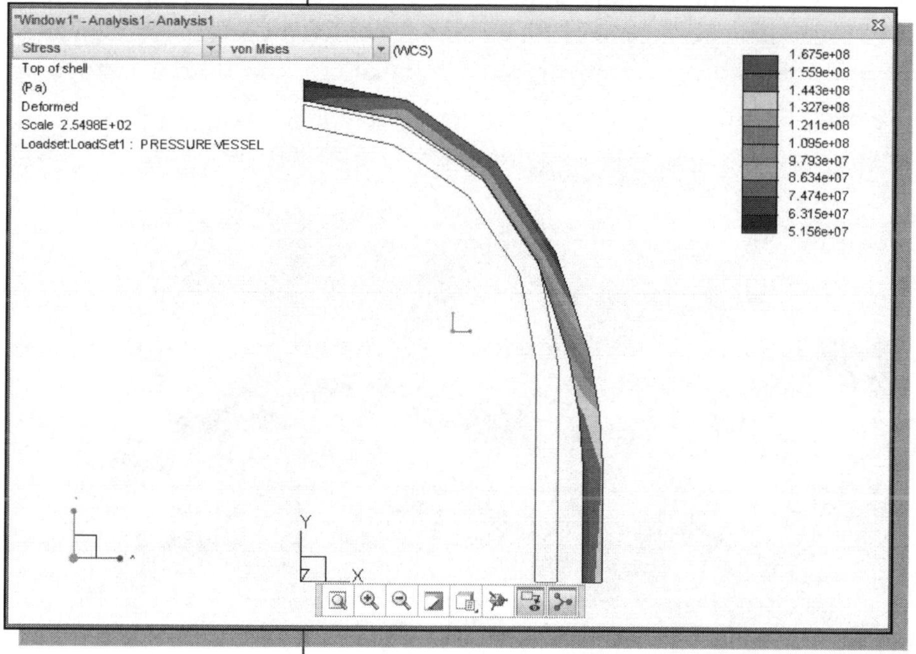

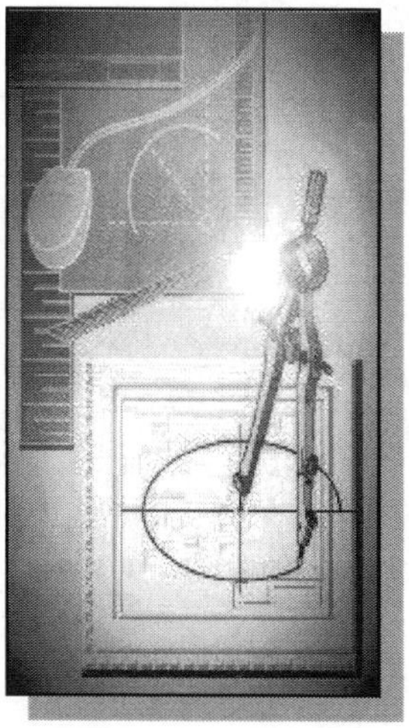

Learning Objectives

- Understand the procedures to create Axisymmetric and Shell FEA models.
- Perform Several FEA Analyses using different element types on the same design.
- Understand the Basic differences of the different 3D FEA elements.
- Use the Insert Window command to display and compare multiple Results.

Introduction

Symmetry is an important characteristic that is often seen in designs. Symmetrical designs are generally more pleasing to the eye and also provide the desirable functionalities in designs. Objects that are symmetrical with respect to an axis of rotation are known as ***axisymmetry***. Objects with axisymmetry are quite common: wheels, containers, drums, pipes, tanks, and rotational devices in general. The main characteristic of axisymmetry objects is that the shape of the object is defined by a 2D planar cross section, which is revolved about the central axis.

In FEA analysis, an ***axisymmetric analysis*** can save a substantial amount of computing time, as a 3D model is simplified to a 2D cross section. To perform an axisymmetric analysis in *Creo Simulate*: (1) the Y axis of the reference coordinate system is used as the axis of rotation; (2) all loads and displacements must be specified in the XY plane; and (3) all the geometry must lie in the positive X and Y portion of the XY plane.

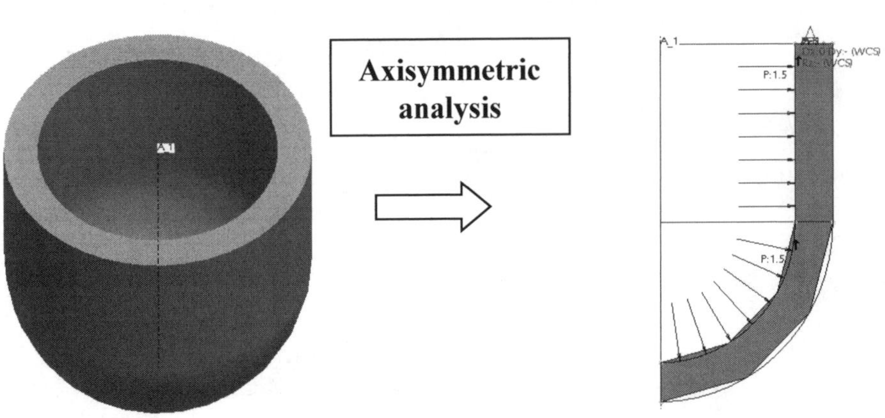

In FEA analysis, if a part is relatively thin compared to its length and width, using ***shell elements*** would be more efficient than using 3D solid elements. In *Creo Simulate,* we can create a shell idealization from an existing solid model. *Creo Simulate* can create a shell pair based on the surfaces of the solid model. *Creo Simulate* will also compress the pair to a midsurface that it uses in the analysis.

Thin Shell analysis

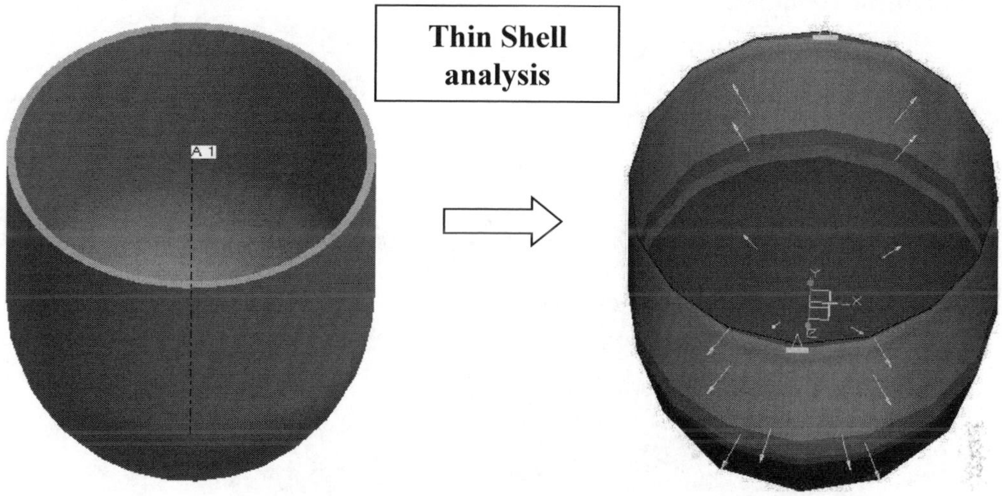

In theory, all designs could be modeled with three-dimensional solid elements. A thin shell design can also be analyzed using ***3D solid elements***. Two of the main considerations in selecting the element type to use are (1) the amount of time it takes to perform the analysis, and (2) the accuracy of the results. One of the main advantages of the P-element method over the H-element method is the relatively small number of nodes and elements used. For designs that involve thin features, the P-element analysis method is much more efficient and generally can reach the convergence much quicker.

3D Solid analysis

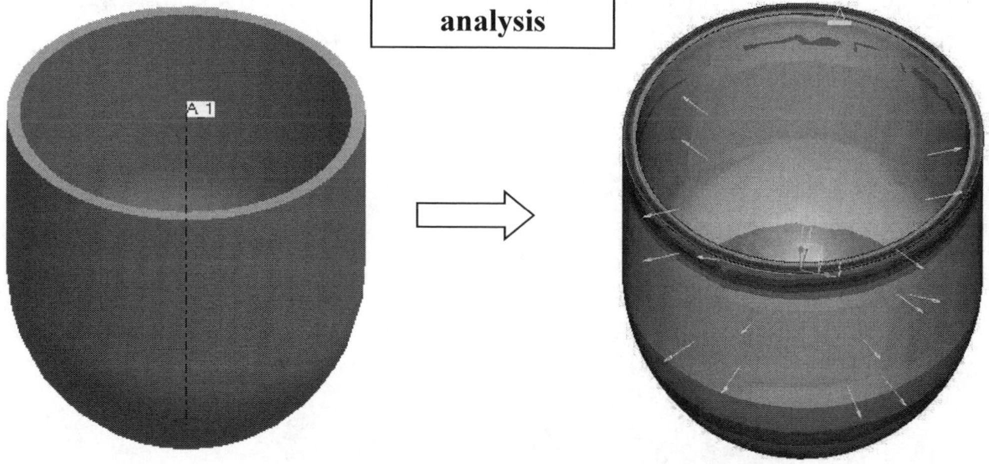

❖ In this chapter, the procedures to perform FEA analyses using the above three types of elements are illustrated. To assure the accuracy of FEA results, performing a second and/or a third analysis can be very practical and effective.

Problem Statement

Determine the **tangential** and **longitudinal stresses** of the thin-wall cylindrical pressure vessel shown in the figure below. The pressure vessel is made of steel and is subject to an **internal pressure of 15MPa**. The dimensions of the vessel are **End radius 300mm, cylindrical length 500mm** and **wall thickness 25mm**.

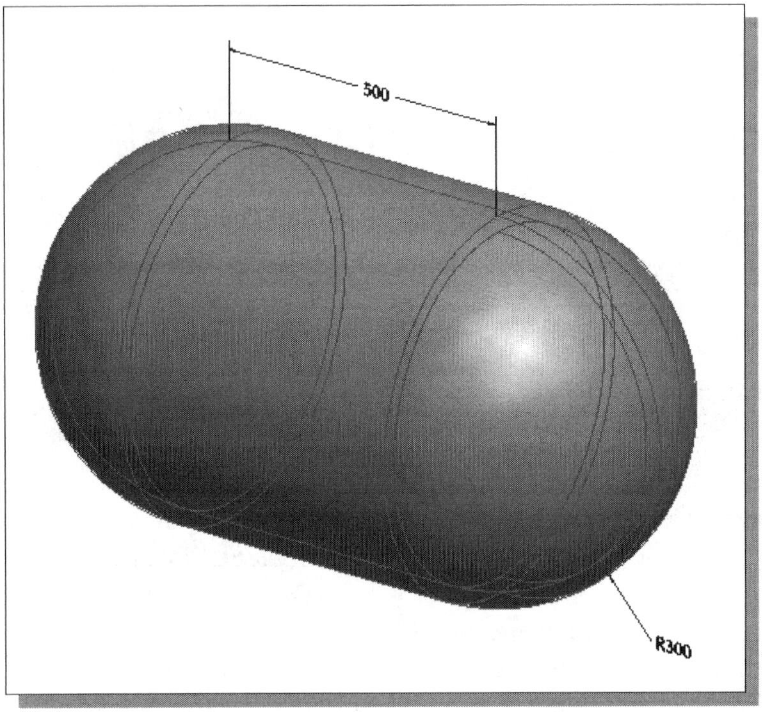

Preliminary Analysis

On the ***cylindrical*** section of the pressure vessel, the two principal stresses are (1) the **tangential** and (2) **longitudinal** stresses. The principal stresses on the ***hemispherical*** ends are **tangential stress**. These primary stresses are identified as shown in the figure below.

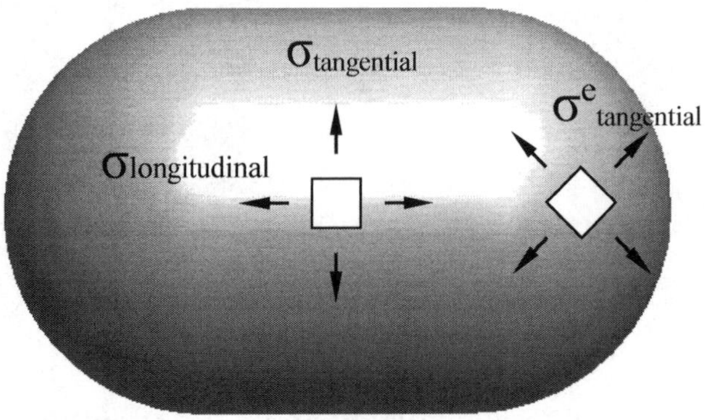

From *Strength of Materials*, the principal stresses on the **cylindrical section** of the pressure vessel walls, which are the tangential and longitudinal stresses, can be determined by

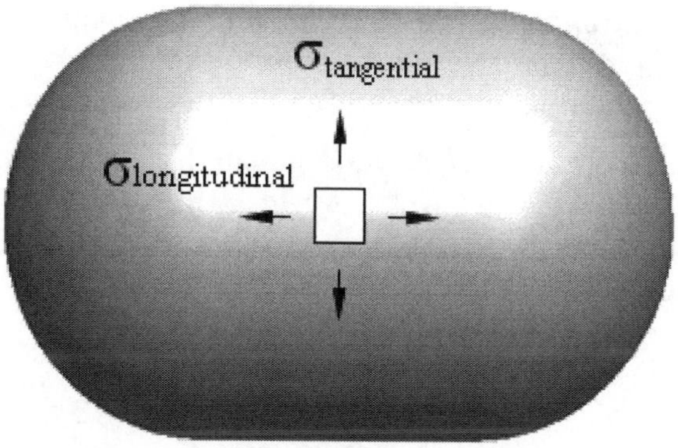

$$\sigma_{tangential} = \frac{Pd_i}{2t} = 15.0 \times 10^6 \times (2 \times 0.275)/(2 \times 0.025)$$

$$= 165 \text{ MPa}$$

$$\sigma_{longitudinal} = \frac{Pd_i}{4t} = 15.0 \times 10^6 \times (2 \times 0.275)/(4 \times 0.025)$$

$$= 82.5 \text{ MPa}$$

The principal stresses on the ***hemispherical*** **section** are tangential stresses, which can be determined by

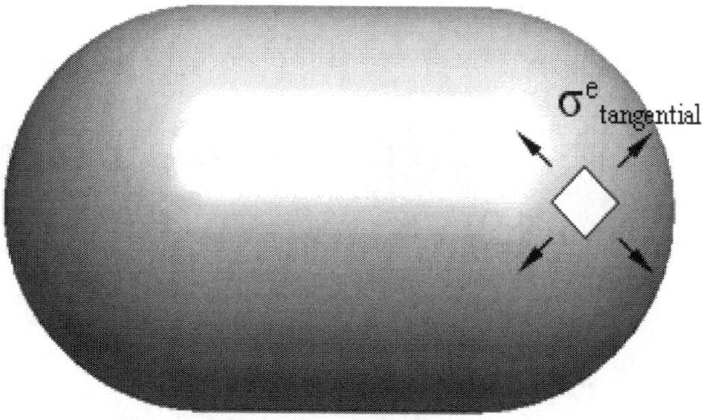

$$\sigma^e_{tangential} = \frac{Pd_i}{4t} = 15.0 \times 10^6 \times (2 \times 0.275)/(4 \times 0.025)$$

$$= 82.5 \text{ MPa}$$

In the following sections, we will perform three FEA analyses, using three different types of FEA elements: **Axisymmetric**, **Shell** and **3D Solid** elements.

Starting Creo Parametric

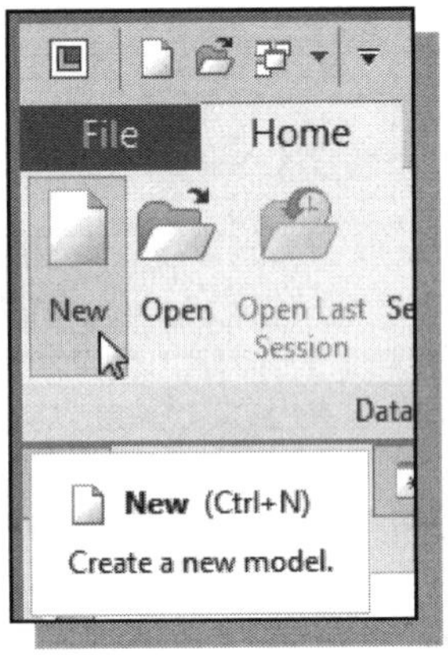

1. Select the **Creo Parametric** option on the *Start* menu or select the **Creo Parametric** icon on the desktop to start *Creo Parametric*. The *Creo Parametric* main window will appear on the screen.

2. Click on the **New** icon, located in the *Ribbon toolbar* as shown.

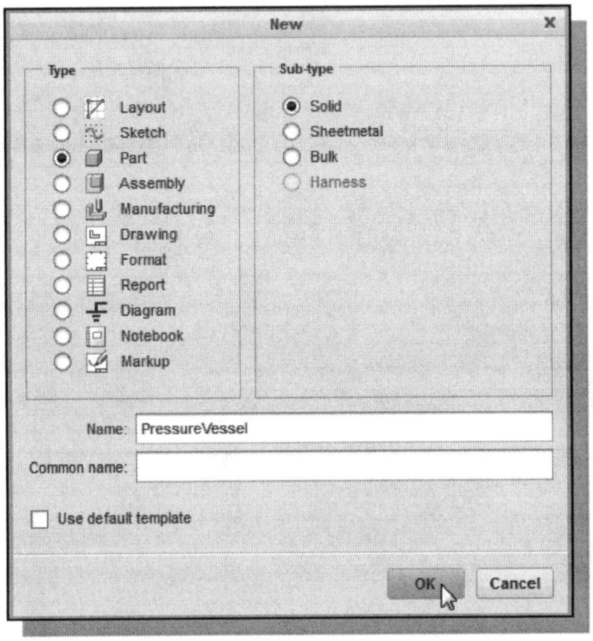

3. In the *New* dialog box, confirm the model's Type is set to **Part** (**Solid** Sub-type).

4. Enter **PressureVessel** as the part Name as shown in the figure.

5. Turn ***off*** the **Use default template** option.

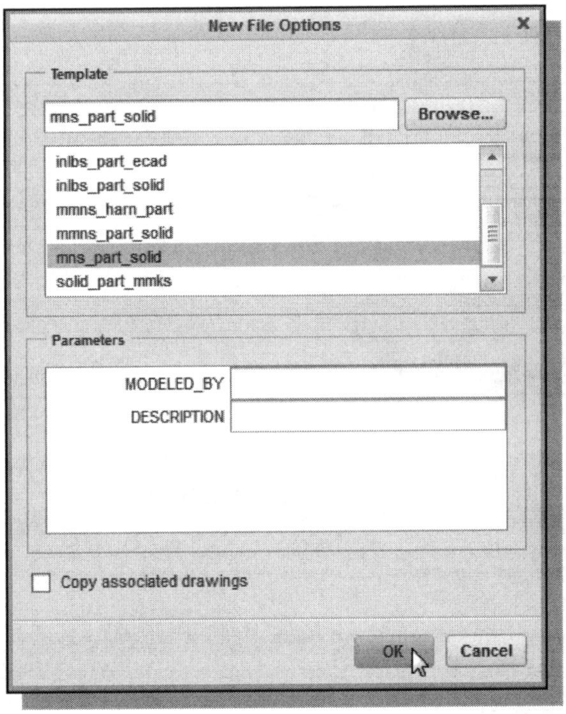

6. In the *New File Options* dialog box, select **mns_part_solid** in the option list as shown.

7. Click on the **OK** button to accept the settings and enter the *Creo Parametric Part Modeling* mode.

Create a CAD Model in Creo Parametric

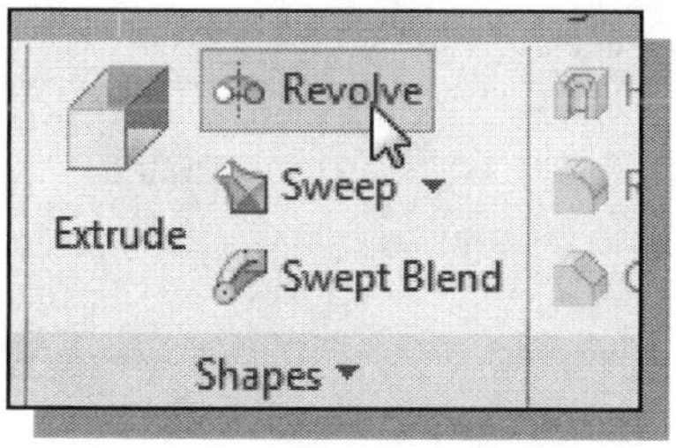

1. In the *Shapes* toolbar, select the **Revolve** tool option as shown.

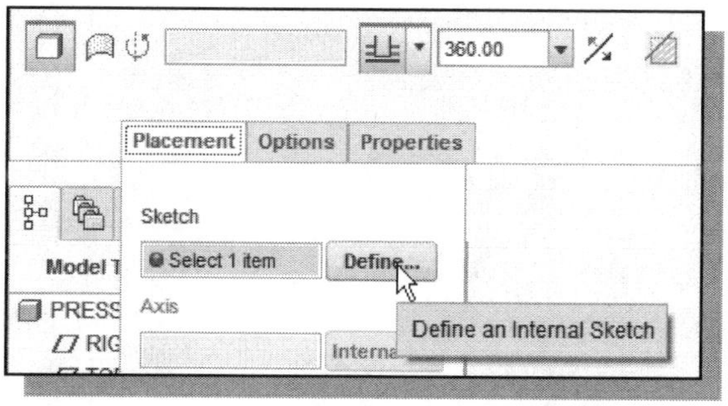

2. Click the **Placement** option and choose **Define** to create a new *internal sketch.*

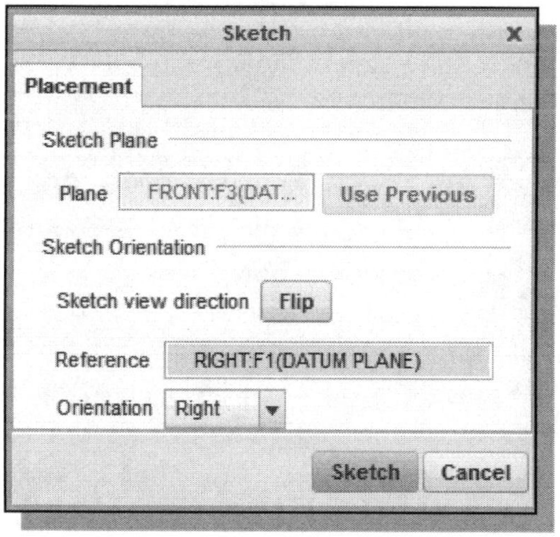

3. On your own, set up the **FRONT** datum plane as the sketch plane with the **RIGHT** datum plane facing the **right** edge of the computer screen, and pick **Sketch** to enter the *Sketcher* mode.

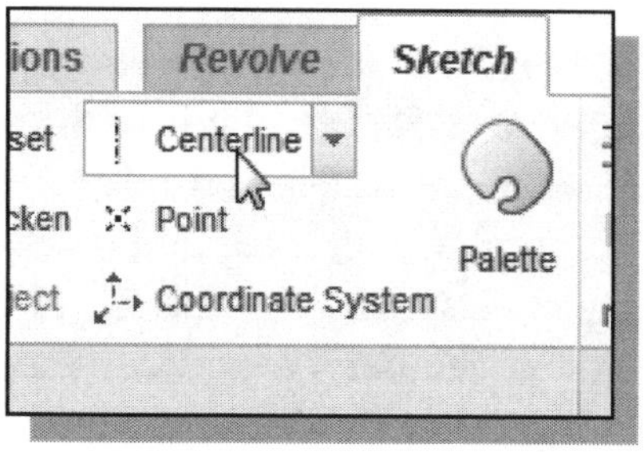

4. In the *Sketching* toolbar, select **Centerline**.

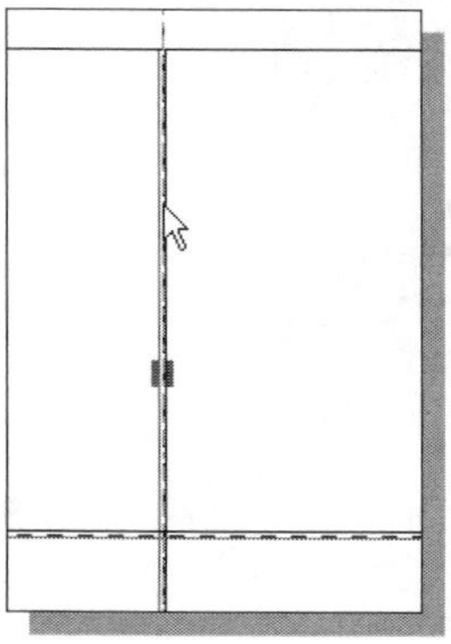

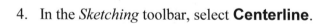

5. On your own, create a **vertical centerline** aligned to datum plane **RIGHT**.

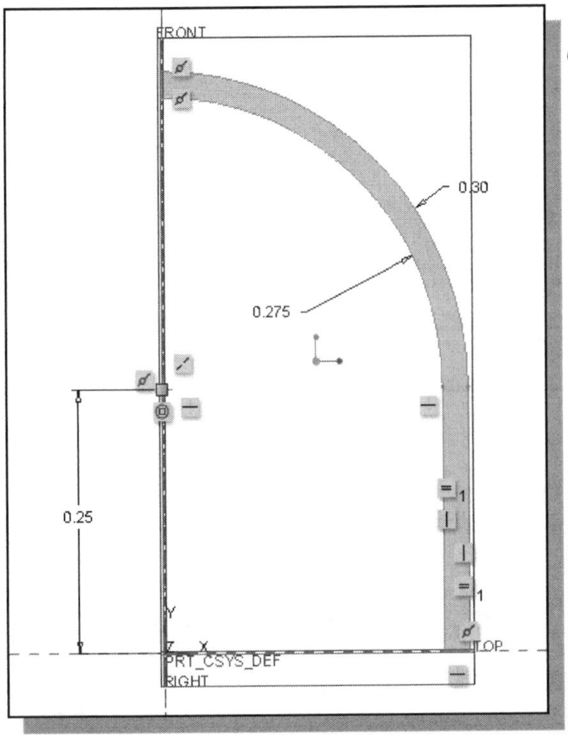

6. On your own, create a sketch with the two sets of parallel curves as shown.

❖ Note that, because of the symmetrical nature of the design, only ¼ of the cross section is needed.

7. On your own, adjust the size of the sketched geometry as shown.

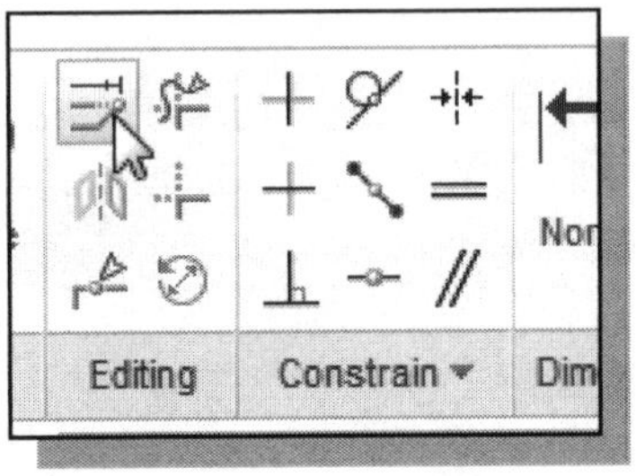

8. Click **OK** to accept the completed sketch.

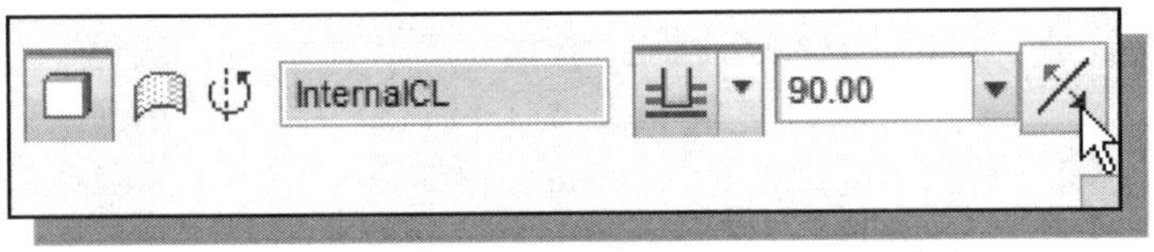

9. Set the *revolve angle* to **90** and click the arrow button once as shown.

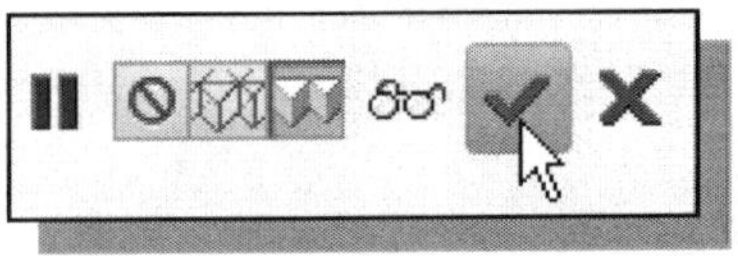

10. Click **Accept** to accept the completed beam model.

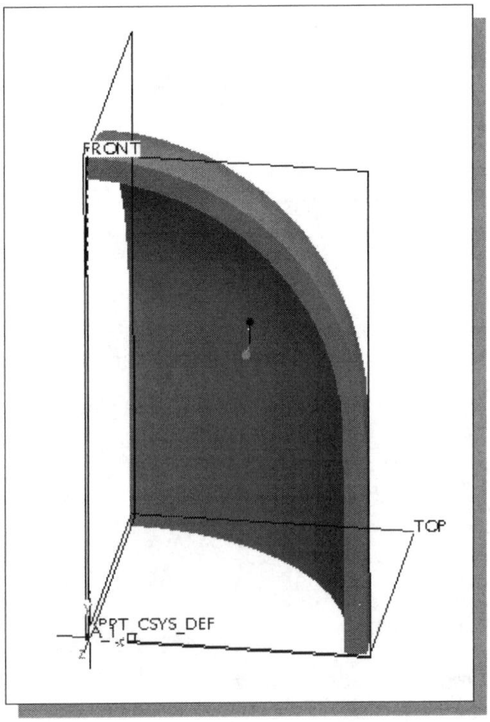

❖ To perform an axisymmetric analysis in *Creo Simulate*: (1) the Y axis of the reference coordinate system is used as the axis of rotation; (2) all loads and displacements must be specified in the XY plane; and (3) all the geometry must lie in the positive X portion of the XY plane.

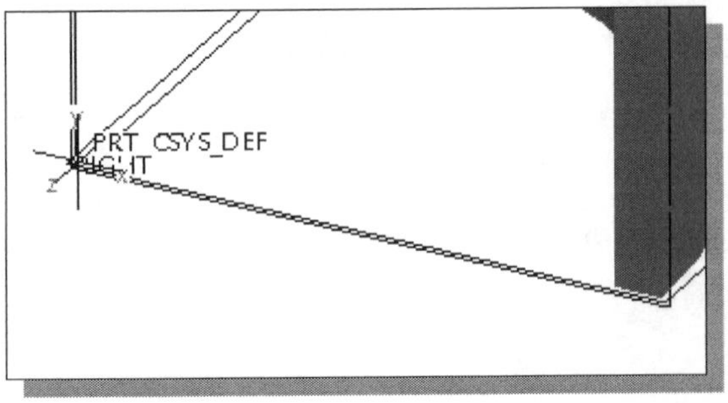

Select and Examine the Part Material Property

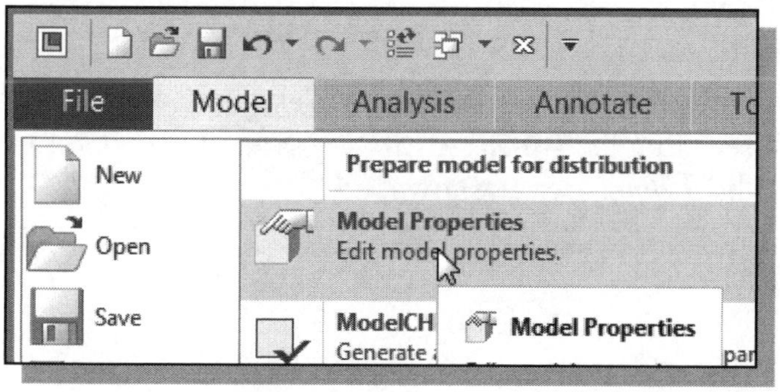

1. Choose **File → Prepare → Model Properties** from the pull-down menu.

2. Select the **Change** option that is to the right of the **Material** option in the ***Model Properties*** window.

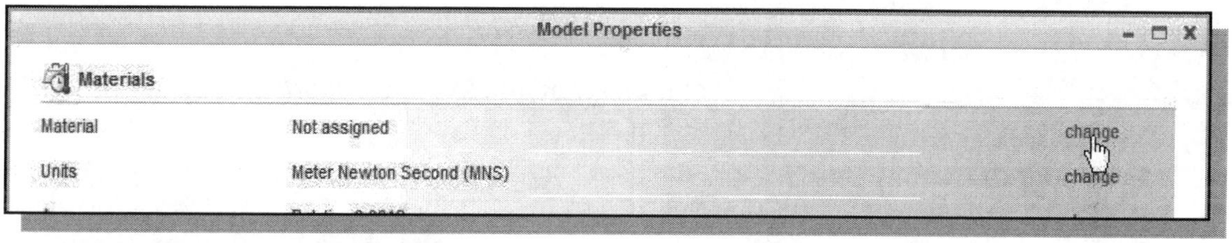

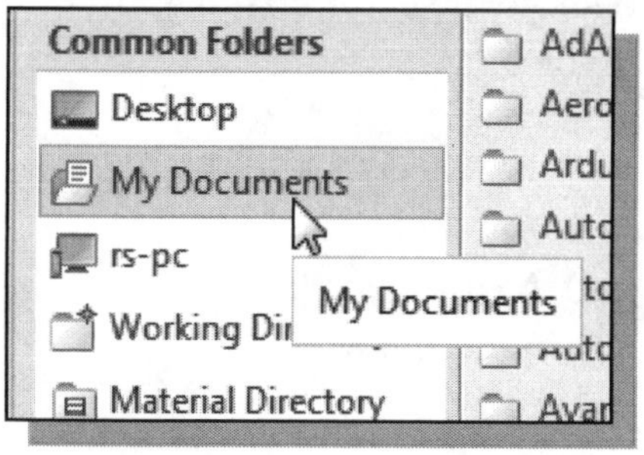

3. On your own, locate the folder where we saved the modified material definition, filename: **Steel_low_Carbon_Modified**.

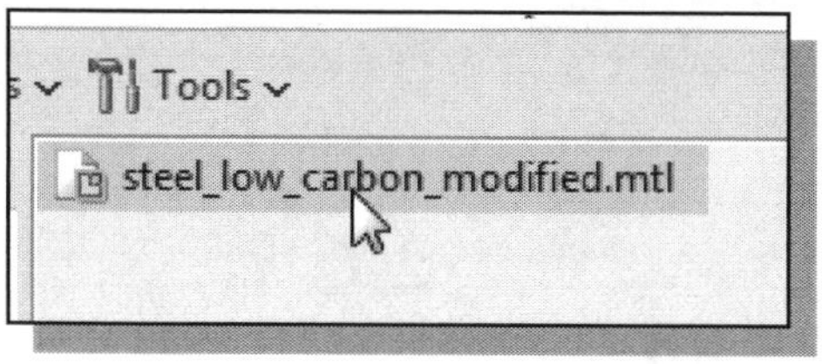

4. **Double-clicking** on the modified material definition for use with the current model.

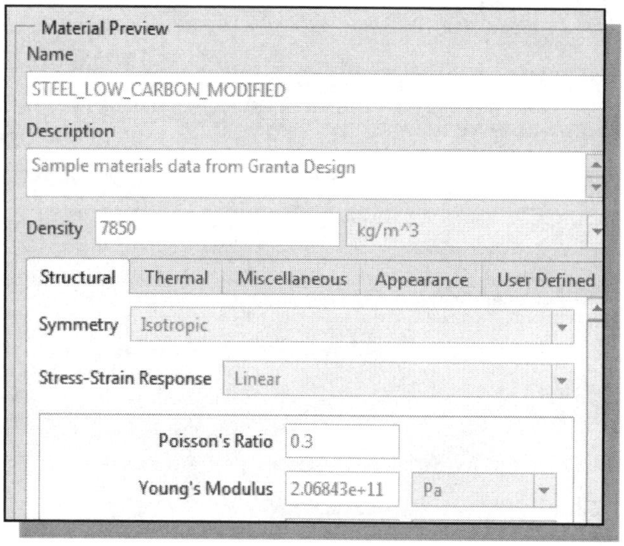

5. Click **Yes** to convert the units to match the model units.

6. Examine the material information in the **Material Preview** area and click **OK** to accept the settings.

The Integrated Mode of Creo Simulate

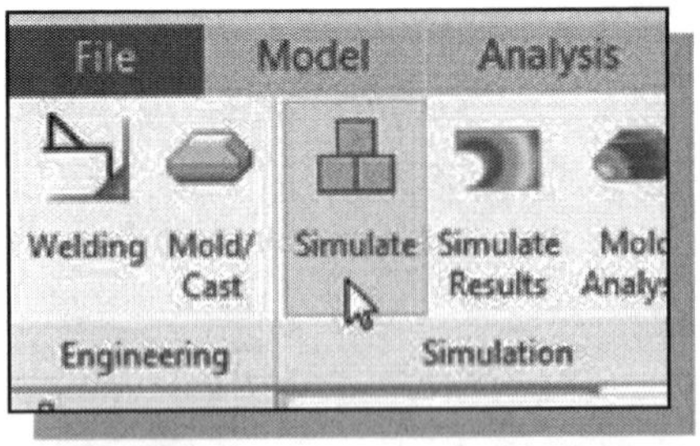

1. Start the *Integrated* mode of *Creo Simulate* by selecting the **Applications → Simulate** option in the *Ribbon* area as shown.

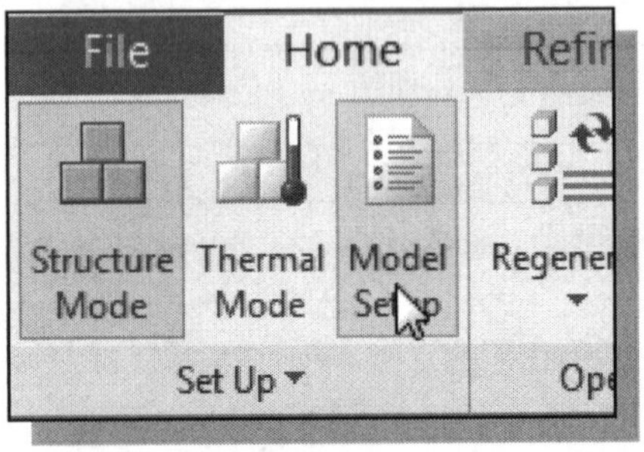

2. In the *Creo Simulate Ribbon* toolbar, confirm **Structure** is set as the analysis type.

3. Click **Model Setup** to open the *setup dialog box* and examine the available *model setup* options.

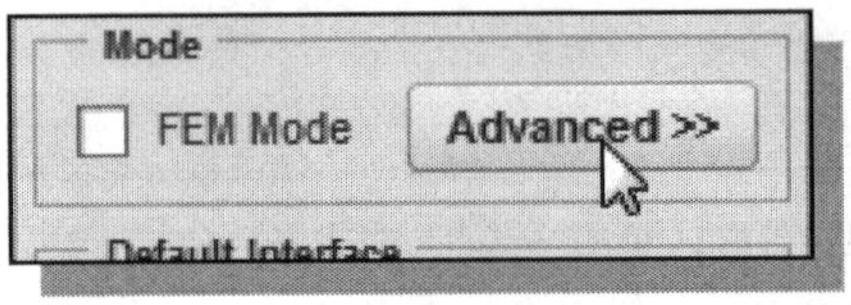

4. Click **Advanced** to show the additional detail settings that are available.

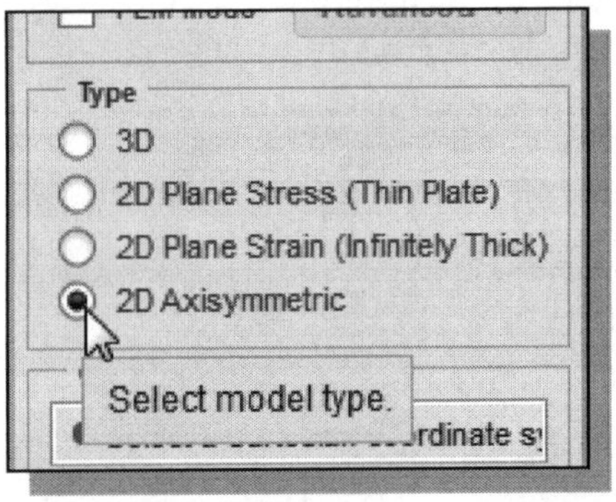

5. Set the selection of **2D Axisymmetric** as the model Type as shown in the figure.

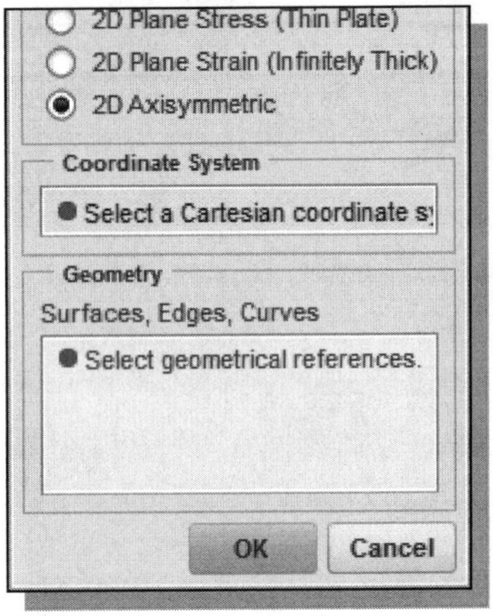

❖ Note that two additional selections are required for using the 2D axisymmetric elements: (1) a reference **Coordinate System** and (2) a reference **Geometry**.

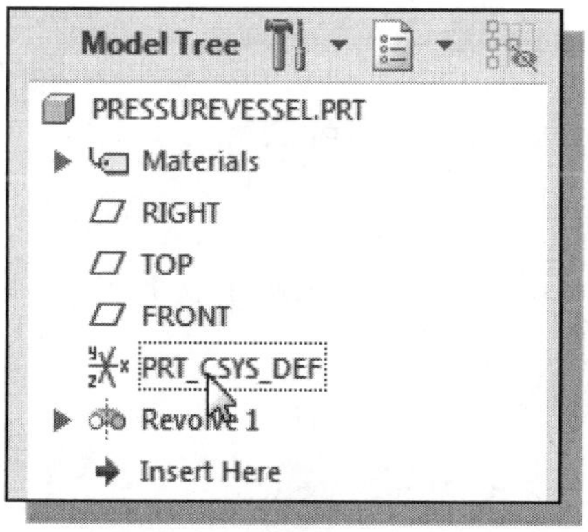

6. Inside the ***Model Tree*** window, select the **default part coordinate system** by clicking once with the left-mouse-button.

7. Select the **surface** that is aligned to the **XY plane** of the part coordinate system as the reference geometry as shown.

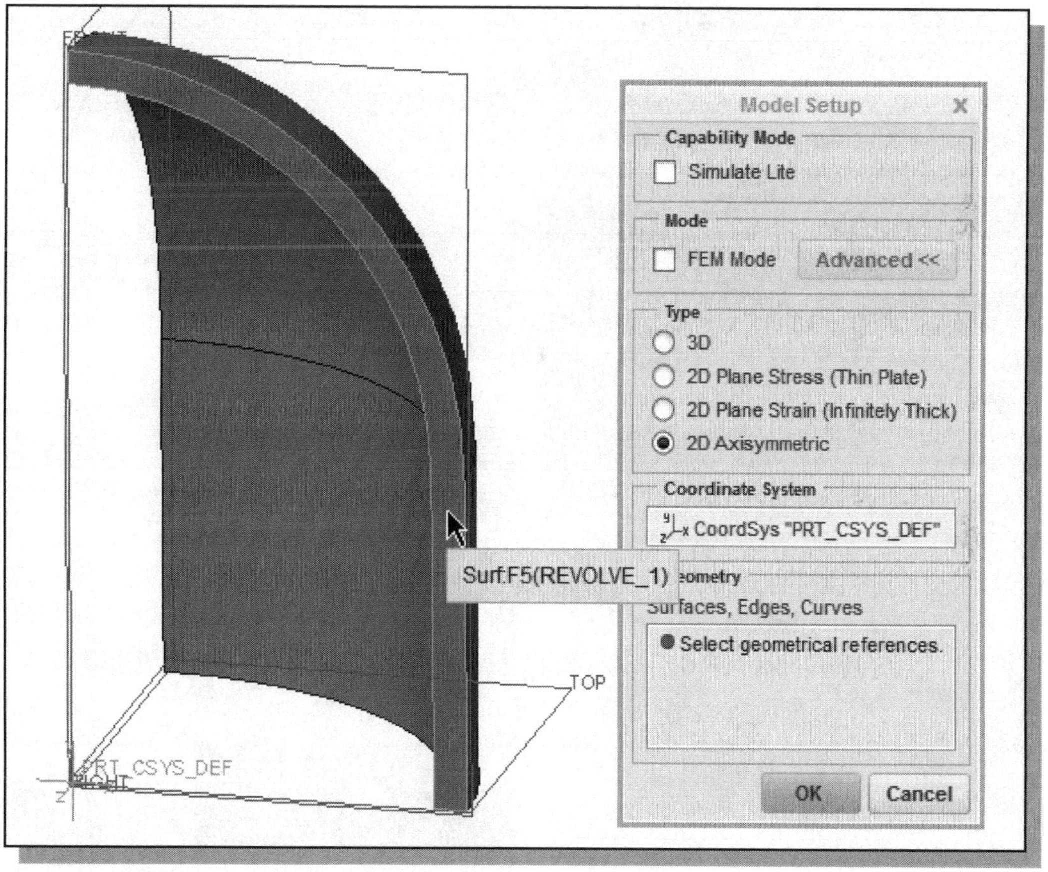

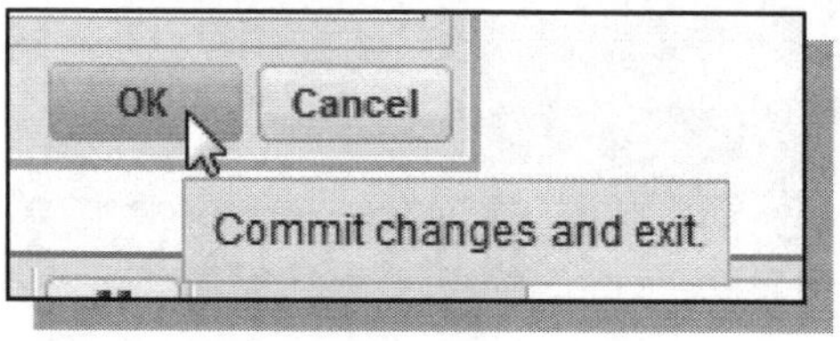

8. In the *Creo Simulate Model Setup* window, click on the **OK** button to accept the settings.

9. Click on the **Confirm** button to proceed to create a new 2D FEA model.

Apply the Boundary Conditions – Constraints

In *Creo Simulate*, the Y axis is set to be the axis of symmetry for axisymmetric model; edges that are aligned to the Y axis will be allowed to move only along the axis of symmetry, so no constraint is needed. For our model, the bottom edge of the FEA model also passes through the horizontal axis of symmetry; we will therefore lock only the Y **Translational** constraint as shown in the figure below.

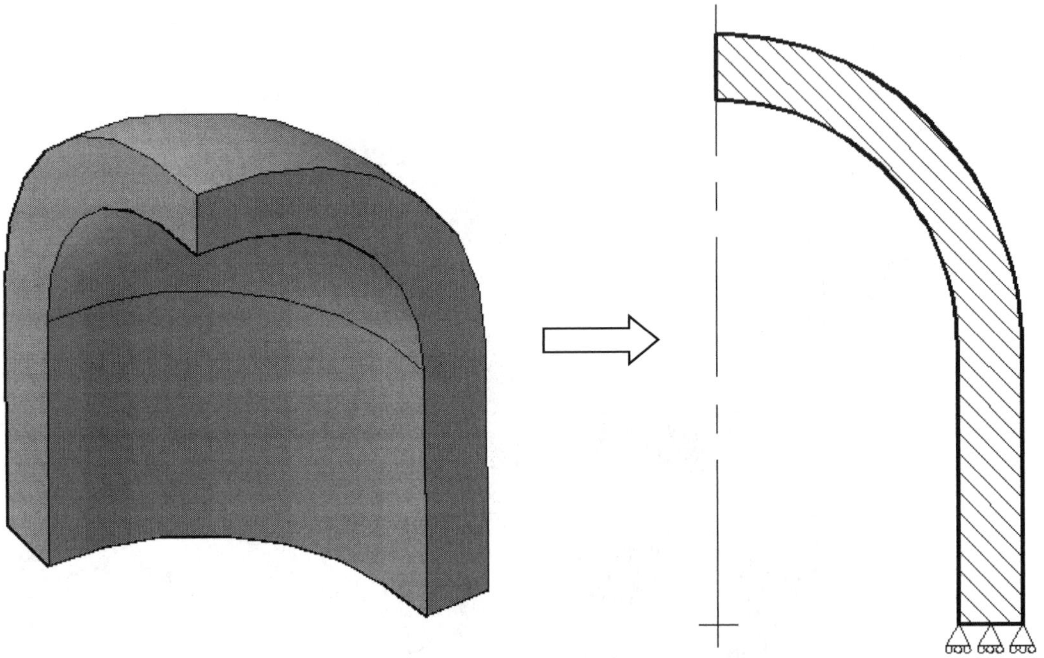

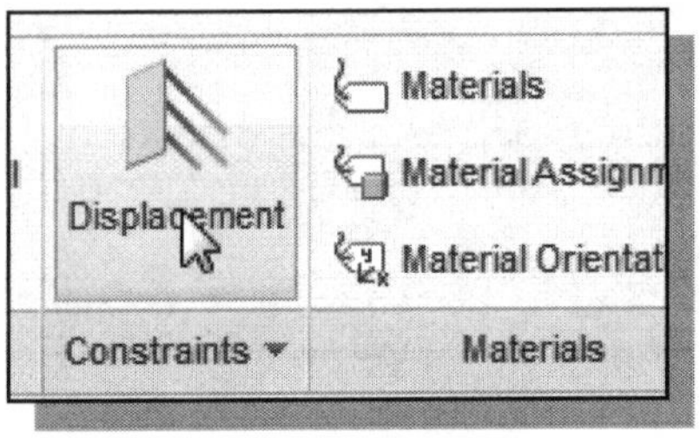

1. Choose **Displacement Constraint** by clicking the icon in the *Constraints* toolbar as shown.

2. Confirm the **References** option is set to **Edges/Curves** and pick the **bottom edge**, aligned to the X axis, of the section as shown.

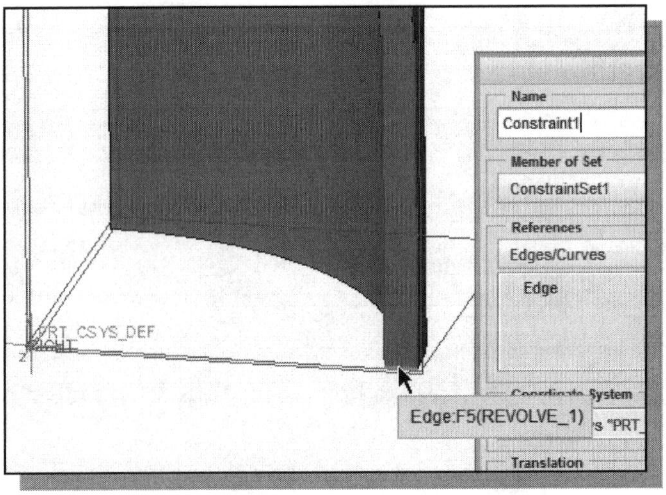

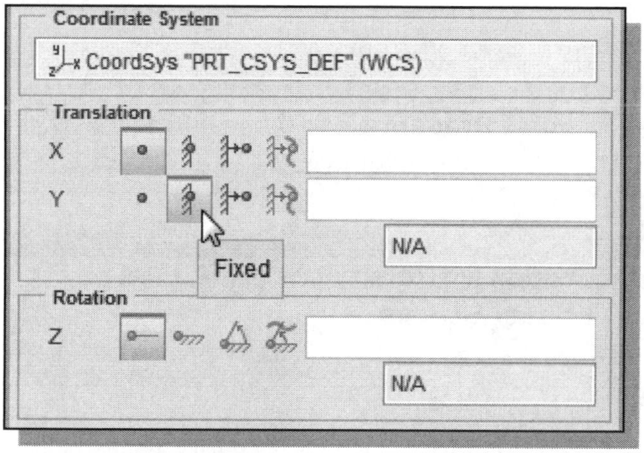

3. Set the Y Translation to **Fixed**, and the X Translation option to **Free** as shown.

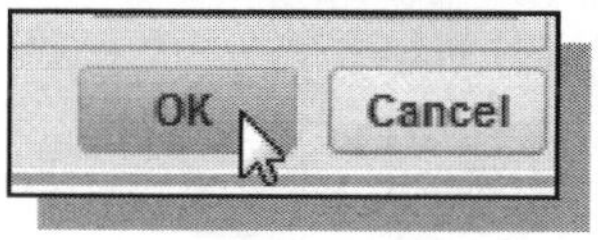

4. Click on the **OK** button to accept the first displacement constraint settings.

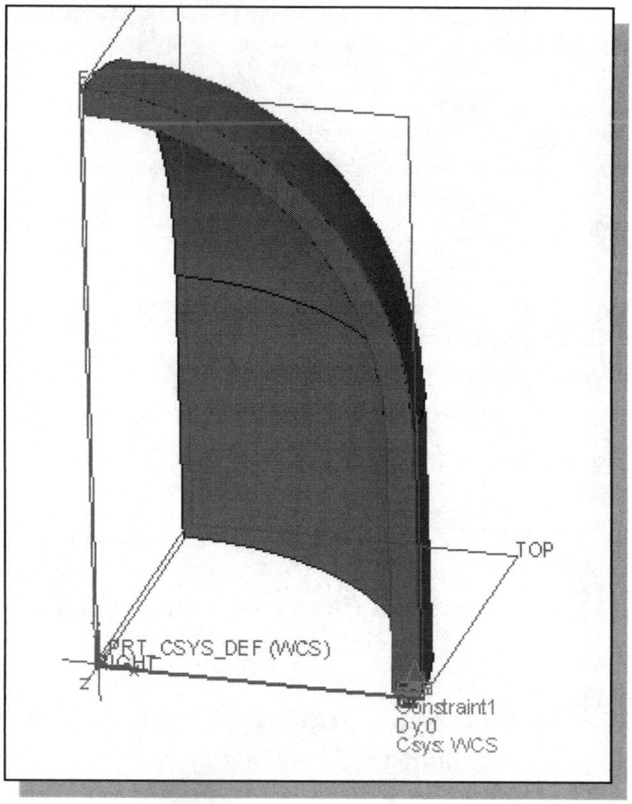

❖ Your model should appear as shown in the figure.

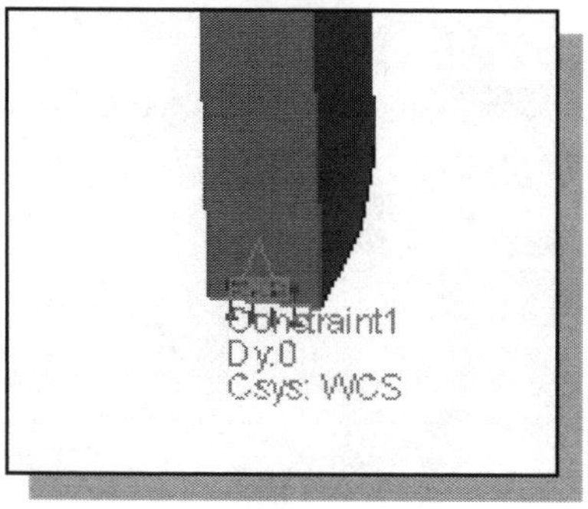

- Note the **applied displacement constraint** can also be identified by examining the small icon and text as shown. **Dy:0** refers to the *Y* Translational being **Fixed**.

Apply the Internal Pressure

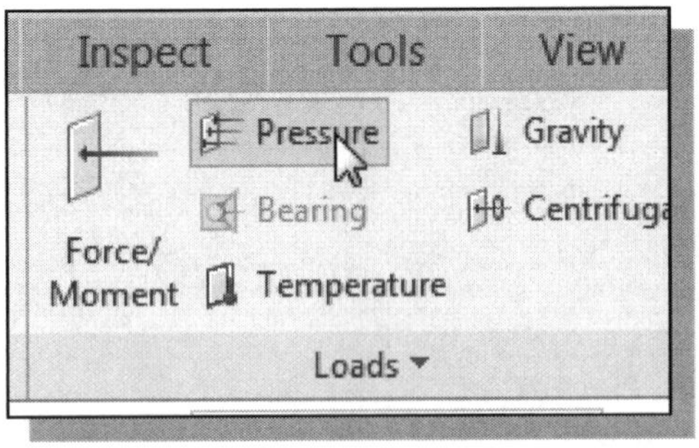

1. Choose **Pressure Load** by clicking the icon in the toolbar as shown.

2. Confirm the References option is set to **Edges/ Curves**.

3. Pick the **two inside edges** of the front surface while holding down the [**Ctrl**] key.

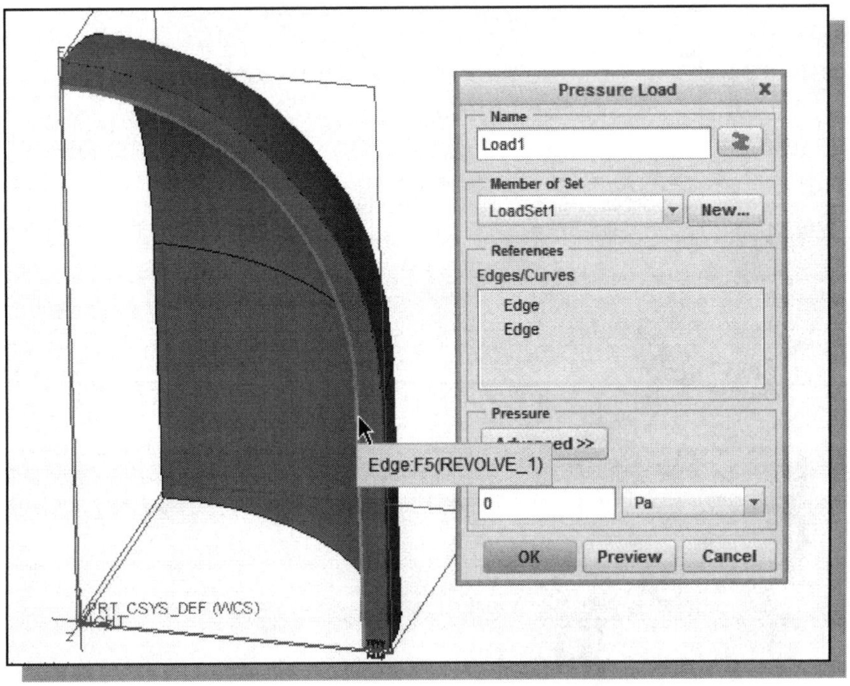

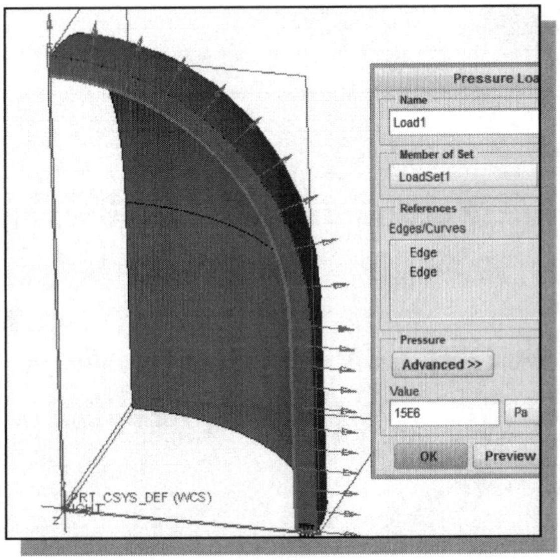

4. Enter **15E6 Pa** in the Pressure Value box as shown.

5. Click on the **Preview** button to confirm the applied load is pointing in the correct direction.

6. Click on the **OK** button to accept the first load settings.

Create the 2D Mesh

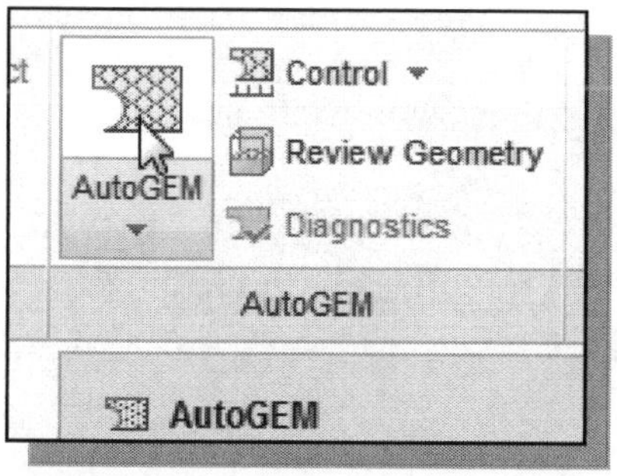

1. Choose **Refine Model** in the *Ribbon* toolbar.

2. Activate the **AutoGEM** command by selecting the icon in the toolbar area as shown.

3. Click **No** to close the retrieve mesh message.

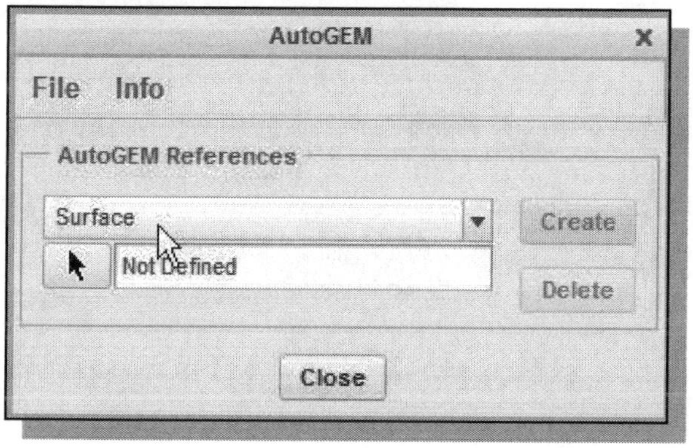

4. Switch the AutoGEM References option to **Surface** as shown.

5. Activate the **Select geometry** option by clicking on the arrow button as shown.

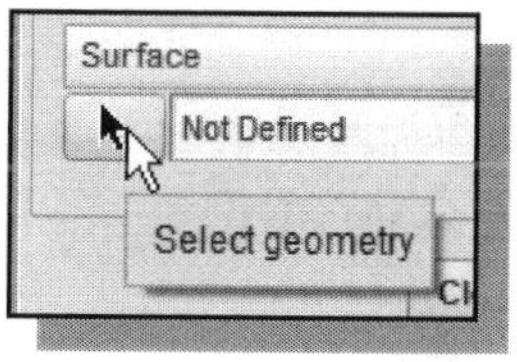

6. On your own, select the **front surface** of the model.

7. Click once with the **right-mouse-button** to accept the selection.

8. Inside the *AutoGEM* dialog box, click **Create** to generate the P-mesh.

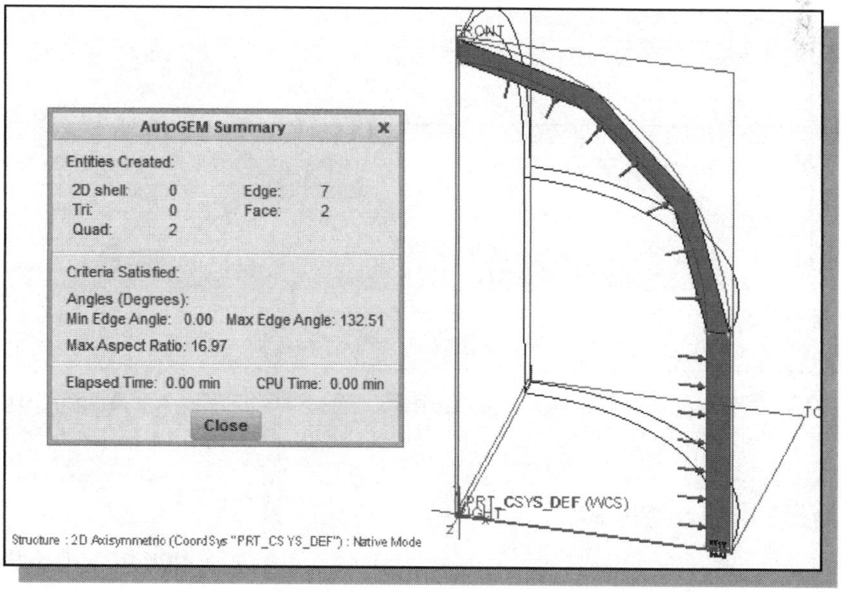

- Note that *Creo Simulate* generated only **two elements** and **seven edges**. This is the main characteristic of the P-element method, using very coarse mesh for the analysis.

Run the FEA Solver

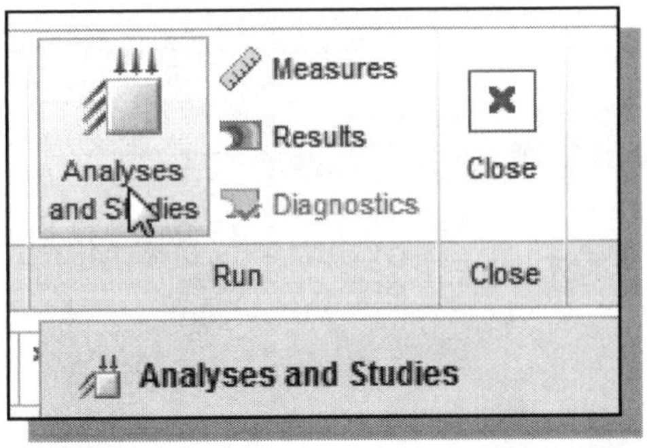

1. Choose **Home** in the *Ribbon* toolbar.

2. Activate the *Creo Simulate* **Analyses and Studies** command by selecting the icon in the toolbar area as shown.

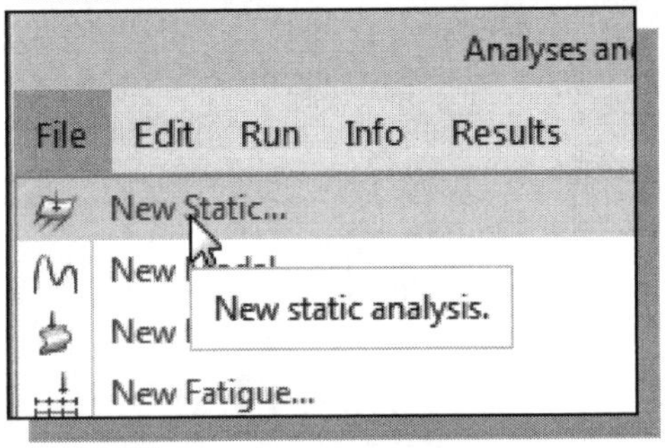

3. In the *Analyses and Design Studies* window, choose **File → New Static** in the pull-down menu. This will start the setup of a basic linear static analysis.

4. Enter **Axisymmetric** as the analysis Name as shown. Also confirm that **ConstraintSet1** and **LoadSet1** are automatically accepted as part of the analysis parameters.

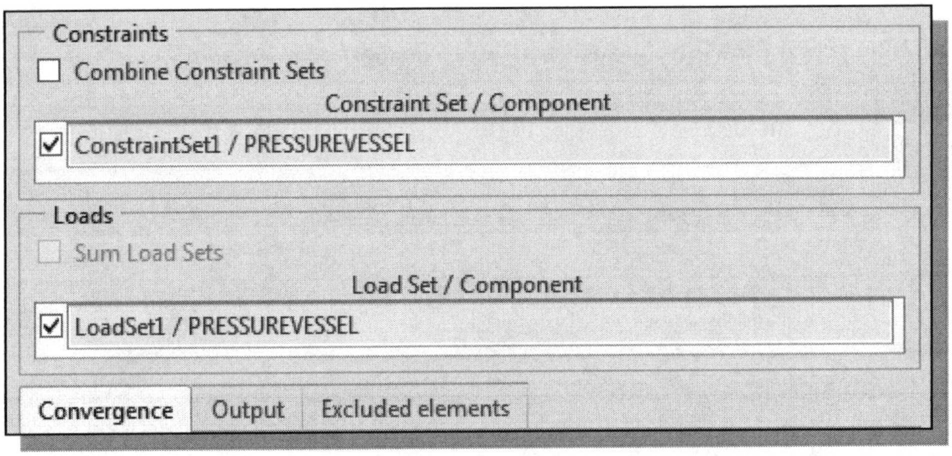

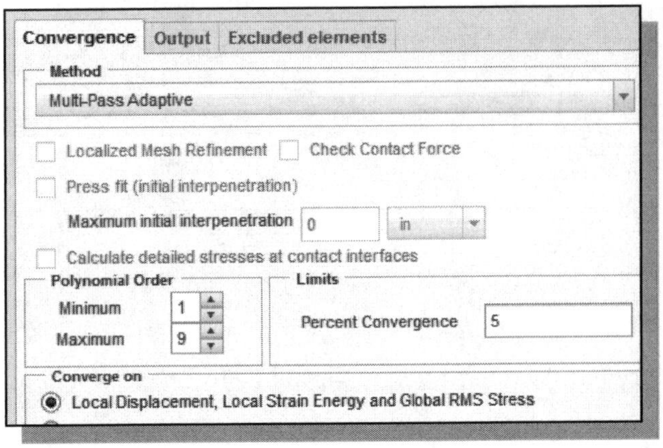

5. Set the Method option to **Multi-Pass Adaptive**, P-order to **9** and Limits to **5** Percent Convergence as shown.

6. Click **OK** to accept the settings.

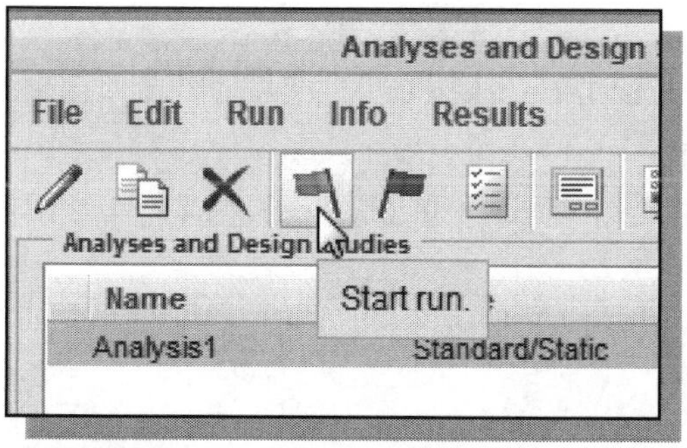

7. Click **Start run** to begin the *Solving* process.

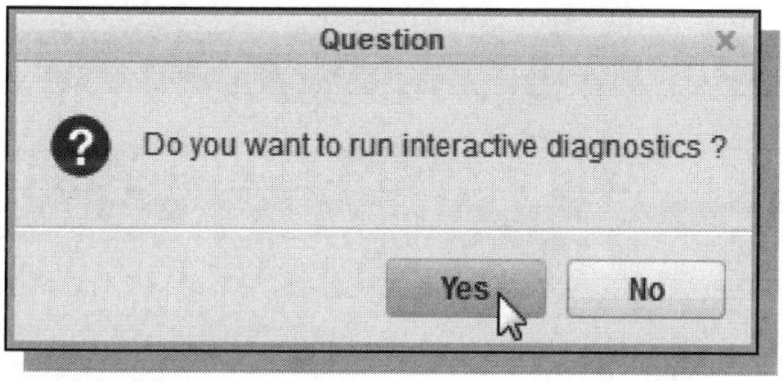

8. Click **Yes** to run interactive diagnostics; this will help us identify and correct problems associated with our FEA model.

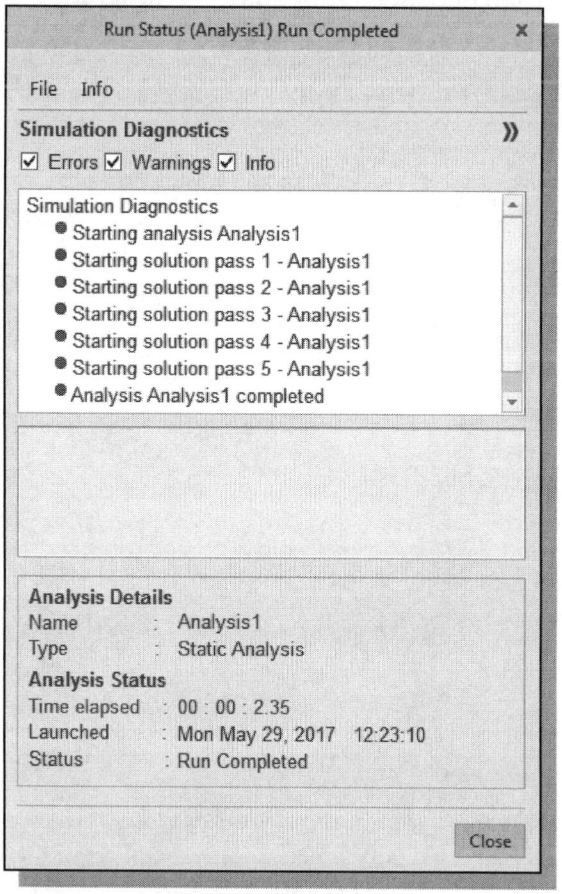

9. In the *Diagnostics* window, *Creo Simulate* will list any problems or conflicts of our FEA model.

10. Click on the **Close** button to exit the *Diagnostics* window.

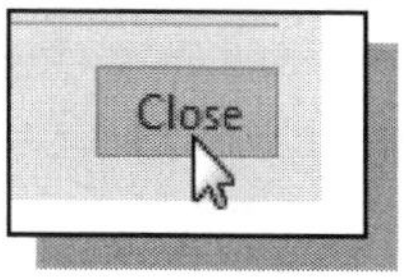

11. In the *Analyses and Design Studies* window, choose **Display study status** as shown.

❖ Note that *Creo Simulate* started with solving 10 equations during the first pass and went up to solving 57 equations to reach the 5% convergence. And it took a very short time for *Creo Simulate* to complete the analysis.

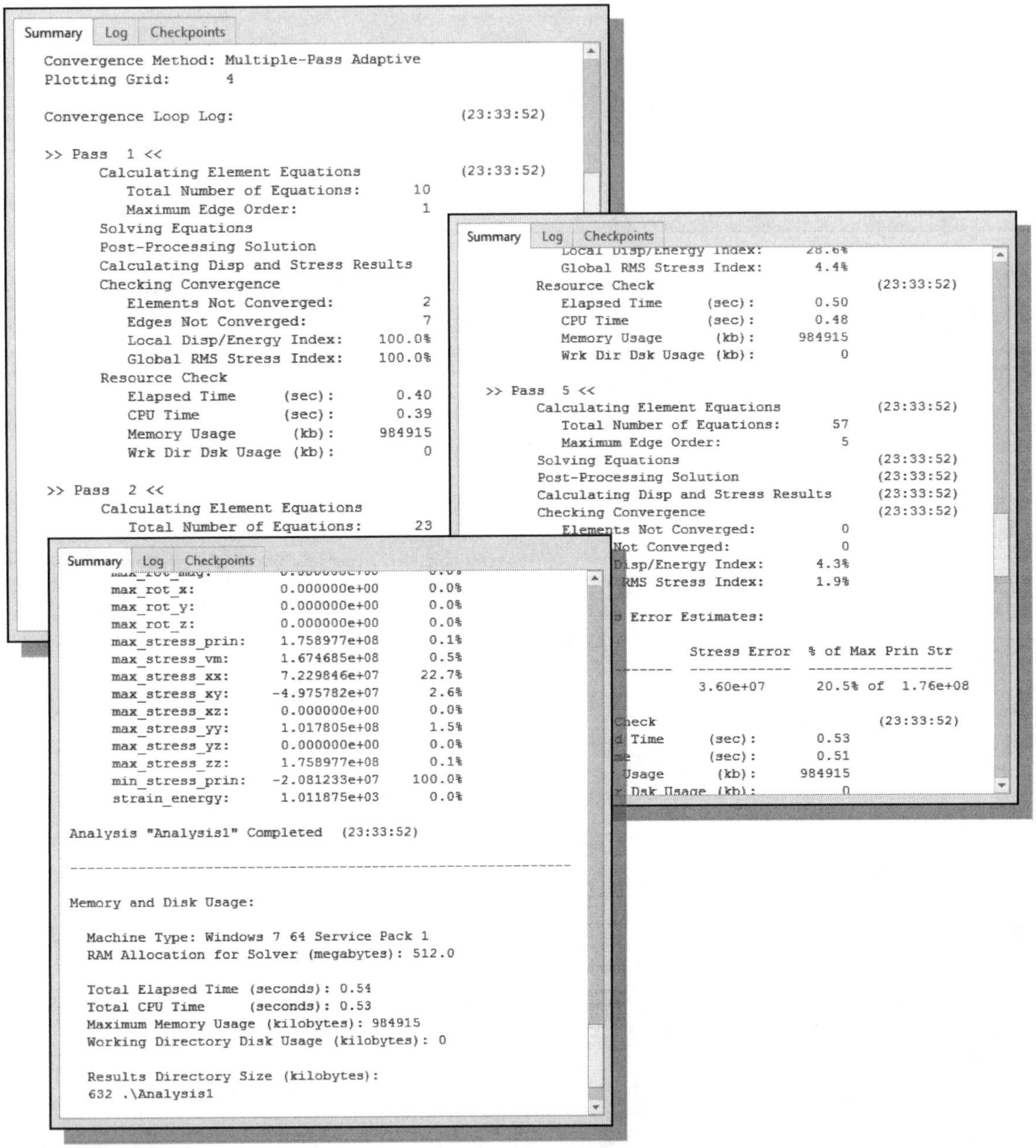

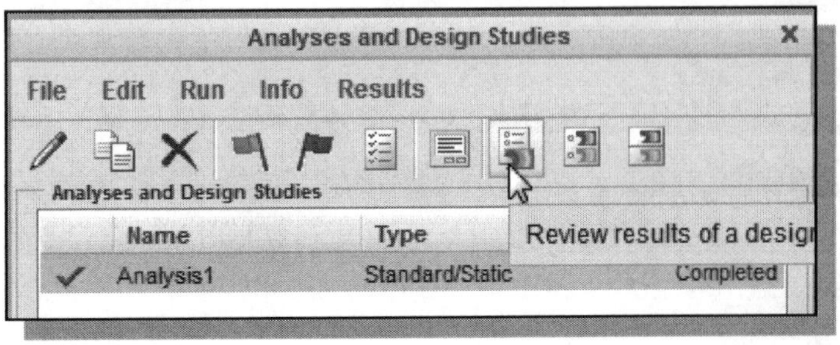

12. In the *Analyses and Design Studies* window, choose **Review results** as shown.

View the Von Mises Stress

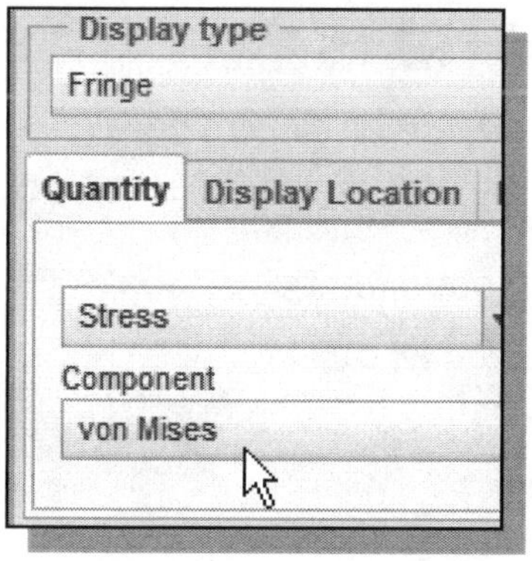

1. On your own, set the **Display type** to **Fringe**.

2. Choose **Von Mises** as the **Stress Component** to be displayed.

3. Click on the **Display Options** tab and switch ***on*** the **Deformed** and **Overlay** options as shown.

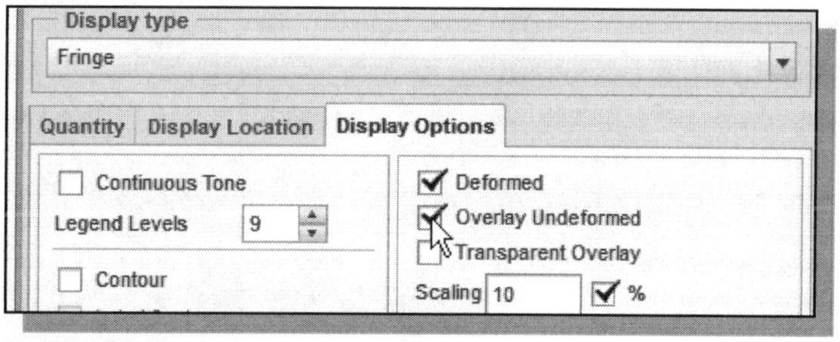

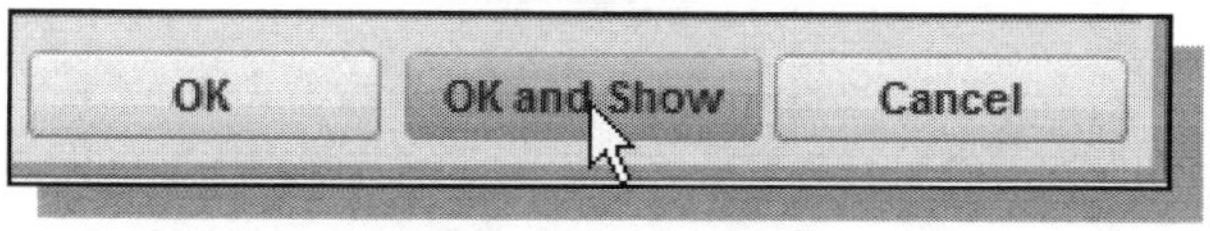

4. Click **OK and Show** to display the results.

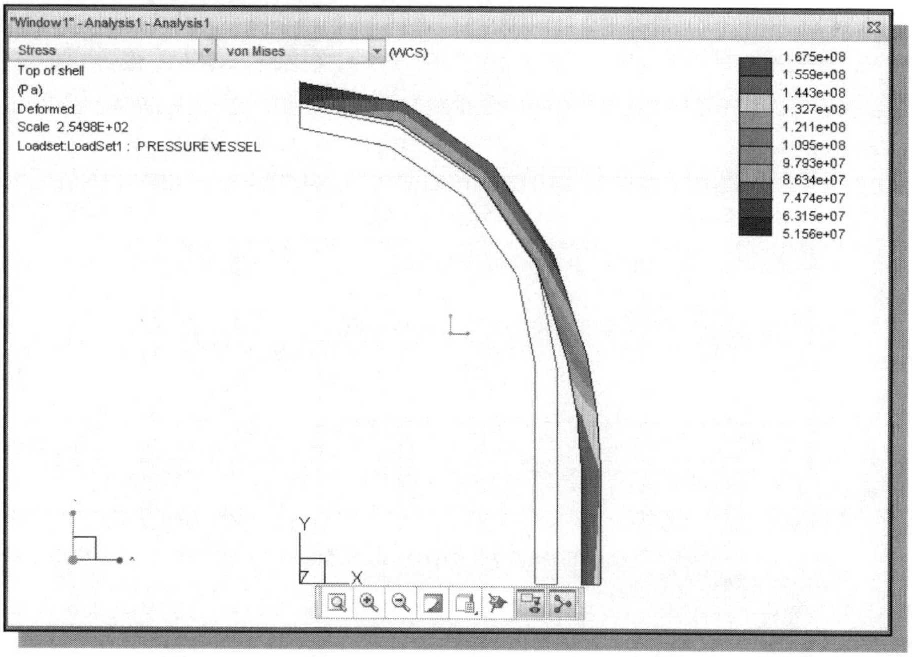

➢ *Creo Simulate* confirms that the maximum stress is **1.675E+8 N/m²**, which occurs on the cylindrical section of the pressure vessel. This number is very close to the **165MPa** we calculated during our preliminary analysis.

Perform a 3D Shell Analysis

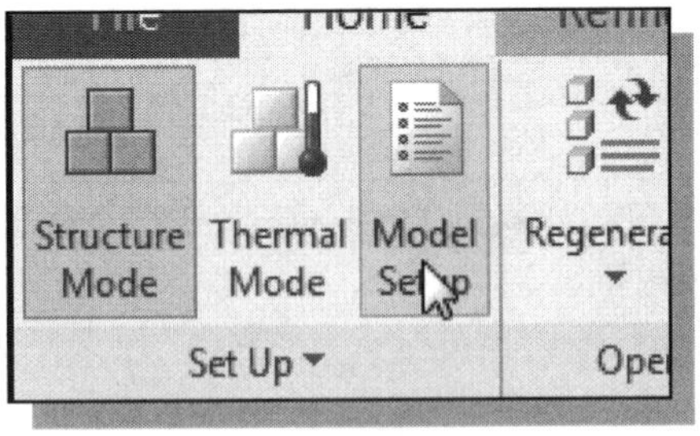

1. On your own, exit the **Review Results** command.

2. In the *Creo Simulate* window, select the *Creo Simulate* **Model Setup** option in the *Ribbon* toolbar as shown.

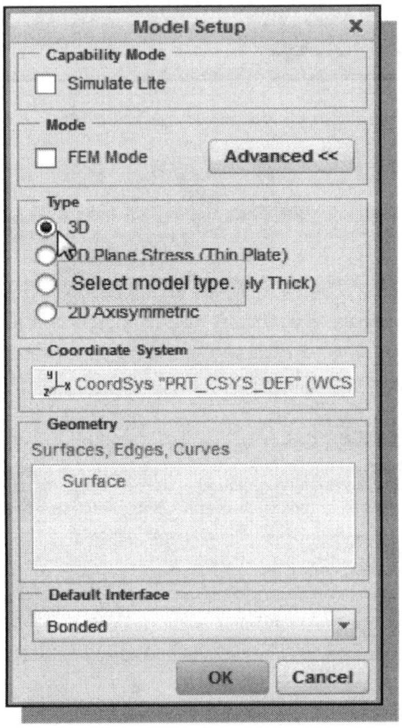

3. In the *Creo Simulate Model Setup* dialog box, set the model **Type** to **3D** as shown.

❖ Note that the 3D FEA model requires no additional references.

4. Click the **OK** button to accept the settings.

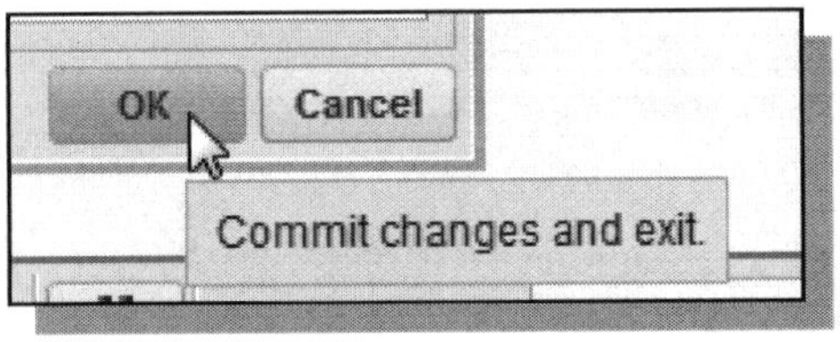

5. The *Confirmation* dialog box appears on the screen, indicating that all simulation modeling entities will be deleted. Only geometric entities and predefined measures will remain. Click **Confirm** to accept the deletion of the axisymmetric constraint set and load set.

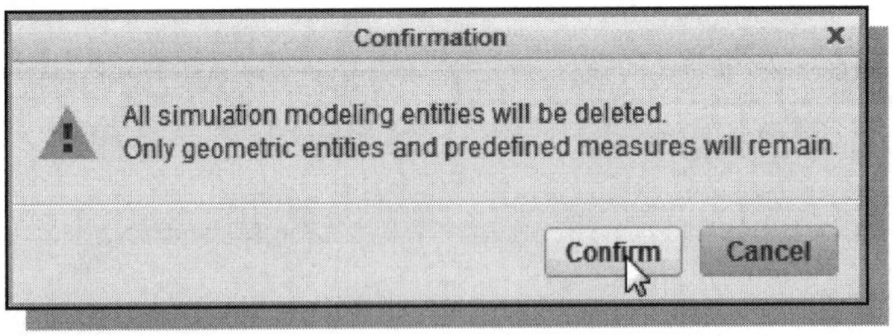

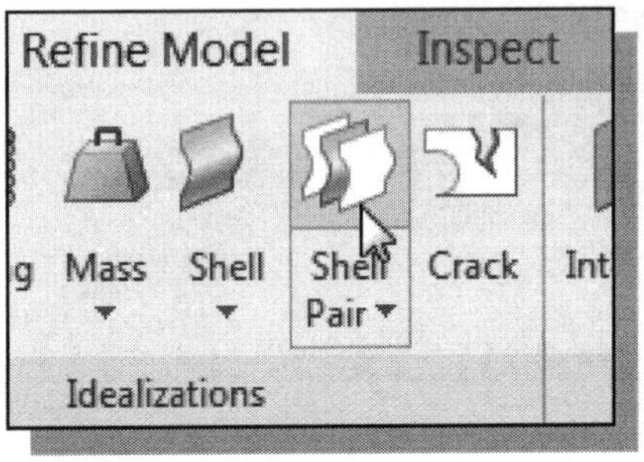

6. Choose **Refine Model** in the *Ribbon* toolbar.

7. Choose **Shell Pair** by clicking the icon in the toolbar as shown.

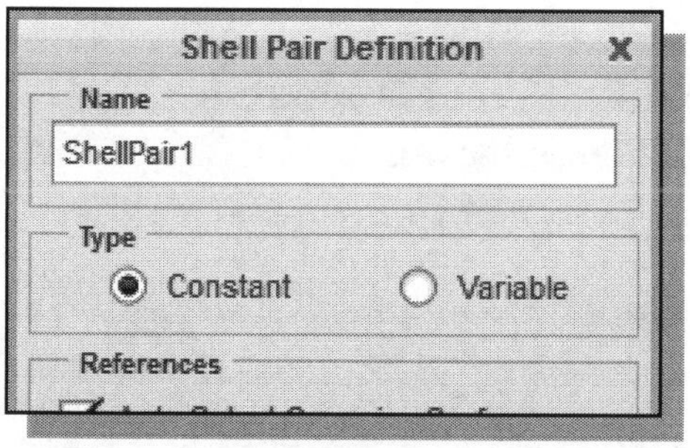

8. In the *Menu Manager*, confirm the shell options are set to **Constant** thickness as shown.

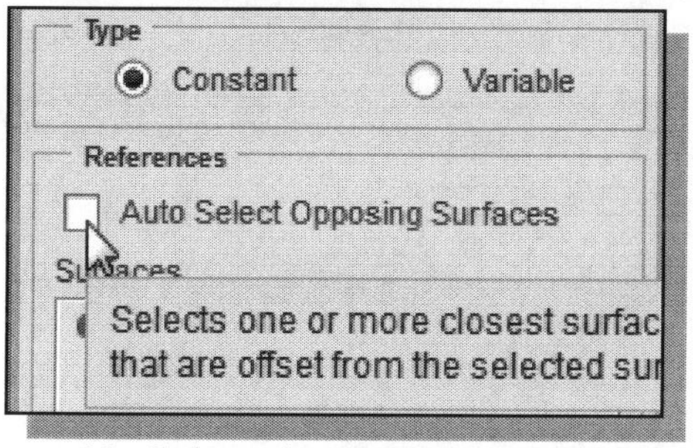

9. Turn ***off*** the **Auto Select Opposing Surfaces** option as shown.

10. Select the **two cylindrical surfaces** while holding down the [**Ctrl**] key.

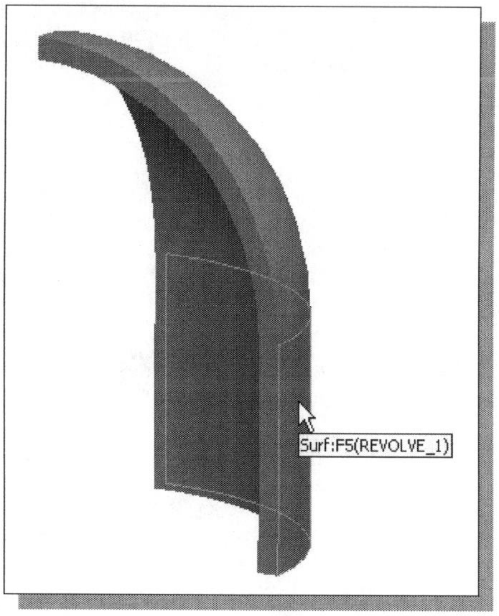

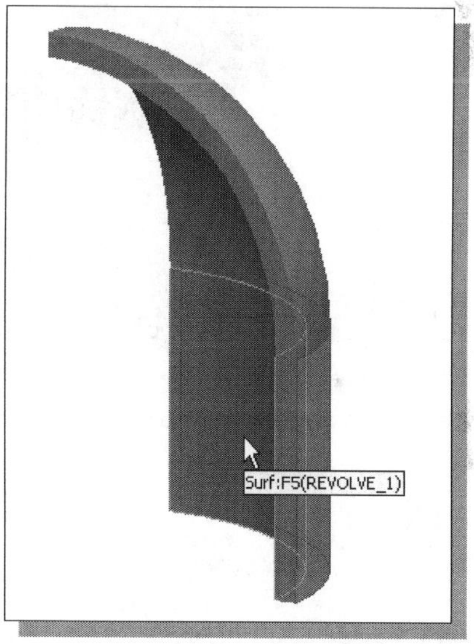

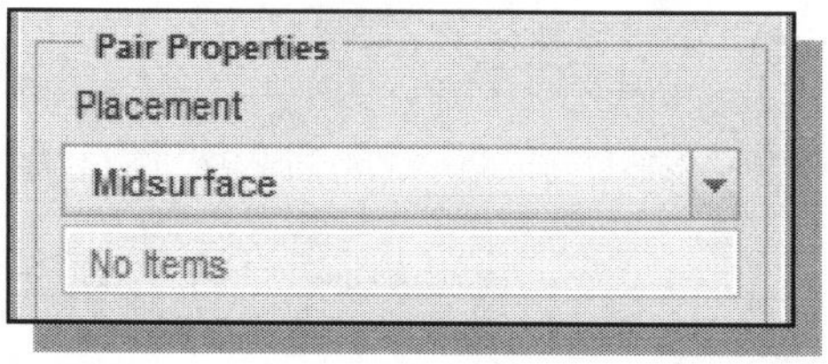

11. Confirm the **Placement** option is set to **Midsurface** in the **Pair Properties** section.

❖ The FEA calculation will be done at the midsurface of the selected pair.

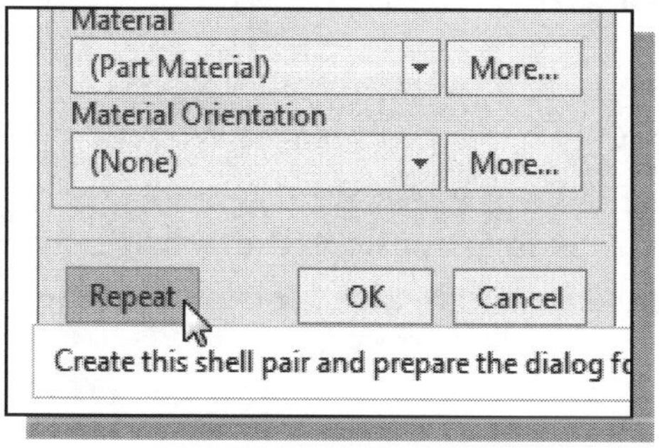

12. Click **Repeat** to create the selected paired surfaces.

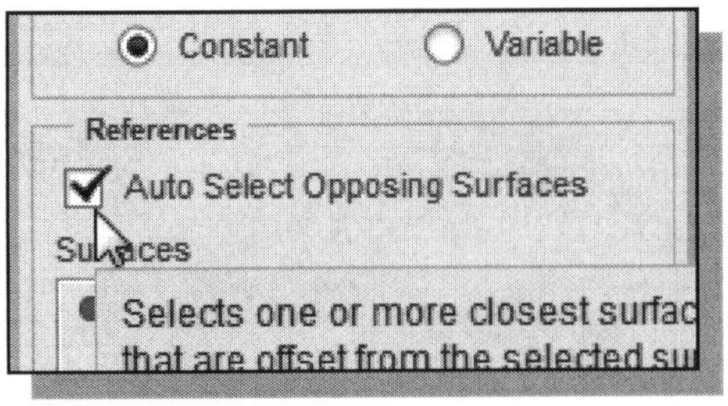

13. Turn ***on*** the **Auto Select Opposing Surfaces** option as shown.

14. Select the **outside hemispherical surface** and notice the corresponding surface is automatically selected with the **Auto Select** option switched ***on***.

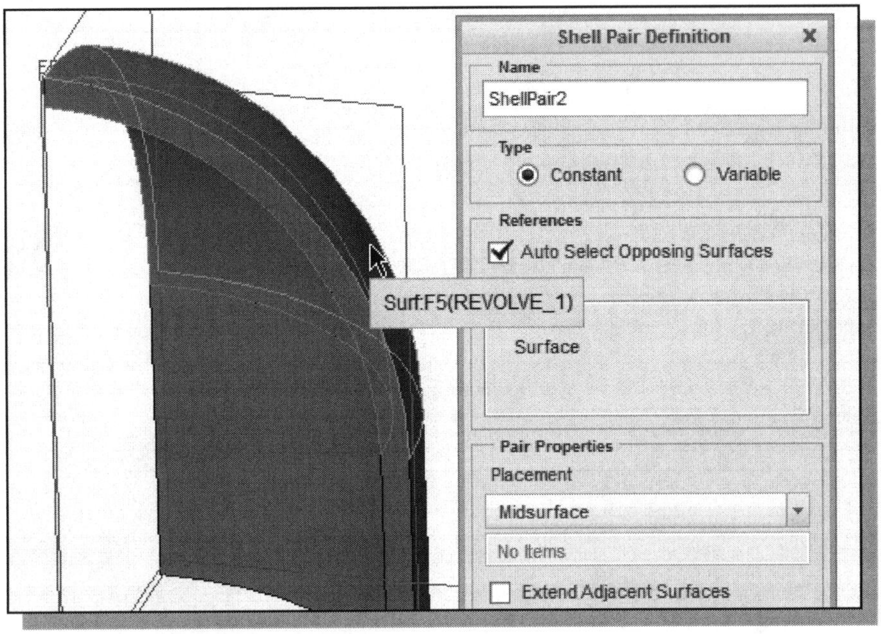

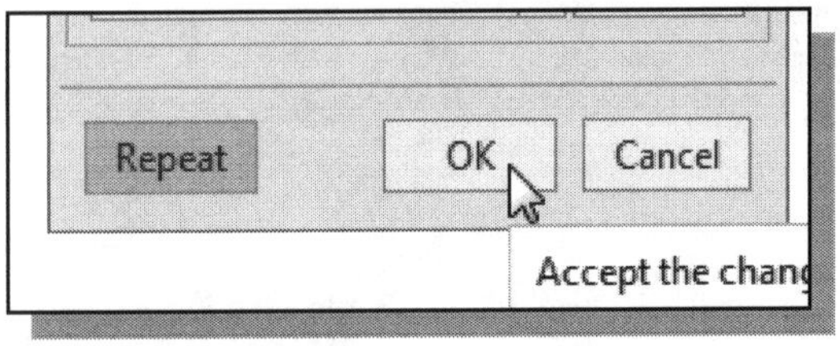

15. Click **OK** to create the paired surface FEA model.

❖ We have created two shell pairs. *Creo Simulate* will generate a new shell model using the two shell pairs we just created. Note that the shell model will have all the properties of the solid model, except the shell pairs are compressed to the mid-surface position.

Apply the Boundary Conditions – Constraints

For our shell analysis, the model we created is a ⅛ solid model of the pressure vessel which is symmetrical about the horizontal mid-plane and two vertical planes. We will need to apply proper constraints to the three symmetry planes. The most important differences between 3D shell analysis and 3D solid analysis is in applying the constraints; **shell elements have rotational degrees of freedom** and **the constraints for shell model must be applied to edges or curves**.

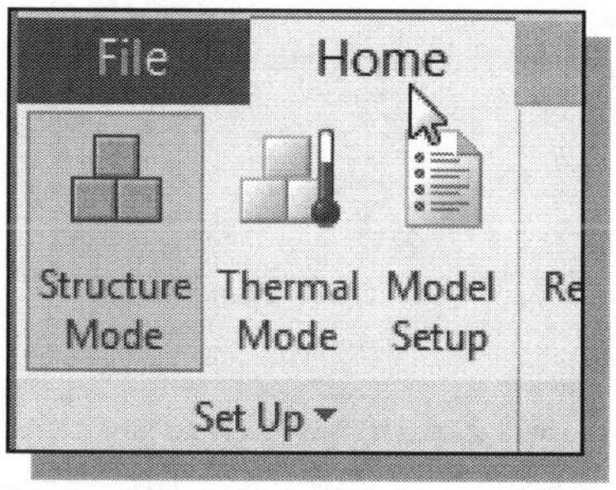

1. Choose **Home** in the *Ribbon* toolbar.

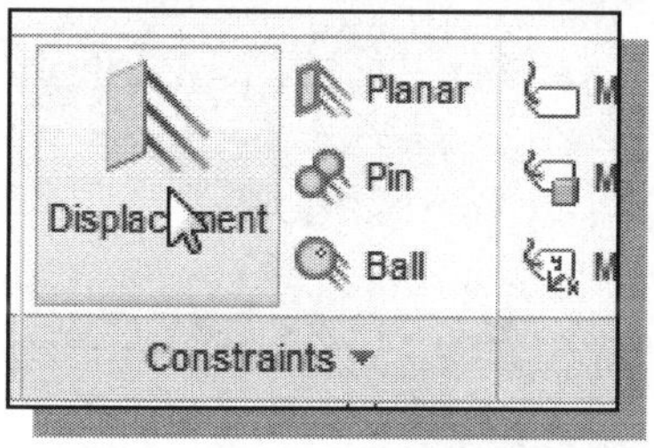

2. Choose **Displacement Constraint** by clicking the icon in the toolbar as shown.

3. Set the *References* option to **Edges/Curves** and pick the **two outside edges** on the XY plane as shown.

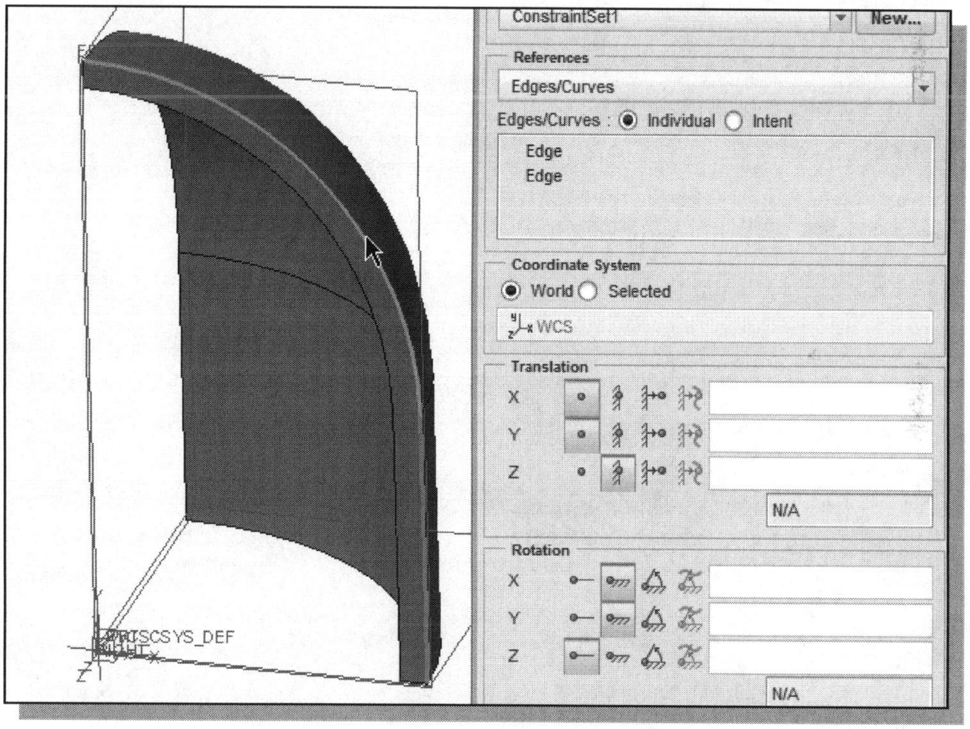

4. **Fix** the Z Translation and **Free** the X and Y Translations.

5. **Free** the Z Rotation and **Fix** the X and Y Rotations.

❖ The above constraints are applied based on the related symmetry; we are restricting all out-of-plane motion, but allowing all in-plane motion.

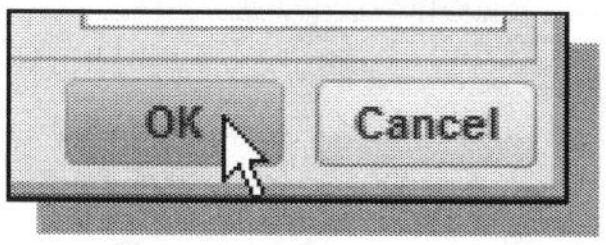

6. Click on the **OK** button to accept the first displacement constraint settings.

7. On your own, repeat the above steps and apply the necessary constraints to the other two edges as shown in the figures below.

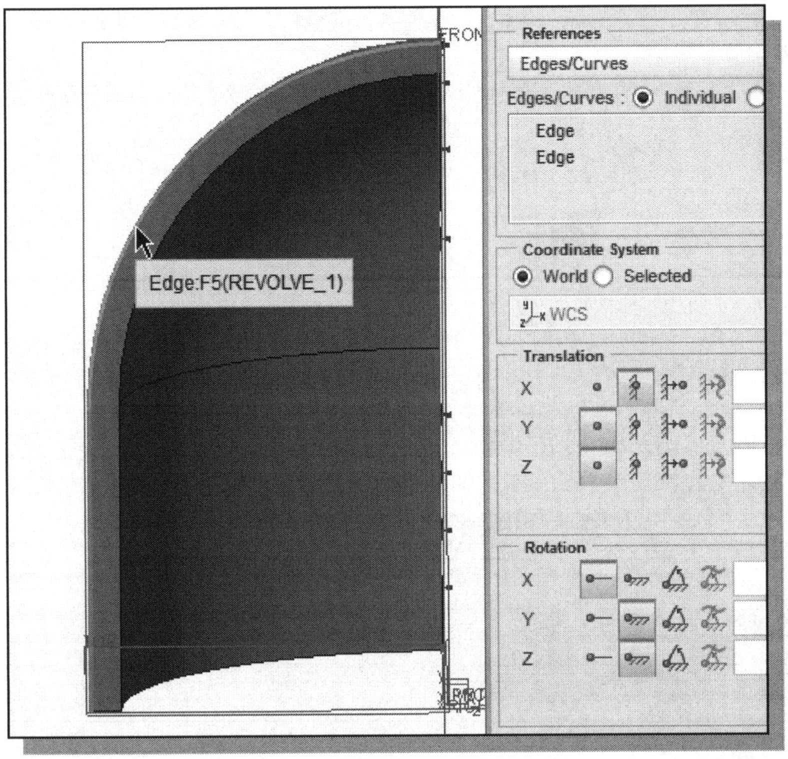

➢ Second vertical plane:

❖ **Fix** the X Translation and **Free** the Z and Y Translations.

❖ **Free** the X Rotation and **Fix** the Z and Y Rotations.

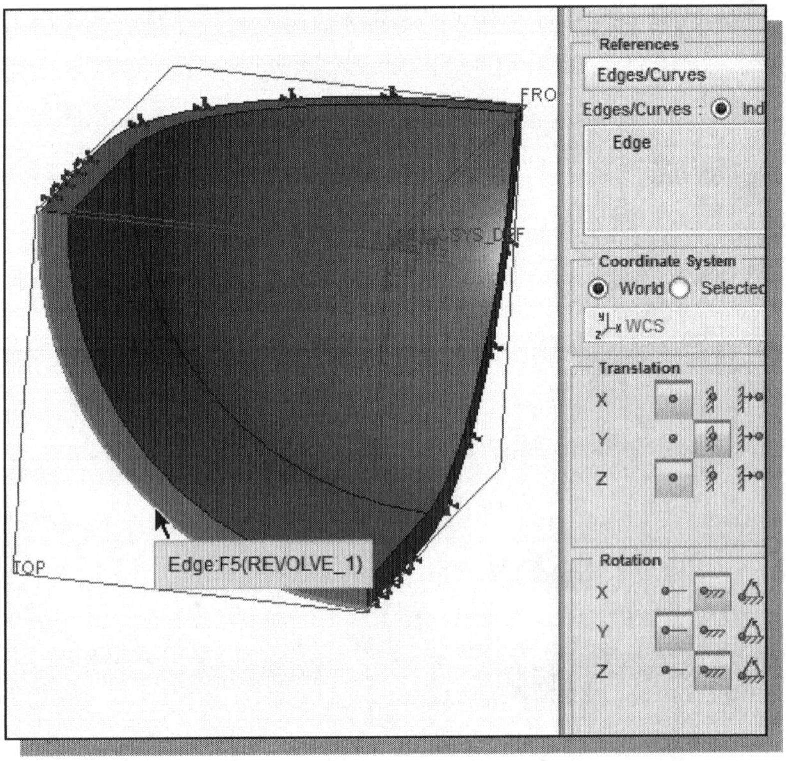

➢ Horizontal plane:

❖ **Fix** the Y Translation and **Free** the Z and X Translations.

❖ **Free** the Y Rotation and **Fix** the Z and X Rotations.

Apply the Internal Pressure

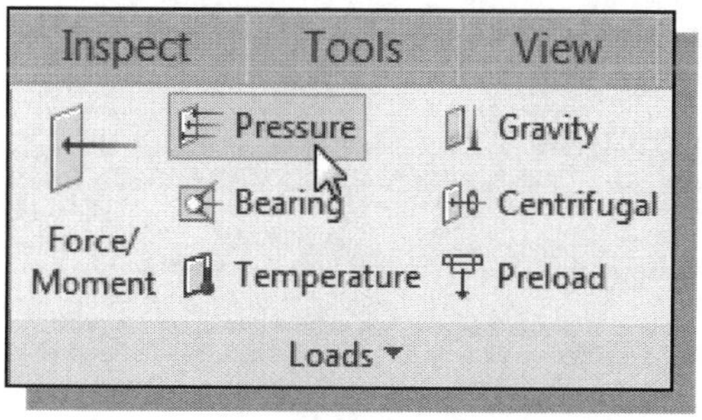

1. Choose **Pressure Load** by clicking the icon in the toolbar as shown.

2. Confirm the References option is set to **Surfaces**.

3. Pick the **two inside surfaces** while holding down the **[Ctrl]** key.

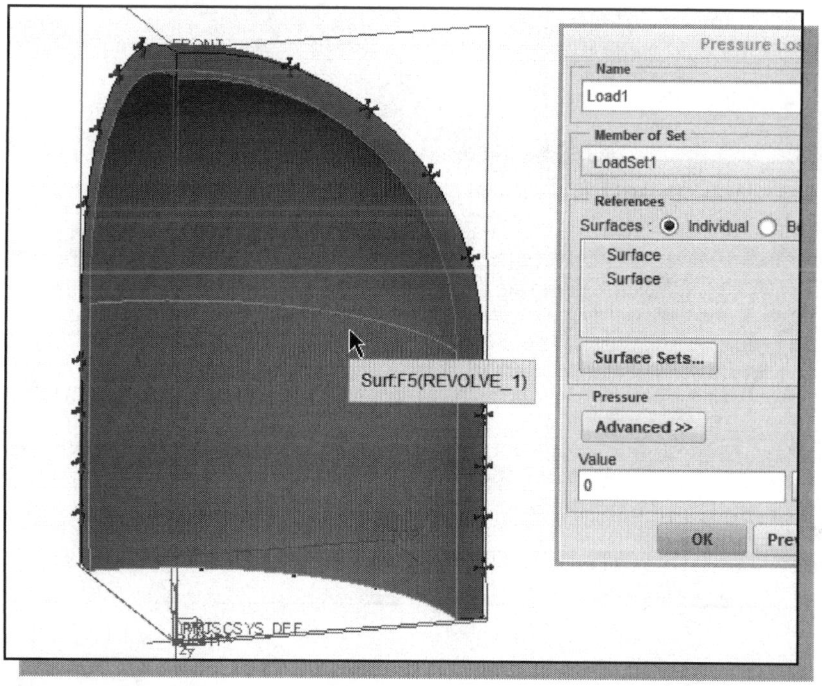

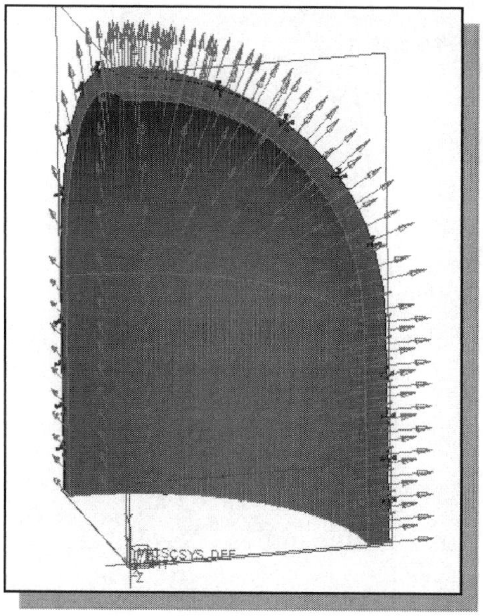

4. Enter **15E6** in the Pressure Value box as shown.

5. Click on the **Preview** button to confirm the applied load is pointing in the correct direction.

6. Click on the **OK** button to accept the first load settings.

Create the 3D Shell Mesh

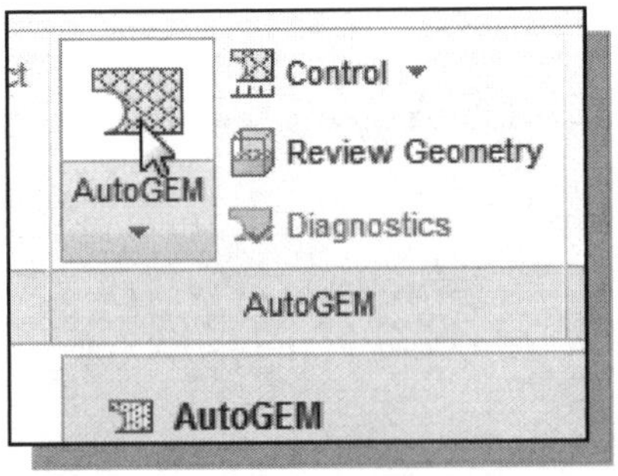

1. Choose **Refine Model** in the *Ribbon* toolbar.

2. Activate the **AutoGEM** command by selecting the icon in the toolbar area as shown.

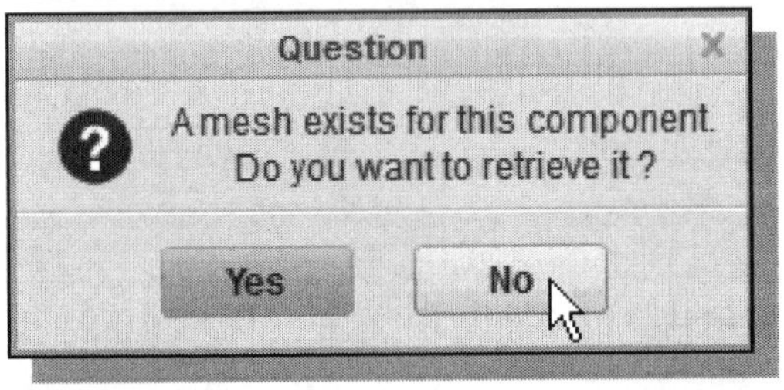

3. Click **No** to decline the use of the old mesh.

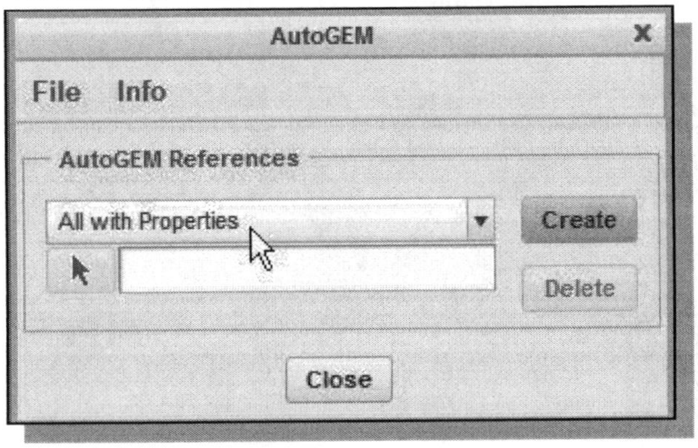

4. Confirm the AutoGEM References option is set to **All with Properties** as shown.

5. Inside the *AutoGEM* dialog box, click **Create** to generate the P-mesh.

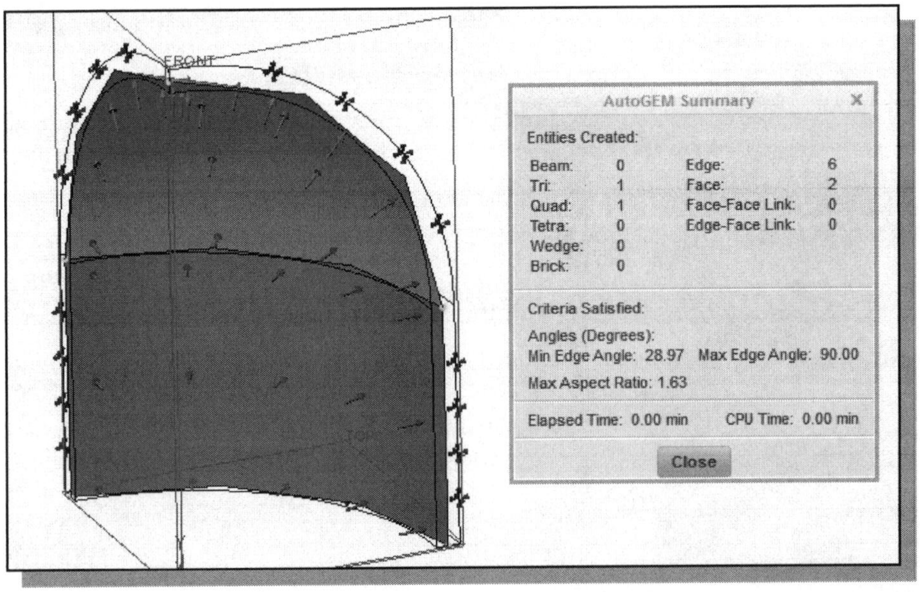

➢ Note that *Creo Simulate* generated only **two elements** and **five nodes**. This is very similar to the mesh of the axisymmetric analysis.

Run the FEA Solver

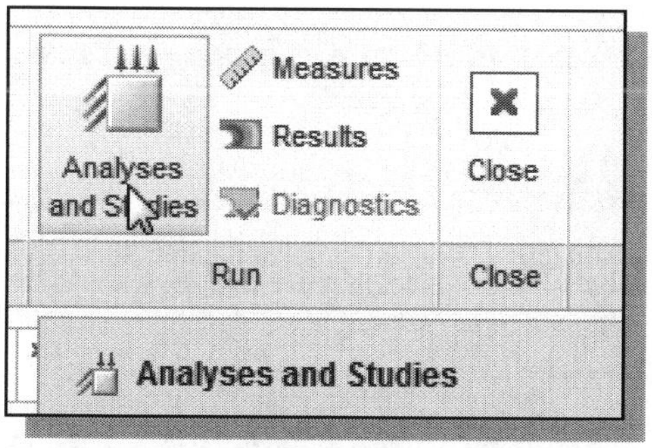

1. Activate the *Creo Simulate* **Analyses and Studies** command by selecting the icon in the toolbar area as shown.

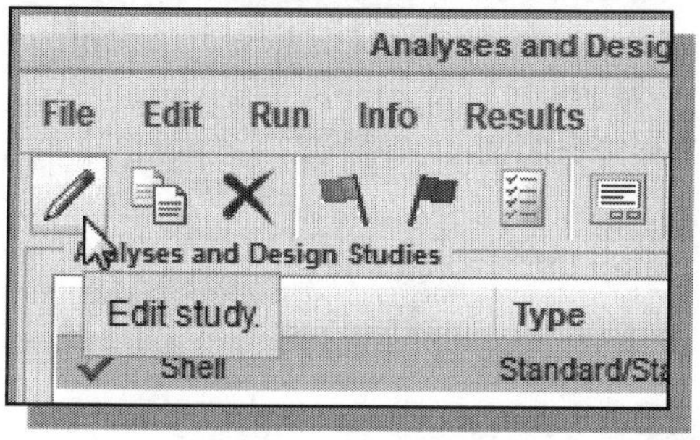

2. In the *Analyses and Design Studies* window, choose **Edit Study** in the icon toolbar. This will allow us to edit the setup of existing linear static analysis.

3. Enter **Shell** as the analysis **Name** as shown. Also confirm that **ConstraintSet1** and **LoadSet1** are automatically accepted as part of the analysis parameters.

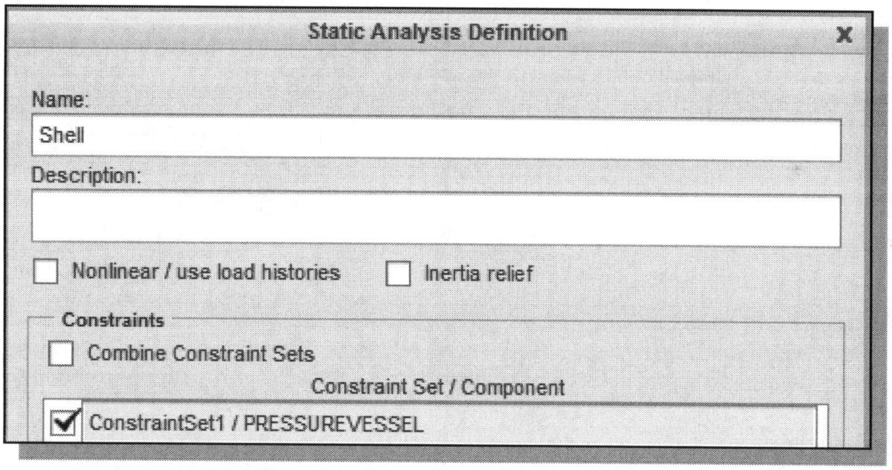

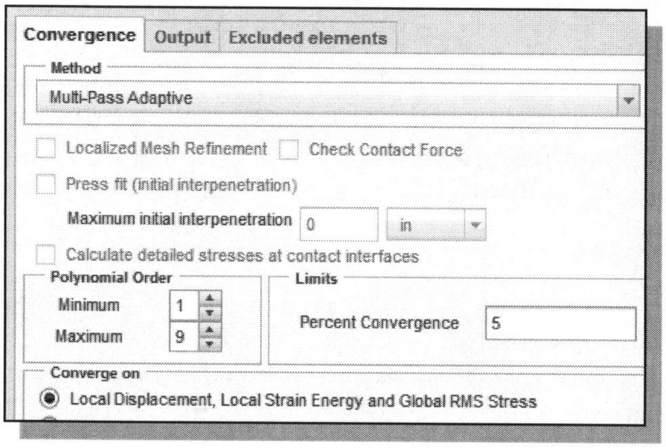

4. Set the Method option to **Multi-Pass Adaptive**, P-order to **9** and Limits to **5** Percent Convergence as shown.

5. Click **OK** to accept the settings.

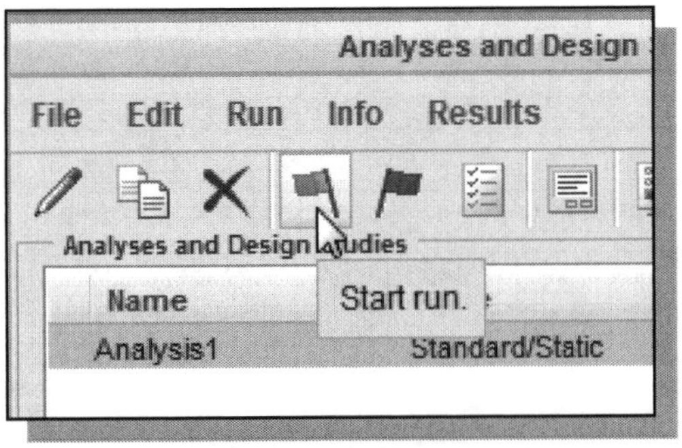

6. Click **Start run** to begin the *Solving* process.

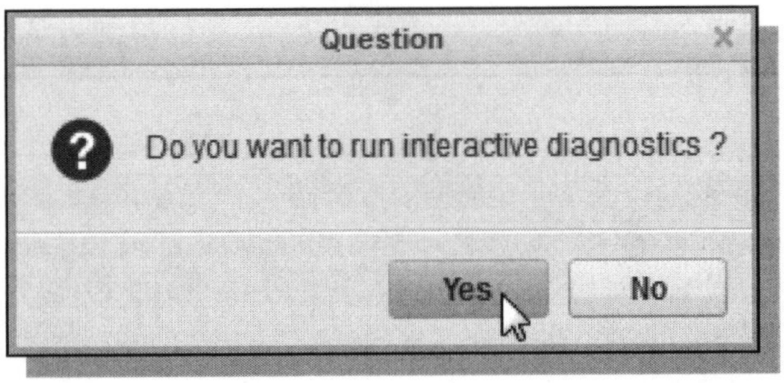

7. Click **Yes** to run interactive diagnostics; this will help us identify and correct problems associated with our FEA model.

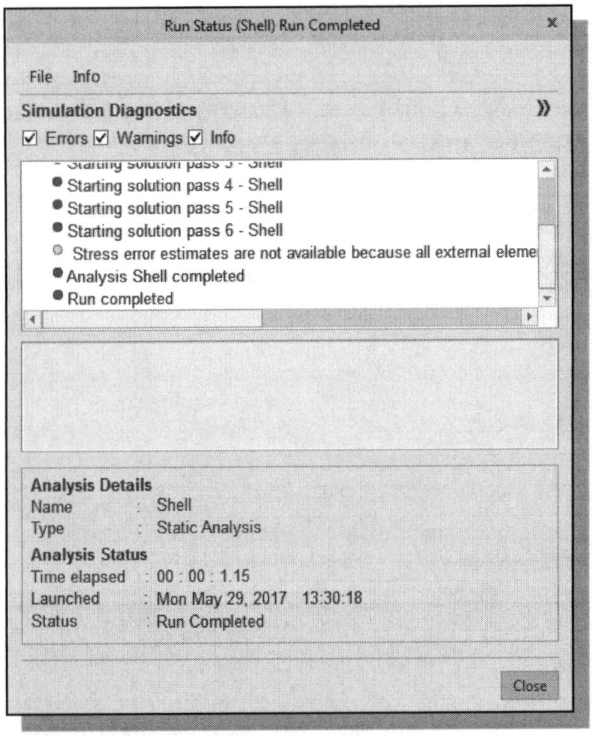

8. In the *Diagnostics* window, *Creo Simulate* will list any problems or conflicts of our FEA model.

9. Click on the **Close** button to exit the *Diagnostics* window.

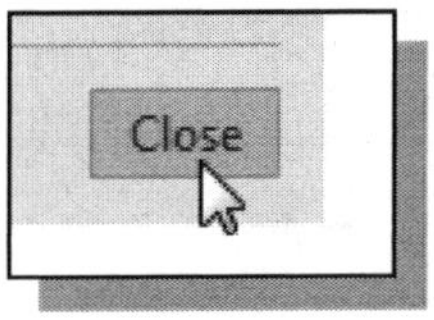

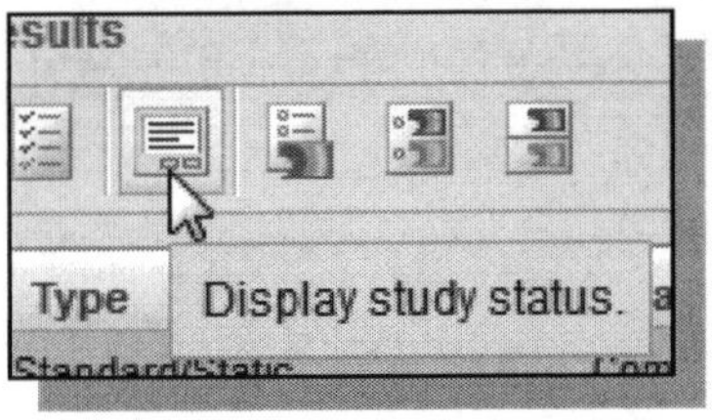

10. In the *Analyses and Design Studies* window, choose **Display study status** as shown.

11. On your own, examine the details of the analysis. How long did it take for *Creo Simulate* to complete this shell analysis?

View the Von Mises Stress

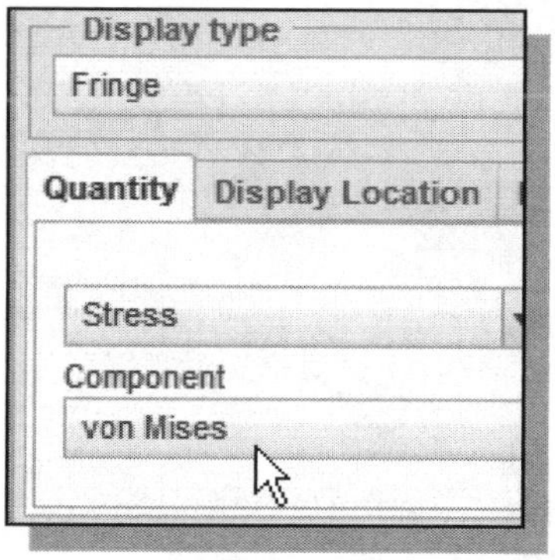

1. In the *Analyses and Design Studies* window, choose **Review results**.

2. On your own, set the Display type to **Fringe**.

3. Choose **Von Mises** as the Stress Component to be displayed.

4. Click on the **Display Options** tab and switch ***on*** the **Deformed** option as shown.

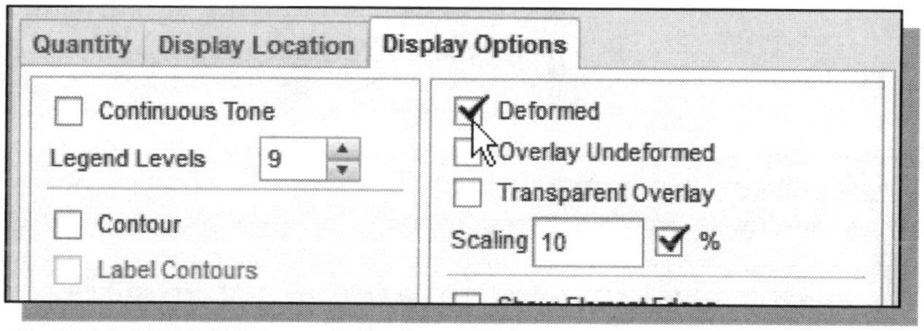

5. Click **OK and Show** to display the results.

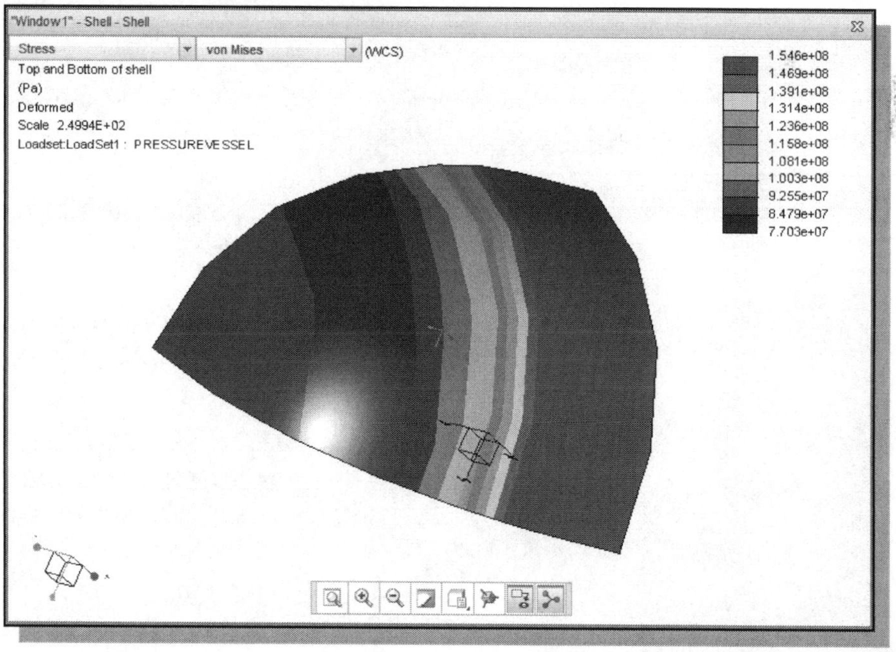

➢ *Creo Simulate* calculated the maximum stress to be 1.545E+8 N/m^2. This number is a bit lower than the 165MPa we calculated during our preliminary analysis. Note the shell model is based on the shell pairs defined in the original solid model.

Perform a 3D Solid Element Analysis

For our 3D solid analysis, the constraints need to be applied to the **surfaces** of the three symmetry planes. Note that the constraints for shell models must be applied to edges or curves.

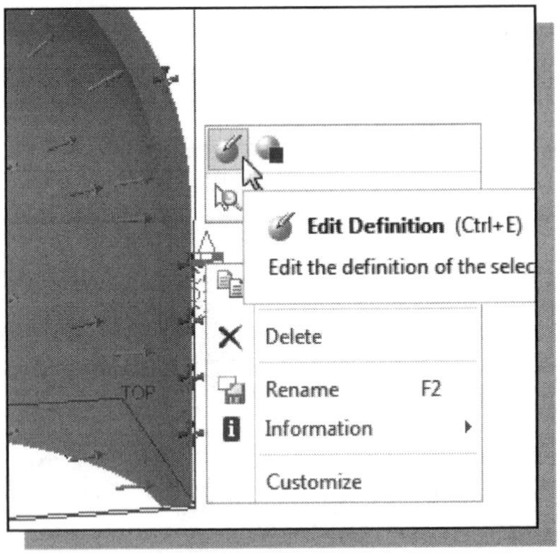

1. On your own, exit the *Review Results* window.

2. Select the **constraint** we applied on the two edges on the **XY plane**.

3. **Right-click** to bring up the option menu and select **Edit Definition** as shown.

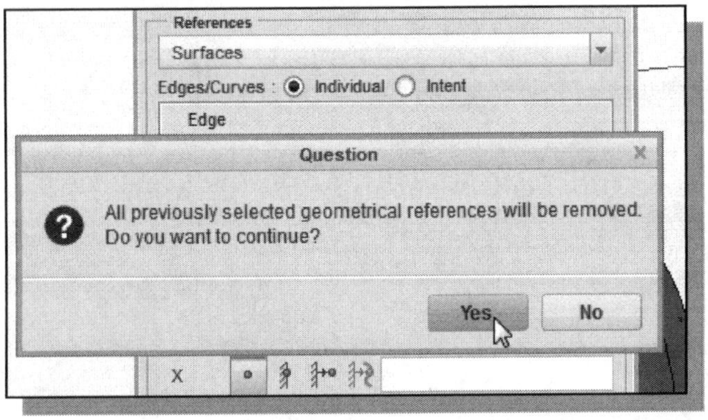

4. Switch the References option to **Surfaces** and pick **Yes** to remove the old constraint applied.

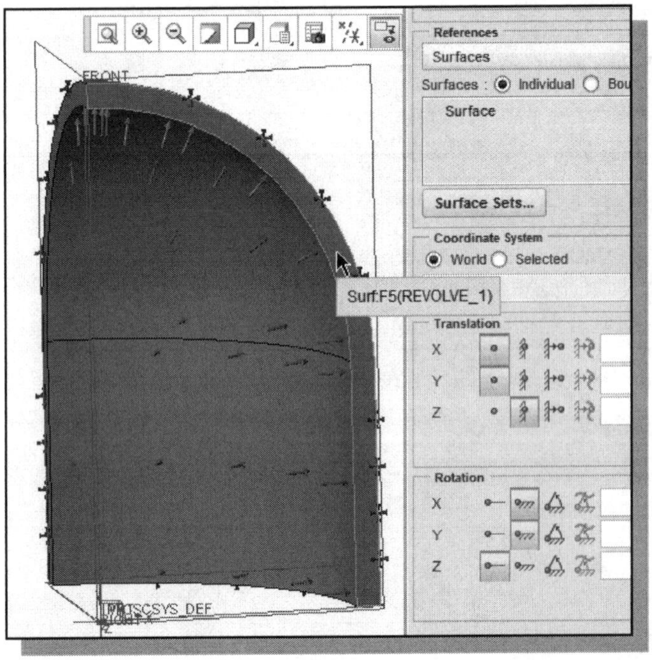

5. Pick the **surface** that is on the XY plane as shown.

6. Confirm the Translational constraints are still applied based on the symmetry conditions.

❖ Note the ***rotational constraints have no effect on solid elements.***

7. Click on the **OK** button to accept the first displacement constraint settings.

8. On your own, repeat the above steps and apply the necessary constraints to the other two surfaces as shown in the figures below.

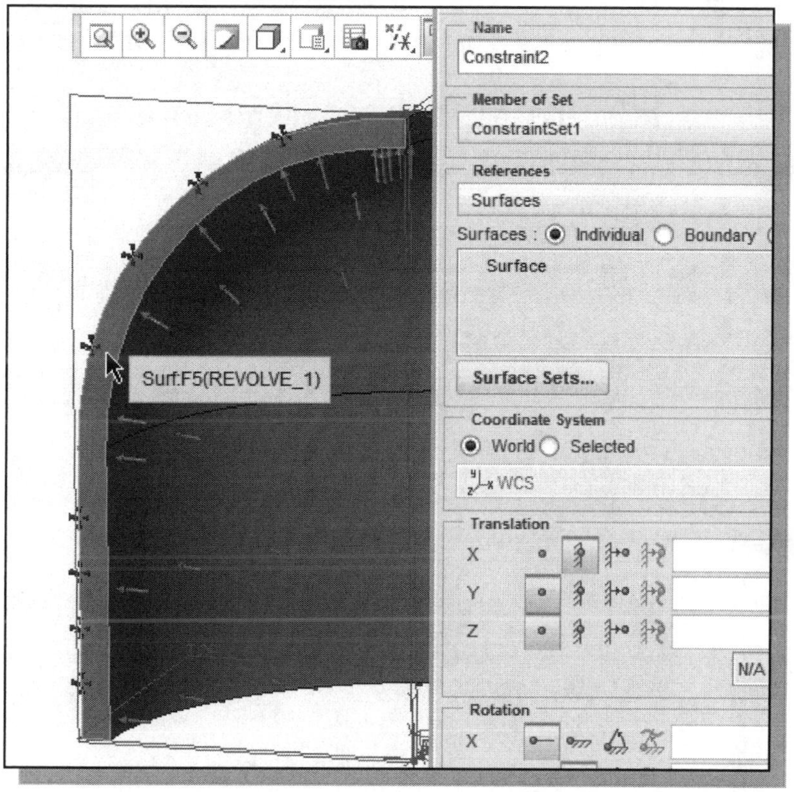

- The surface on the YZ plane.

❖ **Fix** the X Translation and **Free** the Z and Y Translations.

❖ Note the *rotational constraints* have no effects *on solid elements*.

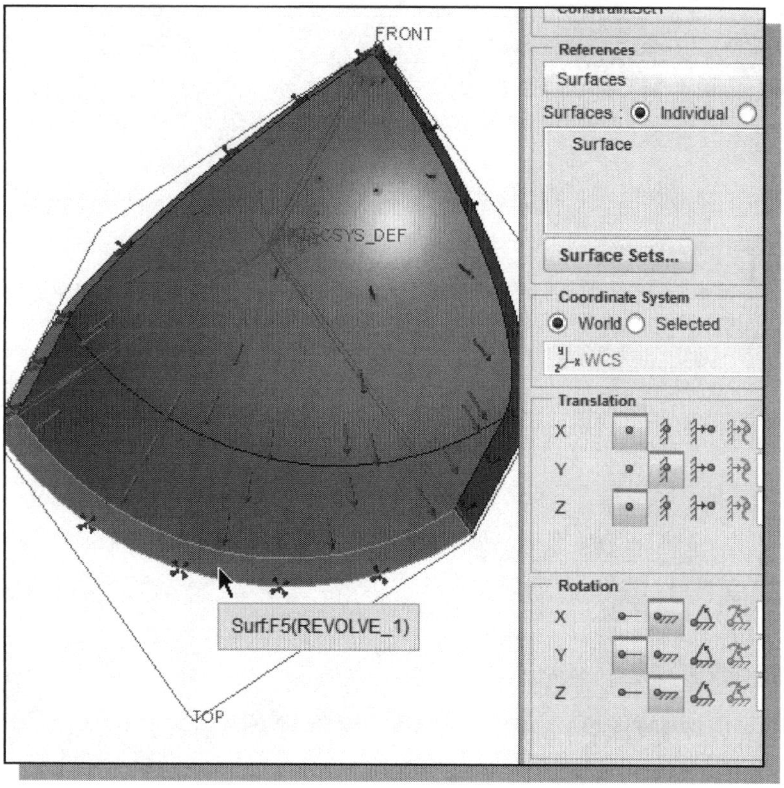

- The surface on the ZX plane:

❖ **Fix** the Y Translation and **Free** the Z and X Translations.

❖ Note the *rotational constraints* have no effect *on solid elements*.

Create the 3D Solid Mesh

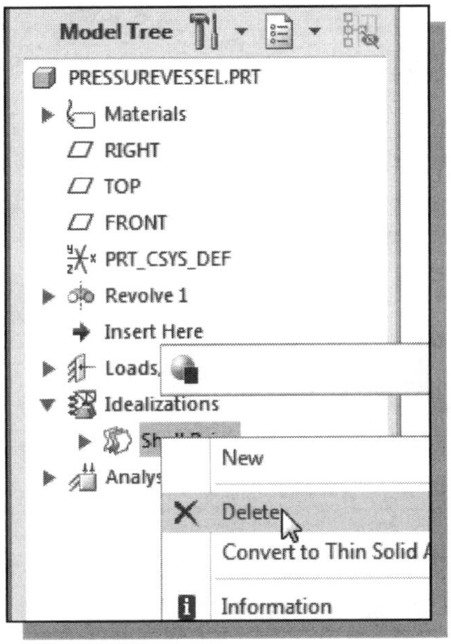

1. Inside the *Model Tree* window, expand the **Idealizations** list by clicking on the [>] symbol.

2. **Right-click** once on the **ShellPair** item to bring up the option menu and select **Delete** as shown.

3. Click **Yes** to delete the shell pairs.

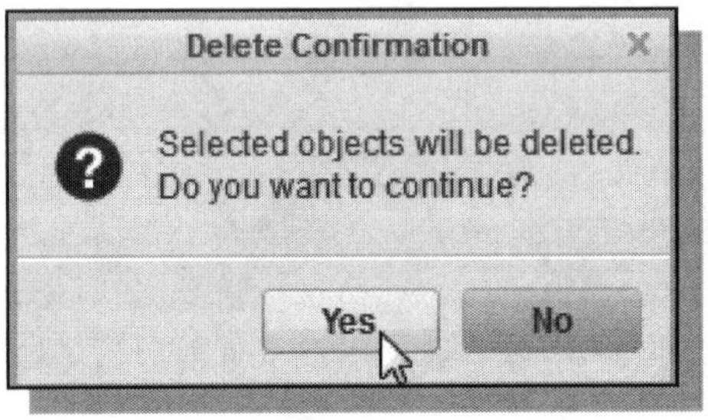

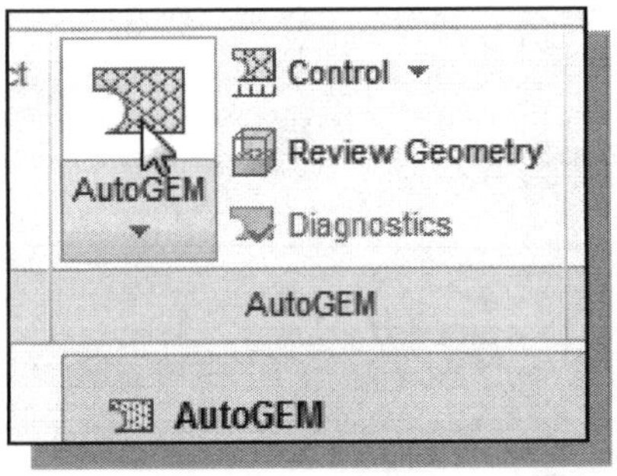

4. Activate the **AutoGEM** command by selecting the icon in the toolbar area as shown.

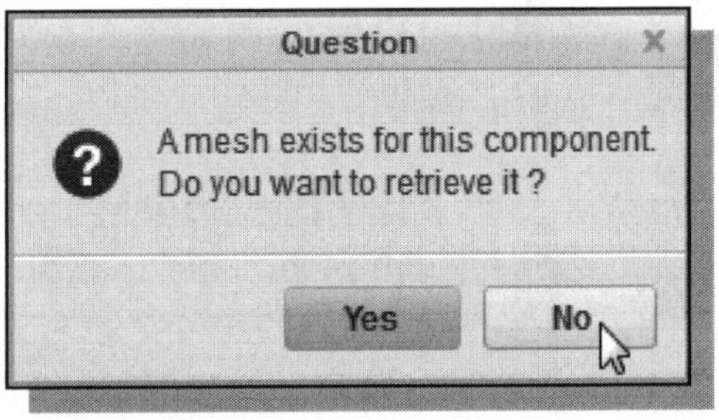

5. Click **No** to decline the use of the old mesh.

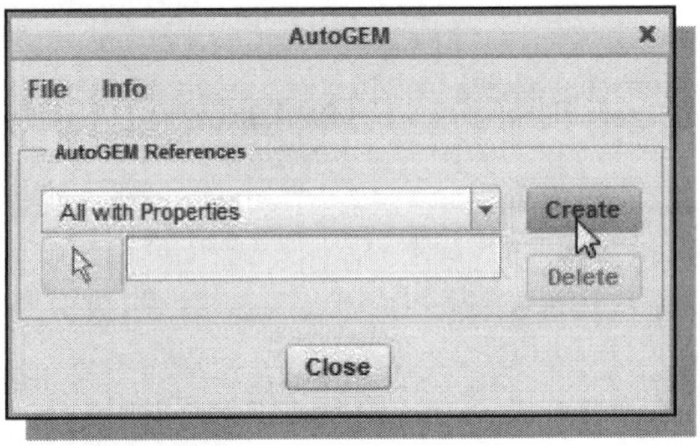

6. Confirm the AutoGEM References option is set to **All with Properties** as shown.

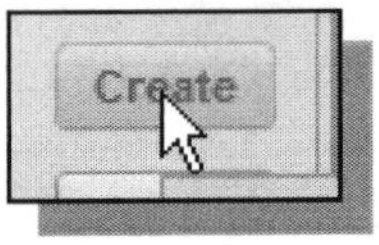

7. Inside the *AutoGEM* dialog box, click **Create** to generate the P-mesh.

➤ Note that *Creo Simulate* generated **155** elements and **70** nodes. These are much higher numbers in comparison to the meshes generated for the shell analysis and the axisymmetric analysis.

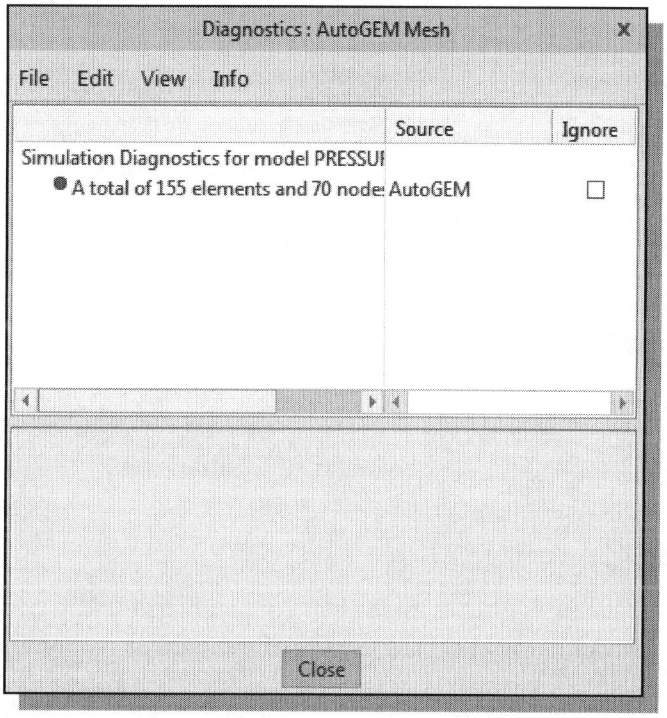

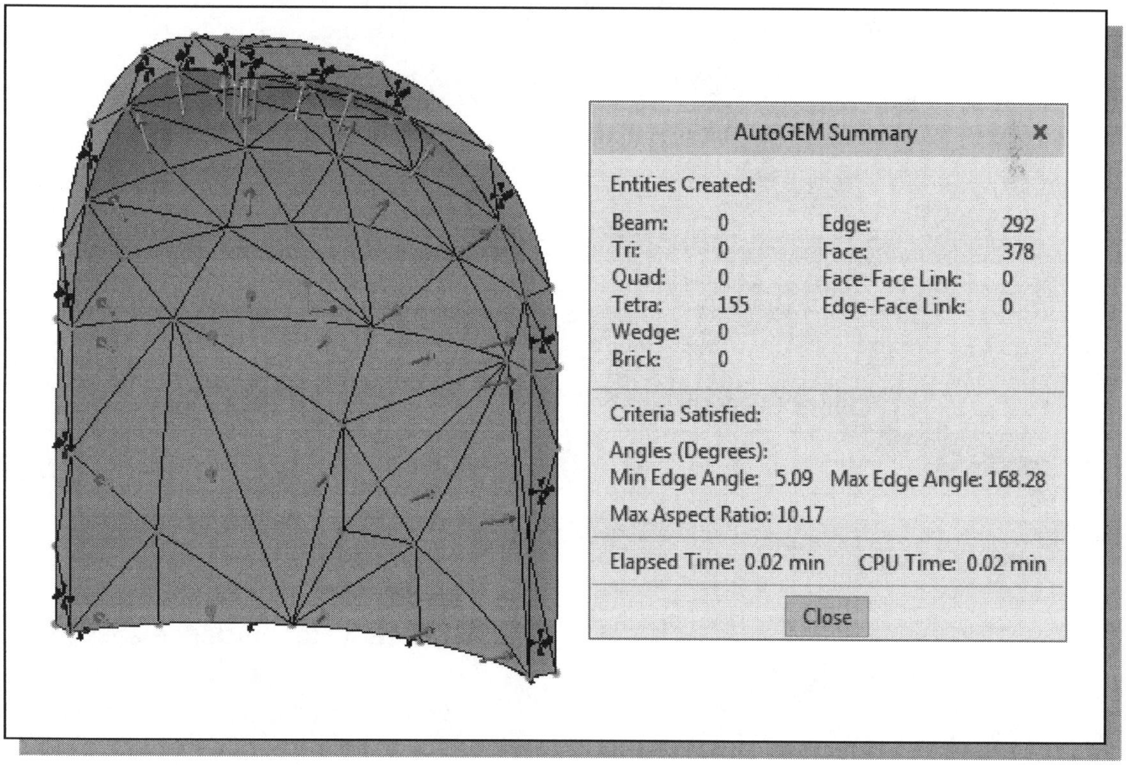

Run the FEA Solver

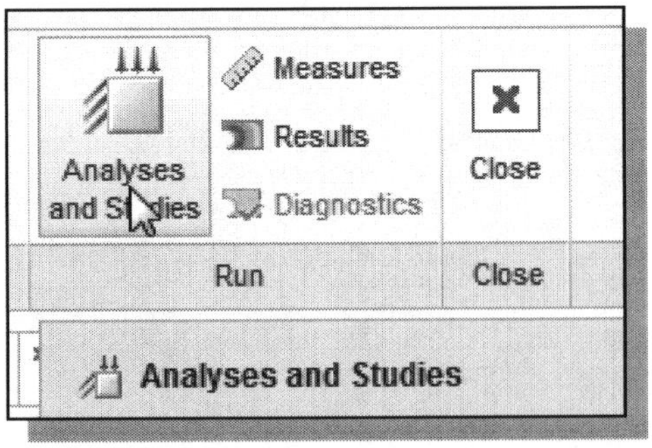

1. Choose **Home** in the *Ribbon* toolbar.

2. Activate the *Creo Simulate* **Analyses and Studies** command by selecting the icon in the toolbar area as shown.

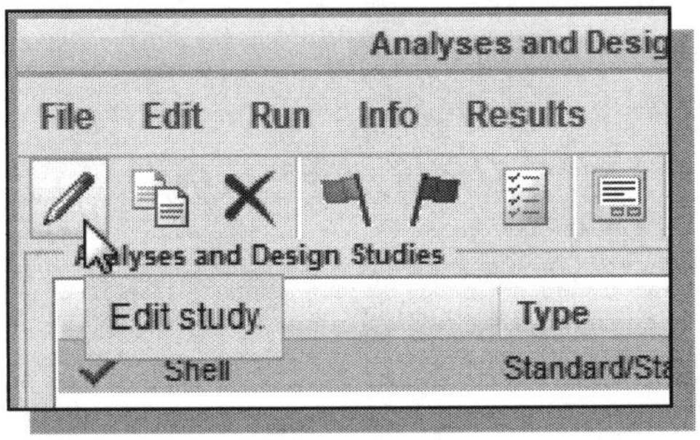

3. In the *Analyses and Design Studies* window, choose **Edit Study** in the toolbar. This will bring up the setup window for the current basic linear static analysis.

4. Enter **Solid** as the analysis Name as shown. Also confirm that **ConstraintSet1** and **LoadSet1** are automatically accepted as part of the analysis parameters.

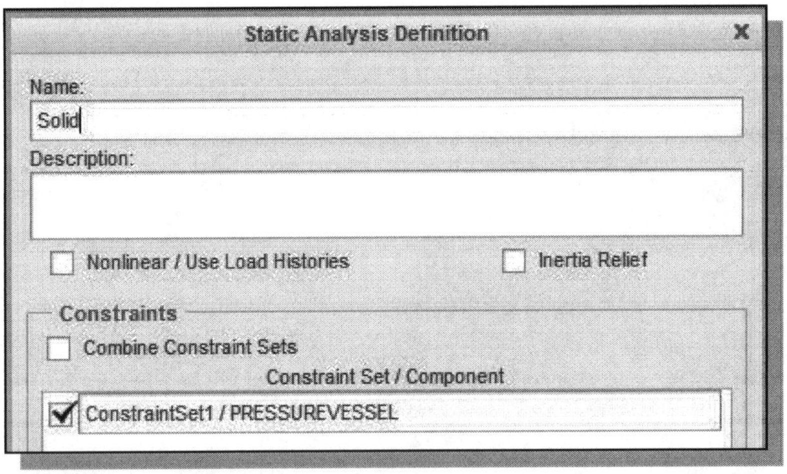

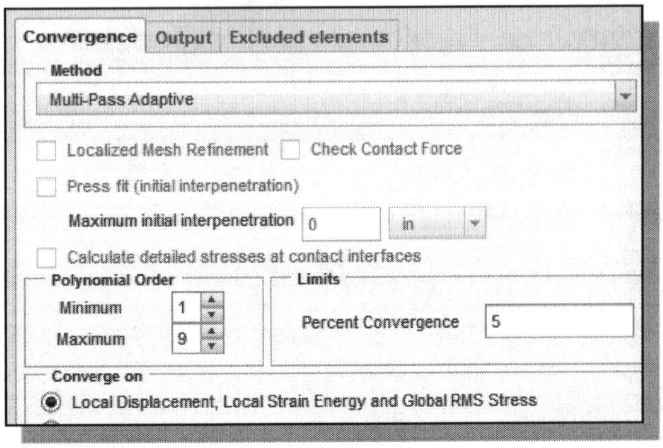

5. Confirm the Method option is set to **Multi-Pass Adaptive**, P-order to **9** and Limits to **5** Percent Convergence as shown.

6. Click **OK** to accept the settings.

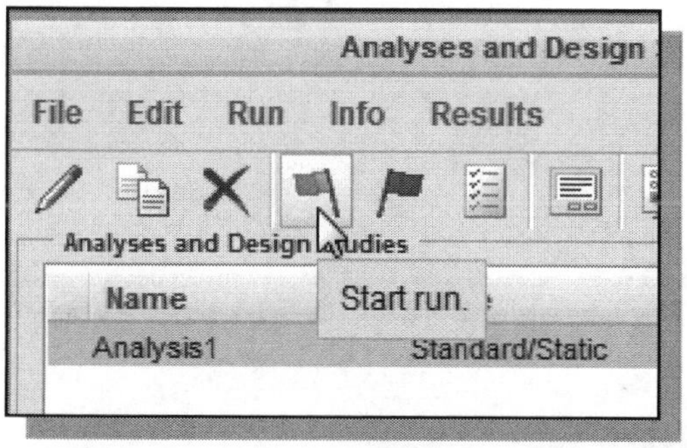

7. Click **Start run** to begin the *Solving* process.

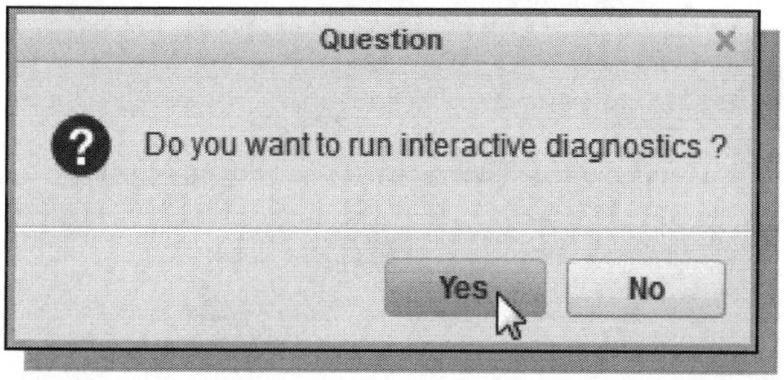

8. Click **Yes** to run interactive diagnostics; this will help us to identify and correct problems associated with our FEA model.

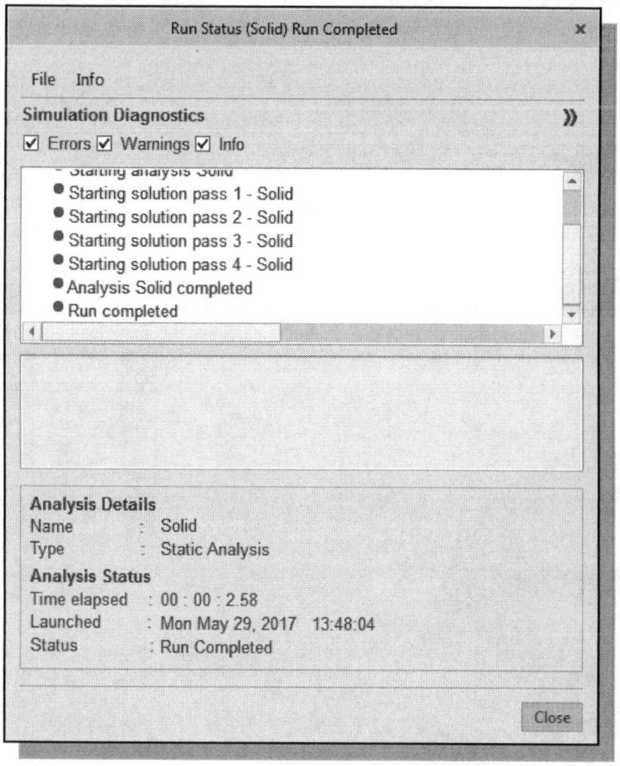

9. In the *Diagnostics* window, *Creo Simulate* will list any problems or conflicts in our FEA model.

10. Click on the **Close** button to exit the *Diagnostics* window.

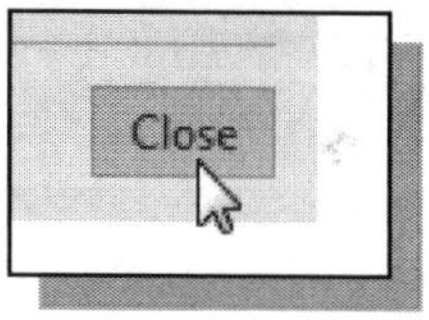

11. In the *Analyses and Design Studies* window, choose **Display study status** as shown.

❖ Note it took four times longer for *Creo Simulate* to complete the 3D solid analysis than the shell analysis.

View the Von Mises Stress

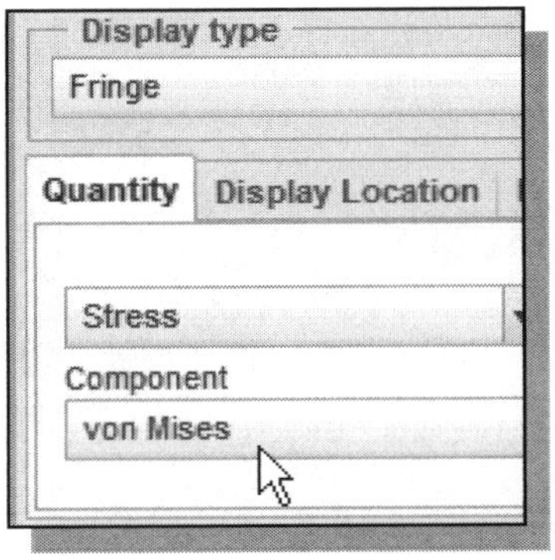

1. On your own, set the Display type to **Fringe**.

2. Choose **Von Mises** as the Stress Component to be displayed.

3. Click on the **Display Options** tab and switch ***on*** the **Deformed** option as shown.

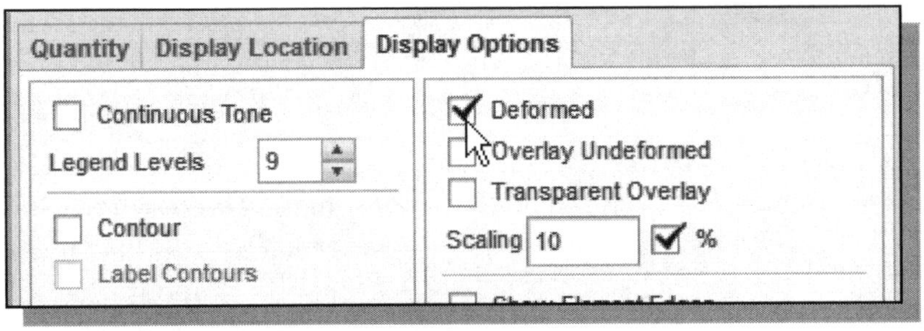

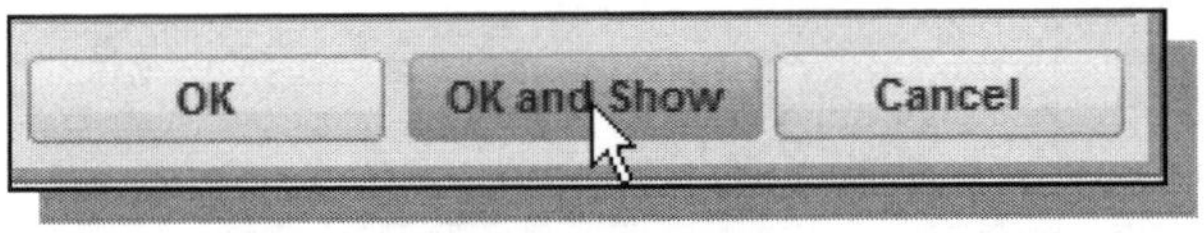

4. Click **OK and Show** to display the results.

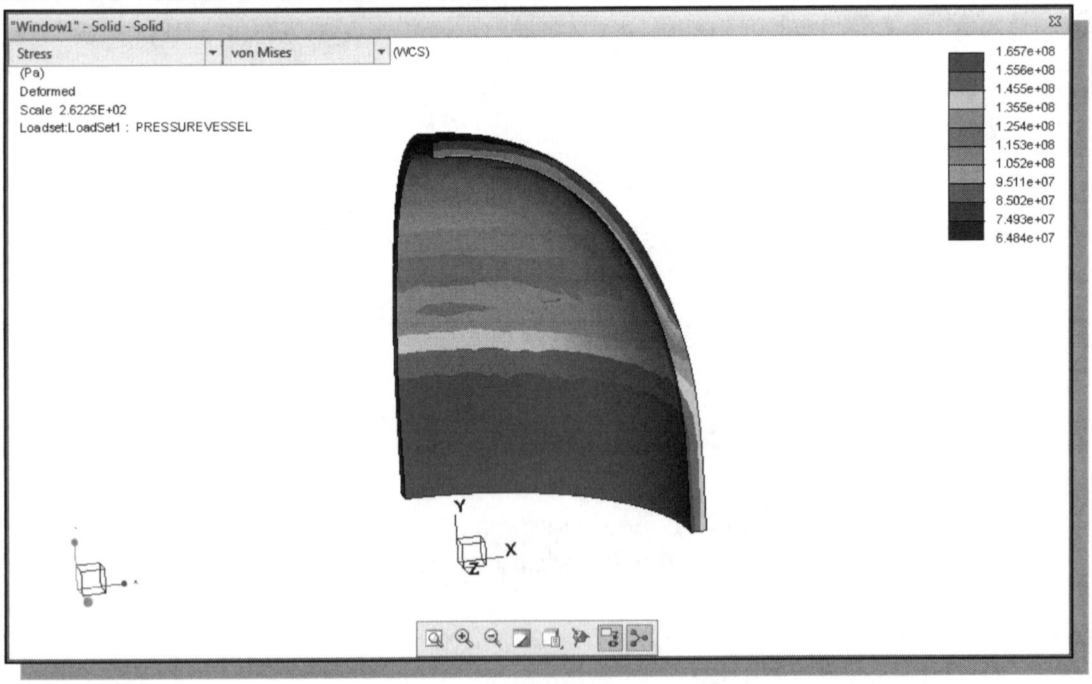

➢ *Creo Simulate* calculated the maximum stress to be **1.657E+8 N/m²**. This number is almost the same as the **165MPa** we calculated during our preliminary analysis. Note the number of solid elements used is much more than the other two analyses.

View Multiple Analyses Results

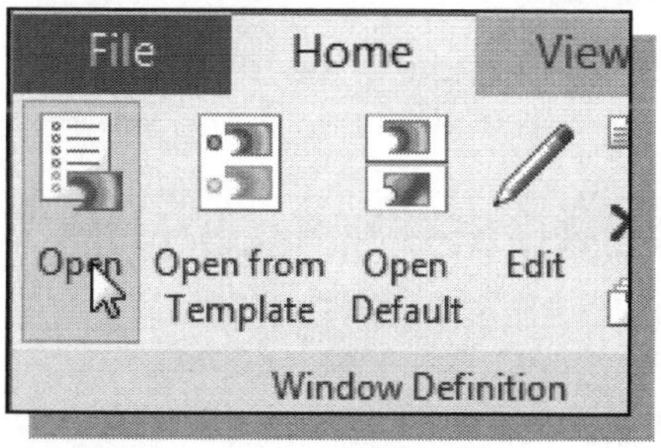

1. Select the **Open** option in the toolbar area as shown.

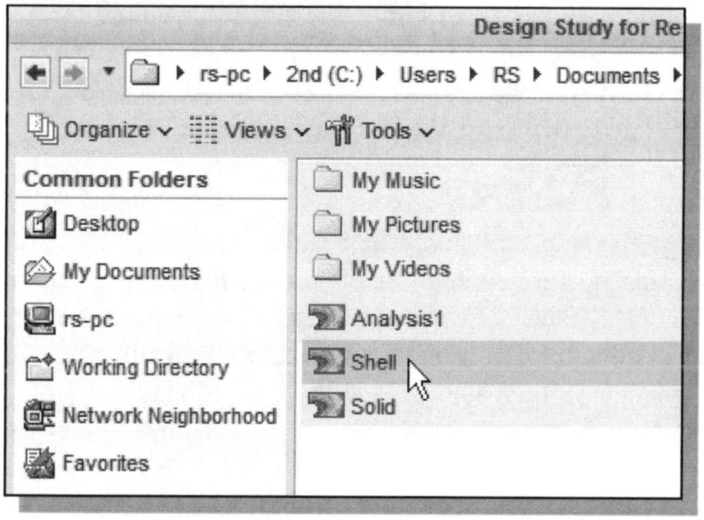

2. Choose **Shell** as shown.

3. Click **Open** to continue.

4. Click **OK and Show** to accept the settings and insert the second result window for side by side viewing.

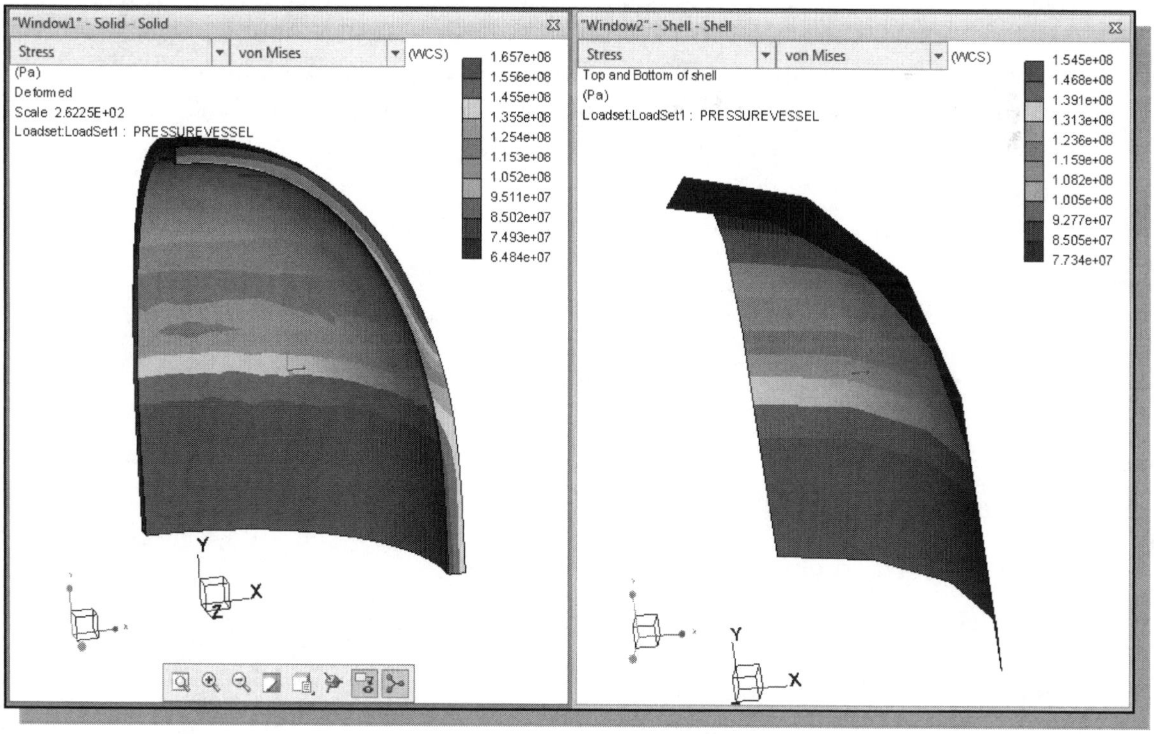

Notes on FEA Linear Static Analyses

In examining the FEA results, one should first examine the deformed shape to check for proper placement of boundary conditions and to determine if the calculated deformation of the model is reasonable. It is always important to perform a *convergence study* to obtain more accurate results. Besides using hand calculations to check the results of the FEA analyses, it is also quite feasible to check the results by using other element types. For example, the curved-beam problem illustrated in this chapter can also be analyzed using both the 1D beam elements and the 2D plane stress elements. We do need to realize that different elements have different limitations; for example, the stress concentration effects are not present with the beam elements. But the purpose of performing a second and/or a third analysis is to assure the results of the first analysis, so it is not necessary to expect all elements types to produce exactly the same results. This approach is further illustrated in the next chapter.

It should be emphasized that, when performing FEA analysis, besides confirming that the systems remain in the elastic regions for the applied loading, other considerations are also important, for example, large displacements and buckling of beams, which can also invalidate the *linear statics* analysis results. In performing finite element analysis, it is also necessary to acquire some knowledge of the theory behind the method and understand the restrictions and limitations of the software. There is no substitution for experience.

Finite element analysis has rapidly become a vital tool for design engineers. However, use of this tool does not guarantee correct results. The design engineer is still responsible for doing approximate calculations, using good design practice, and applying good engineering judgment to the problem. It is hoped that FEA will supplement these skills to ensure that the best design is obtained.

Review Questions

1. For designs that are thin and symmetrical about an axis, what are the different FEA analyses available in *Creo Simulate*?

2. What are the advantages and disadvantages of performing an axisymmetric analysis over a 3D solid analysis?

3. To perform an axisymmetric analysis in *Creo Simulate*, on which plane must the surface be created?

4. In *Creo Simulate*, how do you set up the FEA model to be an axisymmetric model?

5. To perform an axisymmetric analysis in *Creo Simulate*, which axis is the default axis of symmetry?

6. To perform an axisymmetric analysis in *Creo Simulate*, can the surface be created in the negative X portion of the associated coordinate system?

7. What are the requirements to apply constraints on thin shell models?

8. How do you specify the thickness of a thin shell model?

9. How do you display multiple analyses results on the same screen?

Exercises

1. Determine the **maximum Von Mises stress** of the thin-wall round-bottom cylindrical pressure vessel shown in the figure below; dimensions are in mm. The pressure vessel is made of steel and is subject to an **internal pressure of 10MPa**. For the FEA analyses, set all degrees of freedom to **Fixed** at the top edge of the opening. (The Radius 225 and Radius 200 arcs share the center point, which is measured 200 mm along the vertical axis. The *Radius* 265 and *Radius* 235 arcs also share the same center point at the origin.)

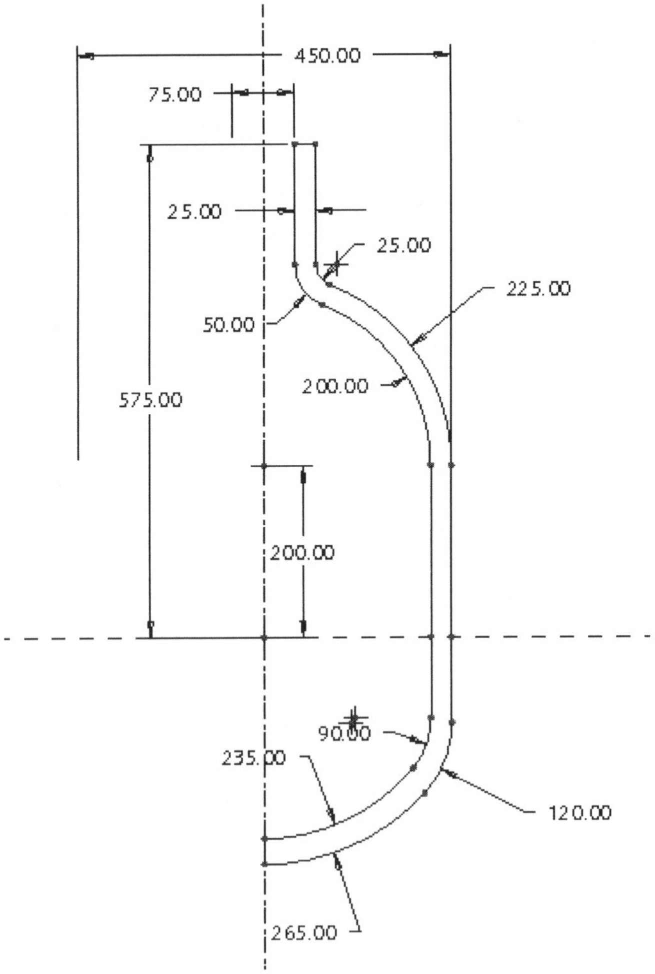

2. Determine the **maximum Von Mises stress** of the thin-wall flat-bottom cylindrical pressure vessel shown in the figure below; dimensions are in inches. The pressure vessel is made of steel and is subject to an **internal pressure of 45Psi**. For the FEA analyses, set all degrees of freedom to **Fixed** at the top edge of the opening.

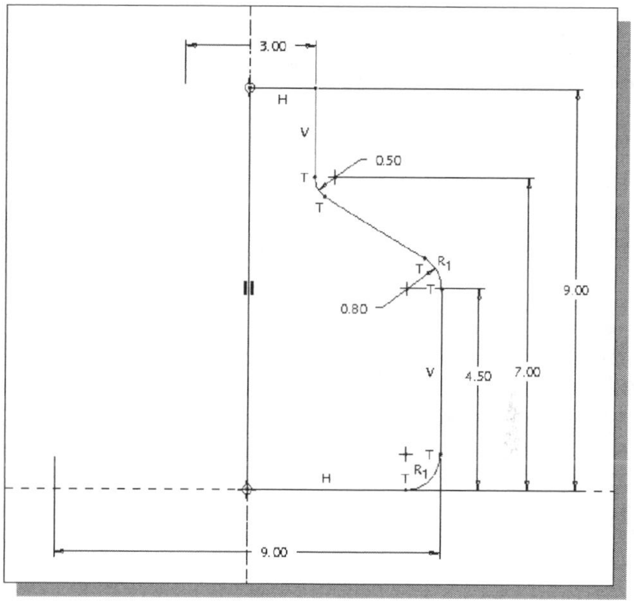

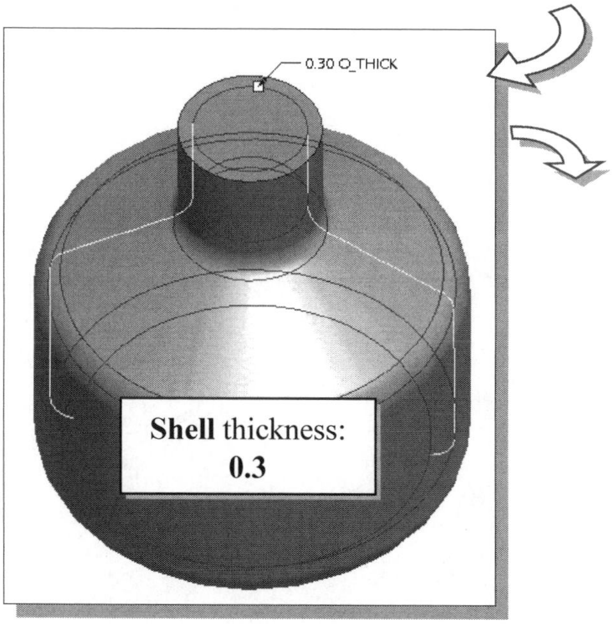

Notes:

Chapter 12
Dynamic Modal Analysis

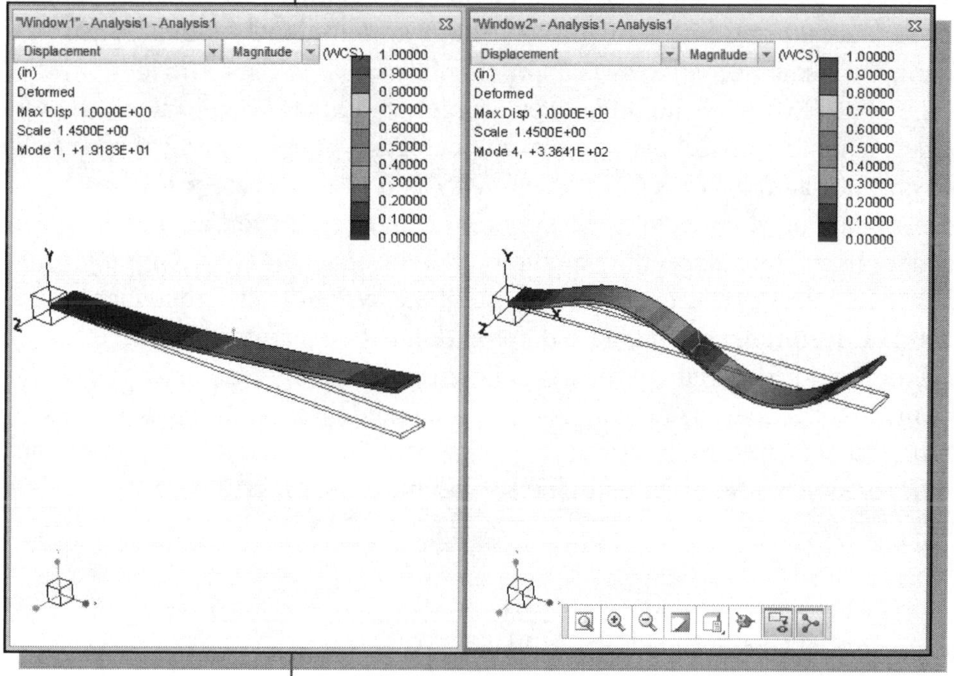

Learning Objectives

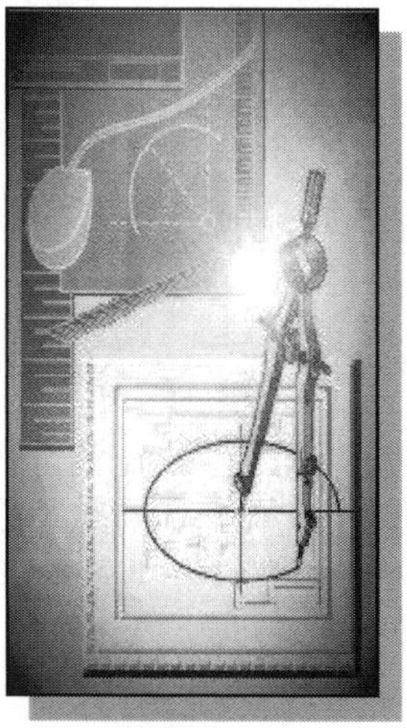

- **Understand the Basic concepts of Natural Frequencies.**
- **Perform FEA Modal Analysis on simple Systems.**
- **Use the Equivalent Mass Spring System approach in performing Modal Analysis.**
- **Use the Insert Window command to display and compare multiple Results.**

Introduction

All *objects* will *vibrate* when subjected to impact, noise or *vibration.* And many systems can **resonate**, where small forces can result in large deformation, and damage can be induced in the systems. **Resonance** is the tendency of a system to oscillate at maximum amplitude at certain frequencies, known as the system's resonant frequencies. At these frequencies, even small driving forces can produce large amplitude vibrations. When damping is small, the resonant frequency is approximately equal to the **natural frequency** of the system, which is the frequency of free vibrations. Any physical structure can be modeled as a number of springs, masses, and dampers. The multitude of spring-mass-damper systems that make up a *Creo Simulate* system are called **degrees of freedom**; and the vibration energy put into a system will distribute itself among the degrees of freedom in amounts depending on their natural frequencies and damping, and on the frequency of the energy source. The response of the system is different at each of the different **natural frequencies**; and these deformation patterns are called **mode shapes**. **Modal analysis** is the study of the dynamic properties of systems under excitation. Detailed modal analysis determines the fundamental vibration mode shapes and corresponding frequencies. Both the natural frequencies and mode shapes can be used to help design better systems for noise and vibration applications.

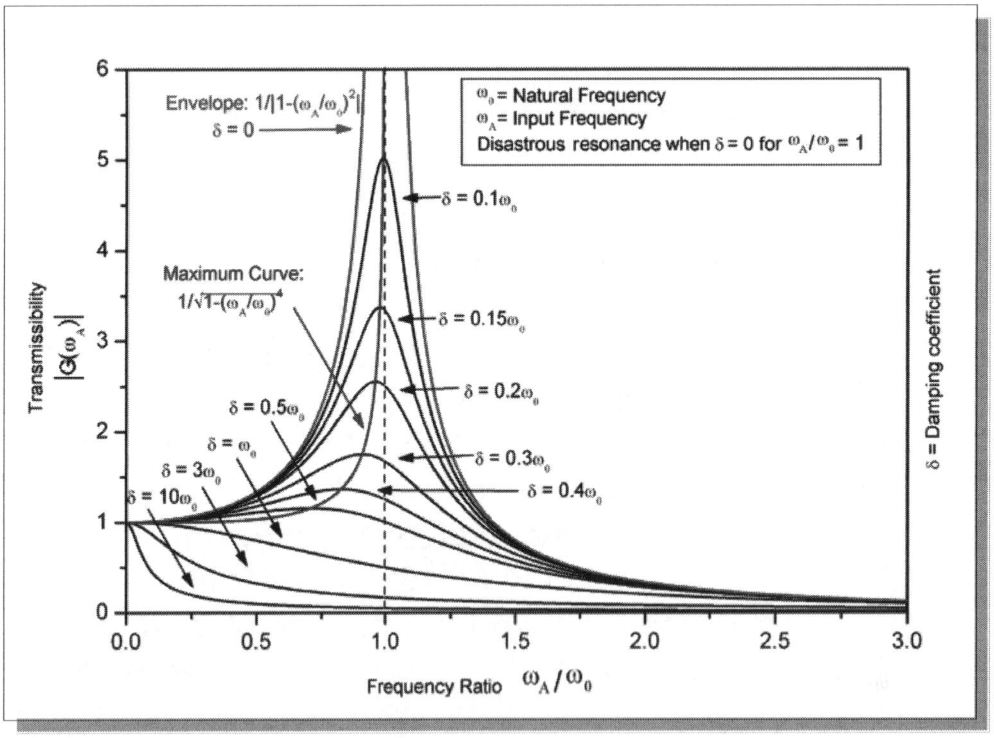

Modal analysis can be relatively simple for basic components of a simple system, and extremely complicated for a complex system, such as a structure exposed to periodic wind loading or during seismic activities. With the advancements of computers, the accurate determination of natural frequencies and mode shapes are best suited to using special techniques such as Finite Element Analysis.

Problem Statement

1) Determine the **natural frequencies** and **mode shapes** of a cantilever beam. The beam is made of steel and has the dimensions of 14.5″ x 1/8″ x 1″.

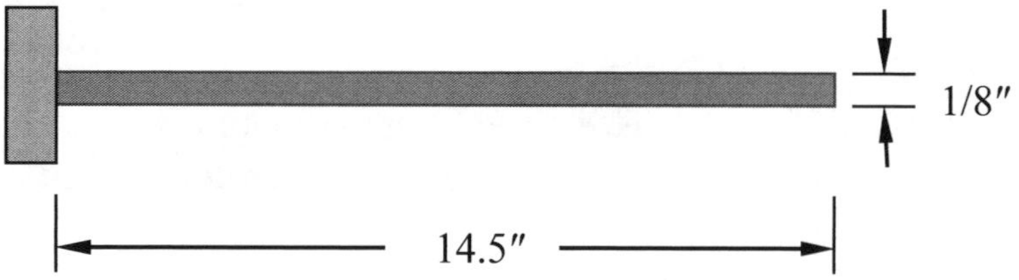

2) Determine the first **natural frequency** and **mode shapes** of a cantilever beam with a small half a pound object attached to the free end of the cantilever beam.

Preliminary Analysis

The cantilever beam is an example of a system, which can be modeled as a simple spring-mass system. In order to model the vibration of the cantilever beam, the end of the beam is chosen as a reference point at which the characteristics and response of the beam are measured. An equivalent system is then built so that the natural frequency of the system can be determined. The equivalent spring constant can be calculated using beam deflection formulae. Calculation of an equivalent mass is necessary because all points along the beam's length do not have the same response as the end of the beam. This means that the equivalent mass, m_e, cannot be determined simply by using the masses of the beam, but must be found by equating the energy of the system as it vibrates.

From *Strength of Materials*, the deflection at the tip of the cantilever beam can be determined by

$$y = \frac{PL^3}{3EI}$$

The deflection equation of the mass spring system is

$$F = ky$$

So, the equivalent spring constant can be expressed as

$$k = \frac{3EI}{L^3}$$

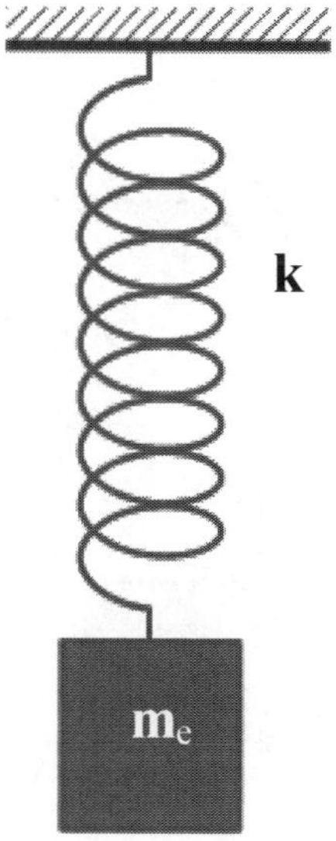

1) For the simple mass-spring system, the natural frequency can be expressed as

$$\omega_n = \sqrt{\frac{k}{m_e}} \quad \text{(in rad/sec)}$$

The equivalent mass (m_e) can be determined from the continuous system analytical approach, and the first three natural frequencies of the cantilever beam can then be expressed as

$$\omega_n = \alpha_n^2 \sqrt{\frac{EI}{m_b L^3}} \quad \text{where} \quad \alpha_n = 1.875, 4.694, 7.855$$

The first three natural frequencies and mode shapes of a cantilever beam are as shown in the figure below.

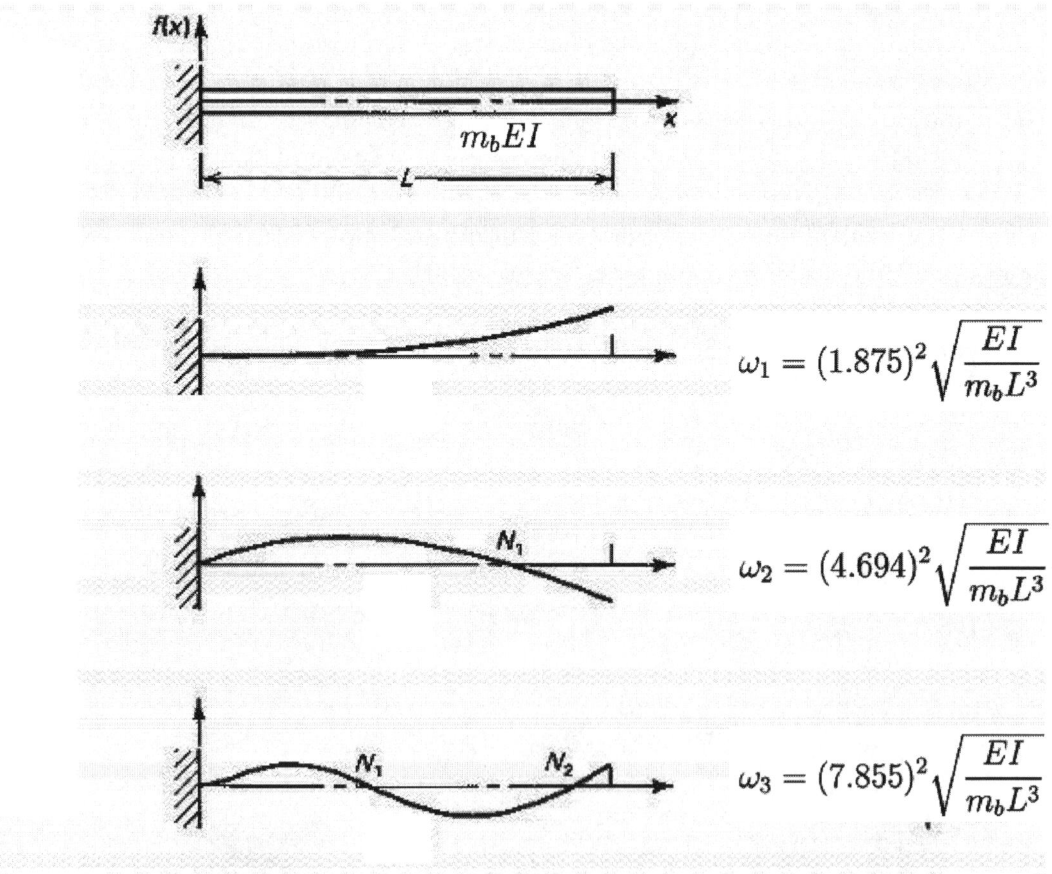

Frequencies and Mode Shapes for Cantilever Beam

For our preliminary analysis, the first three natural frequencies calculations are as follows

$$\omega_n = \alpha_n^2 \sqrt{\frac{EI}{m_b L^3}} \quad \text{where} \quad \alpha_n = 1.875, 4.694, 7.855$$

E = 3.0 x 10^7 psi **(Modulus of Elasticity for Steel)**

I = $bh^3/12$ = **1** x **$(0.125)^3$ /12** = 1.627 x 10^{-4} in^4

L = 14.5 in

m_b = Volume x Density = **1** x **0.125** x **14.5** x **0.0007324** = 0.001326 **lbsec^2**

Therefore,

ω_1 = $(1.875)^2$ x (3.0X10^7 x 1.627 x 10^{-4} / (0.001326 x $(14.5)^3)^{1/2}$
= 122.16 (rad/sec)
= 19.71 (Hz)

ω_2 = $(4.694)^2$ x (3.0 x 10^7 x 1.627 x 10^{-4} / (0.001326 x $(14.5)^3)^{1/2}$
= 765.62 (rad/sec)
= 121.85 (Hz)

ω_3 = $(7.855)^2$ x (3.0 x 10^7 x 1.627 x 10^{-4} / (0.001326 x $(14.5)^3)^{1/2}$
= 2143.99 (rad/sec)
= 341.22 (Hz)

2) The equivalent mass (m_e) for the cantilever beam with a mass attached at the end has a similar format, so the first natural frequency can be expressed as

$$\omega_1 = \sqrt{\frac{3EI}{(m_a + 0.236 m_b) L^3}}$$

ω_1 = (3 x 3.0 x 10^7 x 1.627 x 10^{-4}/((0.5/(32.2 x 12) + 0.236 x 0.001326) x $(14.5)^3)^{1/2}$
= 54.67 (rad/sec)
= 8.70 (Hz)

The Cantilever Beam Modal Analysis program

The ***Cantilever Beam Modal Analysis*** program is a custom-built *MS Windows* based computer program that can be used to perform simple modal analysis on uniform cross section cantilever systems; rectangular and circular cross section calculations are built-in, but other cross sections can also be used. The program is based on the analytical methods described in the previous section. Besides calculating the first six natural frequencies of a cantilever beam, the associated modal shapes are also constructed and displayed. The program is very compact, 75 Kbytes in size, and it will run in any *Windows* based computer systems.

- First download the CBeamModal.zip file, which contains the *Cantilever Beam Modal Analysis* program, from the *SDC Publications* website.

1. Launch your internet browser, such as the *MS Internet Explorer* or *Mozilla Firefox* web browsers.

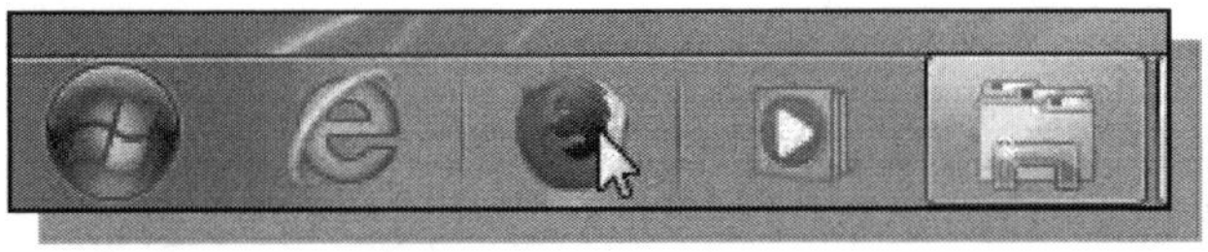

2. In the *URL address* box, enter www.SDCpublications.com/downloads/978-1-63057-108-5.

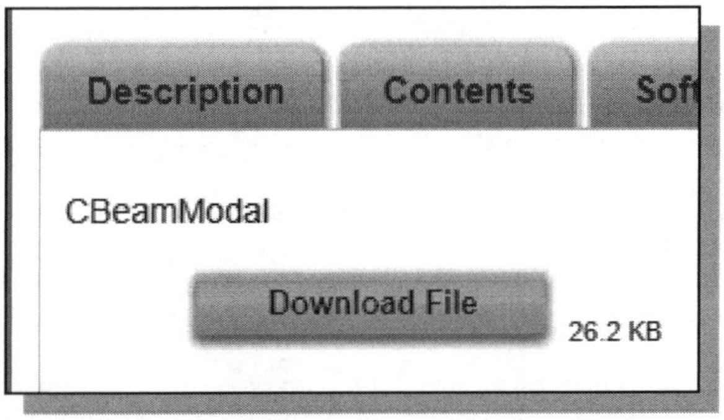

3. On your own, download the file to any folder.

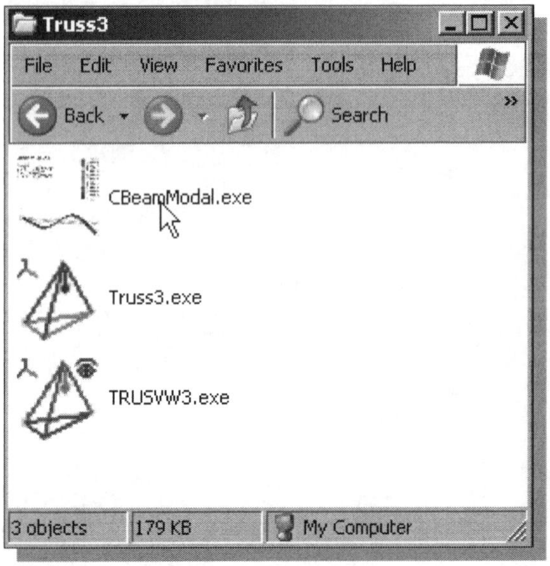

4. On your own, extract the content of the ZIP file to any folder

5. Start the ***Cantilever Beam Modal Analysis*** program by double-clicking on the **CBeamModal.exe** icon.

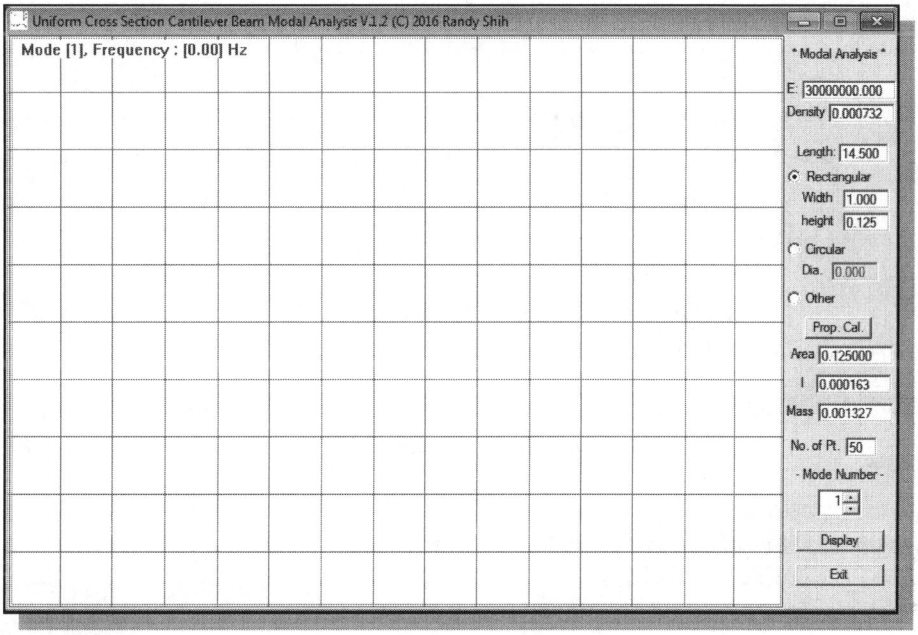

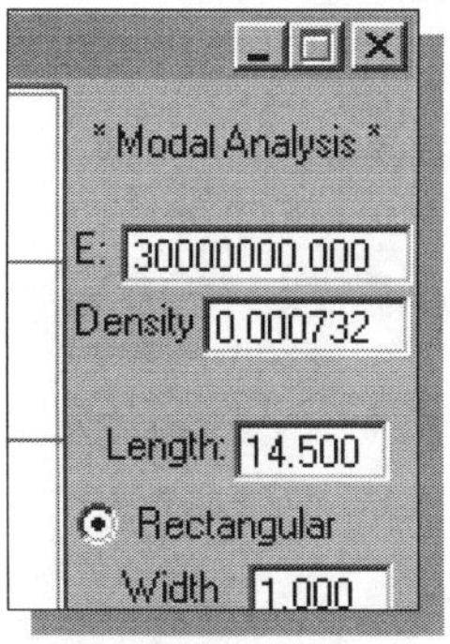

❖ The controls to the inputs and the display options are listed toward the right edge of the program's main window.

6. The first two items in the control panel are the required material information: the **Modulus of Elasticity** (**E**) and the **Density** of the material. Note that it is critical to use the same units for all values entered.

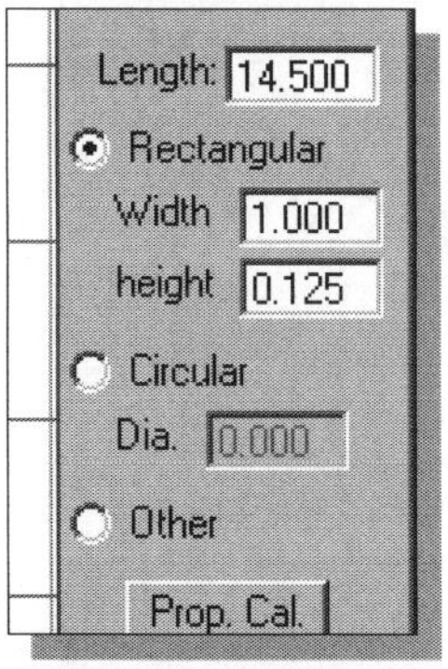

7. Enter the **physical dimensions** of the cantilever beam in the mid-section of the control panel:

- For rectangular shapes, select the **Rectangular** option and enter the **Width** and **Height** dimensions of the cross section.
- For circular shapes, select the **Circular** option and enter the **Diameter** value.
- For any other cross section shapes, select the **Other** option and enter the **Area** and **Area Moment Of Inertia** information in the next section.

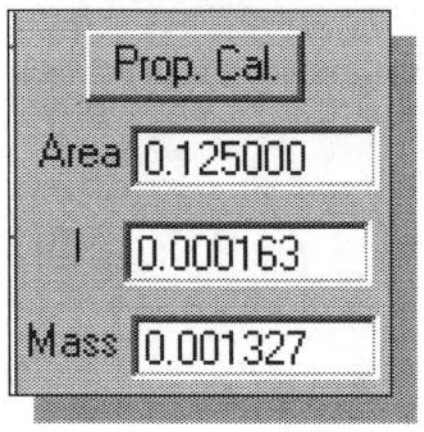

8. Press the **Property Calculation** button to calculate the associated **Area**, **Area Moment Of Inertia** and **Mass** information for the rectangular or the circular cross sections. For any other shapes, enter the information in the edit boxes. Note that the mass information is always calculated based on the density value.

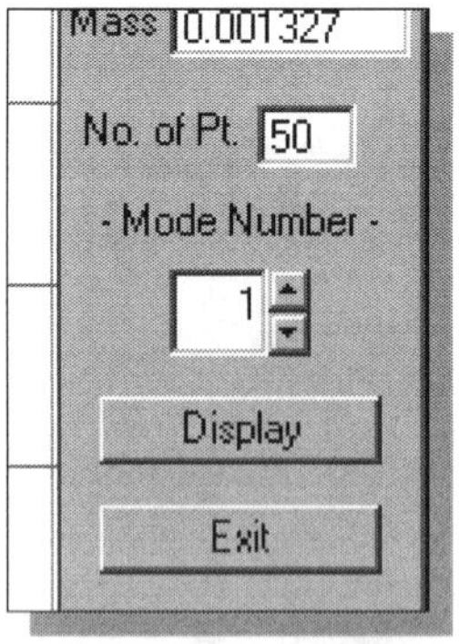

9. The **Number of Point** option is used to control the smoothness of the modal shapes. Enter a number between **5** and **80**; the larger the number, the smoother the curve.

10. The **Mode Number** option is used to examine the different mode shapes of the beam. The first six natural frequencies modal shapes are performed by the program.

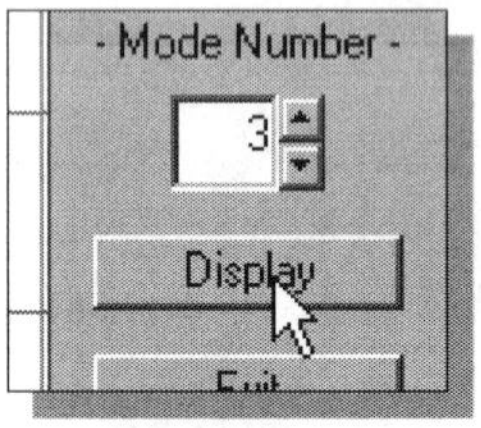

11. Click the **Display** button to show the results of the modal analysis.

➢ The displayed graph shows the modal shape with the horizontal axis corresponds to the length direction of the cantilever beam.

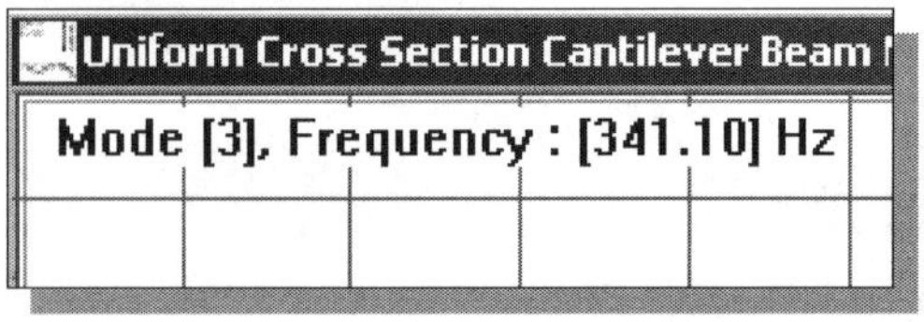

➢ The natural frequency of the selected mode number is calculated and displayed near the upper left corner of the main program window.

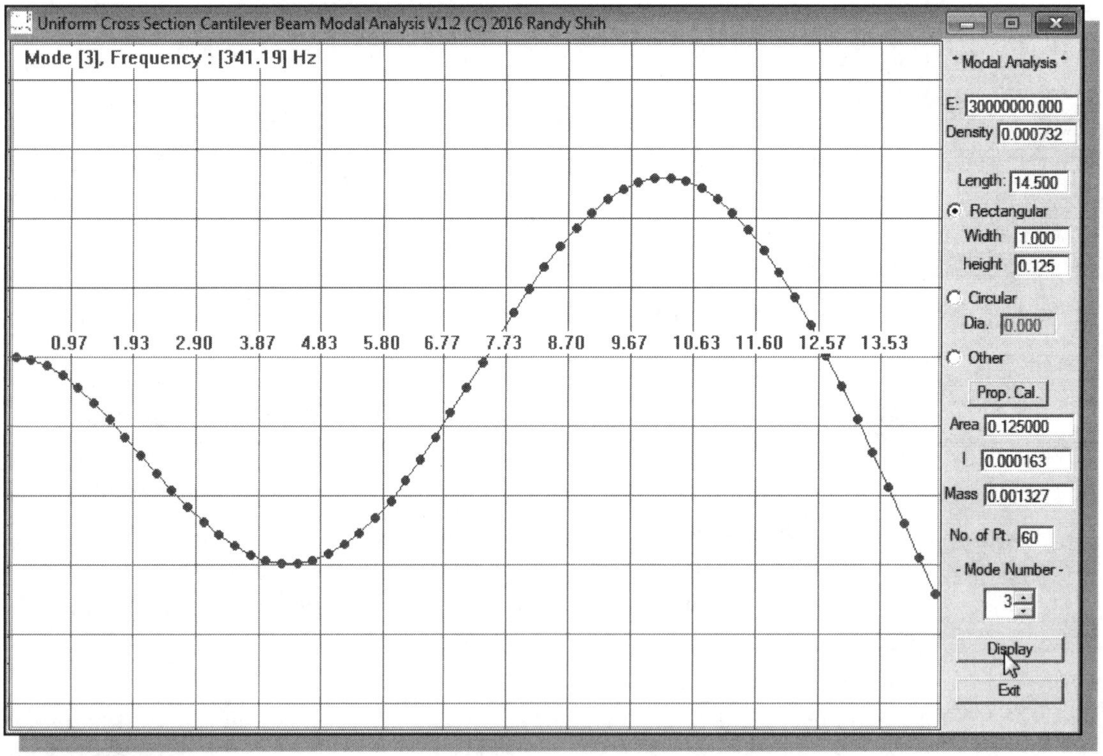

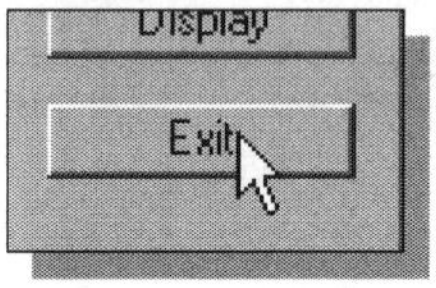

12. Click the **Exit** button to end the program.

Starting Creo Parametric

1. Select the **Creo Parametric** option on the *Start* menu or select the **Creo Parametric** icon on the desktop to start *Creo Parametric*. The *Creo Parametric* main window will appear on the screen.

2. Click on the **New** icon located in the *Standard* toolbar as shown.

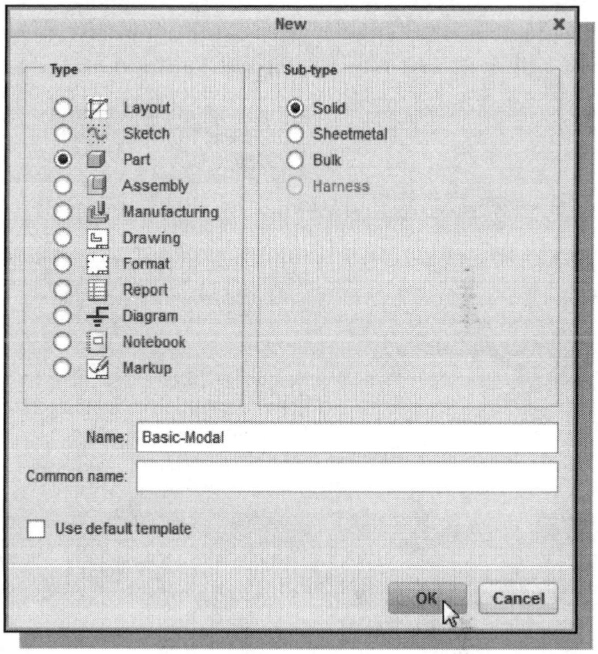

3. In the *New* dialog box, confirm the model's Type is set to **Part** (**Solid** Sub-type).

4. Enter **Basic-Modal** as the part Name as shown in the figure.

5. Turn ***off*** the **Use default template** option.

6. Click on the **OK** button to accept the settings.

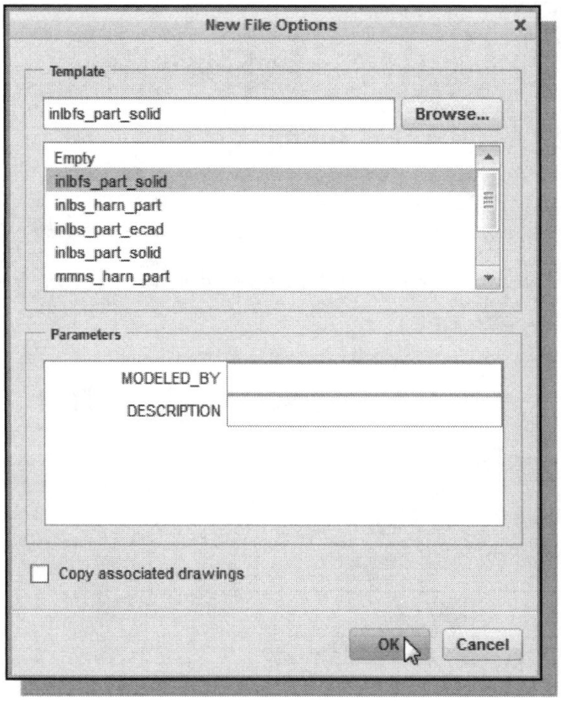

7. In the *New File Options* dialog box, select **inlbfs_part_solid** in the option list as shown.

8. Click on the **OK** button to accept the settings and enter the *Creo Parametric Part Modeling* mode.

Create a CAD Model in Creo Parametric

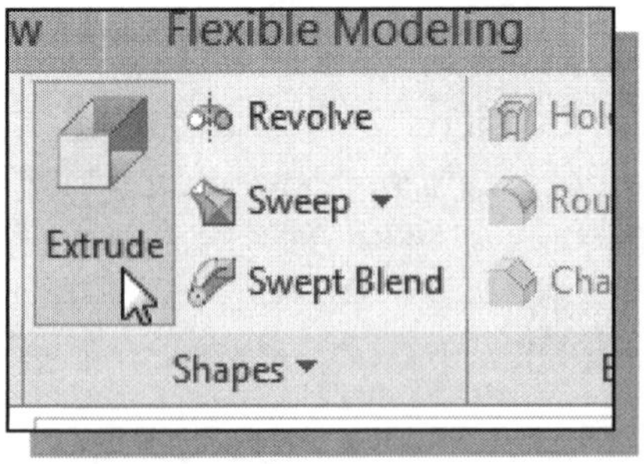

1. In the *Ribbon*, select the **Extrude** tool option as shown.

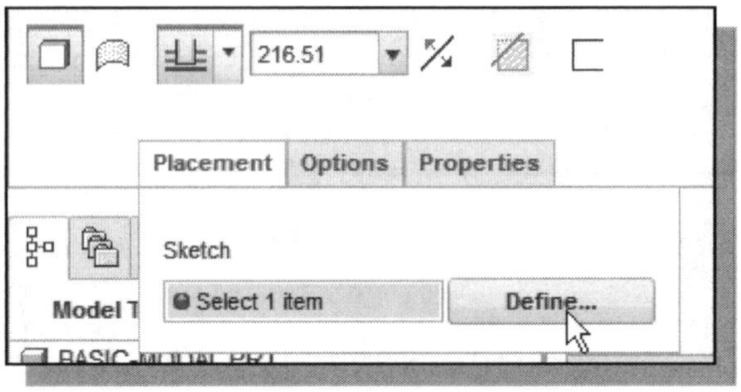

2. Click the **Placement** option and choose **Define** to begin creating a new *internal sketch.*

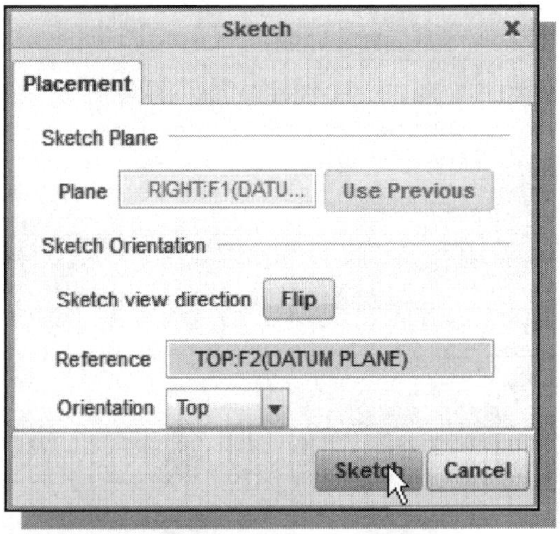

3. On your own, set up the **RIGHT** datum plane as the sketch plane with the **TOP** datum plane facing the **top edge** of the computer screen and pick **Sketch** to enter the *Sketcher* mode.

4. Click **Sketch** to exit the *Sketch* window and proceed to enter the *Creo Parametric Sketcher* mode.

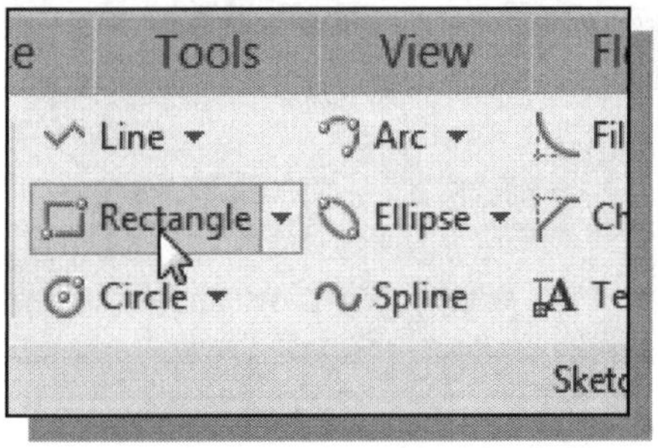

5. In the *Sketching* toolbar, select **Rectangle**.

6. On your own, click on **Sketch View** to switch to the 2D view.

7. On your own, create a rectangle with the lower-left corner aligned to the intersection of the two references as shown.

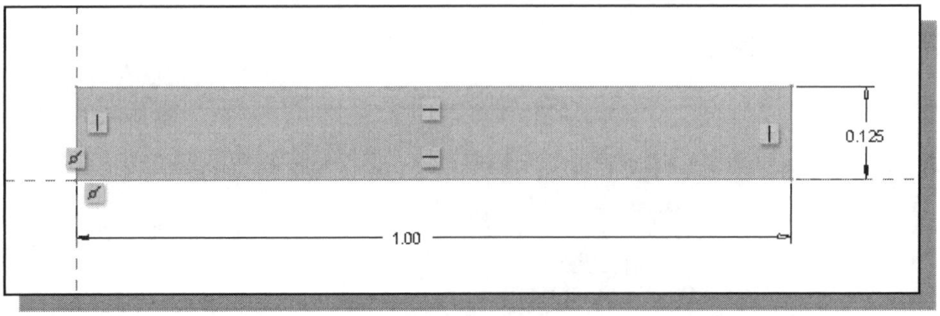

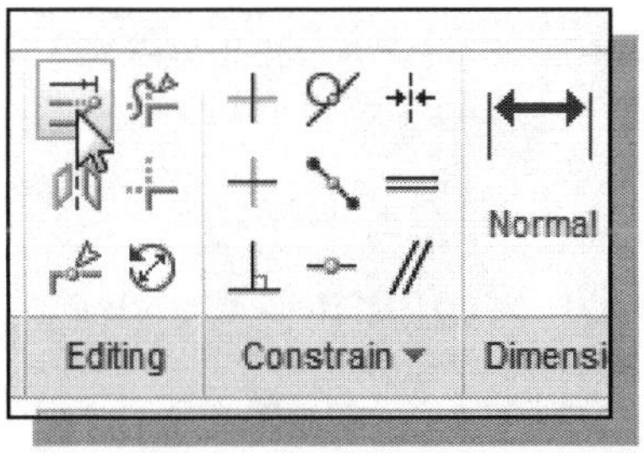

8. On your own, adjust the size of the rectangle to **1″** x **1/8″**.

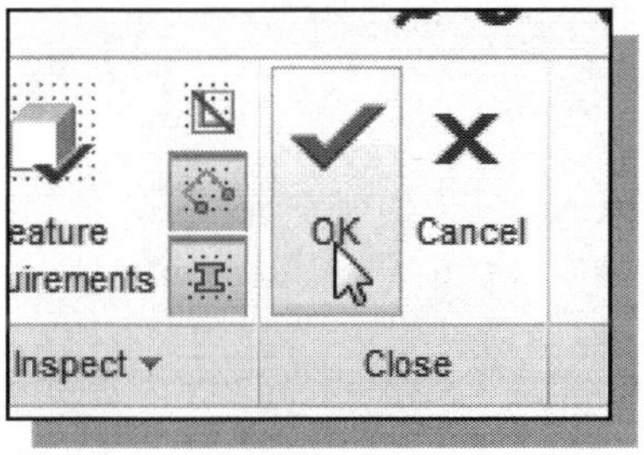

9. Click **OK** to accept the completed sketch.

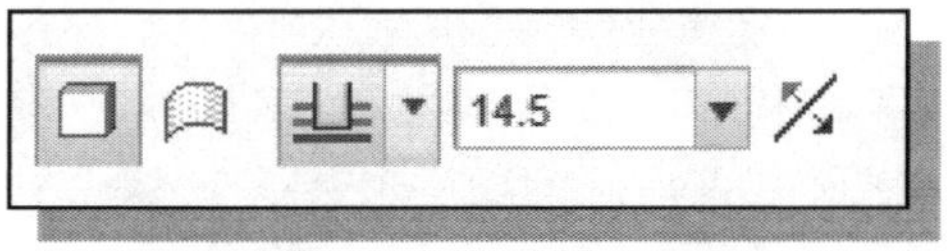

10. Set the *extrusion distance* to **14.5** as shown.

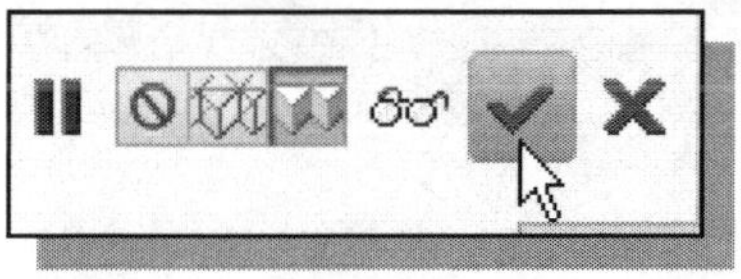

11. Click **Accept** to accept the completed beam model.

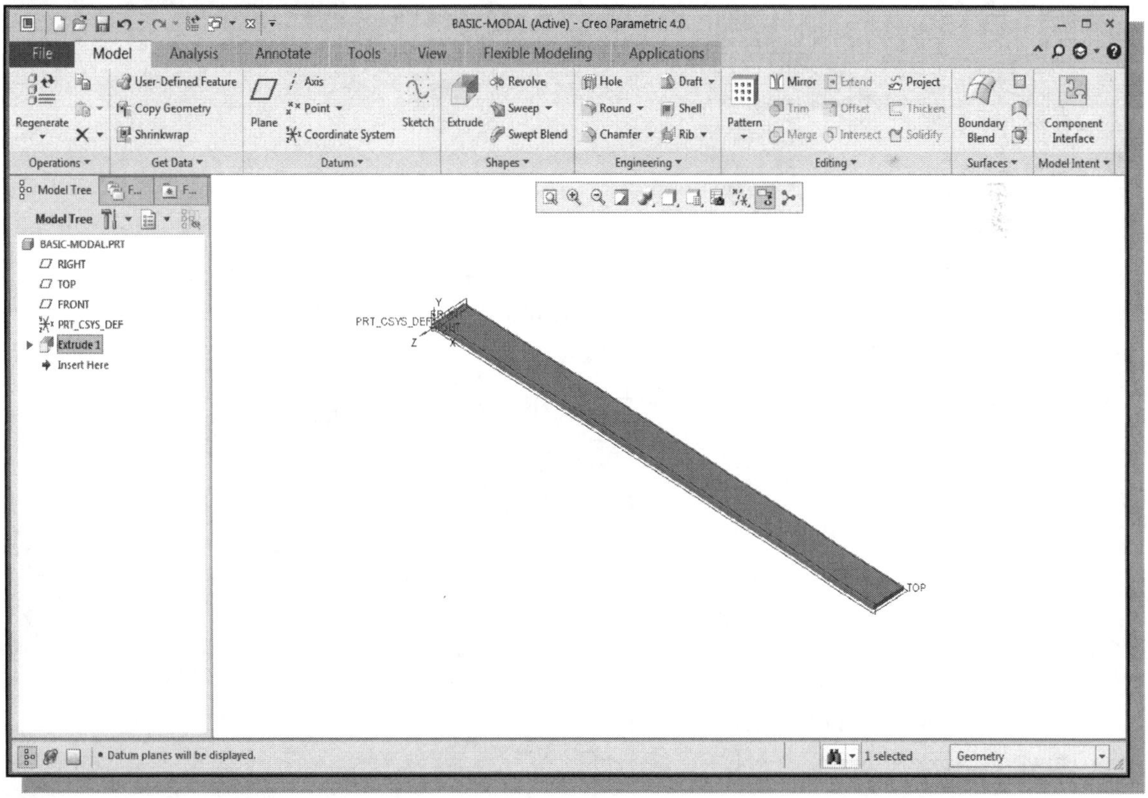

12. Click **Save** and save the CAD model to disk.

Select and Examine the Part Material Property

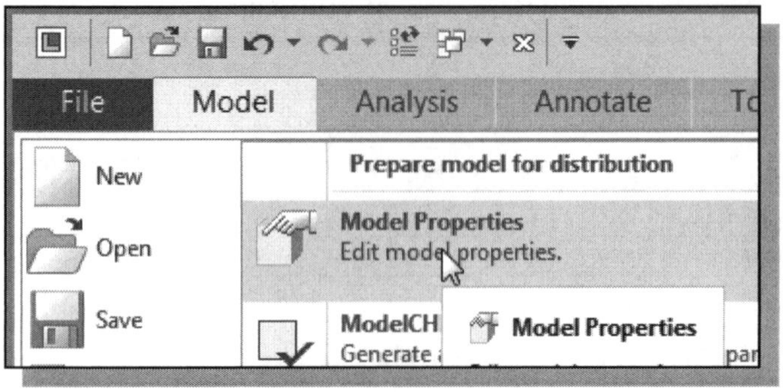

1. Choose **File → Prepare → Model Properties** from the pull-down menu.

2. Select the **Change** option that is to the right of the **Material** option in the *Model Properties* window.

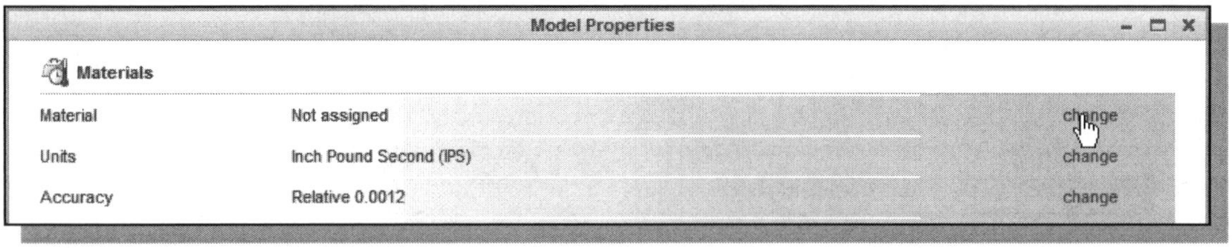

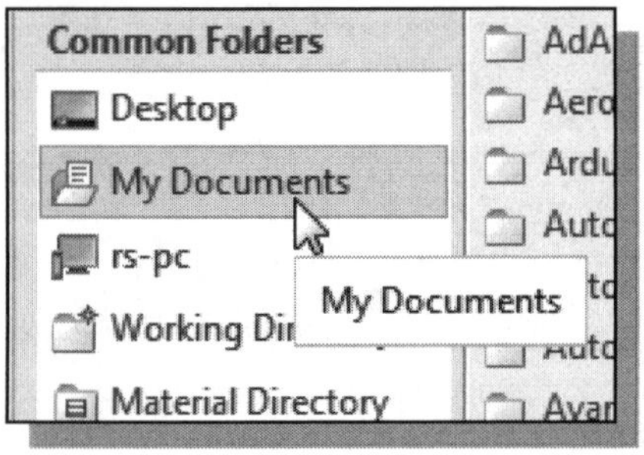

3. On your own, locate the folder where we saved the modified material definition, filename: **Steel_low_Carbon_Modified**.

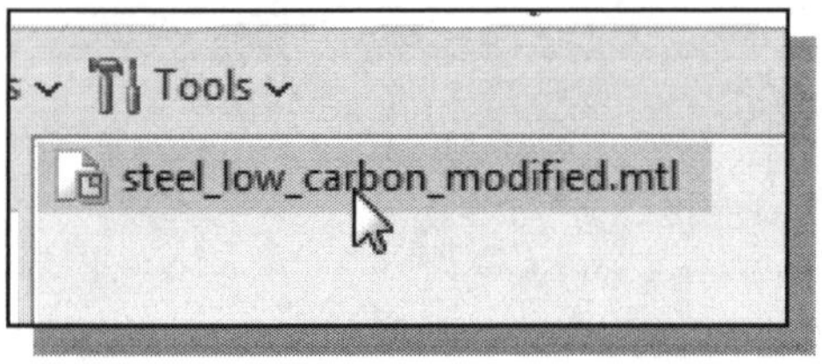

4. **Double-clicking** on the modified material definition for use with the current model.

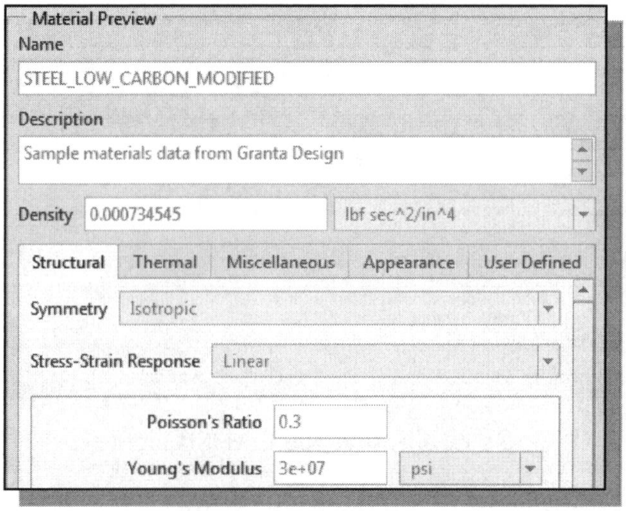

5. Click **Yes** to convert the units to match the model units.

6. Examine the material information in the **Material Preview** area and click **OK** to accept the settings.

The Integrated Mode of Creo Simulate

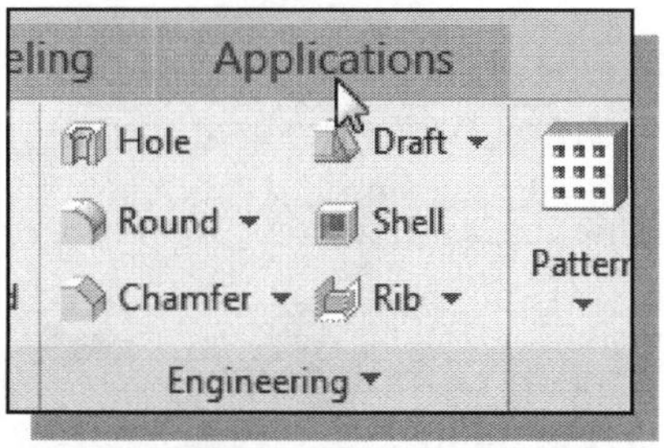

1. Switch to the **Applications** tab in the *Ribbon* area as shown.

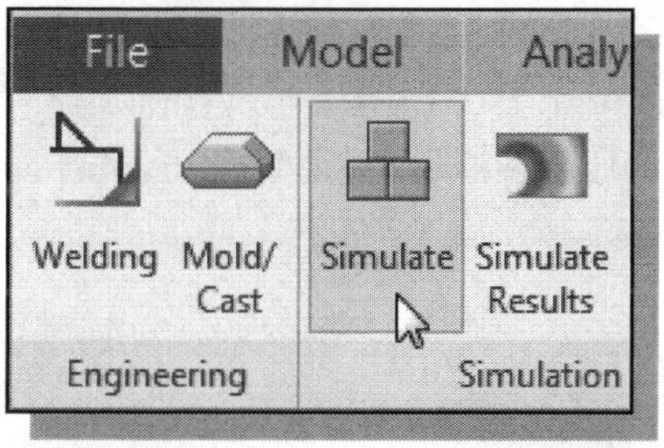

2. Start the *Integrated* mode of *Creo Simulate* by selecting the **Simulate** option as shown.

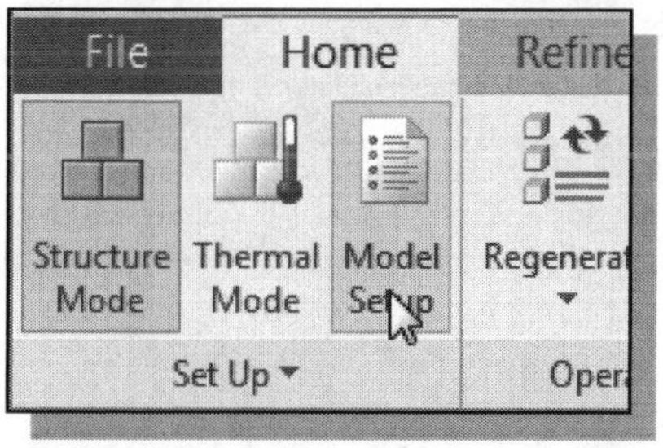

3. In the *Creo Simulate Ribbon* toolbar, confirm **Structure** is set as the analysis type.

4. Click **Model Setup** to examine the *Simulate* model settings.

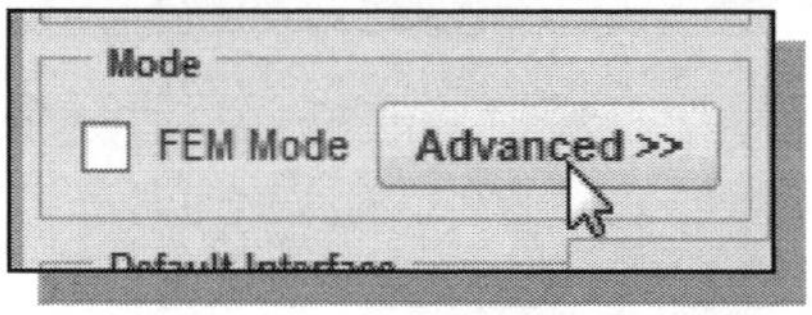

5. Click **Advanced** to show the additional settings that are available.

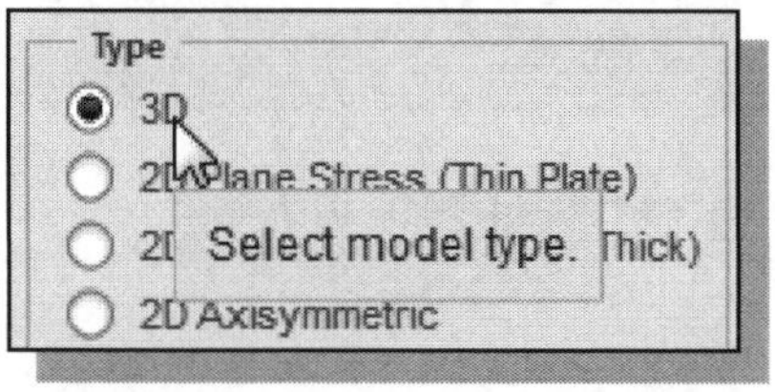

6. Confirm the selection of **3D** as the model **Type** as shown in the figure.

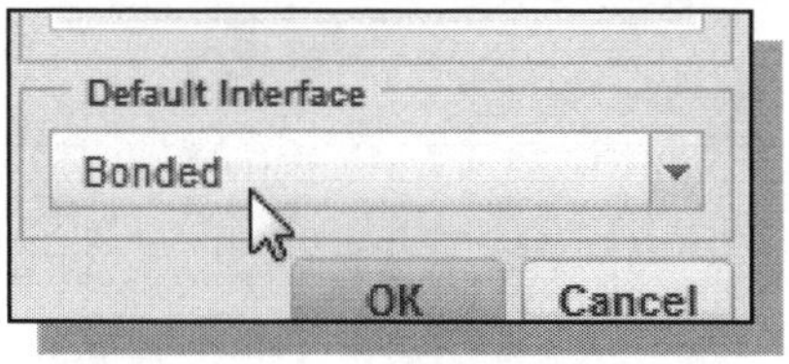

7. Confirm the default interface is set to **Bonded** as shown.

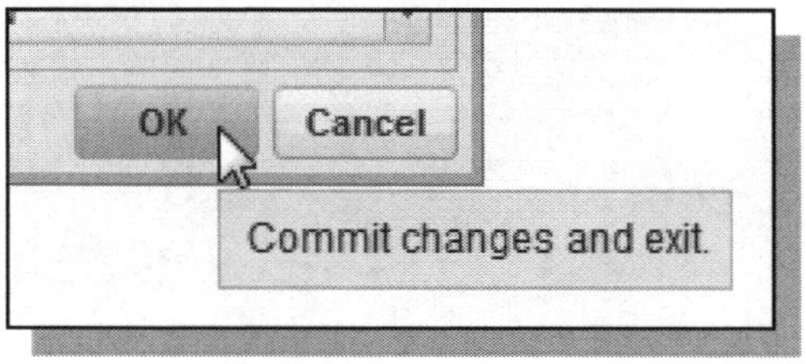

8. In the *Creo Simulate Model Setup* window, click on the **OK** button to accept the settings and proceed to create a 3D FEA model.

Apply the Boundary Conditions – Constraints

For *modal analysis*, we do not need to apply any external load to the system. The natural frequencies of physical systems are dynamic properties, which can be determined through the equivalent mass-spring system. With *Creo Simulate*, unconstrained modal analysis can also be performed.

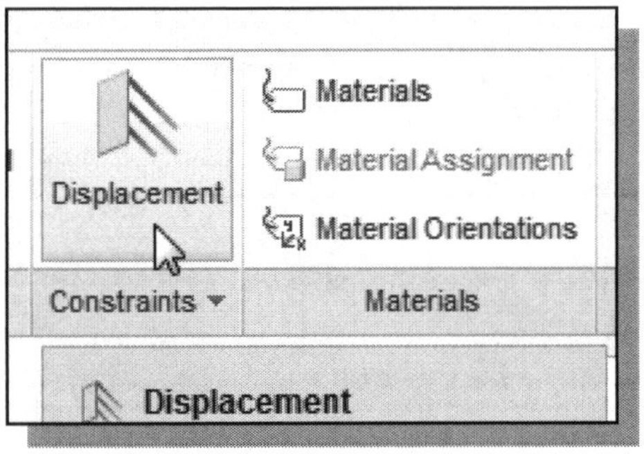

1. Choose **Displacement Constraint** by clicking the icon in the toolbar as shown.

2. Set the References to **Surfaces** and pick the **circular surface** of the top section as shown.

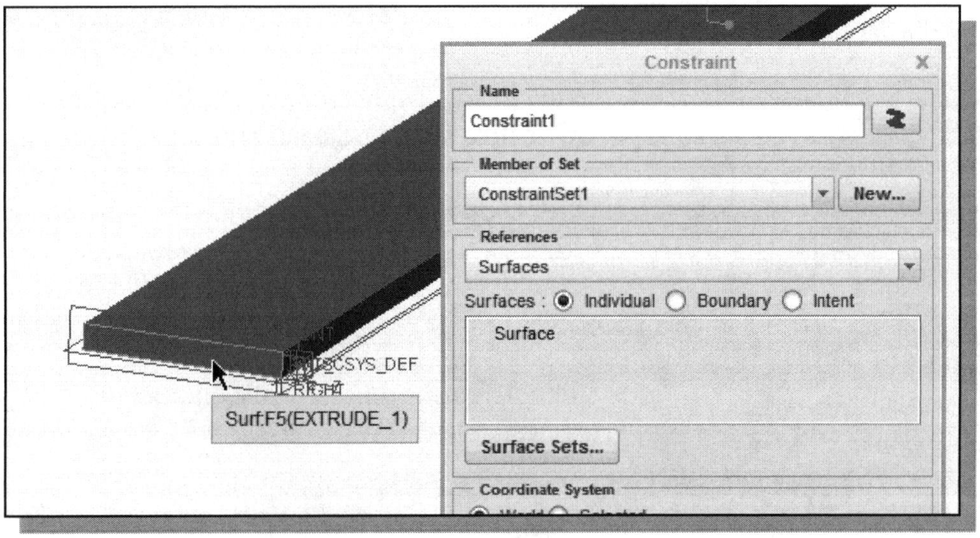

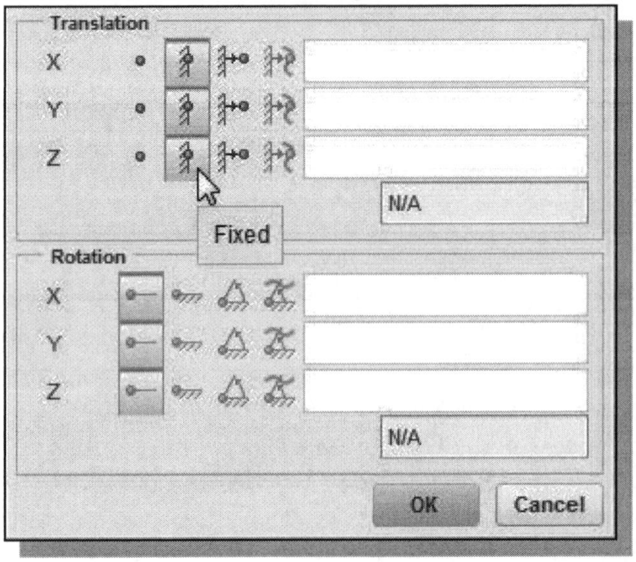

3. Confirm the Translational constraints are set to **Fixed** as shown.

4. Click on the **OK** button to accept the first displacement constraint settings.

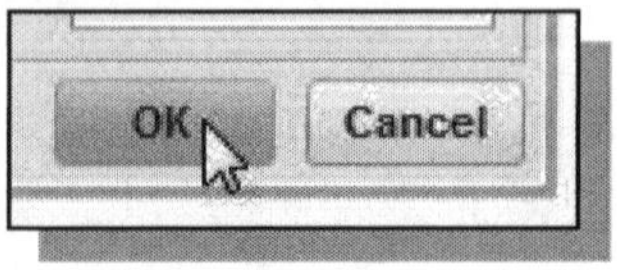

Create the 3D Mesh

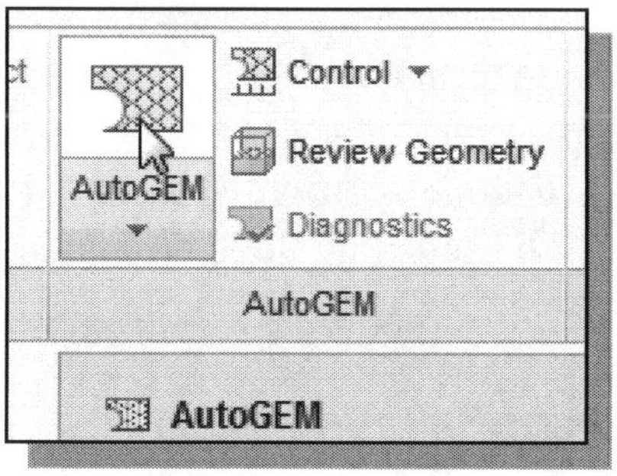

1. Choose **Refine Model** in the *Ribbon* toolbar.

2. Activate the **AutoGEM** command by selecting the icon in the toolbar area as shown.

3. Click **No** to decline the use of the old mesh.

4. Confirm the AutoGEM References option to **All with Properties** as shown.

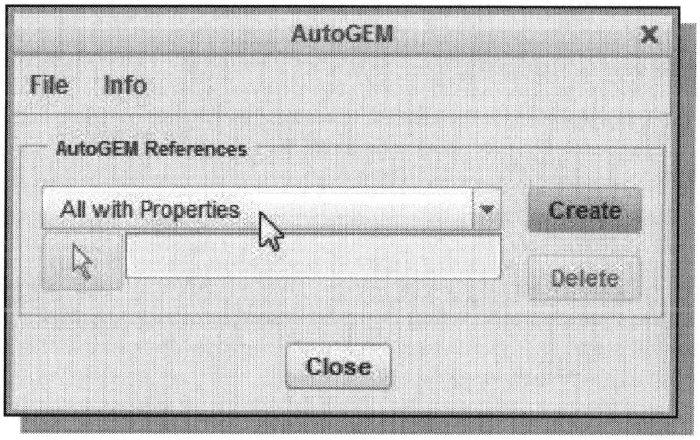

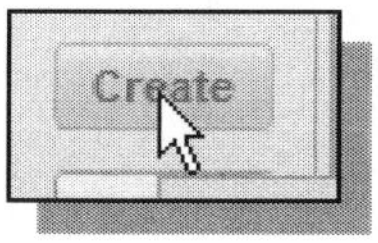

5. Inside the *AutoGEM* dialog box, click **Create** to generate the P-mesh.

➢ Note that *Creo Simulate* generated relatively small numbers of elements and nodes.

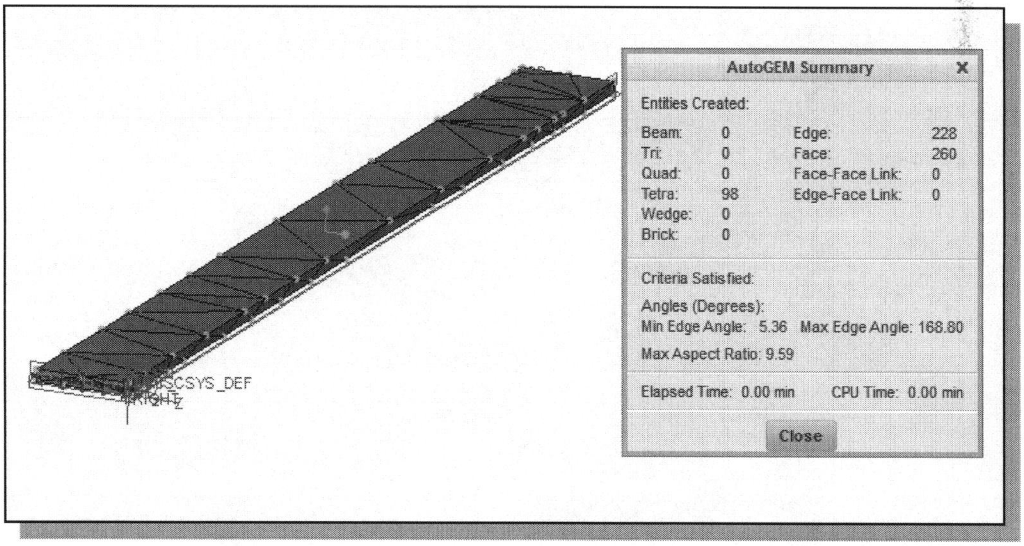

6. On your own, exit the *AutoGEM* dialog box by clicking on the **Close** button.

7. Click **Yes** to save the mesh.

Run the FEA Solver

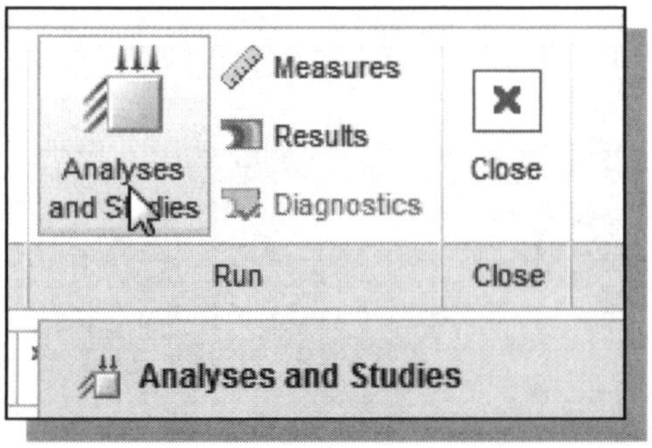

1. Choose **Home** in the *Ribbon* toolbar.

2. Activate the *Creo Simulate* **Analyses and Studies** command by selecting the icon in the toolbar area as shown.

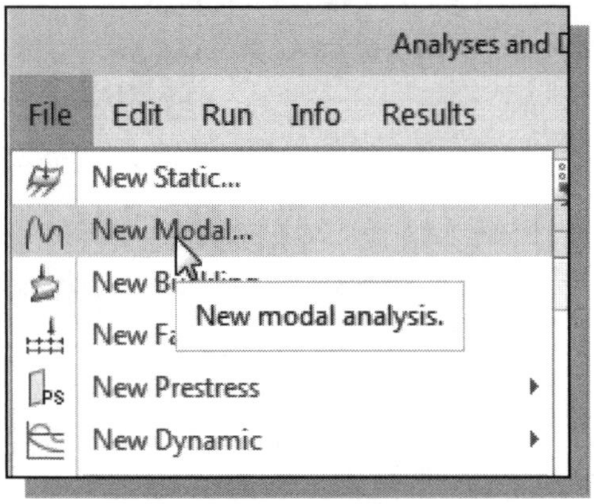

3. In the *Analyses and Design Studies* window, choose **File → New Modal** in the pull-down menu. This will start the setup of a basic modal analysis.

4. Note that **ConstraintSet1** is automatically accepted as part of the analysis parameters.

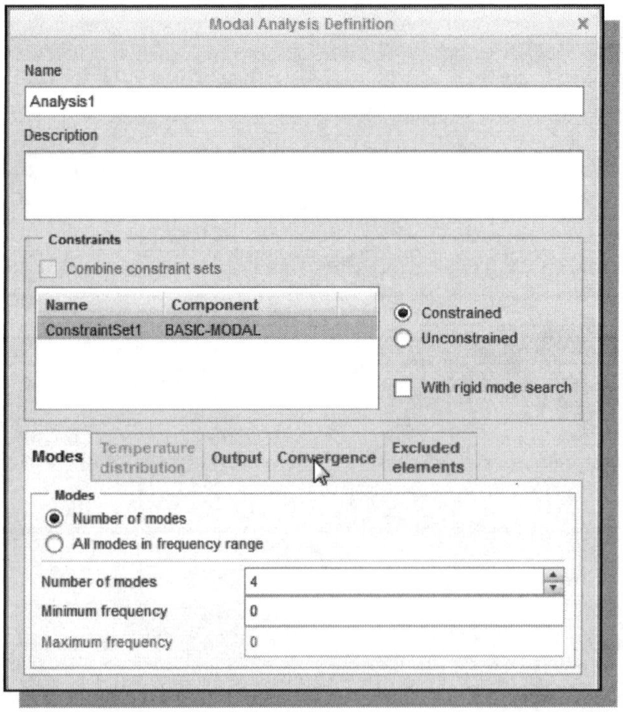

5. Confirm the **Number of Modes** option is set to **4** as shown.

6. Click the **Convergence** tab as shown.

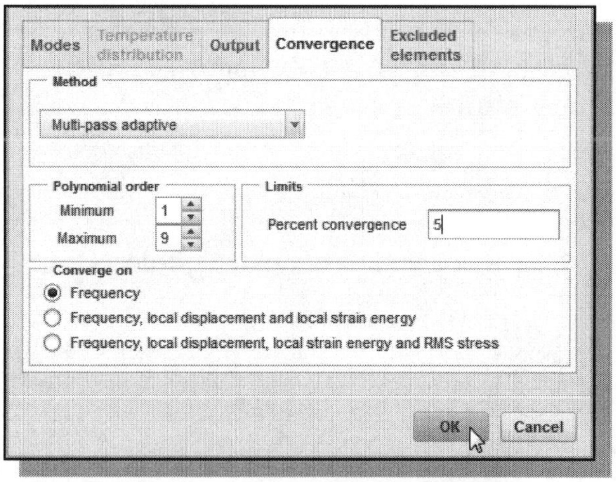

7. Set the Method option to **Multi-Pass Adaptive**, P-order to **9** and Limits to **5** Percent Convergence as shown.

8. Click **OK** to accept the settings.

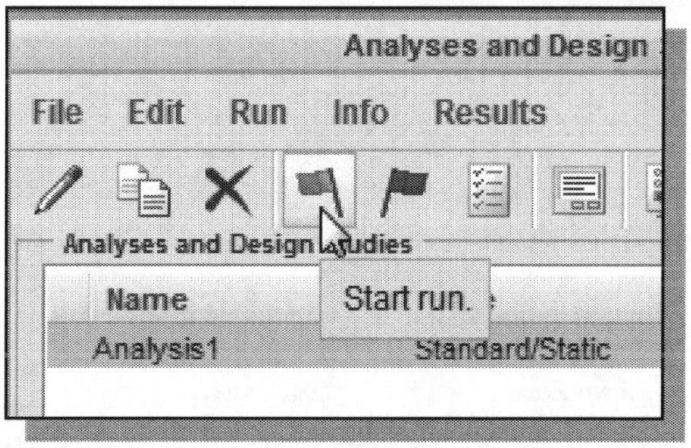

9. Click **Start run** to begin the *Solving* process.

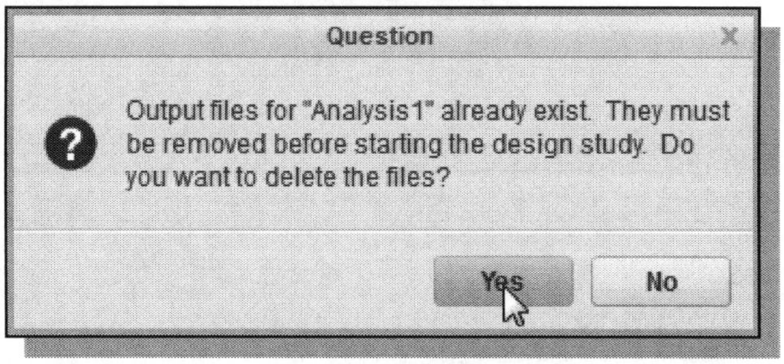

10. Click **Yes** to delete the old output files.

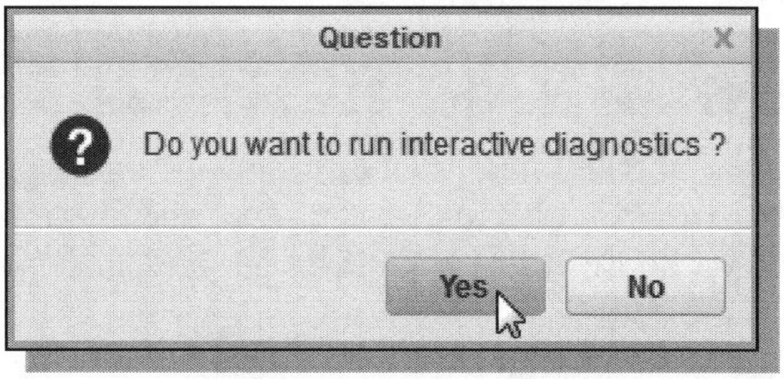

11. Click **Yes** to run interactive diagnostics; this will help us identify and correct problems associated with our FEA model.

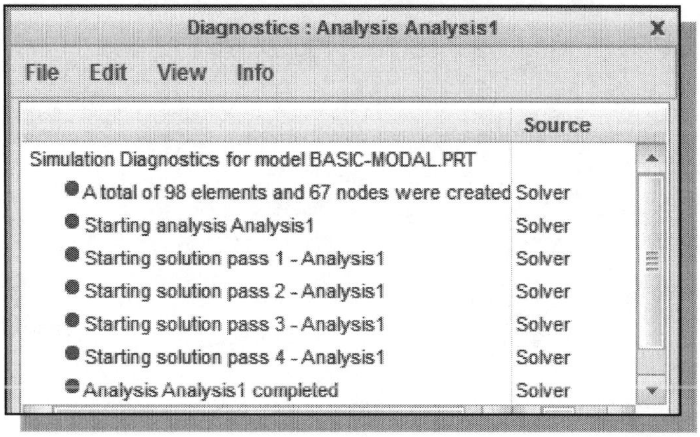

12. In the *Diagnostics* window, *Creo Simulate* will list any problems or conflicts of our FEA model.

13. Click on the **Close** button to exit the *Diagnostics* window.

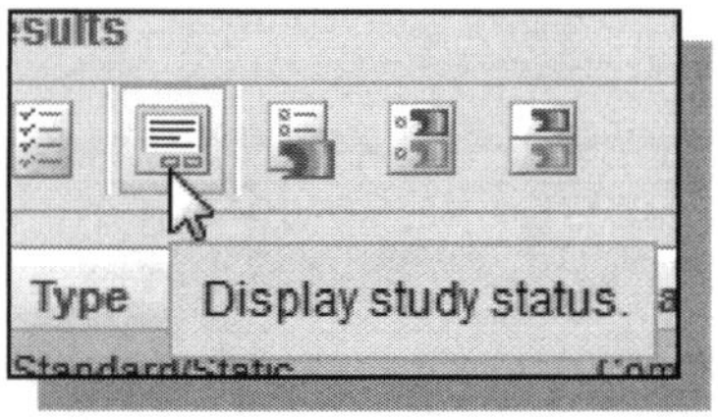

14. In the *Analyses and Design Studies* window, choose **Display study status** as shown.

❖ Note that we started with 98 elements not converging and went up to 4th order polynomials to reach the 5% convergence.

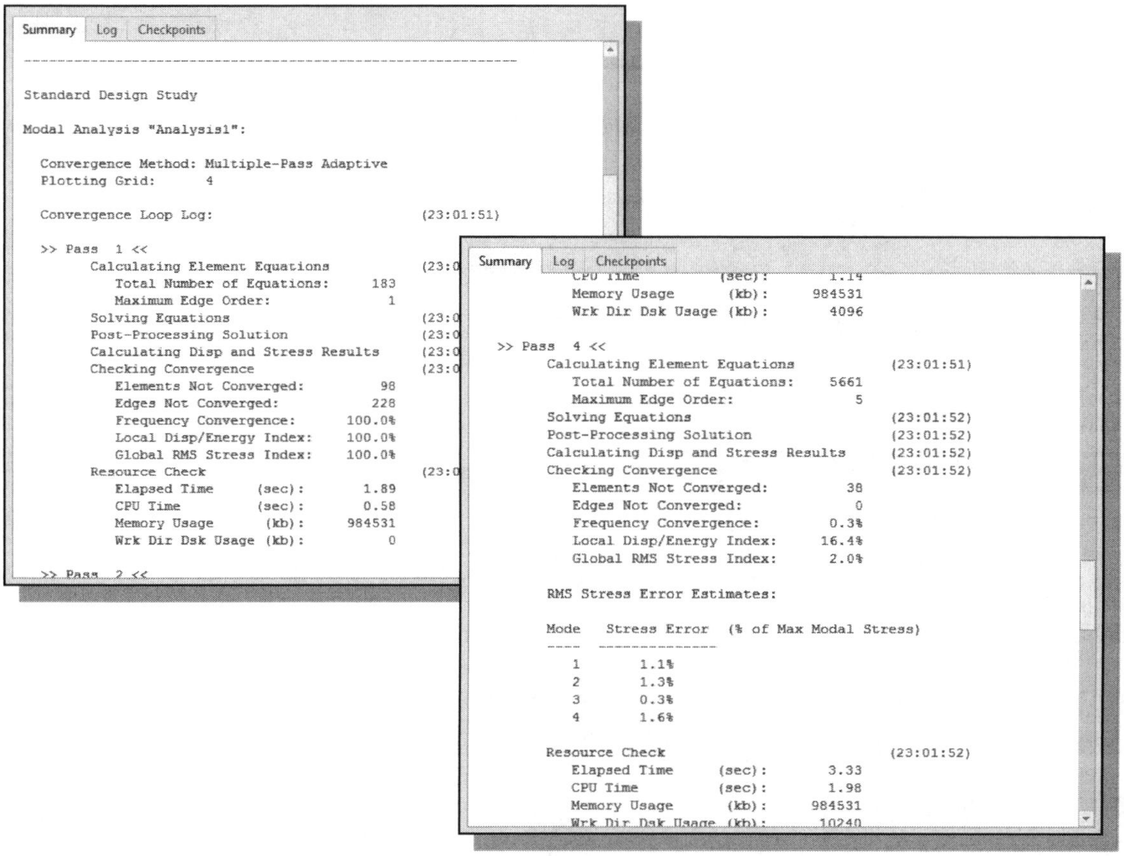

15. On your own, examine the detailed information regarding the analysis and the design.

16. In the *Analyses and Design Studies* window, choose **Review results** as shown.

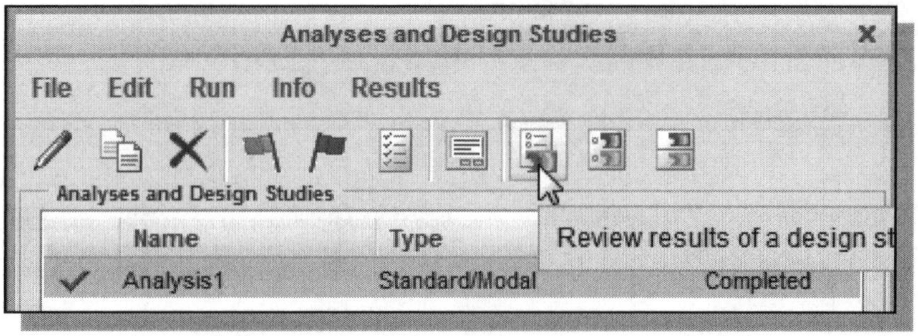

View the FEA Results

Note that the first four natural frequencies of the FEA solutions are shown in the *Result Window Definition* dialog box. The three natural frequencies from our analytical calculations are very similar to the frequencies found by *Creo Simulate*.

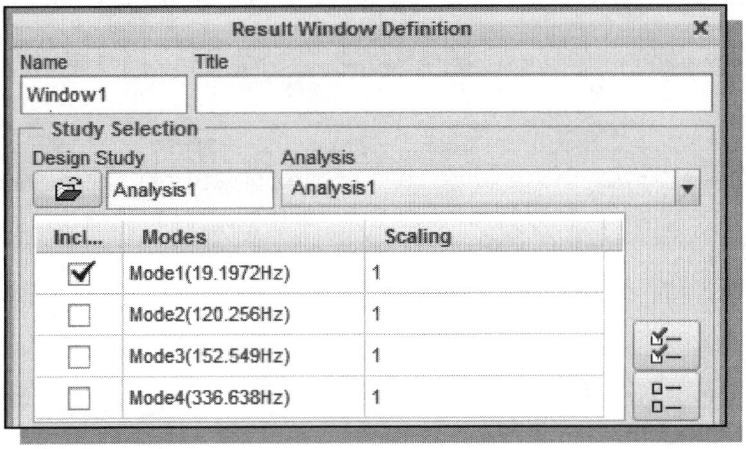

1. Confirm the **Study Selection** settings are as shown in the above figure; **Mode1** is pre-selected.

2. Accept the **default settings** in the **Quantity** option list as shown.

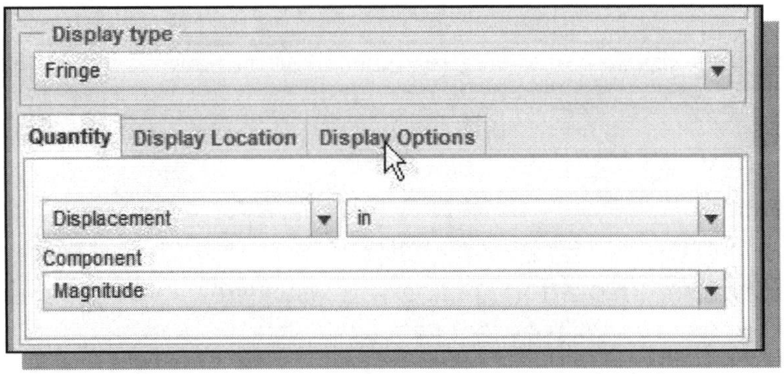

3. Choose **Display Options** to examine the display options.

4. Switch ***on*** the **Deformed** and **Overlay** options as shown.

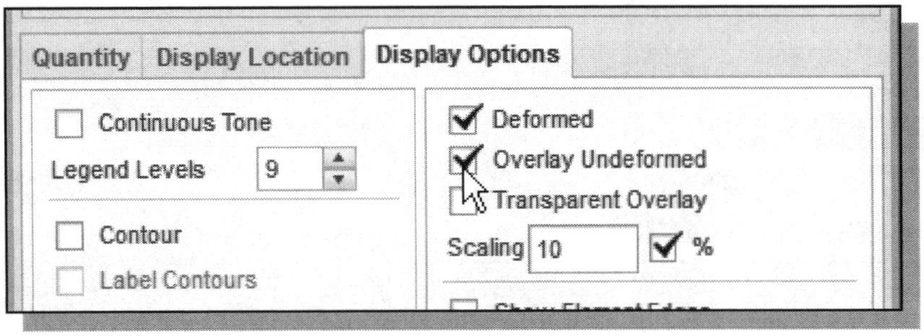

5. Click **OK and Show** to display the results.

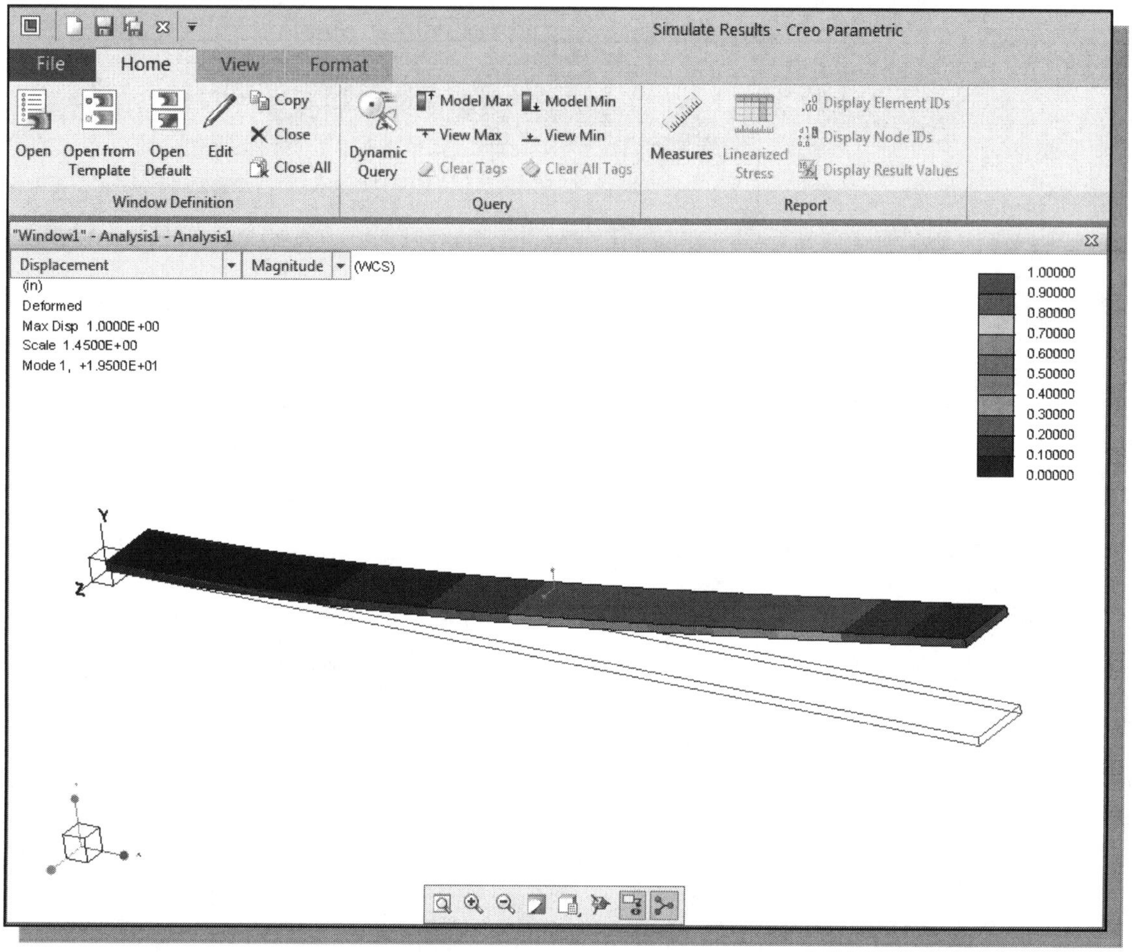

❖ Note that the **Dynamic Rotation** command is also available while the animation is running.

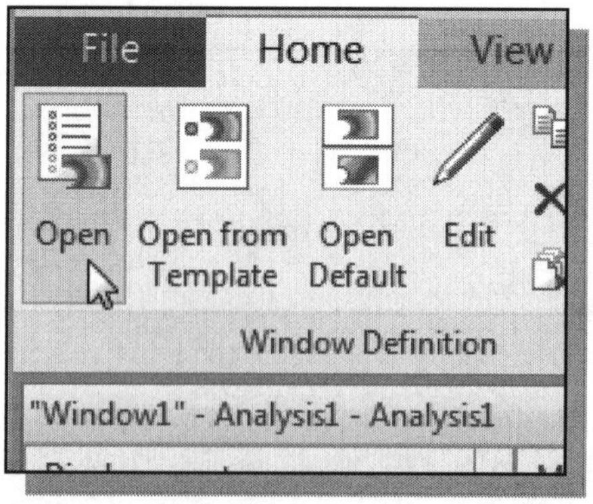

6. Select **Open** in the toolbar area as shown.

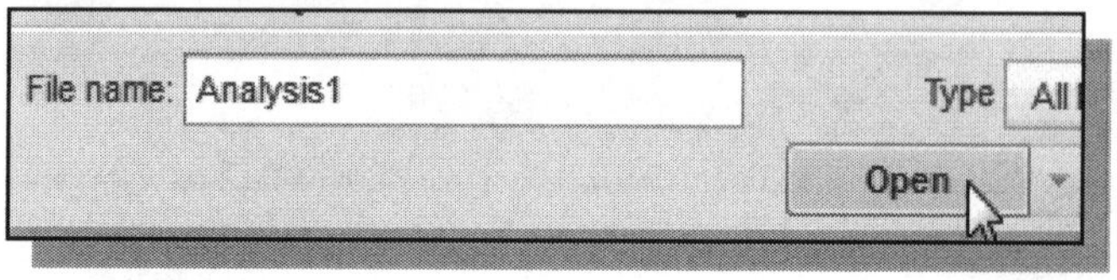

7. Select **Analysis1** from the *Design Study Window* option list.

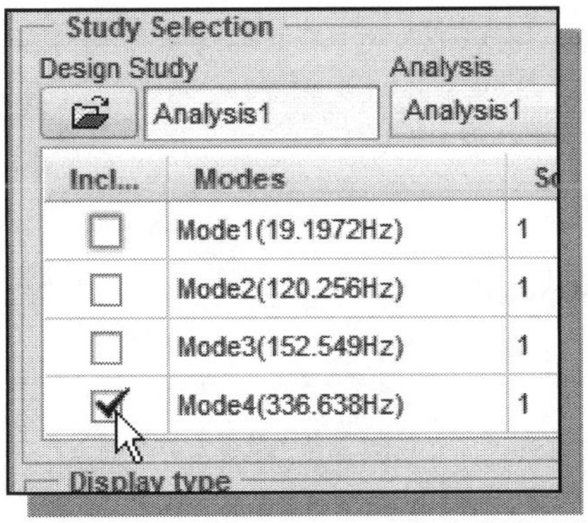

8. Set the Study Selection to **Mode4** as shown.

9. Choose **Display Options** to examine the display options.

10. Switch ***on*** the **Deformed** and **Overlay** options as shown.

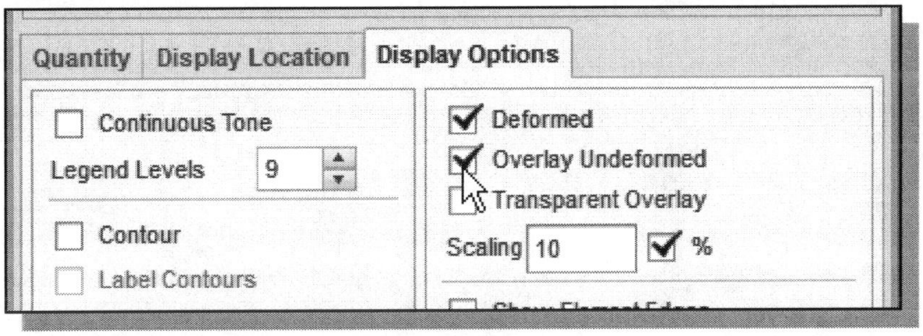

11. Click **OK and Show** to display the results.

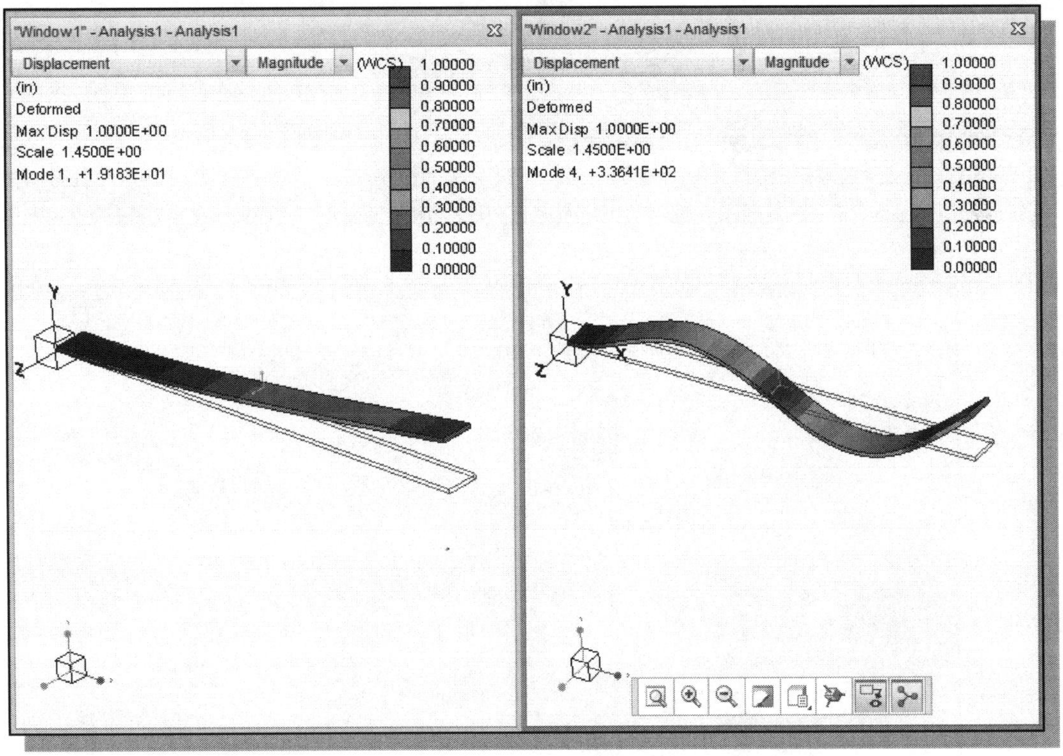

❖ Note that multiple windows can be used to set up displays on the same screen to compare the FEA results.

Adding an Additional Mass to the System

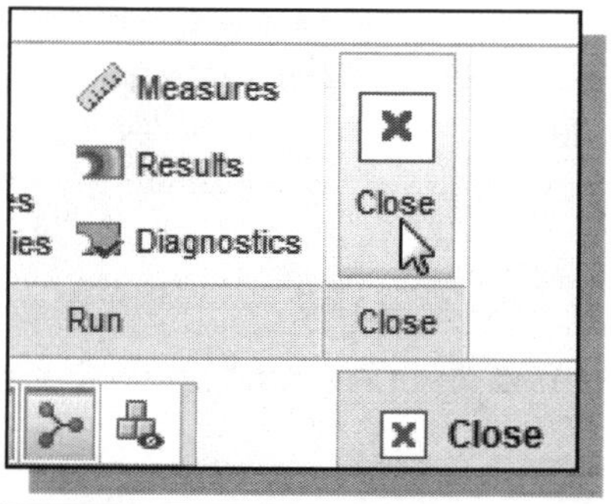

1. On your own, close the **Review Results** windows.

2. Exit the *Integrated* mode of *Creo Simulate* by selecting **Close** in the *Ribbon* toolbar as shown.

❖ We have returned to the standard *Creo Parametric Part Modeling* mode.

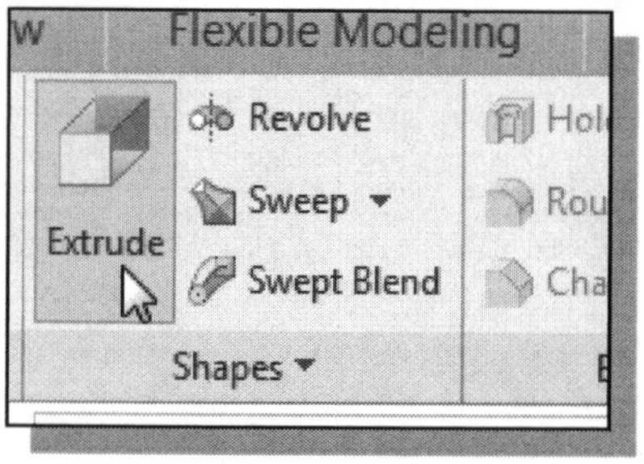

3. In the *Ribbon* toolbar, select the **Extrude** tool option as shown.

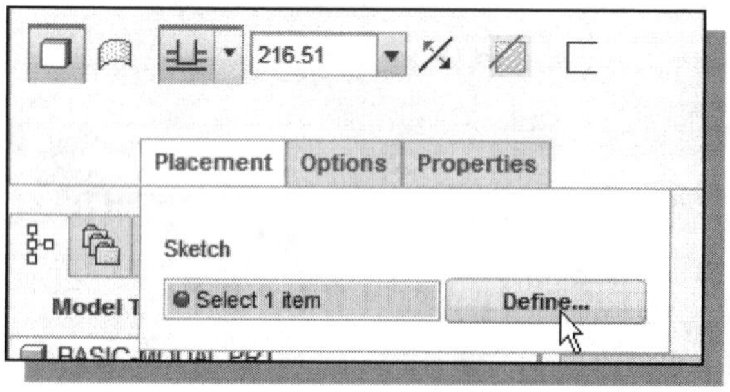

4. Click the **Placement** option and choose **Define** to begin creating a new *internal sketch.*

5. On your own, set up the **TOP** surface of the beam model as the sketch plane with the **RIGHT** datum plane facing the right edge of the computer screen and pick **Sketch** to enter the *Sketcher* mode.

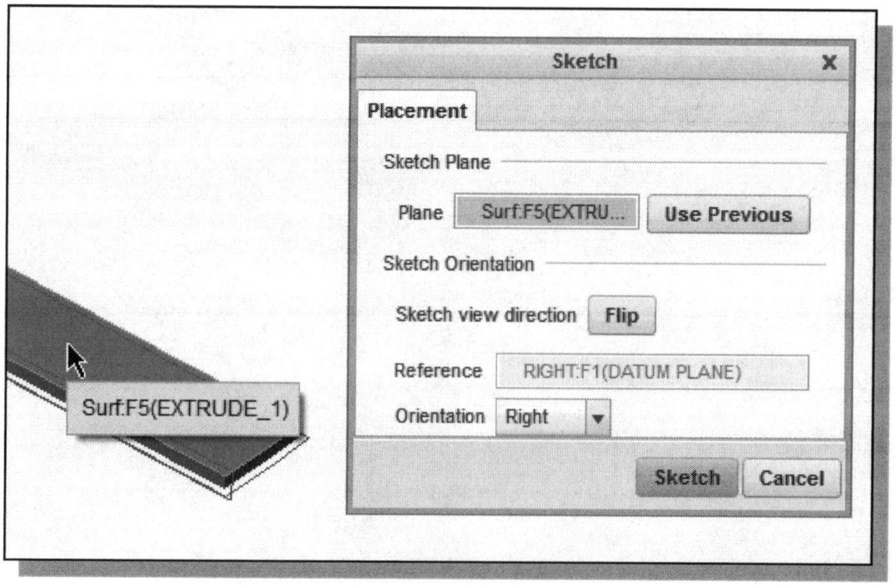

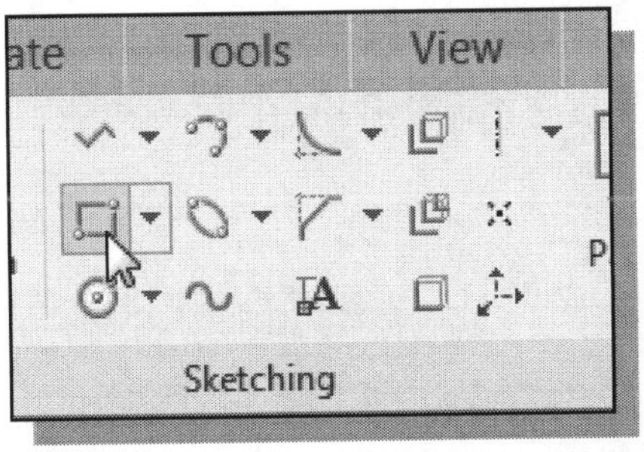

6. In the *Sketching* toolbar, select **Rectangle**.

❖ On your own, create a rectangle with the top, bottom and right edges aligned to the corresponding edges of the solid model as shown.

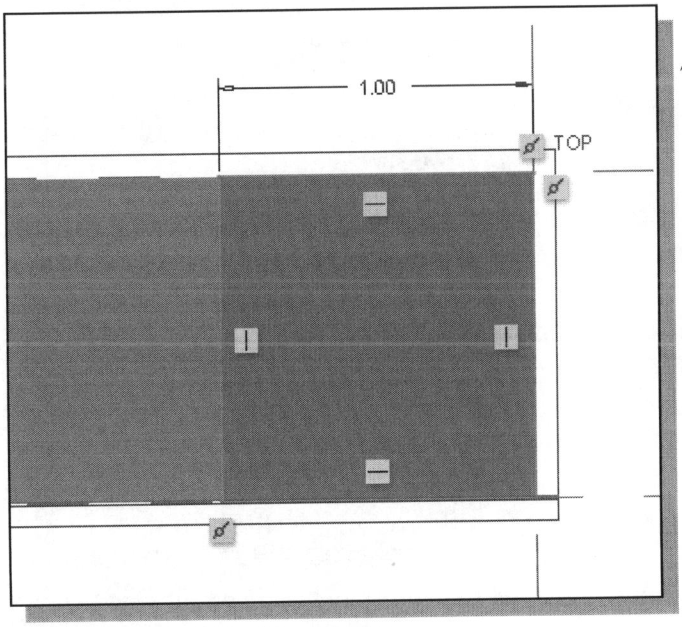

7. On your own, adjust the size of the rectangle to **1″**.

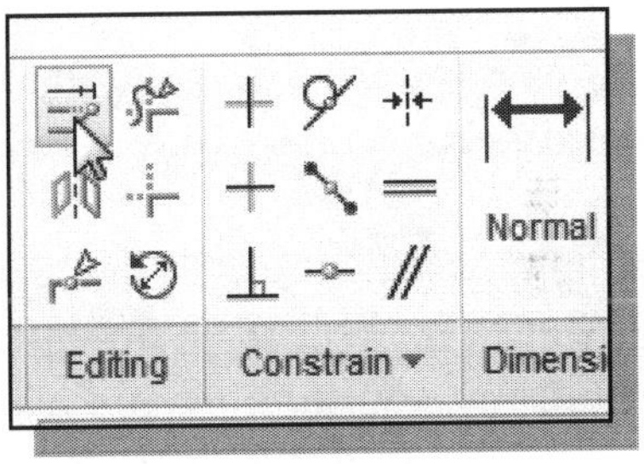

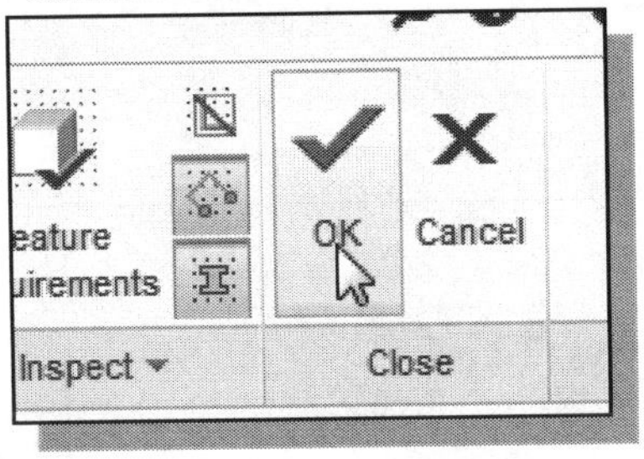

8. Click **OK** to accept the completed sketch.

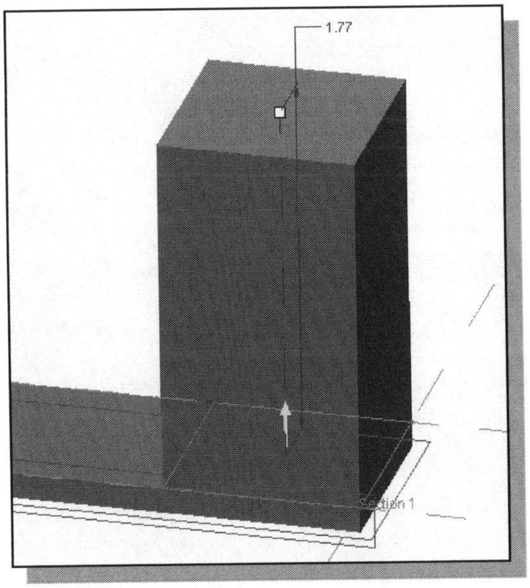

9. Set the *extrusion distance* to **1.77** as shown.

10. Click **Accept** to accept the completed beam model.

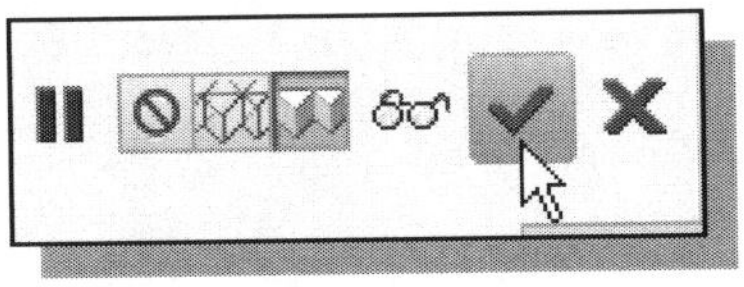

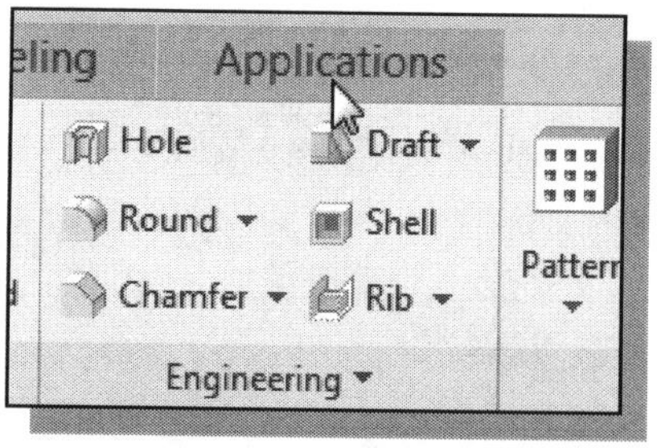

11. Switch to the **Applications** tab in the *Ribbon* area as shown.

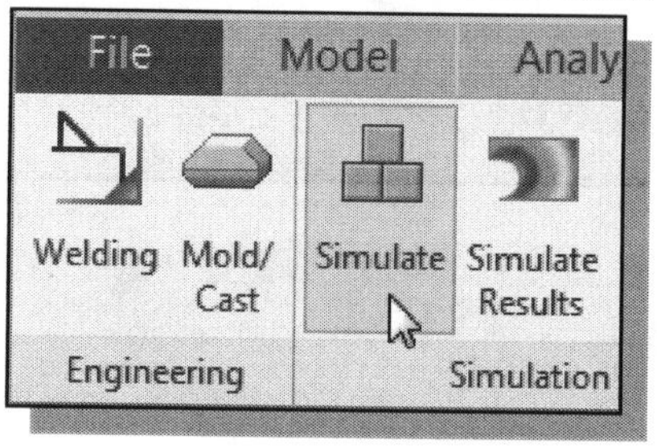

12. Start the *Integrated* mode of *Creo Simulate* by selecting the **Simulate** option as shown.

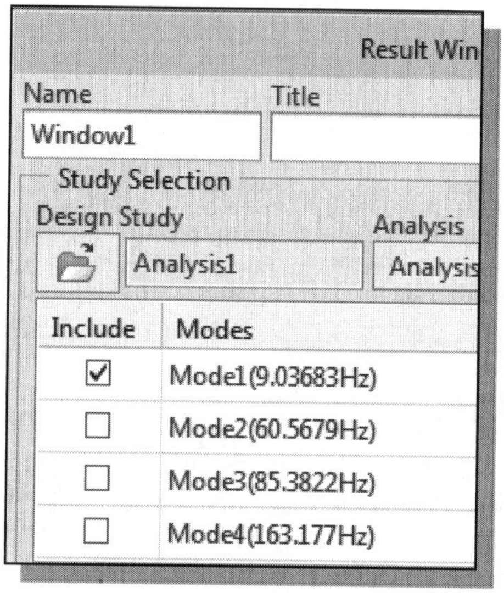

13. On your own, perform another **Modal Analysis**.

➢ Note the first natural frequency is calculated as **9.03 Hz**, which is a little higher than the result from our preliminary calculation. This is due to the additional mass being not centered at the right edge of the beam.

14. On your own, modify the 2D section and update the additional mass as shown.

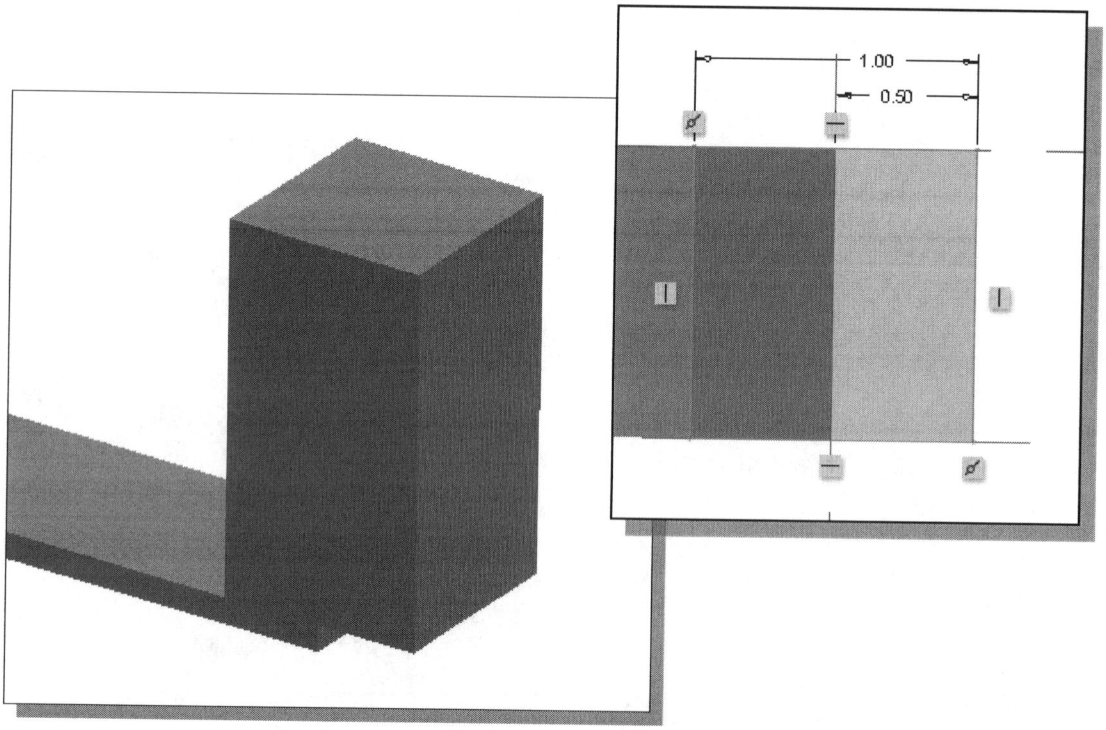

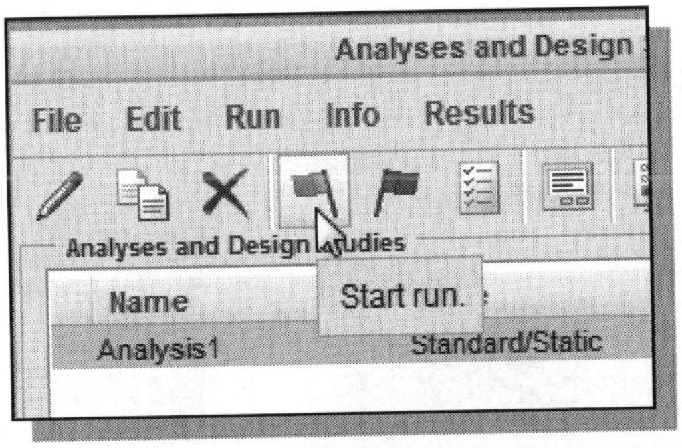

15. On your own, perform another **Modal Analysis**.

➢ Note the calculated first natural frequency is very close to the result of our preliminary analysis.

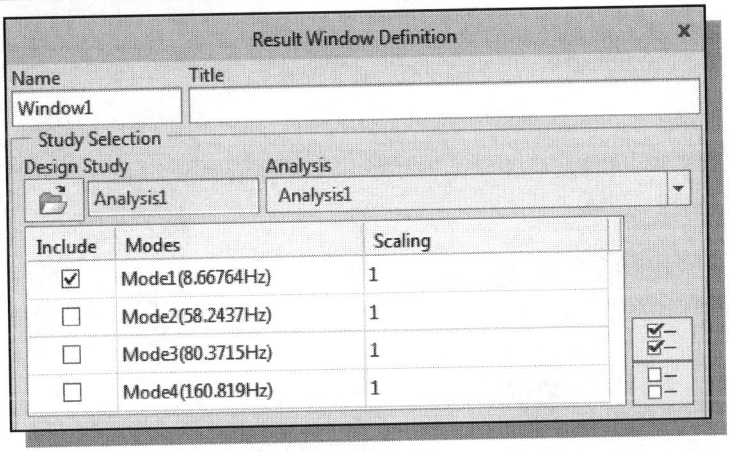

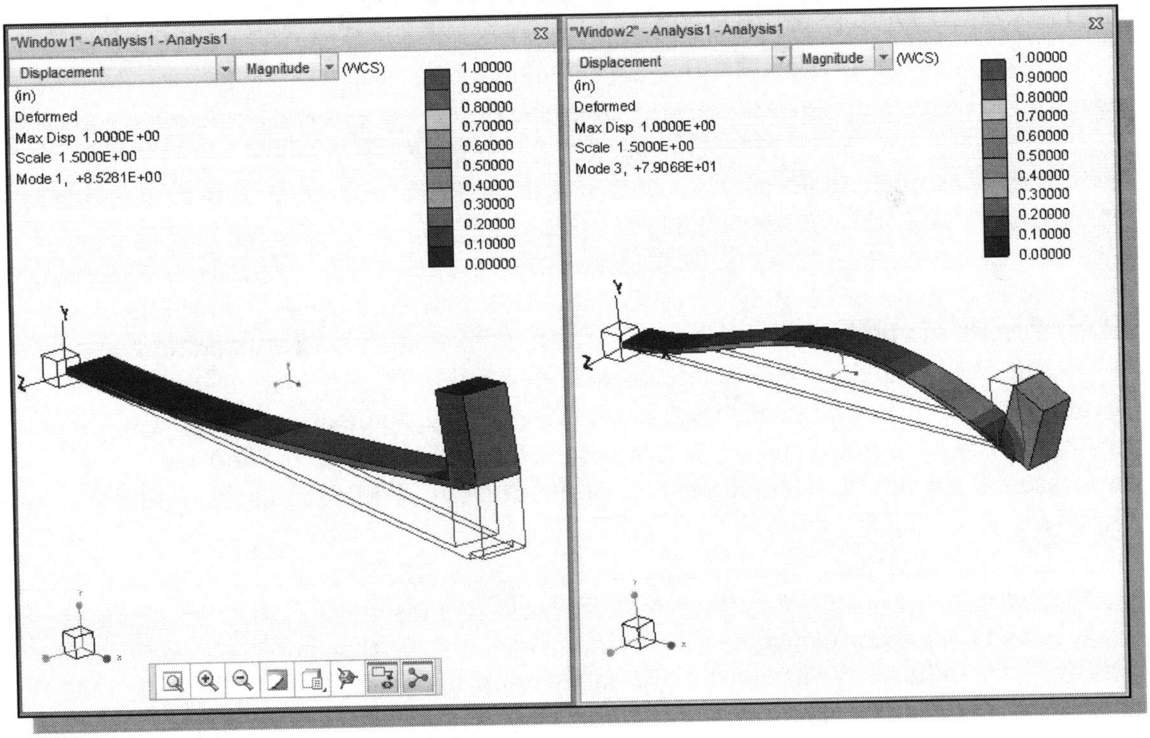

➢ You are also encouraged to create and compare FEA modal analyses of the same problem using one-dimensional beam elements.

Conclusions

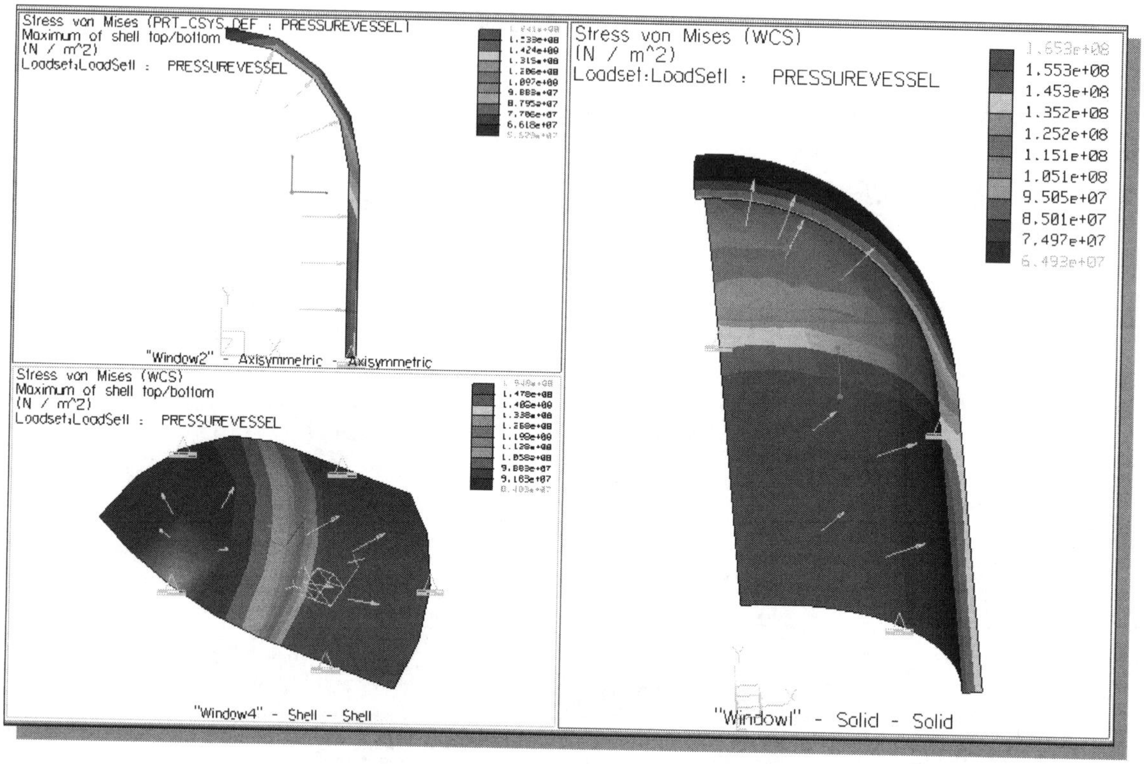

Finite element analysis has rapidly become a vital tool for design engineers. However, use of this tool does not guarantee correct results. The design engineer is still responsible for doing approximate calculations, using good design practice, and applying good engineering judgment to the problem. It is hoped that FEA will supplement these skills to ensure that the best design is obtained.

It should be emphasized that, when performing FEA analysis, besides confirming that the systems remain in the elastic regions for the applied loading, other considerations are also important; for example, large displacements and buckling of beams, which can also invalidate the *linear statics* analysis results. In performing finite element analysis, it is also necessary to acquire some knowledge of the theory behind the method and understand the restrictions and limitations of the software. There is no substitution for experience.

Throughout this text, various basic techniques have been presented. The goals are to make use of the tools provided by *Creo Simulate* and to successfully enhance the DESIGN capabilities. We have gone over the fundamentals of FEA procedures. In many instances, only a single approach to the stress analysis tasks was presented; you are encouraged to repeat any of the chapters and develop different ways of thinking in order to accomplish the goal of making better designs. We have only scratched the surface of *Creo Simulate* functionality. The more time you spend using the system, the easier it will be to perform *Computer Aided Engineering* with *Creo Simulate.*

Review Questions

1. What is the main purpose of performing a modal analysis on a system?

2. What are the relations between resonant frequencies and the natural frequencies of a system?

3. In vibration analysis, any physical structure can be modeled as a combination of what objects?

4. A cantilever beam can be modeled as a simple mass-spring system by examining the deflection at the tip of the beam, what is the equivalent spring constant for such a system?

5. In performing a modal analysis of a cantilever beam, would changing the material properties affect the natural frequencies of the system?

6. In *Creo Simulate*, is it required to have a load applied to the system in order to perform a Modal analysis?

7. In *Creo Simulate*, can we perform a modal analysis without applying a constraint?

8. For the cantilever beam with a mass attached at the end, when the attached mass is relatively big, we can ignore the weight of the beam as a simplified approximation. Calculate the first natural frequency of the tutorial problem using this approximation and compare the results obtained.

Exercises

The first four natural frequencies and mode shapes of a uniform cross section beam, with different types of supports at the ends, are shown in the below table:

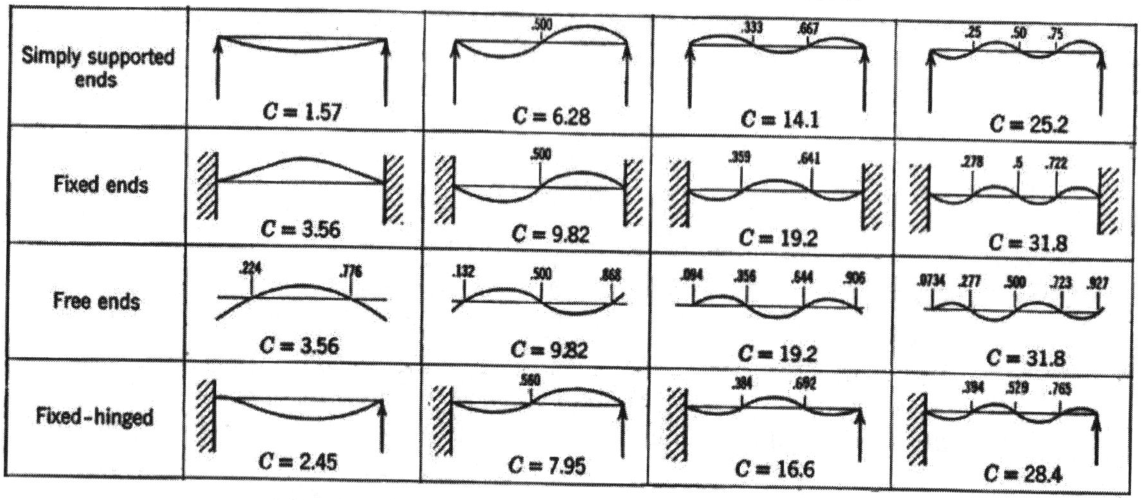

$$f_n = C\sqrt{\frac{EI}{m_b L^3}}$$

f_n = natural frequency in cycles/sec
C = constant from above table

1. Determine the **natural frequencies** and **mode shapes** of a **Fixed-hinged** beam. The beam is made of steel and has the dimensions of 14.5″ x 1/8″ x 1″.

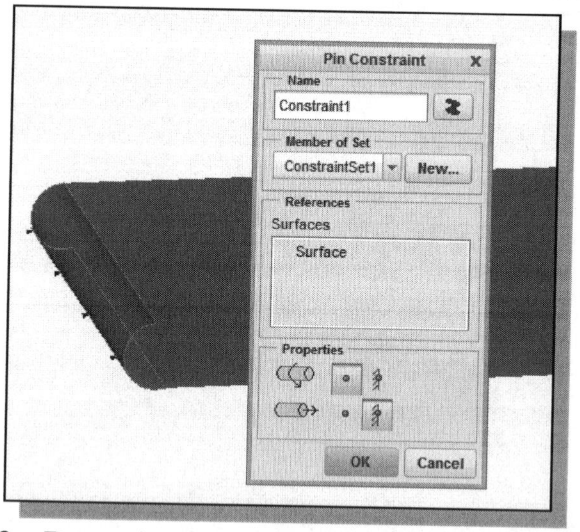

- To apply a **Pin constraint**, first create a cylindrical surface at the Pin location.

2. Determine the **natural frequencies** and **mode shapes** of a **fixed-ends** beam. The beam is made of steel and has the dimensions of 14.5″ x 1/8″ x 1″.

3. Modal analyses can also be performed using 1D beam elements; for the above two problems perform beam modal analyses and compare the results.

INDEX

Notes: